VOLUME FIVE HUNDRED

METHODS IN ENZYMOLOGY

Methods in Systems Biology

METHODS IN ENZYMOLOGY

Editors-in-Chief

JOHN N. ABELSON AND MELVIN I. SIMON

Division of Biology
California Institute of Technology
Pasadena, California

Founding Editors

SIDNEY P. COLOWICK AND NATHAN O. KAPLAN

VOLUME FIVE HUNDRED

METHODS IN ENZYMOLOGY

Methods in Systems Biology

EDITED BY

DANIEL JAMESON, MALKHEY VERMA, AND
HANS V. WESTERHOFF

Manchester Centre for Integrative Systems Biology
The University of Manchester
Manchester, UK

AMSTERDAM • BOSTON • HEIDELBERG • LONDON
NEW YORK • OXFORD • PARIS • SAN DIEGO
SAN FRANCISCO • SINGAPORE • SYDNEY • TOKYO
Academic Press is an imprint of Elsevier

Academic Press is an imprint of Elsevier
525 B Street, Suite 1900, San Diego, CA 92101-4495, USA
225 Wyman Street, Waltham, MA 02451, USA
32 Jamestown Road, London NW1 7BY, UK

First edition 2011

For information on all Academic Press publications
visit our website at elsevierdirect.com

ISBN: 978-0-12-385118-5
ISSN: 0076-6879

Printed and bound in United States of America
11 12 13 14 10 9 8 7 6 5 4 3 2 1

Contents

Contributors

Malgorzata Adamczyk
Warsaw University of Technology, Faculty of Chemistry, Institute of Biotechnology ul.Noakowskiego 3, Warsaw, Poland

James Adjaye
Department of Vertebrate Genomics, Max Planck Institute for Molecular Genetics (Molecular Embryology and Aging group), Ihnestrasse 63-73, Berlin, Germany

J. William Allwood
Manchester Interdisciplinary Biocentre, School of Chemistry, University of Manchester, Manchester, United Kingdom

Barbara M. Bakker
Department of Pediatrics, Center for Liver, Digestive and Metabolic Diseases, University Medical Center Groningen, University of Groningen, Groningen, and Department of Molecular Cell Physiology, Faculty of Earth and Life Sciences, VU University, Amsterdam, The Netherlands

Jennifer A. Bartell
Department of Biomedical Engineering, University of Virginia, Charlottesville, Virginia, USA

Philippe I. H. Bastiaens
Department of Systemic Cell Biology, Max Planck Institute for Molecular Physiology, and Technical University of Dormtund, Dortmund, Germany

Rebekka Biedendieck
Institute of Microbiology, Technische Universität Braunschweig, Spielmannstrasse 7, Braunschweig, Germany

Nils Blüthgen
Institute of Pathology, Charité—Universitätsmedizin Berlin, Charitéplatz 1, and Institute for Theoretical Biology, Humboldt University of Berlin, Invalidenstraße 43, Berlin, Germany

Joost Boele
Amsterdam Institute for Molecules, Medicines and Systems/NISB, VU University Amsterdam, De Boelelaan 1085, Amsterdam, The Netherlands

Claudia Borgmeier
Institute for Molecular Microbiology and Biotechnology, Westfälische Wilhelms-Universität Münster, Corrensstrasse 3, Münster, Germany

Tatiana Borodina
Max Planck Institute for Molecular Genetics, Department of Vertebrate Genomics, Ihnestrasse 63-73, Berlin, Germany

Jildau Bouwman
Biosciences, TNO Quality of Life Zeist, The Netherlands

F. J. Bruggeman
Life Sciences, Centre for Mathematics and Computer Science (CWI), and Netherlands Institute for Systems Biology (NISB), and Molecular Cell Physiology, VU University; Swammerdam Institute for Life Sciences, University of Amsterdam, Amsterdam, The Netherlands

Boyke Bunk
Institute of Microbiology, Technische Universität Braunschweig, Spielmannstrasse 7, Braunschweig, Germany

Kathleen M. Carroll
Manchester Centre for Integrative Systems Biology, Manchester Interdisciplinary Biocentre, and School of Chemistry, The University of Manchester, Manchester, United Kingdom

Kian-Kai Cheng
Department of Biochemistry, University of Cambridge, Cambridge, United Kingdom, and Department of Bioprocess Engineering, Faculty of Chemical and Natural Resources, Engineering, Universiti Teknologi Malaysia, 81310 UTM Skudai, Johor, Malaysia

Ric C. H. De Vos
Plant Research International, Wageningen University and Research Centre (Wageningen UR), and Centre for BioSystems Genomics, Wageningen, The Netherlands

Catherine Deborde
INRA, UMR1332 Biologie du Fruit & Pathologie Centre INRA de Bordeaux, INRA—Université de Bordeaux, and Plateforme Métabolome-Fluxome Bordeaux, Génomique Fonctionnelle Bordeaux, IBVM, BP 81, Villenave d'Ornon, France

M. Dobrzyński
Netherlands Institute for Systems Biology (NISB), Amsterdam, The Netherlands

Franco du Preez
Manchester Interdisciplinary Biocentre, University of Manchester, Manchester, United Kingdom

Warwick B. Dunn
Manchester Centre for Integrative Systems Biology, and Department of Chemistry, Manchester Interdisciplinary Biocentre; Centre for Advanced Discovery and Experimental Therapeutics, School of Biomedicine, University of Manchester, Manchester, United Kingdom

Elizabeth A. Elliot
Centre for Systems Biology at Edinburgh, The University of Edinburgh, Edinburgh, United Kingdom

Alexander Erban
Max Plank Institute of Molecular Plant Physiology, Am Mühlenberg 1, Golm, Germany

Claire E. Eyers
Michael Barber Centre for Mass Spectrometry, School of Chemistry, University of Manchester, Manchester Interdisciplinary Biocentre, Manchester, United Kingdom

Sven Fengler
Department of Systemic Cell Biology, Max Planck Institute for Molecular Physiology, Dortmund, Germany

Carole Goble
School of Computer Science, University of Manchester, Manchester, United Kingdom

Royston Goodacre
Manchester Interdisciplinary Biocentre, School of Chemistry, and Manchester Centre for Integrative Systems Biology, Manchester Interdisciplinary Biocentre, University of Manchester, Manchester, United Kingdom

Hernán E. Grecco
Department of Systemic Cell Biology, Max Planck Institute for Molecular Physiology, Dortmund, Germany

Julian Griffin
Department of Biochemistry, and The Cambridge Systems Biology Centre, University of Cambridge, Cambridge, United Kingdom

Charles R. Haggart
Department of Biomedical Engineering, and Robert M. Berne Cardiovascular Research Center, University of Virginia, Charlottesville, Virginia, USA

Robert D. Hall
Plant Research International, Wageningen University and Research Centre (Wageningen UR), and Centre for BioSystems Genomics, Wageningen, The Netherlands

Neil W. Hayes
Manchester Centre for Integrative Systems Biology, Manchester Interdisciplinary Biocentre, The University of Manchester, Manchester, United Kingdom

Sef J. Heijnen
Biotechnology Department, Delft University of Technology, Julianalaan 67, 2628 BC Delft, The Netherlands

Jan-Hendrik S. Hofmeyr
Department of Biochemistry, and Centre for Studies in Complexity, University of Stellenbosch, Private Bag X1, Stellenbosch, South Africa

Dieter Jahn
Institute of Microbiology, Technische Universität Braunschweig, Spielmannstrasse 7, Braunschweig, Germany

J. Jozefczuk
Department of Vertebrate Genomics, Max Planck Institute for Molecular Genetics (Molecular Embryology and Aging group), Ihnestrasse 63-73, Berlin, Germany

Christoph Kaleta
Department of Bioinformatics, Friedrich Schiller University Jena, Jena, Germany

Joachim Kopka
Max Plank Institute of Molecular Plant Physiology, Am Mühlenberg 1, Golm, Germany

Klaas Krab
Department of Molecular Cell Physiology, IMC, Faculty of Earth and Life Sciences, Vrije Universiteit, De Boelelaan 1085, Amsterdam, The Netherlands

Falko Krause
Max Planck Institute for Molecular Genetics, Department of Computational Molecular Biology, Berlin, Germany

Olga Krebs
Heidelberg Institute for Theoretical Studies (Hits), gGmbH, Schloss-Wolfsbrunnenweg 35, Heidelberg, Germany

Joshua LaBaer
Virginia G. Piper Center for Personalized Diagnostics, Biodesign Institute, Arizona State University, Tempe, Arizona, USA

Karin Lanthaler
Faculty of Life Sciences, University of Manchester, Manchester, United Kingdom

Francesco Lanucara
Michael Barber Centre for Mass Spectrometry, School of Chemistry, University of Manchester, Manchester Interdisciplinary Biocentre, Manchester, United Kingdom

Wolfram Liebermeister
Weizmann Institute of Science, Department of Plant Sciences, Rehovot, Israel

Naglis Malys
Manchester Centre for Integrative Systems Biology, Manchester Interdisciplinary Biocentre, and Faculty of Life Sciences, Manchester Interdisciplinary Biocentre, The University of Manchester, Manchester, United Kingdom

John E. G. McCarthy
Manchester Centre for Integrative Systems Biology, and Faculty of Life Sciences, Manchester Interdisciplinary Biocentre; School of Chemical and Analytical Engineering, The University of Manchester, Manchester, United Kingdom

Friedhelm Meinhardt
Institute for Molecular Microbiology and Biotechnology, Westfälische Wilhelms-Universität Münster, Corrensstrasse 3, Münster, Germany

Hanan L. Messiha
Manchester Centre for Integrative Systems Biology, Manchester Interdisciplinary Biocentre, and School of Chemistry, The University of Manchester, Manchester, United Kingdom

Annick Moing
INRA, UMR1332 Biologie du Fruit & Pathologie Centre INRA de Bordeaux, INRA—Université de Bordeaux, and Plateforme Métabolome-Fluxome Bordeaux, Génomique Fonctionnelle Bordeaux, IBVM, BP 81, Villenave d'Ornon, France

Wolfgang Mueller
Heidelberg Institute for Theoretical Studies (Hits), gGmbH, Schloss-Wolfsbrunnenweg 35, Heidelberg, Germany

Maria Nardelli
Doctoral Training Centre for Integrative Systems Biology, The University of Manchester, Manchester, United Kingdom

Sanjay M. Nilapwar
Manchester Interdisciplinary Biocentre, The University of Manchester, Manchester, United Kingdom

Stephen G. Oliver
Manchester Centre for Integrative Systems Biology, and Faculty of Life Sciences, Michael Smith Building, The University of Manchester, Manchester, United Kingdom

Stuart Owen
School of Computer Science, University of Manchester, Manchester, United Kingdom

Jason A. Papin
Department of Biomedical Engineering, University of Virginia, Charlottesville, Virginia, USA

Ji Qiu
Virginia G. Piper Center for Personalized Diagnostics, Biodesign Institute, Arizona State University, Tempe, Arizona, USA

Pedro Roda-Navarro
Department of Systemic Cell Biology, Max Planck Institute for Molecular Physiology, Dortmund, Germany

Johann M. Rohwer
Department of Biochemistry, University of Stellenbosch, Private Bag X1, Stellenbosch, South Africa

Sergio Rossell
Center for Molecular and Biomolecular Informatics, Nijmegen Center for Molecular Life Sciences, Radboud University Nijmegen Medical Center, Nijmegen, The Netherlands

K. Rybakova
Netherlands Institute for Systems Biology (NISB), and Molecular Cell Physiology, VU University, Amsterdam, The Netherlands

Reza Salek
Department of Biochemistry, University of Cambridge, Cambridge, United Kingdom

Filipe Santos
Amsterdam Institute for Molecules, Medicines and Systems/NISB, and Kluyver Centre for Genomics of Industrial Fermentations/Netherlands Consortium Systems Biology, VU University Amsterdam, De Boelelaan 1085, Amsterdam, The Netherlands

Jeffrey J. Saucerman
Department of Biomedical Engineering, and Robert M. Berne Cardiovascular Research Center, University of Virginia, Charlottesville, Virginia, USA

Sascha Schäuble
Department of Bioinformatics, Friedrich Schiller University Jena, Jena, Germany

Christian Scherling
Institute of Bioinformatics and Biochemistry, Technische Universität Braunschweig, Langer Kamp 19b, Braunschweig, Germany

Pascal Schulthess
Institute of Pathology, Charité—Universitätsmedizin Berlin, Chariteplatz 1, and Institute for Theoretical Biology, Humboldt University of Berlin, Invalidenstraße 43, Berlin, Germany

Marvin Schulz
Max Planck Institute for Molecular Genetics, Department of Computational Molecular Biology, Berlin, Germany

Stefan Schuster
Department of Bioinformatics, Friedrich Schiller University Jena, Jena, Germany

A. Schwabe
Life Sciences, Centre for Mathematics and Computer Science (CWI), and Netherlands Institute for Systems Biology (NISB), Amsterdam, The Netherlands

Evangelos Simeonidis
Manchester Centre for Integrative Systems Biology, The University of Manchester, Manchester, United Kingdom

Kieran Smallbone
Manchester Centre for Integrative Systems Biology, The University of Manchester, Manchester, United Kingdom

Jacky L. Snoep
Manchester Interdisciplinary Biocentre, University of Manchester, Manchester, United Kingdom, and Department of Biochemistry, Stellenbosch University, Private Bag X1, Matieland, South Africa; Molecular Cell Physiology, VU Amsterdam, The Netherlands

Simon Stammen
Institute of Microbiology, Technische Universität Braunschweig, Spielmannstrasse 7, Braunschweig, Germany

Marc Sultan
Max Planck Institute for Molecular Genetics, Department of Vertebrate Genomics, Ihnestrasse 63-73, Berlin, Germany

Neil Swainston
Manchester Centre for Integrative Systems Biology, University of Manchester, United Kingdom

Bas Teusink
Amsterdam Institute for Molecules, Medicines and Systems/NISB, and Kluyver Centre for Genomics of Industrial Fermentations/Netherlands Consortium Systems Biology, VU University Amsterdam, De Boelelaan 1085, Amsterdam, The Netherlands

Karen van Eunen
Department of Pediatrics, Center for Liver, Digestive and Metabolic Diseases, University Medical Center Groningen, University of Groningen, Groningen, and Department of Molecular Cell Physiology, Faculty of Earth and Life Sciences, VU University, Amsterdam, The Netherlands

Peter J. T. Verheijen
Biotechnology Department, Delft University of Technology, Julianalaan 67, 2628 BC Delft, The Netherlands

Malkhey Verma
Manchester Interdisciplinary Biocentre, The University of Manchester, Manchester, United Kingdom

P. Verschure
Netherlands Institute for Systems Biology (NISB), and Swammerdam Institute for Life Sciences, University of Amsterdam, Amsterdam, The Netherlands

Hans V. Westerhoff
Manchester Interdisciplinary Biocentre, and Doctoral Training Centre for Integrative Systems Biology, and Manchester Centre for Integrative Systems Biology, The University of Manchester, Manchester, United Kingdom, and Netherlands Institute for Systems Biology; Department of Molecular Cell Physiology, Faculty of Earth and Life Sciences, VU University, Amsterdam, The Netherlands

Catherine L. Winder
Manchester Centre for Integrative Systems Biology, and Department of Chemistry, Manchester Interdisciplinary Biocentre, University of Manchester, Manchester, United Kingdom

Jill A. Wishart
Manchester Centre for Integrative Systems Biology, and Faculty of Life Sciences, Michael Smith Building, The University of Manchester, Manchester, United Kingdom

Christoph Wittmann
Institute of Biochemical Engineering, Technische Universität Braunschweig, Gaußstrasse 17, Braunschweig, Germany

Katy Wolstencroft
School of Computer Science, University of Manchester, Manchester, United Kingdom

PREFACE

This volume of *Methods in Enzymology* covers many of the tools and techniques used in systems biology research. Systems biology attempts to understand biological function not as the behavior of individual components (e.g., metabolic enzymes or kinases) but as a complex system, the function of which is largely determined by the interactions of these components. This shift away from what some might consider to be the more traditional paradigms of biological research has necessitated the honing of old techniques and the development of new ones.

Of particular importance for systems biology is the assimilation of acquired data into meaningful mathematical models of biological systems. The meaning is attained by inserting actual parameter values into these models and actual mechanisms. This requires not only a plethora of precise experimental techniques but also mechanistic modeling methods, as well as ways to integrate the two and to convert the data avalanches to dynamically stored understanding. The resulting models can then be used to understand existing behavior and to predict new behaviors, for instance, for when one wishes to cure a disease or produce an important commodity more efficiently than in a chemical factory.

Model-building activities require input from a broad array of individuals with disparate, but complementary, expertise. Part of the massive challenge of delivering successful and meaningful results from systems biology research is having an overview and understanding of the diverse methods that these individuals implement in building up a complete picture of a biological system; it is with this in mind that we have put this volume together.

This volume is divided into several sections. The first is an examination of different strategies in systems biology. This is then followed by a set of chapters examining some of the instruments used.

Sections 3–6 contain chapters examining the techniques required to quantify the properties of biological elements that go toward building models—nucleic acids, proteins, metabolites—in a variety of organisms from bacteria to mammals.

Sections 7 and 8 cover the methods used to assimilate gathered data into meaningful models and interpret the results of running these models. These chapters cover explicit examples of how to build models, the pitfalls that may be encountered en route.

Finally, Section 9 examines the management of systems biology: first, methodology for managing the data associated with the research, and second, the issues surrounding the management of the research itself. Both of these areas present their own unique problems due to the interdisciplinary nature of the field.

This volume contains techniques and methodologies written by experts in their fields that we, as the Manchester Centre for Integrative Systems Biology, use successfully on a regular basis. We consider systems biology to be one of the most exciting and promising areas of biological research that can only become more prolific over time.

We would like to thank all of the authors who have contributed to the volume, particularly Neil Hayes who helped initiate the project. Additionally our thanks are due to Zoe Kruze and Paul Prasad Chandramohan from Elsevier for their help and patience during the publication process.

Daniel Jameson, Malkhey Verma, and Hans V. Westerhoff

Methods in Enzymology

Volume XVI. Fast Reactions
Edited by Kenneth Kustin

Volume XVII. Metabolism of Amino Acids and Amines (Parts A and B)
Edited by Herbert Tabor and Celia White Tabor

Volume XVIII. Vitamins and Coenzymes (Parts A, B, and C)
Edited by Donald B. McCormick and Lemuel D. Wright

Volume XIX. Proteolytic Enzymes
Edited by Gertrude E. Perlmann and Laszlo Lorand

Volume XX. Nucleic Acids and Protein Synthesis (Part C)
Edited by Kivie Moldave and Lawrence Grossman

Volume XXI. Nucleic Acids (Part D)
Edited by Lawrence Grossman and Kivie Moldave

Volume XXII. Enzyme Purification and Related Techniques
Edited by William B. Jakoby

Volume XXIII. Photosynthesis (Part A)
Edited by Anthony San Pietro

Volume XXIV. Photosynthesis and Nitrogen Fixation (Part B)
Edited by Anthony San Pietro

Volume XXV. Enzyme Structure (Part B)
Edited by C. H. W. Hirs and Serge N. Timasheff

Volume XXVI. Enzyme Structure (Part C)
Edited by C. H. W. Hirs and Serge N. Timasheff

Volume XXVII. Enzyme Structure (Part D)
Edited by C. H. W. Hirs and Serge N. Timasheff

Volume XXVIII. Complex Carbohydrates (Part B)
Edited by Victor Ginsburg

Volume XXIX. Nucleic Acids and Protein Synthesis (Part E)
Edited by Lawrence Grossman and Kivie Moldave

Volume XXX. Nucleic Acids and Protein Synthesis (Part F)
Edited by Kivie Moldave and Lawrence Grossman

Volume XXXI. Biomembranes (Part A)
Edited by Sidney Fleischer and Lester Packer

Volume XXXII. Biomembranes (Part B)
Edited by Sidney Fleischer and Lester Packer

Volume XXXIII. Cumulative Subject Index Volumes I-XXX
Edited by Martha G. Dennis and Edward A. Dennis

Volume XXXIV. Affinity Techniques (Enzyme Purification: Part B)
Edited by William B. Jakoby and Meir Wilchek

Volume XXXV. Lipids (Part B)
Edited by John M. Lowenstein

Volume XXXVI. Hormone Action (Part A: Steroid Hormones)
Edited by Bert W. O'Malley and Joel G. Hardman

Volume XXXVII. Hormone Action (Part B: Peptide Hormones)
Edited by Bert W. O'Malley and Joel G. Hardman

Volume XXXVIII. Hormone Action (Part C: Cyclic Nucleotides)
Edited by Joel G. Hardman and Bert W. O'Malley

Volume XXXIX. Hormone Action (Part D: Isolated Cells, Tissues, and Organ Systems)
Edited by Joel G. Hardman and Bert W. O'Malley

Volume XL. Hormone Action (Part E: Nuclear Structure and Function)
Edited by Bert W. O'Malley and Joel G. Hardman

Volume XLI. Carbohydrate Metabolism (Part B)
Edited by W. A. Wood

Volume XLII. Carbohydrate Metabolism (Part C)
Edited by W. A. Wood

Volume XLIII. Antibiotics
Edited by John H. Hash

Volume XLIV. Immobilized Enzymes
Edited by Klaus Mosbach

Volume XLV. Proteolytic Enzymes (Part B)
Edited by Laszlo Lorand

Volume XLVI. Affinity Labeling
Edited by William B. Jakoby and Meir Wilchek

Volume XLVII. Enzyme Structure (Part E)
Edited by C. H. W. Hirs and Serge N. Timasheff

Volume XLVIII. Enzyme Structure (Part F)
Edited by C. H. W. Hirs and Serge N. Timasheff

Volume XLIX. Enzyme Structure (Part G)
Edited by C. H. W. Hirs and Serge N. Timasheff

Volume L. Complex Carbohydrates (Part C)
Edited by Victor Ginsburg

Volume LI. Purine and Pyrimidine Nucleotide Metabolism
Edited by Patricia A. Hoffee and Mary Ellen Jones

Volume LII. Biomembranes (Part C: Biological Oxidations)
Edited by Sidney Fleischer and Lester Packer

Volume LIII. Biomembranes (Part D: Biological Oxidations)
Edited by Sidney Fleischer and Lester Packer

Volume LIV. Biomembranes (Part E: Biological Oxidations)
Edited by Sidney Fleischer and Lester Packer

Volume LV. Biomembranes (Part F: Bioenergetics)
Edited by Sidney Fleischer and Lester Packer

Volume LVI. Biomembranes (Part G: Bioenergetics)
Edited by Sidney Fleischer and Lester Packer

Volume LVII. Bioluminescence and Chemiluminescence
Edited by Marlene A. DeLuca

Volume LVIII. Cell Culture
Edited by William B. Jakoby and Ira Pastan

Volume LIX. Nucleic Acids and Protein Synthesis (Part G)
Edited by Kivie Moldave and Lawrence Grossman

Volume LX. Nucleic Acids and Protein Synthesis (Part H)
Edited by Kivie Moldave and Lawrence Grossman

Volume 61. Enzyme Structure (Part H)
Edited by C. H. W. Hirs and Serge N. Timasheff

Volume 62. Vitamins and Coenzymes (Part D)
Edited by Donald B. McCormick and Lemuel D. Wright

Volume 63. Enzyme Kinetics and Mechanism (Part A: Initial Rate and Inhibitor Methods)
Edited by Daniel L. Purich

Volume 64. Enzyme Kinetics and Mechanism
(Part B: Isotopic Probes and Complex Enzyme Systems)
Edited by Daniel L. Purich

Volume 65. Nucleic Acids (Part I)
Edited by Lawrence Grossman and Kivie Moldave

Volume 66. Vitamins and Coenzymes (Part E)
Edited by Donald B. McCormick and Lemuel D. Wright

Volume 67. Vitamins and Coenzymes (Part F)
Edited by Donald B. McCormick and Lemuel D. Wright

Volume 68. Recombinant DNA
Edited by Ray Wu

Volume 69. Photosynthesis and Nitrogen Fixation (Part C)
Edited by Anthony San Pietro

Volume 70. Immunochemical Techniques (Part A)
Edited by Helen Van Vunakis and John J. Langone

VOLUME 71. Lipids (Part C)
Edited by JOHN M. LOWENSTEIN

VOLUME 72. Lipids (Part D)
Edited by JOHN M. LOWENSTEIN

VOLUME 73. Immunochemical Techniques (Part B)
Edited by JOHN J. LANGONE AND HELEN VAN VUNAKIS

VOLUME 74. Immunochemical Techniques (Part C)
Edited by JOHN J. LANGONE AND HELEN VAN VUNAKIS

VOLUME 75. Cumulative Subject Index Volumes XXXI, XXXII, XXXIV–LX
Edited by EDWARD A. DENNIS AND MARTHA G. DENNIS

VOLUME 76. Hemoglobins
Edited by ERALDO ANTONINI, LUIGI ROSSI-BERNARDI, AND EMILIA CHIANCONE

VOLUME 77. Detoxication and Drug Metabolism
Edited by WILLIAM B. JAKOBY

VOLUME 78. Interferons (Part A)
Edited by SIDNEY PESTKA

VOLUME 79. Interferons (Part B)
Edited by SIDNEY PESTKA

VOLUME 80. Proteolytic Enzymes (Part C)
Edited by LASZLO LORAND

VOLUME 81. Biomembranes (Part H: Visual Pigments and Purple Membranes, I)
Edited by LESTER PACKER

VOLUME 82. Structural and Contractile Proteins (Part A: Extracellular Matrix)
Edited by LEON W. CUNNINGHAM AND DIXIE W. FREDERIKSEN

VOLUME 83. Complex Carbohydrates (Part D)
Edited by VICTOR GINSBURG

VOLUME 84. Immunochemical Techniques (Part D: Selected Immunoassays)
Edited by JOHN J. LANGONE AND HELEN VAN VUNAKIS

VOLUME 85. Structural and Contractile Proteins (Part B: The Contractile Apparatus and the Cytoskeleton)
Edited by DIXIE W. FREDERIKSEN AND LEON W. CUNNINGHAM

VOLUME 86. Prostaglandins and Arachidonate Metabolites
Edited by WILLIAM E. M. LANDS AND WILLIAM L. SMITH

VOLUME 87. Enzyme Kinetics and Mechanism (Part C: Intermediates, Stereo-chemistry, and Rate Studies)
Edited by DANIEL L. PURICH

VOLUME 88. Biomembranes (Part I: Visual Pigments and Purple Membranes, II)
Edited by LESTER PACKER

Volume 89. Carbohydrate Metabolism (Part D)
Edited by Willis A. Wood

Volume 90. Carbohydrate Metabolism (Part E)
Edited by Willis A. Wood

Volume 91. Enzyme Structure (Part I)
Edited by C. H. W. Hirs and Serge N. Timasheff

Volume 92. Immunochemical Techniques (Part E: Monoclonal Antibodies and General Immunoassay Methods)
Edited by John J. Langone and Helen Van Vunakis

Volume 93. Immunochemical Techniques (Part F: Conventional Antibodies, Fc Receptors, and Cytotoxicity)
Edited by John J. Langone and Helen Van Vunakis

Volume 94. Polyamines
Edited by Herbert Tabor and Celia White Tabor

Volume 95. Cumulative Subject Index Volumes 61–74, 76–80
Edited by Edward A. Dennis and Martha G. Dennis

Volume 96. Biomembranes [Part J: Membrane Biogenesis: Assembly and Targeting (General Methods; Eukaryotes)]
Edited by Sidney Fleischer and Becca Fleischer

Volume 97. Biomembranes [Part K: Membrane Biogenesis: Assembly and Targeting (Prokaryotes, Mitochondria, and Chloroplasts)]
Edited by Sidney Fleischer and Becca Fleischer

Volume 98. Biomembranes (Part L: Membrane Biogenesis: Processing and Recycling)
Edited by Sidney Fleischer and Becca Fleischer

Volume 99. Hormone Action (Part F: Protein Kinases)
Edited by Jackie D. Corbin and Joel G. Hardman

Volume 100. Recombinant DNA (Part B)
Edited by Ray Wu, Lawrence Grossman, and Kivie Moldave

Volume 101. Recombinant DNA (Part C)
Edited by Ray Wu, Lawrence Grossman, and Kivie Moldave

Volume 102. Hormone Action (Part G: Calmodulin and Calcium-Binding Proteins)
Edited by Anthony R. Means and Bert W. O'Malley

Volume 103. Hormone Action (Part H: Neuroendocrine Peptides)
Edited by P. Michael Conn

Volume 104. Enzyme Purification and Related Techniques (Part C)
Edited by William B. Jakoby

VOLUME 105. Oxygen Radicals in Biological Systems
Edited by LESTER PACKER

VOLUME 106. Posttranslational Modifications (Part A)
Edited by FINN WOLD AND KIVIE MOLDAVE

VOLUME 107. Posttranslational Modifications (Part B)
Edited by FINN WOLD AND KIVIE MOLDAVE

VOLUME 108. Immunochemical Techniques (Part G: Separation and Characterization of Lymphoid Cells)
Edited by GIOVANNI DI SABATO, JOHN J. LANGONE, AND HELEN VAN VUNAKIS

VOLUME 109. Hormone Action (Part I: Peptide Hormones)
Edited by LUTZ BIRNBAUMER AND BERT W. O'MALLEY

VOLUME 110. Steroids and Isoprenoids (Part A)
Edited by JOHN H. LAW AND HANS C. RILLING

VOLUME 111. Steroids and Isoprenoids (Part B)
Edited by JOHN H. LAW AND HANS C. RILLING

VOLUME 112. Drug and Enzyme Targeting (Part A)
Edited by KENNETH J. WIDDER AND RALPH GREEN

VOLUME 113. Glutamate, Glutamine, Glutathione, and Related Compounds
Edited by ALTON MEISTER

VOLUME 114. Diffraction Methods for Biological Macromolecules (Part A)
Edited by HAROLD W. WYCKOFF, C. H. W. HIRS, AND SERGE N. TIMASHEFF

VOLUME 115. Diffraction Methods for Biological Macromolecules (Part B)
Edited by HAROLD W. WYCKOFF, C. H. W. HIRS, AND SERGE N. TIMASHEFF

VOLUME 116. Immunochemical Techniques
(Part H: Effectors and Mediators of Lymphoid Cell Functions)
Edited by GIOVANNI DI SABATO, JOHN J. LANGONE, AND HELEN VAN VUNAKIS

VOLUME 117. Enzyme Structure (Part J)
Edited by C. H. W. HIRS AND SERGE N. TIMASHEFF

VOLUME 118. Plant Molecular Biology
Edited by ARTHUR WEISSBACH AND HERBERT WEISSBACH

VOLUME 119. Interferons (Part C)
Edited by SIDNEY PESTKA

VOLUME 120. Cumulative Subject Index Volumes 81–94, 96–101

VOLUME 121. Immunochemical Techniques (Part I: Hybridoma Technology and Monoclonal Antibodies)
Edited by JOHN J. LANGONE AND HELEN VAN VUNAKIS

VOLUME 122. Vitamins and Coenzymes (Part G)
Edited by FRANK CHYTIL AND DONALD B. MCCORMICK

VOLUME 123. Vitamins and Coenzymes (Part H)
Edited by FRANK CHYTIL AND DONALD B. MCCORMICK

VOLUME 124. Hormone Action (Part J: Neuroendocrine Peptides)
Edited by P. MICHAEL CONN

VOLUME 125. Biomembranes (Part M: Transport in Bacteria, Mitochondria, and Chloroplasts: General Approaches and Transport Systems)
Edited by SIDNEY FLEISCHER AND BECCA FLEISCHER

VOLUME 126. Biomembranes (Part N: Transport in Bacteria, Mitochondria, and Chloroplasts: Protonmotive Force)
Edited by SIDNEY FLEISCHER AND BECCA FLEISCHER

VOLUME 127. Biomembranes (Part O: Protons and Water: Structure and Translocation)
Edited by LESTER PACKER

VOLUME 128. Plasma Lipoproteins (Part A: Preparation, Structure, and Molecular Biology)
Edited by JERE P. SEGREST AND JOHN J. ALBERS

VOLUME 129. Plasma Lipoproteins (Part B: Characterization, Cell Biology, and Metabolism)
Edited by JOHN J. ALBERS AND JERE P. SEGREST

VOLUME 130. Enzyme Structure (Part K)
Edited by C. H. W. HIRS AND SERGE N. TIMASHEFF

VOLUME 131. Enzyme Structure (Part L)
Edited by C. H. W. HIRS AND SERGE N. TIMASHEFF

VOLUME 132. Immunochemical Techniques (Part J: Phagocytosis and Cell-Mediated Cytotoxicity)
Edited by GIOVANNI DI SABATO AND JOHANNES EVERSE

VOLUME 133. Bioluminescence and Chemiluminescence (Part B)
Edited by MARLENE DELUCA AND WILLIAM D. MCELROY

VOLUME 134. Structural and Contractile Proteins (Part C: The Contractile Apparatus and the Cytoskeleton)
Edited by RICHARD B. VALLEE

VOLUME 135. Immobilized Enzymes and Cells (Part B)
Edited by KLAUS MOSBACH

VOLUME 136. Immobilized Enzymes and Cells (Part C)
Edited by KLAUS MOSBACH

VOLUME 137. Immobilized Enzymes and Cells (Part D)
Edited by KLAUS MOSBACH

VOLUME 138. Complex Carbohydrates (Part E)
Edited by VICTOR GINSBURG

Volume 139. Cellular Regulators (Part A: Calcium- and Calmodulin-Binding Proteins)
Edited by Anthony R. Means and P. Michael Conn

Volume 140. Cumulative Subject Index Volumes 102–119, 121–134

Volume 141. Cellular Regulators (Part B: Calcium and Lipids)
Edited by P. Michael Conn and Anthony R. Means

Volume 142. Metabolism of Aromatic Amino Acids and Amines
Edited by Seymour Kaufman

Volume 143. Sulfur and Sulfur Amino Acids
Edited by William B. Jakoby and Owen Griffith

Volume 144. Structural and Contractile Proteins (Part D: Extracellular Matrix)
Edited by Leon W. Cunningham

Volume 145. Structural and Contractile Proteins (Part E: Extracellular Matrix)
Edited by Leon W. Cunningham

Volume 146. Peptide Growth Factors (Part A)
Edited by David Barnes and David A. Sirbasku

Volume 147. Peptide Growth Factors (Part B)
Edited by David Barnes and David A. Sirbasku

Volume 148. Plant Cell Membranes
Edited by Lester Packer and Roland Douce

Volume 149. Drug and Enzyme Targeting (Part B)
Edited by Ralph Green and Kenneth J. Widder

Volume 150. Immunochemical Techniques (Part K: *In Vitro* Models of B and T Cell Functions and Lymphoid Cell Receptors)
Edited by Giovanni Di Sabato

Volume 151. Molecular Genetics of Mammalian Cells
Edited by Michael M. Gottesman

Volume 152. Guide to Molecular Cloning Techniques
Edited by Shelby L. Berger and Alan R. Kimmel

Volume 153. Recombinant DNA (Part D)
Edited by Ray Wu and Lawrence Grossman

Volume 154. Recombinant DNA (Part E)
Edited by Ray Wu and Lawrence Grossman

Volume 155. Recombinant DNA (Part F)
Edited by Ray Wu

Volume 156. Biomembranes (Part P: ATP-Driven Pumps and Related Transport: The Na, K-Pump)
Edited by Sidney Fleischer and Becca Fleischer

Volume 157. Biomembranes (Part Q: ATP-Driven Pumps and Related Transport: Calcium, Proton, and Potassium Pumps)
Edited by Sidney Fleischer and Becca Fleischer

Volume 158. Metalloproteins (Part A)
Edited by James F. Riordan and Bert L. Vallee

Volume 159. Initiation and Termination of Cyclic Nucleotide Action
Edited by Jackie D. Corbin and Roger A. Johnson

Volume 160. Biomass (Part A: Cellulose and Hemicellulose)
Edited by Willis A. Wood and Scott T. Kellogg

Volume 161. Biomass (Part B: Lignin, Pectin, and Chitin)
Edited by Willis A. Wood and Scott T. Kellogg

Volume 162. Immunochemical Techniques (Part L: Chemotaxis and Inflammation)
Edited by Giovanni Di Sabato

Volume 163. Immunochemical Techniques (Part M: Chemotaxis and Inflammation)
Edited by Giovanni Di Sabato

Volume 164. Ribosomes
Edited by Harry F. Noller, Jr., and Kivie Moldave

Volume 165. Microbial Toxins: Tools for Enzymology
Edited by Sidney Harshman

Volume 166. Branched-Chain Amino Acids
Edited by Robert Harris and John R. Sokatch

Volume 167. Cyanobacteria
Edited by Lester Packer and Alexander N. Glazer

Volume 168. Hormone Action (Part K: Neuroendocrine Peptides)
Edited by P. Michael Conn

Volume 169. Platelets: Receptors, Adhesion, Secretion (Part A)
Edited by Jacek Hawiger

Volume 170. Nucleosomes
Edited by Paul M. Wassarman and Roger D. Kornberg

Volume 171. Biomembranes (Part R: Transport Theory: Cells and Model Membranes)
Edited by Sidney Fleischer and Becca Fleischer

Volume 172. Biomembranes (Part S: Transport: Membrane Isolation and Characterization)
Edited by Sidney Fleischer and Becca Fleischer

VOLUME 173. Biomembranes [Part T: Cellular and Subcellular Transport: Eukaryotic (Nonepithelial) Cells]
Edited by SIDNEY FLEISCHER AND BECCA FLEISCHER

VOLUME 174. Biomembranes [Part U: Cellular and Subcellular Transport: Eukaryotic (Nonepithelial) Cells]
Edited by SIDNEY FLEISCHER AND BECCA FLEISCHER

VOLUME 175. Cumulative Subject Index Volumes 135–139, 141–167

VOLUME 176. Nuclear Magnetic Resonance (Part A: Spectral Techniques and Dynamics)
Edited by NORMAN J. OPPENHEIMER AND THOMAS L. JAMES

VOLUME 177. Nuclear Magnetic Resonance (Part B: Structure and Mechanism)
Edited by NORMAN J. OPPENHEIMER AND THOMAS L. JAMES

VOLUME 178. Antibodies, Antigens, and Molecular Mimicry
Edited by JOHN J. LANGONE

VOLUME 179. Complex Carbohydrates (Part F)
Edited by VICTOR GINSBURG

VOLUME 180. RNA Processing (Part A: General Methods)
Edited by JAMES E. DAHLBERG AND JOHN N. ABELSON

VOLUME 181. RNA Processing (Part B: Specific Methods)
Edited by JAMES E. DAHLBERG AND JOHN N. ABELSON

VOLUME 182. Guide to Protein Purification
Edited by MURRAY P. DEUTSCHER

VOLUME 183. Molecular Evolution: Computer Analysis of Protein and Nucleic Acid Sequences
Edited by RUSSELL F. DOOLITTLE

VOLUME 184. Avidin-Biotin Technology
Edited by MEIR WILCHEK AND EDWARD A. BAYER

VOLUME 185. Gene Expression Technology
Edited by DAVID V. GOEDDEL

VOLUME 186. Oxygen Radicals in Biological Systems (Part B: Oxygen Radicals and Antioxidants)
Edited by LESTER PACKER AND ALEXANDER N. GLAZER

VOLUME 187. Arachidonate Related Lipid Mediators
Edited by ROBERT C. MURPHY AND FRANK A. FITZPATRICK

VOLUME 188. Hydrocarbons and Methylotrophy
Edited by MARY E. LIDSTROM

VOLUME 189. Retinoids (Part A: Molecular and Metabolic Aspects)
Edited by LESTER PACKER

Volume 190. Retinoids (Part B: Cell Differentiation and Clinical Applications)
Edited by Lester Packer

Volume 191. Biomembranes (Part V: Cellular and Subcellular Transport: Epithelial Cells)
Edited by Sidney Fleischer and Becca Fleischer

Volume 192. Biomembranes (Part W: Cellular and Subcellular Transport: Epithelial Cells)
Edited by Sidney Fleischer and Becca Fleischer

Volume 193. Mass Spectrometry
Edited by James A. McCloskey

Volume 194. Guide to Yeast Genetics and Molecular Biology
Edited by Christine Guthrie and Gerald R. Fink

Volume 195. Adenylyl Cyclase, G Proteins, and Guanylyl Cyclase
Edited by Roger A. Johnson and Jackie D. Corbin

Volume 196. Molecular Motors and the Cytoskeleton
Edited by Richard B. Vallee

Volume 197. Phospholipases
Edited by Edward A. Dennis

Volume 198. Peptide Growth Factors (Part C)
Edited by David Barnes, J. P. Mather, and Gordon H. Sato

Volume 199. Cumulative Subject Index Volumes 168–174, 176–194

Volume 200. Protein Phosphorylation (Part A: Protein Kinases: Assays, Purification, Antibodies, Functional Analysis, Cloning, and Expression)
Edited by Tony Hunter and Bartholomew M. Sefton

Volume 201. Protein Phosphorylation (Part B: Analysis of Protein Phosphorylation, Protein Kinase Inhibitors, and Protein Phosphatases)
Edited by Tony Hunter and Bartholomew M. Sefton

Volume 202. Molecular Design and Modeling: Concepts and Applications (Part A: Proteins, Peptides, and Enzymes)
Edited by John J. Langone

Volume 203. Molecular Design and Modeling: Concepts and Applications (Part B: Antibodies and Antigens, Nucleic Acids, Polysaccharides, and Drugs)
Edited by John J. Langone

Volume 204. Bacterial Genetic Systems
Edited by Jeffrey H. Miller

Volume 205. Metallobiochemistry (Part B: Metallothionein and Related Molecules)
Edited by James F. Riordan and Bert L. Vallee

Volume 223. Proteolytic Enzymes in Coagulation, Fibrinolysis, and Complement Activation (Part B: Complement Activation, Fibrinolysis, and Nonmammalian Blood Coagulation Factors)
Edited by Laszlo Lorand and Kenneth G. Mann

Volume 224. Molecular Evolution: Producing the Biochemical Data
Edited by Elizabeth Anne Zimmer, Thomas J. White, Rebecca L. Cann, and Allan C. Wilson

Volume 225. Guide to Techniques in Mouse Development
Edited by Paul M. Wassarman and Melvin L. DePamphilis

Volume 226. Metallobiochemistry (Part C: Spectroscopic and Physical Methods for Probing Metal Ion Environments in Metalloenzymes and Metalloproteins)
Edited by James F. Riordan and Bert L. Vallee

Volume 227. Metallobiochemistry (Part D: Physical and Spectroscopic Methods for Probing Metal Ion Environments in Metalloproteins)
Edited by James F. Riordan and Bert L. Vallee

Volume 228. Aqueous Two-Phase Systems
Edited by Harry Walter and Göte Johansson

Volume 229. Cumulative Subject Index Volumes 195–198, 200–227

Volume 230. Guide to Techniques in Glycobiology
Edited by William J. Lennarz and Gerald W. Hart

Volume 231. Hemoglobins (Part B: Biochemical and Analytical Methods)
Edited by Johannes Everse, Kim D. Vandegriff, and Robert M. Winslow

Volume 232. Hemoglobins (Part C: Biophysical Methods)
Edited by Johannes Everse, Kim D. Vandegriff, and Robert M. Winslow

Volume 233. Oxygen Radicals in Biological Systems (Part C)
Edited by Lester Packer

Volume 234. Oxygen Radicals in Biological Systems (Part D)
Edited by Lester Packer

Volume 235. Bacterial Pathogenesis (Part A: Identification and Regulation of Virulence Factors)
Edited by Virginia L. Clark and Patrik M. Bavoil

Volume 236. Bacterial Pathogenesis (Part B: Integration of Pathogenic Bacteria with Host Cells)
Edited by Virginia L. Clark and Patrik M. Bavoil

Volume 237. Heterotrimeric G Proteins
Edited by Ravi Iyengar

Volume 238. Heterotrimeric G-Protein Effectors
Edited by Ravi Iyengar

VOLUME 256. Small GTPases and Their Regulators (Part B: Rho Family)
Edited by W. E. BALCH, CHANNING J. DER, AND ALAN HALL

VOLUME 257. Small GTPases and Their Regulators (Part C: Proteins Involved in Transport)
Edited by W. E. BALCH, CHANNING J. DER, AND ALAN HALL

VOLUME 258. Redox-Active Amino Acids in Biology
Edited by JUDITH P. KLINMAN

VOLUME 259. Energetics of Biological Macromolecules
Edited by MICHAEL L. JOHNSON AND GARY K. ACKERS

VOLUME 260. Mitochondrial Biogenesis and Genetics (Part A)
Edited by GIUSEPPE M. ATTARDI AND ANNE CHOMYN

VOLUME 261. Nuclear Magnetic Resonance and Nucleic Acids
Edited by THOMAS L. JAMES

VOLUME 262. DNA Replication
Edited by JUDITH L. CAMPBELL

VOLUME 263. Plasma Lipoproteins (Part C: Quantitation)
Edited by WILLIAM A. BRADLEY, SANDRA H. GIANTURCO, AND JERE P. SEGREST

VOLUME 264. Mitochondrial Biogenesis and Genetics (Part B)
Edited by GIUSEPPE M. ATTARDI AND ANNE CHOMYN

VOLUME 265. Cumulative Subject Index Volumes 228, 230–262

VOLUME 266. Computer Methods for Macromolecular Sequence Analysis
Edited by RUSSELL F. DOOLITTLE

VOLUME 267. Combinatorial Chemistry
Edited by JOHN N. ABELSON

VOLUME 268. Nitric Oxide (Part A: Sources and Detection of NO; NO Synthase)
Edited by LESTER PACKER

VOLUME 269. Nitric Oxide (Part B: Physiological and Pathological Processes)
Edited by LESTER PACKER

VOLUME 270. High Resolution Separation and Analysis of Biological Macromolecules (Part A: Fundamentals)
Edited by BARRY L. KARGER AND WILLIAM S. HANCOCK

VOLUME 271. High Resolution Separation and Analysis of Biological Macromolecules (Part B: Applications)
Edited by BARRY L. KARGER AND WILLIAM S. HANCOCK

VOLUME 272. Cytochrome P450 (Part B)
Edited by ERIC F. JOHNSON AND MICHAEL R. WATERMAN

VOLUME 273. RNA Polymerase and Associated Factors (Part A)
Edited by SANKAR ADHYA

Volume 328. Applications of Chimeric Genes and Hybrid Proteins (Part C: Protein–Protein Interactions and Genomics)
Edited by Jeremy Thorner, Scott D. Emr, and John N. Abelson

Volume 329. Regulators and Effectors of Small GTPases (Part E: GTPases Involved in Vesicular Traffic)
Edited by W. E. Balch, Channing J. Der, and Alan Hall

Volume 330. Hyperthermophilic Enzymes (Part A)
Edited by Michael W. W. Adams and Robert M. Kelly

Volume 331. Hyperthermophilic Enzymes (Part B)
Edited by Michael W. W. Adams and Robert M. Kelly

Volume 332. Regulators and Effectors of Small GTPases (Part F: Ras Family I)
Edited by W. E. Balch, Channing J. Der, and Alan Hall

Volume 333. Regulators and Effectors of Small GTPases (Part G: Ras Family II)
Edited by W. E. Balch, Channing J. Der, and Alan Hall

Volume 334. Hyperthermophilic Enzymes (Part C)
Edited by Michael W. W. Adams and Robert M. Kelly

Volume 335. Flavonoids and Other Polyphenols
Edited by Lester Packer

Volume 336. Microbial Growth in Biofilms (Part A: Developmental and Molecular Biological Aspects)
Edited by Ron J. Doyle

Volume 337. Microbial Growth in Biofilms (Part B: Special Environments and Physicochemical Aspects)
Edited by Ron J. Doyle

Volume 338. Nuclear Magnetic Resonance of Biological Macromolecules (Part A)
Edited by Thomas L. James, Volker Dötsch, and Uli Schmitz

Volume 339. Nuclear Magnetic Resonance of Biological Macromolecules (Part B)
Edited by Thomas L. James, Volker Dötsch, and Uli Schmitz

Volume 340. Drug–Nucleic Acid Interactions
Edited by Jonathan B. Chaires and Michael J. Waring

Volume 341. Ribonucleases (Part A)
Edited by Allen W. Nicholson

Volume 342. Ribonucleases (Part B)
Edited by Allen W. Nicholson

Volume 343. G Protein Pathways (Part A: Receptors)
Edited by Ravi Iyengar and John D. Hildebrandt

Volume 344. G Protein Pathways (Part B: G Proteins and Their Regulators)
Edited by Ravi Iyengar and John D. Hildebrandt

Volume 345. G Protein Pathways (Part C: Effector Mechanisms)
Edited by Ravi Iyengar and John D. Hildebrandt

Volume 346. Gene Therapy Methods
Edited by M. Ian Phillips

Volume 347. Protein Sensors and Reactive Oxygen Species (Part A: Selenoproteins and Thioredoxin)
Edited by Helmut Sies and Lester Packer

Volume 348. Protein Sensors and Reactive Oxygen Species (Part B: Thiol Enzymes and Proteins)
Edited by Helmut Sies and Lester Packer

Volume 349. Superoxide Dismutase
Edited by Lester Packer

Volume 350. Guide to Yeast Genetics and Molecular and Cell Biology (Part B)
Edited by Christine Guthrie and Gerald R. Fink

Volume 351. Guide to Yeast Genetics and Molecular and Cell Biology (Part C)
Edited by Christine Guthrie and Gerald R. Fink

Volume 352. Redox Cell Biology and Genetics (Part A)
Edited by Chandan K. Sen and Lester Packer

Volume 353. Redox Cell Biology and Genetics (Part B)
Edited by Chandan K. Sen and Lester Packer

Volume 354. Enzyme Kinetics and Mechanisms (Part F: Detection and Characterization of Enzyme Reaction Intermediates)
Edited by Daniel L. Purich

Volume 355. Cumulative Subject Index Volumes 321–354

Volume 356. Laser Capture Microscopy and Microdissection
Edited by P. Michael Conn

Volume 357. Cytochrome P450, Part C
Edited by Eric F. Johnson and Michael R. Waterman

Volume 358. Bacterial Pathogenesis (Part C: Identification, Regulation, and Function of Virulence Factors)
Edited by Virginia L. Clark and Patrik M. Bavoil

Volume 359. Nitric Oxide (Part D)
Edited by Enrique Cadenas and Lester Packer

Volume 360. Biophotonics (Part A)
Edited by Gerard Marriott and Ian Parker

Volume 361. Biophotonics (Part B)
Edited by Gerard Marriott and Ian Parker

Volume 362. Recognition of Carbohydrates in Biological Systems (Part A)
Edited by Yuan C. Lee and Reiko T. Lee

Volume 363. Recognition of Carbohydrates in Biological Systems (Part B)
Edited by Yuan C. Lee and Reiko T. Lee

Volume 364. Nuclear Receptors
Edited by David W. Russell and David J. Mangelsdorf

Volume 365. Differentiation of Embryonic Stem Cells
Edited by Paul M. Wassauman and Gordon M. Keller

Volume 366. Protein Phosphatases
Edited by Susanne Klumpp and Josef Krieglstein

Volume 367. Liposomes (Part A)
Edited by Nejat Düzgüneş

Volume 368. Macromolecular Crystallography (Part C)
Edited by Charles W. Carter, Jr., and Robert M. Sweet

Volume 369. Combinational Chemistry (Part B)
Edited by Guillermo A. Morales and Barry A. Bunin

Volume 370. RNA Polymerases and Associated Factors (Part C)
Edited by Sankar L. Adhya and Susan Garges

Volume 371. RNA Polymerases and Associated Factors (Part D)
Edited by Sankar L. Adhya and Susan Garges

Volume 372. Liposomes (Part B)
Edited by Nejat Düzgüneş

Volume 373. Liposomes (Part C)
Edited by Nejat Düzgüneş

Volume 374. Macromolecular Crystallography (Part D)
Edited by Charles W. Carter, Jr., and Robert W. Sweet

Volume 375. Chromatin and Chromatin Remodeling Enzymes (Part A)
Edited by C. David Allis and Carl Wu

Volume 376. Chromatin and Chromatin Remodeling Enzymes (Part B)
Edited by C. David Allis and Carl Wu

Volume 377. Chromatin and Chromatin Remodeling Enzymes (Part C)
Edited by C. David Allis and Carl Wu

Volume 378. Quinones and Quinone Enzymes (Part A)
Edited by Helmut Sies and Lester Packer

Volume 379. Energetics of Biological Macromolecules (Part D)
Edited by Jo M. Holt, Michael L. Johnson, and Gary K. Ackers

Volume 380. Energetics of Biological Macromolecules (Part E)
Edited by Jo M. Holt, Michael L. Johnson, and Gary K. Ackers

VOLUME 399. Ubiquitin and Protein Degradation (Part B)
Edited by RAYMOND J. DESHAIES

VOLUME 400. Phase II Conjugation Enzymes and Transport Systems
Edited by HELMUT SIES AND LESTER PACKER

VOLUME 401. Glutathione Transferases and Gamma Glutamyl Transpeptidases
Edited by HELMUT SIES AND LESTER PACKER

VOLUME 402. Biological Mass Spectrometry
Edited by A. L. BURLINGAME

VOLUME 403. GTPases Regulating Membrane Targeting and Fusion
Edited by WILLIAM E. BALCH, CHANNING J. DER, AND ALAN HALL

VOLUME 404. GTPases Regulating Membrane Dynamics
Edited by WILLIAM E. BALCH, CHANNING J. DER, AND ALAN HALL

VOLUME 405. Mass Spectrometry: Modified Proteins and Glycoconjugates
Edited by A. L. BURLINGAME

VOLUME 406. Regulators and Effectors of Small GTPases: Rho Family
Edited by WILLIAM E. BALCH, CHANNING J. DER, AND ALAN HALL

VOLUME 407. Regulators and Effectors of Small GTPases: Ras Family
Edited by WILLIAM E. BALCH, CHANNING J. DER, AND ALAN HALL

VOLUME 408. DNA Repair (Part A)
Edited by JUDITH L. CAMPBELL AND PAUL MODRICH

VOLUME 409. DNA Repair (Part B)
Edited by JUDITH L. CAMPBELL AND PAUL MODRICH

VOLUME 410. DNA Microarrays (Part A: Array Platforms and Web-Bench Protocols)
Edited by ALAN KIMMEL AND BRIAN OLIVER

VOLUME 411. DNA Microarrays (Part B: Databases and Statistics)
Edited by ALAN KIMMEL AND BRIAN OLIVER

VOLUME 412. Amyloid, Prions, and Other Protein Aggregates (Part B)
Edited by INDU KHETERPAL AND RONALD WETZEL

VOLUME 413. Amyloid, Prions, and Other Protein Aggregates (Part C)
Edited by INDU KHETERPAL AND RONALD WETZEL

VOLUME 414. Measuring Biological Responses with Automated Microscopy
Edited by JAMES INGLESE

VOLUME 415. Glycobiology
Edited by MINORU FUKUDA

VOLUME 416. Glycomics
Edited by MINORU FUKUDA

VOLUME 417. Functional Glycomics
Edited by MINORU FUKUDA

VOLUME 418. Embryonic Stem Cells
Edited by IRINA KLIMANSKAYA AND ROBERT LANZA

VOLUME 419. Adult Stem Cells
Edited by IRINA KLIMANSKAYA AND ROBERT LANZA

VOLUME 420. Stem Cell Tools and Other Experimental Protocols
Edited by IRINA KLIMANSKAYA AND ROBERT LANZA

VOLUME 421. Advanced Bacterial Genetics: Use of Transposons and Phage for Genomic Engineering
Edited by KELLY T. HUGHES

VOLUME 422. Two-Component Signaling Systems, Part A
Edited by MELVIN I. SIMON, BRIAN R. CRANE, AND ALEXANDRINE CRANE

VOLUME 423. Two-Component Signaling Systems, Part B
Edited by MELVIN I. SIMON, BRIAN R. CRANE, AND ALEXANDRINE CRANE

VOLUME 424. RNA Editing
Edited by JONATHA M. GOTT

VOLUME 425. RNA Modification
Edited by JONATHA M. GOTT

VOLUME 426. Integrins
Edited by DAVID CHERESH

VOLUME 427. MicroRNA Methods
Edited by JOHN J. ROSSI

VOLUME 428. Osmosensing and Osmosignaling
Edited by HELMUT SIES AND DIETER HAUSSINGER

VOLUME 429. Translation Initiation: Extract Systems and Molecular Genetics
Edited by JON LORSCH

VOLUME 430. Translation Initiation: Reconstituted Systems and Biophysical Methods
Edited by JON LORSCH

VOLUME 431. Translation Initiation: Cell Biology, High-Throughput and Chemical-Based Approaches
Edited by JON LORSCH

VOLUME 432. Lipidomics and Bioactive Lipids: Mass-Spectrometry–Based Lipid Analysis
Edited by H. ALEX BROWN

SECTION ONE

CHAPTER ONE

Systems Biology Left and Right

Hans V. Westerhoff

Contents

Abstract

Systems biology has come of age. In most scientifically active countries, significant research programs are funded. Various scientific journals, standards, repositories, and Web sites are devoted to the topic. Systems biology has spun off new subdisciplines such as synthetic biology and systems medicine. There are training courses at the M.Sc. and Ph.D. level at various Universities. And various industries are engaging systems biology in their R&D. Systems biology has also developed numerous new methodologies. This chapter attempts to organize these methodologies from the perspectives of the unique aims of systems biology, and by comparing with one of its parents, molecular biology.

1. A Fundamental Definition of Systems Biology

Definitions of systems biology abound. This may explain why some see systems biology as vague, others see it as nothing new, while yet others see it as hype. Some vehemently disagree with any definition that has been published, or refuse to take note of the literature that has been explicit on defining it (Alberghina and Westerhoff, 2005; Boogerd *et al.*, 2007; Klipp *et al.*, 2005). The discipline of molecular biology has passed through a similar history, now, many years ago. First, tenets of molecular biology were deemed unimportant, then they were accepted to be potentially

Manchester Centre for Integrative Systems Biology, The University of Manchester, Manchester, United Kingdom
Netherlands Institute for Systems Biology, VU University Amsterdam, Amsterdam, The Netherlands

Methods in Enzymology, Volume 500
ISSN 0076-6879, DOI: 10.1016/B978-0-12-385118-5.00001-3

important but wrong, and then they were understood to be right but judged to be nothing new. Ultimately, molecular biology changed the way both biology and medicine were practiced. It was a paradigm shift. And it did bring a scientific revolution (Kuhn, 1962), more recently leading to functional genomics.

Systems biology is another such new discipline (Westerhoff *et al.*, 2009). It is useful to demarcate what it is, because that will explain why much of what is present in this book on methods in systems biology is similar to what could be done in any of the other sciences and yet is different because of a different perspective.

Systems biology is not just an item on the list of biomedical disciplines (Fig. 1.1). It is much more the in-between. Where molecular biology deals with the structure of macromolecules, and biochemistry with the chemical conversions in biology, systems biology attempts to understand how in the interactions between components new properties arise that give the pathways their functional properties (Kolodkin *et al.*, 2010; Westerhoff *et al.*, 1984). While cell biology looks at the compartments in living cells and describes how these function and how molecules are moving within and between them, systems biology attempts to understand how the functions of

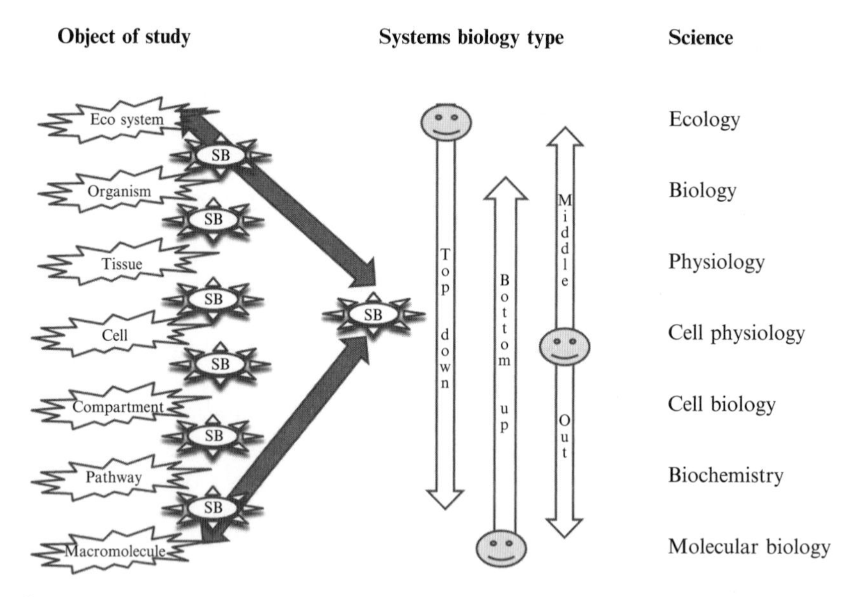

Figure 1.1 The position of systems biology and its different forms relative to its topics of interest and the underpinning other sciences. Systems biology is the in-between: it examines how nonlinear interactions between components lead to functional properties of a system that are not present in the components themselves. It does this either downward, from system to components, or upward, that is, from components to system.

compartments come about in the nonlinear interactions between their pathways and their macromolecules (Bakker *et al.*, 2000; Haanstra *et al.*, 2008; Westerhoff *et al.*, 1981). And where biology looks at the functioning of an organism such as a rabbit, systems biology tries to understand how much of that function comes about in the interactions between its organs/tissues.

In short, systems biology focuses on the emergence of biological function from interactions. Since systems biology is a science, it aims at understanding, not only in descriptive but also in mechanistic terms (Alberghina and Westerhoff, 2005). For interactions to lead to properties that are not yet present in the components themselves, they need to be nonlinear, in any out of a number of ways. The nonlinearity can reside in nonlinear kinetic equations such as those that give rise to oscillations (Reijenga *et al.*, 2005; Tyson *et al.*, 2001), in vectorial reactions such as the ones that lead to ion pumping (Mitchell, 1961; Westerhoff and Van Dam, 1987), in catalytic hierarchies giving rise to amplification in regulation (Bruggeman *et al.*, 2005; Hellingwerf *et al.*, 1995; Kholodenko, 2006; Koshland *et al.*, 1982; Rhee *et al.*, 1989), in network motifs such as the ones that give rise to robust detection (Alon, 2007), and in differences in concentrations of conserved moieties giving rise to temporal organization (Hardin *et al.*, 2009; Hofmeyr *et al.*, 1986).

As a consequence, systems biology depends on the understanding of nonlinear interactions. This has three implications: First, systems biology cannot reduce its objects of study to simplicity: at least part of their complexity is essential. Indeed, systems biology may aim at making its analyses as simple as possible, but not simpler (Einstein, 1934; Westerhoff *et al.*, 2009). Second, systems biology requires mathematics of some sort in order to be able to deal with the necessary complexity. And third, systems biology requires "detail" and hence accurate experimentation *in vivo*, or under *in vivo* conditions (*ex vivo*).

2. The Importance of the Integration of *In Vivo* and *In Vitro* Analyses

The reason for the requirement of *in vivo* conditions is the phenomenon that the nonlinearity is diverse. First, if one is interested in how a function f depends on a molecular property x, then that function may be nonlinear in x itself, such as in

$$f(x) = \alpha x + \beta x^2$$

This implies that the dependence of the function on the molecular property x (which may be quantified as $\mathrm{d}f/\mathrm{d}x$) depends on the magnitude of x, that is,

$$\frac{df}{dx} = \alpha + 2\beta x$$

Therefore, one needs to determine the value of the molecular property (x) experimentally before one analyzes the functional role x plays in the system.

The function may also depend on other molecules, such as y. And therefore one also has to determine the concentration of y to be able to understand the function of y. The situation is more complex, however, for the function of x is likely to depend on the magnitude of the molecule y, that is,

$$f(x) = \alpha x + \beta x^2 + \gamma y + \delta y^2 + \varepsilon xy$$

so that

$$\frac{\mathrm{d}f}{\mathrm{d}x} = \alpha + 2\beta x + \varepsilon y$$

Consequently, in any assay that aims to determine the dependence of function on a particular molecule, one needs to have all other molecules that affect this dependence present at their actual concentrations.

It may seem that this pleads for doing all further experiments exclusively *in vivo*. Here, there is an issue with validation, however, of the understanding that would be achieved. Suppose one measures *in vivo* the parameters α, β, ε, the actual magnitudes of x and y, as well as $\mathrm{d}f/\mathrm{d}x$, and that the above equation is found to hold. This would suggest that the system is understood. Yet, it would be possible (and in view of biological complexity, likely) that there is a hidden dependence on yet another molecule, z, for instance, such that

$$\alpha = \alpha(z) = \alpha_0 z$$

with z *in vivo* equalling α/α_0. Then we would not have understood that z determines the role of x; hence, we would not have understood the system completely. Moreover, if for some reason z would vary between conditions (e.g., because of a mutation) for which x and y would be the same, then one would find the determination of the function of x ($\mathrm{d}f/\mathrm{d}x$) to be irreproducible. How reminiscent of biological practice this is!

If one were to determine all parameter values *in vitro*, one would find absence of correspondence between the predictions based on the determination of the parameter values *in vitro* and the function *in vivo*. This would then lead to the discovery of the role of z and, ultimately, to complete understanding (Snoep *et al.*, 2006).

The issue we raise here is profound and has wide implications. Intracellular networks are strongly connected, to the extent that (almost) any molecule is connected with any other molecule (Jeong *et al.*, 2000). Consequently, *in vitro* assays would have to consider the roles of all other gene products in determining the function of any single one of them. For an organism of 25,000 genes, this would require more than 625 million accurate studies. Luckily, there are arguments that suggest the scaling is linear (Westerhoff *et al.*, 2010), such that the number may be closer to 50,000. And, more strategic thinking will reduce the amount of work needed (Westerhoff *et al.*, 2010). Recently, a tip of this iceberg was lifted when an "*in vivo*-like" medium was designed for the *in vitro* determination of enzyme kinetic parameters: various components of this medium that are usually thought of as "neutral" had significant effects on these parameters (van Eunen *et al.*, 2010).

3. Alternative Definitions of Systems Biology

The above may serve to appreciate why there are multiple definitions of systems biology and may provide place for these definitions in perspective. Systems biology is often defined as the application of mathematical modeling to biology. This definition accords with the fact that systems biology requires mathematical modeling because it depends on precision, and it requires biology because it must deal with actual function. This definition has some disadvantages, however. First, it is secondary rather than primary, that is, it can be deduced from the more fundamental definition given above. Second, there are many fields outside systems biology that would fall under this definition, including the extensive field of structural biology. X-ray or NMR data are input to extensive computations that then lead to the understanding of structures of proteins. Also enzyme kinetics combines computation with experimentation and biology.

Another definition of systems biology is the utilization of genome-wide datasets in biology. This definition also derives from our more fundamental definition because the role that every type of molecule plays in function may depend on all other molecules in the system. Hence all analyses should be genome-wide or at least be wary of the entire genome.

A fourth definition of systems biology is "the interface between biology and medicine on the one hand and physics, chemistry, and mathematics on the other hand." Although systems biology has to be at this interface due to its fundamental definition, there are many other activities at this interface that do not study emergence of function from interactions. Genome sequencing, for instance, requires analytical chemistry, molecular biology, and computation but is not systems biology.

4. Different Types of Systems Biology

Figure 1.1 shows that there are different types of systems biology. These different types can be distinguished in multiple ways. One distinction is whether they work in an upward or a downward direction. The former is called bottom-up systems biology. It considers components and their interactions and then tries to understand how new properties emerge when the components are allowed to interact (Novak *et al.*, 1998; Teusink *et al.*, 1998). The latter, top-down systems biology looks for patterns of behavior in the whole and tries to find empirical laws for correlations and ultimately cause–effect relations (Lauffenburger, 2000). Top-down and bottom-up systems biology can begin and end at any points in biological organization (see Fig. 1.1): they may try to explain ecological phenomena on the basis of interactions of the organisms within an ecosystem (e.g., Getz *et al.*, 2003; Roling *et al.*, 2007), or the production of ATP or DNA structure at the biochemical level from detailed molecular biology (e.g., Westerhoff *et al.*, 1981, 1988).

Top-down systems biology is not the same as genome-wide molecular biology. In Fig. 1.2, we illustrate this further. One may look at biology from different perspectives, such as the molecular perspective, the chemical perspective, or the structure perspective. The molecular perspective tries to understand how interactions of molecules lead to the dynamic behavior of all molecules together. This is only one aspect of function however. A different aspect is that of Physics, where one wonders how the forces that reign between components lead to forces or displacement at high levels of organization. From the chemical perspective, one may see detailed biochemical reactions such as the anabolic chemical pathways that lead to the synthesis of new individuals of the organism. And then there is the genetic perspective, which connects molecular genetics to Mendelian genetics and ultimately to population genetics.

There is yet another perspective on systems biology, that is, the historical one. Part of it has emerged from physical chemistry and mathematical biology, and part from molecular biology through genomics (Westerhoff and Palsson, 2004).

We have drawn up Fig. 1.2 because we think that at present, these different perspectives operate like silos in which different types of systems biology are being professed. We note, however, that there is much excitement in trying to connect the silos, for instance, when asking how the structure and composition of the membranes of mammalian cells depend on the activity of multiple chemical and physical processes (e.g., Maeder *et al.*, 2007). We call the systems biology that attempts to integrate such different perspectives left–right systems biology. In this sense, there are

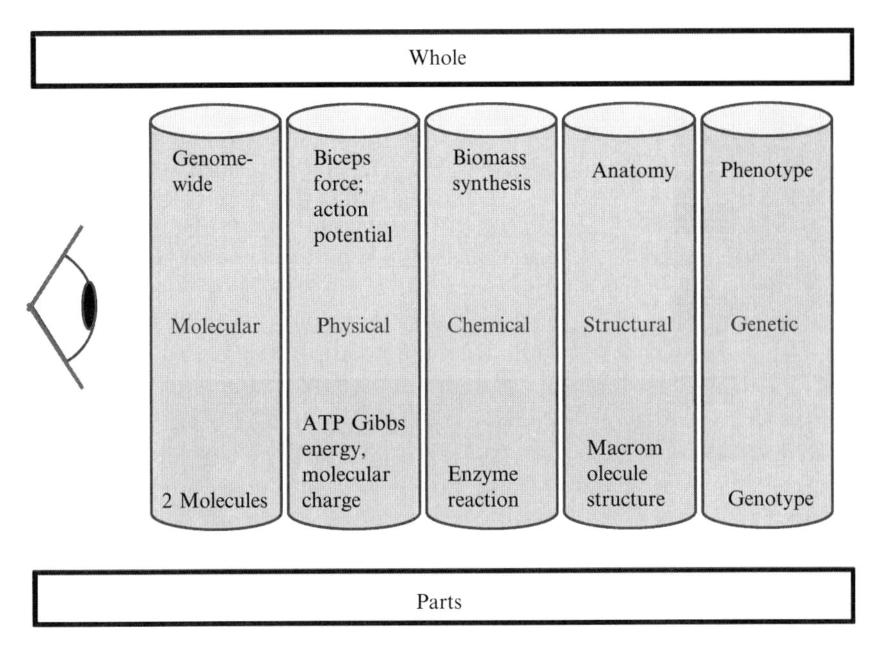

Figure 1.2 Left–right consideration of systems biology. The various perspectives of systems biology may all work downward (top-down) from more intact systems to interacting components, or upward (bottom-up). One of the upcoming challenges for systems biology is not only to integrate top-down or bottom-up but also left–right, that is, for example, to integrate the molecular with the physical perspective, or the chemical with the structural perspective.

different types of systems biology, therefore. In another, there may be only one, as all perspectives described in Fig. 1.2 are needed to understand the functioning of living organisms and the role of each perspective therein.

ACKNOWLEDGMENTS

This work reflects discussions with many of my colleagues, too many to mention: most of what I wrote here I owe to the insights of others and to funding by various organizations—STW, the NGI-Kluyver Centre, NWO-SysMo, BBSRC-MCISB, SysMO, ERASysBio and BRIC, EPSRC, AstraZeneca—and by the EU grants BioSim, NucSys (and extensions), ECMOAN, and UniCellSys.

REFERENCES

Alberghina, L., and Westerhoff, H. V. (eds.), (2005). Systems Biology: Definitions and Perspectives (Topics in Current Genetics), Springer-Verlag/GmbH & Co. K, Berlin/Heidelberg.

Alon, U. (2007). Network motifs: Theory and experimental approaches. *Nat. Rev. Genet.* **8,** 450–461.

Bakker, B. M., Mensonides, F. I. C., Teusink, B., van Hoek, P., Michels, P. A. M., and Westerhoff, H. V. (2000). Compartmentation protects trypanosomes from the dangerous design of glycolysis. *Proc. Natl. Acad. Sci. USA* **97,** 2087–2092.

Boogerd, F. C., Bruggeman, F. J., Hofmeyr, J. H. S., and Westerhoff, H. V. (eds.), (2007). Systems Biology: Philosophical Foundations, Elsevier Science, Amsterdam.

Bruggeman, F. J., Boogerd, F. C., and Westerhoff, H. V. (2005). The multifarious short-term regulation of ammonium assimilation of *Escherichia coli*: Dissection using an in silico replica. *FEBS J.* **272,** 1965–1985.

Einstein, A. (1934). On the method of theoretical physics. *Philos. Sci.* **1,** 163–169.

Getz, W. M., Westerhoff, H. V., Hofmeyr, J. H. S., and Snoep, J. L. (2003). Control analysis of trophic chains. *Ecol. Modell.* **168,** 153–171.

Haanstra, J. R., van Tuijl, A., Kessler, P., Reijnders, W., Michels, P. A. M., Westerhoff, H. V., Parsons, M., and Bakker, B. M. (2008). Compartmentation prevents a lethal turbo-explosion of glycolysis in trypanosomes. *Proc. Natl. Acad. Sci. USA* **105,** 17718–17723.

Hardin, H. M., Zagaris, A., Krab, K., and Westerhoff, H. V. (2009). Simplified yet highly accurate enzyme kinetics for cases of low substrate concentrations. *FEBS J.* **276,** 5491–5506.

Hellingwerf, K. J., Postma, P. W., Tommassen, J., and Westerhoff, H. V. (1995). Signal transduction in bacteria: Phospho-neural network(s) in Escherichia coli? *FEMS Microbiol. Rev.* **16,** 309–321.

Hofmeyr, J. H., Kacser, H., and van der Merwe, K. J. (1986). Metabolic control analysis of moiety-conserved cycles. *Eur. J. Biochem.* **155,** 631–641.

Jeong, H., Tombor, B., Albert, R., Oltvai, Z. N., and Barabasi, A. L. (2000). The large-scale organization of metabolic networks. *Nature* **407,** 651–654.

Kholodenko, B. N. (2006). Cell-signalling dynamics in time and space. *Nat. Rev. Mol. Cell Biol.* **7,** 165–176.

Klipp, E., Herwig, R., Kowald, A., Wierling, C., and Lehrach, H. (2005). *Systems Biology in Practice: Concepts, Implementation, and Application.* Wiley-VCH. Verlag GmbH & Co. KGaA, Weinheim.

Kolodkin, A. N., Bruggeman, F. J., Plant, N., Mone, M. J., Bakker, B. M., Campbell, M. J., van Leeuwen, J. P., Carlberg, C., Snoep, J. L., and Westerhoff, H. V. (2010). Design principles of nuclear receptor signaling: How complex networking improves signal transduction. *Mol. Syst. Biol.* **6,** 446.

Koshland, D. E., Jr., Goldbeter, A., and Stock, J. B. (1982). Amplification and adaptation in regulatory and sensory systems. *Science* **217,** 220–225.

Kuhn, T. S. (1962). The Structure of Scientific Revolutions. University of Chicago Press, Chicago.

Lauffenburger, D. A. (2000). Cell signaling pathways as control modules: Complexity for simplicity? *Proc. Natl. Acad. Sci. USA* **97,** 5031–5033.

Maeder, C. I., Hink, M. A., Kinkhabwala, A., Mayr, R., Bastiaens, P. I. H., and Knop, M. (2007). Spatial regulation of Fus3 MAP kinase activity through a reaction-diffusion mechanism in yeast pheromone signalling. *Nat. Cell Biol.* **9,** 1319–1326.

Mitchell, P. (1961). Coupling of phosphorylation to electron and hydrogen transfer by a chemi-osmotic type of mechanism. *Nature* **191,** 144–148.

Novak, B., Csikasz-Nagy, A., Gyorffy, B., Chen, K., and Tyson, J. J. (1998). Mathematical model of the fission yeast cell cycle with checkpoint controls at the G1/S, G2/M and metaphase/anaphase transitions. *Biophys. Chem.* **72,** 185–200.

Reijenga, K. A., van Megen, Y. M., Kooi, B. W., Bakker, B. M., Snoep, J. L., van Verseveld, H. W., and Westerhoff, H. V. (2005). Yeast glycolytic oscillations that are not controlled by a single oscillophore: A new definition of oscillophore strength. *J. Theor. Biol.* **232,** 385–398.

Rhee, S. G., Chock, P. B., and Stadtman, E. R. (1989). Regulation of Escherichia coli glutamine synthetase. *Adv. Enzymol. Relat. Areas Mol. Biol.* **62,** 37–92.

Roling, W. F. M., van Breukelen, B. M., Bruggeman, F. J., and Westerhoff, H. V. (2007). Ecological control analysis: Being(s) in control of mass flux and metabolite concentrations in anaerobic degradation processes. *Environ. Microbiol.* **9,** 500–511.

Snoep, J. L., Bruggeman, F., Olivier, B. G., and Westerhoff, H. V. (2006). Towards building the silicon cell: A modular approach. *Biosystems* **83,** 207–216.

Teusink, B., Walsh, M. C., van Dam, K., and Westerhoff, H. V. (1998). The danger of metabolic pathways with turbo design. *Trends Biochem. Sci.* **23,** 162–169.

Tyson, J. J., Chen, K., and Novak, B. (2001). Network dynamics and cell physiology. *Nat. Rev. Mol. Cell Biol.* **2,** 908–916.

van Eunen, K., Bouwman, J., Daran-Lapujade, P., Postmus, J., Canelas, A. B., Mensonides, F. I., Orij, R., Tuzun, I., van den Brink, J., Smits, G. J., *et al.* (2010). Measuring enzyme activities under standardized in vivo-like conditions for systems biology. *FEBS J.* **277,** 749–760.

Westerhoff, H. V., and Palsson, B. O. (2004). The evolution of molecular biology into systems biology. *Nat. Biotechnol.* **22,** 1249–1252.

Westerhoff, H. V., and Van Dam, K. (1987). Thermodynamics and Control of Biological Free-Energy Transduction. Elsevier Science Publishers B.V. (Biomedical Division), Amsterdam.

Westerhoff, H. V., Simonetti, A. L. M., and Van Dam, K. (1981). The hypothesis of localized chemiosmosis is unsatisfactory. *Biochem. J.* **200,** 193–202.

Westerhoff, H. V., Melandri, B. A., Venturoli, G., Azzone, G. F., and Kell, D. B. (1984). A minimal-hypothesis for membrane-linked free-energy transduction—The role of independent, small coupling units. *Biochim. Biophys. Acta* **768,** 257–292.

Westerhoff, H. V., Odea, M. H., Maxwell, A., and Gellert, M. (1988). DNA supercoiling by DNA gyrase—A static head analysis. *Cell Biophys.* **12,** 157–181.

Westerhoff, H. V., Winder, C., Messiha, H., Simeonidis, E., Adamczyk, M., Verma, M., Bruggeman, F. J., and Dunn, W. (2009). Systems Biology: The elements and principles of life. *FEBS Lett.* **583**(24), 3882–3890.

Westerhoff, H. V., Verma, M., Nardelli, M., Adamczyk, M., van Eunen, K., Simeonidis, E., and Bakker, B. M. (2010). Systems biochemistry in practice: Experimenting with modelling and understanding, with regulation and control. *Biochem. Soc. Trans.* **38,** 1189–1196.

SECTION TWO

MACHINES FOR SYSTEMS BIOLOGY

CHAPTER TWO

Mass Spectrometry in Systems Biology: An Introduction

Warwick B. Dunn

Contents

Abstract

The qualitative detection, quantification, and structural characterization of analytes in biological systems are important requirements for objectives to be fulfilled in systems biology research. One analytical tool applied to a multitude of systems biology studies is mass spectrometry, particularly for the study of proteins and metabolites. Here, the role of mass spectrometry in systems biology will be assessed, the advantages and disadvantages discussed, and the instrument configurations available described. Finally, general applications will be briefly reviewed.

Manchester Centre for Integrative Systems Biology, Manchester Interdisciplinary Biocentre, University of Manchester, Manchester, United Kingdom
Department of Chemistry, Manchester Interdisciplinary Biocentre, University of Manchester, Manchester, United Kingdom
Centre for Advanced Discovery and Experimental Therapeutics, School of Biomedicine, University of Manchester, Manchester, United Kingdom

Methods in Enzymology, Volume 500
ISSN 0076-6879, DOI: 10.1016/B978-0-12-385118-5.00002-5

1. Introduction

The experimental aspect of systems biology studies focuses on the components, and more importantly, the interactions between components, which describe the emergent properties of biological systems. There are many thousands of single components in biological systems (including genes, RNA, proteins, and metabolites), and some components are modified depending on the biological status of the organism (e.g., posttranslational modifications (PTMs) of proteins). The number of potential binary interactions is significantly larger, and in many examples, multiple species interact and operate (e.g., the coupling of multiple proteins in an enzyme complex). These studies require powerful experimental tools.

One powerful analytical tool applied in the study of metabolites and proteins in systems biology is mass spectrometry. The history of mass spectrometry has been reviewed (El-Aneed *et al.*, 2009; Griffiths, 2008) with the first mass spectrometer operating at a vacuum pressure being constructed by J. J. Thomson in 1913. The technology has grown from strength to strength over the past 100 years and has impacted significantly on biological research. Five Nobel prizes have been awarded to Nobel laureates working in the field of mass spectrometry (Thomson, Aston, Paul, Fenn, and Tanaka). Today, a wide range of instrument configurations are available which provide a multitude of potential applications, many of which would not be achievable without the mass spectrometer.

2. What is Mass Spectrometry?

As the name implies, mass spectrometry is focused on the measurement of mass. More specifically, mass spectrometry determines the mass-to-charge (m/z) ratio of electrically charged molecules. A molecule with a nominal molecular weight of 1000 and coupled to a single proton will be detected with an m/z of 1001 [mass of molecule (1000) + mass or proton (1)/charge (1)]. A molecule with a molecular weight of 1000 and coupled with two protons will be detected with an m/z of 501 [mass of molecule (1000) + mass or proton (2)/charge (2)].

Mass spectrometry is most frequently applied to the study of metabolites and peptides in biological systems. However, peptides are created from the enzymatic or chemical digestion of proteins during sample preparation. The majority of metabolites in biological systems are of low molecular weight and are detected as a single-charged species, as the ability to carry multiple charges is limited. However, peptides and proteins are typically of higher molecular weights and are normally detected as multiply charged species.

Mass is an appropriate property of matter to measure so to provide discrimination of different analyte species. All analytes are constructed by the combination of different elements and different numbers of each element. Each element has a different atomic mass, and therefore, the wide array of different combinations of elements observed in metabolites and proteins have different molecular masses. For example, the two amino acids phenylalanine and tryptophan have molar masses of 165 and 204 g mol^{-1}, respectively, and can be detected with high specificity in the same sample without prior separation. However, some chemical species have identical molar masses as they have the same type and number of elements in their molecular formula (e.g., fructose and glucose, $C_6H_{12}O_6$) or in the case of peptides/proteins are composed of identical amino acid compositions, that is, isomeric peptides (rearrangement of amino acid sequence) and isobaric peptides where isoleucine and leucine are substituted, for example, TFAEA*L*R and TFAEA*I*R. These species cannot be directly differentiated by mass. However, they can be separated from a complex mixture using chromatography before mass spectrometric detection. Alternatively, these species will have the same elemental or amino acid composition but a different structural conformation. The application of gas-phase fragmentation will produce different fragmentation patterns which are specific for each species and can be employed to differentiate chemical species of the same molecular weight and different chemical structures. In proteomics, proteolysis with specific enzymes (e.g., trypsin) followed by mass spectrometry detection can be applied to differentiate between proteins with the same amino acid composition but different primary sequence.

Mass spectrometry has evolved from early work in the late nineteenth and early twentieth century. From these origins have developed a large range of mass spectrometers with different instrumental configurations, as will be discussed below. Today, mass spectrometers are applied in diverse fields from biological research and drug development (Feng, 2004) to space exploration (Petrie and Bohme, 2007).

3. Mass Spectrometer Configurations

The mass spectrometer is constructed with four basic units as shown in Fig. 2.1. These are the sample introduction system, ionization or ion source, mass analyzer, and detection system. Powerful computers are required to control the many electronic and physical components of modern mass spectrometers and to acquire and process the complex raw analytical data acquired. Today, the size of mass spectrometers varies greatly from miniaturized systems (Ouyang and Cooks, 2009) to "bench-top" systems to instruments which are too large for normal analytical

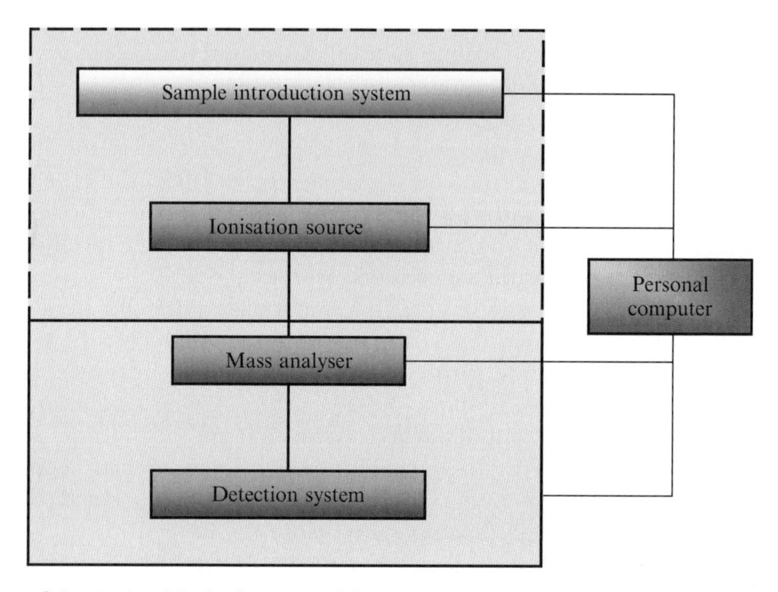

Figure 2.1 A simplified schematic of the components of the mass spectrometer. The mass analyzer and detection system operate at a vacuum pressure as shown by the box with fixed edges. The sample introduction system and ionization source can operate at atmospheric or vacuum pressure as shown by the box with the dashed edges. Samples are introduced into the ionization source where ion formation is performed, followed by separation of ions of different m/z in the mass analyzer, and detection. A personal computer controls the complex operation of the mass spectrometer and stores and processes acquired data.

chemistry laboratories (e.g., Fourier transform ion cyclotron resonance mass analyzers (FTICR-MS)).

In the majority of applications, the mass analyzer and detection system operate in a vacuum at pressures of 10^{-6} torr or lower. Mass separation is typically performed by creating different trajectories for ions of different m/z. If the gas pressure in the mass analyzer was high, then collisions between ions and gas molecules would occur during mass analysis which would disrupt ion trajectories and reduce mass resolution but could also lead to loss of ions and a reduction in sensitivity. The ion source can operate at either low pressures (e.g., electron impact ion source) or atmospheric pressures (e.g., electrospray ion source (ESI)).

Figure 2.1 shows a simplistic view of the construction of mass spectrometers. In reality, the number of components and their operation are significantly more complex. For example, focussing lenses are applied in systems to improve high ion transmission efficiencies so as to improve sensitivity. These lenses are especially important when interfacing atmospheric pressure ion sources to mass analyzers operating at vacuum pressures, where ions are focussed through multiple stages to reduce instrument operating pressures.

Today, a range of instrument configurations are available commercially. The components most frequently applied in biological research and systems biology will be discussed here.

3.1. Sample introduction

The objective of sample introduction systems is to introduce samples to the ion source in a representative and in most cases nondestructive approach. Some systems directly introduce the complete sample (e.g., direct insertion probes) whereas other systems provide prefractionation or separation methods prior to the ion source (e.g., chromatography systems). In systems biology, separation prior to detection is a common approach for complex sample systems, whereas direct introduction is applied for samples fractionated offline to produce pure samples containing a single biochemical species. For example, matrix assisted laser desorption ionization (MALDI) is applied, among others, to analyze purified proteins. Chromatographic separations are essential in the analysis of complex mixtures composed of 100–1000 s of different analytes or matrix components. This is to provide detection of individual analytes without interference from other analytes or matrix components. This can provide improved limits of detection and avoid the nondetection of analytes present at low concentrations caused by masking from high concentration analytes. A second reason to apply chromatographic separations is to differentiate between isomers, which have the same molecular weight and therefore are detected with the same m/z. Quantification or identification of the separate isomers requires separation before detection.

Chromatography systems act to separate components of a mixture based on equilibration between two phases, a mobile phase and a stationary phase (Robards *et al.*, 1997). In gas chromatography (GC), the two phases are a solid (or immobilized liquid) stationary phase and a gaseous mobile phase. In liquid chromatography (LC), the two phases are a liquid mobile phase and a stationary phase composed of a solid or a liquid covalently coated onto a solid surface. The stationary phase is contained in a chromatography column through which the mobile phase passes. For GC, these are narrow and hollow silica capillaries with the stationary phase coated on the inner surface. For LC, these are metal columns with the stationary phase located inside the column on small silica spheres or in porous monolithic layers. Hollow silica capillary LC columns with the stationary phase packed into the capillary are also applied in proteomics and are described as nano-LC.

As the sample in the mobile phase passes through the column, analytes continuously transfer between mobile and stationary phases. The period of time the analyte is present in each phase defines the time to elute through the column. Analytes of different properties (e.g., molecular formula and structure) will have different elution or retention times, and this partition can be optimized to provide appropriate chromatographic separations. For GC–MS, the stationary phase composition, column dimensions, and oven temperature ramp in which the column operates are varied during the optimization process. For LC–MS, the stationary phase and mobile phase compositions are varied. The mobile phase composition is typically binary, and the relative composition of the two solvents is changed during a chromatographic run.

A typical GC–MS chromatogram for a yeast intracellular extract is shown in Fig. 2.2. This describes the separation of two components based on the boiling point of the trimethylsilyl ether derivatives.

GC is frequently applied to study metabolites, as shown in Fig. 2.2. LC is frequently applied to study metabolites, peptides, and proteins. High-performance liquid chromatography (HPLC) is the most routinely applied

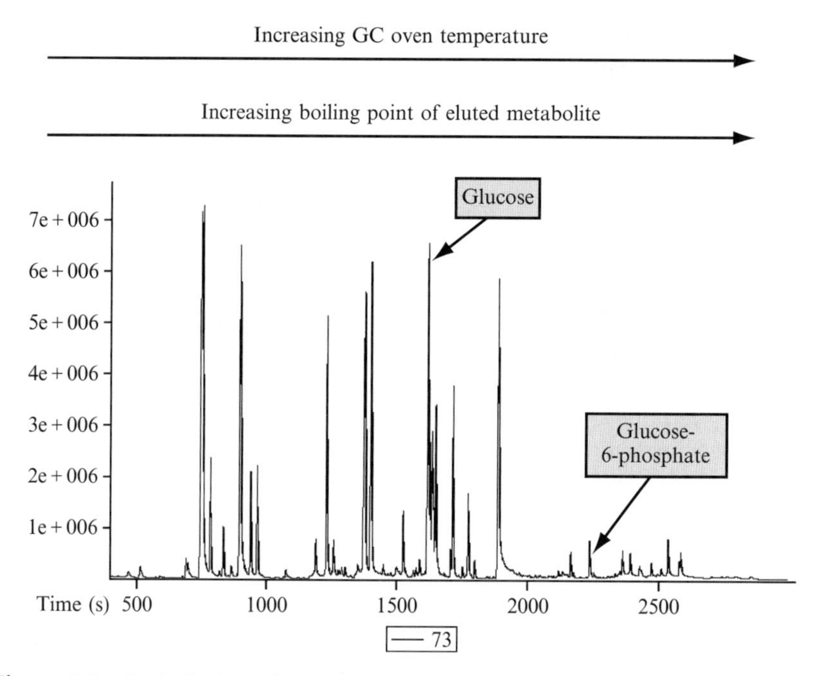

Figure 2.2 A single ion chromatogram (*m/z* 73) of a yeast intracellular extract acquired on a gas chromatography–mass spectrometry system. The abscissa and ordinate are the retention time and measured ion current, respectively. The chromatogram shows the high level of chromatographic resolving power available in profiling of metabolites and proteins/peptides. Two metabolites, which are trimethylsilyl (TMS) derivatized, are included to define the chromatographic separation principles. For gas chromatography, separation is primarily based on boiling point for metabolites of similar chemical structure. The boiling point of the TMS derivatization product of glucose-6-phosphate is greater than that for glucose. Therefore glucose-6-phosphate elutes at a later retention time than for glucose.

tool, though the recent introduction of Ultra-Performance Liquid Chromatography (UPLC; Swartz, 2005) and associated variants has provided the ability to perform separations with greater chromatographic resolution and sensitivity and in shorter analysis times. The column dimensions and flow rates define the type of LC system. For example, column internal diameters of less than 100 μm and flow rates less than 10 μL min^{-1} are defined as nano-LC, whereas internal diameters greater than 1.0 mm and flow rates of 50–1000 μL min^{-1} are defined as analytical LC. Nano-LC has the advantage of lower solvent and sample consumption and is applied in proteomics.

In metabolomics, reversed phase chromatography is most frequently applied where the stationary phase is nonpolar and the mobile phase begins with a polar mobile phase (e.g., water) with the organic mobile phase (methanol or acetonitrile) increasing during the chromatographic run. This provides elution of polar metabolites first followed by elution of nonpolar metabolites at later retention times. Hydrophilic interaction chromatography (HILIC) is becoming more prominent as it provides a complementary separation process, where nonpolar metabolites elute before polar metabolites in a mobile phase operating from a low to high aqueous composition (Gika *et al.*, 2008). In proteomics, reversed phase chromatography is frequently applied, sometimes in combination with a second level of separation such as ion-exchange or a second reverse phase (RP/RP). An example of the former is the case of 2D-LC which is routinely applied in proteomics [e.g., when applying multidimensional protein identification technology (MudPIT) for complex samples (Washburn *et al.*, 2001)] where two columns applying complementary separation techniques are coupled offline or online. The two separation techniques are commonly strong cation exchange and reversed phase.

Other methods for separation are also employed, to a significantly lower frequency. These include capillary electrophoresis (Ramautar *et al.*, 2009) and comprehensive GC × GC–MS (Welthagen *et al.*, 2005) in metabolomics.

3.2. Ionization sources

Ionization, or ion, sources operate with a single goal to create a charged species for each analyte for subsequent mass analysis. There are a variety of different charged species which are created depending on the species which is added to or removed from an analyte. Electron impact ionization removes a single electron to produce a radical cation. Positive ion electrospray and MALDI operate by addition of a cation (H^+, NH_4^+, Na^+, K^+, and many others) to produce a positively charged "adduct" ion.

A multitude of ion sources are available, though the most prominent in biological research are electrospray ionization, electron impact ionization (EI), and MALDI.

3.2.1. Electrospray ionization

ESI (El-Aneed *et al.*, 2009; Gaskell, 1997) operate at atmospheric pressure and were developed separately by Fenn and Tanaka in the 1970s–1980s. Their invention solved a large problem in coupling LC to mass spectrometry, where the vaporization of a liquid stream can overwhelm the instrument vacuum if performed in the vacuum region of the mass spectrometer. ESI provides ion formation at atmospheric pressures from an "electrosprayed" liquid solution followed by introduction to the vacuum of the mass spectrometer. Today, ESI are one of the most frequently applied in biological research. Before ESI and MALDI, the mass analysis of large biomolecules was not routinely feasible because other methods could not provide the ionization and vaporization of large biomolecules. This highlights the requirement for technological advances before biological discoveries, which has been reviewed in a systems biology context (Kell, 2005). The sources provide soft ionization of small and large biomolecules with high sensitivity and limited fragmentation. The process of ion formation is not fully resolved and two methods are reported, the ion evaporation model (IEM; Iribarne and Thomson, 1976) and the charged residue model (CRM; Dole *et al.*, 1968). In reality, both mechanisms are believed to be in operation. In simple terms, a large voltage is applied to a metal needle through which the liquid solution flows. This causes the creation of a Taylor Cone at the end of the needle which produces droplets of the same charge as the needle. These droplets are repelled and drawn toward the sampling orifice and cone of the mass spectrometer, which is held at the opposite charge, and into the vacuum region of the mass spectrometer. As droplets traverse the region between needle and ion orifice at atmospheric pressure, with assistance from nebulizing and desolvation gas flows, solvent evaporation occurs along with coulombic explosions when the charge of the droplet exceeds the Rayleigh point and coulombic repulsion occurs. These explosions produce droplets of smaller size and desolvated charged molecules. This continuous process produces charged molecules for mass analysis. Nanoelectrospray (or nanospray) is applied in proteomic research and operates at low $\mu L\ min^{-1}$ or $nL\ min^{-1}$ flow rates without desolvation/nebulization gases. The low flow rates and needle diameter produce droplets of a narrower diameter than conventional electrospray which aids in ion release and improved sensitivity.

3.2.2. Electron impact ionization

EI is applied with gases (Gross, 2004; Mirsaleh-Kohan *et al.*, 2008). These gases can be directly introduced, produced from thermal heating of samples, or be an eluent from a GC system. Therefore, only analytes which are volatile or whose volatility can be increased can be analyzed. These are typically low molecular weight analytes including metabolites, but excluding proteins. The ion source operates at low vacuum pressures and so low flow rates of gas can be analyzed without a reduction in the operating pressure. GC–MS is the most frequent application of EI in biological research. The gaseous eluent of the GC (typically 0.5–2.0 mL min^{-1} flow rate) is introduced through a heated transfer line to the ion source. Here, an electron beam is accelerated at a specific and reproducible energy of 70 eV and is continuously focused through the region containing a flow of column eluent. Electrons are produced by thermoionic emission of a metal filament which has an electrical current running through it. A quantum mechanical mechanism results in ejection of a single electron from biomolecules to produce a positively charged radical ion which is passed forward for mass analysis.

EI is described as a hard ionization mechanism. The intramolecular energy after ionization is high. As operating under vacuum pressures, ion–molecule collisions to release this energy are not available. The mechanism to release energy is covalent bond fission. The molecular ion formed during ionization rapidly fragments to produce a range of fragment ions of lower m/z. The fragmentation pattern and resulting fragment ions are dependent on the chemical structure and this pattern is routinely applied to identify metabolites. The process of ionization and fragmentation is reproducible between different instruments, and mass spectral libraries which are transferable between instruments are available to undertake identification. This is not yet fully realized for ESI instruments.

3.2.3. Matrix-assisted laser desorption ionization

MALDI (El-Aneed *et al.*, 2009; Gross, 2004) is an ionization technique applied frequently in proteomics but infrequently in metabolomics, though imaging is performed (Chaurand *et al.*, 2006; Fletcher, 2009; van Hove *et al.*, 2010). The sample, normally a relatively simple composition, is mixed with an excess of matrix and spotted onto a metal sample plate. The matrix is chosen based on a number of reasons. It should absorb strongly in the UV or visible range so that laser radiation can be efficiently absorbed which assists in efficient ionization. Chemicals in the matrix should be of low molecular weight to provide vaporization during ionization but should not be too volatile so to be removed during sample preparation where the sample–matrix mixture is dried at room temperatures. The matrix chemicals should form crystals once dried and often are acidic so to provide efficient protonation of proteins or peptides. The sample is mixed with an excess of matrix which is dissolved in a water–organic solvent mixture to provide solubilization of hydrophilic and hydrophobic proteins and peptides followed by drying. Cocrystals of analyte and matrix are formed in this process which aids the ionization and vaporization process.

MALDI applies a UV laser (e.g., nitrogen at 337 nm) which is fired at a sample–matrix dried spot on a MALDI plate. The energy is absorbed by the matrix and produces sublimation and ionization of molecules. Typically, ionization is performed by protonation of molecules from the acidic matrix, though other adducts can be detected. Deprotonation can also be applied to detect negatively charged ions, used to study phosphoproteins or phosphopeptides or acidic metabolites.

3.3. Mass analyzers

3.3.1. Single quadrupole mass analyzers

Quadrupole mass analyzers (Villas-Boas *et al.*, 2007) are the cheapest to manufacture and purchase though only offer low mass resolution and accuracy in their application. They are also very robust and simple to use. The system is constructed with four circular metal rods which are assembled into a symmetrical and parallel construction. Opposite rods are electrically connected to each other with DC and RF voltages. If the RF is the only voltage applied, a wide range of m/z ions will traverse through the quadrupole and this can be applied as an ion-focusing component in many instruments. When a DC voltage is superimposed onto the RF voltage, ions of a specific m/z can be tuned to traverse the quadrupole, while ions of higher or lower m/z values will be lost by collisions with the quadrupoles through unstable trajectories. By increasing the DC and RF voltages, while keeping their ratio constant, the mass range can be scanned to transmit ions of increasing m/z to acquire a mass spectrum. Quadrupole mass analyzers are therefore described as scanning instruments.

Instrument sensitivity is lower in full-scan mode, where not all ions of different m/z reach the detector at all times. Increased scan times in modern instruments has reduced this problem. To overcome this problem and increase selectivity and sensitivity, continuous detection of a single m/z ion can be performed; this is described as single ion monitoring (SIM). The triple quadrupole (QQQ) mass spectrometer, discussed below, applies this strategy to provide greater sensitivity in quantitation studies.

3.3.2. Triple quadrupole mass analyzers

Tandem mass spectrometry is applied to either (a) provide lower limits of detection and increased selectivity in quantification studies or (b) provide gas-phase fragmentation for analyte identification. The QQQ mass spectrometer (Gross, 2004) is constructed with three sets of quadrupoles in series, defined as Q1, Q2, and Q3. Q1 and Q3 operate as mass analyzers. Q2 functions as a collision cell which operates under vacuum but at a higher pressure with a collision gas such as nitrogen or argon introduced. Ions from Q1 are accelerated through Q2 where ion–molecule collisions create transfer of potential energy to the ions which is released by covalent bond fission. The m/z of fragments ions is determined in Q3. This type of common fragmentation mechanism is defined as collision-induced dissociation (CID). The level of fragmentation is dependent on the collision gas pressure and acceleration potential. QQQ instruments are applied for qualitative and quantitative studies. These instruments are described as "in-space" systems as fragmentation processes are separately performed in different components (or spaces) of the instrument. This is in comparison with "in-time" instruments, such as the ion trap, where experiments are performed in the same "space" but are separated in time.

The QQQ instrument can be operated in four processes: product ion scanning, precursor ion scanning, neutral loss scanning, and selected reaction monitoring (SRM) or multiple reaction monitoring (MRM). The product ion scan selects a single m/z in Q1 and transmits to Q2 followed by m/z scanning of all fragment ions in Q3. This is appropriate for structural elucidation work of unknowns. SRM or MRM operates in a similar way but Q3 is set to scan for a single m/z for SRM analysis of a single biochemical or alternates between chosen but different m/z in Q1 and Q3 for analysis of multiple biochemicals. The use of SRM and MRM options increases sensitivity by reducing the chemical noise in Q3 and ensuring that ions of a specific m/z are continuously detected. SRM and MRM methods are applied for quantitation studies.

3.3.3. Time-of-flight mass analyzers

Time-of-flight (TOF) mass analyzers (Gross, 2004) determine m/z through the measurement of flight time. Ions are accelerated at the start of a flight tube so all ions have the same kinetic energy. Kinetic energy, and resulting velocity, is related to m/z and ions will traverse the flight tube to the detector with different flight times, lower m/z ions have higher velocities and reach the detector before higher m/z ions with lower velocities. TOF mass analyzers can acquire multiple mass spectra per second, and depending on the electronics, higher acquisition frequencies may reduce the sensitivity. However, as all ions are detected simultaneously, sensitivity can be higher than quadrupole instruments. As all ions are detected simultaneously TOF mass analyzers are not scanning instruments. As acquisition rates are high, TOF instruments are regularly applied with high resolution chromatographic systems such as GC, comprehensive GC $\times$ GC, and UPLC. For these reasons, TOF instruments have become popular for profiling experiments in metabolomics and proteomics where samples are complex. TOF mass analyzers can also offer other advantages. In well-designed systems, the mass resolution can be significantly higher than quadrupole instruments, which depending on the instrument can be as high as 40,000 FWHM (Full Width Half Maximum) at m/z 1000 Da. However, it must be noted that mass resolution is m/z dependant. High mass resolution allows m/z ions of similar but not identical monoisotopic mass to be separated by m/z without the requirement of prior physical or chemical methods of separation. This is important in complex samples which are typical in metabolomics and proteomics research. When combined with accurate mass calibration, this also allows the *m/z* of ions to be accurately determined to mass accuracies of generally less than 5 ppm. The potential molecular formula of metabolites or amino acid composition of peptides and proteins related to the *m/z* can be calculated from these data.

The flight time and mass resolution/mass accuracy are highly dependent on the initial ion energy and ion position at the start of the flight tube. To overcome diffuse ion energies and spatial position, orthogonal acceleration systems and reflectrons are employed. Orthogonal acceleration systems are employed to couple continuous ion sources (e.g., electrospray) to TOF mass analyzers and operate to allow pulses of ions to be mass analyzed. The system allows orthogonal introduction of ions compared to the ion flight direction in the TOF mass analyzer and focusing of spatial resolution provides higher mass resolutions. This is because the position of the ion on acceleration before mass analysis will determine its flight time; ions nearer the detector in the flight tube will reach the detector faster. Delayed extraction from the ion source is commonly applied where the switching on of the electrical field for ion extraction and acceleration from the source is delayed in respect to ion formation. This method is particularly common in MALDI applications, where the ion formation is pulsed and provides increased mass resolutions. Reflectrons, or ion mirrors, reduce the kinetic energy distribution of ions of the same *m/z* which increases the mass resolution. The system uses a constant electrostatic field, and ions of slightly greater velocity will penetrate further than ions of lower velocity. On exiting the reflectron, the distribution of kinetic energies of ions of the same *m/z* is narrower than before entering the reflectron. Reflectrons are located at the opposite end of the flight tube to the source and detector in single-reflectron systems so that ions are transmitted from the source and reflected back to the detector by the reflectron.

3.3.4. Quadrupole–time-of-flight hybrid systems

A hybrid instrument applied regularly in metabolomics and proteomics is the quadrupole–time-of-flight (Q-TOF, or QqTOF) instrument (El-Aneed *et al.*, 2009). This provides increased capabilities compared to the separate quadrupole and TOF instruments. The system can be operated in TOF-only mode where the quadrupole operates to transmit all *m/z* ions. However, the system can be operated as a tandem mass spectrometer also where ions of specific *m/z* can be isolated in the quadrupole for gas-phase fragmentation in a collision cell (similar to Q2 in QQQ instruments) followed by full mass scans in the TOF mass analyzer. Q-TOF instruments operate to detect product ions to aid in identification, for example. However, all ions are transmitted and detected with post-acquisition isolation of the product ion(s) of interest, whereas for QQQ instruments, the product ions are isolated in Q3. Scanning modes which are feasible with other instrument designs are not feasible with the Q-TOF system. However, quantification can be performed post-acquisition via plotting of single *m/z* ions. In metabolomics and proteomics, this instrument configuration offers an advantage in that full mass scans and product ion scans can be acquired sequentially to acquire both accurate *m/z* data and fragmentation data for quantitation and identifications processes. For example, a full-scan mass spectrum can be acquired from which the top *n*th ions of highest response can then be chosen by the acquisition software for MS/MS analysis with product ion scanning in the next *n*th scans. This is feasible in part because of the rapid acquisition times of the TOF where multiple acquisition scans are achievable per second so that multiple mass spectra can be acquired even when applied with chromatographic systems operating at high resolutions. High mass accuracy measurement of molecular and fragment ions is routinely achievable which aids identification. This is not achievable with QQQ instruments.

3.3.5. Ion trap

Ion traps (March, 1997), also referred to as quadrupole ion traps, trap and store ions in an orbital motion within the ion trap and eject ions for detection. Storage is performed by collecting ions in a potential energy well, and ejection provides energy for ions to escape the potential energy well. This is in comparison to quadrupole mass analyzers which provide continuous transmission of ions. The systems are cheap to manufacture and operate as for quadrupole mass analyzers, though mass resolution is lower than for TOF, Orbitrap, and FTICR-MS instruments. However, multiple MS/MS experiments (or MS^n) can be performed which is a benefit of the system.

The first available ion trap was the 3D system which is composed of two hyperbolic electrodes (end caps) facing each other and separated by a hyperbolic ring electrode. A damping gas, such as helium, is introduced to cool energetic ions, and RF and DC voltages are applied to the three electrodes. Mass scanning is performed by increasing the amplitude of the voltages to eject ions of increasing *m/z* for detection. The ion capacity of 3D traps is limited and gain control systems are typically employed to eliminate space-charge effects which are detrimental to sensitivity, mass accuracy, and mass resolution. To reduce this effect and increase sensitivity, linear ion traps are today more commonly applied in research laboratories. These systems allow a greater number of ions to be trapped. The linear ion trap (Gross, 2004) is composed of three sets of quadrupole rods in which DC and RF voltages are applied to radially constrain ions in the second set of rods and on which potentials are applied to the end electrodes to constrain ions axially. These systems are easier to construct, operate at a faster scan rate, and can contain more ions without space-charge effects.

Ion trap instruments provide a significant advantage over many other instruments. Ions can be stored indefinitely, and multiple gas-phase fragmentation (MS^n) experiments can be performed "in-time" to provide structural information not only on the molecular ion but also on resulting fragment ions to construct mass spectral trees. QQQ and Q-TOF instruments can only perform MS/MS experiments. CID is a common fragmentation mechanism.

3.3.6. Orbitrap mass analyzer

The Orbitrap mass analyzer (Hu *et al.*, 2005) is a more recent addition to the toolbox and was invented by Alexander Makarov. The Orbitrap is constructed with a coaxial central spindle electrode surrounded by a barrel-like electrode and with an electrostatic field operating between the inner and outer electrodes. Ions injected tangentially into the electrostatic field are constrained by two forces; the electrostatic field compensates for the centrifugal force field and results in ions orbiting the central electrode both axially and radially. The frequency of harmonic oscillations axially along the central spindle is applied for m/z measurement, as this oscillatory pattern is independent of ion energy. The image current on the outer electrode is applied for detection.

The mass resolution is dependent on the acquisition time (more specifically the number of harmonic oscillations) and so higher mass resolutions are acquired over longer periods of time, which can be a disadvantage when applied with chromatographic systems operating at high resolutions. Mass resolutions of 200,000 at m/z 400 are achievable which is significantly higher than for most TOF instruments. The mass analyzer can also provide ppm or ppb mass accuracies with mass accuracy not being dependent on dynamic range and so mass accuracy for ions of low and high intensity can be identical.

Commercially, the Orbitrap mass analyzer is coupled to a linear ion trap via a C-trap. The C-trap acts to electrodynamically squeeze ions to convergence at the entrance of the Orbitrap mass analyzer. The linear ion trap and Orbitrap mass analyzers can act separately or in combination. For example, in profiling mode, the Orbitrap can perform mass analysis, and the ion trap is used for interfacing an ion source continuously transmitting ions with a periodic introduction of ions to the Orbitrap. Both can be operated in tandem where mass analysis is performed in the Orbitrap, while MS/MS or MS^n fragmentation experiments with mass analysis are performed in the linear ion trap. This provides the tools for both profiling and identification as discussed for Q-TOF instruments but with a higher mass resolution and accuracy. Finally, gas-phase MS^n experiments in the linear ion trap can be mass analyzed in the Orbitrap.

3.3.7. Fourier transform ion cyclotron resonance mass analyzers

Also known as Fourier transform mass spectrometry (FTMS; Marshall *et al.*, 1998), the system operates by determining the cyclotron frequency of ions in a fixed magnetic field to determine the m/z. Ions are trapped in a magnetic field by two electrical end plates and are excited to orbital radii by an RF electric field applied perpendicular to the magnetic field. Ions of the same m/z orbit in a packet in phase, and the cyclotron frequency is determined by measurement of the free induction decay image current on end plates followed by a mathematical Fourier transform procedure to transform from the frequency domain to a mass spectrum. As for the Orbitrap mass analyzer, instruments are constructed so to store and transmit ions from an ion source continuously producing ions to a mass analyzer requiring periodic introduction of ions.

The system provides the highest mass resolution achievable of all mass spectrometers. Mass resolution is dependent on the magnetic field strength and acquisition time, and systems applying higher magnetic field strengths are favorable and can provide mass resolutions of 1,000,000. Mass accuracies measured in units of ppb are also readily achievable with good mass calibration allowing the determination of molecular formula. However, the operation and maintenance of these systems can be thought of as more difficult than other systems. An array of gas-phase experiments can and have been performed on the trapped ions. Electron transfer dissociation (ETD; Mikesh *et al.*, 2006) and more extensively electron capture dissociation (ECD; Cooper *et al.*, 2005) are applied in proteomic studies. Radical anions such as anthracene (ETD) and free electrons (ECD) are used. Both methods transfer electrons to the molecular ion which results in covalent bond fission. However, the fragmentation mechanism is different to CID, and complementary fragmentation data can be acquired from using both complementary methods. CID generally causes PTM losses during fragmentation, whereas PTMs remain intact after ETD/ECD fragmentation. Infrared multiphoton dissociation (IRMPD; Fukui *et al.*, 2006) and sustained off-resonance irradiation-collision induced dissociation (SORI-CID; Li *et al.*, 2009) can also be performed.

3.3.8. Ion mobility spectrometry

Ion mobility spectrometry (IMS) is not primarily based around separation according to m/z but can play an important role in hybrid instruments (Uetrecht *et al.*, 2010). IMS separations are based on the mass, charge, size, and shape which all influence the ion mobility. Ions are accelerated along a drift tube operating with an applied electric field and a buffer gas flow opposing the ion motion. Ions with different size and shape will have different ion mobilities and therefore different drift times before the detector is reached. Ions of small cross-sectional area will have faster drift times than ions of larger cross-sectional areas. IMS has been coupled to mass analyzers to create hybrid systems. The most significant commercial adaption is the Waters Synapt system (Q-IMS-TOF) where a multitude of applications in protein analysis and protein complex analysis has been applied (Politis *et al.*, 2010).

4. The Benefits of Mass Spectrometry

Mass spectrometry is applied routinely in biological and systems biology research as it is a powerful tool capable of performing multiple processes.

The advantages of some or all mass spectrometers include sensitivity, high mass resolution and accuracy, rapid scan speeds to allow interfacing with chromatographic systems, and the ability to perform gas-phase fragmentation experiments in-space or in-time. Mass spectrometers can typically detect analytes at concentrations of $\mu mol\ L^{-1}$ or lower in routine profiling applications providing the detection of many metabolites and proteins at physiological concentrations. The application of tandem mass spectrometry can lower the limit of detection significantly by 2–5 orders of magnitude.

The high mass resolution and mass accuracies observed with TOF, Orbitrap, and FTICR-MS instruments allow the separation of ions with similar but not identical monoisotopic masses without prior separation. This can result in the detection of a greater number of analytes in direct introduction experiments and also in systems applying chromatographic separations before mass analysis. Accurate mass data also provides information for identification processes.

The ability to acquire many mass spectra every second allows the majority of mass spectrometers to be interfaced with chromatography systems to provide component separation combined with online analysis. This is not achievable with NMR spectroscopy in routine applications.

Finally, mass spectrometers provide multiple mechanisms to aid in the identification of chemicals. This includes accurate mass measurements to define a molecular formula or amino acid composition and the ability to perform gas-phase fragmentation experiments to define a chemical structure or amino acid sequence to metabolite or peptide/protein, respectively,

The disadvantages of mass spectrometry include the destruction of samples and the cost to commercially purchase instruments. The process of sample introduction and ionization results in sample loss, whereas for some other instruments, including NMR spectroscopy, samples can be recovered after data acquisition. The cost of mass spectrometry systems are high from £50,000 for a single quadrupole instrument to greater than £500,000 for Orbitrap instruments and approximately £1million for high magnetic field strength FTICR-MS instruments. In profiling and targeted quantitation of metabolites and peptides where chromatographic systems are interfaced to a mass spectrometer, analysis times are short (typically less than 30 min and in specific examples less than 10 min). This provides a relatively high-throughput of samples which ensures that analysis costs per sample can be reduced in large-scale studies. This is important in studies requiring the analysis of thousands of samples, an example being the study of human populations (Dunn *et al.*, 2011; Kenny *et al.*, 2010).

5. Applications of Mass Spectrometry in Systems Biology

In systems biology research, and in basic terms, the mass spectrometer is applied to (1) determine the accurate mass and structural configuration of ionized biomolecules for qualitative identification, (2) perform profiling of complex samples, and (3) perform quantitation of biomolecules. The majority of these applications are observed for metabolites, peptides, and proteins. There are other applications but whose frequency is significantly lower. The brief of this chapter is not to discuss each application in detail. Other chapters in this book discuss in detail specific applications including the application of QconCAT for protein quantitation (Chapter 7), targeted metabolite quantitation (Chapter 15), and profiling of plant metabolomes (Chapter 16). A number of excellent reviews discussing the role of mass spectrometry in systems biology are available (Aebersold and Mann, 2003; Dunn *et al.*, 2011; Feng *et al.*, 2008; Glinski and Weckwerth, 2006; Kislinger, 2007; Yates *et al.*, 2009).

5.1. Structural characterization and identification

Structural characterization of metabolites and proteins can be performed in complex mixtures. However, isolation of the metabolites and proteins from all other species will provide increased accuracy and confidence of identification. Isolation in metabolomics can involve a range of extraction techniques such as liquid–liquid extraction and solid phase extraction or can be performed by single or multiple fraction collection experiments applying, for example, LC. Isolation or online separation in proteomics is commonly performed applying 2D-LC and fraction collection or applying 2D gel electrophoresis and subsequent excision of spots containing single proteins. These approaches can be applied to identify unknowns in a sample or to confirm identity by comparison to an authentic standard.

Accurate mass is commonly applied, measured to a high accuracy in high mass resolution instruments, to determine the empirical formulae or amino acid composition. However, the presence of metabolites, peptides, or proteins of the same monoisotopic masses is a real possibility in complex samples, and the application of gas-phase fragmentation experiments (MS/MS or MS^n) is a necessity to provide extra confidence to the identification and reduce the number of putative identifications by interpretation to determine the 2D or 3D structural configuration. As a final step of identification, the chromatographic and mass spectrometric properties of an authentic chemical standard should be matched to the experimental-derived data for the unknown.

In protein identification work, a separate process can be performed. For purified or isolated proteins, peptide mass fingerprinting (PMF) can be performed by MALDI–TOF instruments. Here, the isolated protein is passed through a proteolysis step with specific enzymes which cleave the protein at specific linkages (e.g., trypsin, which cleaves C-terminal to lysine and arginine). This is followed by mass measurement of each peptide either directly by MALDI or after chromatographic separation (LC–MALDI), and the resulting data can be used for database searching (in the fairly basic sense matching *m/z* to predicted *m/z* from *in silico* digests of proteins, i.e., PMF Mascot search). For more complex experiments involving structural elucidation of peptides and validation, which require the additional step of tandem MS, typically nano-LC systems coupled to mass spectrometers equipped with a nanospray source are used to reduce sample consumption and provide increased sensitivity. The resulting tandem MS data generated are used for database searching, facilitated through dedicated search engines such as Mascot and Sequest, whereby the experimentally derived masses for precursor *m/z* and product ions are matched to theoretical masses resulting from *in silico* proteolysis.

Protein identification from complex mixtures of proteins is more difficult and applies LC–MS or the more powerful 2D-LC–MS in approaches described as shotgun proteomics or MuDPiT (Washburn *et al.*, 2001). Following proteolysis of the complex protein mixture, the sample is analyzed by LC-MS or 2D-LC–MS leading to fractionation. 2D-LC–MS applies two columns of different functionalities, typically a strong cation exchange column followed by a reversed phase column. The eluent can be introduced into a MS or spotted onto a plate for MALDI analysis. There are inherent difficulties interpreting the data from multiple LC–MS runs, which is not trivial.

PTMs have a large influence on protein biological function, are reversible, and are dependent on the biological and environmental conditions (e.g., cell cycle). PTMs are chemical modifications by addition of chemical moieties to the protein (e.g., phosphate) or by structural changes (e.g., disulphide bond formation between two cysteine amino acids). Chemical modifications are wide ranging and include phosphorylation, ubiquitination, glycosylation, and acylation. Mass spectrometry is a powerful tool to define the type and position of PTMs, especially on purified proteins through a combination of accurate mass and gas-phase fragmentation where CID and ETD/ECD are applied as complementary tools.

5.2. Profiling of metabolomes and proteomes

Profiling has the objective to detect all, or more realistically large portions, of the complete proteome or metabolome in a holistic approach. This provides semiquantitative information to define differences in the metabolome or proteome related to a genetic, biological, or environmental perturbation. For example, this can be applied to define the differences in specific metabolites and proteins related to changes in nutrient supply in yeast or related to the pathophysiological mechanism of human diseases. Alternatively, this can be applied to define differences related to human diseases to identify potential biomarkers of disease or drug toxicity or to provide knowledge on molecular pathophysiological mechanisms of mammalian diseases.

The experimental objective is to detect and identify in a reproducible and robust semiquantitative methodology as many proteins, peptides, or metabolites in a single sample (Dunn *et al.*, 2011; Zelena *et al.*, 2009). In large-scale studies, the application of quality assurance (QA) and quality control (QC) samples are essential.

These studies are applied in discovery or hypothesis-generating experiments (Kell and Oliver, 2003) where knowledge regarding changes is not known *a priori* but are induced from the data acquired. Experimental design of the biological study is important as the probability of false positive discoveries is high (Broadhurst and Kell, 2006). The types of instruments applied are high-specification employing high chromatographic resolution systems coupled to mass spectrometers offering high mass resolution and mass accuracy and rapid scan times (e.g., TOF, Q-TOF, Orbitrap, FT). For metabolomics, accurate mass data only is typically applied. For proteomics, accurate mass and MS/MS data are typically acquired to enable better identification tools. The data provides potential avenues for further research, of a targeted nature to test generated hypotheses.

5.3. Quantitation

More targeted approaches determine the absolute quantification of analytes. These target specific proteins, peptides, and metabolites (maybe less than 10) with methods of high specificity and sensitivity. Sample preparation is more intensive so as to separate analytes of interest from other metabolites, proteins and peptides, and the sample matrix. Specific methods are also employed for calibration and quantitation using authentic chemical standards and/or isotopic labeling as is observed in proteomics. Higher specificity is provided by the use of hybrid instruments to provide gas-phase fragmentation experiments and QQQ instruments are the most appropriate for these types of studies as they provide excellent sensitivity and high specificity.

ACKNOWLEDGMENTS

W. B. D. wishes to thank BBSRC for financial support of The Manchester Centre for Integrative Systems Biology (BBC0082191). Include WBD also wished to thank NIHR and NWDA for financial support of CADET.

REFERENCES

Aebersold, R., and Mann, M. (2003). Mass spectrometry-based proteomics. *Nature* **422,** 198–207.

Broadhurst, D. I., and Kell, D. B. (2006). Statistical strategies for avoiding false discoveries in metabolomics and related experiments. *Metabolomics* **2,** 171–196.

Chaurand, P., Norris, J. L., Cornett, D. S., Mobley, J. A., and Caprioli, R. M. (2006). New developments in profiling and imaging of proteins from tissue sections by MALDI mass spectrometry. *J. Proteome Res.* **5,** 2889–2900.

Cooper, H. J., Hakansson, K., and Marshall, A. G. (2005). The role of electron capture dissociation in biomolecular analysis. *Mass Spectrom. Rev.* **24,** 201–222.

Dole, M., Mack, L. L., and Hines, R. L. (1968). Molecular beams of macroions. *J. Chem. Phys.* **49,** 2240–2249.

Dunn, W. B., Broadhurst, D. I., Atherton, H. J., and Goodacre, R.Griffin, J. L. (2011). Systems level studies of mammalian metabolomes: The roles of mass spectrometry and nuclear magnetic resonance spectroscopy. *Chem. Soc. Rev.* **40,** 387–426.

El-Aneed, A., Cohen, A., and Banoub, J. (2009). Mass spectrometry, review of the basics: Electrospray, MALDI, and commonly used mass analyzers. *Appl. Spectrosc. Rev.* **44,** 210–230.

Feng, W. Y. (2004). Mass spectrometry in drug discovery: A current review. *Curr. Drug Discov. Technol.* **1,** 295–312.

Feng, X. J., Liu, X., Luo, Q. M., and Liu, B. F. (2008). Mass spectrometry in systems biology: An overview. *Mass Spectrom. Rev.* **27,** 635–660.

Fletcher, J. S. (2009). Cellular imaging with secondary ion mass spectrometry. *Analyst* **134,** 2204–2215.

Fukui, K., Takada, Y., Sumiyoshi, T., Imai, T., and Takahashi, K. (2006). Infrared multiphoton dissociation spectroscopic analysis of peptides and oligosaccharides by using Fourier transform ion cyclotron resonance mass spectrometry with a midinfrared free-electron laser. *J. Phys. Chem. B* **110,** 16111–16116.

Gaskell, S. J. (1997). Electrospray: Principles and practice. *J. Mass Spectrom.* **32,** 677–688.

Gika, H. G., Theodoridis, G. A., and Wilson, I. D. (2008). Hydrophilic interaction and reversed-phase ultra-performance liquid chromatography TOF-MS for metabonomic analysis of Zucker rat urine. *J. Sep. Sci.* **31,** 1598–1608.

Glinski, M., and Weckwerth, W. (2006). The role of mass spectrometry in plant systems biology. *Mass Spectrom. Rev.* **25,** 173–214.

Griffiths, J. (2008). A brief history of mass spectrometry. *Anal. Chem.* **80,** 5678–5683.

Gross, J. H. (2004). Mass Spectrometry, A Textbook. Springer-Verlag, Berlin.

Hu, Q. Z., Noll, R. J., Li, H. Y., Makarov, A., Hardman, M., and Cooks, R. G. (2005). The Orbitrap: A new mass spectrometer. *J. Mass Spectrom.* **40,** 430–443.

Iribarne, J. V., and Thomson, B. A. (1976). Evaporation of small ions from charged droplets. *J. Chem. Phys.* **64,** 2287–2294.

Kell, D. B. (2005). Metabolomics, modelling and machine learning in systems biology: Towards an understanding of the languages of cells. The 2005 Theodor Bücher lecture. *FEBS J.* **273,** 873–894.

Kell, D. B., and Oliver, S. G. (2003). Here is the evidence, now what is the hypothesis? The complementary roles of inductive and hypothesis-driven science in the post-genomic era. *Bioessays* **26,** 99–105.

Kenny, L. C., Broadhurst, D. I., Dunn, W., Brown, M., North, R. A., McCowan, L., Roberts, C., Cooper, G. J. S., Kell, D. B., and Baker, P. N. Screening for Pregnancy Endpoints Consortium (2010). Robust early pregnancy prediction of later preeclampsia using metabolomic biomarkers. *Hypertension* **56,** 741–749.

Kislinger, T. (2007). Mass spectrometry-based systems biology. *In* "Cancer Treatment and Research," (I. Jurisica, D. A. Wigle, and B. Wong, eds.), pp. 59–83. Springer, New York.

Li, H. L., Song, F. R., Xing, J. P., Tsao, R., Liu, Z. Q., and Liu, S. Y. (2009). Screening and structural characterization of alpha-glucosidase inhibitors from hawthorn leaf flavonoids extract by ultrafiltration LC-DAD-MSn and SORI-CID FTICR MS. *J. Am. Soc. Mass Spectrom.* **20,** 1496–1503.

March, R. E. (1997). An introduction to quadrupole ion trap mass spectrometry. *J. Mass Spectrom.* **32,** 351–369.

Marshall, A. G., Hendrickson, C. L., and Jackson, G. S. (1998). Fourier transform ion cyclotron resonance mass spectrometry: A primer. *Mass Spectrom. Rev.* **17,** 1–35.

Mikesh, L. M., Ueberheide, B., Chi, A., Coon, J. J., Syka, J. E. P., Shabanowitz, J., and Hunt, D. F. (2006). The utility of ETD mass spectrometry in proteomic analysis. *Biochim. Biophys. Acta* **1764,** 1811–1822.

Mirsaleh-Kohan, N., Robertson, W. D., and Compton, R. N. (2008). Electron ionization time-of-flight mass spectrometry: Historical review and current applications. *Mass Spectrom. Rev.* **27,** 237–285.

Ouyang, Z., and Cooks, R. G. (2009). Miniature mass spectrometers. *Annu. Rev. Anal. Chem.* **2,** 187–214.

Petrie, S., and Bohme, D. K. (2007). Ions in space. *Mass Spectrom. Rev.* **26,** 258–280.

Politis, A., Park, A. Y., Hyung, S.-J., Barsky, D., Ruotolo, B. T., and Robinson, B. T. (2010). Integrating ion mobility mass spectrometry with molecular modelling to determine the architecture of multiprotein complexes. *PLoS One* **5,** e12080.

Ramautar, R., Somsen, G. W., and de Jong, G. J. (2009). CE-MS in metabolomics. *Electrophoresis* **30,** 276–291.

Robards, K., Haddad, P. R., and Jackson, P. E. (1997). Principles and Practice of Modern Chromatographic Methods. Academic Press, London.

Swartz, M. E. (2005). UPLC (TM): An introduction and review. *J. Liq. Chromatogr. Relat. Technol.* **28,** 1253–1263.

Uetrecht, C., Rose, R. J., van Duijn, E., Lorenzen, K., and Heck, A. J. R. (2010). Ion mobility mass spectrometry of proteins and protein assemblies. *Chem. Soc. Rev.* **39,** 1633–1655.

van Hove, E. R. A., Smith, D. F., and Heeren, R. M. A. (2010). A concise review of mass spectrometry imaging. *J. Chromatogr. A* **1217,** 3946–3954.

Villas-Boas, S. G., Roessner, U., Hansen, M. A. E., Smedsgaard, J., and Nielsen, J. (2007). Metabolome Analysis: An Introduction. John Wiley and Sons Inc., New York.

Washburn, M. P., Wolters, D., and Yates, J. R. (2001). Large-scale analysis of the yeast proteome by multidimensional protein identification technology. *Nat. Biotechnol.* **19,** 242–247.

Welthagen, W., Shellie, R. A., Spranger, J., Ristow, M., Zimmermann, R., and Fiehn, O. (2005). Comprehensive two-dimensional gas chromatography-time-of-flight mass spectrometry (GC x GC-TOF) for high resolution metabolomics: Biomarker discovery on spleen tissue extracts of obese NZO compared to lean C57BL/6 mice. *Metabolomics* **1,** 65–73.

Yates, J. R., Ruse, C. I., and Nakorchevsky, A. (2009). Proteomics by mass spectrometry: Approaches, advances, and applications. *Annu. Rev. Biomed. Eng.* **11,** 49–79.

Zelena, E., Dunn, W. B., Broadhurst, D., Francis-McIntyre, S., Carroll, K. M., Begley, P., O'Hagan, S., Knowles, J. D., Halsall, A., Wilson, I. D., Kell, D. B., and Husermet, C. (2009). Development of a robust and repeatable UPLC-MS method for the long-term metabolomic study of human serum. *Anal. Chem.* **81,** 1357–1364.

CHAPTER THREE

High-Throughput Quantification of Posttranslational Modifications *In Situ* by CA-FLIM

Hernán E. Grecco,[*,1] Pedro Roda-Navarro,[*,1] Sven Fengler,[*] *and* Philippe I. H. Bastiaens[*,†]

Contents

Abstract

Signal transduction is mediated by posttranslational modifications (PTMs) of proteins in complex signaling networks. Quantifying PTM levels of multiple network components in response to a stimulus is therefore the key to understand how their concerted activities give rise to cellular function. We have shown that fluorescence lifetime imaging microscopy (FLIM) on cell arrays (CA-FLIM) provides a method to accurately quantify PTM levels of many proteins *in situ*. Herein, we describe the detailed protocol for CA-FLIM. Less than 2 days are needed from cell array preparation to data analysis, where the main limiting step is the 24 h needed for transfection. After generating a single cell array containing 384 spots, it can be imaged and analyzed in less than 2 h.

[*] Department of Systemic Cell Biology, Max Planck Institute for Molecular Physiology, Dortmund, Germany
[†] Technical University of Dormtund, Dortmund, Germany
[1] These authors contributed equally to this work.

Methods in Enzymology, Volume 500
ISSN 0076-6879, DOI: 10.1016/B978-0-12-385118-5.00003-7

1. Introduction

Posttranslational modifications (PTMs) of proteins are ubiquitous in intracellular signaling networks as they transmit information quickly in a highly regulated manner (Tyson *et al.*, 2003). Understanding how a given stimulus is relayed within the cell requires the accurate determination of PTMs in the whole set of proteins that might be involved in signaling (Verveer and Bastiaens, 2008). An effective strategy to address such a variety of targets is to use universal reagents that bind the PTM, such as antibodies or protein domains that are labeled or fused to fluorophores. However, how can a specific protein containing a PTM be detected by a generic readout? Förster resonance energy transfer (FRET; Förster, 1948) quantified by fluorescence lifetime imaging microscopy (FLIM; Verveer *et al.*, 2000a) provides the specificity required for such a task. By monitoring the excited state lifetime of fluorescent proteins that are fused to the protein targets, the binding of the acceptor-labeled generic reagent to the protein can be deduced by FRET. The short range of FRET (~6 nm) guarantees that the signal will be highly specific for this interaction yielding a very low false positive rate. FRET measured by FLIM goes beyond intensity based read-outs as it provides a map of the absolute fraction of donor in complex with the acceptor without an external calibration. Using FLIM on cell arrays (CA-FLIM; Grecco *et al.*, 2010) with a generic reagent that recognizes a PTM provides a high-throughput approach to quantify PTM of many proteins *in situ* with subcellular resolution. In this way, the molecular state of a protein in a network can be correlated with other molecular or phenotypic traits in single cells. Here, we described the method and protocols for the high-throughput quantification of tyrosine phosphorylation by CA-FLIM using generic antiphosphotyrosine antibodies labeled with the FRET acceptor.

2. Methods

2.1. Sample preparation

2.1.1. Arrays of plasmids coding for fluorescent fusion proteins

The first step of the CA-FLIM workflow (Fig. 3.1) is the generation of a library of plasmids that encodes fluorescent fusion proteins. Initially, a cDNA pool is obtained from purified mRNA of human tissues by reverse transcription. Open reading frames (ORFs) of the cDNA pool are amplified by PCR and inserted by the gateway system (Walhout *et al.*, 2000) in an expression vector containing the DNA coding for the fluorescent protein that will act as FRET donor.

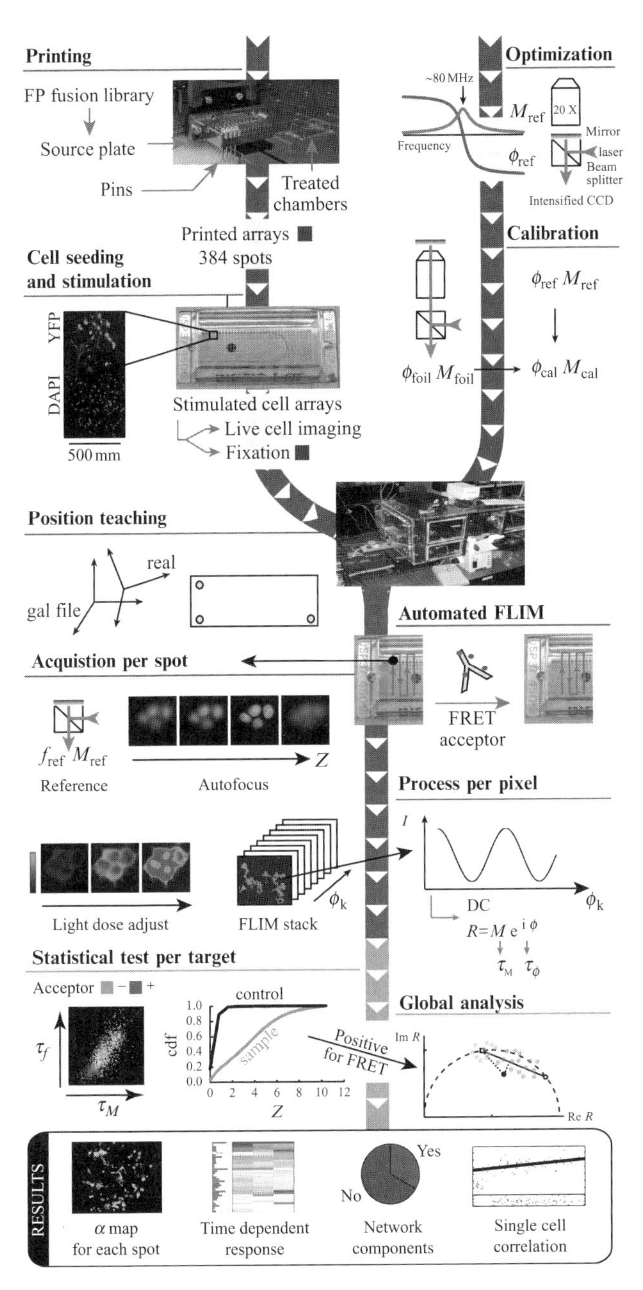

Figure 3.1 CA-FLIM visual workflow. The blue lane indicates the sample preparation steps to obtain a cell array. Green squares indicate pause points as indicated in the procedure section. The red lane indicates the microscopy preparation workflow. Both lanes converge in the automated FLIM violet lane, where after teaching the positions the spots are sequentially scanned. In each spot, optimization preparation steps can be

Plasmids are printed on imaging compatible LabTek chambered coverglasses (Nalge Nunc International; see Appendix 3 for a full list of materials) with a robot spotter Qarray2 (Genetix) to generate arrays. These chambers are pretreated with 1 *M* NaOH for 15 min at room temperature (RT) and washed twice with H_2O. This treatment dramatically improves the round shape of the spots (Fig. 3.2A). As source plates for the spotter, we use low volume 384-well plates (Nalge Nunc International) filled with transfection mixtures containing gelatine (Sigma–Aldrich), fibronectin (Sigma–Aldrich), Lipofectamine 2000 (Invitrogen), and the different plasmids as previously described for siRNA (Erfle *et al.*, 2007). Source plates are loaded with reagents manually or by using the Microlab Star Line Liquid Handling Workstation (Hamilton). Several replicates of each fluorescent fusion protein are included in arrays in randomized positions to decrease the error due to within-array biases and estimate experimental error. We also include reference spots in three of the four corners of the array with a plasmid that encodes fluorescent proteins such as EGFP or EYFP, which are good indicators of the transfection efficiency due to their slow turnover in cells.

Before the 384-well plates are used in the spotter, they are centrifuged for 30 s at 145 rcf to ensure that the transfection mixture is at the bottom of the well. The Qarray2 system is used as follows:

1. Switch on the vacuum system to immobilize chambers and source plates to the holders in the bed of the spotter.
2. Adjust the humidity and temperature during the spotting procedure to 60% and 20 °C, respectively.
3. Use four solid microarray steel pins (Arrayit corporation) in horizontal position in the "microarraying head" of the spotter. These pins produce spots of around 500 μm of diameter (Fig. 3.2A). Set the source order by columns.
4. Set the "slide design" to print spots in grids of 32 rows and 12 columns with a 1125 μm square unit cell (distance between the center of neighboring spots; Fig. 3.2A). This yields 384 samples per LabTek chamber.
5. Set the "number of stamps" per ink and spot to 3.
6. Set the "inking time" (the time that the pin stays in the well) to 110 ms.
7. Set the "stamp time" (the time that the pin is in contact with the glass in the LabTek chamber) to 110 ms.
8. Set the "print depth" to 180 μm.
9. Set the "multistamp timing" to immediate (the stamps will be done consecutively.)

---

performed before the FLIM stack is acquired. The yellow lane indicates the postprocessing of the data, in which targets showing FRET are selected using an statistical test and the fraction of posttranslationally modified protein per pixel is obtained. The resulting dataset can be used for different analysis. (See Color Insert.)

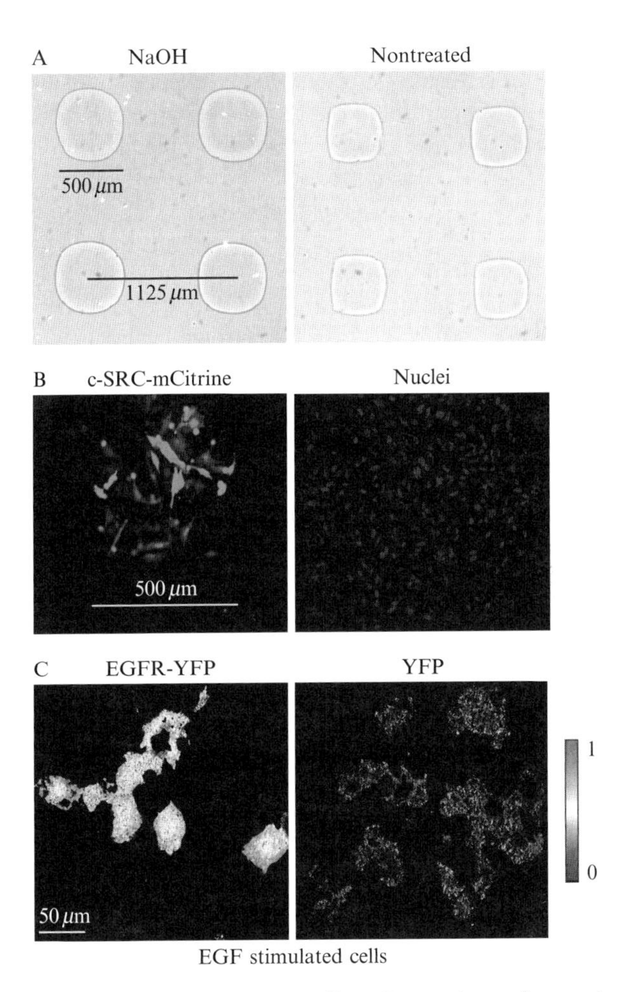

Figure 3.2 Quality control. (A) Spot quality. Comparison of spot shape and size between NaOH treated and nontreated LabTek chambers. (B) Transfection efficiency. Representative example of the transfection efficiency. Total number of cells is counted by staining nuclei (left panel). (C) Spots containing plasmids encoding for EGFR-YFP and YFP alone are used as a positive and negative controls respectively in cell arrays stimulated with EGF. (See Color Insert.)

10. Set the "wash station" to the following protocol: 3 s in H_2O followed by 1 s of drying time, 3 s in EtOH followed by 1 s of drying time, and 3 s in H_2O followed by 5 s of drying time and 1 s waiting step.
11. Prepare an Excel sheet with all the positions of wells in the source plate and the name of the sample that each of them contains. Import this file in the QSoft Data Tracking software (included with the Qarray2 system). The system will use this information for generating the Galfile (see next step).

12. Make a run in "data tracking spot only" to get the Galfile that contains the position of each sample in the array. This file is used as the input for the software of the automated FLIM microscope (see below).
13. Make a run in "normal" to print the array using all the wells in the source plate.

If the spot size and shape are not good, check that temperature and humidity in the Qarray2 spotter are set to 20 °C and 60%, respectively, and replace the LabTek chambers by new ones pretreated with NaOH. It is also necessary to inspect the pins using a reflective microscope or a magnifying glass. If the surface shows any sign of damage, replace the pins with new ones (Table 3.1).

Once the LabTek chamber is printed, label the first corner of the array with a marker pen that corresponds to the position A1 of the source plate. Printed LabTek chambers can be stored for several months in a box containing silica Gel Orange (Carl Roth GmbH, Germany). However, best results are obtained by minimizing the storage time.

2.2. Cell arrays for automated microscopy

2.2.1. Preparation of cell arrays

MCF7, MDA-MB-231, or HeLa cells (ATCC) cells are grown in DMEM (PAN Biotech GmbH) supplemented with 10% heat-inactivated fetal calf serum (FCS; PAN Biotech GmbH), 2 mM L-glutamine (PAN Biotech GmbH), and 100 U ml^{-1} penicillin and 100 μg ml^{-1} streptomycin (GIBCO) at 37 °C and 5% (MCF7 and HeLa) or no added (MDA-MB-231) CO_2. To determine the cell density, cells are detached by trypsinization, centrifuged at 290 rcf for 5 min, and resuspended in growth medium. Before seeding on cDNA arrays, we heat cells and LabTek chambers to 37 °C. Chambers are kept at this temperature with a hot plate controller (Labotect) during the seeding process to ensure an even distribution of cells on the array. The amount of cells seeded depends on the duplication time of the cell line. We use 2.5×10^5 MCF7 or HeLa cells, and 3×10^5 MDA-MB-231 cells per array. Cells are dispensed onto the array in a total volume of 3 ml by using a 5-ml pipette. Arrays are tilted to ensure that the cell suspension reaches all four corners of the chamber. Contact of the pipette to the chamber should be avoided as this could damage the array. In order to get a strong attachment of those cell lines that grow in suspension or show low adhesion to glass, we coat LabTek chambers with extracellular matrix components or poly-L-lysine (Sigma–Aldrich). For example, we use type I collagen (Serva) to favor a strong attachment of MDA-MB-231 cells.

Cell arrays constituted in this way are incubated at 37 °C and 5% CO_2 for 24 h to allow reverse transfection, protein expression, and fluorophore maturation. If low transfection efficiencies are observed, print new arrays and always use them shortly after they are printed,

Table 3.1 Troubleshooting common problems

Problem	Possible reason	Solution
Spot size and shape are not good	The environmental conditions are not properly adjusted	Set temperature and humidity in the Qarray2 spotter to 20 °C and 60%, respectively. Replace the LabTek chambers by new ones pretreated with NaOH
	The printing pins are damaged	Inspect the pins using a reflective microscope or a magnifying glass. If the surface shows any sign of damage, replace the pins with new ones
Low transfection efficiency	Old arrays are used	Print new arrays and always use them shortly after they are printed
	Presence of contaminants in the plasmid samples	Prepare endonuclease free plasmids
	Poor quality of the transfection reagent	Check the quality of the transfection reagent in liquid phase experiments
Low modulation	Signal generators are not synchronized	Lock the signal generators to a common external clock
	The AOM is misaligned	Check the modulation of the laser using a fast photodiode and oscilloscope. Realign to reach the Bragg condition
The reference phase and modulation change	The temperature of the AOM is not stable	Improve the temperature stabilization of the room. Couple the AOM to a thermostatic water bath/circulator to keep the temperature of the AOM within ±0.1 °C the set temperature (e.g., 2° above RT)

prepare endotoxin free plasmids, and check the quality of the transfection reagent in liquid phase experiments (Table 3.1).

2.2.2. Cell array stimulation

Different extracellular stimuli and stimulation times can be used along with CA-FLIM. For example, we stimulate cells with 1 mM pervanadate (PV) for 30 min in order to inhibit protein tyrosine phosphatases (Huyer *et al.*, 1997) and to promote phosphorylation of proteins on tyrosine residues, or with 100 ng/ml epidermal growth factor (EGF; cell signaling) to determine the structure and dynamics of signaling networks downstream of EGFR activation. PV is freshly prepared by mixing orthovanadate (Sigma–Aldrich) with H_2O_2 (Calbiochem) at 2:1 molar ratio for 10 min at RT. In order to reduce the basal levels of tyrosine phosphorylation, cells are starved in low serum medium (DMEM containing 0.5% FCS) at 37 °C for 4 h prior to stimulation.

A fusion protein known to be modified by the stimulation can be included as positive control in each cell array. We typically use the epidermal growth factor receptor (EGFR; Ullrich *et al.*, 1984) as a positive control for tyrosine phosphorylation in response to PV and EGF. In the same way, at least one protein that does not contain the PTM under study should be included as a negative control (Fig. 3.2C). In spots containing this control, the fluorescence lifetime of the FRET donor before and after the addition of the FRET acceptor should match.

At this step, live cell quantitative imaging can be performed. For *in situ* experiments, refer to the next section.

2.2.3. Fixation of cell arrays

After the stimulation time, aspirate the medium and incubate the cell array with the fixative solution. Fixatives that preserve the fluorescence lifetime of the FRET donor should be used. For example, cell fixation with 4% paraformaldehyde (PFA; Sigma–Aldrich) by 5 min incubation at RT preserves the monoexponential fluorescence decay of the YFP. Remove fixative solution and wash away by incubating the cell array 10 min in TBS (20 mM Tris–HCl, pH 7.5, 150 mM NaCl). At this step fixed cell arrays can be stored for several weeks at 4 °C in PBS containing 0.01% (w/v) sodium azide (Sigma–Aldrich).

Cell arrays are permeabilized for 5 min at RT by incubation with PBS containing 0.1% Triton X-100 and washed once with PBS. The cell nuclei are stained with 0.5 μg ml^{-1} of Hoechst solution (Sigma–Aldrich) in PBS and cell arrays are washed with PBS to make them ready for frequency domain (FD) FLIM. We measure the fluorescence lifetimes τ_ϕ and τ_M of the FRET donor (YFP) before and after 20 min incubation at RT with the PY72 antibody labeled with the FRET acceptor, the fluorolink Cy3.5 reactive dye (GE Healthcare) as explained below.

2.2.4. Labeling of the PY72 antibody

We label antibodies at free amino groups by 30 min incubation at RT with a 20-fold molar excess of the fluorolink Cy3.5 reactive dye in 0.1 *M* bicine/NaOH, pH 9.0, 3–5% dimethylformamide. Alkaline pH promotes deprotonation of the amino groups of lysines and therefore assists in the efficient coupling. We then purify the Cy3.5-labeled antibody by gel filtration chromatography with Econo-Pac 10 DG columns (Biorad), concentrate the antibody solution to 1 mg/ml with Microcon YM-10 (Amicon Bioseparation), and measure the labeling ratio (dye/protein) with a spectrophotometer. Labeling ratios around 3–4 are required in order to obtain high FRET efficiencies. Labeled antibody can be stored at -20 °C (optionally the antibody can be diluted in one volume of glycerol).

2.3. Image acquisition and analysis

2.3.1. FLIM optimization and calibration

Initialize the FLIM system (described in Appendix 1) by setting the intensifier, mode to "RF" the voltage to 700 V, and the "RF Gain" to 100. Set the voltage of both signal generators to -13 dBM. These values might have to be optimized for a particular system in order to obtain high modulation.

To find the frequency of the system that provides the maximum modulation, acquire FLIM stacks of the mirror located in the filtercube while scanning the modulation frequency of the intensifier in 40 steps with a range of 4 MHz around 80 MHz. The ratio between the frequencies of the signals applied to the intensifier and the acousto-optic modulator (AOM) has to be kept at 2:1. In addition to the FLIM stack, acquire a dark image by keeping the same exposure time but blocking the laser. Calculate for each frequency the phase (ϕ_{ref}) and modulation (M_{ref}) as described in Section 2.3.3. To further refine the frequency selection, repeat the previous step for a narrower frequency range around one of the peaks and select the frequency at which the modulation is maximal. M_{ref} should be above 0.9 to obtain robust FLIM data.

After the modulation depth has been optimized, acquire $I_{\text{foil}}(i, j, k)$, a FLIM stack, and a dark image of a mirror located in the sample plane (e.g., a metallic foil in LabTek chamber) to calibrate the dephase and demodulation introduced by the objective. Calculate the calibration phase and modulation as

$$\phi_{\text{cal}} = \phi_{\text{foil}} - \phi_{\text{ref}}^{\text{startup}} \tag{3.1a}$$

$$M_{\text{cal}} = M_{\text{foil}}/M_{\text{ref}}^{\text{startup}} \tag{3.1b}$$

Notice that this value is only dependent on the objective and labware used. As it does not show a significant temperature dependency, it can be considered as constant during the time of the experiment. To verify the proper calibration of the system, acquire a FLIM stack of a diluted sample of Alexa 488 ($\tau_{\phi} = \tau_M = \tau = 4.06$ ns). In addition, perform a dilution series to estimate the range of intensities at which the measured fluorescence lifetime is accurate.

2.3.2. Automated FLIM

As a first step, the coordinates as given by the spotter have to be transformed to the coordinated system of the stage. For this purpose, move the motorized stage to focus three or more reference spots in the array and record their

3D positions in the coordinate system of the stage ($\vec{w}$) and the spotter ($\vec{v}$). The rotation–translation transformation from the spotter coordinate system to the stage coordinate system can be obtained by minimizing the sum of the distances between the transformed N reference points to the measured positions in the stage:

$$\chi^2_{\min}\left(\vec{\Omega},\vec{D}\right) = \underset{\vec{\Omega},\vec{v}}{\arg\min} \sum_{n=1}^{N} \left\| \begin{pmatrix} \vec{w} \\ 1 \end{pmatrix} - \left[\begin{array}{c|c} \mathbb{R}\left(\vec{\Omega}\right) & \vec{D} \\ \hline 0 & 1 \end{array} \right] \begin{pmatrix} \vec{v} \\ 1 \end{pmatrix} \right\|^2 \tag{3.2}$$

where $\mathbb{R}$ is the usual 3D rotation matrix dependent on three angles ($\vec{\Omega}$) and $\vec{D}$ is a translation vector. As starting values for $\vec{D}$, use the shift in the centroid of the system. For Ω, use the angle of the average normal to the array. Use the transformation matrix obtained to convert all spotter coordinates to the stage coordinate system.

Once that the positions have been given to the system, the software should go over the whole slide performing the following actions in each spot:

- A refocusing step might be necessary dependent on the stability of the system. Due to its brightness, the DAPI image of the nuclei can be used to perform a fast software autofocus.
- Acquire a set of images at different excitation powers and exposure times to find the light dose that delivers optimal images.
- Acquire a FLIM phase stack and a dark image.
- A reference acquisition step might be necessary dependent on the stability of the system.
- Acquire a fluorescence image of the nuclei staining.
- Optionally, acquire fluorescence to track other organelles or molecules.

The FLIM stacks acquired in the absence of antibody provides a measurement of the fluorescence lifetime of the donor and allows quantifying the uncertainty of the measurement. Add the PY72–Cy3.5 and incubate for 20 min at RT and acquire for each spot the FLIM stack again. Washing is not necessary as the specificity of the assay is given by the donor and the stringent distance requirement of FRET.

The outcome of these steps is the following set of images for each spot:

- $I(i, j, k)^{\mathrm{no}ab}$: FLIM stack in the absence of the acceptor-labeled antibody.
- $I(i, j, k)^{ab}$: FLIM stack in the presence of the acceptor-labeled antibody.
- Other fluorescence images.

As described in Appendix 1, i, j are the indices of the pixels and k corresponds to the phase ϕ_k. There should be a number of $I_{\mathrm{ref}}(i, j, k)$ taken throughout the screening process to compensate for thermal drift in the AOM.

2.3.3. FLIM image processing

The following image processing can be performed using freeware tools such as ImageJ (Abramoff *et al.*, 2004) or CellProfiler (Carpenter *et al.*, 2006). For all acquired FLIM stacks, calculate the maximum projection intensity image and segment it into foreground and background using Otsu's method (Otsu, 1979). Define a background region of interest (ROI) by selecting a small region in the background close to the center of the image. For a homogeneous sample such as the reference, the foil, or dye solution, use the dark image to calculate the background.

For each image in the phase stack, subtract the mean intensity of the background ROI. Obtain for each pixel (i, j) the mean value (DC) of the FLIM stack and the real and imaginary part of the Fourier coefficients R using single value decomposition to fit:

$$\begin{aligned} I(i,j,k) = \mathrm{DC}(i,j) + \mathbb{R}\mathrm{e}\,R(i,j)\frac{\cos(\phi_k - \phi_{\mathrm{ref}} - \phi_{\mathrm{cal}})}{M_{\mathrm{ref}}M_{\mathrm{cal}}} \\ + \mathbb{I}\mathrm{m}\,R(i,j)\frac{\sin(\phi_k - \phi_{\mathrm{ref}} - \phi_{\mathrm{cal}})}{M_{\mathrm{ref}}M_{\mathrm{cal}}} \end{aligned} \quad (3.3)$$

Finally, calculate the apparent frequency-dependent fluorescence lifetime images as

$$\tau_\phi(i,j) = \frac{1}{w}\frac{\mathbb{I}\mathrm{m}\,R(i,j)}{\mathbb{R}\mathrm{e}\,R(i,j)} \quad (3.4a)$$

$$\tau_M(i,j) = \frac{1}{w}\sqrt{\frac{1}{\mathbb{R}\mathrm{e}\,R(i,j)^2 + \mathbb{I}\mathrm{m}\,R(i,j)^2} - 1} \quad (3.4b)$$

2.3.4. Assessment of positives

Segment the foreground of $\mathrm{DC}(i, j)^{\mathrm{no}ab}$ into cells by performing a connected component analysis. If an image of the nuclei is available, a two-step process can be performed by finding first the center of all nuclei (primary objects) and then the cells in $\mathrm{DC}(i, j)^{\mathrm{no}ab}$ (secondary objects). After segmentation, calculate for each cell the Z-score defined as

$$Z_{\mathrm{cell}} = \sqrt{\left(\frac{\tau_\phi^{ab} - \tau_\phi}{\sigma\tau_\phi}\right)^2 + \left(\frac{\tau_M^{ab} - \tau_M}{\sigma\tau_M}\right)^2} \quad (3.5)$$

To determine if FRET was detected for a particular target, compare the Z_{cell} distribution to the control distribution using a Kolmogorov–Smirnov Test (Massey, 1951).

2.3.5. Global analysis

For each target found positive for FRET, pool all pixels from all cells from all conditions and all treatments. Fit a straight line to the data

$$\mathbb{I}\mathrm{m}\, R(i,j) = u + v\mathbb{R}\mathrm{e}\, R(i,j) \tag{3.6}$$

and find R_D and R_F, the intersection points of the line with the "mono-exponential semicircle" (Fig. 3.4, Appendix 2). From these intersection points, calculate the lifetime of the donor τ_D and donor-in-complex τ_F.

From the slope v and the offset u of 6, the fluorescence lifetimes can be directly obtained:

$$\tau_D \tau_F = \frac{1 \pm \sqrt{1 - 4u(u + v)}}{2\omega u} \tag{3.7}$$

The FRET efficiency E can be calculated as

$$E = 1 - \frac{\tau_F}{\tau_D} \tag{3.8}$$

Calculate the fraction of donor in complex with acceptor α by projecting the R value of each pixel into the straight line:

$$\alpha(i,j) = \frac{\omega(\tau_F + \tau_D)\mathbb{R}\mathrm{e}\, R(i,j) + (\omega^2 \tau_F \tau_D - 1)\mathbb{I}\mathrm{m}\, R(i,j) - \omega\tau_F}{\omega(\tau_D - \tau_F)} \tag{3.9}$$

3. Results and Discussion

3.1. Arrays of plasmids coding for fluorescent fusion proteins

The main advantage of cloning by recombination is that it avoids restriction site incompatibilities and also allows fusion proteins to be generated with the fluorophore at either end of the cDNA. This enhances the chance to exclude fluorescent fusion proteins that do not localize to the proper compartment. The subcellular distribution of the fluorescent fusion proteins used in our screening has been previously described (http://www.dkfz-heidelberg.de; Simpson *et al.*, 2000). However, in the case that this information is missing, the images obtained by CA-FLIM allow the subcellular distribution of each fluorescent fusion protein to be directly analyzed.

We have improved the printing procedure by pretreatment of LabTek chambers with 1 *M* NaOH. This pretreatment increased the size and improved the shape of spots (Fig. 3.2A).

3.2. Cell arrays

Transfection efficiencies were optimal (around 40% of transfected cells per spot) with 1–1.5 μg of DNA and 1.5–3.5 μl of Lipofectamine 2000 in a final volume of transfection mixture per well of 13 μl (Fig. 3.2B). We did not find substantial transfection efficiency variations in different batches of Lipofectamine 2000. Other transfection reagents, including Effectene, Fugene, as well as transfection mediated by precipitates of calcium phosphate did not improve the results obtained with Lipofectamine 2000.

Cell arrays as opposed to multiwell plates allow homogeneous stimulation. The high density of samples on cell arrays is optimal for automated microscopy and reduces the sources of random errors associated with the large dimensions of multiwell plates such as well location, liquid dispensing, or signal intensity (Malo *et al.*, 2006).

An established general inductor of a PTM allows the assessment of its detectability on modified proteins. For example, we have used sodium PV in order to inhibit protein tyrosine phosphatases (Huyer *et al.*, 1997) and thereby ubiquitously increase the phosphorylation on tyrosine residues. In this way, one is able to detect potential phosphoproteins that belong to a signaling network, decreasing the occurrence of false negatives.

3.3. FLIM and FLIM image processing

In CA-FLIM, it is important to select a donor whose fluorescence decay profile can be characterized by a single lifetime which is not significantly modified upon fusion with another protein. Otherwise, the complexity of resolving more fluorescence lifetimes (Appendix 2) hinders the ability to accurately quantify the fraction of donor in complex, making the analysis less robust. We have found empirically that the enhanced yellow fluorescent protein (EYFP; Clontech) fulfills these conditions.

A substantial overlap between the emission spectrum of the donor and the absorption spectrum of the acceptor is required in order to obtain a high FRET efficiency. Detailed information about selecting fluorophore pairs suitable for FRET pairs can be found elsewhere (Shaner *et al.*, 2005). As a good FRET acceptor for EYFP, we used the Cy3.5 dye coupled to a generic antiphosphotyrosine antibody. FRET will not occur in donor–acceptor complexes when the spatial orientation is not optimal. To minimize the chances of obtaining such false negatives, coupling of multiple labels to the antibody increases the effective FRET efficiency. We labeled the generic antiphosphotyrosine antibody PY72 with ~3–4 Cy3.5 molecules per antibody when screening for tyrosine phosphorylation patterns (Grecco *et al.*, 2010; Verveer *et al.*, 2000a).

For assessing FRET occurrence, we calculated a *Z*-score per cell for each fluorescent fusion protein. This score quantifies the decrease of τ_{ϕ} and

τ_M after the addition of the FRET acceptor. The cumulative distribution of this indicator for each target is statistically compared with the negative control (typically the donor alone) using a nonparametric test. Only targets that show p-values < 0.05 are considered positive for FRET and further analyzed by global analysis.

Global analysis enables the determination of the FRET efficiency in the complex (E) and the quantification of the fraction of modified protein per pixel (α map) by using *a priori* knowledge (Appendix 2; Grecco *et al.*, 2009; Verveer and Bastiaens, 2003). An important feature of this analysis is that it can simultaneously analyze countless numbers of datasets, therefore yielding more accurate values for E and α.

4. Conclusion: Why Use CA-FLIM?

The most important feature of this protocol is that it allows quantification *in situ* of the extent of PTM using generic reagents in high-throughput screening platforms. We have applied CA-FLIM in our laboratory for the quantification of phosphotyrosine levels *in situ*. Moreover, we can constrain the scope of the readout by using more reagents that recognize modifications within specific sequences. For example, SH2 domains can be used to bind certain phosphorylated tyrosines (Pawson *et al.*, 2001). However, any PTM can be quantified provided that a reagent that binds to the modified amino acid is available.

A major advantage of CA-FLIM is that PTMs are quantified with single cell resolution, which allows PTM levels to be correlated with other molecular or phenotypic traits. For example, by inhibiting protein tyrosine phosphatases with PV, and based on Michaelis–Menten kinetics, we have shown that the correlation between the phosphorylated fraction and the expression level of a fluorescent fusion protein conveys information about network motifs. We have also obtained information about the structure and dynamics of the signaling network downstream of EGFR stimulation. For each target, we calculated the change in the phosphorylated fraction upon stimulation with EGF at different time points. Hierarchical clustering using an Euclidean distance of the response profile provided information about the similitude of the different targets in strength and response dynamics. This analysis classifies proteins according to their biological functionality and can be used to predict the function of unknown proteins (Grecco *et al.*, 2010). Due to the fact that FLIM measures PTM with spatial resolution, CA-FLIM also provides the means for studying spatially regulated signaling processes.

The ability to quantify the PTM state of many different proteins in a short time allows the causality in the signaling network to be studied by perturbation analysis. This task requires that the expression levels or activity of each component of a given network are perturbed, while quantifying the induced response in all other components. This can be accomplished by using a quantitative, generic, and high-throughput technique such as CA-FLIM. We expect that CA-FLIM will constitute a widely used tool for untangling the intricate signaling networks that alter the cell's phenotype in response to a great variety of extracellular stimuli.

Appendix 1. Frequency Domain Fluorescence Lifetime Imaging Microscopy

A1. General principles

In FD-FLIM, the sample is excited with an intensity-modulated light source inducing an intensity-modulated fluorescence emission. The dephase and demodulation between excitation and emission is a direct measurement of the lifetime of the excited state (Lakowicz, 2008; Fig. 3.3A). The optimal frequency f for measuring a fluorescence lifetime τ is such that $2\pi f\tau \sim 1$, resulting in a range of frequencies from 10 to 100 MHz for most fluorophores used in cell biology.

The current inability to acquire images at GHz frame rate makes it impossible to directly measure the dephase and demodulation of the fluorescence emission using a CCD camera. Therefore, most wide field homodyne FD-FLIM implementations recover these parameters from a stack of images (FLIM stack) acquired using a microchannel plate (MCP) intensifier to provide the necessary bandwidth. This intensifier is placed in front of the camera to effectively modulate the gain at the frequency of the excitation modulation but phase shifted in ϕ_k (Gadella *et al.*, 1993). The FLIM stack $I(i, j, k)$ is acquired by taking multiple images at different phase shifts ϕ_k over 2π (Fig. 3.3B). The phase ϕ and modulation M at each pixel or alternative their complex representation $R = Me^{i\phi}$ can be obtained using singular value decomposition. It is common to express these quantities as their lifetime counterparts τ_ϕ and τ_M. These apparent frequency-dependent fluorescence lifetimes are only equal to each other when the fluorescence decay can be characterized by a single exponential.

It is necessary to calibrate the phase and modulation of the excitation beam at the sample plane due to the delays introduced by the electronics and optics. This can be achieved by acquiring a FLIM stack of a reflective metallic foil at the sample position $I_{\text{foil}}(i, j, k)$. A periodic recalibration is needed to account for the drift in environmental and alignment conditions, but replacing the sample by a reflective one is impractical in screening applications. A more convenient method is to register periodically a reference FLIM stack of a mirror located in a filter cube $I_{\text{ref}}(i, j, k)$, which can be traced to the sample plane with a single measurement of $I_{\text{foil}}(i, j, k)$. In this

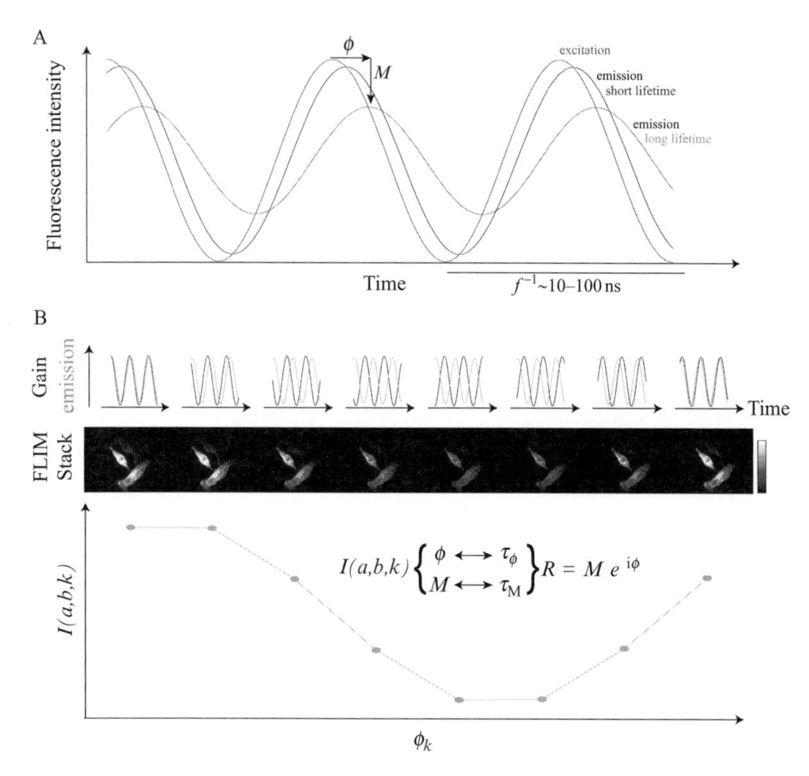

Figure 3.3 FD-FLIM. (A) Sinusoidally modulated excitation and emission of two fluorophores with two different lifetimes (short and long). The phase shift, demodulation, and the period of the wave are indicated. (B) Acquisition of the FLIM stack by homodyne detection. Upper panels show the modulated emission along with the different phases of the modulated gain of the intensifier. Middle panel shows the different phases of the acquired FLIM stack. Plot in the lower panel shows the change in $I(i, j, k)$ in the different phases of the FLIM stack. (See Color Insert.)

way a calibration phase (ϕ_{cal}) and modulation (M_{cal}) are obtained at startup of the system and the reference values ϕ_{ref} and M_{ref} are measured periodically during the screening.

A2. FLIM instrumentation

In our general purpose microscopic screening setup, we use an argon laser (Coherent, Innova 305) as a light source, as it provides enough power (>100 mW) in the lines (457.0, 488.0, and 514.5 nm), which are used to excited commonly used fluorescent proteins (CFP/TFP, GFP, and YFP/mCitrine). The desired excitation wavelength and power is selected with an acousto-optic tunable filter (AOTF, AA, AOTFnC-VIS-TN). We also used the AOTF as a fast shutter by connecting its blanking input to the exposure out TTL of the camera. Intensity modulation is achieved by creating an oscillating diffraction grating with a standing wave acousto-optic modulator (AOM; IntraAction, SWM-804AE1-1) powered with a sinusoidal wave of frequency $f/2$, which deflects the light from the zero order at a frequency f. The AOM is mounted in a rotating unit (Thorlabs, PR01) for alignment in the Bragg condition. The modulated zeroth order beam is selected by placing an iris diaphragm (Thorlabs, ID25/M) in the optical path. The modulation of the laser can be checked using a fast photodiode (Thorlabs, DET10A) and oscilloscope (Tektronix, TDS 820). Depending on the wavelength or power requirement of a particular application, a more compact but less flexible modulated solid state laser with 1–100 MHz bandwidth could be used to replace these devices.

The laser is coupled into a vibrationally isolated inverted microscope (Olympus, IX81) using a multimode fiber (Schäfter + Kirchhoff GmbH, #46688-03). The spatial coherence of the laser is disrupted by vibrating the fiber using a rotating eccentric wheel attached to the fiber. This results in a randomly moving speckle pattern, which averages out during detection. Homogeneous (Koehler) illumination at the sample plane is achieved by imaging the fiber core in the backfocal plane of the objective (for CA-FLIM: Olympus, UPSLAPO 20X). Dichroic and emission filters (for YFP: Chroma, 530LP and 538/25) are used to deliver the excitation light (514 nm) into the sample and collect the fluorescence. The emitted fluorescence is detected using an Intensified CCD camera (LaVision, Picostar HRI 12). The sample plane is imaged into a photocathode from where photoelectrons are ejected onto the front face of the MCP. The electron image is intensified as it travels through the MCP and converted into photons by a phosphorous screen at its rear face. These photons are detected by a 12-bit CCD camera using a 0.5× magnification lens to match the CCD chip and the phosphorous screen sizes. The gain of the system is controlled by the voltage across the MCP and the photocathode voltage. By applying a biased sinusoidal voltage with frequency f to the photocathode, the effective gain of the intensifier can be modulated.

To perform phase-sensitive homodyne detection, it is important that (1) the frequencies of the modulation and detection are precisely matched and (2) the phase between the two signals can be shifted. We achieve (1) by using the same clock reference applied to a pair of signal generators (Aeroflex, 2023A). We generate the dephase ϕ_k by directly commanding the initial phase of the signal generator. Alternatively, a single signal generator at $f/2$ and a frequency doubler can be used to feed both the AOM and the photocathode. In this case, a delay unit with subnanosecond resolution is used to achieve the dephase. To minimize the impact of photobleaching in the calculation of the lifetimes the dephased images are not acquired in a linear sequence of ϕ_k over 2π, but rather in pseudorandom order (van Munster and Gadella, 2004).

Appendix 2. Global Analysis of FRET–FLIM Data

The phase and modulation parameterization of the fluorescence decay profile can be superseded by a parameterization in terms of three biophysical relevant quantities. The first two are the fluorescence lifetime of the donor (τ_D) alone and the donor in complex with the acceptor (τ_F), which together define the FRET efficiency $E = 1 - \tau_F/\tau_D$. The third parameter is the fraction of donor in complex with the acceptor (α).

When assessing interactions, the fluorescence lifetimes are only dependent on the photophysical properties of the donor alone and in complex. Therefore, for a given FRET pair, these values can be considered constant and globally linked across all datasets. In contrast, the biologically relevant parameter $\alpha(i, j)$ quantifies in each pixel the fraction of interacting protein.

In the complex plane representation of the data (Grecco *et al.*, 2009; Verveer and Bastiaens, 2003; Verveer *et al.*, 2000b), points corresponding to a monoexponential fluorescence decay lay in a semicircle with radius and center 0.5 (Fig. 3.4).

$$R_\tau = \frac{1}{1 - i\omega\tau}$$

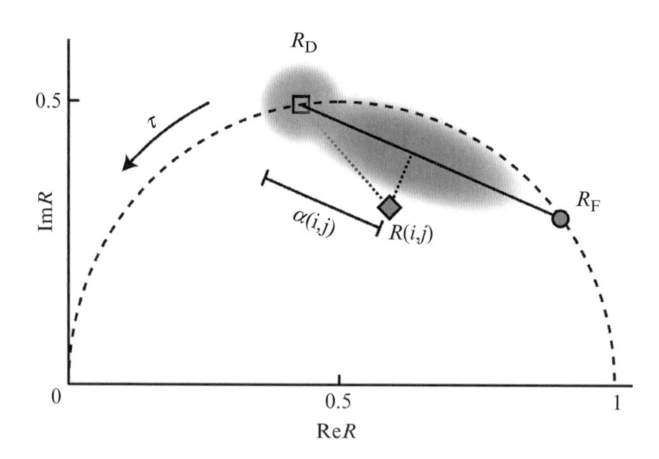

Figure 3.4 Global analysis in the complex plane. Complex plane schematic representation of phase shift and modulation data before (green cloud) and after (red cloud) the addition of the FRET acceptor. The dashed semicircle represents those points characterized by a single fluorescence lifetime (τ). R_D and R_F are the values for the donor alone and in complex with the acceptor. A mixture will be a linear combination of those. The fraction of donor in complex with the acceptor (α) is the length of the projection of $R - R_D$ onto the vector $R_F - R_D$. (See Color Insert.)

A biexponential decay (τ_D and τ_F) is just the linear combination of the complex quantities for each component and the prefactors are the relative fractions of each component

$$R(i,j) = (1 - \alpha(i,j))R_D + \alpha(i,j)R_F$$

If there is enough dispersion in the α values, a straight line can be fitted to the cloud of $R(i, j)$. The intersection points with the "monoexponential semicircle" are RD and RF from which the corresponding lifetimes can be derived. The value of $\alpha(i, j)$ can be obtained by projecting the vector $R(i, j) - R_D$ into the vector $R_F - R_D$ (Fig. 3.4).

The fact that recovering these parameters constitutes a linear problem is of special interest in screening applications a large number of datasets can be analyzed together (Clayton *et al.*, 2004).

Appendix 3. Materials

Reagents

- ORF-YFP clones previously generated (see Section 3; Simpson *et al.*, 2000).
- Lipofectamine 2000 (Invitrogen, cat. no. 11668–019)
- Sucrose (Sigma–Aldrich, cat. no. S-0389)
- Sodium hydroxide (J. T. Baker, cat. no. UN1823)
- Gelatine (Sigma–Aldrich, cat. no. G-9391)
- Fibronectin, human (Sigma–Aldrich, cat. no. F0895)
- Collagen R solution (Serva, cat. no. 47254)
- Poly-L-lysine (Sigma–Aldrich, cat. no. P6516)
- DMEM (PAN Biotech GmbH, cat. no. P04-03600)
- OptiMEM I + GlutaMAX I (GIBCO, cat. no. 51985–026)
- PBS (PAN Biotech GmbH, cat. no. P04-36500)
- Fetal calf serum (PAN Biotech GmbH, cat. no. P30-0402)
- L-Glutamine (PAN Biotech GmbH, cat. no. P04-80100)
- Penicillin/streptomycin (GIBCO, cat. no. 15070)
- Trypsin (PAN Biotech GmbH, cat. no. P10-023100)
- Hydrogen peroxide (H_2O_2; Calbiochem, cat. no. 386790)

! CAUTION: H_2O_2 causes burns. Handle it with protective clothing, gloves, and eye/face protections.

- Sodium orthovanadate (Sigma–Aldrich, cat. no. 6508)
- Epidermal growth factor (EGF) (Cell signaling, cat. no. 8916SC)
- Silica Gel Orange (Carl Roth GmbH, cat. no. P077.1)
- Sodium azide (Sigma–Aldrich, cat. no. S-8032)

- Paraformaldehyde (PFA; Sigma–Aldrich, cat. no. P6148)
- Triton X-100 (Sigma–Aldrich, cat. no. 234729)
- Generic antiphosphotyrosine antibody (PY72; *in vivo* Biotech Services)
- Fluorolink Cy3.5 reactive dye (GE Healthcare, cat. no. PA23501)
- Alexa Fluor 488 Dye (Invitrogen)
- Hoechst solution (Sigma–Aldrich, cat. no. H6024)
- Cell lines:
  - Mammary gland-derived adenocarcinoma, MCF7 (ATCC, cat. no. HTB-22)
  - Mammary gland-derived adenocarcinoma, MDA-MB-231 (ATCC, cat. no. HTB-26)

Equipment

- 384-well low volume plates (Nalge Nunc International, cat. no. 264360)
- 96-well cell culture plates (cell star, cat. no. 651180)
- Sterile filter (0.45 μm; Millipore, cat. no. 83.1826)
- Reservoir (Nalge Nunc International, cat. no. 370905)
- Econo-Pac 10 DG chromatography columns (Biorad, cat. no. 732–2010)
- Microcon centrifuge filter devices YM-10 (Amicon Bioseparation, cat. no. 42407)
- Centrifuge for plates 5430 (Eppendorf, cat. no. 5427 000.216)
- Hot plate controller 062 (Labotect)
- Solid microarray pins (Array-It corporation, cat. no. 946NS15)
- LabTek chambered coverglass (Nalge Nunc International, cat. no. 155361)
- Nanodrop spectrophotometer (Peqlab Biotechnologie GmbH, cat. no. ND-1000)
- Cell counter (Beckman Coulter, Vi-Cell-XR)
- Microlab Star Line Liquid Handling Workstation (Hamilton)
- Robot spotter Qarray2 (Genetix)
- Fully motorized microscope (Olympus, IX81)
- Computer controlled stage (Marzhauser, SCAN IM 120 × 100)
- 20× objective (Olympus, UPSLAPO 20×)
- Acousto-optic modulator (IntraAction, SWM-804AE1-1)
- RF Amplifier (IntraAction, PA)

! CAUTION: Do not turn on the amplifier if the AOM is disconnected.

- Dichroic filter (Chroma, 530LP)
- Emission filter (Chroma, 538/25)
- Intensified CCD and controller (LaVision, Picostar HRI 12)
- Two signal generators (Aeroflex, 2023A)
- Laser (Coherent, Innova 305)

- Acousto-optic tunable filter (AA, AOTFnC-VIS-TN)
- Acousto-optic tunable filter Controller (AA,MDS8C-D66-22-80.153)

! CAUTION: Do not turn on the controller if the AOTF is disconnected.

- Fiber (Schäfter + Kirchhoff GmbH, #46688-03)
- Mode scrambler (home built, see Appendix 1)
- Rotation mount (Thorlabs, PR01)
- Mirrors and mirror holders (Thorlabs, KM100-E02)
- Iris diaphragm (Thorlabs, ID25/M)
- High frequency oscilloscope (Tektronix, TDS 820)
- Fast photodiode (Thorlabs, DET10A)

REFERENCES

Abramoff, M. D., Magelhaes, P., and Ram, S. (2004). Image processing with imagej. *Biophotonics Int.* **11**(7), 36–42.

Carpenter, Anne E., Jones, T. R., Lamprecht, M. R., Clarke, C., Kang, I. H., Friman, O., Guertin, D. A., Chang, J. H., Lindquist, R. A., Moffat, J., Golland, P., and Sabatini, D. (2006). Cellprofiler: Image analysis software for identifying and quantifying cell phenotypes. *Genome Biol.* **7**(10), R100R100.

Clayton, A. H. A., Hanley, Q. S., and Verveer, P. J. (2004). Graphical representation and multicomponent analysis of single-frequency fluorescence lifetime imaging microscopy data. *J. Microsc.* **213,** 1–5.

Erfle, H., Neumann, B., Liebel, U., Rogers, P., Held, M., Walter, T., Ellen-berg, J., and Pepperkok, R. (2007). Reverse transfection on cell arrays for high content screening microscopy. *Nat. Protoc.* **2**(2), 392–399.

Förster, T. (1948). Zwischenmolekulare energiewanderung und fluoreszenz. *An-nalen der Physik* **437**(1–2), 55–75.

Gadella, T. W. J., Jovin, T. M., and Clegg, R. M. (1993). Fluorescence lifetime imaging microscopy (flim)—Spatial-resolution of microstructures on the nanosecond time-scale. *Biophys. Chem.* **48**(2), 221–239.

Grecco, H. E., Roda-Navarro, P., and Verveer, P. J. (2009). Global analysis of time correlated single photon counting fret-flim data. *Opt. Express* **17**(8), 6493–6508.

Grecco, H. E., Roda-Navarro, P., Girod, A., Hou, J., Frahm, T., Truxius, D. C., Pepperkok, R., Squire, A., and Bastiaens, P. I. (2010). In situ analysis of tyrosine phosphorylation networks by flim on cell arrays. *Nat. Methods* **7**(6), 467–472.

Huyer, G., Liu, S., Kelly, J., Moffat, J., Payette, P., Kennedy, B., Tsaprailis, G., Gresser, M. J., and Ramachandran, C. (1997). Mechanism of inhibition of protein-tyrosine phosphatases by vanadate and pervanadate. *J. Biol. Chem.* **272**(2), 843–851.

Lakowicz, J. (2008). Principles of Fluorescence Spectroscopy. 3rd edn. Spring Street, New York, NY, USA.

Malo, N., Hanley, J. A., Cerquozzi, S., Pelletier, J., and Nadon, R. (2006). Statistical practice in high-throughput screening data analysis. *Nat. Biotechnol.* **24**(2), 167–175.

Massey, F. (1951). The kolmogorov–smirnov test for goodness of fit. *J. Am. Stat. Assoc.* **46**(253), 68–78.

Otsu, N. (1979). A threshold selection method from gray-level histograms. *IEEE Trans. Syst. Man Cybern.* **9**(1), 62–66.

Pawson, T., Gishb, G. D., and Nashb, P. (2001). Sh2 domains, interaction modules and cellular wiring. *Trends Cell Biol.* **11**(12), 504–511.
Shaner, N. C., Steinbach, P. A., and Tsien, R. Y. (2005). A guide to choosing fluorescent proteins. *Nat. Methods* **2**(12), 905–909.
Simpson, J. C., Wellenreuther, R., Poustka, A., Pepperkok, R., and Wiemann, S. (2000). Systematic subcellular localization of novel proteins identified by large-scale cDNA sequencing. *EMBO Rep.* **1**(3), 287–292.
Tyson, J. J., Chen, K. C., and Novak, B. (2003). Sniffers, buzzers, toggles and blinkers: Dynamics of regulatory and signaling pathways in the cell. *Curr. Opin. Cell Biol.* **15**(2), 221–231.
Ullrich, A., Coussens, L., Hayflick, J. S., Dull, T. J., Gray, A., Tam, A. W., Lee, J., Yarden, Y., Libermann, T. A., Schlessinger, J., *et al.* (1984). Human epidermal growth factor receptor cDNA sequence and aberrant expression of the amplified gene in a431 epidermoid carcinoma cells. *Nature* **309**(5967), 418–425.
van Munster, E. B., and Gadella, T. W. J. (2004). Suppression of photobleaching-induced artifacts in frequency-domain flim by permutation of the recording order. *Cytometry A* **58A**(2), 185–194.
Verveer, P. J., and Bastiaens, P. I. H. (2003). Evaluation of global analysis algorithms for single frequency fluorescence lifetime imaging microscopy data. *J. Microsc.* **209,** 1–7.
Verveer, P. J., and Bastiaens, P. I. H. (2008). Quantitative microscopy and systems biology: Seeing the whole picture. *Histochem. Cell Biol.* **130**(5), 833–843.
Verveer, P. J., Wouters, F. S., Reynolds, A. R., and Bastiaens, P. I. H. (2000a). Quantitative imaging of lateral erbb1 receptor signal propagation in the plasma membrane. *Science* **290** (5496), 1567–1570.
Verveer, P. J., Squire, A., and Bastiaens, P. I. H. (2000b). Global analysis of fluorescence lifetime imaging microscopy data. *Biophys. J.* **78**(4), 2127–2137.
Walhout, A. J., Sordella, R., Lu, X., Hartley, J. L., Temple, G. F., Brasch, M. A., Thierry-Mieg, N., and Vidal, M. (2000). Protein interaction mapping in *c. elegans* using proteins involved in vulval development. *Science* **287**(5450), 116–122.

CHAPTER FOUR

Absorption Spectroscopy

Sanjay M. Nilapwar,* Maria Nardelli,† Hans V. Westerhoff,*,†,‡,§ *and* Malkhey Verma*

Contents

Abstract

Absorption spectroscopy is one of the most widely used techniques employed for determining the concentrations of absorbing species (chromophores) in solutions. It is a nondestructive technique which biologists and biochemists and now systems biologists use to quantify the cellular components and characteristic parameters of functional molecules. This quantification is most relevant in the context of systems biology. For creating a quantitative depiction of a metabolic pathway, a number of parameters and variables are important

* Manchester Interdisciplinary Biocentre, The University of Manchester, Manchester, United Kingdom
† Doctoral Training Centre for Integrative Systems Biology, The University of Manchester, Manchester, United Kingdom
‡ Manchester Centre for Integrative Systems Biology, The University of Manchester, Manchester, United Kingdom
§ Netherlands Institute for Systems Biology, VU University Amsterdam, Amsterdam, The Netherlands

Methods in Enzymology, Volume 500
ISSN 0076-6879, DOI: 10.1016/B978-0-12-385118-5.00004-9

and these need to be determined experimentally. This chapter describes the UV–visible absorption spectroscopy used to produce experimental data for bottom-up modeling approaches of systems biology which uses concentrations and kinetic parameters (K_m and V_{max}) of enzymes of metabolic/signaling pathways, intracellular concentrations of metabolites and fluxes. It also briefly describes the application of this technique for quantification of biomolecules and investigating biomolecular interactions.

1. Introduction

Absorption spectroscopy is widely used to obtain the absorbance spectra of specific molecules in solution and as solids. In the previous century, it has evolved as the preferred method for qualitative and quantitative determination of molecules present in solution. Absorption spectroscopy is also used as collective term for describing various spectroscopic techniques, such as UV–visible, fluorescence, circular dichroism (CD), and infrared spectroscopy. This chapter briefly covers the theoretical details and experimental aspects of UV-visible absorption spectroscopy for quantitative and qualitative measurement of biomolecules.

2. Theory of Absorption Spectroscopy

2.1. Origins of spectra

Light visible to the human eye can be defined as electromagnetic radiation with wavelength between 400 and 780 nm (Threlfall, 1993). The apparent color of a material is complementary to the color of the incident light absorbed (see Table 4.1). So when white light is irradiated on a sample, the light will be partially reflected giving out a white color to the sample. If the light is fully absorbed, the substance will appear as black. Selective absorption of a color such as yellow results in the remaining light lacking the color yellow being reflected and the object in this example appearing blue (see Table 4.1). Blue is then referred to as the complementary color. Table 4.1 also describes the relationship between absorbed radiation (nm) and apparent color. The UV-visible spectral region (200–780) is normally expressed in terms of nanometers (nm) or angstroms (Å) of the corresponding wavelengths. The units are interrelated as

$$1\,\text{nm} = 10\ \text{Å}$$

It is customary to use nanometers rather than the pre-SI Angstrom unit.

Table 4.1 Apparent color and the complementary absorbed color

Wavelength range (nm)[a]	Apparent color	Absorbed color
400–465	Violet	Yellow–green
465–482	Blue	Yellow
482–487	Greenish-blue	Orange
487–493	Blue–green	Red–orange
493–498	Bluish-green	Red
498–530	Green	Red–purple
530–559	Yellowish-green	Reddish-purple
559–571	Yellow–green	Purple
571–576	Greenish-yellow	Violet
576–580	Yellow	Blue
580–587	Yellowish-orange	Blue
587–597	Orange	Greenish-blue
597–617	Reddish-orange	Blue–green
617–780	Red	Blue–green

[a] Wavelength range given is approximate (Burns, 1993).

Some insight into quantum theory is useful for appreciating practical UV-visible absorption spectroscopy. First, light of a specific color is not just a collection of waves with various wavelengths, it is also a collection of particles with a precise kinetic energies (nonrest masses in relativity theory), called photons. The wavelength and energy are related by

$$E = h\frac{c}{\lambda}$$

where, h is Planck's constant and c the velocity of light *in vacuo*.

The energies of photons in the region of 200–780 nm allow excitation of outer valence electrons and inner shell d–d transitions with associated vibrational levels (Burns, 1993). When molecules are closely but irregularly packed together, as they are in spectroscopic measurements of solutions, they influence each other's energy levels which become broadened blurring the sharp spectral lines present in the vapor states into wide spectral bands. These effects can be seen as a difference between the spectrums of benzene as a vapor and in the liquid state (Burns, 1993).

The color of a substance (Table 4.1) is related to its electronic structure. The absorption of ultraviolet or visible light by the substance will be accompanied by changes in the electronic state of the molecules in the sample, but only if such change is possible; light absorption by substances is quantal.

Only certain energy changes are possible, because the structure of the molecule in its ground state only allows for certain changes to take place. The energy provided should enable electrons in an orbital in the molecule to transit to an orbital of higher energy. Because photon absorption is too fast for heat to escape from the absorbing molecule, the difference in energy between the ground state and the excited state of the molecule should be (almost, see above) precisely equal to the energy of the photon. In most stable molecules, all bonding orbitals are fully populated by electrons already and transition needs to be very fast; they do not allow the electrons to change the position and must therefore between overlapping orbitals. Consequently, the bonding electrons jump mostly to the corresponding antibonding orbitals. The ground state orbitals involved are σ, π, and n (nonbonding) orbitals, while antibonding orbitals are σ*(sigma star) and π* (pi star). Transition of an electron from a bonding to the corresponding antibonding σ orbital is referred as σ–σ* transition, and the transition of one electron from a π orbital to an antibonding π* orbital is referred as π–π* transition (see Fig. 4.1). Other electronic transitions also occur due to the absorption of UV–visible light and include n to σ* and n to π*. Both σ to σ* and n to σ* transitions require high energy and occur in the far ultraviolet region of 150–250 nm. These transitions are relatively weak absorbers. Most of the UV–visible spectra involve n to π* and the more intense π to π* transitions (Table 4.2); it is these orbitals that overlap most.

A spectrum is the absorbance of the substance as a function of the wavelength of the light. The spectrum exhibits peaks at the wavelengths that correspond to the energy differences between n and π* and between π and π* orbitals. Every single covalent bond in a molecule will have a σ and a corresponding σ* orbital. Every double bond will have a π and a corresponding π* orbital. Every lone electron pair in a molecule has an n orbital. Two lone pairs typically occur on oxygen atoms, because these have six electrons and at most two covalent bonds and one lone pair occurs on nitrogen atoms, because these have five electrons and mostly three covalent bonds.

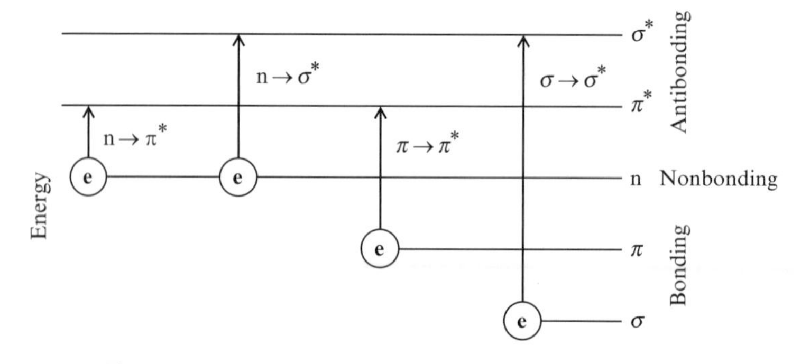

Figure 4.1 Electronic transition of π, σ, and n electrons.

Table 4.2 Orbital transitions and λ_{max} of common organic molecules

Molecule name	Chemical structure	Transition	λ_{max} (nm)	ε
Ethane	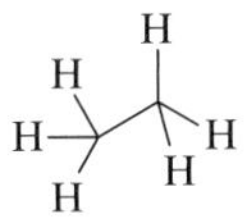	σ to σ*	135	
Methanol	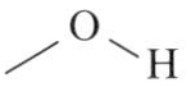	σ to σ* n to σ*	150 183	
Acetone	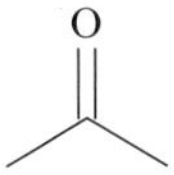	n to π*	290	
Benzene		π to π*	254	200
Ethylene	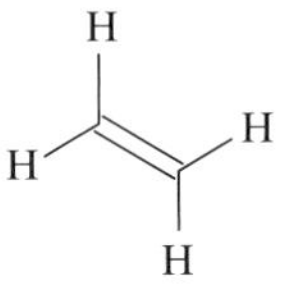	π to π*	180–200	1000
Nitromethane	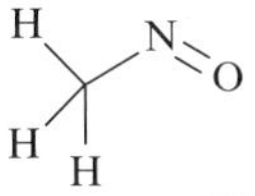	n to π* π to π*	275 200	17 5000

Here ε is molar absorptivity.
(Thermo Spectronic, 2011)

Clearly, already a small molecule will exhibit an absorption spectrum in the UV region with many peaks, which because of their broadening (see above) tends to become a number of connecting broad absorbance bands.

2.2. The Beer–Lambert law

In absorption spectroscopy, the Beer–Lambert law, also known as Beer's law or Beer–Lambert–Bouguer law, relates the absorption of light to the properties of the material through which the light passes. The Beer's law states that the absorbance of a beam of collimated monochromatic radiation

Table 4.3 Standard parameters used in UV–visible spectrophotometry

Parameter	Symbol	Synonym	Mathematical representation
Absorbance	A	OD (optical density), D, E	$\log I_0/I$
Extinction value	$A_{1\text{ cm}}^{1\%}$	$E_{1\text{ cm}}^{1\%}$	$10\varepsilon/M$
Molar extinction coefficient	ε	Molar absorptivity	A/lM $M = \text{mol/dm}$
Path length	l	Cell length	b or d
Transmittance	T	Transmission	I_0/I

in a homogenous isotropic medium is proportional to the absorption path length l, and to the concentration c, or, in the gas phase, to the pressure of the absorbing species (McNaught and Wilkinson, 2006). Beer's law can be expressed as

$$A = \log_{10}\frac{I_0}{I} = \varepsilon lc$$

where I_0 and I are intensities of incident and transmitted light, respectively. The portionality constant ε is called molar absorption coefficient. For l in cm and c in mol/dm^3 or M in mol/l, ε will result in dm^3/mol/cm or M/cm is the commonly used unit. Although the SI unit of ε is m^2/mol. The Beer–Lambert law holds only if the spectral bandwidth of the light is narrow compared to spectral line widths, otherwise the effective ε decreases with increasing bandwidth of the radiation. Some of the standard parameters used in UV–visible spectroscopy are listed in Table 4.3.

3. Hardware

3.1. Instrumentation for absorption spectroscopy

An absorption spectrophotometer is a device used to measure absorbed light intensity as a function of wavelength. In UV–visible spectrophotometers, a beam of light from a suitable UV and/or visible light source is passed through a prism or diffraction grating monochromator. The light then passes through the sample to be analyzed before reaching the detector (Fig. 4.2). UV–visible spectrophotometers have five main components: the light source, monochromator, sample holder, detector, and interpreter. The standard light source consists of a deuterium arc (190–330 nm) and a

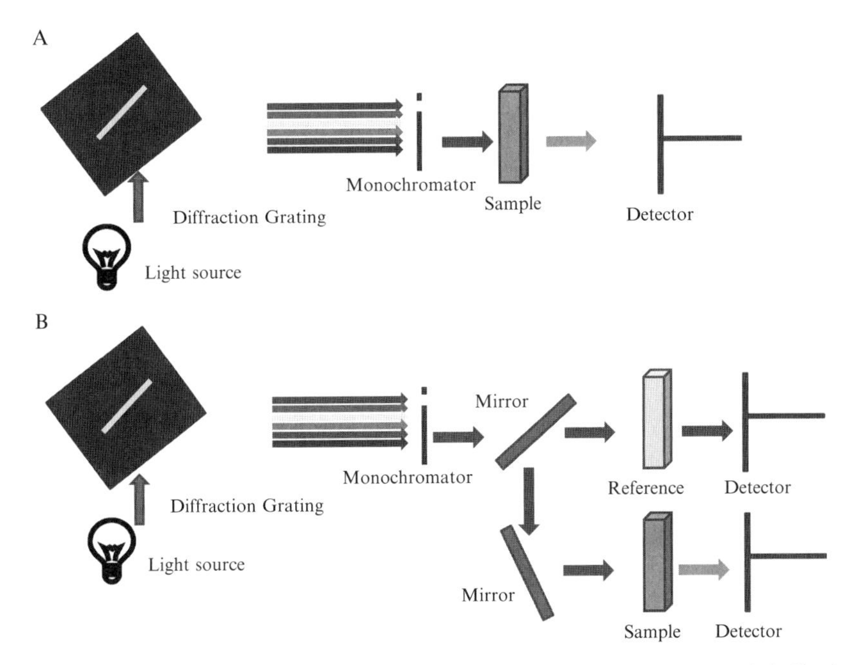

Figure 4.2 Typical optical layout of an absorption spectrophotometer. (A) Single beam spectrophotometer and (B) double beam spectrophotometer.

tungsten filament lamp (330–800 nm), which together generates a light beam across the 190–800 nm spectral range. The monochromator produces a compact optical path and reduces optical aberrations. Modern instruments use grating monochromators in reflection mode as the dispersing element.

There are two classes of spectrophotometers: single and double beam. In single beam configuration, the components are in single beam sequence; this instrument is cheaper to build and easy to use and maintain. A single beam spectrometric instrument requires a reference sample to be measured separately from the test sample. In a double beam spectrophotometer, the light from the source is split into two separate beams after passing through the monochromator (Fig. 4.2B). One beam is used for the sample, while the other one is used for reference determination. This configuration is advantageous because sample and reference reading can be conducted simultaneously so that the measurement becomes independent from variations in the intensity and spectral composition of the light source; a single beam spectrophotometer should always be switched on some time before use to allow the lamp to reach a constant temperature. Some double beam spectrophotometers employ a beam chopper, in which by blocking one beam at a time, measurement of sample and reference beams by a single photomultiplier at essentially the same time becomes possible.

3.2. Choice of materials for sample holders and buffer/solvents

Sample cells are available in a wide variety of materials and pathlengths, with most routine measurements being made using 10 mm rectangular cell. These are conventionally being fabricated using materials such as plastic, glass, fused quartz, and synthetic silica. Quartz and silica have an advantage over the other materials in that they can be used to measure absorbance down to 190 nm, compared to 300 nm for glass and plastic cuvettes. For specialized experiments, numerous designs and sizes of sample (0.2–4.0 ml) cuvettes/cells are available.

The solvent or medium used for estimation of UV–visible absorbance needs to be selected carefully as every solvent has a specific UV absorbance wavelength cutoff. The solvent cutoff can be defined as the wavelength below which the solvent itself will absorb all the light. If the solvent cutoff wavelength is close to the solute detection wavelength, a different solvent should be used. Table 4.4 provides some of the examples of commonly used solvents.

4. Applications of UV–Visible Spectrometry

Spectrometry has been extensively used in wide variety of applications. UV-visible spectroscopy can provide qualitative and quantitative information of compounds, examples of which are discussed below.

4.1. Estimation of protein concentration at 280 nm (A_{280})

UV–visible absorbance is used widely for determining protein concentrations in sample solutions and can be carried out by a simple spectrophotometer. Proteins in solution absorb light, resulting in absorbance maxima at

Table 4.4 Cutoff wavelengths for various solvents

Solvent	UV absorbance cutoff (nm)
Acetone	329
Benzene	278
Dimethylformamide (DMF)	267
Ethanol	205
Toluene	285
Water	180

(Thermo Spectronic, 2011)

280 and 205 nm. Absorption of light in the near-UV range is largely dependent on amino acids, tryptophan (Trp), and tyrosine (Tyr), while phenylalanine (Phe) and the disulfide bonds present in protein solutions may also have minor effect. A_{280} for various proteins ranges from 0 to 4 for 1 mg/ml solution of protein, although most of the proteins give 0.5–1.5 (Kirschenbaum, 1975). The protein solution can also be measured in a wide range of buffers. The advantages of absorbance assays include fast and convenient operation and no additional reagents or incubations being required.

Some of the disadvantages of this assay result from the fact that it is not strictly quantitative as it is based on strong absorbance of Phe, Tyr, and Trp residues. Different proteins may therefore have varying extinction coefficients, resulting in proteins containing neither Phe nor Tyr or Trp remaining undetected by UV–visible spectroscopy. Secondary, tertiary, and quarternary structure can also affect absorbance, along with factors such as pH and ionic strength. Other disadvantages include strong interference by nucleic acids and other chromophores. The extinction coefficient of nucleic acid in the near-UV (280 nm) region is as much as 10 times higher than that of proteins, resulting in a small amount of nucleic acid greatly influencing the total absorption (Aitken and Learmonth, 2002).

4.2. Estimation of DNA melting temperature by absorption spectroscopy

DNA consists of two long polymers of simple units called nucleotides, with backbones made of sugars and phosphate groups joined by ester bonds. The nucleotides within DNA are bonded together such that the sugar of one nucleotide is always attached to the phosphate group of the next nucleotide. These two polymeric chains of DNA are bonded by two hydrogen bonds between adenine and thymine and three hydrogen bonds between guanine and cytosine from opposite chains. These hydrogen bonds are responsible for providing thermodynamic stability to the double-stranded DNA in solution.

When DNA in solution is exposed to extremes of heat, pH, or solutes, such as urea or amides, the double-stranded helical structure of DNA shifts into a randomly unfolded single-stranded form called denatured DNA. During denaturation, the interactions between complementary base pairs are disrupted and this results in significant changes in a number of physical properties, such as an increase in UV absorption at 260 nm, an increase in buoyant density, and a decrease in viscosity. The effect of increase in UV absorption provides an appropriate method for monitoring the denaturation of DNA.

In this method, the temperature of the DNA solution is increased in stepwise manner so that the DNA gradually denatures by strand separation. A temperature control accessory strapped to a UV–visible spectrophotometer is perfectly suited for these studies, as temperature can readily be controlled and monitored and the denaturation can simultaneously be recorded by observing changes at 260 nm. The absorbance versus temperature (thermal denaturation) curve obtained, after fitting with nonlinear curve fitting algorithm, gives the melting temperature (T_m) of DNA.

UV absorption of nucleic acids related to the formation of double-stranded and/or single-stranded conformations has been used to examine the interaction with PNA or oligonucleotide probes with DNA. The determination of the thermodynamic stability and the kinetics of duplex formation are important. Analyses include the measurement of UV-absorption changes when nucleic acid targets form duplexes with antigene probes (Kushon *et al.*, 2001).

4.3. Biomolecular interaction analysis

The ability of a biomolecule (protein or nucleic acid) to absorb UV–visible light may allow the monitoring and characterization of complexes formed with ligands with concomitant alteration in the structure and function. One example is the protein serum albumin, which is involved in maintaining the colloid osmotic pressure and pH of blood and transport of hydrophobic substances by blood. Serum albumin has a characteristic absorbance spectrum with two peaks at 225 and 275 nm. An alteration of the spectrum after interaction with drugs has given useful information on the nature of the interaction and on the possible variation of protein activity as a consequence (Chi *et al.*, 2010; Hu *et al.*, 2006). DNA constitutes another example. Drugs targeting DNA molecules tend to alter the DNA absorbance at 260 nm, an event that can be easily observed and monitored (Ahmadi *et al.*, 2011; Devi and Singh, 2011).

4.4. Enzyme kinetics

The kinetic parameters of enzymes can be determined by measuring absorption of chromophore substrates or products of enzymatic reactions as discussed elsewhere in this volume. Low throughput assays are often carried out using single or double beam UV–visible spectrophotometers and a 1.0-ml cuvette. Medium throughput kinetic assays can be conducted using UV–visible plate readers using 96- and 384-well plates and manual pipetting. High-throughput assays can also be performed in microtitre plates containing up to 3456 wells using modern robotic liquid handling instruments (Bonowski *et al.*, 2010; King *et al.*, 2009).

Measurement of kinetic parameters of hexokinase (HK), an enzyme which phosphorylates glucose (Glc) in the glycolytic pathway, is achieved using absorption spectroscopy. The V_{max} and K_m of HK were measured through coupled enzyme assay using glucose-6-phosphate dehydrogenase (G6PDH) as a coupling enzyme (Sheel and Neet, 1975). To measure the kinetic parameters for glucose, the assay was conducted in 100 mM Tris–HCl at pH 7.0, 5 mM $MgCl_2$, 5 mM NAD, 50 U/ml G6PDH, 2 mM ATP at various glucose concentrations (0–20 mM). The kinetic parameters with respect to ATP can be measured under similar reaction conditions keeping the glucose concentration fixed (20 mM) and varying the ATP concentrations (0–2.5 mM). The activity was monitored at 30 °C in a BMG POLARStar microplate reader by measuring change in the absorbance of cofactor NADH at 340 nm. The spectrograms for the HK assay at various concentrations of glucose and determination of the kinetic parameters by fitting the initial rates to the Michaelis–Menten equation are shown in the Fig. 4.3A. The spectrogram and progressive curve fitting for determination of kinetic parameters are shown in Fig. 4.3B.

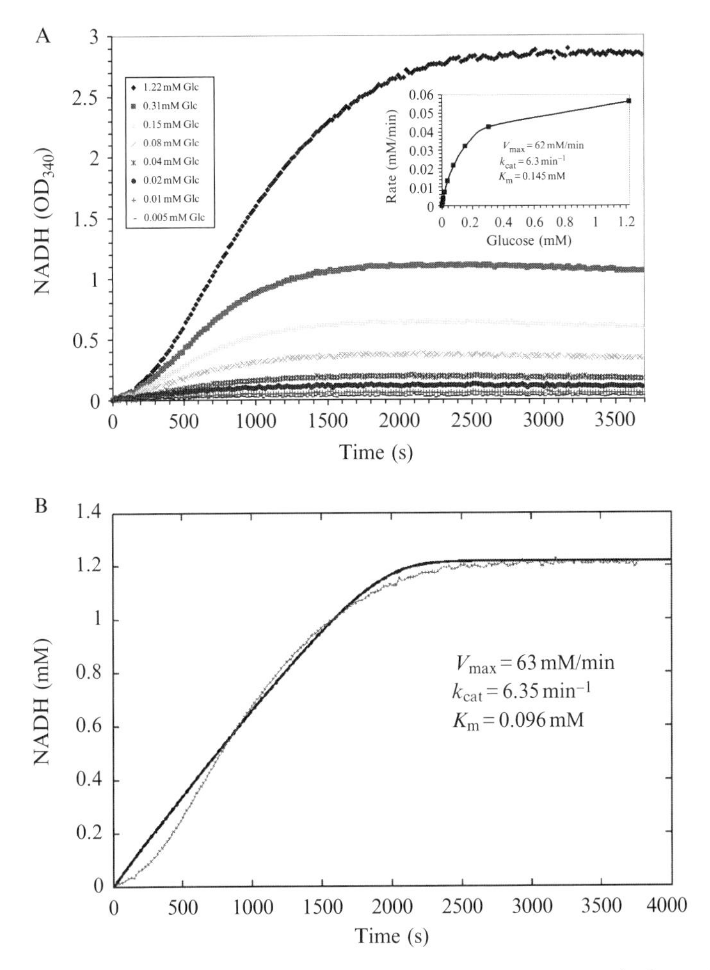

Figure 4.3 Absorption spectrograms of a hexokinase assay and determination of kinetic parameters by fitting the data using (A) Michaelis–Menten equation and (B) progressive curve fitting based on that equation.

The activities of glycolytic enzymes and enzymes in other pathways can be measured in freshly prepared cell-free extract under defined assay conditions. The standard protocol for the preparation of cell-free extract for *Saccharomyces cerevisiae* is described below and see also the other Chapters on enzyme assays in this volume.

4.4.1. Cell-free extract preparation

Fifty milliliter of cells at an $OD_{600\ nm}$ of 1.0 were harvested by centrifugation in a precooled centrifuge at 4000×*g* for 5 min at 4 °C. The cell pellet was washed three times with 10 ml of ice cold (4 °C) freeze buffer (10 mM potassium phosphate buffer, 2 mM Na_2H_2–EDTA at pH 7.5) to remove the growth medium completely. The final cell pellet was resuspended in 2.5 ml of sonication buffer (100 mM potassium phosphate buffer, 2 mM $MgCl_2$ at pH 7.5) to obtain a final $OD_{600\ nm}$ of about 20, which will give enough intracellular proteins to measure the activity of all the glycolytic enzymes using a cell extract volume of 5-10 μl. One milliliter of this cell suspension was transferred into a precooled screw top tube containing 0.7-0.8 g of glass beads (e.g., 425–600μm, Sigma). The cells were homogenized in the Mini bead-beater (8–10 bursts of 10 s each at a shaking speed of 3450 oscillations/min each burst followed by an incubation of 1 min on ice). Homogenization can also be performed by vortexing the cells at maximum speed for 30 s 10 times and cooling the samples between two rounds of vortexing for 1 min on ice. The samples were then centrifuged at 4 °C and 13000 rpm for 10 min. The supernatant (cell-free extract) was transferred to a new 1.5-ml eppendorf tube and stored on ice. The protein concentration in this cell-free extract was then measured, and 5–10 μl of cell extract was used for each enzyme assay under uniform/defined conditions.

4.5. Measurement of intracellular metabolites

Intracellular metabolites with chromophore characteristics can be measured directly by UV–visible spectroscopy in tandem with HPLC (Teusink *et al.*, 2000). Another method to measure the intracellular metabolite concentrations is through enzymatic assay using UV-visible absorption spectroscopy (Teusink *et al.*, 2000).

4.5.1. Quenching and extraction of intracellular metabolites

Intracellular metabolism is dynamic in nature and the flux through a pathway changes quickly when cells are disconnected from their extracellular environment. This affects the intracellular metabolites concentrations at a timescale equal to the ratio of their concentration to the flux through them. For major pathways such as glycolysis in yeast, these times are on the order of a second (Teusink *et al.*, 2000). When cells are separated from their external environment, a very rapid (<1 s) inhibition of all metabolism is required to provide a sample that is representative of the initial culture condition. This process of universal metabolic inhibition is called quenching. There are various methods reported in the literature for quenching and extraction of intracellular metabolites. Here, we describe a general protocol for quenching and extraction of metabolites for cultures from various modes of culture cultivation.

4.5.1.1. *Materials and instruments*

(i) Methanol, ethanol, ammonium hydrogen carbonate, and water of analytical grade.
(ii) 2 and 50 ml centrifuge tubes made of inert and high-quality plastic.
(iii) Dry ice (solid carbon dioxide).
(iv) Temperature controlled microcentrifuge and centrifuge capable of operating with 50 ml centrifuge tubes.
(v) Vortex mixer.
(vi) Quenching solution composed of 60:40 (v/v) methanol:water containing ammonium hydrogen carbonate at a final concentration of 0.85% (w/v). The solution should be adjusted, if required, to pH 5.5 and cooled to −48 °C in a dry ice bath or −80 °C freezer.
(vii) Extraction buffer; 70% ethanol, and 30% 50 mM Tris buffer, pH 7.0.

4.5.1.2. *Quenching protocol*

(i) Chill 25 ml of quenching buffer in a 50 ml centrifuge tube to about −40 °C for an hour.
(ii) Rapidly transfer 5.0 ml of culture into the quenching solution; expel the culture to the center of the tube by dipping the pipette into the quenching solution down to half its depth of quenching solution.
(iii) Rapidly mix the culture and quenching solution by inverting the tube.
(iv) Separate the cells by centrifugation for 2 min at a temperature of −9 °C and 4000×*g*.
(v) Rapidly transfer all supernatant into a new 50-ml centrifuge tube. Store the cell pellet on ice for metabolite extraction.

4.5.1.3. *Metabolite extraction protocol*

(i) Add 400 μl of extraction buffer to the cell pellet.
(ii) Suspend the cell pellet by vortex mixing.
(iii) Transfer the cell suspension to a 2-ml microcentrifuge tube and store on ice.
(iv) Add another 100 μl of extraction buffer to the 50-ml centrifuge tube and vortex mix in order to suspend any remaining cell

biomass. Transfer the cell suspension to the previous 2-ml centrifuge tube.

(v) Heat the cell suspension at 80–85 °C in a water bath for 5 min then place the centrifuge tubes on ice for condensation for 15 min.

(vi) Centrifuge at −9 °C and 13,000×*g* for 10–15 min to separate the cell debris from extracted metabolites.

(vii) Transfer the extracted supernatant to a new 1.5-ml centrifuge tube and store on dry ice.

(viii) Lyophilize the supernatant.

(ix) Suspend the lyophilized matter in a volume of 50 mM Tris–HCl buffer of pH 7.0 corresponding to the extraction volume before the lyophilization.

(x) Centrifuge it and transfer the soluble fraction to a new eppendorf tube for enzymatic quantification of metabolites.

4.5.2. Enzymatic rate assay for metabolite concentrations

Enzymatic measurement of intracellular metabolites is demonstrated here for ADP and AMP. The ADP in the metabolite extract may be measured by a system of two coupled enzymes, that is, pyruvate kinase (PK) and lactate dehydrogenase (LDH) by monitoring the decrease of absorbance of NADH at 340 nm (Negelein, 1944). The traditional way of doing this (see elsewhere in this volume) is to add an amount of ADP that is considerably less than the amount of NADH in the assay and then to follow the decrease in NADH absorbance. At some point, that decrease will stop because ADP has been depleted. By measuring the total decrease in absorbance and dividing this by the molar absorption coefficient of NADH, one then obtains the concentration of ADP added to the cuvette. This method has the advantage that it is independent of any factors in the sample that may affect the activities of the enzymes in the sample. The "end-point method" method can have the disadvantage that it is slow especially if the K_m of the enzyme consuming the ADP is relatively high. Also, it is less suitable for robotization and parallelization.

In such cases, one may switch to a rate assay of the concentration of ADP, using the same coupled enzyme system. Here, we illustrate this method: The assay was performed in 50 mM imidazole–HCl, pH 7.6, 50 mM KCl, 15 mM $MgCl_2$, 1 mM NADH, 5 mM phosphoenol pyruvate, 2 U/ml PK, and 15 U/ml LDH. When preparing a buffer mixture like this, it should always be checked that the pH of the final mixture is the intended pH. Particularly, when a $MgCl_2$ solution of a certain pH is added to the solution of a phosphorylated compound of the same pH, the final pH will be more acidic.

A standard curve for the ADP assay was produced by conducting the assay for various ADP concentrations (0-0.2 mM). The concentration of ADP was calibrated in an end-point assay using stoichiometry of enzymatic

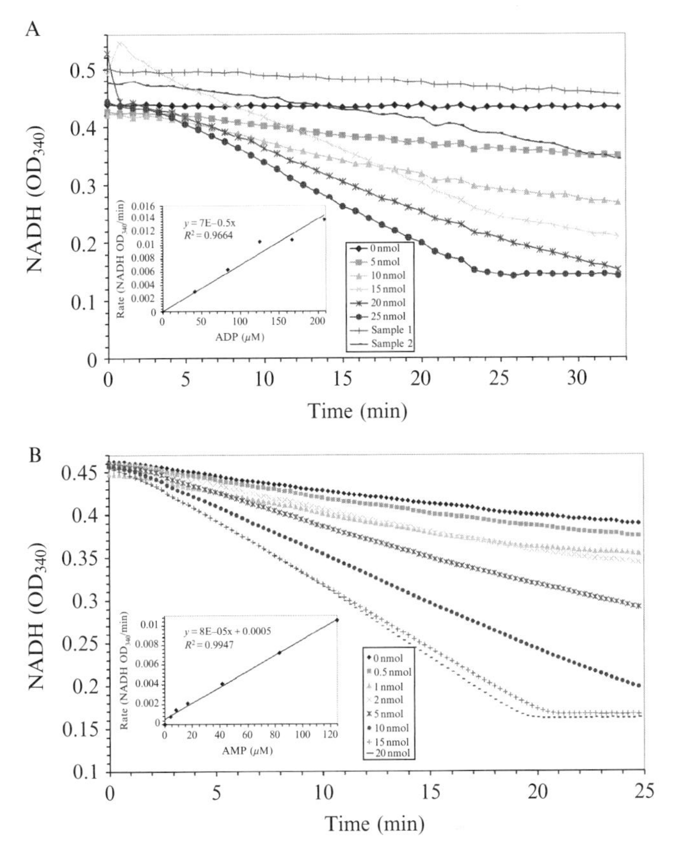

Figure 4.4 Absorption spectrogram and calibration curves of enzymatic quantification of intracellular metabolites (A) ADP and (B) AMP.

reactions and an extinction coefficient ($\varepsilon_{340\ \mathrm{nm}}$) of 6.620 mM^{-1} cm^{-1} for NADH. The spectrograms for various ADP concentrations and calibration curve are shown in Fig. 4.4A.

AMP in the metabolite extract was measured by three enzymes, that is, adenylate kinase (AK), PK, and LDH, coupled assay by monitoring the decrease in NADH absorbance at 340 nm. This AMP rate assay was developed in our lab by extending Negelein's protocol for the ADP assay. The assay was performed in 50 mM imidazole–HCl, pH 7.6, 50 mM KCl, 15 mM $MgCl_2$, 1 mM NADH, 5 mM phosphoenol pyruvate, 2 U/ml PK, 15 U/ml LDH, 10 U/ml AK (myokinase), and 5 mM ATP. An AMP standard curve assay was produced by performing the assay at various AMP concentrations (0–0.2 mM). The concentration of AMP was calculated by considering the additional stoichiometry of AK to ADP assay. The spectrograms for various concentrations of AMP and the calibration curve for AMP are shown in the Fig. 4.4B.

4.6. Measurement of fluxes

Living cells take up micronutrients (carbon, nitrogen, and inorganic phosphate sources) from the extracellular environment and metabolize them by various pathways in the cell for Gibbs-free energy extraction and anabolism. The carbon source, once entered into a cell, distributes over various branches of metabolism. The time-course measurement of extracellular concentration of the carbon source and of the concentrations of metabolites accumulated extracellularly provides a measure for the flow of carbon flux through the metabolic pathways. In this way, for example, the carbon flux through the glycolytic pathway in anaerobic *S. cerevisiae* can be determined by measuring the change in time of extracellular glucose, the glycogen inside the cytoplasm, and extracellular glycerol and alcohol accumulated in time. Glucose, glycogen, glycerol, and alcohol as metabolites can all be measured by enzymatic assay using the UV–visible spectroscopy as described in Section 4.4.

5. Perspective

Systems biology research aims to understand how biological function emerges from the nonlinear interactions of biological macromolecules. Enzymes communicate with each other through the metabolites. They "talk" by producing and consuming them. They "listen" by adjusting the rates at which they change the concentrations of the metabolites. These concentration changes are caused by the joint talking of all the enzymes. Together, this produces the symphony that leads to biological functions such as Gibbs energy transfer from catabolism to useful processes, cell cycling, and differentiation.

UV–visible absorption spectroscopy measuring changes in absorption of NAD(P)H while monitoring a reaction catalyzed by an enzyme in a test tube can provide useful and accurate information about the talking and listening that enzyme can engage in. It can also reveal the concentrations of metabolites and enzymes in the cell, as well as the fluxes through the latter, for simple pathways: These applications of UV–visible spectroscopy can be used to measure kinetome, metabolome, and fluxome alone or in combination with other tools and techniques. Using mathematical models to integrate this information, we may begin to understand how the talking and listening of molecules lead to Life.

Acknowledgments

We are grateful to our colleagues at The University of Manchester, UK, for their help and support during this work. We also thank the BBSRC for funding through the grants for SysMO projects and the MCISB grant. H. V. Westerhoff is and has been supported by a Roslind Franklin Fellowship to B. M. Bakker; by STW, the NGI-Kluyver Centre,

NWO-SysMo, BBSRC-MCISB, SysMO, BBSRC-EPSRC-DTC, ERASysBio and BRIC, EPSRC, AstraZeneca; and by the EU grants BioSim, NucSys (and extensions), ECMOAN, and UniCellSys.

REFERENCES

Ahmadi, S. M., Dehghan, G., Hosseinpourfeizi, M. A., Dolatabadi, J. E., and Kashanian, S. (2011). Preparation, characterization, and DNA binding studies of water-soluble quercetin-molybdenum(VI) complex. *DNA Cell Biol.* **30,** 517–523.

Aitken, A., and Learmonth, M. P. (2002). Protein determination by UV absorption. *In* "The Protein Protocols Handbook," (J. M. Walker, ed.), pp. 3–6. Humana Press Inc, Totowa NJ.

Bonowski, F., Kitanovic, A., Ruoff, P., Holzwarth, J., Kitanovic, I., Bui, V. N., Lederer, E., and Wolfl, S. (2010). Computer controlled automated assay for comprehensive studies of enzyme kinetic parameters. *PLoS One* **5,** e10727.

Burns, D. T. (1993). Principles of spectrophotometric measurements with particular reference to the UV-visible region. *In* "UV Spectroscopy: Technique, Instrumentation, Data Handling," (B. J. Clark, T. Frost, and M. A. Rusell, eds.), pp. 1–15. Chapman and Hall, London.

Chi, Z., Liu, R., Teng, Y., Fang, X., and Gao, C. (2010). Binding of oxytetracycline to bovine serum albumin: Spectroscopic and molecular modeling investigations. *J. Agric. Food Chem.* **58,** 10262–10269.

Devi, C. V., and Singh, N. R. (2011). Absorption spectroscopic probe to investigate the interaction between Nd (III) and calf-thymus DNA. *Spectrochim. Acta A Mol. Biomol. Spectrosc.* **78,** 1180–1186.

Hu, Y. J., Yu, H. G., Dong, J. X., Yang, X., and Liu, Y. (2006). Spectroscopic studies on the interaction between 3,4,5-trimethoxybenzoic acid and bovine serum albumin. *Spectrochim. Acta A Mol. Biomol. Spectrosc.* **65,** 988–992.

King, R. D., Rowland, J., Oliver, S. G., Young, M., Aubrey, W., Byrne, E., Liakata, M., Markham, M., Pir, P., Soldatova, L. N., Sparkes, A., Whelan, K. E., *et al.* (2009). The automation of science. *Science* **324,** 85–89.

Kirschenbaum, D. M. (1975). Molar absorptivity and A-1% 1 cm values for proteins at selected wavelengths of the ultraviolet and visible regions. X. *Anal. Biochem.* **64,** 186–213.

Kushon, S. A., Jordan, J. P., Seifert, J. L., Nielsen, H., Nielsen, P. E., and Armitage, B. A. (2001). Effect of secondary structure on the thermodynamics and kinetics of PNA hybridization to DNA hairpins. *J. Am. Chem. Soc.* **123,** 10805–10813.

McNaught, A. D., and Wilkinson, A. (2006). IUPAC. Compendium of Chemical Terminology. The "Gold Book" http://goldbook.iupac.org (2006-) created by M. Nic, J. Jirat, B. Kosata; updates compiled by A. Jenkins. [2]. 1997. Oxford, Blackwell Scientific Publications. Ref Type: Electronic Citation.

Negelein, E. (1944). Unpublished observations; cf. F. Kubowitz and P. Ott. *Biochem. Z.* 193.

Sheel, J. P., and Neet, K. E. (1975). Allosteric properties and the slow transition of yeast hexokinase. *J. Biol. Chem.* **250,** 2259–2268.

Teusink, B., Passarge, J., Reijenga, C. A., Esgalhado, E., van der Weijden, C. C., Schepper, M., Walsh, M. C., Bakker, B. M., van, D. K., Westerhoff, H. V., and Snoep, J. L. (2000). Can yeast glycolysis be understood in terms of in vitro kinetics of the constituent enzymes? Testing biochemistry. *Eur. J. Biochem.* **267,** 5313–5329.

Thermo Spectronic (2011). *Basic UV-Vis Theory, Concepts and Applications. 9* Thermo Spectronic. 5-4-2011.

Threlfall, T. L. (1993). Colour. *In* "UV Spectroscopy: Technique, Instrumentation, Data Handling," (B. J. Clark, T. Frost, and M. A. Rusell, eds.), pp. 88–96. Chapman and Hall, London.

SECTION THREE

NUCLEIC ACIDS AND SYSTEMS BIOLOGY

CHAPTER FIVE

A Strand-Specific Library Preparation Protocol for RNA Sequencing

Tatiana Borodina, James Adjaye, *and* Marc Sultan

Contents

Abstract

The analysis of transcriptome, which was over the past decade based mostly on microarray technologies, is now being superseded by so-called next generation sequencing (NGS) systems that changed the way to explore entire transcriptome. RNA sequencing (RNA-Seq), one application of NGS, is a powerful tool, providing information not only about the expression level of genes but also further about the structure of transcripts as it enables to unequivocally identify splicing events, RNA editing products, and mutations in expressed coding sequences within a single experiment. Herein, we describe step by step the deoxy-UTP (dUTP) strand-marking protocol [Parkhomchuk, D., Borodina, T., Amstislavskiy, V., Banaru, M., Hallen, L., Krobitsch, S., Lehrach, H., Soldatov, A. (2009). Transcriptome analysis by strand-specific sequencing of complementary DNA. *Nucleic Acids Res.* **37**(18), e123], which has been recently reviewed as the leading protocol for strand-specific RNA-Seq library preparation [Levin, J. Z., Yassour, M., Adiconis, X., Nusbaum, C., Thompson, D. A., Friedman, N., Gnirke, A., Regev, A. (2009). Comprehensive comparative analysis of strand-specific RNA sequencing methods. *Nat. Methods* **7**(9), 709–715]. The procedure starts with the isolation of the polyA fraction (mRNA) within a pool of total RNA, followed by its fragmentation. Then double-stranded (ds) cDNA synthesis is performed with

Max Planck Institute for Molecular Genetics, Department of Vertebrate Genomics, Ihnestrasse 63-73, Berlin, Germany

Methods in Enzymology, Volume 500
ISSN 0076-6879, DOI: 10.1016/B978-0-12-385118-5.00005-0

the incorporation of dUTP in the second strand. The ds cDNA fragments are further processed following a standard sequencing library preparation scheme tailored for the Illumina sequencing platform: end polishing, A-tailing, adapter ligation, and size selection. Prior to final amplification, the dUTP-marked strand is selectively degraded by Uracil-DNA-Glycosylase (UDG). The remaining strand is amplified to generate a cDNA library suitable for sequencing.

1. Introduction

1.1. NGS

The next generation sequencing (NGS) has revolutionized molecular biology and is currently one of the most promising and fastest developing technologies. Since the first instrument from Roche appeared on the market in 2004, enabling the sequencing of thousands of DNA fragments of 100 base pair (bp) in parallel, the throughput and quality of the generated sequences has continuously increased. Today's Genome Sequencer (FLX Titanium) from Roche is able to generate about 1 million of 400 bp sequences per run. This system belongs to the so-called long-read technologies. Though it is "next generation" in terms of throughput (0.4–0.6 Gb per run), it is only ~20 times cheaper than the conventional Sanger sequencing approach. The major breakthrough came from the so-called short-read sequencing technologies, which widely broadened sequencing applications and significantly reduced the sequencing costs. In 2006, Solexa (acquired by Illumina in 2007) was the first company to introduce a short-read sequencing system to the market, based on sequencing by synthesis approach. A year later, the ligation-based SOLiD system from ABI appeared on the market. Both systems have considerably evolved over the passed years, increasing the throughput, optimizing the machine operation, and constantly decreasing the reagents costs (up to twofold per year so far). Currently, both the Illumina and SOLiD systems are comparable in terms of sequencing costs and throughput: ~25 Gb of mappable reads per day at ~70€ per Gb for the Illumina HiSeq2000 and ~6 Gb of mappable reads per day at ~45€ per Gb for SOLiD4 systems[1] (180 and 280 times cheaper than Sanger sequencing, respectively). Both systems have similar applications spectrum and are now widely used for genomic *de novo* and resequencing, transcriptome analysis, epigenetics studies, genotyping, etc. The two platforms have made NGS an affordable tool for routine laboratory research and rendered challenging large-scale projects like the 1000 genomes project, possible.

[1] The throughput and costs information are derived from the official announcements of the respective companies.

Recently, authors began to distinguish second generation sequencing (Roche, Illumina and ABI platforms) and third generation single molecule sequencing technologies. The single molecule sequencer Heliscope from Helicos Bioscience is already on the market. Several platforms are being developed and should be available soon on the market (e.g., sequencers from Pacific Biosciences and Oxford Nanopore).

1.2. RNA sequencing

Among all NGS applications, high-throughput complementary DNA (cDNA) sequencing (RNA-Seq) is probably one of most challenging. RNA-Seq is a powerful tool for whole-transcriptome analysis, providing information not only on the expression level of genes but also on the structure of transcripts as it enables the identification of splicing events and RNA editing products, mutations in expressed coding sequences, and allele-specific transcription. The RNA-Seq procedure itself is straightforward and has a large dynamic range as well as high sensitivity. RNA-Seq has clear advantages over previous high-throughput approaches, such as microarray hybridization, gene-specific and tiling arrays, or SAGE analyses (Stoughton, 2005; Velculescu *et al.*, 1995). In contrast to SAGE, RNA-Seq does not depend on the presence of specific restriction sites within the cDNA. Unlike hybridization-based approaches, RNA-Seq does not require prior information on transcript sequences, allowing the detection of novel transcripts. It has a very low background noise and provides a dynamic range ($\sim 10^5$) typically an order of magnitude higher than one can achieve with hybridization-based arrays. The digital character of RNA-Seq data allows easily comparing and integrating results obtained from different laboratories. The applications and advantages of RNA-Seq have already been shown in several studies (e.g., Denoeud *et al.*, 2008; Marioni *et al.*, 2008; Mortazavi *et al.*, 2008; Nagalakshmi *et al.*, 2008; Sultan *et al.*, 2008; Wilhelm *et al.*, 2008). An exhaustive review about RNA sequencing was recently published by Costa *et al.* (2010).

The RNA-Seq procedure can be divided into four main steps: (i) preparation of the sequencing library, (ii) clonal amplification of library molecules on the sequencing support (glass surface in the case of the Illumina platform; magnetic beads in Roche and ABI schemes), (iii) the sequencing itself, and (iv) data analysis. The second and the third steps are sequencer-dependent and are therefore highly standardized. Library preparation and data analysis allow more freedom in adjusting the protocol for a particular purpose leading to a great variety of methods, where standards and guidelines are still evolving. In this chapter, we discuss the first step of the RNA-Seq procedure, which is the preparation of the cDNA library.

Until recently, the common strategy of an RNA-Seq library preparation was to convert single-stranded RNA molecules into double-stranded (ds)

DNA fragments of a certain size flanked by platform-specific adapter sequences. Originally, RNA-Seq was not strand specific, that is, reads corresponding to original RNA molecule or complementary to it were indistinguishable, which created difficulties for the data analysis. Transcription can occur in both directions, genes being located on either DNA strand, sometimes overlapping. In eukaryotes, a complementary RNA molecule to a given mRNA can also be transcribed: this has been described as antisense transcription and these molecules are involved in regulatory mechanism (He *et al.*, 2008; Kapranov *et al.*, 2007). Knowing from which DNA strand the RNA molecule originates from is an important piece of information, which helps resolving annotation ambiguities for known and novel genes, provides hints to the function of the studied RNA, and helps to correctly predict the expression levels of a given transcript.

Several strand-specific RNA-Seq (ssRNA-Seq) protocols based either on the ligation of adapters to the RNA molecules or on modifications of the ds cDNA synthesis procedure were suggested recently. Schemes that involve adapter ligation directly to single-stranded RNA molecules (Lister *et al.*, 2008) have no restrictions in the RNA length and are the only choice for the analysis of short RNA molecules like micro-RNA. This principle is used in many commercial preparation kits,[2] but these methods are laborious and are sensitive to ribosomal RNA (rRNA) contamination, so the RNA fraction of interest (e.g., mRNA, micro-RNA, or short transcripts) must be preselected.

In alternative directional transcriptome profiling protocols, the ds cDNA synthesis procedure is modified. Cloonan and colleagues used a tagged random hexamer primer in the synthesis of the first cDNA strand and DNA–RNA template-switching primers during the synthesis of the second cDNA strand (Cloonan *et al.*, 2008). Another approach is based on substituting all cytidine residues in RNA molecules to uridines by bisulfite treatment prior to cDNA synthesis (He *et al.*, 2008). Thus, the first strand contains no guanines, while the second strand has no cytidines. This method requires a tailored data analysis scheme and leads to the loss of $\sim$30% of sequences that can be mapped uniquely to the genome, as the transformation of a four-bases code to a three-bases code partly results in skewing of the complexity of the genome under investigation. The directional RNA-Seq procedure described herein is based on the incorporation of deoxy-UTP (dUTP) during either the first- or second-strand cDNA synthesis and the subsequent selective destruction of this strand (Parkhomchuk *et al.*, 2009).

Recently, Levin and coworkers compared seven ssRNA-Seq protocols in terms of strand specificity, library complexity, evenness and continuity of coverage, agreement with known annotations, and accuracy for expression

[2] For example, DGE small RNA Sample Preparation Kit (Illumina); directional mRNA-Seq library preparation kit (Illumina); SOLiD small RNA expression kit (Applied Biosystems); total RNA-Seq kit (Applied Biosystems).

level analysis (Levin *et al.*, 2009). Based on their conclusion, the dUTP approach (Parkhomchuk *et al.*, 2009) provided the most compelling overall balance across criteria, followed closely by the Illumina RNA ligation protocol (Directional mRNA-Seq Library Preparation Kit, Illumina). Despite a wide range of strand-specific protocols, many researchers still often rely on nonstrand-specific RNA-Seq library preparation, where RNA is first converted into ds cDNA, and then processed into a sequencing library, mostly because it is simple, straightforward, and stable (Wilhelm *et al.*, 2010). In this respect, the dUTP strand-marking method is a good alternative, as it adds only minor modifications to the standard cDNA synthesis workflow, and is compatible with commercially available kits.

2. ssRNA-Seq Protocol

In this chapter, we present the detailed protocol for the preparation of an ssRNA-Seq library from mRNA suitable for Illumina-based sequencing. The protocol relies on the incorporation of dUTP during the second-strand synthesis, allowing subsequent selective destruction of this strand by Uracil-DNA-Glycosylase (UDG; Fig. 5.1).

The initial step of this part is to generate a cDNA molecule to the initial mRNA, in a reaction known as first-strand cDNA synthesis. The results of this reaction are DNA–RNA hybrids, which will serve as templates for the generation of the second cDNA strand. During the latter reaction, the RNA is degraded and the second cDNA strand is synthesized using a nucleotide mixture in which dTTP has been replaced by dUTP. The next steps enable the ligation of adaptor sequences on both ends of the ds cDNAs. The adaptor sequences have a Y-shaped structure, which preserves the directionality of the molecules. Finally, the second strand is being degraded using an enzyme called UDG that selectively removes the incorporated uracil from the DNA. The remaining intact first-strand cDNAs complemented with Y-shaped adapter sequences enable the directional sequencing of all molecules while preserving the strand information of the RNA.

The protocol proceeds over 3.5 days and includes several steps, which are detailed below and in Fig. 5.2.

Day 1:. mRNA purification from total RNA, DNase treatment, RNA fragmentation.

Day 2:. Synthesis of ds cDNA with incorporation of dUTP within the second strand.

Day 3:. Preparation of the Illumina Paired End sequencing library (end polishing, A-Tailing, adapter ligation, size selection), degradation of the dUTP-marked strand.

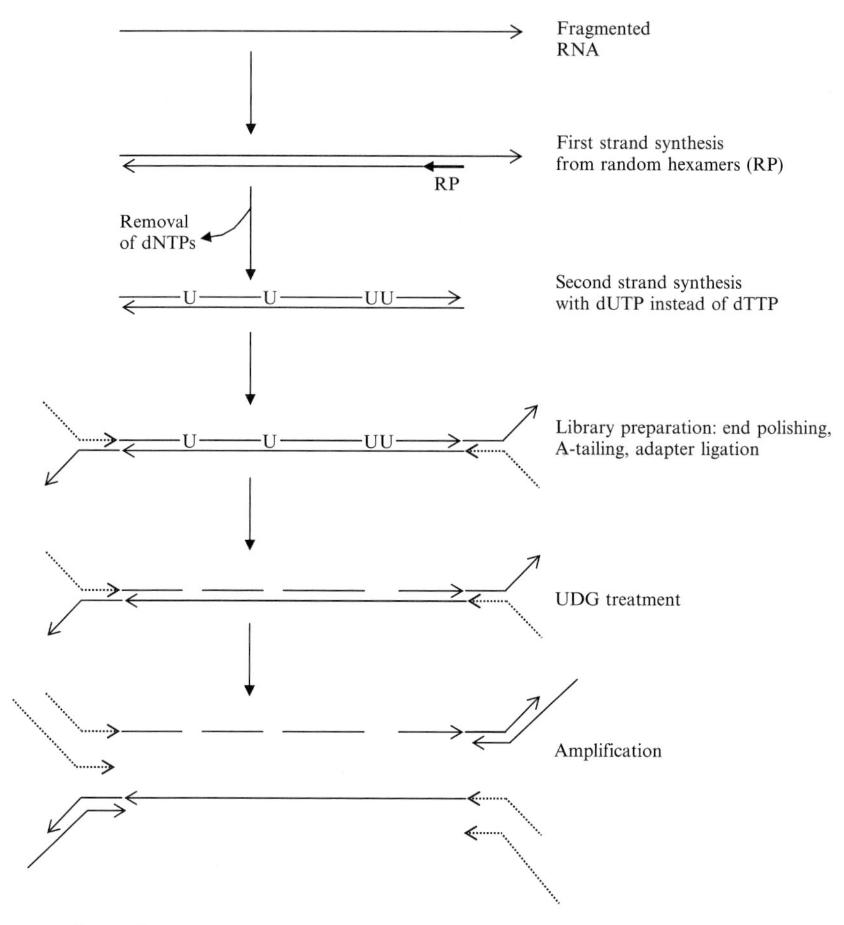

Figure 5.1 Principle of the dUTP strand-marking ssRNA-Seq method.

Day 4:. Amplification of the cDNA library.

All protocols presented below with a possibility of automatic calculation for different volumes of reagents are available online (Zbio-wiki, 2010).

2.1. General issues

2.1.1. Selection of fragmentation approach

The presented protocol exploits chemical RNA fragmentation before cDNA synthesis. Alternatively, the ds cDNA can be first synthesized, and then sheared (preferably with ultrasound). Both procedures work well with the dUTP strand-marking method.

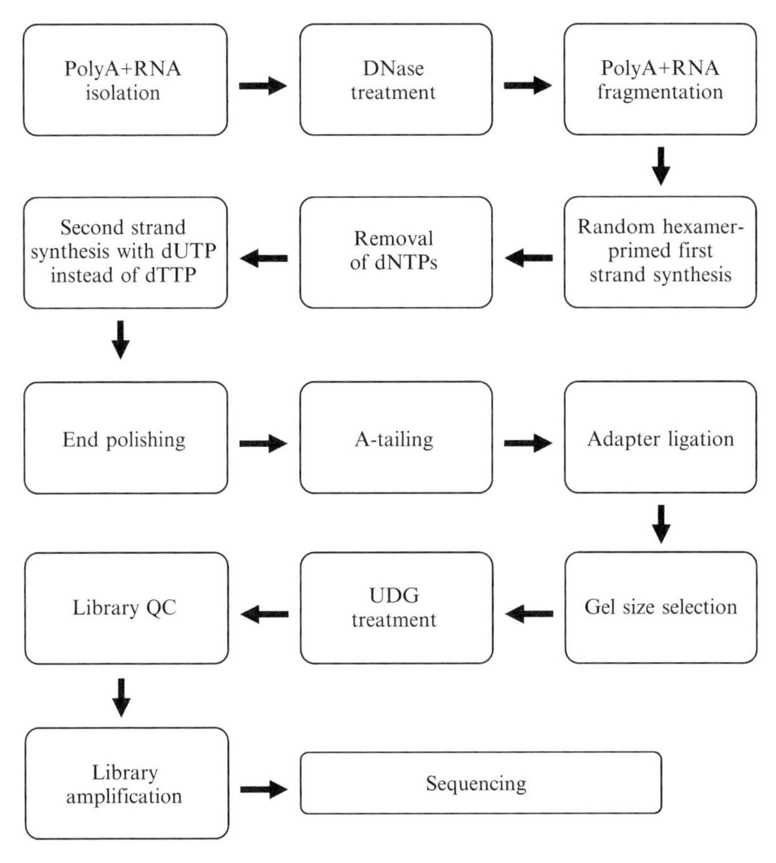

Figure 5.2 Workflow of the Illumina PE sequencing library preparation integrating the dUTP strand-marking method.

The advantage of the RNA fragmentation approach is a significantly more even reads distribution along the transcript. In the cDNA shearing scheme, RNA secondary structures distort locally the cDNA synthesis, which leads to nonuniform reads distribution along the RNA and, if oligo (dT) primer was used for the first-strand synthesis together with random primers, to a 3′-bias. Also, the RNA fragmentation is much quicker.

The cDNA shearing scheme is advantageous for samples where rRNA is highly represented (e.g., preselection of polyA$^+$RNA fraction is not possible like in prokaryotic transcriptomes). The rRNA has strong secondary structures which suppress cDNA synthesis. For example, if cDNA is synthesized from total RNA, where the rRNA fraction constitutes 85–95% of the material, ribosomal reads would be ~50% of the sequencing output. After one round of polyA$^+$RNA isolation on Oligo $d(T)_{25}$ Dynabeads, about 50% of the RNA washed off from the Dynabeads is rRNA. Using the cDNA shearing approach would result in just a small percent of rRNA reads in the sequencing data. For RNA fragmentation, it is obligatory to perform an additional round of polyA$^+$RNA isolation to get rid of rRNA.

If compared to RNA chemical fragmentation, ultrasonic shearing of cDNA is a safer procedure in terms of over digestion. Ultrasound gives 150–250 bp fragments size distribution. This range would not change if the shearing time is increased. Chemical over digestion, or fragmentation of partly degraded samples, would lead to formation of smaller than necessary fragments (below 100 bp instead of ~170 bp expected).

When performing the protocol, it is necessary to consider that shorter RNA fragments require higher concentration of random primers for an effective primer extension: 2.4 per 0.5 μg of RNA for the RNA fragmentation approach and 40 ng per 0.5 μg for the cDNA shearing scheme.

2.1.2. Required materials

Here, the items are listed which are necessary throughout the whole protocol.

- Water (Ambion, e.g., # AM9930)
- DNA LoBind 1,5-ml tubes (Eppendorf, #0030.108.051)
- DNA LoBind 0.5-ml tubes (Eppendorf, #0030.108.035)
- 0.2 ml thin-wall tubes (e.g., Applied Biosystems, #N8010540)
- Filtered tips (e.g., Axygen, Biozym)

Note 1: We recommend to avoid using DEPC-treated reagents for the protocol: DEPC may slightly inhibit the cDNA synthesis.

Note 2: Powder-free gloves should be worn when performing the protocol.

2.2. Purification of polyA$^+$RNA

The first step of an RNA-Seq experiment is to extract good quality intact RNA from the tissue or cells of interest. For this purpose, one of many standard protocols for total RNA extraction (Trizol, Qiagen Kits, etc.) can be employed. Depending on the task, the sequencing library can be prepared either from total RNA or from a distinct RNA fraction (e.g., mRNA, noncoding RNA, small RNA, etc.). Here, we focus on the selective extraction of the polyadenylated RNA fraction (polyA$^+$RNA or mRNA) as it is the most commonly used RNA fraction for gene expression studies. This protocol was also successfully applied for the sequencing ribo minus[3] RNA samples.

[3] Ribo minus RNA is total RNA treated with, for example, RiboMinus™ Eukaryote Kit for RNA-Seq (Invitrogen, SKU# A10837-08).

2.2.1. Required materials

Equipment

- Water bath: 65, 70, 37 °C
- Magnetic stand for 1.5-ml Eppendorf tubes (e.g., Applied Biosystems, #AM10055)
- Rotator for 1.5-ml Eppendorf tubes (room temperature, ~10 rpm required)
- PCR thermal cycler for 0.2-ml tubes
- Centrifuge for 1.5–2 ml tubes (4 °C, 14,000 rpm required)

Reagents/enzymes

- Dynabeads® mRNA Purification Kit for mRNA purification from total RNA preps (Invitrogen SKU# 610–06)
- TURBO™ DNase 2 u/μL (Ambion, #AM2238)
- 10× Fragmentation Reagent and Stop Solution (Ambion, #AM8740) or, alternatively, 5× Fragmentation Buffer and Fragmentation Stop Solution (included in the Illumina mRNA-Seq 8-Sample Prep Kit, #RS-100-0801). Stop Solution is 0.2 M EDTA, pH 8.0
- Glycogen, 10 μg/μl
- 3 M Sodium acetate, pH 5.2
- Ethanol, 100%
- 10 mM Tris–Cl, pH 8.5
- RNeasy MinElute Cleanup Kit (Qiagen, #74204)

Additional material

- Intact total RNA sample of high quality (RIN ≥8 measured on an Agilent Bioanalyzer). We recommend starting with 30–50 μg of total RNA. It is possible to go down to 10 μg.

2.2.2. Isolation of polyA$^+$RNA

Preparation of Oligo d(T)$_{25}$ Dynabeads:

1. Resuspend Dynabeads by vortexing. Transfer 0.5 μl of suspension per 1 μg of total RNA (minimum 50 μl of Oligo d(T)$_{25}$ Dynabeads) to a 1.5-ml LoBind tube.
2. Vortex the tube for 30 s and then keep the tube on the magnetic stand for 1 min. Remove the supernatant.
3. Resuspend the beads in Binding Buffer—take the same volume as the volume of Dynabeads suspension.
4. Repeat steps 2 and 3.

Preparation of the total RNA sample and binding polyA$^+$RNA to Dynabeads:

1. Transfer your RNA to the LoBind tube. Place the sample at 65 °C for 2 min to denature the secondary structures, then place the sample immediately on ice.
2. If necessary, add Tris–HCl, 10 mM pH7.5 or Binding Buffer.

   Note: Optimal hybridization conditions are obtained when $V_{RNA} = V_{Dynabeads}$, where V is the volume. If the RNA sample volume is smaller than the volume of the Dynabeads suspension, fill it up with Tris–HCl, 10 mM pH7.5. If the RNA sample volume is larger, then add Binding Buffer.
3. Combine the Dynabeads and the total RNA sample, mix carefully by inverting the tube several times.
4. Rotate the tube for 30–45 min at room temperature at ~10 rpm.
5. Put the tube on the magnetic stand for 1 min. Transfer the supernatant to a new LoBind tube.

   Note: We advise the retention of the non-polyA$^+$RNA fraction at least until it is clear that the polyA$^+$RNA fraction has been successfully isolated.
6. Wash the Dynabeads twice with Washing Buffer (volume $= V_{RNA} + V_{Dynabeads}$).
7. Resuspend beads in 50 μl of Tris–HCl, 10 mM pH 7.5. Place the tube in a 70 °C water bath for 2 min (to wash off RNA from beads), then put immediately on ice. Add 50 μl of Binding Buffer and mix carefully by inverting the tube. Repeat steps 4–6.

   Note: Second round of mRNA purification on Dynabeads is obligatory for the RNA fragmentation approach and optional for cDNA shearing approach.

polyA$^+$RNA elution from Dynabeads:

1. Resuspend the Dynabeads in 10 μl of 10 mM Tris–HCl, pH7.5.
2. Place the tube in the 70 °C water bath for 2 min.
3. Heat the magnetic stand in the 70 °C water bath for ~5 min, take it out shortly before taking out the tube with Dynabeads.

   Note: Be careful not to heat the magnetic stand to 80 °C or above, it would loose its magnetic properties.
4. Place the tube with Dynabeads on the preheated magnetic stand for ~15 s, collect the eluate into a new 0.5-ml LoBind tube.
5. Repeat steps 1–4.
6. Combine the eluates from step 3. Measure the concentration of the polyA$^+$RNA.

   Note: Typically polyA$^+$RNA constitutes about 1–5% of total RNA.

7. For safe storage, snap freeze the tube with polyA$^+$RNA in liquid nitrogen, place at −20 °C for short-term storage, otherwise store the tube at −70 °C.

2.2.3. DNase treatment

DNase treatment should be performed systematically for all RNA-Seq samples to minimize nonspecific sequencing reads, which might also interfere with strand specificity and increase background noise. We recommend you to perform DNase treatment after mRNA purification because the RNA might be partly fragmented during the DNase treatment, which would lead to an underrepresentation of 5′ ends of mRNA. We further recommend the use of TURBO™ DNase (Ambion), as it is much more efficient for low concentration of DNA in solution, than DNase from other vendors.

1. Prepare the reaction mix: RNA (up to ~0.5 μg/μl), 1× TURBO™ DNase Buffer, TURBO™ DNase (take 0.2 U of TURBO™ DNase per 1 μg of RNA, minimum—0.2 U per reaction).
2. Incubate the reaction at 37 °C (water bath) for 30 min.
3. Proceed immediately to the RNA cleanup. We recommend the RNeasy MinElute Cleanup Kit.
4. Elute the RNA in RNase-free water. Check the sample concentration (optional).
5. For safe storage, snap freeze the tube with polyA$^+$RNA in liquid nitrogen, place at −20 °C for short-term storage, otherwise store the tube at −70 °C.

2.2.4. RNA fragmentation

Chemical RNA fragmentation yields ~ 170 nt RNA fragments

1. Preparation of the reaction mix: RNA (up to ~2.5 μg/μl) in 1× Fragmentation Buffer.
2. Incubate the reaction at 70 °C (in a preheated PCR thermal cycler) for 5 min. Then place the tube on ice and add 2 μl of Stop Solution.

   Note: If the Illumina Fragmentation Buffer is used, incubate at 94 °C for exactly 5 min.
3. Transfer the reaction into a LoBind tube.
4. Perform RNA precipitation with salt/2.5 V ethanol. We use 0.3 M NaOAc, pH 5.2 as a salt component. If the amount of RNA is less than 1 μg, add 20 μg of glycogen as a coprecipitant. Incubate ~30 min at −20 °C before centrifugation.

5. Elute the RNA in RNase-free water. Check the sample concentration (optional).

2.3. cDNA synthesis

2.3.1. Required materials

Equipment

- PCR machine for 0.2-ml tubes
- Water bath: 16 °C

Reagents/enzymes

- SuperScript III First-Strand Synthesis System for RT-PCR (Invitrogen, #18080-051)
- Actinomycin D (Sigma, #A1410), 120 ng/μl
- 1 mM Tris–HCl, pH7.0
- dNTP Set, 100 mM each A,C,G,T (GE Healthcare, #28-4065-51)
- dUTP, 100 mM (GE Healthcare, #28-4065-41)
- Second-Strand Buffer (Invitrogen, #10812-014)
- *Escherichia coli* DNA ligase (New England BioLabs, # M0205)—Optional
- DNA Polymerase I (*E. coli*) (New England BioLabs, #M0209)
- RNase H (Invitrogen Cat #18021–071)
- Illustra ProbeQuant G-50 Micro Columns (GE Healthcare, #28-9034-08)
- QIAquick PCR Purification Kit (Qiagen, #28106)

2.3.2. First-strand synthesis

First-strand synthesis is performed in 1× reverse transcription buffer, 0.5 mM dNTPs, 5 mM $MgCl_2$ and 10 mM DTT. $(dN)_6$ primers (for fragmented polyA$^+$RNA: 40 ng per 0.5 μg of mRNA, min. 40 ng; for full-length polyA$^+$RNA: 2.4 μg per 0.5 μg of mRNA, min. 2.4 μg), RNase OUT (20 U per 1 μg of RNA, min. 20 u) and SuperScript III polymerase (100 U per 1 μg of RNA, min. 100 U) are added to the reaction. In case cDNA is synthesized from nonfragmented mRNA, oligo(dT) are added to the mix (25 pmol per 0.5 μg of mRNA, min. 25 pmol). Usually we start with 0.2–0.5 μg of mRNA and perform the first-strand synthesis in 20 μl.

1. Combine in a 0.2 ml thin-wall tube, all reaction components except enzymes.
2. Add mRNA (up to 50 ng/μl).
3. Place the reaction in PCR machine and run the following program:
   - 70 °C for 5 min
   - cooling down to 15 °C, at 0.1 °C/s rate

- 15 °C—pause

  Note 1: In case cDNA is synthesized from nonfragmented mRNA, we recommend to hold mRNA at 98 °C for 1 min to melt RNA secondary structures, and then go down to 70 °C.

  Note 2: Slow cooling allows annealing of primers being as reproducible as possible.

4. Add Actinomycin D at a concentration of 6 ng/μl while the tube is still within the PCR machine (optional step: if missed, proceed further to step 5).

   Note: The addition of Actinomycin D specifically inhibits DNA-dependent, but not RNA-dependent, DNA synthesis (Ruprecht *et al.*, 1973). This prevents the synthesis of artifact second-strand cDNA, which is an important source of false antisense transcription (Perocchi *et al.*, 2007).
5. Add RNase OUT and SuperScript III polymerase to the reaction while the tube is still within the PCR machine.

   Note: It is convenient to have a premix of these enzymes, stored at −20 °C.
6. Run the following program:
   - Heating from 15 to 25 °C, 0.1 °C/s
   - 25 °C for 10 min
   - Heating from 25 to 42 °C, 0.1 °C/s
   - 42 °C for 45 min
   - Heating from 42 to 50 °C, 0.1 °C/s
   - 50 °C for 25 min
   - 75 °C for 15 min to inactivate the SuperScript III polymerase

   Note: The temperature of the reverse transcription reaction is increased gradually as a compromise between the stability of primer annealing (for 6-mers — ~25 °C), biological activity of the enzyme (active up to 60 °C, optimal temperature ~42 °C), and denaturation of RNA secondary structures.

2.3.3. Removal of dNTPs

The labeling of the second cDNA strand with uridines requires the removal of dTTP from the reaction. To remove dNTPs, we recommend the use of Illustra ProbeQuant G-50 Micro Columns. It is also possible to clean up the first-strand synthesis reaction by applying a PCIA extraction followed by ethanol precipitation (Levin *et al.*, 2009). We recommend gel filtration because it is faster.

1. Change buffer in the gel filtration column: pass 1.5 ml of 1 mM Tris–HCl, pH 7.0 through the column under gravity.

   Note: This step is required to change the buffer in the column. Illustra ProbeQuant G-50 Micro Columns are originally filled with the STE Buffer (100 mM NaCl, 10 mM Tris–HCl, 1 mM EDTA). For the second-strand synthesis, the cDNA has to be eluted in 1 mM Tris–HCl, pH 7.0.

2. Add 10 mM Tris–HCl, pH 8.5, to the first-strand synthesis reaction up to 50 μl. Load the reaction on the column and spin down at 2000 rpm for 1 min.
3. Transfer the eluate to a new 1.5-ml LoBind tube, put the tube on ice.

2.3.4. Second-strand synthesis

Second-strand synthesis reaction is performed in 0.13× reverse transcription buffer, 0.26 mM dNTPs with dTTP replaced by dUTP, 0.67 mM $MgCl_2$ and 1.33 mM DTT and 1× second-Strand Synthesis Buffer.[4] *E. coli* DNA polymerase I (0.27 U/μl) and RNase H (0.013 U/μl) are added to the reaction. If ds cDNA synthesis is performed on nonfragmented polyA$^+$RNA, *E. coli* ligase (0.07 U/μl) is added to the reaction.

Note: The substitution of dTTP by dUTP during second-strand synthesis does not change DNA synthesis rate.

1. Combine in a LoBind tube all reaction components. Usually, we prepare 25 μl of the reaction mixture: 15 μl of 5× second-Strand Synthesis Buffer, 1 μl of 10× reverse transcription buffer, 2 μl of 10 mM dNTPs with dTTP replaced by dUTP, 0.5 μl of 100 mM $MgCl_2$, 1 μl of 100 mM DTT, 2 μl of *E. coli* DNA polymerase I (10 U/μl), and 0.5 μl of RNase H (2 U/μl).
2. Add the reaction mixture to the cDNA (50 μl).
3. Incubate the reaction at 16 °C for 2 h.
4. Perform the reaction cleanup with Qiagen PCR Purification Kit.
5. Check the sample concentration (optional).

Note: typical outcome of ds cDNA synthesis is 70–90% of the starting mRNA amount.

2.4. ssRNA-Seq library preparation

The procedure of sequencing library preparation for Illumina sequencing platform that we describe below is based on the original Illumina protocol of library construction from genomic DNA. But additionally, uridine digestion by UDG is performed before the final library amplification. Thus, the uridine-containing strand is destroyed and is not amplified. We, further, routinely perform a check of the library concentration by real-time PCR to accurately determine the number of cycles for the final amplification.

[4] We use the Invitrogen system for cDNA synthesis, in which the first-strand synthesis reaction is diluted in 7.5 times in the second-strand synthesis reaction.

2.4.1. Required materials

Equipment

- PCR machine for 0.2-ml tubes
- Real-time PCR machine
- Water bath: 20, 37 °C
- Power supply
- Agarose electrophoresis chamber

Reagents/enzymes

- 10× T4 DNA Ligase Reaction Buffer (New England BioLabs, # B0202)
- T4 DNA Polymerase (New England BioLabs, # M0203)
- DNA Polymerase I, Klenow large fragment (New England BioLabs, # M02103)
- T4 Polynucleotide Kinase (New England BioLabs, # M0201)
- Klenow fragment (3′→5′ exo-) (New England BioLabs, # M0212)
- NEBuffer 2 (New England BioLabs, #B7002S)
- 10 mM dATP
- Quick Ligation™ Kit (New England BioLabs, #M2200)
- Low range ultra agarose 125 g (Bio-Rad, #161-3107)
- Phusion™ High-Fidelity PCR Master Mix with HF Buffer (New England BioLabs, #F-531S)
- PE genomic adapters, 12 μM (sequences are available)
- PE PCR primers (sequences are available)

  Note: all enzymes and oligos mentioned above are included in the Illumina Genomic DNA Sample Prep Kit (e.g., #PE-102-1001).

- SYBR® Green PCR Core Reagents (Applied Biosystems, #4304886)
- UDG (New England BioLabs, #M0280)
- QIAquick PCR Purification Kit (Qiagen, #28106)
- MinElute Reaction Cleanup Kit (Qiagen, #28206)
- QIAquick Gel Extraction Kit (Qiagen, #28706)
- Oligos for real-time PCR:
  - # PE adapter primer 2:
    5′-CTCGGCATTCCTGCTGAACCGCTCTTCCGATCT-3′
  - # adapter primer 1:
    5′-ACACTCTTTCCCTACACGACGCTCTTCCGATCT-3′
- 50 bp DNA Ladder (New England BioLabs, #N3236)

Disposables

- Scalpels
- Optical plastic (tubes, strips or plates) for the real-time PCR machine

2.4.2. End-repair

1. Prepare the reaction mix in a 1.5-ml LoBind tube: 30 μl of ds cDNA, 1× T4 DNA Ligase Buffer with 1 mM ATP, 0.25 mM dNTPs, T4 DNA Polymerase (0.1 U/μl), DNA Polymerase I, Klenow large fragment (0.03 U/μl), T4 Polynucleotide kinase (0.3 U/μl).

   Note: We perform the end-repair reaction in 50 μl. We recommend a final concentration of ds cDNA up to 20 ng/μl. The volume of the enzymes added should not exceed 10% of the volume of the reaction.
2. Incubate the reaction at 20 °C for 30 min.

   Note: It is important not to increase the temperature as this would otherwise increase the exonuclease activity of T4 polymerase.
3. Perform the reaction cleanup with Qiagen PCR Purification Kit, elute in 30 μl EB.

   Note: The presence of artificial hybrid molecules in the library poses a problem when searching for rare structural rearrangements. Such molecules may appear during adapter ligation when some fragments may ligate to each other. To avoid *dimers*, a gel size selection should be performed after end-repair. It is necessary to select the cut-out range of this first size selection (before adapter ligation) in such a way that, during the second size selection (after adapter ligation), the doubled smallest fragments reliably differ from the largest fragments: $2 \times L_{small} \gg L_{large}$.

2.4.3. A-Tailing

1. Prepare the reaction mix in a 1.5-ml LoBind tube: 30 μl of ds cDNA, 1× NEBuffer 2, 1 mM dATP, Klenow fragment; 3′ → 5′ exo- (1 U per ~160 ng DNA, min. 2.5 I per reaction).

   Note: The A-tailing reaction is performed in a volume of 50 μl. The volume of the enzymes added should not exceed 10% of the volume of the reaction.
2. Incubate the reaction at 37 °C for 30 min.
3. Perform the reaction cleanup using the MinElute Reaction Cleanup Kit and following the manufacturer instructions. Elute in 10 μl EB.

2.4.4. Adapter ligation

1. Prepare the reaction mix in a 1.5-ml LoBind tube: 10 μl of ds cDNA, PE genomic adapter (10× molar excess over the amount of ds cDNA, min. 0.5 μl), 1× Quick Ligase Buffer, Quick Ligase (10% of the volume of the reaction).

   Note: To minimize nonspecific ligation, we first mix the cDNA with the adapters, then mix thoroughly with the buffer, and then add ligase.

2. Incubate the reaction at 20 °C for 30 min.
3. Perform the reaction cleanup with Qiagen PCR Purification Kit, elute in 30 μl EB.

2.4.5. Size selection

1. Load the sample on a 2% Low Range Ultra Agarose 1× TAE gel along with an appropriate marker (e.g., 50 bp marker, NEB). Run the gel until the marker bands in the range 100–400 bp are well separated (the 100 bp band should be about 5 cm from the well).

   Note: There should be at least one empty well between sample and marker and at least two empty wells between two samples. To minimize cross contaminations, we recommend avoiding loading samples differing more than 10-fold in concentration. For precious samples, it is worth running one gel per sample.
2. Cut out the gel band corresponding to a 180–220 nt fragment.

   Note 1: The selected size range corresponds to a 110–150 bp insertion (PE adapters add ~70 bp). We use this size range for the transcriptome analysis of human samples. For the analysis of transcript structure, it is desirable that PE reads belong to different exons. We have chosen the range comparable with the average size of an exon (~170 bp).

   Note 2: If the library is prepared exclusively for expression profiling, insert size variation is not important. If structural analysis would be performed, the size range should be as small as possible.

   Note 3: We recommend cutting out neighboring band (220–250 bp) and to freeze it in liquid nitrogen and store at −20 °C. Such a reserve might be helpful especially for precious samples.
3. Perform the reaction cleanup with Qiagen Gel Extraction Kit and following the manufacturer instructions. Elute the library in 30 μl EB.

2.4.6. UDG treatment

1. Add 3.4 μl of 10× UDG Buffer to the library. Then add UDG (1 U per 1 μg of DNA, min. 0.5 U per reaction).
2. Incubate the reaction tube at 37 °C for 15 min. No cleanup is required.

2.4.7. Real-time quality check

1. Prepare the 20 μl real-time PCR reaction. Combine in an appropriate tube 12.7 μl of water, 2 μl of 10× SYBRGreen Buffer, 2 μl of 12.5 nM dNTPs with dUTP, 2 μl of 25 mM $MgCl_2$, 0.4 μl of each of 20 μM

oligos #PE adapter primer 2 and #adapter primer 1, and 0.1 μl of AmpliTaq Gold (5 U/μl). Add 0.5 μl of library per reaction.
2. Place the reaction in real-time PCR machine and run the following program:
   - 95 °C for 10 min
   - 40 cycles: 95 °C for 15 s, 65 °C for 30 s, 72 °C for 30 s
3. Determine the number of cycles A, which corresponds to the middle of the logarithmic phase.
   Note: Over-amplification leads to the distortion of the original proportion of different DNA fragments.
4. Calculate the number of cycles for the final library amplification N, taking into account the volume of final amplification V_{amp} and volume of library taken for amplification

$$V_{library} : N = A - \log_2\{(V_{library}[\mu l]/V_{amp}[\mu l])/(0.5\,\mu l/20\,\mu l)\}$$

2.4.8. Amplification of the library

1. Prepare the amplification reaction mix in a 0.2 ml thin-wall tube: 1× Phusion HF Master Mix, 0.4 μM oligos #PE PCR primer 1.1 and #PE PCR primer 2.1. Take 10–30 μl of the library for the reaction.
   Note:We recommend the use of 1/3 to 1/2 of the material for the final amplification. Taking less would result in an unreasonable reduction of the complexity of the library. If all the material is used, there would be no reserve to repeat the amplification in case of an accident (like unsuccessful amplification, broken well in the gel, etc.).
   Note: The Phusion DNA polymerase used for the library amplification is strongly suppressed by uridines in the template (Hogrefe *et al.*, 2002). So, even in case that UDG occasionally would not remove the uridine base efficiently, the molecule will still not be amplified.
2. Place the reaction in a PCR machine and run the following program:
   - 95 °C for 30 s
   - Selected (by real-time library check) number of cycles: 95 °C for 15 s, 65 °C for 30 s, 72 °C for 30 s.
3. Load the entire PCR reaction on an agarose gel and excise cDNAs ranging in size between ~230 and 270 bp.
   Note: this additional gel selection step results in a higher quality library giving a better fragment size distribution and eliminating PCR by-products like primer dimers.
4. Determine the library concentration.

Proceed to the flowcell preparation, cluster generation, and sequencing reaction on the Illumina Cluster Station and Genome Analyzer according to the manufacturer's instructions

REFERENCES

Cloonan, N., Forrest, A. R., Kolle, G., Gardiner, B. B., Faulkner, G. J., Brown, M. K., Taylor, D. F., Steptoe, A. L., Wani, S., Bethel, G., *et al.* (2008). Stem cell transcriptome profiling via massive-scale mRNA sequencing. *Nat. Methods* **5,** 613–619.

Costa, V., Angelini, C., De Feis, I., and Ciccodicola, A. (2010). Uncovering the complexity of transcriptomes with RNA-Seq. *J. Biomed. Biotechnol.* 2010: 853916.

Denoeud, F., Aury, J. M., Da Silva, C., Noel, B., Rogier, O., Delledonne, M., Morgante, M., Valle, G., Wincker, P., Scarpelli, C., Jaillon, O., and Artiguenave, F. (2008). Annotating genomes with massive-scale RNA sequencing. *Genome Biol.* **9**(12), R175.

He, Y., Vogelstein, B., Velculescu, V. E., Papadopoulos, N., and Kinzler, K. W. (2008). The antisense transcriptomes of human cells. *Science* **322,** 1855–1857.

Hogrefe, H. H., Hansen, C. J., Scott, B. R., and Nielson, K. B. (2002). Archaeal dUTPase enhances PCR amplifications with archaeal DNA polymerases by preventing dUTP incorporation. *Proc. Natl. Acad. Sci. USA* **99,** 596–601.

Kapranov, P., Cheng, J., Dike, S., Nix, D. A., Duttagupta, R., Willingham, A. T., Stadler, P. F., Hertel, J., Hackermuller, J., Hofacker, I. L., *et al.* (2007). RNA maps reveal new RNA classes and a possible function for pervasive transcription. *Science* **316,** 1484–1488.

Levin, J. Z., Yassour, M., Adiconis, X., Nusbaum, C., Thompson, D. A., Friedman, N., Gnirke, A., and Regev, A. (2009). Comprehensive comparative analysis of strand-specific RNA sequencing methods. *Nat. Methods* **7**(9), 709–715.

Lister, R., O'Malley, R. C., Tonti-Filippini, J., Gregory, B. D., Berry, C. C., Millar, A. H., and Ecker, J. R. (2008). Highly integrated single-base resolution maps of the epigenome in Arabidopsis. *Cell* **133,** 523–536.

Marioni, J., Mason, C., Mane, S., Stephens, M., and Gilad, Y. (2008). RNA-seq: An assessment of technical reproducibility and comparison with gene expression arrays. *Genome Res.* **18,** 1509–1517.

Mortazavi, A., Williams, B. A., McCue, K., Schaeffer, L., and Wold, B. (2008). Mapping and quantifying mammalian transcriptomes by RNA-Seq. *Nat. Methods* **5,** 621–628.

Nagalakshmi, U., Wang, Z., Waern, K., Shou, C., Raha, D., Gerstein, M., and Snyder, M. (2008). The transcriptional landscape of the yeast genome defined by RNA sequencing. *Science* **320,** 1344–1349.

Parkhomchuk, D., Borodina, T., Amstislavskiy, V., Banaru, M., Hallen, L., Krobitsch, S., Lehrach, H., and Soldatov, A. (2009). Transcriptome analysis by strand-specific sequencing of complementary DNA. *Nucleic Acids Res.* **37**(18), e123.

Perocchi, F., Xu, Z., Clauder-Münster, S., and Steinmetz, L. M. (2007). Antisense artifacts in transcriptome microarray experiments are resolved by actinomycin D. *Nucleic Acids Res.* **35,** e128.

Ruprecht, R. M., Goodman, N. C., and Spiegelman, S. (1973). Conditions for the selective synthesis of DNA complementary to template RNA. *Biochim. Biophys. Acta* **294,** 192–203.

Stoughton, R. B. (2005). Applications of DNA microarrays in biology. *Annu. Rev. Biochem.* **74,** 53–82.

Sultan, M., Schulz, M. H., Richard, H., Magen, A., Klingenhoff, A., Scherf, M., Seifert, M., Borodina, T., Soldatov, A., Parkhomchuk, D., *et al.* (2008). A global view of gene activity and alternative splicing by deep sequencing of the human transcriptome. *Science* **321,** 956–960.

Velculescu, V. E., Zhang, L., Vogelstein, B., and Kinzler, K. W. (1995). Serial analysis of gene expression. *Science* **270,** 484–487.

Wilhelm, B. T., Marguerat, S., Watt, S., Schubert, F., Wood, V., Goodhead, I., Penkett, C. J., Rogers, J., and Bähler, J. (2008). Dynamic repertoire of a eukaryotic transcriptome surveyed at single-nucleotide resolution. *Nature* **453,** 1239–1243.

Wilhelm, B. T., Marguerat, S., Goodhead, U., and Bähler, J. (2010). Defining transcribed regions using RNA-Seq. *Nat. Protoc.* **5,** 255–266.

Zbio-wiki (2010). Next-generation_sequencing. Available at:http://www.zbio.net/wiki/Next-generation_sequencing(accessed 11 November 2010).

CHAPTER SIX

Quantitative Real-Time PCR-Based Analysis of Gene Expression

J. Jozefczuk *and* J. Adjaye

Contents

Abstract

Quantitative real-time polymerase chain reaction (QRT-PCR) has become an extensively applied technique. It enables quantitative analyses of gene expression applicable to basic molecular biology, medicine, and diagnostics. Nowadays, it is broadly used to describe messenger RNA (mRNA) expression patterns and to compare the relative levels of mRNA within distinct biological samples. The scope of the QRT-PCR technique makes it applicable across a wide range of experimental conditions and allows experimental comparison between normal and abnormal tissue. Most importantly, this technique enables additional independent confirmation of microarray or next generation sequencing (NGS)-based results. An inherent advantage of QRT-PCR is the large dynamic range, remarkable sensitivity, and sequence-specificity. We provide a detailed step by step guide to the principles underlying a successful QRT-PCR experiment.

Department of Vertebrate Genomics, Max Planck Institute for Molecular Genetics (Molecular Embryology and Aging group), Ihnestrasse 63-73, Berlin, Germany

Methods in Enzymology, Volume 500
ISSN 0076-6879, DOI: 10.1016/B978-0-12-385118-5.00006-2

1. Introduction

Quantitative real-time polymerase chain reaction (QRT-PCR) enables the continuous monitoring of the amplification process as it occurs and uses fluorescent reporter dyes to merge the amplification and detection steps. The assay is based on measuring the increase in fluorescent signal, which corresponds to the amount of DNA produced during each PCR cycle. A single PCR reaction is characterized by the PCR cycle at which fluorescence first rises above threshold background levels (threshold cycle, Ct). Hence, higher the messenger RNA (mRNA) concentration of a target gene, lower is the Ct value. The process of QRT-PCR consists of three steps: reverse transcriptase-based conversion of RNA to cDNA, the amplification of cDNA by PCR, and the detection and quantification of amplified products-referred as amplicons. Several concerns exist, with regard to consistency as the results depend on the quantity and quality of the template RNA. In order to minimize inconsistencies and maximize reproducibility, a quality control of each component of the QRT-PCR assay is advisable.

The extensive use of this technique has resulted in the development of various protocols that enable the generation of quantitative data applying (i) fresh, frozen, or archived FFPE (formalin-fixed, paraffin-embedded) samples, (ii) whole-tissue biopsies, microdissected samples, single cells, tissue cultured cells, (iii) total or mRNA, (iv) different cDNA priming strategies, (v) different enzymes, (vi) assays of distinct efficiencies, sensitivities, and robustness, (vii) diverse detection chemistry, reaction conditions, thermal cyclers, (viii) individual analysis and reporting methods (Nolan *et al.*, 2006). The sensitivity of QRT-PCR has enabled the investigation of gene expression during oogenesis and embryogenesis in mouse, bovine, and human (Adjaye, 2005, 2007; Adjaye *et al.*, 2005; Kues *et al.*, 2008; Zuccotti *et al.*, 2008). Essential prerequisites for a successful reverse transcription (RT) reaction are (i) good quality RNA and (2) an mRNA priming method that enables the synthesis of full length corresponding cDNAs. First-strand cDNA synthesis can be primed using either one of the three distinct methods; (1) random primers (hexamers), (2) oligo(dT), or (3) gene-specific primers. Each approach has its advantages and disadvantages. Both random priming and oligo(dT) priming lead to the production of a representative pool of distinct cDNAs within a single reaction. However, it has been shown that priming with random hexamers does not result in equal efficiencies of RT for all target genes within the sample, nonetheless, it increases the probabilities of generating pools of cDNAs with a high representation of 5′-end sequences. Gene-specific primers are the most specific and sensitive method and are recommended when the quantity of RNA is not a limiting factor and when one aims at only amplifying a single gene rather than generating a pool of distinct cDNAs. Despite these features, there can be differences in the efficiencies at which individual RT reactions proceed (Nolan *et al.*, 2006).

As QRT-PCR is now a common method for quantifying gene expression, it is imperative to be aware of the various options available to enable the execution of a successful reaction (Wong and Medrano, 2005). A well-designed experiment carried out with the appropriate controls can be the most sensitive, efficient, and reproducible assay to measure gene expression.

2. Primer Design

Well-designed gene-specific primers are essential for the successful analysis of expression. The design of primers for QRT-PCR using SYBR® Green should fulfil important parameters like: 30–80% G/C content, amplicon length between 50 and 150 bases (the shorter the better), T_m 58–60 °C, minimum 18-bases length, 3′-instability (max. 2 (Gs/Cs) within the last 3′-bases). Primers can be designed using one of many available tools such as Primer Express 2.0 from Applied Biosystems, or Primer 3 (The Whitehead Institute, http://frodo.wi.mit.edu/primer3/). Most importantly, the forward and reverse primers should span an intron/exon boundary in order to avoid the amplification of contaminating genomic DNA.

3. Primer Validation Experiments

For comparative gene expression analyses by QRT-PCR, the efficiency of the primer pairs has to be tested prior to comparing differential gene expression between distinct biological samples. Meaningful comparative expression will only be attained if the different genes are amplified with nearly the same efficiency by the designed primers. In general, a primer validation experiment has to be carried out for each gene. To determine the priming efficiency of a primer pair, each gene should be amplified using four different dilutions of the RT reaction, that is, 1:1, 1:10, 1:100, and 1:1000. In a typical primer efficiency validation, all samples (distinct genes and dilutions) should be analyzed in triplicate, and for each gene, two NTCs (nontemplate controls) should be analyzed in parallel.

The results of a real-time plot for a gene without NTCs are shown in Fig. 6.1A. When threshold is specified it is evident that different dilutions result in distinct numbers of cycles. The mean of these threshold cycles is then plotted against the corresponding dilutions and the slope of the linear trend established (Fig. 6.1B). Applying this slope, the efficiency of the primers can then be calculated using the following equation

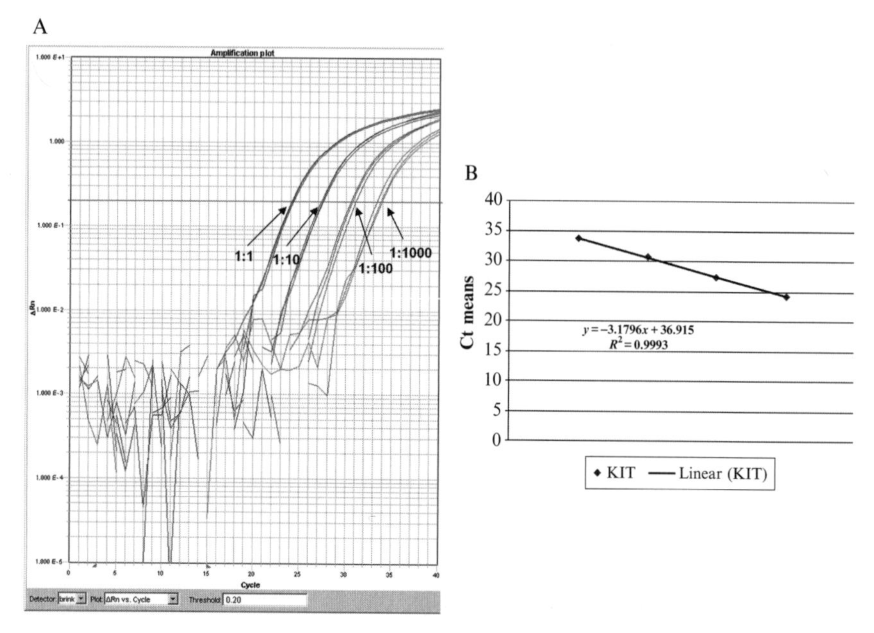

Figure 6.1 Primers validation. (A) An illustration of an amplification plot of a primer validation experiment. The plot shows the amplification of four distinct cDNA dilutions. These are from left to right: 1:1, 1:10, 1:100, and 1:1000. To obtain the threshold cycles, the threshold is set by the horizontal red line. (B) Determination of the efficiency of primers designed to amplify the gene *KIT*. The threshold cycles of the four dilutions are displayed in the graph. The dilutions are from left to right: 1:1000, 1:100, 1:10, and 1:1. The equation of the trend-line is shown. The efficiency can be calculated by using the slope (−3.1796) and the Eq. (6.1) described in the text. (For interpretation of the references to color in this figure legend, the reader is referred to the Web version of this chapter.)

$$E = 10^{(1/-\text{slope})}, \tag{6.1}$$

where E is the efficiency of the amplification.

4. Cell Culture-Based Isolation and Quantification of RNA

Prior to RNA isolation, cells have to be washed with Ca^{2+}- and Mg^{2+}-free PBS (Gibco, Invitrogen) and lysed directly in the culture dish. Total RNA can be isolated using any of the available isolation kits such as the RNeasy® Mini Kit (Qiagen, Hilden, Germany) following the instructions suggested by the manufacturer. In addition, DNase I (RNase-free DNase set, Qiagen) digest should be included. The concentration and quality of the isolated RNA can be determined using a Nanodrop spectrophotometer (Thermo Fischer Scientific, Waltham, MA) or a Bioanalyzer (Agilent Technologies, Santa Clara, CA). An example of a quality control picture produced by a Bioanalyzer is shown in Fig. 6.2.

The quality of isolated RNA can also be determined by using agarose gel electrophoresis and ethidium bromide staining. A typical 1.5% agarose gel (Life Technologies, Paisley, Scotland) is made up by melting in 100 ml of 1× TAE (Tris-acetate-EDTA) buffer. One microliter of ethidium bromide (10 mg/ml; Invitrogen) is then added directly to the molten gel (final concentration, 100 ng/ml), mixed, and poured into an appropriate electrophoresis chamber. RNA samples to be analyzed should be mixed with 1/3 volume of 6× loading buffer (Fermentas, St. Leon-Rot, Germany) loaded into the wells of the agarose gel and then resolved in an appropriate electrophoresis chamber. The length of an amplicon can be determined using as a reference DNA size marker such as the GeneRuler™ 1-kb DNA ladder (Fermentas). An image analyzer such as the AlphaImager™ (Alpha Innotech, San Leandro, CA) can be used to visualize nucleic acids under UV light.

5. Reverse Transcription

RT is carried out in a total volume of 25 μl as follows: RNA (500 ng–2 μg, equal concentrations for all samples under investigation, adjust to a final volume of 9.5 μl with water), and 0.5 μl of oligo(dT_{15}) (1 μg/μl; Invitek, Berlin, Germany) should be incubated for 3 min at 70 °C then

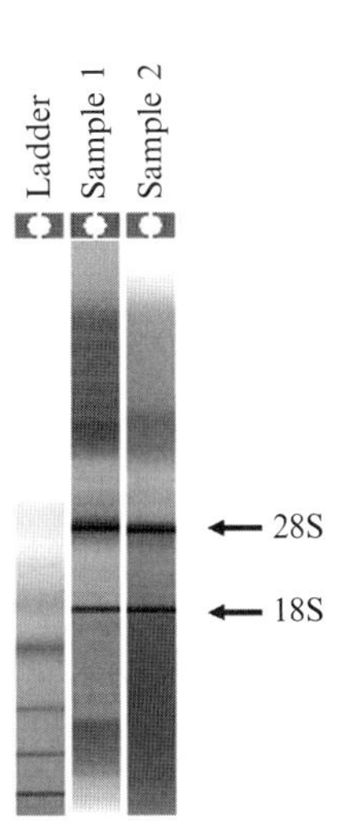

Figure 6.2 A Bioanalyzer-based image of intact RNA isolated from cells in culture. The figure shows the typical 28s and 18s ribosomal RNA bands with an intensity ratio of 2:1.

cooled on ice to attain annealing of the primer to the polyA tail within the mRNA. Then supplement with a Master Mix (15 μl) consisting of the following components: 5.0 μl of 5× reaction buffer (Promega), 0.5 μl of (25 mM) dNTP, 0.1 μl of M-MLV reverse transcriptase, (Moloney Murine Leukemia Virus reverse transcriptase; 200 U/μl; USB), and 9.4 μl of dH_2O. After 1-h incubation at 42 °C, the reaction is stopped at 65 °C for 10 min (Jozefczuk *et al.*, 2010).

M-MLV reverse transcriptase can be used in cDNA synthesis with long mRNA templates (>5 kb), and their RNase H (endoribonuclease H) activity is weaker than that of the commonly used Avian Myeloblastosis Virus (AMV) reverse transcriptase.

5.1. Quantitative real-time polymerase chain reaction

QRT-PCR can be carried out on the Applied Biosystems 7900 instrument, in 96-Well Optical Reaction Plates (Applied Biosystems). The following program can be employed:

Stage 1 (activation)	50 °C for 2 min
Stage 2 (initial denaturing)	95 °C for 10 min
Stage 3 (40 repeats)	
Denaturing	95 °C for 15 s
Elongation	60 °C for 1 min
Stage 4 (dissociation curve)	95 °C for 15 s
	60 °C for 15 s
	95 °C for 15 s

In addition, dissociation curves of the products should be created (SDS 2.2.1 software, Applied Biosystems). The final reaction volume of 20 μl consists of 10 μl of SYBR™ Green PCR Master Mix (Applied Biosystems), 2.5 μM of each primer (1.5 μl of each primer), and 7 μl of cDNA (not less than 1% of RT reaction). Each gene should be analyzed in triplicate. Relative mRNA levels are calculated using the comparative ΔΔCt method (Livak and Schmittgen, 2001) and presented as a percentage of the controls. mRNA levels of both *β-ACTIN* and *GAPDH* genes are used as controls for normalization.

In the best possible scenario, the efficiency of a primer pair is 2 which would imply that the number of amplicons is doubled in each cycle.

5.2. Comparative gene expression experiments

Since in comparative experiments all results are relative, a consistent reference is needed and the results have to be normalized. As reference, the housekeeping genes like *GAPDH* and *β-ACTIN* should be used. Each gene should be analyzed in triplicate and two NTCs should be included in the analysis.

In an example of a comparative experiment, data for two definitive endoderm (DE) marker genes (*SOX17*, *FOXA2*) was generated. For each gene, threshold cycles for undifferentiated human embryonic stem cells (hESCs) and derived DE cells were obtained. An amplification plot of such a comparative experiment is shown in Fig. 6.3. Threshold cycles of DE marker genes were compared against the corresponding cycles in the reference sample-undifferentiated hESCs.

Using the average threshold cycles in the QRT-PCR run, the amount of target DNA can be calculated. This amount, normalized to the endogenous reference (*GAPDH* or *β-ACTIN*) and relative to a calibrator (hESCs), is given by:

$$2^{-\Delta\Delta \mathrm{Ct}}$$

Details of the derivation of this formula can be found in Applied Biosystems (2001) and Livak and Schmittgen (2001).

In essence, the exponential amplification of amplicons is described by:

$$X_n = X_0(1 + E_X)^n$$

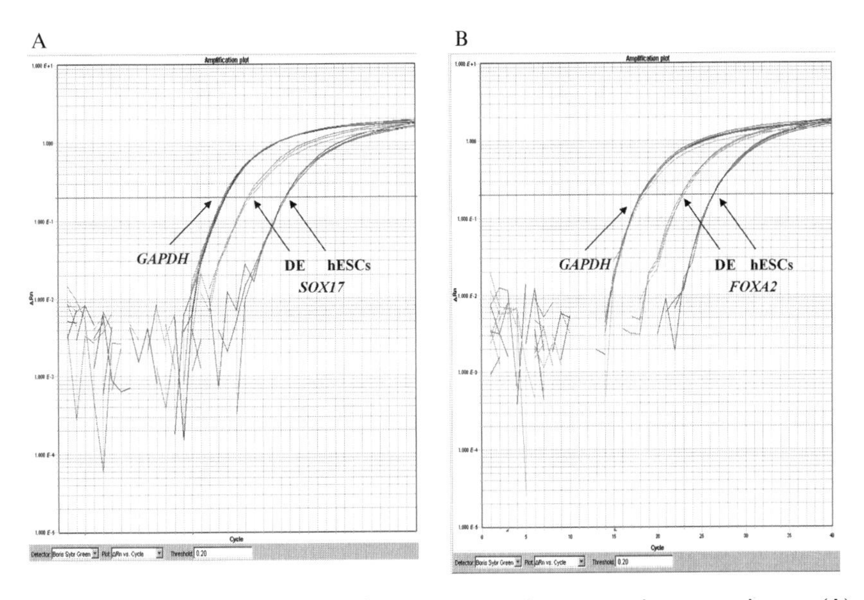

Figure 6.3 Amplification plots of a gene expression comparison experiment. (A) Amplification plot of a QRT-PCR analysis for the expression of *SOX17* plus the reference (*GAPDH*). (B) Amplification plot of a QRT-PCR analysis for the expression of *FOXA2* plus the reference (*GAPDH*). Both plots present results for DE-differentiated cells and undifferentiated hESCs.

where X_n = number of target molecules at cycle n, X_0 = initial number of target molecules, E_X = efficiency of target amplification, and n = number of cycles.

The threshold cycle (Ct) indicates the fractional cycle number at which the amount of amplified target reaches a fixed threshold. Thus,

$$X_T = X_0(1 + E_X)^{\mathrm{Ct},x} = K_X$$

where X_T = threshold number of target molecules, Ct, x = threshold cycle for target amplification, and K_X = constant.

A similar equation for the endogenous reference reaction is

$$R_T = R_0(1 + E_R)^{\mathrm{Ct},r} = K_R$$

where R_T = threshold number of reference molecules, R_0 = initial number of reference molecules, E_R = efficiency of reference amplification, Ct, r = threshold cycle for reference amplification, and K_R = constant.

Dividing X_T by R_T gives the following expression:

$$\frac{X_T}{R_T} = \frac{X_0(1 + E_X)^{\mathrm{Ct},x}}{R_0(1 + E_R)^{\mathrm{Ct},r}} = \frac{K_X}{K_R} = K$$

Assuming efficiencies of the target and the reference are the same:

$$E_X = E_R = E$$

$$\frac{X_0}{R_0}(1 + E)^{\mathrm{Ct},x-\mathrm{Ct},r} = K$$

or

$$X_N(1 + E)^{\Delta \mathrm{Ct}} = K$$

where $X_N = \frac{X_0}{R_0}$, the normalized amount of target

ΔCt = Ct, x − Ct, r, the difference in threshold cycles for target and reference.

Rearranging gives the following expression:

$$X_N = K(1 + E)^{-\Delta \mathrm{Ct}}$$

The final step is to divide the X_N for DE samples by the X_N for the calibrator (hESCs):

$$\frac{X_{\mathrm{N,DE}}}{X_{\mathrm{N,hESCs}}} = \frac{K(1+E)^{-\Delta \mathrm{Ct,DE}}}{K(1+E)^{-\Delta \mathrm{Ct,hESCs}}} = (1+E)^{-\Delta\Delta \mathrm{Ct}}$$

where $\Delta\Delta\mathrm{Ct} = (\Delta\mathrm{Ct},\ \mathrm{DE}) - (\Delta\mathrm{Ct},\ \mathrm{hESCs})$

Assuming that the primers have efficiency close to 1, and then the amount of amplicon is defined by the equation;

$$2^{-\Delta\Delta \mathrm{Ct}}$$

One should bear in mind that the efficiency values are not the same as that calculated for the primer validation experiments. The optimal efficiency of 1 in this case is equal to an efficiency of 2 in the primer validation experiments.

5.3. Drawbacks associated with QRT-PCR

The inclusion of assessing dissociation curve is crucial for a successful outcome of an SYBR® Green based QRT-PCR reaction. Because a fluorescent dye is incorporated in the SYBR® Green Master Mix, this intercalates with double stranded DNA and as such for short amplicons between 50 and 100 nucleotides, primer dimers (up to 50 nucleotides) can be falsely interpreted as genuine amplicons. To verify if primer dimerization has occurred, dissociation curves are obtained by a very slow stepwise increase in temperatures from 60 to 95 °C. Figure 6.4 shows typical curves for a perfect QRT-PCR reaction lacking primer dimerization. Contamination of the amplicons by primer dimerization would be presented as an additional peak with a lower temperature, whereas contamination with genomic DNA will result in a peak with a higher melting temperature.

Another problem associated with the use of SYBR® Green based QRT-PCR is that the annealing of the primers and the subsequent elongation occurs simultaneously hence by default there will be a tendency for primer dimerization but, however, this can be bypassed by the use of RNA templates at concentrations ranging from 500 ng to 2 μg.

5.4. Required materials

Devices

Nanodrop spectrophotometer (Thermo Fischer Scientific)
Bioanalyzer (Agilent Technologies)

AlphaImager™ (Alpha Innotech)

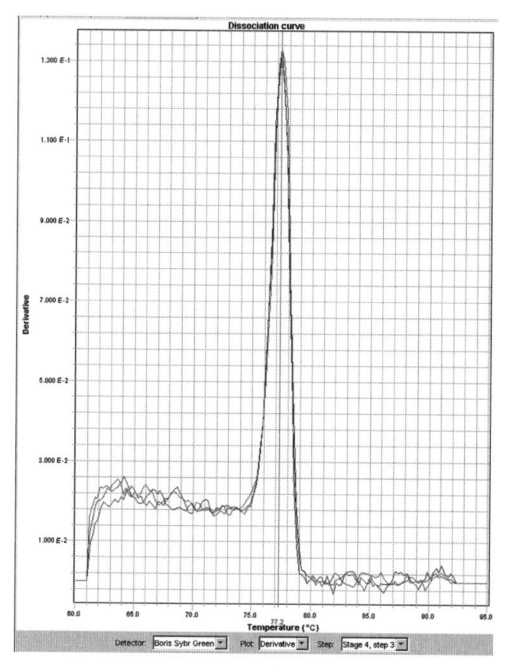

Figure 6.4 A typical dissociation curve showing the absence of primer dimerization. The figure shows dissociation curves for a QRT-PCR reaction, with a unique peak for the melting temperature of the produced amplicon.

Applied Biosystems 7900 instrument (Applied Biosystems)
SDS 2.2.1 software (Applied Biosystems)

Additional materials

oligo(dT15), dNTPs, primers
96-well optical reaction plates (Applied Biosystems)

Other reagents

SYBR Green PCR Master Mix (Applied Biosystems)
PBS (Gibco, Invitrogen)
RNeasy® Mini Kit (Qiagen)
DNase I–RNase-free DNase set (Qiagen)
Agarose (Life Technologies)
TAE buffer
Ethidium bromide (Invitrogen)
GeneRuler™ 1 kb DNA ladder (Fermentas)
6× loading buffer (Fermentas)
M-MLV reverse transcriptase 5× Reaction Buffer (Promega)

M-MLV reverse transcriptase (USB)
Disposables
1.5-ml (Eppendorf) tubes
Tissue culture plates

REFERENCES

Adjaye, J. (2005). Whole-genome approaches for large-scale gene identification and expression analysis in mammalian preimplantation embryos. *Reprod. Fertil. Dev.* **17,** 37–45.

Adjaye, J. (2007). Generation of amplified RNAs and cDNA libraries from single mammalian cells. *Methods Mol. Med.* **132,** 117–124.

Adjaye, J., Huntriss, J., Herwig, R., BenKahla, A., Brink, T. C., Wierling, C., Hultschig, C., Groth, D., Yaspo, M. L., Picton, H. M., Gosden, R. G., and Lehrach, H. (2005). Primary differentiation in the human blastocyst: Comparative molecular portraits of inner cell mass and trophectoderm cells. *Stem Cells* **23,** 1514–1525.

Applied Biosystems: ABI PRISM 7700 Sequence Detection System. User Bulletin #2. 11–13, 2001.

Jozefczuk, J., Stachelscheid, H., Chavez, L., Herwig, R., Lehrach, H., Zeilinger, K., Gerlach, J. C., and Adjaye, J. (2010). Molecular characterization of cultured adult human liver progenitor cells. *Tissue Eng. Part C Methods* **16**(5), 821–834.

Kues, W. A., Sudheer, S., Herrmann, D., Carnwath, J. W., Havlicek, V., Besenfelder, U., Lehrach, H., Adjaye, J., and Niemann, H. (2008). Genome-wide expression profiling reveals distinct clusters of transcriptional regulation during bovine preimplantation development in vivo. *Proc. Natl. Acad. Sci. USA* **105,** 19768–19773.

Livak, K. J., and Schmittgen, T. D. (2001). Analysis of relative gene expression data using real-time quantitative PCR and the 2(-Delta Delta C(T)) method. *Methods* **25,** 402–408.

Nolan, T., Hands, R. E., and Bustin, S. A. (2006). Quantification of mRNA using real-time RT-PCR. *Nat. Protoc.* **1,** 1559–1582.

Wong, M. L., and Medrano, J. F. (2005). Real-time PCR for mRNA quantitation. *Biotechniques* **39,** 75–85.

Zuccotti, M., Merico, V., Sacchi, L., Bellone, M., Brink, T. C., Bellazzi, R., Stefanelli, M., Redi, C. A., Garagna, S., and Adjaye, J. (2008). Maternal Oct-4 is a potential key regulator of the developmental competence of mouse oocytes. *BMC Dev. Biol.* **8,** 97.

SECTION FOUR

PROTEIN PRODUCTION AND QUANTIFICATION FOR SYSTEMS BIOLOGY

CHAPTER SEVEN

Quantification of Proteins and Their Modifications Using QconCAT Technology

Kathleen M. Carroll,* Francesco Lanucara,† *and* Claire E. Eyers†

Contents

* Manchester Centre for Integrative Systems Biology, University of Manchester, Manchester Interdisciplinary Biocentre, Manchester, United Kingdom
† Michael Barber Centre for Mass Spectrometry, School of Chemistry, University of Manchester, Manchester Interdisciplinary Biocentre, Manchester, United Kingdom

Methods in Enzymology, Volume 500
ISSN 0076-6879, DOI: 10.1016/B978-0-12-385118-5.00007-4

Abstract

Building a mathematical model of a biological system requires input of experimental data for each networked component, ultimately generating a model that can be used to test scientific hypotheses. A fundamental requirement in the computation of these systems is that the total amount of each component can be specified precisely. An added level of complexity occurs because a vast number of protein posttranslational modifications modulate protein function. Each of these modified forms therefore needs to be considered as a separate system component, and must therefore be quantified individually. In this chapter, we describe how designer QconCAT proteins can be used to determine the absolute amounts of both the polypeptide components and their covalently modified derivatives in both yeast and mammalian extracts derived from living cell populations.

ABBREVIATIONS

M9	type of minimal media
MeCN	acetonitrile
MS	mass spectrometry
SOC	super optimal broth with catabolite repression
XIC	extracted ion chromatogram

1. INTRODUCTION

1.1. Isotope-labeled peptides for absolute protein quantification

Quantitative rather than purely qualitative proteomics is becoming increasingly *de rigueur*. Such data can be used both to help understand and define how the protein content of cells and tissues becomes altered as cellular conditions change and to provide essential information for mathematical modeling of specified biological systems (Eyers, 2009). While relative protein quantification between samples is often sufficient to characterize condition specific changes, defining absolute amounts of the same protein components is critical if those pathways are to be computationally modeled. Mass spectrometry (MS) is an extremely sensitive analytical tool; unfortunately, it is an inherently nonquantitative technique. At present, determining the total amount of individual proteins by MS optimally requires the concurrent analysis of known quantities of internal standards, chemically identical analytes, for reference purposes. In order to distinguish these reference analytes from their native counterparts, stable isotope labeling is employed. Addition of known amounts of isotope-labeled peptide to an unknown quantity of native peptide can be used to calculate the amount of native peptide (and thus the protein from which it was derived) with accuracy and precision (Brun *et al.*, 2009). These isotope-labeled peptides can be generated either using chemical synthesis (AQUA) (Gerber *et al.*, 2003) or by taking advantage of the protein production machinery in *Escherichia coli* employed during the synthesis of recombinant proteins (Beynon *et al.*, 2005). When absolute quantification data are required for more than a handful of proteins, extracted from a number of differentially treated samples, production of these reference standards is most readily achieved using "designer" QconCAT (Quantification concatamer) proteins, whose proteolysis yields a stoichiometric mixture of the reference peptides (Q-peptides) incorporated (Fig. 7.1; Pratt *et al.*, 2006; Rivers *et al.*, 2007).

Rather than synthesizing and purifying reference quantification peptides individually, often at great cost, the coding sequences of the peptides selected for quantification can alternatively be concatenated and inserted

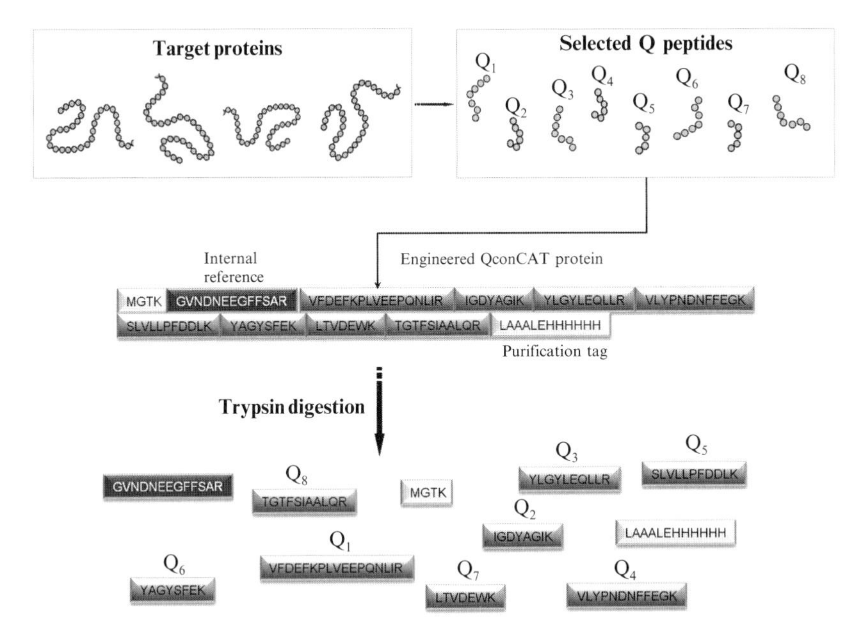

Figure 7.1 Designer "Quantification concatamer" (QconCAT) protein for the generation of a stoichiometric mixture of Q-peptides following proteolysis with trypsin. The schematic indicates selection of a subset of tryptic peptides (Q1–Q8) from a selection of target proteins, which together with an internal reference peptide (GVNDNEEGFF-SAR) and the purification His6-tag are used to design, express, and purify a recombinant protein by heterologous gene expression. Trypsin digestion of the engineered protein yields a stoichiometric mixture of peptide standards as indicated.

into a bacterial expression vector, generating a designer QconCAT construct. The number of Q-peptides that can be simultaneously produced is thus theoretically limited only by polypeptide stability after expression. Expression of QconCAT proteins in *E. coli* in [$^{13}C_6$]-Arg/Lys-containing medium ensures that each of the reference tryptic peptides can be differentially observed by MS (Beynon *et al.*, 2005; Fig. 7.2). However, the QconCAT technique relies on high levels of stable isotope amino acid incorporation, the maximum for which is 99%, as this in turn is dependent on the level present in the commercial stable amino acid isotopes used in the growth media (readily obtainable where the label is [$^{13}C_6$], though not necessarily the case where the stable isotope [^{15}N] is used. For this reason, and due to practical difficulties associated with deconvoluting spectra from [^{15}N]-labeled peptides, we thus recommend [$^{13}C_6$]).

A comparison of the distinct AQUA and QconCAT techniques for the production of Q-peptides for absolute quantification concluded that while all of the peptides under investigation could be readily obtained by digestion of expressed QconCATs, not all of the individual peptides could be generated using standard synthesis techniques (Mirzaei *et al.*, 2008). The issues that arose during QconCAT-based peptide (Q-peptide) generation were due to lack of protein expression thought to arise from problems at the stage

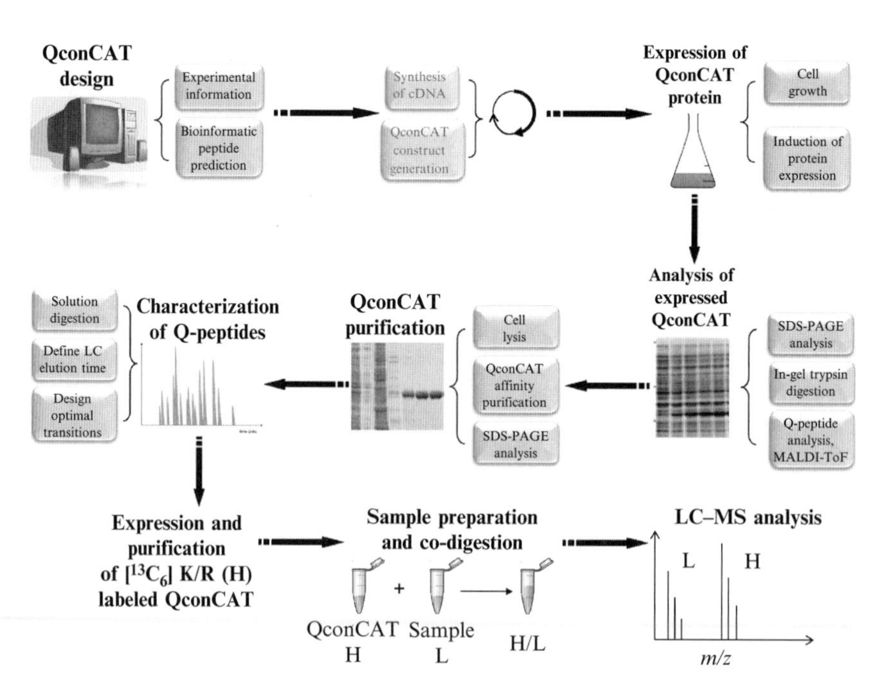

Figure 7.2 Workflow for the generation and use of designer QconCAT proteins for absolute protein quantification.

of protein translation. These issues were easily overcome by reordering of the peptides sequences within the construct (Mirzaei *et al.*, 2008), and indeed optimization of the codon sequence to minimize the formation of stable inhibitory RNA secondary structures is something that is usually taken into consideration during the initial stages of QconCAT design (Pratt *et al.*, 2006).

Irrespective of the method used for Q-peptide production, selection of the optimal reference peptide is critical for accurate calculation of absolute protein levels in cell extracts (Bronstrup, 2004). Once these peptides have been selected and, in the case of QconCAT, used to design the required artificial protein, quantification is achieved by spiking defined quantities of isotope-labeled QconCAT into a protein-containing extract, subjecting the sample to proteolysis (most often with trypsin) and then performing peptide directed mass spectrometric analysis, having first separated the peptides by reverse phase chromatography. Labeling the peptides with ^{13}C ensures that there is no difference in peptide elution from C_{18} reverse phase medium, which might complicate analyses as observed in the case of ^{2}H-labeling (Zhang *et al.*, 2002). A number of mass spectrometric methods are described for analysis of these types of samples; however, the lowest limits of detection to date have been achieved using multiple selective reaction monitoring (SRM) experiments (Lange *et al.*, 2008). Here, the triple quadrupole (or Q-trap) mass spectrometer is set up to selectively transmit a specific peptide ion through the first quadrupole (Q1) into the collision cell, while setting the third quadrupole (Q3) to transmit two (or more) defined product ions from that precursor (Fig. 7.3). Such targeted experiments improve the sensitivity and the specificity of quantitative analyses, with more SRM experiments improving the specificity of analysis, albeit often at the expense of the limit of quantification. The limit of quantification in experiments such as these is also governed by a number of other factors, including the "detectability" of the peptide(s) that are selected to represent the protein of interest, and the amount of proteolyzed extract loaded onto the reversed phase column used for peptide separation in-line with the MS analysis.

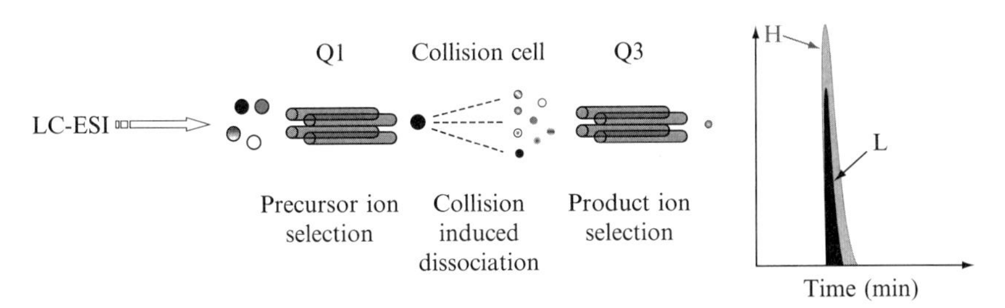

Figure 7.3 Schematic of a selective reaction monitoring (SRM) experiment using a triple quadrupole mass spectrometer.

This chapter describes the methods and considerations required for the quantitative analysis of proteins and their modified peptides from yeast and mammalian cells.

1.2. QconCAT design

To increase the accuracy of protein quantification, a minimum of two Q-peptides should be selected per protein to be quantified. FASTA files are generated for each of the proteins of interest and using CONSeQuence (http://king.smith.man.ac.uk/CONSeQuence/) CONS4, Q-peptides are predicted for each of the proteins (Eyers *et al.*, 2011). For each of the peptides listed, the presence of potential modifications and/or polymorphisms is ascertained using available databases including UniProt Knowledgebase (http://www.uniprot.org/), PhosphoSitePlus® (http://www.phosphosite.org/), PhosphosELM (http://phospho.elm.eu.org/index.html), and any relevant species-specific databases. Mass spectral databases (e.g., PeptideAtlas, http://www.peptideatlas.org/) are also interrogated to ascertain if the peptides selected have previously been observed by MS. The likelihood of incomplete proteolysis of the native protein and thus generation of the limit peptides are also assessed using the incorporated missed cleavage predictor (Siepen *et al.*, 2007a). If appropriate Q-peptides are not listed in CONS4, then CONS3 is used. Whenever possible, adjacent peptides in the native protein should not be selected as representative Q-peptides, because should these peptides fail to undergo complete proteolysis, neither peptide can be used for accurate determination of total protein amount.

1.2.1. Design for posttranslational modification quantification

Having selected Q-peptides for absolute quantification of the protein of interest, the tryptic peptide sequences that include the potential site(s) of modification are then selected. This of course requires that all sites of modification on a protein be mapped prior to quantification (Eyers and Gaskell, 2008). Unlike the Q-peptides used for calculating total protein levels, the reference peptides for quantification of protein modification cannot be optimally selected based on their ability to (i) generate limit peptides (i.e., undergo complete proteolysis from the native protein) and (ii) be readily observable in an LC–MS experiment. In this case, in an attempt to mimic proteolysis of analog peptides, three amino acids from the native sequence at both the C-terminal (P1′–P3′) and N-terminal positions (P1–P3) are maintained in the QconCAT protein sequence (Johnson *et al.*, 2009), as these residues are often the culprits when miscleavage occurs (Siepen *et al.*, 2007a).

Peptides are ordered in the QconCAT such that the surrounding sequence resembles the native primary structure where possible. Ideally the artificial QconCAT proteins should be smaller than ~65 kDa to decrease potential issues with bacterial protein expression (e.g., degradation, lower expression levels). Depending on the number of Q-peptides required, it may therefore be necessary to split the QconCAT into multiple plasmids. Under these circumstances, the Q-peptides for quantification of an individual protein should automatically be included in the same construct. For normalization of spiked reference between constructs, the "Fib" peptide, GVNDNEEGFFSAR (or GGVNDNEEGFFSAR) (Eyers *et al.*, 2008) can be included in each QconCAT construct. Differentially labeled synthesized peptides (accurately quantified) can also be used as an additional step for QconCAT quantification (Fig. 7.1). Having generated a complete, ordered list of Q-peptides for quantification, codon usage is optimized for bacterial transcription such that mRNA secondary structure is avoided where possible, as this limits translation. Finally, the cDNA sequence encoding the QconCAT is synthesized by piecing together primer sequences and inserted into the bacterial expression plasmid pET21a, for expression of a His_6-tagged protein using a T7-driven promoter (Pratt *et al.*, 2006). Design and optimization of the QconCAT cDNA can also be outsourced to one of a number of specialist companies, for example, PolyQuant GmbH who have their own proprietary software, Leto 1.0 for the optimization of DNA sequences for maximum expression yields (see http://www.entelechon.com/synthetic-products-2/bioinformatics/leto/open-source-libraries/).

2. Expression and Purification of QconCAT Proteins

The protocol detailed is amended from Pratt *et al.* (2006).

2.1. Transformation of expression *E. coli* strain with QconCAT plasmid

1. Dissolve the QconCAT plasmid in Tris–EDTA buffer to a final concentration of 0.5 ng/μl. Store at −20 °C.
2. Add 0.5 μl of DNA to 10 μl of chemically competent BL21 (DE3) pLysS cells (Novagen). Pipette to mix and leave on ice for 30 min. Heat shock by placing in a water bath at 42 °C for 60 s and return to ice for a further 2 min.
3. Add 200 μl ± SOC medium and incubate with shaking at 37 °C for 1 h.
4. Spread ∼100 μl on an LB agar plate containing 50 μg/ml ampicillin, invert, and leave overnight at 37 °C.

2.2. Protein expression

1. Using a sterile loop, inoculate 15 ml of LB with antibiotic with a singly colony from the agar plate (or with frozen glycerol stock; see step 2). Incubate between 8 h—O/N with shaking at 37 °C to generate a starter culture.
2. Prepare 15% glycerol stocks for future use by adding 150 μl sterile 50% (v/v) glycerol to 350 μl overnight culture in sterile microcentrifuge tubes. Store at −80 °C.
3. Inoculate 100 μl of starter culture into 10 ml of prewarmed M9 minimal medium with 50 μg/ml ampicillin. 5 × M9 minimal medium: 56.4 g/l M9 salts, 5 mM magnesium sulfate, 0.5 mM calcium chloride, autoclaved prior to addition of sterile 0.25% (w/v) thiamine and 1% (w/v) glucose.
4. Transfer 2–5 ml of the overnight culture to 200 ml of prewarmed M9 minimal medium with ampicillin and amino acids using either $^{13}C_6$-labeled or unlabeled arginine and lysine. Non-autoclavable amino acids (His/Phe/Pro/Trp/Tyr) added at 200 mg/l (filter sterilized), all other amino acids at 100 mg/l.
5. Incubate with shaking at 37 °C until OD_{600} ~0.8 and add 100 μl 1 M isopropyl-β-D-thiogalactoside (IPTG) to induce protein expression (final concentration 500 μM). Transfer to 30 °C and continue vigorous shaking for 5 h.
6. To monitor protein induction, collect samples hourly for analysis by SDS-PAGE and stain with a MS compatible staining solution (e.g., Colloidal Coomassie). For low expressing constructs, it may be necessary to immunoblot using an anti-polyHis-horseradish peroxidase (HRP) conjugated antibody (Sigma).
7. Harvest cells by centrifugation at 4200 rpm for 20 min, 4 °C in pre-weighed centrifuge tubes.

2.3. Analysis of expressed QconCAT

1. Using a clean scalpel blade, remove the Coomassie stained band corresponding to the expressed QconCAT protein (unlabeled) in the bacterial extract, cut each slice in 1 mm^3 pieces, and transfer to a clean tube.
2. Add 30 μl of a 50% (v/v) 100 mM NH_4HCO_3 50% MeCN solution to remove the stain from the gel pieces and incubate at 37 °C for 10 min with shaking.
3. Remove the solvent with a gel loading tip and add a further 30 μl of a 50% (v/v) 100 mM NH_4HCO_3 50% (v/v) MeCN solution and incubate at 37 °C for 10 min.
4. Remove the solvent. The gel pieces should be completely destained. If not, add a further 30 μl of a 50% (v/v) 100 mM NH_4HCO_3, 50% (v/v) MeCN solution and incubate at 37 °C for 15 min.

5. Remove the solvent and add 150 μl of a 10 mM solution of dithiothreitol (DTT) in 50 mM NH_4HCO_3, and incubate at 37 °C for 30 min.
6. Remove the solvent and add 150 μl of a 55 mM solution of iodoacetamide in 50 mM NH_4HCO_3, and incubate at 37 °C for 60 min in the dark.
7. Remove the solvent, add 30 μl of pure MeCN, and incubate at 37 °C for 15 min.
8. Remove the solvent. The gel will turn completely white and dehydrated; if not, add a further 30 μl of pure MeCN and incubate at 37 °C for 15 min.
9. Reconstitute sequence grade trypsin in HCl 1 mM (0.20 mg/ml). Acidic stock solutions can be aliquoted and frozen at −20 °C for up to 6 months. Add 136 μl of incubation solution (40 mM NH_4HCO_3, 5% (v/v) MeCN) to 8 μl of the acidic trypsin solution. Incubate at 37 °C for 1 h.
10. Add 20 μl of the trypsin solution to the gel and wait 5 min. Add the incubation solution in 10 μl increments and wait 5 min, until the gel pieces are covered by at least 2.0 mm excess solution. Incubate overnight at 37 °C with gentle agitation. N.B.: Trypsin proteolysis is much less efficient if the gel pieces are not completely covered during digestion.
11. Add MeCN (volume equal to that used for the overnight incubation) and leave shaking for a further 20 min. Centrifuge briefly to spin the gel pieces at 13,000×*g* for 1 min and collect the supernatant.
12. Extract peptides by adding 50 μl of 1% (v/v) trifluoroacetic acid (TFA), 80% (v/v) MeCN. Spin at 13,000×*g* for 1 min and combine the supernatant with the fraction from step 11.
13. Dry the peptide mixture in a speed vacuum centrifuge to 10 μl and reconstitute by adding 20 μl of 0.1% (v/v) TFA depending on protein concentration.
14. Spot 0.5 μl of digest on a MALDI target and overlay with 0.5 μl of saturated solution of α-cyano-4-hydroxycinnamic acid dissolved in 50% (v/v) MeCN, 0.1% (v/v) TFA.
15. Analyze the samples by MALDI-TOF MS to confirm expression of the QconCAT. Confirm that the Q-peptide ions are present and that digestion has proceeded to completion, that is, no missed cleavage peptide ions are observed.

2.4. Extraction and purification of QconCAT protein

1. Add 2 ml BugBuster® protein extraction reagent (Merck) per 500 mg wet cell pellet and resuspend by pipetting up and down and vortexing.
2. To degrade viscous genomic material, add Benzonase to 50 U/ml and incubate with shaking at room temperature for ~15 min.

3. To separate soluble proteins from inclusion bodies (cell organelles and debris), centrifuge at 18,000×g for 20 min, 4 °C. Retain supernatant for analysis by SDS-PAGE.
4. Resuspend the pellet (containing the inclusion bodies) in the same volume of BugBuster as used in step 1.
5. Add lysozyme to 500 μg/ml. Vortex briefly and leave shaking at room temperature for 5 min.
6. Add a further 6 volumes of BugBuster diluted 1:10 in deionized water and mix by vortexing for 1 min.
7. Pellet the inclusion bodies by centrifugation at 5000×g for 15 min at 4 °C and remove the supernatant by careful pipetting.
8. Resuspend the inclusion bodies in 0.5 volumes of 1:10 diluted BugBuster. Mix by vortexing and centrifuge as in step 7. Retain as wash 1.
9. Repeat the wash detailed in step 8 twice. The final centrifugation step should be performed at 18,000×g for 15 min, 4 °C.
10. Resuspend the purified inclusion bodies in 50 mM sodium phosphate buffer (pH 7.5), 300 mM NaCl, 5 mM imidazole, 6 M guanidine–HCl.
11. Analyze an equivalent amount of the soluble fraction, each wash fraction and the resuspended inclusion bodies by SDS-PAGE to localize the expressed QconCAT protein. N.B.: Samples in 6 M guanidine–HCl will need to be cleaned prior to SDS-PAGE using StrataClean resin (Stratagene) according to manufacturer's instructions.
12. Load 0.5 ml TALON beads into a disposable column. Wash with 5 ml deionized water and equilibrate with 5 ml of 50 mM sodium phosphate buffer (pH 7.5), 300 mM NaCl, 5 mM imidazole (containing 6 M guanidine–HCl if the QconCAT is purified from the inclusion body fraction).
13. Add the QconCAT containing fraction to the beads and allow to flow through under gravity. Collect the flow through from the column, retain a fraction for analysis by SDS-PAGE and store at −20 °C.
14. Wash the column with 10 ml of 50 mM sodium phosphate buffer (pH 7.5), 300 mM NaCl, 5 mM imidazole containing 6 M guanidine–HCl as appropriate, retaining the wash fraction as for step 13.
15. Elute bound proteins with 5 × 0.5 ml of 50 mM sodium phosphate buffer (pH 7.5), 300 mM NaCl, 200 mM imidazole containing 6 M guanidine–HCl, if appropriate.
16. Analyze by SDS-PAGE (after StrataClean treatment, if necessary) and combine eluant fractions containing purified QconCAT. Buffer exchange into 10 mM NH_4HCO_3, 1 mM DTT at 4 °C using a dialysis cassette or spin filter cartridge. N.B.: It may be necessary to include 4 M urea if the QconCAT protein precipitates during buffer exchange.

2.5. Calculation of protein concentration

The Coomassie Plus Protein Assay (Pierce) can be used to determine the protein concentration. However, more accurate assessment of protein concentration can be achieved by MS analysis (XIC or SRM) of the "Fib" reference peptide included in the QconCAT design with reference to a known amount accurately quantified differentially labeled "Fib" peptide.

2.6. Determining the extent of isotope labeling

For accurate quantification, it is critical that the extent of labeling is properly characterized. Unlabeled (light) peptides derived from the heavy labeled QconCAT construct will contribute to the native peptide ion signal intensity and contribute to the amount of light peptide quantified. To determine the extent of labeling, a tryptic digest of heavy labeled QconCAT should be analyzed by MALDI-TOF MS. Comparison of the area under the monoisotopic peak for the "light" and "heavy" peptide ions is then used to assess the ratio of light:heavy peptides in the labeled QconCAT.

3. Preparation of Samples for Analysis

For any experiment where absolute quantitative information is required, it is critical that procedures are employed that enable complete cell lysis. Unless the protein content of the cells are extracted completely, there will be error in calculation of the protein copy number and will enable only relative, rather than absolute, protein levels to be determined (assuming reproducible protein extraction). The methods described should provide a starting point for sample preparation, although care should be taken to determine that complete protein extraction has been achieved in the hands of the end user.

3.1. Yeast

Yeast cells (*Saccharomyces cerevisiae*) are grown in the required media and the cell number determined prior to lysis. This can be done quite simply using an hemocytometer and a light microscope, or a digital cell counting device, if available. Once the cell number has been determined, the cells can be harvested by centrifugation of the culture. Protein can be extracted from the resulting pellet mechanically using glass beads and a mini-bead beater. It is important to ensure that all protein is extracted, and this should be confirmed experimentally by assessing subsequent extractions for protein content, by analyzing them independently. Once this is confirmed, the extracts can either be analyzed individually, with the data summed at the end, or pooled to reduce the number of analyses required. In our hands, we find that five rounds of extraction results in complete lysis. A standard protein assay can be used to ascertain if sufficient rounds of extraction have been carried out.

The protocol is detailed below:

1. Culture yeast cells as required in suitable media depending on the experimental aims, and once the required optical density (OD) is reached (e.g., OD_{600} 0.7), sediment the yeast cells by centrifugation.
2. Resuspend the pellet in an equal volume of extraction buffer (50 mM Tris–HCl, pH 7.5, 750 mM NaCl, 4 mM $MgCl_2$, 5 mM DTT, and 10% (v/v) glycerol) and add approximately the same volume of glass beads (Sigma-Aldrich, Poole UK).
3. Mechanically disrupt the cells using a mini-bead beater (Biospec Products, Inc., Bartlesville, OK) (3 × 1 min on ice). Remove the soluble fraction, and add a further volume of extraction buffer to the glass beads/ yeast mixture as per step 2. Repeat several times, until it is established that sufficient extractions have been carried out such that there is no soluble protein remaining.
4. Estimate the protein content of the independent fractions using a standard assay such as Bicinchoninic acid (BCA) (QuantiPro™ BCA assay, Sigma-Aldrich, Dorset, UK) (Brown *et al.*, 1989; Wiechelman *et al.*, 1988).
5. The fractions can either be analyzed independently and the final numbers summed, or a combined mixture used representing a known percentage of the whole lysate.
6. For trypsin digest, lysates should be reduced by the addition of DTT to a final concentration of 20 mM (from a 1 M stock prepared in 50 mM ammonium bicarbonate) at 56 °C for 1 h.
7. Iodoacetamide should then be added to a final concentration of 10 mM from a 1 M stock prepared in 50 mM ammonium bicarbonate for 30 min at room temperature with light exclusion.
8. Add trypsin at an approximate ratio of 1:50 (trypsin:total cell protein). Leave overnight at 37 °C or until the analyte and standard have digested to completion.

3.2. Mammalian cells

Lysis is performed in SDS and the sample prepared for analysis using a slightly modified version of the filter aided sample preparation (FASP) method (Wisniewski *et al.*, 2009) to maximize protein extraction and solubilization and minimize proteolysis and changes to modification status.

1. Lyse cells by addition of 0.5 ml per 1×10^7 cells of 4% (w/v) SDS, 100 mM DTT in 100 mM Tris–HCl, pH 7.6. Heat at 95 °C for 5 min.
2. Sonicate briefly (3 × 30 s) to shear DNA and clarify by centrifugation (16,000×*g* for 5 min at room temperature).
3. Add protein extract equivalent of 2×10^5 cells (~100 μg of protein extract) and 20–500 fmol of isotope-labeled QconCAT to 200 μl of 8 M urea in 100 mM Tris–HCl, pH 8.5, in a 10-kDa cutoff centrifugal filter unit (Amicon Ultra-0.5 Centrifugal Filter Unit with Ultracel-10 membrane). Centrifuge at 14,000 *g* for 40 min.
4. Add a further 200 μl of 8 M urea in 100 mM Tris–HCl, pH 8.5. Centrifuge as in step 3 and discard the flow through.
5. Add 100 μl of 50 mM iodoacetamide in 8 M urea, 100 mM Tris–HCl, pH 8.5, for 1 min at 800 rpm and leave for 5 min without mixing. Centrifuge at 14,000×*g* for 40 min.
6. Add 100 μl of 8 M urea in 100 mM Tris–HCl (pH 7.9) and centrifuge at 14,000×*g* for 40 min. Repeat twice, discarding flow through.
7. Add 40 μl of 8 M urea in 100 mM Tris–HCl (pH 7.9) with 1% (w/w) Lys-C and mix at 800 rpm for 1 min at 20 °C. Leave overnight without shaking.
8. Transfer filter units to new collection tubes and add 120 μl of 50 mM NH_4HCO_3 containing 2% (w/w) trypsin. Mix for 1 min and leave for 4 h at 20 °C. Centrifuge at 14,000×*g* for 40 min. Retain filtrate.
9. Add 50 μl of 500 mM NaCl and centrifuge the filter units at 14,000 *g* for 20 min collect the filtrate. Add a further 20 μl of 10% (v/v) MeCN and collect the filtrate.
10. Acidify filtrate by addition of TFA to 0.2% (w/v) and desalt using C_{18} solid phase extraction (SPE) cartridges (Empore cartridges C_{18}-SD 1 ml/4 mm) as per manufacturer's instructions, washing in 0.1% (v/v) TFA and eluting in 70% (v/v) MeCN.

3.3. Considerations for the quantification of posttranslational modifications

Most posttranslational modifications (PTMs) are dynamic, enabling the reversible regulation of protein function. Cell extracts therefore need to be made with care, preserving the state of protein modification so that it resembles that in the cell under the specified conditions. Samples that are lysed in SDS and heated are unlikely to retain functional enzymatic activity. However, this step must be avoided for the analysis of heat labile modifications (e.g., histidine phosphorylation). As a general rule, heating of samples should be minimized wherever possible, decreasing the likelihood of introducing sample preparation artifactual modifications such as deamidation and oxidation. If SDS sample lysis is not used, the composition of the lysis buffer should be such that modifying enzymes are inhibited. In the case of phosphopeptide analysis, this requires the inclusion of inhibitors of protein kinase and phosphatase activity (e.g., ethylenediaminetetraacetic acid (EDTA), ethylene glycol tetra-acetic acid (EGTA), microcystin LR, sodium orthovanadate, β-glycerophosphate, sodium fluoride, calyculin A) to maintain phosphorylation status. Commercial preparations of such mixtures (e.g., PhosphoStop) are also available.

4. Mass Spectrometric Analysis

While the QconCAT method is itself platform independent, we have found the following analytical systems to work well in our hands.

4.1. Triple quadrupole instruments—Multiple reaction monitoring

SRM, the process of monitoring specific fragment ions derived from a target precursor ion (defined as transitions), usually in a defined LC elution time window, is an extremely sensitive method for targeted analyte identification and quantification. These experiments utilize the unique capabilities of triple quadrupole (QqQ) mass spectrometers for targeted analyte studies. In these types of experiments, the first quadrupole is set to act as a filter to specifically allow transmission of predefined precursor ions (m/z values), while the third quadrupole filters for the fragment ion(s) of interest; the second quadrupole meanwhile acts as the collision cell (Fig. 7.3).

Dependent on the exact type and number of transitions, these types of experiments can be uniquely sensitive and specific for the analytes of interest; coeluting ions outside of the transmission window are not transferred to the collision cell, and only those precursor ions yielding fragment ions of interest will generate a signal. Multiple reaction monitoring (MRM), namely the analysis of multiple fragment ions from a single defined precursor ion, thus improves the specificity of the assay, since the likelihood of multiple defined ions being generated from a coeluting ion of the same precursor m/z decreases significantly. However, increasing the number of transitions also decreases the lower limit of quantification, as lower intensity fragment ions will by necessity be used for quantification. The type and number of transitions selected for these MRM assays is thus critical to the outcome of the quantification experiments with two to four transitions being used optimally for quantification (reviewed in Lange *et al.*, 2008).

4.1.1. Defining transitions

Following proteolysis with trypsin, the C-terminal most residue in the peptide contains the isotope-labeled amino acid (Lys or Arg). The fragment ion used in the transition should therefore be a y-ion to ensure differentiation of the labeled Q-peptide from the unlabeled native peptide. Transitions can be defined either by analysis of MS/MS spectra acquired from proteolyzed QconCAT (ideally on the instrument which is being used for the MRM assays) or by prediction with software (Cham *et al.*, 2010) such as MRMaid (Mead *et al.*, 2009), MaRiMba (Sherwood *et al.*, 2009), Skyline (MacLean *et al.*, 2010), or Pinpoint (ThermoFisher Scientific). Using experimental data to define transitions is, of course, the preferred choice whenever possible. To maximize the sensitivity of the assay, an intense observed fragment ion should be selected; to maximize the specificity of the assay, the selected y-ion should be as large as possible, ideally with an m/z value greater than that of the precursor ion, and three transitions per peptide are generally sufficient. To facilitate the design of these types of experiments, a compendium of transitions for yeast peptides has been compiled based on acquired tandem MS data (http://www.mrmatlas.org/) (Picotti *et al.*, 2008). As experiments of this nature become more commonplace, repositories of successful SRM transitions will undoubtedly be made available to the research community.

4.2. High-resolution instruments—Extracted ion chromatogram

High-resolution instruments such as the Orbitrap XL, which have very low signal-to-noise, make it possible to extract ion chromatograms for a specific m/z with minimal baseline interference, even for complex samples such as unfractionated yeast cell lysates. There is also the advantage that tandem MS data can be acquired at the same time, which is useful for database searching and verification of sequence authenticity. The data obtained can also be used for method transfer to a triple quadrupole instrument, once a chromatographic retention time window has been determined.

4.2.1. Software for data analysis

Manually analyzing data derived from quantification experiments is a laborious task. A variety of software tools have been generated to automate and expedite data analysis including the following: MAXQuant (Cox and Mann, 2008; which was designed for Orbitrap data, and is therefore specific to the ThermoScientific Platform), SILACAnalyzer (available through OpenMS; Nilse *et al.*, 2010), which does not require prior MS/MS identification, and our own in-house software (which also facilitates visualization of results; Swainston *et al.*, 2011), and in addition describes a pipeline for automated database searching, quantification, data storage and dissemination. The latter is tailored to QconCAT experiments and is an extension to PrideWizard developed and described previously for iTRAQ applications (Siepen *et al.*, 2007b). This utility provides a user interface to which batches of spectra (in PRIDE XML format) may be submitted.

4.3. Data analysis

Quantification is achieved by first determining the area under the curve of the integrated peaks for each transition if a triple quadrupole is used. Other methods for data obtained on high-resolution instruments involve extracted ion chromatograms for both light and heavy peptides, usually involving vendor supplied software capable of peak detection and area calculation. For analysis of Orbitrap data, we use the ICIS peak detection algorithm component of Qualbrowser (supplied with Xcalibur). This is then used to calculate the ratio of the light (unlabeled) peptide transition to the heavy (labeled) transition for each of the peptide ions, from which the mean and relative standard deviation of the ratios can be calculated for each peptide sequence.

Significant differences in signal intensity of the light and heavy Q-peptides may require that the experiment be repeated by spiking different amounts of QconCAT to the cell lysate, to increase the accuracy and precision of quantification. Attempting to quantify over a large dynamic range, or when the signal intensity from either the light or heavy Q-peptides is near the threshold of quantification, will markedly decrease the analytical precision and should be overcome wherever possible. In the case of low levels of native peptide, this may require prefractionation of proteolyzed samples prior to LC–MS analysis, or employing a reverse-phase column with a greater capacity.

Having experimentally determined the ratio of heavy:light, the number of moles of analyte can be calculated, because the number of spiked moles of the peptide standard is accurately known. When the total cell number is also known, the number of analyte moles can be easily converted to number of molecules per cell.

5. Quantification of Posttranslational Modifications

Calculation of the stoichiometry of modification (e.g., phosphorylation) is performed in a subtractive manner, having first determined the amount of nonmodified (e.g., nonphosphorylated) peptide present in the sample (Fig. 7.4), according to the equation stoichiometry (%) = 100 − $[100 \times (L/H)_{mod}/(L/H)_{ref}]$.

When multiple sites of modification are identified on the same peptide, it may not be possible to determine the stoichiometry of each of the individual sites of modification in this manner. Under these circumstances, the peptide sequence containing the modification of interest could be synthesized in a labeled form and added exogenously to the cell lysate/ QconCAT digest. Alternatively, it may be possible to define the percentage

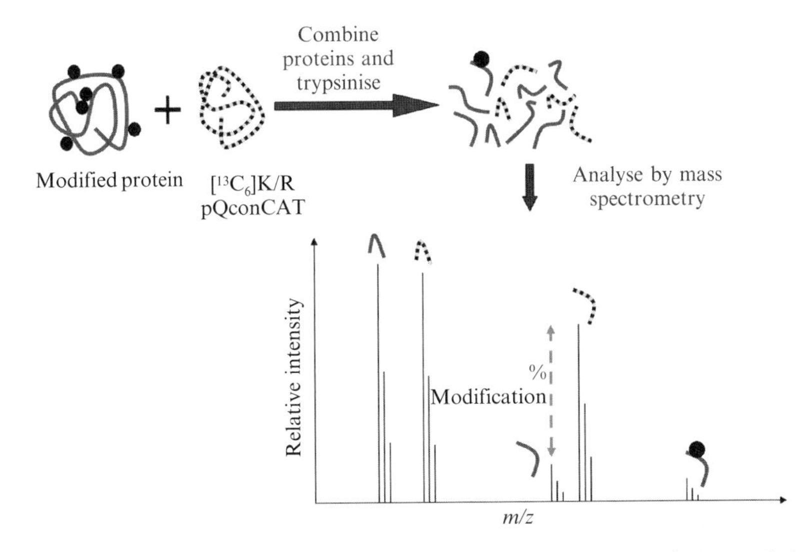

Figure 7.4 QconCAT for determining the absolute stoichiometry of posttranslationally modified peptides.

modification of each individual site by employing an enzyme other than trypsin, thus effectively separating the sites of interest. However, this requires that the original QconCAT be redesigned with this alternative protease in mind.

6. Replicate Analysis—Biological Versus Technical Replicates

As with any type of quantitative analysis, replicates should be factored into the experimental design to assess the statistical precision of the data. Technical replicates (repeat analysis of the same sample) give a measure of the repeatability of the technology (LC–MS), while biological replicates, namely the preparation of different biological samples, will determine the robustness of the biological system. Biological replicates by nature will therefore include the technical variability present in the system. The number of replicates required will largely depend on the type of samples undergoing analysis and the variation observed. For example, cell culture systems might not exhibit as much biological variation as patient derived samples; fewer biological repeats might therefore be required. Where sample is limited, sample pooling may be considered, thus reducing the number of biological replicates required to determine biologically relevant changes (reviewed in Karp *et al.*, 2005). As a rule of thumb, therefore, for quantification of proteins extracted from cell culture, a *minimum* of two biological replicates and three technical replicates (total of $n = 6$) should be used in the first instance to assess variability.

7. Summary and General Considerations

Methods for absolute protein quantification, including QconCAT, facilitate calculation of the number of molecules of analyte protein per sample. When the total cell number is accurately calculated, this information can then be converted to give an estimate of the number of molecules per cell. However, for this data to be used in molecular modeling, these values need to be converted to analyte concentrations (for V_{max} calculation), requiring knowledge of the intracellular volume. If this cannot be determined experimentally, then a value may need to be taken from the literature, which may have its own associated errors.

The reliance on isotope labeling for determining absolute levels of (particularly low abundance) analytes by MS is ongoing, at least until alternative reliable methodology is developed, which will most likely exploit peptide ion signal intensity. Undoubtedly, development of alternative methods for absolute protein/peptide quantification will require better understanding of the factors that contribute to peptide ionization efficiency, enabling the normalization of signal intensity to analyte quantity. Rapid improvements are being made in the application of label-free methods for absolute quantification. However, even the most advanced of these methods is not currently as reliable as the employment of isotope labeling for peptide quantification. 'Novel' methods can be used only routinely for the most abundant proteins, and there are additional problems associated with converting the relative quantification data generated to absolute numbers. Isotope labeling is therefore likely to remain the method of choice for the precise quantification of proteins on the global level in the foreseeable future.

REFERENCES

Beynon, R. J., *et al.* (2005). Multiplexed absolute quantification in proteomics using artificial QCAT proteins of concatenated signature peptides. *Nat. Methods* **2**(8), 587–589.

Bronstrup, M. (2004). Absolute quantification strategies in proteomics based on mass spectrometry. *Expert Rev. Proteomics* **1**(4), 503–512.

Brown, R. E., *et al.* (1989). Protein measurement using bicinchoninic acid—Elimination of interfering substances. *Anal. Biochem.* **180**(1), 136–139.

Brun, V., *et al.* (2009). Isotope dilution strategies for absolute quantitative proteomics. *J. Proteomics* **72**(5), 740–749.

Cham, J. A., *et al.* (2010). Free computational resources for designing selected reaction monitoring transitions. *Proteomics* **10**(6), 1106–1126.

Cox, J., and Mann, M. (2008). MaxQuant enables high peptide identification rates, individualized p.p.b.-range mass accuracies and proteome-wide protein quantification. *Nat. Biotechnol.* **26**(12), 1367–1372.

Eyers, C. E. (2009). Quantitative proteomics for systems biology. *Eur. Pharm. Rev.* **3,** 16–22.

Eyers, C. E., and Gaskell, S. J. (2008). Mass Spectrometry to Identify Post-Translational Modifications. Wiley Encyclopedia of Chemical Biology, pp. 1–34.

Eyers, C. E., *et al.* (2008). QCAL—A novel standard for assessing instrument conditions for proteome analysis. *J. Am. Soc. Mass Spectrom.* **19**(9), 1275–1280.

Eyers, C. E., *et al.* (2011). CONSeQuence: prediction of reference peptides for absolute quantitative proteomics using consensus machine learning approaches. *Mol. Cell. Proteomics,* doi: 10.1074/mcp.M110.003384.

Gerber, S. A., *et al.* (2003). Absolute quantification of proteins and phosphoproteins from cell lysates by tandem MS. *Proc. Natl. Acad. Sci. USA* **100**(12), 6940–6945.

Johnson, H., *et al.* (2009). Rigorous determination of the stoichiometry of protein phosphorylation using mass spectrometry. *J. Am. Soc. Mass Spectrom.* **20**(12), 2211–2220.

Karp, N. A., *et al.* (2005). Impact of replicate types on proteomic expression analysis. *J. Proteome Res.* **4**(5), 1867–1871.

Lange, V., *et al.* (2008). Selected reaction monitoring for quantitative proteomics: A tutorial. *Mol. Syst. Biol.* **4,** 222.

MacLean, B., *et al.* (2010). Skyline: An open source document editor for creating and analyzing targeted proteomics experiments. *Bioinformatics* **26**(7), 966–968.

Mead, J. A., *et al.* (2009). MRMaid, the Web-based tool for designing multiple reaction monitoring (MRM) transitions. *Mol. Cell. Proteomics* **8**(4), 696–705.

Mirzaei, H., *et al.* (2008). Comparative evaluation of current peptide production platforms used in absolute quantification in proteomics. *Mol. Cell Proteomics* **7,** 813–823.

Nilse, L., *et al.* (2010). SILACAnalyzer—A tool for differential quantitation of stable isotope derived data. *Lect. Notes Comput. Sci.* **6160,** 45–55.

Picotti, P., *et al.* (2008). A database of mass spectrometric assays for the yeast proteome. *Nat. Methods,* **5**(11), 913–914.

Pratt, J. M., *et al.* (2006). Multiplexed absolute quantification for proteomics using concatenated signature peptides encoded by QconCAT genes. *Nat. Protoc.* **1**(2), 1029–1043.

Rivers, J., *et al.* (2007). Absolute multiplexed quantitative analysis of protein expression during muscle development using QconCAT. *Mol. Cell. Proteomics* **6**(8), 1416–1427.

Sherwood, C. A., *et al.* (2009). MaRiMba: A software application for spectral library-based MRM transition list assembly. *J. Proteome Res.* **8**(10), 4396–4405.

Siepen, J. A., *et al.* (2007a). Prediction of missed cleavage sites in tryptic peptides aids protein identification in proteomics. *J. Proteome Res.* **6**(1), 399–408.

Siepen, J. A., *et al.* (2007b). An informatic pipeline for the data capture and submission of quantitative proteomic data using iTRAQ (TM). *Proteome Sci.* **5**(4).

Swainston, N., *et al.* (2011). A QconCAT informatics pipeline for the analysis, visualisation and sharing of absolute quantitative proteomics data. *Proteomics* **11**(2), 329–333.

Wiechelman, K. J., *et al.* (1988). Investigation of the bicinchoninic acid protein assay—Identification of the groups responsible for color formation. *Anal. Biochem.* **175**(1), 231–237.

Wisniewski, J. R., *et al.* (2009). Universal sample preparation method for proteome analysis. *Nat. Methods* **6**(5), 359–362.

Zhang, R. J., *et al.* (2002). Controlling deuterium isotope effects in comparative proteomics. *Anal. Chem.* **74**(15), 3662–3669.

CHAPTER EIGHT

Mass Spectrometric-Based Quantitative Proteomics Using SILAC

Francesco Lanucara *and* Claire E. Eyers

Contents

Abstract

One of the main goals of comparative cell signaling analyses is the characterization of protein changes between different biological samples, either globally or by targeting specific proteins of interest. Highly accurate and precise strategies are thus required for the relative quantification of proteins extracted from two or more different cell populations. Stable isotope labeling with amino acids in cell culture (SILAC) is a general method for mass spectrometric quantitative proteomics based on metabolic incorporation of stable isotope-labeled amino acids into the cellular protein pool. This method has been applied with great

Michael Barber Centre for Mass Spectrometry, School of Chemistry, University of Manchester, Manchester Interdisciplinary Biocentre, Manchester, United Kingdom

Methods in Enzymology, Volume 500
ISSN 0076-6879, DOI: 10.1016/B978-0-12-385118-5.00008-6

success to a variety of quantitative proteomics problems aimed at gaining further insight into cell signaling pathways. In this chapter, we describe how SILAC can be used for the elucidation of cellular mechanisms, including temporal proteome profiling and the quantitative analysis of the extent of specific posttranslational modifications.

1. Introduction

Mass spectrometry (MS)-based proteomics analysis is becoming an indispensable tool complementing molecular and cellular biology research. In particular, MS is now almost obligatory for the expanding field of systems biology. The high degree of complexity of cellular proteomes and the low abundance of many of the proteins of interest represents a challenge to the proteomics community. As a highly sensitive analytical technique, MS is now one of the most powerful tools for the analysis of complex protein samples.

Protein analysis encompasses a variety of aspects, including protein primary sequence identification, analysis of posttranslational modifications (PTMs), protein–protein interactions, and absolute and relative quantification of target proteins. Once a protein has been formally identified, it is often useful to determine how much of that protein is present in the sample, either as an absolute value or relatively with respect to the amount of the same protein in a different sample. In this context, quantitative proteomics aims to identify specific changes between control samples and specific experimental conditions, often focusing on a particular subset of proteins, so-called expression proteomics (Ong and Mann, 2005).

Quantitative data are also essential in order to build robust and reliable models of biological systems, which in turn can be used to gain insights into the principles that shape intracellular signaling pathways. More interestingly, these models reveal how pathways can be manipulated to yield a particular cellular output. Absolute and relative quantification approaches allow researchers to gain both specific and complementary information on the system being studied. Absolute quantification determines changes in protein expression in terms of the exact amount or concentration (nanograms or nanomoles per gram of tissue, lysate, or plasma) of each protein present (refer to Chapter 7), whereas relative quantification determines the relative changes in protein levels under particular experimental conditions, relative to the control sample. Both approaches yield valuable information on the system being studied, although if mathematical models have to be derived, absolute values of protein concentration, often reported as copy number (number of molecules of protein per cell), are preferred.

Quantitative proteomics methods can be classified into two main groups. First, methods based on the use of stable isotope labeling, used either to introduce a mass difference between two proteomes or to provide an internal standard for relative quantification. Second, label-free approaches, which can be based either on the integration of the MS signal of the peptides of interest or on spectral counting (Schulze and Usadel, 2010). The main advantage afforded through stable isotope labeling for peptide quantification is that the two forms of the peptide can be processed and analyzed concurrently, generating more accurate results at a lower limit of detection. The earlier in the workflow that these differentially labeled analytes can be combined, the higher the accuracy of quantification, because the ratio of the analytes will be unaffected by potential differences in the efficiency of sample handling and processing (protein extraction, separation, proteolysis, etc.).

1.1. Stable isotope labeling in MS-based quantitative proteomics

MS is not a quantitative technique *per se*, largely because different molecules ionize with different relative efficiency. Internal standards or external references are therefore necessary to determine the amount of a given analyte, either as an absolute or relative value. Stable isotope analogs of the target protein(s) that need to be quantified represent the common choice for accurate quantitative MS of proteins. The isotopic label can be introduced in the protein either synthetically or by exploiting native cellular metabolic pathways and, in a label-dependent manner, directly alter the mass of the polypeptide without affecting either analytical, biochemical, or chemical properties (Iliuk *et al.*, 2009). Isotope labels can be introduced metabolically, enzymatically, or chemically and at either the peptide or the protein level.

Chemical and enzymatic labeling has the advantage of being applicable to all types of protein samples, including extracts from tissues, blood, and biological fluids. Conversely, metabolic labeling requires a population of metabolically competent cells for protein labeling. The main advantage of SILAC over synthetic labeling procedures is the potential for labeling all proteins in a cell in a uniform manner, irrespective of exogenous enzyme activity or the efficiency of chemical derivatization.

1.2. Metabolic labeling of the cellular proteome

In a metabolic labeling experiment, stable isotopes are incorporated into the cellular protein pool during normal cell growth and division, either by growing cells in media containing heavy ^{15}N isotopes (Conrads *et al.*, 2000; Gao *et al.*, 2000; Oda *et al.*, 1999) or amino acids labeled with combinations of ^{13}C, ^{15}N, and ^{3}H. This technique is termed stable isotope labeling with amino acids in cell culture (SILAC) (Ong and Mann, 2006; Ong *et al.*, 2002). One of the main advantages with using metabolic labeling is that samples grown (and thus labeled) in the presence of heavy and light isotopes can be pooled and mixed together before being processed, thus avoiding the inevitable errors resulting from experimental manipulation of each sample.

^{15}N labeling, which was the first method to be introduced for metabolic labeling for MS proteomics analysis, is particularly useful when dealing with organisms that biochemically synthesize amino acids, as incorporation of stable isotope-labeled amino acids would yield partially labeled proteins, upsetting the accuracy and precision of subsequent quantification. One downside of using a labeled nitrogen source, rather than a specifically labeled amino acid, is that the mass difference between the light and heavy forms of a peptide is not consistent; rather it is dependent on the number of N-atoms. Mass difference cannot therefore be predicted in the absence of sequence information. SILAC relies on the incorporation of isotope-labeled (heavy) amino acids during protein synthesis, resulting in peptides of predictable mass difference.

In this chapter, the principles behind the SILAC technique, along with a general experimental procedure and its applications to relative and absolute quantitative proteomics will be presented.

2. SILAC: The Method

2.1. General overview

SILAC is a metabolic strategy that relies on the incorporation of different isotope-labeled amino acids which have been added to the growth medium and incorporated during normal protein synthesis in cells grown in culture. Labeling results in a specific mass difference in all the peptides containing the SILAC amino acids, and this difference is dependent on the isotopes and amino acids used as well as the proteolysis strategy. Comparison of the mass spectrometric signal intensities of the differentially labeled peptides from two or more cell populations permits relative quantification of the proteins from which those peptides were derived. SILAC is a simple and powerful approach to MS-based quantitative proteomics and it can provide accurate relative quantification with no need for further chemical derivatization or excessive sample manipulation. Additionally, because there is no chemical difference between the light and heavy amino acids used during the labeling procedure other than the isotopic composition, comparative cell populations will exhibit identical biochemical and cellular properties. It is often therefore the method of choice when reliable, high-throughput data generation is required. Due to its relative ease, the quality of the data generated and compatibility with the majority of proteomics protocols, SILAC is becoming increasingly popular within the proteomics community (Jiang and English, 2002; Ong *et al.*, 2002, 2003; Zhu *et al.*, 2002). For SILAC, any metabolically active cells in culture can be labeled, and this methodology has been applied to quantification of proteins from mammalian cells, yeast, bacteria, and plants (Gruhler *et al.*, 2005a,b; Jiang and English, 2002; Kerner *et al.*, 2005; Ong *et al.*, 2002).

Mammalian cells are not able to synthesize the eight essential amino acids; therefore, they depend on the supply of these components in the cell culture media. Analogs of amino acids labeled with stable isotopes (^{13}C, ^{2}H, ^{15}N) are readily available from commercial sources. To ensure that cells only incorporate the added, labeled amino acids, only essential amino acids are exploited, and cells are grown in the presence of dialyzed serum, as appropriate. After a sufficient number of cell doublings, at least five in mammalian cells (Ong and Mann, 2006), the whole proteome of the cells theoretically contains the heavy form of the amino acid present in the culture medium (Fig. 8.1). The number of cell doublings required for complete protein labeling is dependent on the relative rates of *de novo* protein synthesis, protein degradation, and cell division; the extent of labeling therefore needs to be carefully assessed prior to quantification. Because SILAC employs highly sensitive MS for analysis of the extracted digested proteins,

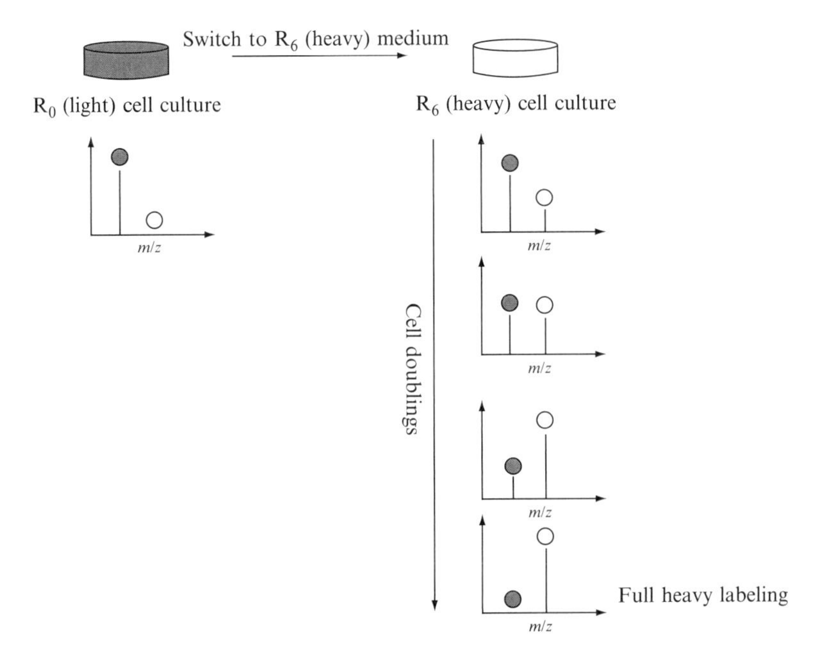

Figure 8.1 Flow chart for a general SILAC experiment: the adaptation phase.

even a small amount of unlabeled amino acid in the heavy protein population can be detected. Unfortunately, this contributes to the unlabeled peptide ion signal, interfering with the heavy to light peptide ratio observed in the resulting mass spectrum and potentially resulting in erroneous quantification.

When complete labeling of the proteins has been achieved, the cell populations are experimentally manipulated, and then combined prior to subsequent cell lysis, protein extraction, and proteolysis. Sets of peptide ions observed during MS analysis will be separated by a residue-specific mass difference, dictated by the isotope difference of the light and heavy versions of the amino acid and the number of labeled amino acids present in the peptide. The quantitative relationship between the two sets of proteins in the mix is determined by the ratio between the MS signal intensity of the heavy and the light peptides (H/L) (Fig. 8.2). Interestingly, metabolic conversion of arginine to proline (and vice versa) has been observed in cells, when levels of arginine in the media are not correctly controlled (Ong and Mann, 2006; Ong *et al.*, 2003), resulting in inaccurate quantification. Conversion of labeled arginine to other amino acids can be overcome by deletion of the arginase genes of the cell line of interest, as has been demonstrated in the fission yeast *Schizosaccharomyces pombe* (Bicho *et al.*, 2010).

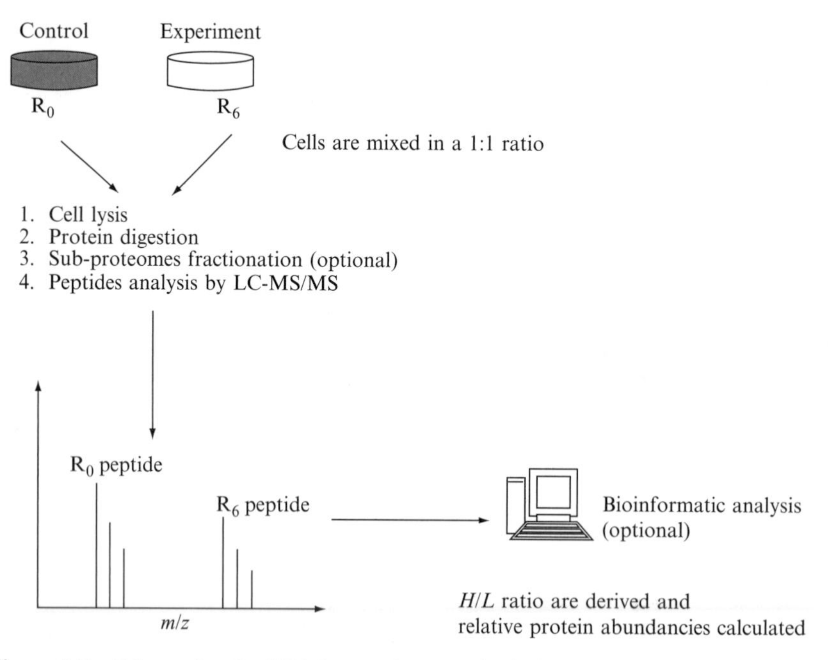

Figure 8.2 Schematic of a SILAC experiment. The light and heavy cell populations are differentially treated, combined, and processed together to measure the heavy to light peptide signal ratios.

It should also be noted that the absence of chemical derivatization procedures downstream of the SILAC protocol avoids any loss resulting from incomplete or nonspecific reactions, thus making SILAC comparable to nonquantitative MS techniques in terms of both sensitivity and throughput. Importantly, SILAC has also recently been exploited for the quantitative analysis of tumor samples from biopsies in a procedure termed Super-SILAC (Geiger *et al.*, 2010): SILAC labeling of a number of related cell lines served as internal standards for quantification of the unlabeled tissue samples.

2.2. Experimental design

The quantitative information in a SILAC experiment is given by the incorporation of the labeled amino acid; therefore, selecting the right label is critical and this parameter can be modified for the specific purposes of the experiment (Doherty *et al.*, 2009; Geiger *et al.*, 2010). As stated above, stable isotope-labeled amino acids are preferred to metabolites or amino acid precursors, because the mass shift introduced by an amino acid is readily predictable. Moreover, using a metabolite could potentially lead to a scrambling of the label, caused by the different anabolic and catabolic pathways through which the metabolite can be fed. The metabolic stability of the amino acid, and in particular the stability of the chemical group bearing the label, is crucial for the success of the experiment. Leucine is known to be metabolically labile, because deamidation and transamination reactions followed by oxidation of the alpha carbon can result in transfer of deuterium from [$^{2}H_{10}$]Leu to another amino acid (Pratt *et al.*, 2002a,b). The initial isotopic composition of the label is also depleted. This is observed as a decreased mass shift in SILAC experiments when using [$^{2}H_{10}$]Leu as the source of the label, due to incorporation of only partially deuterated Leu. The choice of the isotope is related to the chemical stability of the element and to the effect of extensive labeling on the chromatographic behaviour of the peptides. In the majority of applications, SILAC is followed by trypsin digestion, so labeled Arg and Lys are generally preferred for labeling. The resulting peptides (except those derived from the protein carboxy-terminus) should therefore be readily quantifiable (Steen and Mann, 2004). As SILAC relies on the mass difference between the peptides in an isotopic peptide ion pair, the labeled amino acid should induce spacing in the SILAC doublet of at least 4 Da to avoid any overlapping between the two isotopic envelopes. In order to ensure a predictable and reproducible mass shift, the isotopic enrichment of the heavy amino acid should also be $\geq 98\%$. Because coelution of the SILAC peptides during reverse-phase liquid chromatography (LC) greatly simplifies the analysis, ^{13}C labeling is preferential, since deuterium (^{2}H) labeling can induce a shift in the peptide LC retention time (Zhang *et al.*, 2002).

When the amino acids for a SILAC experiment have been chosen, the experimental design is relatively straight forward and very similar to a purely qualitative proteomics analysis. The SILAC work flow comprises a phase of adaptation, during which cells are grown in the labeled and unlabeled media until full incorporation of the heavy amino acids has been achieved (Fig. 8.1). The two cell populations are grown under identical conditions during this phase in order to avoid any difference in protein expression. The evaluation of the degree of incorporation of the SILAC amino acids is performed by means of LC-MS/MS: the area under the curve (AUC) of the MS peaks for the heavy and light peptide pairs are used to evaluate the extent of labeling, while the peptide sequences are derived by performing tandem MS. The heavy population is harvested, cells are lysed, and the proteins extracted and digested into peptides. The degree of labeling is thus evaluated at the peptide level, normally using a small fraction of the total protein content (a few bands from a sodium dodecyl sulfate (SDS) polyacrylamide gel or an in-solution digest) and when full incorporation has been confirmed, the two cell populations can be submitted to differential treatment, then mixed together and analyzed (Fig. 8.2).

3. General SILAC Protocol

3.1. Preparation of SILAC media

3.1.1. SILAC amino acids

Stock solutions of SILAC amino acids are prepared in small volumes (0.5 ml) of phosphate buffered saline (PBS) and can be stored at −20 °C. Arg and Lys are prepared at 84 and 146 mg/ml in PBS and sterile filtered through 0.22 μm syringe filters before use. *Note*: to avoid loss of expensive labeled amino acid, directly add the correct amount of solvent to the vial rather than trying to precisely weigh the amino acid.

3.1.2. SILAC media

The following is a general recipe for SILAC media for mammalian cell lines and can be modified and adapted to particular cell lines.

Add 50 ml of dialyzed fetal bovine serum (FBS) to 444 ml of customized SILAC Dulbecco modified Eagle's minimal essential medium (DMEM) or Roswell Park Memorial Institute medium (RPMI) 1640 without Arg and Lys (or other SILAC amino acids of choice). Add 0.5 ml of Arg or Lys stock solutions (prepared as above). The final concentration of Lys and Arg will be 0.4 and 0.8 mM, respectively. Add 5.5 ml of 1% Streptomycin/Penicillin as appropriate. Mix well and filter through 0.22 μm sterile filters under vacuum. The filtrate can be stored at 4 °C for up 2 months.

3.2. Cell growth in SILAC media

Note: SILAC medium is normally prepared using dialyzed FBS as a source of essential proteins and growth factors, although the dialysis process may remove some of these components. It is therefore important to test the growth of your cell lines in a non-labeled SILAC medium before starting the adaptation phase. It is also possible to customize the FBS by using dialysis membranes with larger pore diameters (10 kDa cut off).

1. Depending on the cell line, harvest cells from a flask containing $1–2 \times 10^5$ cells (80–90% confluency) by washing the flask with 10 ml of PBS at 37 °C. Do not use trypsin to detach the cells, as it can be a source of light amino acids. Split the cells 1–10 into two cell culture flasks, one containing 9 ml of heavy and one containing 9 ml of light SILAC media.
2. Incubate the two cell populations and split them as appropriate (normally every 2 or 3 days) maintaining the confluence between 30% and 90% so that cells are actively growing in log phase. Check the viability of the cells (cells in heavy and light media should show similar growth rates).
3. After five cell doublings, the incorporation of the heavy amino acid should be complete (~98%). Harvest 10^6–10^7 cells from the heavy cell population, making sure to maintain the heavy culture, and use these to evaluate the incorporation of the label.

3.3. Determining the degree of incorporation of SILAC amino acids

This analysis should be done each time a new SILAC cell line is started.

1. Remove SILAC medium from the flask.
2. Gently wash cells with an equivalent volume of PBS.
3. Detach cells from the flask surface by harshly washing with PBS. Scrape the dish with a cell scraper, if necessary.
4. Spin down the cells by centrifuging at 1500 rpm for 10 min.
5. Wash twice with 10 ml of PBS, centrifuge at 1500 rpm for 10 min and discard supernatant.
6. Add 250 μl of complete, Mini ethylenediaminetetraacetic acid (EDTA)-free Protease Inhibitor Cocktail solution [Roche] (or any other type of inhibitors as convenient).
7. Vortex the pellet intermittently for 5 min.
8. Lyse cells in a sonication bath for 15 min or with a suitable lysis buffer (e.g., 6 M urea and 2 M thiourea).
9. Pellet the debris by centrifugation for 10 min at 16,000×*g*, 4 °C. Collect the supernatant in a new tube.
10. Estimate protein concentration using a Bradford assay or equivalent.

11. Make up 25–50 μg of protein in 50 μl NH_4HCO_3 (50 mM).
12. Add dithiothreitol (DTT) (stock solution of 100 mM in 50 mM NH_4HCO_3) to a final concentration of 4 mM.
13. Vortex and incubate at 60 °C for 10 min or at room temperature for 45 min while shaking.
14. Cool to room temperature and add iodoacetamide (stock solution of 100 mM in 50 mM NH_4HCO_3) to a final concentration of 14 mM.
15. Vortex and incubate at room temperature for 45 min in the dark.
16. Quench the excess iodoacetamide by addition of DTT to a final concentration of 7 mM.
17. Add 10–20 μl of 0.1 μg/μl trypsin in water (enzyme/protein ratio from 1–50 to 1–100)
18. Incubate overnight at 37 °C with gentle shaking.
19. Acidify the sample by addition of formic acid to 0.1% (v/v). The peptide mixture can be stored at −80 °C for 6 months to 1 year.
20. Analyze the sample by LC-nanospray MS/MS or matrix-assisted laser desorption/ionization (MALDI)-MS (Fig. 8.3).
21. Search the tandem MS data using the search algorithm(s) of choice (e.g., Mascot) and the appropriate database. N.B.: The heavy SILAC labels should be set as variable modifications in the search parameters.

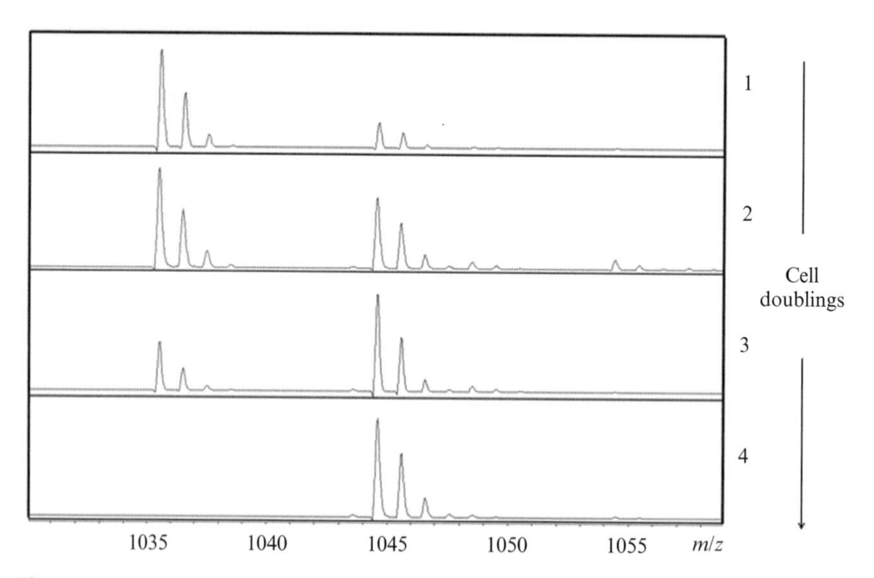

Figure 8.3 MALDI-Time of Flight (TOF) MS analysis to determine the incorporation of isotope label during the adaptation phase. The four mass spectra show the sequential increase in the intensity of the heavy peptide (*m/z* 1044.56) and decrease of the light peptide (*m/z* 1035.56) until full incorporation of the label ($^2HLeu_9$) is reached. The peptide sequence in this case is LFGPEK(Ac)FR and the difference between the heavy and light versions of the same peptide is 9 Da, as expected for singly charged ions.

Other modifications should be set as appropriate: that is, carbamido-methyl Cys (static), oxidized Met (variable), deamidated Asn (variable).

22. Locate a SILAC-labeled peptide in the chromatographic run using the appropriate retention time and *m/z* value.
23. Confirm that the mass difference between a heavy and light peptide pair is as expected; this should correspond with the mass difference of the light/heavy SILAC label and the number of those amino acids in the peptide sequence (see Fig. 8.3 as an example of a SILAC peptide pair). There are some considerations to take into account when analyzing SILAC peptides. First, the peptide identified in the search output must match the raw MS and MS/MS data (precursor *m/z* and charge state). Manually inspect the MS/MS spectra and confirm the identity of the peptide. Once the sequence of the heavy peptide has been confirmed, verify the absence of the light peptide ion in the MS spectrum. Refer to Ong and Mann (2006) for guidelines in the manual interpretation of MS/MS of peptides.
24. If full incorporation has been confirmed with a sufficient number of peptides (and proteins), proceed to the differential treatment.

3.4. Treatment of heavy SILAC cells

1. Either the light or heavy cell population can be used as a control. Induce a differential response in one of the two cell populations with an appropriate stimulus/treatment for the correct amount of time.
2. Wash the control and stimulated cells with PBS and lyse them appropriately (this will be dependent on the cell line and any fractionation employed).
3. Mix the heavy and light lysate in a 1:1 ratio taking into account relative protein concentration of the samples, vortex the mixture intermittently for 5 min and spin the lysate at 15,000×*g* for 10 min at 4 °C. Collect the supernatant and transfer to a new tube.

3.5. Analysis of SILAC-labeled cell extracts

Samples can be processed for MS analysis using a number of different strategies: these can either involve protein digestion in solution, or in a polyacrylamide gel, either with or without protein prefractionation. The degree of protein fractionation will depend on the depth of analysis required of the cell lysate. The manner of protein separation will depend largely on the facilities available and may include (but are not limited to) separation by size, charge, or isoelectric point. Detailed here is a relatively straightforward method of protein separation by size (SDS-PAGE) combined with in-gel proteolysis. N.B.: To minimize contamination with keratin, fresh SDS and SDS-sample buffer should be used and gels polymerized in a laminar flow hood.

1. Resolve proteins by SDS-PAGE and stain with a MS-compatible staining solution (e.g., Colloidal Coomassie).
2. Destain the gel with 40% (v/v) MeOH, 10% (v/v) AcOH destaining solution to visualize protein bands.
3. Using a new scalpel blade, cut the gel lanes in ~40 slices/lane.
4. Cut each slice in 1 mm^3 pieces and transfer them to a 1.5 ml tube.
5. Add 30 μl of a 50% (v/v) 100 mM NH_4HCO_3, 50% MeCN solution to destain the gel pieces and incubate at 37 °C for 10 min.
6. Remove the solvent with a gel loading tip, add a further 30 μl of a 50% (v/v) 100 mM NH_4HCO_3, 50% (v/v) MeCN solution, and incubate at 37 °C for 10 min.
7. Remove the solvent. The gel pieces should now be completely destained. If not, add a further 30 μl of a 50% (v/v) 100 mM NH_4HCO_3, 50% (v/v) MeCN solution, and incubate at 37 °C for 15 min.
8. Remove the solvent, add 150 μl of a 10 mM solution of DTT in 50 mM NH_4HCO_3, and incubate at 37 °C for 30 min.
9. Remove the solvent, add 150 μl of a 55 mM solution of iodoacetamide in 50 mM NH_4HCO_3, and incubate at 37 °C for 60 min in the dark.
10. Remove the solvent, add 30 μl of pure MeCN, and incubate at 37 °C for 15 min.
11. The gel will turn completely white and dehydrated, if not add a further 30 μl of pure MeCN and incubate at 37 °C for 15 min. Remove the solvent and leave the remaining MeCN to evaporate.
12. Reconstitute sequence grade trypsin (20 μg) in 100 μl of HCl 1 M. Acidic stock solutions can be aliquoted and frozen at −20 °C for up to 6 months. Add 136 μl of incubation solution (40 mM NH_4HCO_3, 5% (v/v) MeCN) to 8 μl of the acidic trypsin solution.
13. Add 20 μl of the trypsin solution to the gel. Add the incubation solution in 10 μl increments and wait 5 min, until the gel pieces are covered by at least 2.0 mm excess solution. Incubate at 37 °C for 1 h. If necessary, add 20 μl of 50 mM NH_4HCO_3. Incubate overnight at 37 °C with gentle agitation. N.B.: Trypsin proteolysis is much less efficient if the gel pieces are not completely covered during digestion.
14. Add an equal volume of MeCN to that used for incubation and leave shaking for a further 20 min.
15. Centrifuge briefly to spin the gel pieces at 13,000×*g* for 1 min and collect the supernatant.
16. Extract peptides by adding 50 μl of 1% (v/v) trifluoroacetic acid (TFA), 80% (v/v) MeCN and incubating at 37 °C for 15 min.
17. Spin at 13,000×*g* for 1 min and combine the supernatant with the fraction from 15.

18. Dry the peptide mixture in a speed vacuum centrifuge to 10 μl.
19. Reconstitute by adding from 10 to 20 μl of 0.1% (v/v) formic acid depending on protein concentration.
20. Analyze the samples by LC-nanospray MS/MS to identify the peptides and quantify the heavy to light peptide ratios.

3.6. Quantification of SILAC peptides

1. Identify peptides and proteins using database search software, making sure the SILAC labeling is included as variable modification in the search settings.
2. Using the retention times and the m/z ratio of the identified peptides, locate them in the LC run.
3. Manually confirm the identity of the peptide: the mass must match the nominal mass of the peptide as given in the database search output and the MS/MS must match the amino acid sequence.
4. Extract the ion chromatogram (XIC) associated to the signal of the heavy peptide using an appropriate m/z window around the monoisotopic peak. This value must be optimized depending on the instrument's resolution.
5. Extract the XIC associated to the signal of the light peptide using an appropriate m/z window around the monoisotopic peak (see step 4). This value must be adjusted in order to minimize the overlapping of unrelated peaks to the isotopic distribution of the SILAC peptide.
6. Measure the ratio of the heavy and light SILAC peptides using the XIC signal intensity (area under curve).
7. Repeat the ratio measurement for all the peptides associated to the same protein and obtain an average value from them. This value will be used to relatively quantify the proteins of interest between the two samples.

Data analysis for large scale quantitative proteomics experiments is extremely challenging, because the amount of information (i.e., number of peptides and proteins) generated from a single LC-MS/MS run is enormous. Therefore, manual validation and quantification of all the peptides of interest in a SILAC experiment is a limitation to the rate of throughput of the technique. There are several tools available to the proteomics community to analyze SILAC data. MSQuant is the software for quantitative proteomics and is mainly used to process LC-MS data and extract quantitative information about peptides and hence proteins. The software allows manual inspection of raw data; annotated spectra are used for quantification and product ion mass spectra for manual validation of peptide sequences (Mortensen *et al.*, 2010) can be fully automated. MaxQuant (Cox and Mann, 2008; Cox *et al.*, 2009), software specifically designed for analysis of high-resolution MS quantitative data, identifies and annotates peaks, isotope clusters, and SILAC peptide pairs and correlates them to their m/z, elution times, and signal intensity values. Accuracy in the p.p.b. range is ensured by the integration of multiple mass measurements. MaxQuant automatically identifies and quantifies several hundred thousand peptides per SILAC experiment, and it has been shown to provide accurate quantification of more than 40,000 proteins from a lysate (Cox and Mann, 2008).

SILACAnalyzer is a recently developed algorithm for quantitative analyses of SILAC MS experiments (Nilse *et al.*, 2010). This software is freely available and fully automated; it accepts .xml output files from a variety of MS platforms and requires minimal parameter tuning from the user. SILACAnalyzer determines m/z values and retention times, and through a sequential filtering, clustering, and linear fitting procedure calculates heavy to light peptide ratios without any prior identification of the peptide sequence by tandem MS, thus allowing a substantial reduction of the instrument duty cycle. One of the main appealing capabilities of the software is the possibility of identifying proteins or groups of proteins that undergo quantitative changes between different sets of experiments, thus offering valuable information for targeting protein candidates for further studies.

4. Applications

4.1. Relative quantification

Relative quantification of proteins from two different biological samples (comparative proteomics) represents one of the most useful and widespread applications of SILAC. Once the desired extent of protein labeling is achieved ($\geq$98%), the two cell populations are harvested, mixed together, and the peptides are relatively quantified by means of their heavy to light MS peaks relative isotope abundance (RIA). It is crucial to ensure that complete labeling has been reached before mixing the cells otherwise the RIA measurements will be biased. The time required for individual proteins to be completely labeled is dependent on their turnover rate. The SILAC technique has been successfully applied to monitor changes in protein expression during different phases in the cell cycle and after biological stimulation, thus gaining insight into the role of differential protein turnover in particular biological process, such as muscle cell differentiation (Ong *et al.*, 2002).

As discussed above, the SILAC technique requires growing and metabolically active cells for protein labeling. If relative quantification of proteins from tissue samples extracted under different pathophysiological conditions is required as may be the case during biomarker discovery, the problem of protein labeling in tissue samples has to be overcome. A recent study has demonstrated how SILAC can be used to compare differentially expressed proteins from mammalian tissues in two different states (Xu *et al.*, 2009). A reference human embryonic kidney (HEK) cell culture grown in Leu-d3 medium was used as an internal standard to which proteins obtained from two histological samples could be compared. Proteins extracted from human renal samples (healthy and a cancerous example), whose peptides were not experimentally labeled, were extracted and quantified with respect to the protein levels observed in the Leu-d3 HEK control extract. The two H/L ratios were then used to identify changes in the expression of the same proteins between the normal and diseased sample. This approach expands the SILAC strategy to the quantitative analysis of proteins from tissues, thus allowing one to gain substantial information on the molecular changes associated with the development and progression of diseases (Xu *et al.*, 2009).

The SILAC technique is also currently being expanded, with increasing success, in the area of biomarker discovery, where the analysis of the expression levels of several target proteins under normal and pathological conditions can result in the identification of new candidates for developing novel, and potentially faster, MS-based diagnostic procedures (Lund *et al.*, 2009; Ren *et al.*, 2010; Zamo and Cecconi, 2010).

4.2. Protein turnover

SILAC can also be used to study protein life cycles in mammalian cells (Pratt *et al.*, 2002a). In this type of experiment, cells grown in heavy SILAC medium (and completely labeled) are transferred to the light medium and harvested at different time points. When cultured in the light medium, proteins begin to incorporate light amino acids as a result of protein degradation and new synthesis. By measuring the ratio between the heavy and the light peptides for proteins at different time points, it is possible to define a degradation rate constant. This technique (referred to as dynamic SILAC) has been used to determine the relative stability of proteins in human cell lines and to correlate rates of degradation to potential protein sequence, which could be implicated in modulating the turnover rate of cellular proteomes (Doherty *et al.*, 2009).

4.3. Studies of posttranslational modifications

The SILAC technique has also been extended to the quantitative analysis of a variety of protein posttranslational modifications. Protein phosphorylation is one of the key mechanisms involved in cell signaling, and analysis of protein phosphorylation represents a continuous and important challenge for the proteomics community. Identification of phosphorylation sites *per se* only represents a first step, because there is a real need for quantitative analysis of this modification. SILAC is advantageous over other techniques (e.g., two-dimensional PAGE), because it can focus the quantitative analysis on individual phosphorylation sites, thus providing crucial information linking phosphorylation stoichiometry under diverse signaling conditions. The advantage of using SILAC to monitor dynamic changes in the extent of PTMs on a particular set of peptides, such as phosphopeptides, comes from the independence of the labeling strategy from the PTM under investigation. This technique thus provides a means of performing a comprehensive analysis of dynamic changes of multiple PTMs from the same sample, as has been demonstrated in the case of histone modifications, where methylation, acetylation, and phosphorylation were analyzed and quantified concurrently at different time points throughout the cell cycle (Bonenfant *et al.*, 2007).

A variation of the conventional SILAC approach, known as heavy methyl SILAC, uses a stable isotope-labeled version of methionine which is converted into the "heavy" *S*-adenosylmethionine. Proteins incorporate the heavy methyl group when the methylation reaction involves one of the heavy methylated substrates. This technique has been used in the reliable identification and quantification of sites of protein methylation (Ong *et al.*, 2004). Furthermore, a recent study has demonstrated the application of SILAC to the quantification of methylated and acetylated peptides derived from histones H3 and H4 from normal and disease tissue samples (Cuomo *et al.*, 2011).

5. Conclusion

SILAC is a powerful technique, whose applications cover the main fields of proteomic research, including relative protein quantification, temporal profiling, and posttranslational modification studies. The relative ease of the protocols and the possibility of using different combinations of heavy amino acids for labeling proteomes make this technique suitable for many different purposes. The high volume of SILAC publications that have appeared in the last 8 years clearly demonstrate that an evergrowing number of proteomic studies are including SILAC in their workflows. Stable isotopic labeling of proteins with amino acids in cell culture therefore represents one of the most promising techniques for fundamental biomolecular studies and for important clinical applications such as biomarker discovery.

References

Bicho, C. C., *et al.* (2010). A genetic engineering solution to the "arginine conversion problem" in stable isotope labeling by amino acids in cell culture (SILAC). *Mol. Cell. Proteomics* **9**(7), 1567–1577.

Bonenfant, D., *et al.* (2007). Analysis of dynamic changes in post-translational modifications of human histones during cell cycle by mass spectrometry. *Mol. Cell. Proteomics* **6**(11), 1917–1932.

Conrads, T. P., *et al.* (2000). Utility of accurate mass tags for proteome-wide protein identification. *Anal. Chem.* **72**(14), 3349–3354.

Cox, J., and Mann, M. (2008). MaxQuant enables high peptide identification rates, individualized p.p.b.-range mass accuracies and proteome-wide protein quantification. *Nat. Biotechnol.* **26**(12), 1367–1372.

Cox, J., *et al.* (2009). A practical guide to the MaxQuant computational platform for SILAC-based quantitative proteomics. *Nat. Protoc.* **4**(5), 698–705.

Cuomo, A., *et al.* (2011). SILAC-based proteomic analysis to dissect the "histone modification signature" of human breast cancer cells. *Amino Acids* **41**(5), 387–399.

Doherty, M. K., *et al.* (2009). Turnover of the human proteome: Determination of protein intracellular stability by dynamic SILAC. *J. Proteome Res.* **8**(1), 104–112.

Gao, H. Y., *et al.* (2000). Two-dimensional electrophoretic/chromatographic separations combined with electrospray ionization FTICR mass spectrometry for high throughput proteome analysis. *J. Microcolumn Sep.* **12**(7), 383–390.

Geiger, T., *et al.* (2010). Super-SILAC mix for quantitative proteomics of human tumor tissue. *Nat. Methods* **7**(5), 383–385.

Gruhler, A., *et al.* (2005a). Quantitative phosphoproteomics applied to the yeast pheromone signaling pathway. *Mol. Cell. Proteomics* **4**(3), 310–327.

Gruhler, A., *et al.* (2005b). Stable isotope labeling of Arabidopsis thaliana cells and quantitative proteomics by mass spectrometry. *Mol. Cell. Proteomics* **4**(11), 1697–1709.

Iliuk, A., *et al.* (2009). Playing tag with quantitative proteomics. *Anal. Bioanal. Chem.* **393**(2), 503–513.

Jiang, H., and English, A. M. (2002). Quantitative analysis of the yeast proteome by incorporation of isotopically labeled leucine. *J. Proteome Res.* **1**(4), 345–350.

Kerner, M. J., *et al.* (2005). Proteome-wide analysis of chaperonin-dependent protein folding in Escherichia coli. *Cell* **122**(2), 209–220.

Lund, R., *et al.* (2009). Efficient isolation and quantitative proteomic analysis of cancer cell plasma membrane proteins for identification of metastasis-associated cell surface markers. *J. Proteome Res.* **8**(6), 3078–3090.

Mortensen, P., *et al.* (2010). MSQuant, an open source platform for mass spectrometry-based quantitative proteomics. *J. Proteome Res.* **9**(1), 393–403.

Nilse, L., *et al.* (2010). SILACAnalyzer—A tool for differential quantitaion of stable isotope derived data. *Lect. Notes Comput. Sci.* **6160**(1), 45–55.

Oda, Y., *et al.* (1999). Accurate quantitation of protein expression and site-specific phosphorylation. *Proc. Natl. Acad. Sci. USA* **96**(12), 6591–6596.

Ong, S. E., and Mann, M. (2005). Mass spectrometry-based proteomics turns quantitative. *Nat. Chem. Biol.* **1**(5), 252–262.

Ong, S. E., and Mann, M. (2006). A practical recipe for stable isotope labeling by amino acids in cell culture (SILAC). *Nat. Protoc.* **1**(6), 2650–2660.

Ong, S. E., *et al.* (2002). Stable isotope labeling by amino acids in cell culture, SILAC, as a simple and accurate approach to expression proteomics. *Mol. Cell. Proteomics* **1**(5), 376–386.

Ong, S. E., *et al.* (2003). Properties of C-13-substituted arginine in stable isotope labeling by amino acids in cell culture (SILAC). *J. Proteome Res.* **2**(2), 173–181.

Ong, S. E., *et al.* (2004). Identifying and quantifying in vivo methylation sites by heavy methyl SILAC. *Nat. Methods* **1**(2), 119–126.

Pratt, J. M., *et al.* (2002a). Dynamics of protein turnover, a missing dimension in proteomics. *Mol. Cell. Proteomics* **1**(8), 579–591.

Pratt, J. M., *et al.* (2002b). Stable isotope labelling in vivo as an aid to protein identification in peptide mass fingerprinting. *Proteomics* **2**(2), 157–163.

Ren, F., *et al.* (2010). Comparative serum proteomic analysis of patients with acute-on-chronic liver failure: Alpha-1-acid glycoprotein maybe a candidate marker for prognosis of hepatitis B virus infection. *J. Viral Hepat.* **17**(11), 816–824.

Schulze, W. X., and Usadel, B. (2010). Quantitation in mass-spectrometry-based proteomics. *Annu. Rev. Plant Biol.* **61,** 491–516.

Steen, H., and Mann, M. (2004). The ABC's (and XYZ's) of peptide sequencing. *Nat. Rev. Mol. Cell Biol.* **5**(9), 699–711.

Xu, Y. H., *et al.* (2009). Application of the SILAC (stable isotope labelling with amino acids in cell culture) technique in quantitative comparisons for tissue proteome expression. *Biotechnol. Appl. Biochem.* **54,** 11–20.

Zamo, A., and Cecconi, D. (2010). Proteomic analysis of lymphoid and haematopoietic neoplasms: There's More than biomarker discovery. *J. Proteomics* **73**(3), 508–520.

Zhang, R. J., *et al.* (2002). Controlling deuterium isotope effects in comparative proteomics. *Anal. Chem.* **74**(15), 3662–3669.

Zhu, H. N., *et al.* (2002). Amino acid residue specific stable isotope labeling for quantitative proteomics. *Rapid Commun. Mass Spectrom.* **16**(22), 2115–2123.

CHAPTER NINE

Nucleic Acid Programmable Protein Array: A Just-In-Time Multiplexed Protein Expression and Purification Platform

Ji Qiu *and* Joshua LaBaer

Contents

Abstract

Systematic study of proteins requires the availability of thousands of proteins in functional format. However, traditional recombinant protein expression and purification methods have many drawbacks for such study at the proteome level. We have developed an innovative *in situ* protein expression and capture system, namely NAPPA (nucleic acid programmable protein array), where C-terminal tagged proteins are expressed using an *in vitro* expression system and efficiently captured/purified by antitag antibodies coprinted at each spot. The NAPPA technology presented in this chapter enable researchers to produce and display fresh proteins just in time in a multiplexed high-throughput fashion

Virginia G. Piper Center for Personalized Diagnostics, Biodesign Institute, Arizona State University, Tempe, Arizona, USA

Methods in Enzymology, Volume 500
ISSN 0076-6879, DOI: 10.1016/B978-0-12-385118-5.00009-8

and utilize them for various downstream biochemical researches of interest. This platform could revolutionize the field of functional proteomics with it ability to produce thousands of spatially separated proteins in high density with narrow dynamic rand of protein concentrations, reproducibly and functionally.

1. Introduction

The advancement of proteomics has significantly accelerated the pace of biomedical research. To this end, there is great demand for high-throughput data-driven approaches to study proteins and their functions at the proteome level. Protein microarrays provide a key enabling technology where thousands of proteins are spotted in high spatial density on a microscopic glass slide and enable the assay of protein biochemical properties on a planar surface in a multiplexed fashion(MacBeath and Schreiber, 2000; Zhu *et al.*, 2000). In contrast to DNA microarrays, which only measure the relative concentration of DNA molecules complementary to spotted "oligonucleotides" (Schena *et al.*, 1995), protein microarrays can be employed not only to measure target protein concentrations in biological samples (antibody arrays; Haab *et al.*, 2001) but also to study a broad range of biochemical properties for thousands of proteins in a high-throughput fashion (LaBaer and Ramachandran, 2005). These biochemical properties include protein functions and interactions with other molecules such as small-molecule drugs, DNA, or other proteins.

While holding tremendous potential to meet the pressing need to study thousands of proteins in parallel, high-density recombinant protein microarrays have not been widely deployed into the biomedical research community. The production of protein microarrays requires the expression and purification of proteins at a scale that traditional methodologies cannot deliver. Unlike DNA that can be produced *en masse* using Polymerase Chain Reaction (PCR) by one enzyme—DNA polymerase—or even synthesized chemically, proteins must be produced from cDNA by complex transcription and translation machinery. During and/or after translation, proteins also need suitable environment and machinery to promote and maintain their native functional conformations.

The production and purification of recombinant protein for biological studies typically includes cloning the gene into an expression vector, transforming it into an expression system, inducing the cells to produce protein, and isolating the protein through a laborious set of purification steps on affinity columns. Although commonly employed, and even automated in some circumstance, this approach has serious drawbacks. First, there is variable protein yield during production that can lead to >1000-fold differences in concentrations from one protein to another in preparation.

Because most biochemical reactions are concentration dependent, this can lead to false negatives/positives if proteins of interest are under/overrepresented on the microarray. Second, prepared proteins require storage at −20 °C or even −80 °C to maintain functionality and still have limited shelf life. Third, the recombinant proteins are often expressed in *Escherichia coli*, yeast, or insect cell expression system that lacks the mammalian context sometimes essential for adequate production and proper protein folding. The manipulation during purification and storage of the recombinants may further limit proper protein folding. Lastly, the expression, purification, and spotting of tens of thousands recombinant proteins is a laborious task that elevates its cost, reduces its reproducibility, and limits its utilization.

To address the deficiencies of spotting purified proteins, we have developed a high-throughput just-in-time protein expression and *in situ* purification system, namely nucleic acid programmable protein array (NAPPA; Ramachandran *et al.*, 2004). Several platforms have been developed to use cell-free systems to generate protein microarrays, such as protein *in situ* array (PISA) (He and Taussig, 2001) and DNA array to protein array (DAPA; He *et al.*, 2008). However, to our knowledge, NAPPA is the only nonprotein printing platform that has advanced beyond proof-of-concept to produce arrays of >1000 unique proteins for biomedical studies at the proteome level. NAPPA will not only achieve protein expression and isolation at the proteome level but also deliver the end products in a microarray format for multiplexed high-throughput studies.

In this method, full-length cDNAs corresponding to the proteins of interest are printed on the microarray substrate. The cDNAs are configured to append a common epitope tag to all of the proteins on the N- or C-termini so that they can be captured by a high-affinity capture reagent that recognizes the epitope tag and is immobilized along with the cDNA. To produce the protein at the time of assay, *in vitro* transcription and translation (IVTT)-coupled rabbit reticulocyte lysate is used. This approach offers the following advantages over traditional method:

1. Replaces preparing proteins with the more reliable and less expensive process of preparing DNA.
2. Avoids the need to express, purify, and store individual proteins.
3. Avoids concerns about protein shelf life because the proteins are made fresh at the time of assay.
4. Displays better than 95% of sequence-verified full-length genes, including membrane proteins.
5. Protein display levels are more consistent from protein to proteins; 93% of display levels are within twofold of the mean.
6. Assures protein integrity by using mammalian expression machinery to synthesize and fold proteins.
7. Easy to create custom arrays by simply rearranging plasmids.

Using this approach, ~20,000 different proteins have been expressed including human kinases, transcription factors, G-protein coupled receptors, and various druggable targets. Early studies demonstrated functional proteins by documenting 85% of the known protein interactions in the human DNA prereplication complex. More recently, we have demonstrated that kinases expressed on the array are active enzymes by measuring autophosphorylation activity that can be inhibited selectively by known kinase inhibitors (Festa and LaBaer, unpublished data). Since development, this technology has been effectively used for disease biomarker discovery and functional protein assays and successfully adopted by several other labs (Anderson *et al.*, 2008, 2010; Ceroni *et al.*, 2010; Montor *et al.*, 2009; Ramirez *et al.*, 2010; Wright *et al.*, 2010).

In this chapter, we will detail the methods from plasmid DNA preparation to array production to protein display assessment. Rabbit reticulocyte lysates will be used as an example for the *in vitro* expression system. However, we have demonstrated that this protocol can be easily adapted to other expression systems, such as insect cell or human cell lysates. Our standard expression vector pANT7-cGST is freely available to the research community (Ramachandran *et al.*, 2004). Genes of interest can be cloned into pANT7-cGST through recombinational cloning (Park and Labaer, 2006). Other versions of expression vectors are also available through DNASU (www.dnasu.asu.edu). Furthermore, ~10,000 human gene clones including ~500 human kinase genes together with tens of thousands genes from different pathogens that are already in NAPPA compatible vector are readily available from DNASU (Hu *et al.*, 2007; Labaer *et al.*, 2004; Murthy *et al.*, 2007; Park *et al.*, 2005; Rolfs *et al.*, 2008).

2. Overview of NAPPA Chemistry

The innovation of NAPPA lies in the functional expression of proteins from immobilized plasmid DNA on a solid surface and efficient capture *in situ* by cospotted capture agents. Microscopic slides were treated with 3-aminopropyltriethoxysilane (APS) to attach a functional primary amine group to the surface. Plasmid DNAs and capture antibodies are immobilized on the slide surface with a homobifunctional primary amine cross-linker BS3 without compromise of integrity in terms of expression of cDNAs and binding of antibodies. The addition of Bovine Serum Albumin (BSA) in the printing mixture provides unexplained promoting effects on both effective immobilization and efficient expression (Ramachandran *et al.*, 2008). Expressed proteins are tagged at the C-termini and captured by corresponding polyclonal antibody against the tag. The employment of the C-terminal tag ensures the capture of only full-length proteins (Fig. 9.1).

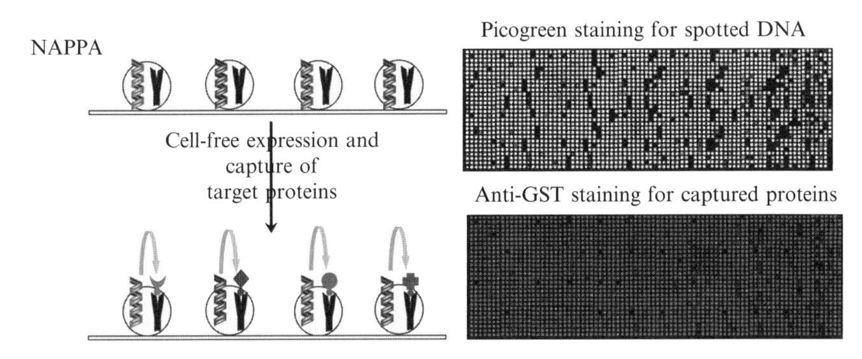

Figure 9.1 On-array protein expression and capture. Left: Plasmid DNA is mixed with BSA, BS3 cross-linker, and the anti-GST capture antibody and arrayed on the aminosilane-coated glass slide. After blocking, cell-free expression mix is applied to the slide, and during a temperature-programmed incubation, the proteins are produced and bind to the capture antibody. Captured proteins can be detected by detecting the GST tag using a monoclonal anti-GST antibody, an HRP-labeled anti-mouse antibody, and Cy3-tyramide (TSA) HRP substrate. Right: Sample NAPPA images for PicoGreen staining of spotted DNA and anti-GST staining for captured proteins.

NAPPA chemistry is robust, reproducible, and versatile. Our standard NAPPA system uses an expression vector that encodes the protein of interest with a C-terminal glutathione-*S*-transferase (GST) fusion protein under the transcriptional control of the T7 promoter. Cospotted polyclonal anti-GST antibody is used to capture the expressed target proteins. Captured proteins can be confirmed by incubating the slides with an antibody that recognizes a different epitope on the tag than the antibody used for capture. Alternative tags such as HA and FLAG, alternative promoters such as SP6, and alternative expression systems such as wheat germ and insect cell lysates have also been demonstrated compatibility with our NAPPA system to accommodate different research needs and clone availability. Moreover, the schemes outlined here can be also altered by the user to accommodate different immobilization chemistries for the plasmid DNA and/or target proteins. Printed slides are stable when stored in a dry environment at room temperature (RT).

3. Array Production

Required materials

Automated liquid handling system
Filtered compressed air
Tabletop centrifuge capable of processing multiwell plates

3.1. Preparation of slides

- Glass slides
- 3-APS (Pierce 80370)
- Stainless steel 30-slide Wheaton rack (VWR 25461–014), handle removed
- Glass staining dish (Wheaton 900201)
- Lock & Lock 1.5 cup boxes (Heritage Mint Ltd., ZHPL810)

1. Prepare 300 ml of aminosilane coating solution (2% 3-APS in acetone) in a Wheaton glass staining dish.
2. Put glass slides in a 30-slide Wheaton metal rack.
3. Immerse slides in the aminosilane coating solution for 15 min at RT on a shaker.
4. Rinse with acetone in a Wheaton glass staining dish by dipping the slides in rack for five times.
5. Rinse with deionized water.
6. Dry with filtered compressed air.
   a. It is important to dry slides quickly and evenly across slide surfaces to prevent water stain that might cause high background during assay.
7. Store at RT in metal rack in Lock & Lock boxes with desiccant packs until use.

3.2. Preparation of nucleobond anion exchange resin plate for DNA preparation

- Nucleobond resin (Machery-Nagel custom order)
- Solution N2 (Equilibration buffer): 100 mM Tris, 15% EtOH, 900 mM KCl, 0.15% Triton X-100, add phosphoric acid until pH is 6.3
- 96-well deep-well block (Marsh AB-0661)
- 800 μl glass fiber MBPP 25 μm filter plate (Whatman 13503–040)

1. In a 1-l glass bottle, add the nucleobond resin to 300 ml, and then add solution N2 to 900 ml.
   a. This step should be done in a hood.
2. Mix the resin with N2 solution until it is homogeneous.
3. Pour the resin solution out into a Wheaton glass staining dish.
4. Place an 800 μl Whatman MBPP filter plate on top of a 96 deep well block (Marsh).
5. Using wide-bore P1000 tips and a multichannel pipette, transfer 450 μl resin slurry to into each well of the Whatman MBPP filter plate.
   a. Gently mix the resin solution with the multichannel pipette laterally several times before each transfer to avoid sedimentation and ensure consistent transfer.

6. Spin the filter plate with the deep well block (Marsh) at the bottom for $131 \times g$ for 5 min.
   a. The resin is ready to use at step 14 in Section 3.3.

3.3. Preparation of plasmid DNA

- Terrific Broth media
- Luria–Bertani media
- Ampicillin stock: 100 mg/ml in H_2O. Store at −20 °C
- Agar
- Omni plate (NUNC 242811)
- 96-pin device (Boekel 140500)
- 96-well deep-well block (Marsh AB-0661)
- Gas permeable plate seal (VWR 47749-924)
- Multitron shaker (Appropriate Technical Resources, Inc.)
- Thermomixer (Eppendorf)
- Matrix WellMate (ThermoFisher)
- Aluminum plate seal (CIC FS-100)
- Solution 1 (Resuspension buffer): 50 mM Tris (pH 8.0), 10 mM EDTA, and 0.1 mg/ml RNAse. Store at 4 °C
- Solution 2 (Lysis buffer): 0.2 N NaOH with 1% SDS
- Solution 3 (Neutralization buffer): 3 M Potassium Acetate (KOAc), add glacial acetic acid until pH is 5.1. Store at 4 °C
- Solution N2 (Equilibration buffer): 100 mM Tris, 15% EtOH, 900 mM KCl, 0.15% Triton X-100, add phosphoric acid until pH is 6.3
- Solution N3 (Wash buffer): 100 mM Tris, 15% EtOH, 1.15 M KCl, add phosphoric acid until pH is 6.3
- Solution N5 (Elution buffer): 100 mM Tris, 15% EtOH, 1 M KCl, add phosphoric acid until pH is 8.5
- 800 μl 96-well block (Abgene AB-0859)
- 350 μl 96-well plate (Greiner 651201)

Note: Liquid transferring between plates can be done either by a multichannel pipette or a robotic liquid handling system.

1. Take out the 96-well glycerol stock plate from −80 °C freezer.
2. Thaw at RT and pulse spin at $233 \times g$ for 3 min to spin down media from the wall and the aluminum foil cover to the bottom of wells.
3. Open carefully to avoid cross contamination.
4. Spot 3 μl from the glycerol stock onto a prewarmed LB/Agar omni plate. Overnight incubation at 37 °C.
5. Sterilize a 96-pin device in 80% ethanol, then flame. Let it cool down before use.

6. Inoculate the culture block from the omni-agar plate to a 96-well deep plate (Marsh) with 1.5 ml TB supplied with 100 μg/ml ampicillin using the 96-pin device.
   a. Sterilize the pin device between each inoculation.
   b. Approximately six inoculates can be done from on a single agar plate.
7. Cover the 96-deep-well plate (Marsh) with a gas permeable seal and put it on an ATR multitron shaker for 24 h at 37 °C, 800 rpm.
   a. Dilute the culture by 1:10 to check OD_{600} periodically.
   b. Ideal OD_{600} should be 0.16–0.25 at the end of incubation.
8. Spin the 96-well deep plate (Marsh) for 20 min at $3724\times g$ on a tabletop centrifuge.
9. Decant supernatant and blot the plate upside down briefly on paper towels.
10. Add 200 μl of solution 1 to each well using wellmate and resuspend by putting on a thermomixer for 1 min at 2000 rpm.
11. Add 200 μl of solution 2 to each well using wellmate, seal with an aluminum plate seal, and mix by inverting the plate for five times.
    a. Solution 2 has to be made fresh.
12. Five minutes later, add 200 μl of solution 3, seal with an aluminum plate seal, and mix by inverting the plate for five times.
    a. Wipe the top of plate with a piece of clean Kimwipes before applying the aluminum foil plate seal may prevent leak and cross contamination among wells.
13. Spin the 96-well deep plate (Marsh) for 20 min at $3724\times g$.
14. Transfer 600 μl supernatant lysates to the nucleobond resin plate using a robotic liquid handling system.
    a. 150 μl each for four times.
    b. If encountering the block problem for lysate transfer using a robotic liquid handling system, use toothpick to remove the pellet and spin again.
15. Spin the plate for 5 min at $21\times g$ using a tabletop centrifuge with slow start.
16. Add 400 μl washing solution N3 to each well using a wellmate, place the plate on the top of a vacuum manifold, and drain by vacuum.
    a. Repeat the washing step four times.
17. Spin for 5 min at $131\times g$ at RT.
18. Place the resin plate onto a clean 800 μl collection plate (Abgene). Add 300 μl elution buffer N5 to each well.
19. Spin the stacked plates for 5 min at $21\times g$ with a slow start followed by 1 min at $233\times g$ with a fast start.
    a. Eluted DNA can be quantified using Nanodrop or Hoechst Dye. >30 ng/μl is desirable before precipitation.
    b. Eluted DNA can be stored -20 °C before precipitation.

20. To precipitate DNA, add 40 μl of 3 M NaOAc, and then add 240 μl of isopropanol to each well.
21. Seal the plate with aluminum plate seal and mix by inverting the plate for three times.
22. Put the plate at −80 °C for 15 min.
23. Warm the plate up at RT for 10 min before spinning at 3724×*g* for 30 min at 20 °C.
24. Decant the supernatant.
25. Add 300 μl of 80% ethanol to each well.
26. Seal with aluminum plate seal and shake at 1200 rpm for 30 min on a Thermomixer.
27. Decant the supernatant.
    a. Spin for 5 min at 233×*g* 20 °C to bring all pellets down to the bottom of well.
28. Air-dry the plate.
    a. Precipitated DNA can now be stored at −20 °C before use.

3.4. Production of arrays

- Contact arrayer with humidity control
- Mater mix: 3.67 mg/ml BSA (Sigma), 1.25 mg/ml BS3 (bis[sulfosuccinimidyl] suberate, Pierce 21580), 50 μg/ml polyclonal anti-GST antibody (Amersham Biosciences 27457701)
- 384-well plate for arraying (Genetix x7020)
- Purified GST protein (Sigma G5663)
- Whole mouse IgG (Pierce 31202)

1. Using P100 multichannel pipette to transfer 25 μl of master mix to each well of 96-well DNA plate prepared as in Section 3.2.
    a. Pre-equilibrate the DNA plate to RT if taking out from −20 °C.
    b. Various control samples such as master mix, purified GST protein, and mouse IgG can be included quality assurance and data analysis purposes.
2. Seal with an aluminum plate seal and shake for 45 min at 1000 rpm on a thermomixer.
3. Spin at 524×*g* for 1 min at 20 °C.
4. Transfer 23 μl resuspending DNA in master mix to a 384-well plate (Genetix x7020) using a robotic liquid handling system.
5. Spin the 384-well plate for 1 min at 524×*g* at 20 °C to remove bubbles.
6. Array, using humidity control at 60%.
7. Put barcode labels at the end of the nonspotting side of printed slides.
8. Store printed slides in a Lock & Lock box with desiccant packs at RT until use.

4. Detection of Protein Expression

4.1. Detection of spotted DNA

- SuperBlock blocking solution in TBS (Pierce 37535)
- Lifterslips (VWR 100499–632)
- PicoGreen (Molecular Probes P11495) stock solution: to the 100 μl/vial that comes, add 200 μl TE buffer. Before use do a 1:600 dilution in SuperBlock
- 4-well dish (NUNC, 73521–424)

1. Block a slide with SuperBlock for 1 h at RT in a 4-well dish. Use ~4 ml per slide. Rock gently.
2. Before applying the PicoGreen reagent, tap the slides gently and wipe the nonspotting side on clean paper towels to remove excess water from the slides, but do not let dry.
   a. If the slides are too wet, it will dilute the PicoGreen reagent.
3. For a single slide: apply 500 μl PicoGreen mix, and apply a lifterslip slowly to avoid trapping bubbles.
4. Incubate for 10 min at RT.
5. Rinse with deionized water for three times.
6. Dry with filtered compressed air- or spin-dry at 131×*g* for 3 min at 20 °C.
7. Scan in microarray scanner at proper settings for Cy3.
   a. Should avoid saturated pixels for quantitative analysis.

4.2. Expression of proteins

- EchoTherm™ IN35 Bench Top Chilling/Heating Incubators with Fully Programmable Controls (Torrey Pines Scientific)
- HybriWell gaskets (Grace Bio-Labs, 44904)
- TNT® Quick Coupled Transcription/Translation Systems (Rabbit Reticulocyte Lysate) (Promega TM045)
- SuperBlock blocking solution in TBS (Pierce 37535)
- Milk blocking solution: 5% Milk in PBS with 0.2% Tween 20
- 4-well dish (NUNC, 73521–424)
- Microarray scanner

1. Block slides for ~1 h at RT with SuperBlock in a 4-well dish. Use ~4 ml per slide. Rock gently.
2. Quickly rinse with deionized water.
3. Dry with filtered compressed air or spin at 131×*g* for 3 min at 20 °C.
4. While blocking, take out Promega T7 Coupled Rabbit Reticulocyte Lysates from −80 °C and thaw on ice.

   a. Good expression of the NAPPA arrays depends on high quality rabbit reticulocyte lysates, which may vary from lot to lot. It is recommended to run one set of experiments with well-tested comparable lots.
5. Prepare IVTT mix. 160 μl is needed for 1 slide.
   a. 128 μl rabbit reticulocyte lysates
   b. 28.8 μl DEPC water
   c. 3.2 μl 1 mM Met
6. Apply a HybriWell gasket to each slide. Use the wooden stick to rub the areas where the adhesive is to make sure it is well stuck all around.
7. Apply IVTT mix through one of the sample application ports on the Hybriwell gasket. Pipette the mix in slowly. Gently massage the HybriWell with both hands to get the IVTT mix to spread out and cover all of the area of the array. Get rid of any bubbles in the center by tapping with the wooden stick. Apply the small round port seals to both ports.
8. Incubate for 1.5 h at 30 °C for protein expression, followed by 30 min at 15 °C for the query protein to bind to the immobilized protein.
9. Remove the HybriWell; rinse with milk blocking buffer five times in a pipette box.
10. Slides are ready for downstreaming assays and should be rinsed with an assay compatible buffer instead of milk blocking buffer.
11. Block with milk in PBST0.2 at RT for 1 h.

4.3. Detection of captured proteins

- Primary AB solution: mouse anti-GST (Cell Signaling 2624) 1:200 in milk blocking solution.
- Secondary AB solution: HRP-conjugated anti-mouse (JacksonImmunoResearch 515-035-062, 1 mg/ml) 1:400 in milk blocking solution.
- Tyramide signal amplification (TSA) stock solution: use TSA reagent (PerkinElmer SAT704B001EA). Prepare per kit directions. Keep this solution at 4 °C.
- PBST0.2: PBS with 0.2% Tween 20
- Milk blocking solution: 5% milk in PBS with 0.2% Tween 20.
- Lifterslips (VWR 100499–632).

1. Incubate with 3 ml primary antibody mouse anti-GST solution in a 4-well dish for 1 h at RT with gentle rocking.
2. Rinse with milk blocking solution for three times.
3. Wash with milk blocking solution for 3 × 5 min.
4. Incubate with 3 ml secondary antibody HRP-labeled goat anti-mouse IgG solution in a 4-well dish for 1 h at RT with gentle rocking.

5. Rinse with PBST0.2 for three times.
6. Wash with PBST0.2 for 3 × 5 min.
   a. While washing, prepare the TSA reagent and keep in dark.
7. Rinse with deionized water for three times.
8. Before applying the TSA reagent, tap the slides gently on clean paper towels to excess water from the slides, but do not let dry.
   a. If the slides are too wet, it will dilute the TSA reagent.
9. Apply the TSA reagent on the slide and gently cover with a lifter slip.
   a. Avoid trapping bubbles under the lifter slip.
10. Incubate for 10 min at RT.
11. Rinse in deionized water; dry with filtered compressed air- or spin-dry at 131×*g* for 3 min at 20 °C.
    a. Do not let slides air-dry slowly as the water marks may form on array.
12. Scan in a microarray scanner at proper settings for Cy3.
    a. Avoid saturated pixels for quantitative analysis.

ACKNOWLEDGMENTS

The research relevant to this chapter was supported by grants from National Cancer Institute (1-R21/R33-CA099191-01), the Early Detection Research Network (5U01CA117374-02), the Juvenile Diabetes Research Foundation (JDRF 17-2007-1045), and a contract from National Institute of Allergy and Infectious Diseases (HHS2662004000053). The authors would like to thank Eliseo Mendoza Garcia for providing sample NAPPA images.

REFERENCES

Anderson, K. S., Ramachandran, N., Wong, J., Raphael, J. V., Hainsworth, E., Demirkan, G., Cramer, D., Aronzon, D., Hodi, F. S., Harris, L., Logvinenko, T., and LaBaer, J. (2008). Application of protein microarrays for multiplexed detection of antibodies to tumor antigens in breast cancer. *J. Proteome Res.* **7,** 1490–1499.

Anderson, K. S., Wong, J., Vitonis, A., Crum, C. P., Sluss, P. M., Labaer, J., and Cramer, D. (2010). p53 autoantibodies as potential detection and prognostic biomarkers in serous ovarian cancer. *Cancer Epidemiol. Biomarkers Prev.* **19,** 859–868.

Ceroni, A., Sibani, S., Baiker, A., Pothineni, V. R., Bailer, S. M., LaBaer, J., Haas, J., and Campbell, C. J. (2010). Systematic analysis of the IgG antibody immune response against varicella zoster virus (VZV) using a self-assembled protein microarray. *Mol. Biosyst.* **6,** 1604–1610.

Haab, B. B., Dunham, M. J., and Brown, P. O. (2001). Protein microarrays for highly parallel detection and quantitation of specific proteins and antibodies in complex solutions. *Genome Biol.* **2**, RESEARCH0004.

He, M., and Taussig, M. J. (2001). Single step generation of protein arrays from DNA by cell-free expression and in situ immobilisation (PISA method). *Nucleic Acids Res.* **29,** e73.

He, M., Stoevesandt, O., Palmer, E. A., Khan, F., Ericsson, O., and Taussig, M. J. (2008). Printing protein arrays from DNA arrays. *Nat. Methods* **5,** 175–177.

Hu, Y., Rolfs, A., Bhullar, B., Murthy, T. V., Zhu, C., Berger, M. F., Camargo, A. A., Kelley, F., McCarron, S., Jepson, D., Richardson, A., Raphael, J., *et al.* (2007). Approaching a complete repository of sequence-verified protein-encoding clones for Saccharomyces cerevisiae. *Genome Res.* **17,** 536–543.

LaBaer, J., and Ramachandran, N. (2005). Protein microarrays as tools for functional proteomics. *Curr. Opin. Chem. Biol.* **9,** 14–19.

Labaer, J., Qiu, Q., Anumanthan, A., Mar, W., Zuo, D., Murthy, T. V., Taycher, H., Halleck, A., Hainsworth, E., Lory, S., and Brizuela, L. (2004). The Pseudomonas aeruginosa PA01 gene collection. *Genome Res.* **14,** 2190–2200.

MacBeath, G., and Schreiber, S. L. (2000). Printing proteins as microarrays for high-throughput function determination. *Science* **289,** 1760–1763.

Montor, W. R., Huang, J., Hu, Y., Hainsworth, E., Lynch, S., Kronish, J. W., Ordonez, C. L., Logvinenko, T., Lory, S., and LaBaer, J. (2009). Genome-wide study of Pseudomonas aeruginosa outer membrane protein immunogenicity using self-assembling protein microarrays. *Infect. Immun.* **77,** 4877–4886.

Murthy, T., Rolfs, A., Hu, Y., Shi, Z., Raphael, J., Moreira, D., Kelley, F., McCarron, S., Jepson, D., Taycher, E., Zuo, D., Mohr, S. E., *et al.* (2007). A full-genomic sequence-verified protein-coding gene collection for Francisella tularensis. *PLoS One* **2,** e577.

Park, J., and Labaer, J. (2006). Recombinational cloning. *Curr. Protoc. Mol. Biol.* 74:3.20.1–3.20.22.

Park, J., Hu, Y., Murthy, T. V., Vannberg, F., Shen, B., Rolfs, A., Hutti, J. E., Cantley, L. C., Labaer, J., Harlow, E., and Brizuela, L. (2005). Building a human kinase gene repository: Bioinformatics, molecular cloning, and functional validation. *Proc. Natl. Acad. Sci. USA* **102,** 8114–8119.

Ramachandran, N., Hainsworth, E., Bhullar, B., Eisenstein, S., Rosen, B., Lau, A. Y., Walter, J. C., and LaBaer, J. (2004). Self-assembling protein microarrays. *Science* **305,** 86–90.

Ramachandran, N., Raphael, J. V., Hainsworth, E., Demirkan, G., Fuentes, M. G., Rolfs, A., Hu, Y., and LaBaer, J. (2008). Next-generation high-density self-assembling functional protein arrays. *Nat. Methods* **5,** 535–538.

Ramirez, A. B., Loch, C. M., Zhang, Y., Liu, Y., Wang, X., Wayner, E. A., Sargent, J. E., Sibani, S., Hainsworth, E., Mendoza, E. A., Eugene, R., Labaer, J., *et al.* (2010). Use of a single-chain antibody library for ovarian cancer biomarker discovery. *Mol. Cell. Proteomics* **9,** 1449–1460.

Rolfs, A., Hu, Y., Ebert, L., Hoffmann, D., Zuo, D., Ramachandran, N., Raphael, J., Kelley, F., McCarron, S., Jepson, D. A., Shen, B., Baqui, M. M., *et al.* (2008). A biomedically enriched collection of 7000 human ORF clones. *PLoS One* **3,** e1528.

Schena, M., Shalon, D., Davis, R. W., and Brown, P. O. (1995). Quantitative monitoring of gene expression patterns with a complementary DNA microarray. *Science* **270,** 467–470.

Wright, C., Sibani, S., Trudgian, D., Fischer, R., Kessler, B., Labaer, J., and Bowness, P. (2010). Detection of multiple autoantibodies in patients with ankylosing spondylitis using nucleic acid programmable protein arrays. *Mol. Cell. Proteomics,* doi:10.1074/mcp. M900384-MCP200.

Zhu, H., Klemic, J. F., Chang, S., Bertone, P., Casamayor, A., Klemic, K. G., Smith, D., Gerstein, M., Reed, M. A., and Snyder, M. (2000). Analysis of yeast protein kinases using protein chips. *Nat. Genet.* **26,** 283–289.

CHAPTER TEN

Systems Biology of Recombinant Protein Production Using *Bacillus megaterium*

Rebekka Biedendieck,* Claudia Borgmeier,† Boyke Bunk,* Simon Stammen,* Christian Scherling,‡ Friedhelm Meinhardt,† Christoph Wittmann,§ *and* Dieter Jahn*

Contents

Abstract

The Gram-negative bacterium *Escherichia coli* is the most widely used production host for recombinant proteins in both academia and industry. The Gram-positive bacterium *Bacillus megaterium* represents an increasingly used alternative for

* Institute of Microbiology, Technische Universität Braunschweig, Spielmannstrasse 7, Braunschweig, Germany

† Institute for Molecular Microbiology and Biotechnology, Westfälische Wilhelms-Universität Münster, Correnssttrasse 3, Münster, Germany

‡ Institute of Bioinformatics and Biochemistry, Technische Universität Braunschweig, Langer Kamp 19b, Braunschweig, Germany

§ Institute of Biochemical Engineering, Technische Universität Braunschweig, Gaußstrasse 17, Braunschweig, Germany

Methods in Enzymology, Volume 500
ISSN 0076-6879, DOI: 10.1016/B978-0-12-385118-5.00010-4

high yield intra- and extracellular protein synthesis. During the past two decades, multiple tools including gene expression plasmids and production strains have been developed. Introduction of free replicating and integrative plasmids into *B. megaterium* is possible via protoplasts transformation or transconjugation. Using His$_6$- and StrepII affinity tags, the intra- or extracellular produced proteins can easily be purified in one-step procedures. Different gene expression systems based on the xylose controlled promoter P_{xylA} and various phage RNA polymerase (T7, SP6, K1E) driven systems enable *B. megaterium* to produce up to 1.25 g of recombinant protein per liter. Biomass concentrations of up to 80 g/l can be achieved by high cell density cultivations in bioreactors. Gene knockouts and gene replacements in *B. megaterium* are possible via an optimized gene disruption system. For a safe application in industry, sporulation and protease-deficient as well as UV-sensitive mutants are available. With the help of the recently published *B. megaterium* genome sequence, it is possible to characterize bottle necks in the protein production process via systems biology approaches based on transcriptome, proteome, metabolome, and fluxome data. The bioinformatical platform (Megabac, http://www.megabac.tu-bs.de) integrates obtained theoretical and experimental data.

1. Introduction

Since many decades, microorganisms are used for the large-scale production of recombinant proteins. Within the group of bacteria, the Gram-negative *Escherichia coli* is the most widely used production host. Nevertheless, some proteins create major problems when heterologously overproduced in this organism. One of the main problems is the formation of inclusion bodies. As a consequence, often a complicated and expensive denaturation process of the inclusion bodies followed by a refolding process is necessary. Moreover, due to the architecture of its Gram-negative cell wall, proteins are not readily secreted into the surrounding growth medium and rather stay in the periplasmatic space between cell wall and outer membrane. However, secretion of proteins directly into the growth medium does avoid the time-consuming and expensive cell disruption prior the protein purification process. Hence, alternative production hosts are of gaining interest within academia and industry (Bunk *et al.*, 2010).

In this context, during the past two decades, the Gram-positive soil bacterium *Bacillus megaterium* was systematically developed for applications in recombinant protein production. With its eponymous size of 1.5 × 4 μm ("megat(h)erium," Greek "big animal"), this microorganism belongs to the larger bacteria. Due to its dimension, *B. megaterium* is also well suited for morphological studies in cell biology approaches (Vary, 1992). Its size also allows for fluorescence-activated cell sorting which is an important tool for investigations in the new upcoming field of individual cell behavior

(Biedendieck *et al.*, 2007c). Finally, the Gram-positive cell structure without outer membrane enables *B. megaterium* to secrete proteins with high yields directly into its surrounding growth medium (Biedendieck *et al.*, 2007a,b; Malten *et al.*, 2005a,b, 2006; Nahrstedt *et al.*, 2004; Vary *et al.*, 2007).

2. Handling *B. megaterium*

2.1. Plasmids for recombinant protein production in *B. megaterium*

Several *B. megaterium* strains are known to carry significant parts of their genetic material (up to 11%) on up to 10 different free replicating plasmids (Vary, 1994; von Tersch and Carlton, 1983). For industrial applications and research, plasmidless strains like *B. megaterium* DSM319 (Stahl and Esser, 1983) are usually used as hosts for plasmid-driven protein production to prevent incompatibilities. *B. megaterium* is known for its ability to stably replicate and maintain recombinant plasmids, even without antibiotic selection (Rygus and Hillen, 1991; Vary, 1994).

Different shuttle vector systems for cloning in *E. coli* and recombinant gene expression in *B. megaterium* were constructed within the last years. Starting in 1991, Rygus and Hillen developed a xylose-inducible gene expression system for recombinant protein production in *B. megaterium* (Rygus and Hillen, 1991). The xylose-inducible promoter (P_{xylA}) is located upstream of an operon consisting of the genes for xylose isomerase XylA, xylulokinase XylB, and xylose permease XylT (Rygus *et al.*, 1991). XylT is responsible for the active transport of xylose into the cell, while XylA and XylB are utilized for the phosphorylation of xylose to form xylose-5-phosphate. The gene encoding the repressor protein XylR is located divergently oriented upstream of this operon. Consequently, the promoter regions of *xylR* and of the *xylABT* operon are overlapping. The regulation of *xylABT* expression occurs at the transcriptional level. In the absence of xylose, the repressor XylR binds to the two tandem overlapping operator sequences spaced by four base pairs located in P_{xylA} and prevents transcription of the *xylABT* operon (Dahl *et al.*, 1994; Gärtner *et al.*, 1988). In the presence of xylose, the sugar binds to the repressor XylR. This results in a conformational change of XylR which in turn prevents operator binding. In this case, the RNA polymerase (RNAP) is able to recognize the promoter and initiates gene expression. This system was the starting point for a stepwise optimization strategy yielding various protein production systems which can be used for high yield synthesis of recombinant proteins and their purification via His_6- or StrepII-tag affinity chromatography (Biedendieck *et al.*, 2007c; Stammen *et al.*, 2010a). The N-terminally fused affinity tag can be cleaved off using various proteases. Amounts of recombinant protein of up to 1.25 g/l were achieved (Stammen *et al.*, 2010a). The system was expanded by introducing different signal peptides for the secretion of recombinant proteins. These proteins can also be purified directly from the growth medium by affinity chromatography (Biedendieck *et al.*, 2007a; Malten *et al.*, 2006; Stammen *et al.*, 2010a; Fig. 10.1).

Further, the system was upgraded by vectors containing strong promoters for various phage RNAPs including the T7-, the SP6-, and the K1E-RNA polymerase (Gamer *et al.*, 2009; Stammen *et al.*, 2010b; Fig. 10.1). Besides the production of high amounts of recombinant proteins, these phage promoter containing vectors also avoid cloning problems in *E. coli*. The xylose-inducible promoter is leaky in the cloning host *E. coli* (Jordan *et al.*, 2007). Hence, genes encoding toxic products are very difficult or not possible to clone in *E. coli*. The phage RNAP promoters are strictly controlled in this host (Terpe, 2006). Therefore, genes toxic for *E. coli* can now easier be cloned for expression in *B. megaterium*.

All expression vectors have the same multiple cloning site which allows easy subcloning. The whole xylose-dependent expression system is also usable for *Bacillus subtilis* (Rygus *et al.*, 1991).

2.1.1. Materials

LB medium: 5 g/l NaCl, 5 g/l yeast extract, 10 g/l tryptone; add 15 g/l agar agar, if necessary

A5-medium for high cell density cultivation (HCDC) experiments: 2 g/l $(NH_4)_2SO_4$, 300 mg $MgSO_4 \times 7H_2O$, 500 mg/l yeast extract, 0.4–3% (w/v) glucose, 40 mg/l $MnCl_2 \times 4H_2O$, 53 mg/l $CaCl_2 \times 2H_2O$, 2.5 mg/l $FeSO_4 \times 7H_2O$, 2.5 mg/l $(NH_4)_6Mo_7O_{24} \times 4H_2O$, 2.5 mg/l $CoCl_2 \times 6H_2O$, 3.52 g/l KH_2PO_4, 7.26 g/l $Na_2HPO_4 \times 2\ H_2O$ (Malten *et al.*, 2005a)

Minimal medium for batch experiments: 3.52 g/l KH_2PO_4, 6.62 g/l $Na_2HPO_4 \times 2H_2O$, 300 mg/l $MgSO_4 \times 7H_2O$, 25 g/l $(NH_4)_2\ SO_4$, 15 g/l fructose, 80 mg/l $MnCl_2 \times 4H_2O$, 106 mg/l $CaCl_2 \times 2H_2O$, 5 mg/l $FeSO_4 \times 7H_2O$, 4 mg/l $(NH_4)_6Mo_7O_{24} \times 4H_2O$, 2.2 mg/l $CoCl_2$ (David *et al.*, 2010; Stammen *et al.*, 2010a)

Feed medium for batch experiments: 9.9 g/l KH_2PO_4, 14.98 g/l Na_2HPO_4, 300 mg/l $MgSO_4 \times 7H_2O$, 25 g/l $(NH_4)_2SO_4$, 40 mg/l $MnCl_2 \times 4H_2O$, 53 mg/l $CaCl_2 \times 2H_2O$, 2.5 mg/l $FeSO_4 \times 7H_2O$, 2 mg/l $(NH_4)_6Mo_7O_{24} \times 4H_2O$, 1.1 mg/l $CoCl_2$, 5 g/l xylose, required antibiotic concentration (David *et al.*, 2010; Stammen *et al.*, 2010a)

Inducer: 50% (w/v) xylose in deionized H_2O

Lysozyme buffer: 16.39 g/l Na_3PO_4, 5 mg/ml lysozyme, 50 mU/ml benzonase (pH 6.5)

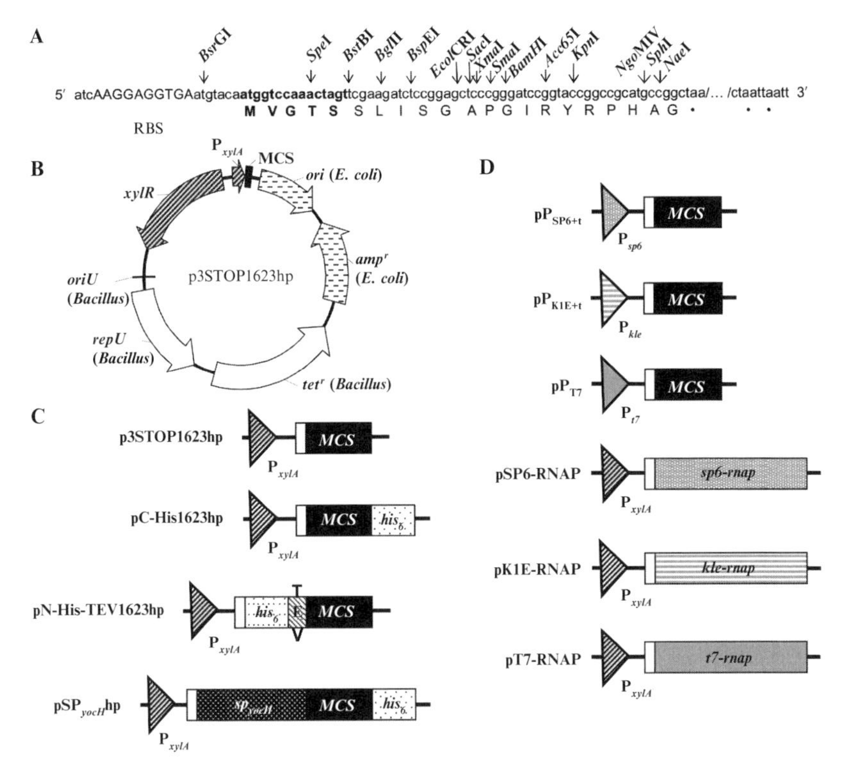

Figure 10.1 Expression and secretion vectors for the intra- and extracellular production of affinity tagged proteins in *B. megaterium*. (A) The DNA sequence of the ribosome binding site (RBS) and multiple cloning site (MCS) of all vectors based on p3STOP1623hp is shown. The RBS is indicated in capital letters. Stop codons downstream of the MCS in all open reading frames are marked with "•." All unique restriction sites of the MCS are indicated. (B) Vector map of the basic vector p3STOP1623hp. (C) Scheme of expression vectors based on p3STOP1623hp. (D) Scheme of expression vectors belonging to the phage-based expression systems for *B. megaterium*. (C and D) All expression plasmids shown allow parallel cloning of the genes of interest into the identical MCS showed in (A). P_{xylA}: promoter of the *xylABT* operon; *xylR*: gene encoding repressor protein XylR; his_6: DNA sequence coding for a 6× histidin tag; TEV: tobacco etch virus protease cleavage site; sp_{YocH}: coding sequence of the signal peptide of YocH; RNAP: RNA polymerase; P_{t7}: T7-RNAP promoter; P_{sp6}: SP6-RNAP promoter; P_{k1e}: K1E-RNAP promoter. For details consult references Biedendieck *et al.* (2007c), Malten *et al.* (2005a), and Vary *et al.* (2007).

2.1.2. Recombinant protein production in *B. megaterium* in shaking flask scale

B. megaterium strain DSM319 is a suitable production host for recombinant proteins using free replicating vectors. This strain does not contain any plasmids naturally. Further, several useful knockout mutants are available (Table 10.1). To avoid the metabolization of the inducer of gene expression

Table 10.1 Selection of mutant strains developed from *B. megaterium* strain DSM319 (MEGABAC, http://www.megabac.tu-bs.de)

Properties	Strain	Description	Reference
Beta-galactosidase negative	WH320	*lacZ–*	Rygus *et al.* (1991)
WH320 + xylose isomerase negative	WH323	*lacZ–*, Δ*xylA*	Rygus and Hillen (1992)
Protease deficient	MS941	Δ*nprM*	Wittchen and Meinhardt (1995)
Protease deficient, amylase negative	MS001	Δ*nprM*, Δ*barM*/*bamM*	Wittchen *et al.* (unpublished)
MS941 + xylose isomerase negative	YYBm1	Δ*nprM*, Δ*xylA*	Yang *et al.* (2006)
MS941 + leucine auxotroph	MS942	Δ*nprM*, Δ*leuC*	Wittchen *et al.* (1998)
MS942 + sporulation negative	MS943	Δ*nprM*, Δ*leuC*, Δ*spoIV*	Wittchen *et al.* (1998)
MS943 + recombination deficient	MS944	Δ*nprM*, Δ*leuC*, Δ*spoIV*, Δ*recA*	Wittchen and Meinhardt (1995)
Amylase negative	MS982	Δ*bamM*	Lee *et al.* (2001)
MS942 + beta-galactosidase negative	MS983	Δ*nprM*, Δ*leuC*, Δ*bgaR*/*bgaM*	Wittchen *et al.* (unpublished)
Recombination deficient	MS991	Δ*recA*1	Nahrstedt *et al.* (2005)
Beta-galactosidase negative	MS021	Δ*bgaR*/*bgaM*	Schmidt *et al.* (2005)
UV sensitive	MS022	Δ*uvrBA::cat*	Nahrstedt and Meinhardt (2004)
UV sensitive	MS033	Δ*uvrB*	Nahrstedt and Meinhardt (2004)
5-Flouro-uracil resistant	MS1004	Δ*upp*	Borgmeier and Meinhardt (unpublished)

xylose, the chemical mutant WH323 was constructed which is *xylA*-negative and does not utilize xylose anymore (Rygus and Hillen, 1991). Further, for the stabilization of secreted recombinant proteins, the mutant MS941 was constructed (Wittchen and Meinhardt, 1995). This strain lacks the major extracellular protease NprM and does only show 1.4% of natural extracellular protease activity (Wittchen and Meinhardt, 1995). Based on this mutant, the *nprM/xylA*-negative strain YYBm1 was developed (Yang *et al.*, 2006).

1. Streak *B. megaterium* plasmid-containing strain onto a LB medium agar plate containing the appropriate antibiotic and incubate for 16 h at 37 °C. If working with the xylose-inducible promoter, one antibiotic (usually 10 μg of tetracycline/ml) is necessary. All the phage RNAP-driven systems are based on a two vector system. Here, two antibiotics (usually 10 μg of tetracycline/ml and 4.5 μg chloramphenicol/ml) are required.
2. Inoculate a 50-ml LB medium liquid starter culture in a 300-ml shaking flask with cell material from the plate and incubate for 16 h at 100 rpm and 37 °C. Cultures must not reach stationary phase.
3. Inoculate a 100-ml LB medium liquid culture in a 500-ml shaking flask in the ratio 1:100 with the preculture and incubate at 250 rpm and 37 °C.
4. When the culture reaches an optical density (OD) at 578 nm (OD_{578nm}) of 0.3–0.4 induce the recombinant gene expression with a final concentration of 0.5% (w/v) of xylose. In case of the xylose-inducible promoter, the xylose directly induces the transcription of the recombinant gene. In a phage RNAP-driven system, the xylose acts indirectly. Here, the xylose induces the formation of phage RNAP in *B. megaterium*, which in turn starts the recombinant target gene expression (Gamer *et al.*, 2009; Stammen *et al.*, 2010b).
5. Samples for OD measurements, SDS-PAGE gel analyses, and protein assays should be taken every hour for a period of 9 h after induction of recombinant gene expression.
6. Harvest the cells by centrifugation (15,000×*g*, 10 min, 4 °C) and store them at −20 °C. The cell-free supernatant should be filtered (pore diameter of 0.2 μm) and afterward stored at 4 °C for further analysis.
7. For analysis, destroy the precipitated cells with lysozyme buffer treatment for 30 min at 37 °C and 1000 rpm. Harvest the soluble protein fraction by centrifugation (15,000 × *g*, 30 min, 4 °C).
8. The extracellular proteins of 1.5 ml of the cell-free supernatant can be precipitated by 70% (w/v) of ammonium sulfate, by 10% (v/v) of trichloracetic acid (TCA) or by four volumes of ice-cold acetone.

2.1.3. Recombinant protein production in *B. megaterium* using high cell density conditions in a bioreactor

The bioreactors used were a Biostat B2 (B. Braun, Melsungen, Germany) with a working volume of 2 l or a RALF Plus 3.7-l bioreactor (Bioengineering, Wald, Switzerland; Biedendieck *et al.*, 2007c; Stammen *et al.*, 2010a). The bioreactor should be connected to an exhaust gas analysis unit (Maihak, Hamburg, Germany). The control unit of the reactor was connected to a computer running the MFCS software (B. Braun). Further, feed and base reservoirs should be stored on scales connected to the control computer.

To control the pH of the culture, regulated addition of either NaOH with a concentration of 200 g/l or of H_3PO_4 (98 g/l) have to take place.

1. Inoculate a starter culture of 50 ml medium for HCDC containing the required antibiotic in a 300-ml shaking flask with 200 μl of a 30% (w/v) glycerol stock and grow the culture for 16 h at 37 °C and 100 rpm. The culture must not reach stationary phase.
2. Add cells of the starter culture to 1 l of cultivation medium in a bioreactor with a total volume of 2 l to reach a final OD_{578nm} of 0.1 and grow them at 37 °C in a batch phase.
3. The end of the batch phase is characterized by a rise in the concentration of dissolved oxygen. Start a feeding profile using the feed medium with a set growth rate (μ_{set}) of 0.12 h^{-1} up to 0.14 h^{-1}.
4. Induce the recombinant gene expression with xylose to a final concentration of 5 g/l. The time point of induction can vary depending on the experimental outline. Usually, it takes place shortly before or after the C-source is exhausted.
5. Samples for the determination of biomass, metabolites, transcriptome, and recombinant protein production should be taken at different time points prior and after induction of recombinant protein production (Fig. 10.2). Cell dry weights of up to 80 g/l and concentrations of recombinant proteins of up to 1.25 g/l can be achieved (Hollmann and Deckwer, 2004; Stammen *et al.*, 2010a).

2.1.4. Purification of recombinantly produced intra- and extracellular proteins

In the cell of a Gram-positive bacterium, there are different possible localizations for produced recombinant proteins, intracellularly in the cytoplasma, bound to the membrane, anchored in the cell wall, or extracellularly in the growth medium. Here, we focused on intracellularly in the cytoplasma and extracellularly in the growth medium localized recombinant proteins. They can be fused to an N- or C-terminal His_6- or StrepII-tag. To purify an intracellularly located recombinant protein, the bacterial cells have to be disrupted and the cell debris has to be removed to allow for the chromatographic purification of a soluble affinity tagged protein. Proteins located extracellularly in the growth medium of *B. megaterium* are much easier to purify. Simply, the cells have to be removed from the medium, and then the proteins can be purified from the cell-free supernatant. The best time point for harvesting the proteins can be determined by SDS-PAGE analysis of the culture at different time points of cultivation.

1. Define the time point after induction of recombinant protein production, where most of the desired protein is accumulated, by SDS-PAGE gel analysis.

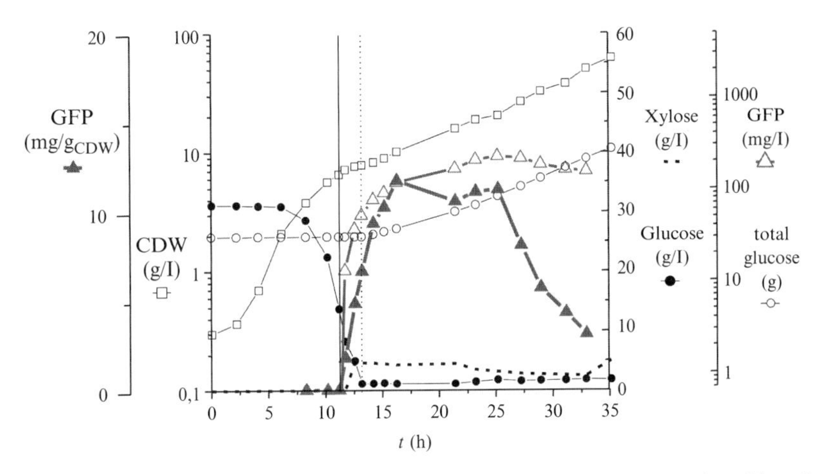

Figure 10.2 Typical diagram of a fed batch cultivation of *B. megaterium* plasmid strains in a bioreactor. The cell dry weight, the concentration of glucose and xylose as well as the cellular and volumetric amount of Gfp are given. The solid line illustrates induction of *gfp* gene expression with xylose. The start of the exponential glucose feeding profile is indicated by a dashed line (adapted from Biedendieck *et al.*, 2007c).

2. Harvest the cells of 50 ml culture medium and keep the precipitated cells for purification of intracellular proteins. For purification of extracellular proteins, recover the supernatant and path it through a filter with a pore diameter of 0.2 μm. The precipitated cells are stored at −20 °C, while the supernatant is kept at 4 °C.

Recovery of intracellular proteins

3. Resuspend the cells in the desired buffer for affinity purification of StrepII- or His_6-tagged proteins (see manufacturer's instructions). Disrupt the cells by sonication (SONOPLUS, Bandelin electronics, Berlin, Germany) three times for 5 min with 50% cycle and 70% power.
4. Centrifuge at 113,000×*g* for 25 min at 4 °C and load the supernatant containing the soluble proteins onto the preequilibrated affinity column.
5. Wash the column and elute the proteins with the appropriate buffers after the manufacturer's instructions.
6. Mix aliquots of the wash and elution fractions with SDS-sample buffer and analyze them via SDS-PAGE gel.

Recovery of extracellular proteins

3. Take the cell-free supernatant and incubate the affinity material with the supernatant (700 μl of affinity material with 50 ml of supernatant) by soft shaking for 1 h at 4 °C.

4. Harvest the affinity material loaded with the recombinant protein by short centrifugation at 2500×*g* and 4 °C. Resuspend the precipitated material in a small volume of the appropriate chromatographic buffer.
5. Wash the chromatography material with the desired buffer and elute the bound recombinant protein from the affinity material after the instructions of the manufacturers.
6. Mix aliquots of the elution fraction with SDS-sample buffer and analyze via SDS-PAGE gel.

2.2. Transformation of *B. megaterium* protoplasts

The transformation of protoplasted *B. megaterium* cells provides an elegant method to introduce foreign plasmid DNA into this organism (von Tersch and Robbins, 1990). For the formation of protoplasts, a suitable osmotic environment is important. Protoplasts are defined as cells without cell wall. This is usually removed by lysozyme treatment. The lacking cell wall makes protoplasts very sensitive against external influences like strong vibration or nonfavorable osmolarity.

For an efficient protoplast transformation, a high-quality plasmid DNA preparation is advised. DNA can be purified by a commercially available extraction kit or by multiple phenol/chloroform extractions. EtOH should not remain in the final DNA solution. Moreover, protoplasts are sensitive against detergents so that the use of new sterile plastic ware is recommended. The tubes containing the very fragile protoplasts should be treated very gently. Further, all solutions used for protoplast preparation and transformation should be warmed up to room temperature. Ready-to-use *B. megaterium* protoplasts are commercially available from the MoBiTec GmbH (Göttingen, Germany).

2.2.1. Materials

2× AB3 Medium (Antibiotic Medium No. 3): 7 g AB3 (Difco) in 200 ml deionized H_2O, sterilize by autoclaving

2× SMM: 1.16 g malic acid, 800 mg NaOH, 2.03 g $MgCl_2 \times 6H_2O$, 85.58 g sucrose, solubilize in 250 ml of deionized H_2O in the given order, sterilize by filtration

SMMP (prepare freshly before each transformation): 2× AB3 and 2× SMM 1:1

SMMP-lysozyme solution: 10 mg lysozyme/ml SMMP, sterilize by filtration

PEG-P: 20 g PEG-6000 in 50 ml of 1× SMM, sterilize by autoclaving

CR5-top-agar (prepare all following solutions seperately):

Solution A: 51.5 g sucrose, 3.25 g MOPS, 300 mg NaOH, solubilize in 250 ml of deionized H_2O, adjust pH to 7.3 with NaOH, sterilize by filtration

Solution B: 2 g agar agar, 100 mg casamino acids, 5 g yeast extract, add deionized H_2O to 142.5 ml, sterilize by autoclaving

8× CR5-salts: 1.25 g K_2SO_4, 50 g $MgCl_2 \times 6H_2O$, 250 mg KH_2PO_4, 11 g $CaCl_2$, solubilize in 625 ml of deionized H_2O, sterilize by autoclaving

12% proline: 3 g proline, solubilize in 25 ml of deionized H_2O, sterilize by filtration

20% glucose: 20 g glucose, solubilize in 100 ml of deionized H_2O, sterilize by filtration

For 2.5 ml portions of CR5-top-agar, mix the following in the given order: 1.25 ml of solution A, 288 μl of CR5-salts, 125 μl of 12% proline, 125 μl of 20% glucose, 713 μl of solution B (after liquefaction)

Buffer 1: 6.06 g/l Tris (pH 8), 2.92 g/l EDTA (pH 8), 100 μg/ml RNase A, 10 mg lysozyme/ml

Buffer 2: 8 g/l NaOH, 10 g/l SDS

Buffer 3: 294.45 g/l potassium acetate (pH 5.5)

2.2.2. Preparation of protoplasts

1. Inoculate a 50-ml *B. megaterium* LB medium starter culture in a 300-ml baffled flask overnight at 100 rpm at 37 °C. It should not reach the stationary phase.
2. Inoculate 50 ml LB medium with 1 ml of preculture in 300 ml baffled flask and grow it at 37 °C and strong shaking (250 rpm) to an OD_{578nm} of 1.
3. Separate cells and growth medium at 4 °C, resuspend the precipitated cells in 5 ml of freshly prepared SMMP, and transfer the cell suspension into a 15-ml sterile plastic reaction tube.
4. Add 150 μl of freshly prepared and filter sterilized SMMP-lysozyme solution and incubate at 37 °C and soft shaking until 80–90% of the cells are protoplasted (keep a reference of untreated cells). Check the stadium of protoplast formation using a microscope. Protoplasts can be recognized as round structures, while cell wall containing *B. megaterium* is a rod. If no protoplast formation is visible after 20 min, increase the lysozyme concentration. Avoid incubation with lysozyme for longer than 40 min.
5. Harvest protoplasted cells carefully for 10 min and 1300×*g* at room temperature and carefully remove the supernatant of the precipitated cells. Do not completely remove the supernatant because the cells only form a soft and fragile precipitate.

6. Carefully resuspend the protoplasts in 5 ml of SMMP by pipetting and spin down again (see step 5.)
7. Carefully resuspend the cells in 5 ml of SMMP and add 750 μl of 87% (w/v) sterile glycerol. Mix softly by rolling and split the protoplasts in portions of 500 μl into sterile reaction tubes.
8. Protoplasts can be stored at −80 °C for up to 6 months, but competence will sink gradually.

2.2.3. Transformation of protoplasts

1. Carefully defrost the protoplasts on ice.
2. Use 10–20 μl of pure plasmid DNA (150 ng/μl) and mix it with 500 μl of protoplasts in a sterile reaction cup.
3. Transfer the DNA–cell mixture into a 15-ml sterile tube which is filled with 1.5 ml of PEG-P and mix softly by rolling the tube.
4. Incubate the mixture for 2 min at room temperature, add 5 ml of freshly prepared SMMP, and mix carefully by rolling the tube.
5. Centrifuge the mixture carefully for 10 min and 1300×*g* at room temperature and carefully discard the supernatant. Usually, the precipitated cells are not visible.
6. Carefully resuspend the cells in 500 μl of SMMP and transfer the mixture into an 1.5-ml reaction tube.
7. Incubate the cells for 45 min without shaking followed by 45 min at 300 rpm.
8. Fifteen minutes before the end of the incubation time, prepare 2.5 ml of CR5-top-agar per transformation in a 15-ml reaction tube and incubate at 43 °C in a waterbath.
9. After the 90 min incubation time (see step 7), pipette the cell suspension to the tube with the CR5-top-agar, mix gently by rolling the tube, and pour the top agar containing the cells on prewarmed (30 °C) LB medium agar plates containing the appropriated antibiotic.
10. Incubate the plates for up to 24 h at 30 °C, separate grown colonies into single colonies, and check the rod shape form of the *B. megaterium* cells microscopically.

2.2.4. Test of protoplast transformation

To control for successful protoplast transformation, the plasmid can be recovered from the *B. megaterium* cells.

1. Inoculate 5.5 ml of LB medium containing the required antibiotic with a single colony of the *B. megaterium* clone of interest and grow the bacteria for 16 h at 200 rpm and 37 °C.

2. Harvest the cells by centrifugation, completely remove the supernatant, and resuspend the precipitated cells in 300 μl of buffer 1 supplemented with 10 mg of lysozyme/ml.
3. Incubate at 37 °C and 800 rpm for 30 min.
4. Add 300 μl of buffer 2, mix by inverting five times, and incubate for 3 min.
5. Add 300 μl of buffer 3, mix by inverting five times, and centrifuge for 20 min at 15,000×*g* and room temperature.
6. Transfer the supernatant in a new 1.5-ml reaction tube, mix with 600 μl of isopropanol by vigorous shaking, and centrifuge for 20 min at 15,000×*g* and room temperature.
7. Discard the supernatant, wash the DNA with 400 μl of 70% (v/v) of EtOH, and dry the precipitated DNA.
8. Resolve the DNA in 50 μl of deionized H_2O.
9. Use 5 μl of this DNA solution to transform *E. coli* and grow the cells for 16 h at 37 °C.
10. Inoculate 5.5 ml of LB medium containing the required antibiotics with a single *E. coli* transformant and grow at 37 °C and 200 rpm for 16 h.
11. Harvest the cells by centrifugation, completely remove the supernatant, and resuspend the precipitated cells in 300 μl of buffer 1.
12. Add 300 μl of buffer 2 and go on with steps 4–8.
13. Perform a restriction enzyme-based analysis of isolated DNA to identify the prepared plasmid.

Another possibility to proof the new plasmid strain can be a colony PCR.

1. For this, take cell material of the new plasmid strain, add it to 30 μl of H_2O, and boil it for 1 min.
2. After centrifugation for 1 min (15,000×*g*), use 2 μl of the supernatant as a template for a PCR reaction using gene or plasmid specific primers.

2.2.5. Storage of *B. megaterium*: Preparation of glycerol stocks

1. Incubate a *B. megaterium* strain carrying a plasmid in 50 ml of the appropriate medium and antibiotics for 14 h and 100 rpm at 37 °C.
2. Thoroughly mix a culture volume of 650 μl with 350 μl of 87% (w/v) of glycerol to achieve a final glycerol concentration of ~30%.
3. Store this glycerol stock at −80 °C until further use. While streaking out the bacteria from the glycerol stock onto agar plates, never defrost the cryo cultures.

2.3. Transconjugation as genetic tool for *B. megaterium*

The direct knockout of genes routinely requires high transformation efficiencies which can be achieved in *B. megaterium* by conjugal DNA transfer from *E. coli* as reported recently (Richhardt *et al.*, 2010).

The procedure is based on a method described previously for *Lysinibacillus sphaericus* (Aquino de Muro and Priest, 2000). For this purposes, *E. coli* S17-1 (Δ*recA*, *end*A1, *hsd*R17, *sup*E44, *thi*-1, *tra*+) carrying a mobilizable vector designed for use in *B. megaterium* (Table 10.2) serves as the donor strain. Different counter selection systems can be applied to eliminate *E. coli* donor cells from the conjugation mixture. Sporogenous strains can easily be selected by pasteurization. Asporogenous strains (Table 10.1; e.g., Δ*spoIV*), preferred in industrial scale production, require an alternative counter selection such as the *sacB* gene of *B. subtilis* (Gay *et al.*, 1985; see Section 2.3.5).

2.3.1. Materials

Holding buffer: 1.7 g/l KH_2PO_4, 2.18 g/l K_2HPO_4, 120.36 mg/l $MgSO_4$ (pH 7.2), sterilize by autoclaving

Sporulation medium: 16 g/l nutrient broth, 2 g/l KCl, 500 mg/l $MgSO_4$; after autoclaving, add 1 ml/l of 100 μM $Ca(NO)_3$, 1 ml/l of 100 μM $MnCl_2 \times 4\ H_2O$, 1 ml/l of 1 mM $FeSO_4$, 2 ml/l of 40% (w/v) glucose (pH 7), all sterilized by filtration; add 15 g/l agar agar, if necessary

Selection medium: 5 g/l yeast extract, 10 g/l tryptone, 15 g/l agar agar, 10 mg/l chloramphenicol, 10% (w/v) sucrose (pH 7.4)

2.3.2. Transformation of *E. coli* competent cells with pJR-derived plasmids

1. Mix 50 μl of competent *E. coli* cells with 100 ng plasmid DNA and incubate the mixture on ice for 20 min.
2. Heat shock them at 42 °C for 45 s, place on ice for additional 2 min, and add 300 μl of LB medium.
3. Mix for 30 min at 750 rpm and 37 °C, add 0.5 μg/ml chloramphenicol, and mix for additional 30 min at 750 rpm.
4. Spread cells on LB medium agar plates containing 10 μg/ml chloramphenicol and incubate at 37 °C for up to 2 days.

2.3.3. Preparation of *B. megaterium* and *E. coli* cells for transconjugation

1. Incubate a 50-ml *B. megaterium* LB medium starter culture in a 300-ml baffled flask overnight at 37 °C and 100 rpm so that it does not reach the stationary phase. In parallel, incubate a 5-ml LB medium starter culture containing the appropriate antibiotic for *E. coli* S17-1 carrying the plasmid to be transferred and grow it overnight at 37 °C at 200 rpm.

Table 10.2 List of plasmids for *B. megaterium* (see also Fig. 10.1)

Applications	Plasmid	Description	Reference
Targeted gene deletion	pUCTV2	Shuttle plasmid; Ap^R, Tc^R, *ori E. coli*, ori^{ts}	Wittchen and Meinhardt (1995)
Targeted gene deletion	pSKE194	Shuttle plasmid; Ap^R, Em^R, *ori E. coli*, ori^{ts}	Nahrstedt *et al.* (2005)
Promoter test studies	pptBm1	Freely replicating promoter test vector; promoterless *bgaM*, Tc^R, Ap^R, *ori E. coli*, ori^{ts}	Schmidt *et al.* (2005)
Promoter test studies	ppts	Integrative promoter test vector; Δ*leuC*::*bgaM*, Tc^R, Ap^R, *ori E. coli*, ori^{ts}	Schmidt *et al.* (2005)
Targeted gene deletion	pE007	Plasmid for single-copy replacement; MCS, Em^R, ori^{ts} (*Bacillus*)	Hoffmann *et al.* (2010)
Targeted gene deletion	pBBRE194	Mobilizable shuttle plasmid; *ori E. coli*, mob, MCS, Tc^R, ori^{ts}, Em^R	Richhardt and Meinhardt (unpublished)
Direct knockout	pJRSu1	Mobilizable suicide vector; Cm^R, mob, ori^T, rep	Richhardt *et al.* (2010)
Targeted gene deletion	pJR1	Mobilizable vector; Cm^R, mob, ori^T, ori^{ts}, sacB, rep	Richhardt *et al.* (2010)
Direct knockout	pJRSu2	Mobilizable suicide vector; optimized MCS, Cm^R, mob, ori^T, rep	Borgmeier *et al.* (unpublished)
Targeted gene deletion	pJR2	Mobilizable vector; optimized MCS, Cm^R, mob, ori^T, ori^{ts}, rep	Borgmeier *et al.* (unpublished)
Recombinant protein production	p3Stop1623hp	Shuttle plasmid; Ap^R, Tc^R, *ori E. coli*, *oriU*, P_{xylA}	Stammen *et al.* (2010a)
Recombinant protein production and protein purification	pC-His1623hp	Shuttle plasmid; Ap^R, Tc^R, *ori E. coli*, *oriU*, P_{xylA}	Stammen *et al.* (2010a)
Recombinant protein production and protein purification	pN-His-TEV1623hp	Shuttle plasmid; Ap^R, Tc^R, *ori E. coli*, *oriU*, P_{xylA}	Stammen *et al.* (2010a)
Recombinant protein secretion	pSP_{yocH}-hp	Shuttle secretion plasmid; Ap^R, Tc^R, *ori E. coli*, *oriU*, P_{xylA}	Stammen *et al.* (2010a)

(*Continued*)

Table 10.2 (*Continued*)

Applications	Plasmid	Description	Reference
T7 RNAP production	pT7-RNAP	Shuttle plasmid; Ap^R, Cm^R, *ori E. coli*, *oriBM100*, P_{xylA}-$rnap_{T7}$	Gamer *et al.* (2009)
Recombinant protein production under control of P_{T7}	pP_{T7}	Shuttle plasmid; Ap^R, Tc^R, *ori E. coli*, *oriU*, P_{T7}	Gamer *et al.* (2009)
SP6 RNAP production	pSP6-RNAP	Shuttle plasmid; Ap^R, Cm^R, *ori E. coli*, *oriBM100*, P_{xylA}-$rnap_{SP6}$	Stammen *et al.* (2010b)
Recombinant protein production under control of P_{SP6}	pP_{SP6+t}	Shuttle plasmid; Ap^R, Tc^R, *ori E. coli*, *oriU*, P_{SP6}	Stammen *et al.* (2010b)
K1E RNAP production	pK1E-RNAP	Shuttle plasmid; Ap^R, Cm^R, *ori E. coli*, *oriBM100*, P_{xylA}-$rnap_{K1E}$	Stammen *et al.* (2010b)
Recombinant protein production under control of P_{K1E}	$pP_{K1E\text{-}1+t}$	Shuttle plasmid; Ap^R, Tc^R, *ori E. coli*, *oriU*, P_{K1E}	Stammen *et al.* (2010b)

2. Inoculate two 50-ml LB medium main cultures in 300-ml baffled flasks with 1 ml of the starter culture of *B. megaterium* and *E. coli* S17-1, respectively, and grow them at 37 °C and strong shaking to an OD_{578} nm of 0.6–0.8.
3. Harvest the cells by centrifugation (15 min, 3220×*g*, 4 °C), wash twice in 15 ml of holding buffer, and resuspend in 30 ml of holding buffer.
4. Store *E. coli* donor cells on ice while heating *B. megaterium* cells for 9 min at 49 °C.
5. Mix *E. coli* donor cells with *B. megaterium* cells. Best results are obtained with a donor/recipient mixing ratio in the range of 2:1 up to 6:1. Close contact between cells is realized by pressing the mixture through a sterile nitrocellulose filter (pore size of 0.45 μm). Cells are collected on the surface of the filter.
6. Place the filter, cells upside, on sporulation agar (Schaeffer *et al.*, 1965) or on LB medium agar plates, depending on the desired counter-selection system (Sections 2.3.4 and 2.3.5).
7. Incubate filter plates at 30 °C for 24 h (LB medium agar) or 48 h (sporulation medium agar).

2.3.4. Counter selection by pasteurization

1. Incubate the mating filter on sporulation medium agar plates (Schaeffer *et al.*, 1965) for 2 days. This should allow for an almost complete sporulation of the *B. megaterium* cells.
2. Subsequently resuspend the cells in 900 μl of holding buffer and pasteurize them (20 min at 80 °C).
3. Spread the cell suspension on LB medium agar plates containing the appropriate antibiotics.

2.3.5. Counter selection by the *sacB* suicide system

1. Recover the cells from the mating filter (incubated on LB medium agar plates) by suspension in 900 μl holding buffer.
2. Plate the cells on sucrose selection medium agar plates containing the appropriate antibiotic.
3. Upon incubation for 20–24 h at 30 °C, the transconjugants usually form a lawn of cells, while the *E. coli* cells do not survive due to detrimental sugar polymer formation via the *sacB* encoded glycosyl transferase. To obtain single *B. megaterium* colonies, resuspend the cells in 3 ml of holding buffer, spread on fresh sucrose selection medium agar plates, and incubate for another 20–24 h at 30 °C.

2.4. Gene knockout/replacement in *B. megaterium*

Another method for single-copy replacement of chromosomally localized genes with inactivated copies and subsequent curing of the mediating vector that carries a temperature-sensitive origin of replication (ori^{ts}; Table 10.2) enables the inactivation of genes in bacterial species not yet amenable to highly efficient transformation. In contrast to classical gene disruption strategies, no selection marker gene remains in the genome. Hence, such technique is advantageous when consecutive rounds of gene inactivation are necessary. Additionally, it meets major biosafety standards. A gene deletion cartridge localized on such a vector, that comprises regions flanking the gene of interest, is integrated into the genome via homologous recombination (Fig. 10.3). Employing a second recombination event, the plasmid can be excised. Two different scenarios are possible. If the second recombination event is mediated via the identical flank as used for the integration event, the wild-type situation is restored. However, if vector excision takes places via the other flank, gene deletion is manifested in the genome. In such case, the target gene is excised along with the vector. This so-called curing process is significantly enhanced, if temperature-sensitive origins of replication are used. They lead to the loss of plasmid when the culture is incubated at the nonpermissive temperature for several generation times.

2.4.1. Protocol

1. Start the screening process with the transformation of *B. megaterium* with the deletion vector (see Section 2.2).
2. Cultivate the transformed *B. megaterium* strain in LB medium containing a high concentration of the appropriate antibiotic (e.g., 5 μg/ml erythromycin for pSKE194 derivatives; Table 10.2) at the permissive temperature of 30 °C overnight to select for the free replication version of the multicopy gene deletion plasmid.
3. To achieve the integration of the gene deletion vector, cells are plated on LB medium agar plates containing a low antibiotic concentration (e.g., 0.5 μg/ml erythromycin for pSKE194-derivatives; Table 10.2), and most importantly incubation is continued at the nonpermissive temperature of 42 °C overnight.
4. Toothpicking of colonies on LB medium agar plates without antibiotic and further incubation at the nonpermissive temperature of 42 °C sustains curing of the excised plasmid. Finally, growth of cells that have lost the plasmid is ensured.
5. To verify the establishment of the desired deletion mutant, chromosomal DNA is isolated from all colonies growing on replica plates by scraping them from the plate and DNA is analyzed by PCR.

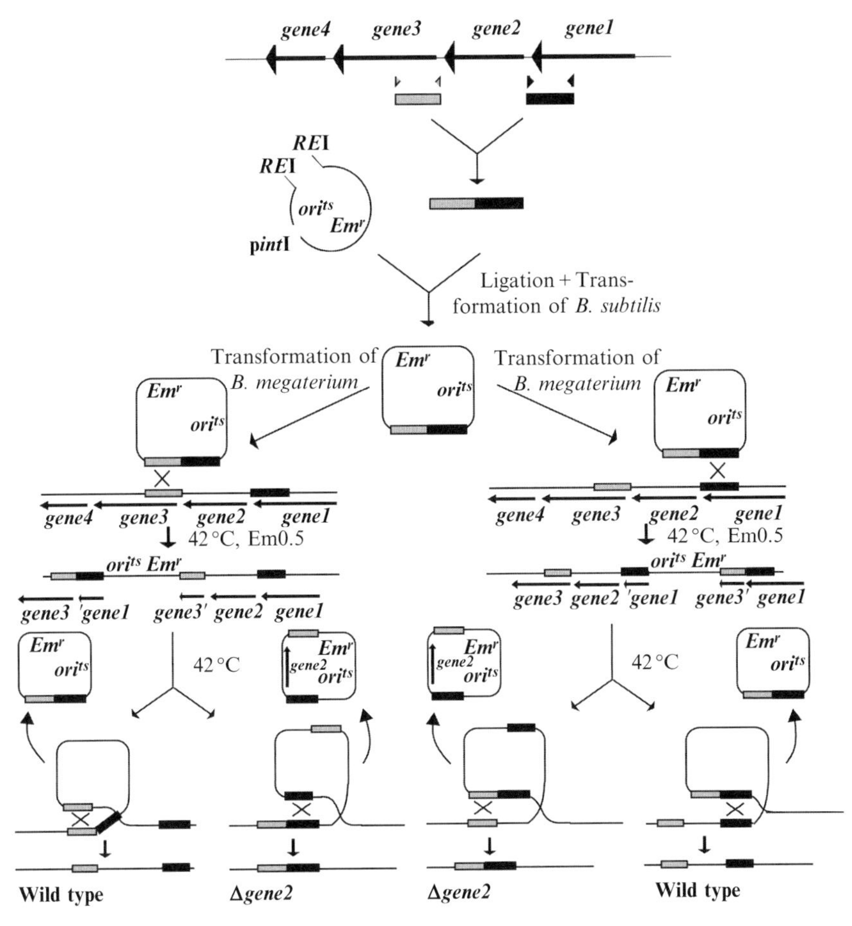

Figure 10.3 Generation of chromosomal gene deletion mutants via single-copy replacement and plasmid curing. The example of deletion of the chromosomal *gene2* is given. *RE*I, restriction enzyme I; p*int*I, integrative plasmid (such as pE007); Em^r, erythromycin resistance gene; *ori*^{ts}, temperature-sensitive origin of replication; Em 0.5, erythromycin concentration of 0.5 μg/ml (modified from Atlagic *et al.*, 2006).

6. Mutant cells are enriched by narrowing down the number of colonies that are pooled in one DNA preparation. In most cases, colonies will harbor wild type as well as mutant cells so that liquid cultures obtained from a single colony have to be streaked on solid medium agar plates repeatedly to separate mutant and wild-type cells testing by PCR.

2.4.2. Counter selection

Plasmid curing with a separation of mutant from wild-type cells is usually a laborious and time-consuming process, especially, if colonies cannot be screened phenotypically. Consequently, a clear-cut counter selection for integrated as well as freely replicating plasmid is desirable. Currently, a very efficient counter-selection system is based on the *upp* gene which encodes the enzyme uracil-phosphoribosyltransferase (UPRTase). It catalyzes the conversion of uracil to uridine monophosphate (UMP) with the help of phosphoribosyl pyrophosphate and enables the cells to use exogenous uracil via conversion to UMP (Neuhard, 1983; Nygaard, 1993). Deletion of the *upp* gene renders the cells resistant to the antimetabolite 5-fluorouracil (5FU), which is normally converted into 5-fluoro-UMP and subsequently metabolized to 5-fluoro-dUMP. The latter compound is a strong inhibitor of the thymidylate synthetase. Consequently, uptake of 5FU leads to bacterial cell death. The *upp* gene can be integrated in replicative as well as suicide vectors thus enabling the separation of plasmid-containing cells from completely cured cells. Consequently, reexcision events of suicide plasmids can be selected. In the case of suicide vectors that can be employed in *B. megaterium* in combination with highly efficient conjugal DNA transfer (Section 2.3), the generation of so-called clean deletions can be achieved with only two consecutive selection steps. Integration of the complete plasmid construct can be selected by antibiotic resistance mediated by the plasmid encoded resistance gene. Subsequent incubation with 5FU selects for the reexcision event. The system also allows direct gene conversion, that is, introduction of point mutations (Fabret *et al.*, 2002).

3. Systems Biology of *B. megaterium*

The systematic investigation of biotechnological processes using *B. megaterium* started in 2005, when first analyses of the intra- as well as of the extracellular proteome were conducted (Wang *et al.*, 2005, 2006a,b). At that time, proteins were primary analyzed via 2D gel electrophoresis. Obtained data gave a first indication about the changes in protein compositions of a *B. megaterium* strain related to the growth under different environmental conditions.

However, without possessing a reliable genome sequence further, whole scale high-throughput analyses toward a systems biology understanding were prohibited (Biedendieck *et al.*, 2010a).

3.1. Genome sequence of *B. megaterium*

The functional analysis of the whole genome of a microorganism of interest is called “genomics”. This technique is the basis for a variety of new techniques all ending with the suffix -omics as transcriptomics, proteomics, metabolomics, and fluxomics. In the case of *B. megaterium*, the genomes of two different strains were sequenced, of the plasmidless isolate DSM319 and of strain QM B1551 which harbors seven plasmids [GeneBank accession numbers CP001982 (DSM319), CP001983 (QM B1551), and CP001984-1990 (plasmids pBM100–pBM700)] (Eppinger *et al.*, 2011). Both genomes now enable the use of all -omics techniques for this microorganism.

Compared to *B. subtilis* (Barbe *et al.*, 2009), the circular bacterial chromosomes of both *B. megaterium* strains with 5.1 Mbp are 900 kbp larger. Nevertheless, the chromosomal gene composition is only partially reflected by the *B. subtilis* genome. Prominent unique features of *B. megaterium* genome exist comparing it to other Bacilli. For example, two vitamin B_{12} biosynthetic gene clusters are encoded in the genome sequences, one of them already biotechnologically utilized (Biedendieck *et al.*, 2010b; Eppinger *et al.*, 2011; Raux *et al.*, 1998). Upon oxygen deprivation, *B. megaterium* is able to generate energy via a mixed acid-type fermentation. Most likely, the organism tends to avoid this situation by motility through peritrichous flagella and gas vesicles. The analysis of the regulation of different genes under different conditions can now easily be analyzed by microarray experiments.

3.2. The database MegaBac

To facilitate comparative genome analysis of *B. megaterium*, a new database called MegaBac v9 (http://megabac.tu-bs.de) has been set up using genomic data from the two *B. megaterium* strains and of other Bacilli. Annotated and manually curated DNA sequences of all chromosomes and plasmids are available. MegaBac v9 is based on the recently developed OGeR system (Klein *et al.*, 2009). In addition to database function, MegaBac v9 offers powerful comparative tools. The intuitive query of genes is possible via the "Genes/Protein" form. Genes and proteins can be identified using a product name, gene symbol, locus tag, or EC number as input. Further, the results can be limited to the organism of interest. Display of the corresponding results can be done by choosing either the gene or the protein result page.

Additionally, MegaBac v9 provides BLAST functionality with a customizable e-value. Intergenic region analysis is also accessible by a direct connection to the Virtual Footprint tool, a central feature of the PRODORIC database (Grote *et al.*, 2009; Münch *et al.*, 2005). Finally, whole proteome analyses are possible based on the precalculated homolog predictions.

3.3. Transcriptomics

As a central technique for transcriptome analysis, DNA microarrays are usually used (Patterson *et al.*, 2006). This method provides a semiquantitative comparison of mRNA profiles for the detection of differentially expressed genes. Using DNA microarrays technology in one-color mode, whole time courses can be recorded measuring the RNA level of a cell at a given time point.

Making use of the final genome sequence of *B. megaterium* DSM319, Agilent microarrays were designed (Agilent Technologies, Boeblingen, Germany). For this purpose, the 8 × 15 K layout was chosen (eight arrays on one slide, each representing 15,000 probes in 60mers). This allows for a representation of up to three different probes per gene of the *B. megaterium* genome, which therefore cannot be treated as technical replicates. Although Agilent's eArray platform (https://earray.chem.agilent.com/earray) took care of the core probe design, some additional bioinformatical steps were necessary.

For submission of data from the *B. megaterium* DSM319 genome to eArray, following preprocessing steps were established:

1. A FASTA file containing all coding sequences of the genome was created with BMD locus-tags in the header.
2. A second FASTA file was provided containing the whole genome sequence in 1 Mbp pieces. This was necessary, as eArray accepts no sequences longer than 1 Mbp for calculation and avoidance of cross-hybridization. Next, 120 bp "patches" over the cut-sites of the 1 Mbp genome pieces were created to generate probes cross-hybridizing exact at these positions. All sequences were provided as reverse complement.

Since gene names and descriptions are usually completely missing in the Agilent grid files, they have to be added for each of the 15,000 probes.

3.3.1. Materials

Killing buffer: 2.42 g/l Tris (pH 7.5), 476 mg/l $MgCl_2$, solve in RNase-free H_2O, autoclave twice, add 5 ml of a RNase-free 2 M NaN_3 stock solution, make aliquots of 25 ml and store at −20 °C

Lysis buffer: 452.64 g/l guanidine thiocyanate, 8.32 ml of 3 M sodium acetate (pH 5.2), 50 ml 10% (w/v) *N*-lauroylsarcosinate, solve in RNase-free H_2O

Acid-phenol solution: aqua-phenol:chloroform:isoamylalcohol 50:48:2

RNA-storage buffer: 2.83 g/l Na_2PO_4 (pH 6.5), 292 mg/l EDTA (pH 8), autoclave twice

10× DNase buffer: 16.41 g/l sodium acetate (pH 4.5), 9.52 g/l $MgCl_2$, 5.84 g/l NaCl, autoclave twice

3.3.2. RNA preparation for microarray analysis of *B. megaterium*

While working with RNA, it is very important to wear gloves all the time to protect the RNA from RNases. All solutions and plastic ware should be autoclaved twice to destroy all RNases. RNase away spray further removes RNases and DNA contaminations (Molecular BioProducts, San Diego, CA, USA).

1. Streak a *B. megaterium* plasmid-containing strain onto a LB medium agar plate containing the required antibiotics and incubate for 24 h at 30 °C.
2. Inoculate a 50-ml liquid starter culture in a 300-ml shaking flask with cell material from the freshly prepared plate and incubate for 15 h at 100 rpm and 37 °C.
3. Inoculate a 100-ml liquid culture in a 500-ml shaking flask with 3×10^9 cells of the starter culture and incubate at 250 rpm and 37 °C.
4. Induce the recombinant gene expression with a final concentration of 0.5% (w/v) of xylose when the culture reaches an OD_{578nm} of 0.3–0.4.

Preparation for cell harvesting and disruption: Provide the acid-phenol solution (Section 3.3.1), defrost killing buffer, and get liquid nitrogen. Further, prepare 2 ml safe-lock reaction tubes with 600 mg of glass beads (diameter of 38–45 μm), lysis buffer, new safe-lock reaction tubes, and RNase-free plastic ware.

5. Thoroughly mix 2.5×10^{10} cells with 25 ml killing buffer. From now on working at 4 °C is necessary.
6. Centrifuge for 3 min at 2600×*g*.
7. Discard the supernatant. Use 1 ml of the supernatant to resuspend the precipitate and transfer the solution into a 2-ml reaction tube.
8. Centrifuge for 1 min at 15,000×*g*. Remove the supernatant and transfer the reaction tube containing the precipitate into liquid nitrogen.
9. Add 600 mg of glass beads (diameter of 38–45 μm) and 1 ml of lysis buffer and disrupt the cells in a FastPrep24 (MP Biomedicals, Solon, OH, USA) for two times 60 s at 6.5 m/s.
10. Centrifuge for 1 min at 15,000×*g*, transfer the supernatant into a 2 ml safe-lock reaction tube containing 1 ml of acid-phenol solution, mix thoroughly, and freeze in liquid nitrogen. Storage at −80 °C is possible.

Preparation for the RNA purification: Provide RNase-free plastic ware, 2 ml reaction tubes containing 1 ml chloroform:isoamylalcohol 48:2, reaction tubes containing 70 μl of 3 M sodium acetate (pH 5.2), and isopropanol.

11. Defrost disrupted cells in acid-phenol solution-mix carefully and open reaction tube continuously. Centrifuge for 5 min at room temperature at 15,000×*g*.
12. Transfer 800 μl of supernatant into a new reaction tube containing 1 ml chloroform:isoamylalcohol 48:2 and mix thoroughly. Centrifuge for 5 min at room temperature at 15,000×*g*.
13. Transfer 700 μl of supernatant into a new reaction tube containing 70 μl of 3 M sodium acetate (pH 5.2), mix thoroughly, add 1 ml of isopropanol, mix thoroughly, and store for at least 1 h at −80 °C. During this time, prepare DNase buffer, RNA-storage buffer, RNase-free plastic

ware, and RNA purification kit innuPrep RNA Mini Kit (Analytik Jena, Jena, Germany), and dilute DNase (RNase-free) 1:5 in RNA-storage buffer.

14. Centrifuge the sample for 5 min at room temperature at 15,000×*g*, completely discard the supernatant and dry precipitated RNA for 5 min at room temperature.
15. Resuspend RNA in 180 μl of RNA-storage buffer (10 min at 50 °C), add 20 μl of 10× DNase buffer and 1 μl of 1:5 diluted DNase.
16. After incubation for 30 min at 37 °C, mix 200 μl of pure RNA solution with 400 μl of Lysis solution RL (innuPrep RNA Mini Kit, Analytik Jena) and 600 μl of 70% (v/v) EtOH. Continue protocol "RNA extraction from bacterial cells" according to the manufacturer's instruction at step 6. Final elution volume is 30 μl.
17. The RNA is ready for performing microarrays according to the manufacturer's instruction.

3.4. Proteomics

First proteome analyses with *B. megaterium* were published before. Wang *et al.* describe the intra- as well as extracellular proteome of a recombinant *B. megaterium* strain (Wang *et al.*, 2005, 2006a,b).

For sample preparation for proteome analysis, see Sections 2.1.2 and 2.1.3.

3.5. Metabolomics

Metabolomic techniques are innovative tools which have been successfully applied to analyze genetic and environmental perturbations and to define phenotypes of organisms. They are increasingly used in gene function annotation and systems biology (Morgenthal *et al.*, 2005; Weckwerth, 2003). For such approaches with *B. megaterium*, a strategy of Krall *et al.* was modified (Krall *et al.*, 2009; Fig. 10.4).

3.5.1. Materials

Wash buffer: 5.84 g/l NaCl

Extraction solution: 2.5:1:0.5 (v/v/v) methanol:chloroform:water solution supplemented with 100 ng/ml of U-^{13}C-sorbitol

40 mg/ml methoxyamine hydrochloride in pyridine

Derivatization solution: *N*-methyl-*N*-treimethylsilyl trifluoro acetamid

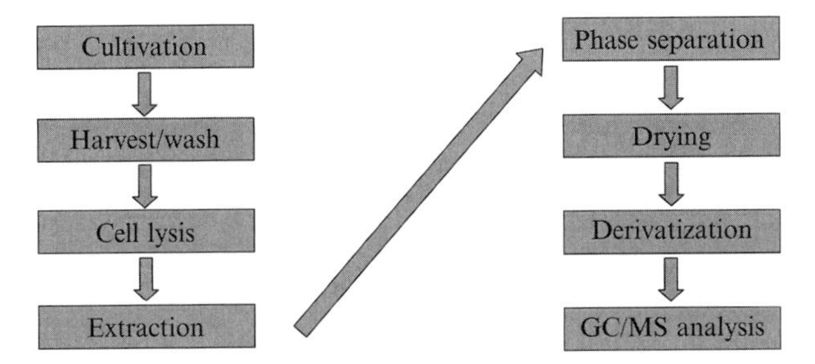

Figure 10.4 Metabolome analysis of *B. megaterium* via GC/MS. The process of sample handling to analyze metabolites of *B. megaterium* is given.

3.5.2. Metabolome analysis via GC/MS of *B. megaterium*

1. For analyzing the metabolome of *B. megaterium*, grow the cells as described in Section 2.1.2.
2. Samples are taken by fast filtering with a vacuum filtration manifold (Millipore, Schwalbach/Ts., Germany). Take 2 ml of cell culture and filter it through a 2.5-cm diameter filter with a pore size of 0.65 μm. Wash cells with 6 ml of wash buffer.
3. Transfer the filter into a 2-ml reaction tube and freeze it in liquid nitrogen.
4. Extract the samples with 600 μl of a cold extraction solution supplemented with 100 ng/ml of U-^{13}C-sorbitol as internal standard in the presence of glass beads (Precellys, France) by vigorously shaking for 1 min and for additional 2 min in a FastPrep24 (MP Biomedicals) for 60 s at 6.5 m/s.
5. Centrifuge at 4 °C for 2 min and 14,000×*g*.
6. Concentrate 500 μl of the upper polar phase to dryness using a vacuum vaporizer.
7. Use a mixture of *n*-alkanes as internal retention index (RI) markers to calculate individual RIs for a clear identification of each target compound.
8. Add 5 μl of 40 mg/ml methoxyamine hydrochloride in pyridine to the samples and shake for 90 min at 30 °C. Then derivatize the samples by trimethylsilylation of acidic protons by addition of 45 μl derivatization solution and incubate for 30 min at 37 °C.
9. The samples are ready for GC-TOF-MS data mining. For this purpose, the raw data were analyzed with the ChromaTof software from LECO (http://www.leco.com/index2.htm), which performed the deconvolution of all mass spectra data, built mass-spectral correction for coeluting metabolites, calculated the RIs, and identified a suitable fragment mass-to-charge ratio for selective quantification. The obtained data were

analyzed by defining a reference chromatogram with the maximum number of detected peaks over a signal/noise threshold of 50. All chromatograms were matched against this reference chromatogram with a minimum match factor of 800. Compounds were annotated by RI and mass spectra comparison to a user-defined spectra library. Selected fragment ions specific for each individual metabolite were used for peak area quantification. Each compound was normalized by the peak area from the internal standard and by the fresh weight of each sample.

3.6. Fluxomics

Fluxomics aims at the quantification of small molecule fluxes through metabolic networks and provides access to the *in vivo* activity of reactions and pathways in intact living cells. In this regard, fluxomics integrates the different cellular components, that is, proteins, transcripts, and metabolites into biological network function and is thus in the core of metabolic engineering (Stephanopoulos, 1999) and systems biology (Kohlstedt *et al.*, 2010). Most of the comprehensive approaches for metabolic flux studies today involve isotopic tracer studies and mass spectrometry for measurement of the labeling pattern of metabolites which are then utilized, together with directly measured fluxes from the cultivation, to estimate the distribution of fluxes (Wittmann, 2007). Recently, these approaches have been adapted and applied to investigation of recombinant protein producing *B. megaterium* (Fürch *et al.*, 2007a,b) to unravel key properties of the underlying metabolic networks upon genetic or environmental perturbation and, in combination with genome-scale network models predict pathway bottle necks for strain optimization.

3.6.1. Materials

99% [1-^{13}C] glucose
NaCl solution: 0.9 g NaOH in 100 ml water
HCl: 6 M HCl
Dimethylformamide (DMF) with 0.1% pyridine
N-Methyl-*N*-*t*-butyl-dimethylsilyltrifluoracetamide (MBDSTFA)

3.6.2. Fluxome analysis of *B. megaterium*

1. Grow cells as described in Section 2.1.2, whereby the natural glucose is replaced by 99% of [1-^{13}C] glucose. To minimize the disturbance by the nonlabeled inoculum on subsequent labeling analysis, the starting cell concentration should be below 1% of the corresponding value at harvesting. Verify that the cells grow in metabolic steady state

(e.g., constant specific growth rate and yields) by online monitoring of concentrations of cells, substrates, and products. Harvest the cells by centrifugation (5 min, 9300×*g*, 4 °C).

2. Hydrolyze the cell protein after careful washing to completely remove any complex medium constituents by centrifugation (5 min, 9300×*g*, 4 °C) and resuspend the precipitated cells in 10 ml of deionized water.
3. Wash again in same conditions.
4. Harvest the washed biomass pellet by centrifugation (5 min, 9300×*g*, 4 °C) and mix with 50 μl of 6 M HCl on a vortex. Incubate the mixture at 105 °C for 24 h for hydrolysis of the cell protein.
5. Evaporate HCl under an N_2 stream. Centrifuge the mixture (5 min, 9300×*g*, 4 °C) through a centrifugal filter device (0.22 μm) to remove insoluble components.
6. Transfer the clear filtrate into a 1-ml glass vial and freeze at −70 °C. Freeze-dry the frozen filtrate.
7. For GC/MS analysis the amino acids are converted into *t*-butyl-tri-methylsilyl (TBDMS) derivates. Add 50 μl DMF containing 0.1% pyridine and 50 μl MBDSTFA on the top of the solid sample.
8. Incubate the mixture at 80 °C for 60 min and transfer the derivatized sample into a 100-μl GC-micro insert.
9. The labeling analysis is carried out with 0.1 μl sample injection on a GC with a HP5MS-capillary column, electron impact ionization at 70 eV, and quadrupole detector. The approach is based on a protocol for the ^{13}C labeling measurement of TBDMS-derivatized amino acids in cell extracts. In contrast to previous work, the temperature gradient was modified to reduce the analysis time. The temperature gradient is 120 °C for 1 min, 10 °C/min up to 310 °C, and 310 °C for 1 min. Further operation temperatures are 320 °C (inlet), 320 °C (interface), and 325 °C (quadrupole). Carrier gas flow is helium at 0.7 ml/min. Measurement is carried out in SIM mode with corresponding masses given elsewhere.
10. Use an isotopomer model, reflecting the intracellular reactions to be studied to compute the fluxes from the ^{13}C labeling patterns, measured fluxes in substrate uptake and product formation as well as on anabolic precursors demand (Wittmann, 2007). Different, user friendly software tools such as OpenFLUX (Quek *et al.*, 2009) or FiatFlux (Zamboni *et al.*, 2005) can be used for this purpose.

ACKNOWLEDGEMENTS

This work was financially supported by "Deutsche Forschungsgemeinschaft (SFB578)." The work was supported by the Federal Ministry of Education and Research (BMBF, Bonn-Bad Godesberg, Germany) grant no. 0315283.

REFERENCES

Atlagic, D., Kilic, A. O., and Tao, L. (2006). Unmarked gene deletion mutagenesis of *gtfB* and *gtfC* in *Streptococcus mutans* using a targeted hit-and-run strategy with a thermosensitive plasmid. *Oral Microbiol. Immunol.* **21,** 132–135.

Aquino de Muro, M., and Priest, F. G. (2000). Construction of chromosomal integrants of *Bacillus sphaericus* 2362 by conjugation with *Escherichia coli*. *Res. Microbiol.* **151,** 547–555.

Barbe, V., Cruveiller, S., Kunst, F., Lenoble, P., Meurice, G., Sekowska, A., Vallenet, D., Wang, T., Moszer, I., Medigue, C., and Danchin, A. (2009). From a consortium sequence to a unified sequence: The *Bacillus subtilis* 168 reference genome a decade later. *Microbiology* **155,** 1758–1775.

Biedendieck, R., Beine, R., Gamer, M., Jordan, E., Buchholz, K., Seibel, J., Dijkhuizen, L., Malten, M., and Jahn, D. (2007a). Export, purification, and activities of affinity tagged *Lactobacillus reuteri* levansucrase produced by *Bacillus megaterium*. *Appl. Microbiol. Biotechnol.* **74,** 1062–1073.

Biedendieck, R., Gamer, M., Jaensch, L., Meyer, S., Rohde, M., Deckwer, W. D., and Jahn, D. (2007b). A sucrose-inducible promoter system for the intra- and extracellular protein production in *Bacillus megaterium*. *J. Biotechnol.* **132,** 426–430.

Biedendieck, R., Yang, Y., Deckwer, W. D., Malten, M., and Jahn, D. (2007c). Plasmid system for the intracellular production and purification of affinity-tagged proteins in *Bacillus megaterium*. *Biotechnol. Bioeng.* **96,** 525–537.

Biedendieck, R., Bunk, B., Fürch, T., Franco-Lara, E., Jahn, M., and Jahn, D. (2010a). Systems biology of recombinant protein production in *Bacillus megaterium*. *Adv. Biochem. Eng. Biotechnol.* **120,** 133–161.

Biedendieck, R., Malten, M., Barg, H., Bunk, B., Martens, J. H., Derry, E., Leech, H., Warren, M. J., and Jahn, D. (2010b). Metabolic engineering of cobalamin (vitamin B_{12}) production in *Bacillus megaterium*. *Microb. Biotechnol.* **3,** 24–37.

Borgmeier, C., and Meinhardt, F. (unpublished).

Borgmeier, C., Hoffmann, K., and Meinhardt, F. (unpublished).

Bunk, B., Biedendieck, R., Jahn, D., and Vary, P. S. (2010). Industrial production by *Bacillus megaterium* and other Bacilli. *In* "Encyclopedia of Industrial Biotechnology: Bioprocess, Bioseparation, and Cell Technology," (M. C. Flickinger, ed.), pp. 1–15. John Wiley & Sons, Hoboken (New Jersey), USA.

Dahl, M. K., Degenkolb, J., and Hillen, W. (1994). Transcription of the xyl operon is controlled in *Bacillus subtilis* by tandem overlapping operators spaced by four base-pairs. *J. Mol. Biol.* **243,** 413–424.

David, F., Westphal, R., Bunk, B., Jahn, D., and Franco-Lara, E. (2010). Optimization of antibody fragment production in *Bacillus megaterium*: The role of metal ions on protein secretion. *J. Biotechnol.* **150,** 115–124.

Eppinger, M., Bunk, B., Johns, M. A., Edirisinghe, J. N., Kutumbaka, K. K., Koenig, S. S. K., Creasy, H. H., Rosovitz, M. J., Riley, D. R., Daugherty, S., Martin, M., Elbourne, L. D. H., *et al.* (2011). Genome sequences of the biotechnologically important *B. megaterium* strains QM B1551 and DSM319. *J. Bacteriol.,* in press.

Fabret, C., Ehrlich, S. D., and Noirot, P. (2002). A new mutation delivery system for genome-scale approaches in *Bacillus subtilis*. *Mol. Microbiol.* **46,** 25–36.

Fürch, T., Hollmann, R., Wittmann, C., Wang, W., and Deckwer, W. D. (2007a). Comparative study on central metabolic fluxes of *Bacillus megaterium* strains in continuous culture using ^{13}C labelled substrates. *Bioprocess Biosyst. Eng.* **30,** 47–59.

Fürch, T., Wittmann, C., Wang, W., Franco-Lara, E., Jahn, D., and Deckwer, W.-D. (2007b). Effect of different carbon sources on central metabolic fluxes and the recombinant production of a hydrolase from *Thermobifida fusca* in *Bacillus megaterium*. *J. Biotechnol.* **132,** 385–394.

Gamer, M., Fröde, D., Biedendieck, R., Stammen, S., and Jahn, D. (2009). A T7 RNA polymerase-dependent gene expression system for *Bacillus megaterium*. *Appl. Microbiol. Biotechnol.* **82,** 1195–1203.

Gärtner, D., Geissendorfer, M., and Hillen, W. (1988). Expression of the *Bacillus subtilis xyl* operon is repressed at the level of transcription and is induced by xylose. *J. Bacteriol.* **170,** 3102–3109.

Gay, P., Le Coq, D., Steinmetz, M., Berkelman, T., and Kado, C. I. (1985). Positive selection procedure for entrapment of insertion sequence elements in Gram-negative bacteria. *J. Bacteriol.* **164,** 918–921.

Grote, A., Klein, J., Retter, I., Haddad, I., Behling, S., Bunk, B., Biegler, I., Yarmolinetz, S., Jahn, D., and Münch, R. (2009). PRODORIC (release 2009): A database and tool platform for the analysis of gene regulation in prokaryotes. *Nucleic Acids Res.* **37,** D61–D65.

Hoffmann, K., Wollherr, A., Larsen, M., Rachinger, M., Liesegang, H., Ehrenreich, A., and Meinhardt, F. (2010). Facilitation of direct conditional knockout of essential genes in *Bacillus licheniformis* DSM13 by comparative genetic analysis and manipulation of genetic competence. *Appl. Environ. Microbiol.* **76,** 5046–5057.

Hollmann, R., and Deckwer, W. D. (2004). Pyruvate formation and suppression in recombinant *Bacillus megaterium* cultivation. *J. Biotechnol.* **111,** 89–96.

Jordan, E., Hust, M., Roth, A., Biedendieck, R., Schirrmann, T., Jahn, D., and Dübel, S. (2007). Production of recombinant antibody fragments in *Bacillus megaterium*. *Microb. Cell Fact.* **6,** 2.

Klein, J., Munch, R., Biegler, I., Haddad, I., Retter, I., and Jahn, D. (2009). Strepto-DB, a database for comparative genomics of group A (GAS) and B (GBS) streptococci, implemented with the novel database platform 'Open Genome Resource' (OGeR). *Nucleic Acids Res.* **37,** D494–D498.

Kohlstedt, M., Becker, J., and Wittmann, C. (2010). Metabolic fluxes and beyond-systems biology understanding and engineering of microbial metabolism. *Appl. Microbiol. Biotechnol.* **88,** 1065–1075.

Krall, L., Huege, J., Catchpole, G., Steinhauser, D., and Willmitzer, L. (2009). Assessment of sampling strategies for gas chromatography-mass spectrometry (GC-MS) based metabolomics of cyanobacteria. *J. Chromatogr. B Analyt. Technol. Biomed. Life Sci.* **877,** 2952–2960.

Lee, J. S., Wittchen, K. D., Stahl, C., Strey, J., and Meinhardt, F. (2001). Cloning, expression and carbon catabolite repression of the *bamM* gene encoding beta-amylase of *Bacillus megaterium* DSM319. *Appl. Microbiol. Biotechnol.* **56,** 205–211.

Malten, M., Hollmann, R., Deckwer, W. D., and Jahn, D. (2005a). Production and secretion of recombinant *Leuconostoc mesenteroides* dextransucrase DsrS in *Bacillus megaterium*. *Biotechnol. Bioeng.* **89,** 206–218.

Malten, M., Nahrstedt, H., Meinhardt, F., and Jahn, D. (2005b). Coexpression of the type I signal peptidase gene *sipM* increases recombinant protein production and export in *Bacillus megaterium* MS941. *Biotechnol. Bioeng.* **91,** 616–621.

Malten, M., Biedendieck, R., Gamer, M., Drews, A. C., Stammen, S., Buchholz, K., Dijkhuizen, L., and Jahn, D. (2006). A *Bacillus megaterium* plasmid system for the production, export, and one-step purification of affinity-tagged heterologous levansucrase from growth medium. *Appl. Environ. Microbiol.* **72,** 1677–1679.

Morgenthal, K., Wienkoop, S., Scholz, M., Selbig, J., and Weckwerth, W. (2005). Correlative GC-TOF-MS based metabolite profiling and LC-MS based protein profiling reveal time-related systemic regulation of metabolite-protein networks and improve pattern recognition for multiple biomarker selection. *Metabolomics* **1,** 109–121.

Münch, R., Hiller, K., Grote, A., Scheer, M., Klein, J., Schobert, M., and Jahn, D. (2005). Virtual Footprint and PRODORIC: An integrative framework for regulon prediction in prokaryotes. *Bioinformatics* **21,** 4187–4189.

Nahrstedt, H., and Meinhardt, F. (2004). Structural and functional characterization of the *Bacillus megaterium uvrBA* locus and generation of UV-sensitive mutants. *Appl. Microbiol. Biotechnol.* **65,** 193–199.

Nahrstedt, H., Schröder, C., and Meinhardt, F. (2005). Evidence for two *recA* genes mediating DNA repair in *Bacillus megaterium*. *Microbiology* **151,** 775–787.

Nahrstedt, H., Wittchen, K., Rachman, M. A., and Meinhardt, F. (2004). Identification and functional characterization of a type I signal peptidase gene of *Bacillus megaterium* DSM319. *Appl. Microbiol. Biotechnol.* **64,** 243–249.

Neuhard, J. (1983). Utilization of preformed pyrimidine bases and nucleosides. *In* "Metabolism of Nucleotides, Nucleosides and Nucleobases in Microorganisms," (A. Munch-Petersen, ed.), pp. 95–148. Academic Press, New York.

Nygaard, P. (1993). Purine and pyrimidine salvage pathways. *In* "*Bacillus subtilis* and Other Gram-Positive Bacteria," (A. L. Sonenshein, J. A. Hoch, and R. Losick, eds.), pp. 359–378. American Society for Microbiology, Washington, DC.

Patterson, T. A., Lobenhofer, E. K., Fulmer-Smentek, S. B., Collins, P. J., Chu, T. M., Bao, W., Fang, H., Kawasaki, E. S., Hager, J., Tikhonova, I. R., Walker, S. J., Zhang, L., *et al.* (2006). Performance comparison of one-color and two-color platforms within the MicroArray Quality Control (MAQC) project. *Nat. Biotechnol.* **24,** 1140–1150.

Quek, L. E., Wittmann, C., Nielsen, L. K., and Kromer, J. O. (2009). OpenFLUX: Efficient modelling software for ^{13}C-based metabolic flux analysis. *Microb. Cell Fact.* **8,** 25.

Raux, E., Lanois, A., Warren, M. J., Rambach, A., and Thermes, C. (1998). Cobalamin (vitamin B_{12}) biosynthesis: Identification and characterization of a *Bacillus megaterium cobI* operon. *Biochem. J.* **335,** 159–166.

Richhardt, J., Larsen, M., and Meinhardt, F. (2010). An improved transconjugation protocol for *Bacillus megaterium* facilitating a direct genetic knockout. *Appl. Microbiol. Biotechnol.* **86,** 1959–1965.

Richhardt, J., and Meinhardt, F. (unpublished).

Rygus, T., and Hillen, W. (1991). Inducible high-level expression of heterologous genes in *Bacillus megaterium* using the regulatory elements of the xylose-utilization operon. *Appl. Microbiol. Biotechnol.* **35,** 594–599.

Rygus, T., and Hillen, W. (1992). Catabolite repression of the *xyl* operon in *Bacillus megaterium*. *J. Bacteriol.* **174,** 3049–3055.

Rygus, T., Scheler, A., Allmansberger, R., and Hillen, W. (1991). Molecular cloning, structure, promoters and regulatory elements for transcription of the *Bacillus megaterium* encoded regulon for xylose utilization. *Arch. Microbiol.* **155,** 535–542.

Schaeffer, P., Millet, J., and Aubert, J. P. (1965). Catabolic repression of bacterial sporulation. *Proc. Natl. Acad. Sci. USA* **54,** 704–711.

Schmidt, S., Wolf, N., Strey, J., Nahrstedt, H., Meinhardt, F., and Waldeck, J. (2005). Test systems to study transcriptional regulation and promoter activity in *Bacillus megaterium*. *Appl. Microbiol. Biotechnol.* **68,** 647–655.

Stahl, U., and Esser, K. (1983). Plasmid heterogeneity in various strains of *Bacillus megaterium*. *Eur. J. Appl. Biotechnol.* **17,** 248–251.

Stammen, S., Müller, B. K., Korneli, C., Biedendieck, R., Gamer, M., Franco-Lara, E., and Jahn, D. (2010a). High-yield intra- and extracellular protein production using *Bacillus megaterium*. *Appl. Environ. Microbiol.* **76,** 4037–4046.

Stammen, S., Schuller, F., Dietrich, S., Gamer, M., Biedendieck, R., and Jahn, D. (2010b). Application of *Escherichia coli* phage K1E DNA-dependent RNA polymerase for *in vitro* RNA synthesis and *in vivo* protein production in *Bacillus megaterium*. *Appl. Microbiol. Biotechnol.* **88,** 529–539.

Stephanopoulos, G. (1999). Metabolic fluxes and metabolic engineering. *Metab. Eng.* **1,** 1–11.

Terpe, K. (2006). Overview of bacterial expression systems for heterologous protein production: From molecular and biochemical fundamentals to commercial systems. *Appl. Microbiol. Biotechnol.* **72,** 211–222.

Vary, P. S. (1992). Development of genetic engineering in *Bacillus megaterium*. *In* "Biology of Bacilli, Applications to Industry," (R. H. Doi and M. McGloughlin, eds.), pp. 251–310. Butterworth-Heinemann, Boston.

Vary, P. S. (1994). Prime time for *Bacillus megaterium*. *Microbiology* **140,** 1001–1013.

Vary, P. S., Biedendieck, R., Fürch, T., Meinhardt, F., Rohde, M., Deckwer, W. D., and Jahn, D. (2007). *Bacillus megaterium*—From simple soil bacterium to industrial protein production host. *Appl. Microbiol. Biotechnol.* **76,** 957–967.

von Tersch, M. A., and Carlton, B. C. (1983). Megacinogenic plasmids of *Bacillus megaterium*. *J. Bacteriol.* **155,** 872–877.

von Tersch, M. A., and Robbins, H. L. (1990). Efficient cloning in *Bacillus megaterium*: Comparison to *Bacillus subtilis* and *Escherichia coli* cloning hosts. *FEMS Microbiol. Lett.* **70,** 305–310.

Wang, W., Hollmann, R., Fürch, T., Nimtz, M., Malten, M., Jahn, D., and Deckwer, W. D. (2005). Proteome analysis of a recombinant *Bacillus megaterium* strain during heterologous production of a glucosyltransferase. *Proteome Sci.* **3,** 4.

Wang, W., Hollmann, R., and Deckwer, W. D. (2006a). Comparative proteomic analysis of high cell density cultivations with two recombinant *Bacillus megaterium* strains for the production of a heterologous dextransucrase. *Proteome Sci.* **4,** 19.

Wang, W., Sun, J., Hollmann, R., Zeng, A.-P., and Deckwer, W. D. (2006b). Proteomic characterization of transient expression and secretion of a stress-related metalloprotease in high cell density culture of *Bacillus megaterium*. *J. Biotechnol.* **126,** 313–324.

Weckwerth, W. (2003). Metabolomics in systems biology. *Annu. Rev. Plant Biol.* **54,** 669–689.

Wittchen, K. D., Lee, J. S., Stahl, C., Strey, J., and Meinhardt, F. (unpublished).

Wittchen, K. D., and Meinhardt, F. (1995). Inactivation of the major extracellular protease from *Bacillus megaterium* DSM319 by gene replacement. *Appl. Microbiol. Biotechnol.* **42,** 871–877.

Wittchen, K. D., Strey, J., and Bültmann, A. (1998). Molecular characterization of the operon comprising the *spoIV* gene of *Bacillus megaterium* DSM319 and generation of a deletion mutant. *J. Gen. Appl. Microbiol.* **44,** 317–326.

Wittmann, C. (2007). Fluxome analysis using GC-MS. *Microb. Cell Fact.* **6,** 6.

Yang, Y., Biedendieck, R., Wang, W., Gamer, M., Malten, M., Jahn, D., and Deckwer, W. D. (2006). High yield recombinant penicillin G amidase production and export into the growth medium using *Bacillus megaterium*. *Microb. Cell Fact.* **5,** 36.

Zamboni, N., Fischer, E., and Sauer, U. (2005). FiatFlux—A software for metabolic flux analysis from ^{13}C-glucose experiments. *BMC Bioinformatics* **6,** 209.

CHAPTER ELEVEN

Protein Production in *Saccharomyces cerevisiae* for Systems Biology Studies

Naglis Malys,[*,†] Jill A. Wishart,[*,‡] Stephen G. Oliver,[*,‡,1] *and* John E. G. McCarthy[*,†,§,2]

Contents

Abstract

Proteins together with metabolites, nucleic acids, lipids, and other intracellular molecules form biological systems that involve networks of functional and physical interactions. To understand these interactions and the many other characteristics of proteins in the context of biochemical networks and systems

* Manchester Centre for Integrative Systems Biology, The University of Manchester, Manchester, United Kingdom
† Faculty of Life Sciences, Manchester Interdisciplinary Biocentre, The University of Manchester, Manchester, United Kingdom
‡ Faculty of Life Sciences, Michael Smith Building, The University of Manchester, Manchester, United Kingdom
§ School of Chemical and Analytical Engineering, The University of Manchester, Manchester, United Kingdom
1 Present address: Cambridge Systems Biology Centre, Department of Biochemistry, Sanger Building, 80 Tennis Court Road, Cambridge CB2 1GA, United Kingdom
2 Present address: School of Life Sciences, Gibbet Hill Campus, University of Warwick, Coventry, United Kingdom

Methods in Enzymology, Volume 500
ISSN 0076-6879, DOI: 10.1016/B978-0-12-385118-5.00011-6

biology, research aimed at studying medium and large sets of proteins is required. This either involves an investigation focused on individual protein activities in the mixture (e.g., cell extracts) or a protein characterization in the isolated form. This chapter provides an overview on the currently available resources and strategies for isolation of proteins from *Saccharomyces cerevisiae*. The use of standardized gene expression systems is discussed, and protein production protocols applied to the data generation pipeline for systems biology are described in detail.

1. Introduction

The necessity for producing comprehensive collections of active proteins is becoming critically important as systematic studies of biological processes and attempts at reconstituting various biological systems are under way. This requires accelerated speed and high efficiency from the protein production pipeline.

In the past few decades, overexpression in *Escherichia coli* has established itself as a mainstream strategy for efficient protein production (Gold, 1990; Junge *et al.*, 2008; Makrides, 1996; Tabor and Richardson, 1985; Zerbs *et al.*, 2009). Yeast, baculovirus-insect, mammalian, and other bacterial or cell-free expression systems have also been studied and used effectively (Cregg *et al.*, 2009; Jarvis, 2009; Junge *et al.*, 2008; Katzen *et al.*, 2005; Malys and McCarthy, 2011; Zerbs *et al.*, 2009). However, the most fundamentally studied eukaryote—*Saccharomyces cerevisiae*—has attracted less attention as a possible tool for protein production. This is due to the following problems: (1) difficulties in disrupting the thick cell wall to successfully extract intracellular proteins; (2) absence of very strong transcriptional and translational systems, which could offer very high protein synthesis levels similar to those available in *E. coli*.

Despite the shortcomings, in the past decade, *S. cerevisiae* has become a reliable source for protein production and its advantages (such as good genetic characterization, ease, and safety of the organism to work with, relatively inexpensive protein production process, tightly regulated and efficient expression systems, appropriate protein folding, and some level of posttranslational modification) have been better appreciated. This chapter reviews commonly used protein production systems for yeast and describes in more detail the comprehensive gene-expression libraries that can be used for the production of proteins in a research environment focused on systems biology. Specific procedures for cell preparation, gene expression, and protein purification and analysis form a major part of this chapter.

2. Comparison of Commonly Used Expression Systems

Choosing an appropriate host for expressing proteins of interest is a decisive step for any biological research project. Selection of an inadequate expression system can result in several undesirable consequences for the protein: poorly expressed, misfolded, or lacking natural posttranslational modifications. A comparison of commonly used expression systems highlighting their positive and negative characteristics is shown in Table 11.1.

For studies that require characterization of whole biological systems and, hence, a relatively large number of proteins, other important factors to consider are speed, reproducibility, and the ability to express and purify all required proteins using a single (or a few very similar) expression system(s). The intended use of the produced proteins is also important in decision making, as different quantities and qualities of the protein sample might be required for *in vitro* activity assays or structural studies, for example.

3. Comprehensive Libraries for Protein Production in *S. cerevisiae*

For protein production in *S. cerevisiae*, a number of comprehensive gene-overexpression libraries have been created by inserting yeast ORFs into vectors in which gene expression is usually controlled by the strong galactose-inducible *GAL1* promoter. Of the currently available libraries, two collections stand out as the most complete and suitable for yeast protein overproduction (Boone *et al.*, 2007).

First, the Yeast ORF Collection has been developed by collaborative work between Eric Phizicky's and Mike Snyder's groups (Gelperin *et al.*, 2005). In this collection, yeast ORFs are placed under the control of the inducible *GAL1* promoter and cloned into individual plasmid vectors designed for tagged protein overexpression. A C-terminal 19-kDa tandem fusion tag is formed of 6xHis and hemagglutination (HA) protein domains, a 3 °C protease cleavage site, and two IgG-binding domains from *Staphylococcus aureus* protein A. This tag configuration enables robust affinity purification, sensitive immunodetection, and partial tag removal (i.e., protein A). Over 4900 *S. cerevisiae* genes can be expressed and proteins purified by using this collection (Gelperin *et al.*, 2005). The Yeast ORF Collection has been used successfully for the identification and characterization of enzymatic activities (Jackman *et al.*, 2007, 2008; King *et al.*, 2009) and protein interaction studies (Kung and Snyder, 2006; Li *et al.*, 2008).

Table 11.1 Characteristics of commonly used expression systems for protein production

Characteristics	Expression system					
	Bacteria	Yeast	Insect	Mammalian	Plant	Cell free
Expression level	Moderate–high Usually several milligrams of protein per liter of culture, gram per liter can be achieved in some instances	Moderate–high Occasionally, up to several milligrams of protein per liter of culture	Low–high Rarely, up to several milligrams of protein per liter of culture	Low–moderate Occasionally, up to few milligrams of protein per liter of culture	Low Several micrograms of protein per gram of cell tissue	Low–high Broad range, depending on cell type used for extract preparation
Time efficiency	High	High	Low	Low	Low	Very high
Cell growth	20–30 min/division	80–120 min/division	18–24 h/division	24 h/division	18–24 h/division	N/A
Time required for protein production	3–4 days	5–7 days	2–4 weeks	3–8 weeks	3–8 weeks	less than 1 day
Cost of resources	Low	Low	High	High	Low	High
Process operation and scalability	Easy	Moderate	Difficult	Difficult	Difficult	Easy and expensive
Density of biomass	High	Moderate–high	Low	Low	Low–high	N/A
Recombinant protein quality	Low	Moderate	High	High	High	Low–moderate

Folding	Frequently improper	Rarely improper	Usually proper	Usually proper	Usually proper	Depends on extract source
Glycosylation	No	O-linked	O-linked	Yes	Yes	Depends on extract source
Phosphorylation, acetylation, and acylation	No	Yes	Yes	Yes	Yes	Depends on extract source
Gamma-carboxylation	No	No	Yes	Yes	–	
Development and reliability level of expression system	Very high	High	Moderate	Moderate	Low	High

Information from the following sources: Brondyk, 2009; Junge *et al.*, 2008; Lindbo, 2007; Sabate *et al.*, 2010; Yin *et al.*, 2007; www.invitrogen.com.

The second major resource is the Yeast GST-Tagged Collection, which has been assembled by the groups of Charles Boone and Brenda Andrews (Sopko *et al.*, 2006) based on the previously developed overexpression plasmid library (Zhu *et al.*, 2001). It includes more than 5000 strains that allow the overexpression of more than 80% of *S. cerevisiae* proteins. In each plasmid, the yeast ORF is under the control of the inducible *GAL1/10* promoter and contains GST-6xHis tag at the N-terminus. This allows reliable protein overexpression under galactose induction, affinity purification (using glutathione sepharose), and immunodetection (by anti-GST antibodies). GST-tagged protein screens have been used in phenotype and pathway mapping (Sopko *et al.*, 2006), transcription factor functional analysis (Chua *et al.*, 2006), and many other applications. It has also been proposed as a metabolite screening tool for systems biology (Kell, 2004).

An additional comprehensive library, which expresses proteins at endogenous levels, is most useful for protein complex purification. The Yeast TAP-Tagged Collection was constructed by the O'Shea and Weissman groups (Ghaemmaghami *et al.*, 2003) and uses the tandem affinity purification (TAP) strategy (Puig *et al.*, 2001; Rigaut *et al.*, 1999). In each strain of this collection, a C-terminal-tagged ORF is under the control of the endogenous promoter at its original chromosomal location. The TAP tag includes a calmodulin-binding peptide, TEV cleavage site, and the IgG-binding domains of protein A (which, similarly to the Yeast ORF Collection, allows affinity purification, immunodetection, and partial tag removal). The Yeast TAP-Tagged Collection has been used for global quantitative analysis of the yeast proteome (Ghaemmaghami *et al.*, 2003). Over 4200 proteins were detected, and absolute protein levels in exponentially growing cells were quantified using Western blot analysis. C-terminal and N-terminal TAP tag strategies have also been used successfully for the identification, purification, and functional characterization of distinct protein complexes (Fatica *et al.*, 2002; Hernández *et al.*, 2006; Malys and McCarthy, 2006; Malys *et al.*, 2004; Panigrahi *et al.*, 2007) and in high-throughput functional analyses (Gavin *et al.*, 2006; Krogan *et al.*, 2006).

4. Protocols for Protein Expression and Purification from Tagged Collections of *S. cerevisiae*

In order to overexpress and purify proteins for systems biology studies, the Yeast ORF Collection (Gelperin *et al.*, 2005) can be used as the primary source. This section describes procedures that have been used to produce a comprehensive collection of enzymes for studying individual biochemical networks and generating data for mathematical modeling. To produce enzyme kinetics parameters, proteins need to be isolated in submicrogram quantities with their full enzymatic activity retained. For example, for a systems analysis of the glycolysis and trehalose biosynthesis pathways, 35 proteins were initially selected as core components of the networks. Thirty-two proteins were identified to be available for expression from the Yeast ORF Collection and, after initial expression and a first round of binding to IgG sepharose (Fig. 11.1), 26 proteins were expressed in quantities sufficient for application in systems biology studies. Table 11.2 lists the successfully purified enzymes and the quantities obtained from single-culture experiments.

In order to produce required proteins that are not available from the Yeast ORF Collection, the Yeast GST-Tagged Collection and Yeast TAP-Tagged Collection can be used as alternatives. The ability to coexpress protein subunits from the Yeast TAP-Tagged Collection under control of their native promoters and with natural stoichiometries allows the purification of enzyme complexes formed of different peptides (e.g., heterooctamer of Pfk1/Pfk2 for phosphofructokinase).

4.1. Cell growth and protein expression

All three *S. cerevisiae* strain collections (Yeast ORF, Yeast GST-Tagged, and Yeast TAP-Tagged) are available from Open Biosystems (www.openbiosystems.com). Normally, strains are stored at −80 °C in 96-well plates.

Yeast ORF and Yeast GST-Tagged libraries include expression vectors pBG1805 and pEGH. Both contain the *URA3* gene, which is required by *S. cerevisiae* strains Y258 (*MAT***a**, *pep4-3, his4-580, ura3-52, leu2-3, 112*)

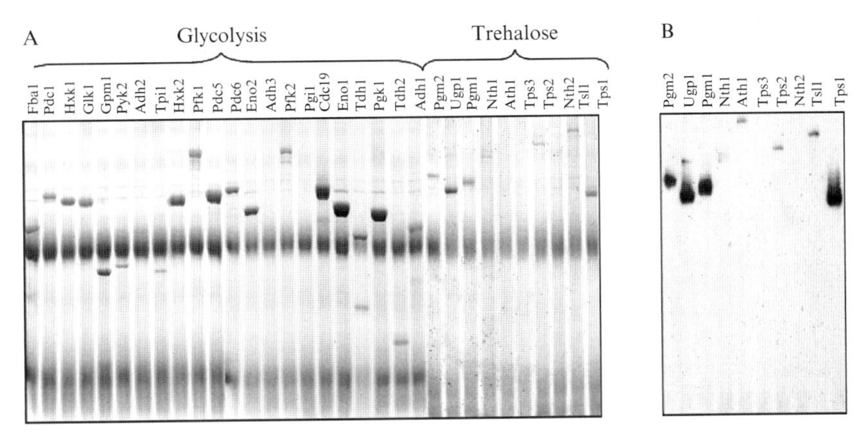

Figure 11.1 Proteins expressed in strains from the Yeast ORF Collection. SDS/PAGE analysis of IgG sepharose-bound enzymes from glycolysis and trehalose biosynthesis pathways (A) and Western blot analysis of trehalose biosynthesis enzymes (B).

Table 11.2 Examples of purified enzymes with quantities and percentages of full-length proteins in sample preparation

Protein	Quantity, mg/culture[a]	Percentage of full-length protein	Protein	Quantity, mg/culture[a]	Percentage of full-length protein
Adh1	0.08	97	Pfk2	0.12	97
Cdc19	0.48	96	Pgi1	0.08	97
Eno1	0.56	93	Pgk1	0.35	97
Eno2	1.62	98	Tdh1	0.11	95
Fba1	0.38	85	Tdh3	0.16	93
Glk1	0.70	98	Tpi1	0.32	95
Gpm1	1.02	99	Tps1	0.80	95
Hxk1	0.54	99	Tsl1	0.08	82
Hxk2	1.56	97	Tps2	0.24	97
Pdc1	0.48	97	Nth1	0.16	98
Pdc5	0.63	96	Pgm1	0.28	98
Pdc6	0.39	94	Ugp1	0.64	97
Pfk1	0.25	88	Pgm2	0.90	99

The majority of proteins were purified from the Yeast ORF Collection. Gpm1 and Eno2 were from the Yeast GST-Tagged Collection. Cdc19, Tpi1, and Tdh3 were from the Yeast TAP-Tagged Collection.
[a] Represents average calculated from values determined by two independent protein quantification techniques as described in Section 5.

and BY4741 (*MAT***a** *his3Δ1 leu2Δ0 met15Δ0 ura3Δ0*), respectively, when strains are cultured in a synthetic medium without uracil (S-Ura medium: 1.7 g/L YNB (w/o) amino acids and (w/o) ammonium sulfate, 5 g/L ammonium sulfate, 20 mg/L adenine, 20 mg/L arginine, 30 mg/L isoleucine, 20 mg/L histidine, 60 mg/L leucine, 30 mg/L lysine, 20 mg/L methionine, 50 mg/L phenylalanine, 20 mg/L tryptophan, 30 mg/L tyrosine, and 150 mg/L valine) to select for the expression plasmid containing the ORF of interest. Strains from the Yeast TAP-Tagged Collection (constructed in the haploid BY4741 genetic background) can be grown in synthetic complete medium without histidine, or in rich YPD medium.

To start with, 96-well stock plates are thawed. The foil seals of the wells containing the strains of interest are pierced, and cells are grown by pipetting 10 μL of cell culture and spreading it for single colonies on S-URA/raffinose agar plates (Yeast ORF and Yeast GST-Tagged Collections) or YPD agar plates (Yeast TAP-Tagged Collection). Cells are grown on agar plates at 30 °C for 3–5 days. Single colonies are then used to inoculate liquid medium, cultures are first grown overnight at 30 °C, 220 rpm in 5-ml volumes of S-Ura with 2% (w/v) raffinose in 25-mL flasks for Yeast ORF and Yeast GST-Tagged Collections, or 50 ml of YPD in 300-mL flasks for Yeast TAP-Tagged Collection. Overnight cultures of strains from the Yeast

TAP-Tagged Collection are diluted into 800-mL YPD in 3- L flasks and grown for 6–8 h to OD_{600} 2–4, followed by cell harvesting and then treated as described later in this section. Overnight cultures of Yeast ORF and Yeast GST-Tagged Collections are diluted into 60-mL S-Ura with 2% (w/v) raffinose in 300-mL flasks and grown for 24 h at 30 °C and 220 rpm, followed by dilution into 760-mL S-Ura/2% raffinose to OD_{600} 0.05 in 3-L flasks and overnight growth to OD_{600} 0.6–0.8. ORF expression is induced by adding 40 mL of 40% (w/v) galactose to each culture bringing the total volume to 800 mL and allowing cells to grow for 5 h. All cultures are harvested by centrifugation at 6000×*g* and washed twice with 50 mL of ice-cold water. Protein purification procedures can be initiated immediately, or cell pellets can be stored at −80 °C. Occasionally, to produce a minimum of 1 mg protein, the volume of culture has to be adjusted by increasing the number of flasks for each strain of interest.

4.2. Cell disruption

Cells from each 800 mL culture are resuspended in 4 mL of disruption buffer (10 mM Tris–Cl, pH 8.0, 150 mM NaCl, 2.5 mM PMSF, 2.5 μg/mL leupeptin, 2.5 μg/mL pepstatin, and 1 Protease inhibitor cocktail tablet (Roche)) and are lyzed mechanically by bead beating in a Mini-BeadBeater-16 (Biospec Inc.). Standard 2 mL screw-cap microcentrifuge tubes containing 0.4–0.5 mL of 0.425–0.600 mm glass beads (Sigma) and 0.8–1.0 mL of cell suspension in disruption buffer are subjected to four cycles of shaking (3450 oscillations/min) for 30 s, each with a 1-min interval on ice between each cycle. Then, tubes are centrifuged at 8000×*g* for 5 min and the supernatant is collected. A new aliquot of disruption buffer is added to the cell/debris pellet, bead beating is repeated, and the supernatant collected as above. Lysis efficiency can be checked microscopically by determining the ratio of the cell debris to intact cells. When the bead-beaten cell suspension is centrifuged, the pellet usually contains two layers: a top lighter layer of cell debris and organelles from disrupted cells, and a bottom darker layer of intact cells. The ratio between the top and bottom layers of 5:1 or higher is a good indicator of efficient cell lysis. The combined cell lysate from both supernatants from the above centrifugations is subjected to a further centrifugation step at 20,000×*g* for 15 min, and the resulting supernatant is used for the protein purification.

4.3. Protein purification

Immediately after proteins are extracted from yeast cells, soluble fractions are applied to the protein purification procedure. First, affinity purification on IgG sepharose is performed for Yeast ORF and Yeast TAP-Tagged Collections, or glutathione sepharose purification is used for Yeast GST-Tagged Collection.

Extracts (10–12 mL) in disruption buffer are combined with 100–120 μL of 10% NP40 detergent and added to 200–400 μL of IgG Sepharose beads (GE Healthcare) for Yeast ORF and Yeast TAP-Tagged Collections, and to 200–400 μL of glutathione sepharose beads for Yeast GST-Tagged Collection, which have both been prewashed three times with 2 mL of washing buffer (10 mM Tris–Cl, pH 8.0, 150 mM NaCl, 0.1% NP40). The suspension is mixed in a 15-mL tube at 4 °C for 2 h and then passed through a 9-cm high Poly-Prep column (Bio-Rad). Unbound proteins are separated by gravity from beads with bound protein of interest. After unbound proteins are removed, both types of beads are subsequently washed six to eight times with 10 mL of washing buffer, twice with 5 mL of protease cleavage buffer (10 mM Tris–Cl, pH 8.0, 150 mM NaCl, 0.5 mM EDTA, 1 mM DTT, 0.1% NP40; for Yeast ORF and Yeast TAP-Tagged Collections only), and transferred to 2-mL microcentrifuge tubes.

IgG Sepharose beads are resuspended in 400 μL of protease cleavage buffer and with 0.5–2 units of PreScission protease (human rhinovirus 3 °C protease–GST fusion, GE healthcare) for the Yeast ORF Collection, or with 2–10 units of AcTEV Protease (Tobacco Etch Virus (TEV) protease, Invitrogen) for Yeast GST-Tagged Collection, and vigorously mixed (600 rpm) in thermomixer (Eppendorf) for 16 h at 4 °C. Then, the PreScission protease can be removed by additional incubation with 20 μL of pre-equilibrated glutathione sepharose 4B beads (GE healthcare) for 1 h. For the Yeast GST-Tagged Collection, glutathione sepharose beads are resuspended in 400 μL of elution buffer (50mM Tris–Cl, pH 7.5; 150 mM NaCl; 10 mM Glutathione, reduced form) and vigorously mixed (600 rpm) in thermomixer for 1 h at 4 °C.

To collect proteins of interest, which have been released from IgG Sepharose by the protease digestion, or from glutathione sepharose by the reduced form of glutathione, the 2-mL tubes with bead suspensions are centrifuged at 300×*g* for 1 min and supernatants with proteins of interest are collected. The beads are washed twice more with 300 μL of either protease cleavage buffer or elution buffer. The resulting 1 mL supernatant, containing the protein of interest, is clarified by passage through a SPIN-X filter centrifuge tube, 0.22 μm cellulose acetate (Corning, Inc.). The buffer in the purified protein sample can be exchanged for any desired storage buffer using Vivaspin 2 centrifugal filters (Sartorius). For enzyme kinetics studies in our systems biology investigations, proteins are stored in 50 mM MES, pH 6.5, 150 mM KCl, 2 mM $MgCl_2$ at 4 °C or −80 °C.

As shown in Fig. 11.2, the success of each step of purification of an individual protein can be assessed by analyzing intermediates and the final protein sample on SDS–PAGE (Laemmli, 1970).

In the case of the Yeast ORF Collection, where the 6xHis motif protein tag remains, the protein can be subjected to further purification using Ni^{2+} affinity chromatography (e.g., Ni-NTA Superflow; Qiagen). If required,

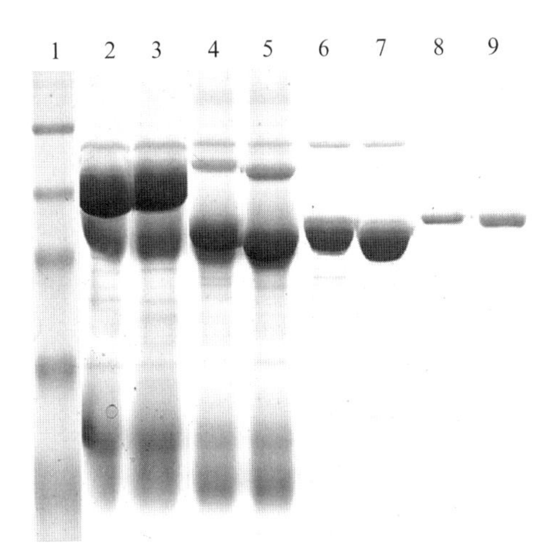

Figure 11.2 SDS–PAGE representing key stages of protein purification from *S. cerevisiae*: enolase isoenzymes Eno1 and Eno2. Lane 1 represents a protein molecular weight marker containing 116, 66, 45, 25, and 18 kDa protein standards. Lanes 2, 4, 6, 8 are for Eno1; lanes 3, 5, 7, 9 are for Eno2. Protein bound to the IgG sepharose (2 and 3), cleaved with 3 °C protease (4 and 5), eluted from the IgG sepharose (6 and 7), and after purification on HisTrap HP and HiTrap desalting columns (GE Healthcare).

1–2 mM $MgCl_2$ and 10% (w/v) glycerol can be used in disruption, washing, elution, and storage buffers. For unstable and environment-sensitive enzymes, it is advisable that proteins are tested for different purification and storage conditions and assayed immediately after purification is complete.

5. Protein Analysis and Quantification

Quantities and concentrations of purified enzymes are determined using protein QuantiPro™ BCA Assay Kit (Sigma–Aldrich; Smith *et al.*, 1985). The quantity of protein of interest and quality of sample preparation can be further evaluated using 2100 Bioanalyzer (Agilent Technologies) as shown for neutral trehalase (Fig. 11.3). Proteins are separated using gel electrophoresis principles and microfluidic capillary technology, which are replicated into the chip format. During separation, proteins are detected by laser-induced fluorescence. The manufacturer's protocol and Protein 230 kit are employed for sample preparation and chip priming. Agilent Expert Software is used for data analysis. Table 11.2 lists some successfully purified enzymes, their quantities, and the percentage yield of the full-length protein of interest. It should be noted that some proteins are unstable or susceptible to proteases resulting in lower percentage yields of full-length proteins in the protein preparation (e.g., Tsl1).

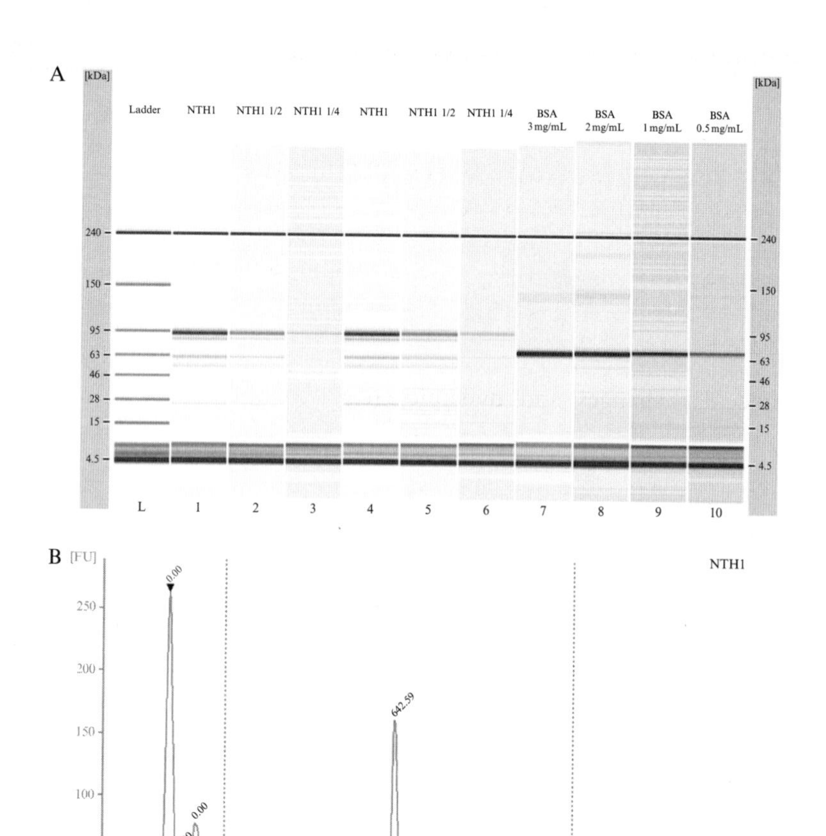

C

Size [kDa]	Rel. Conc. [ng/μl]	Calib. Conc. [ng/μl]	Time corrected area	% Total	Observations	Area
4.5	0	0	750.9	0	Lower marker	136.7
7.1	0	0	163.5	0	System peak	30.9
9.3	0	0	219.5	0	System peak	43.5
25.9	43.5	0	13.7	4.7		3.3
54.5	60.6	0	19.1	6.5		5.4
61.8	101.1	0	31.9	10.8		9.4
85.2	86.6	0	27.3	9.3		8.6
91.8	642.6	674.4	202.8	68.7	Calibrated protein	64.7
240	60	0	18.9	0	Upper marker	8.1

Figure 11.3 Protein quantification and quality of sample preparation evaluated using 2100 Bioanalyzer. (A) Gel-like image of a 14–230 kDa chip run of partially purified neutral trehalase (Nth1) from *S. cerevisiae* and bovine serum albumin (BSA) standard. Protein samples were loaded on the protein chip using Protein 230 kit in the following order: molecular weight ladder (4.5–240 kDa), two repeats of various dilutions (undiluted, two and four times) of Nth1 protein after the first step of purification on IgG Sepharose column, last four lanes are different concentrations of BSA (3, 2, 1, and 0.5 mg/mL). (B) Electropherogram shows chip analysis of the partially purified enzyme (undiluted Nth1) sample that contains some truncated protein fragments. (C) Peak table represents quantitative characteristics of the contents in the first sample (undiluted Nth1). Nth1 was quantified on the basis of internal and BSA standards. Based on the relative quantification with the Bioanalyzer, the overall concentration for full-length Nth1 was estimated to 642.6–674.430 ng/μL accounting for 68.7% of the total protein in the sample.

6. Protein Use in Proteomics and Enzyme Kinetics Measurements

Systems biology requires experimental data that are quantitative, comprehensive, and compatible with mathematical/computational simulation of a biological system (Kell, 2006). This essentially entails establishing: (1) *in vivo* levels of biological molecules (e.g., proteins and metabolites); (2) molecular changes and dynamics (e.g., enzymatic reactions); and (3) spatial distribution of where such changes occur (e.g., in the cellular cytoplasm). Protein production contributes to the collection of these data in at least two areas. First, purified proteins can be used as internal standards for quantification and quality control in establishing protein levels of the biological system (Carroll *et al.*, 2011). Second, proteins play key roles in measuring molecular interactions and changes including signaling, transport, and enzymatic reactions (Franklin and Ullu, 2010; George *et al.*, 2009; Hernández *et al.*, 2006; Malys *et al.*, 2002, 2009; Messiha *et al.*, 2011; Norris and Malys, 2011; Smallbone *et al.*, 2011).

7. Concluding Remarks

Overexpression of tagged proteins brings enormous advantages: it significantly increases solubility and throughput allowing purification of multiple proteins of choice (Malhotra, 2009). It also brings several limitations and problems such as functional and structural issues. Occasionally, the position of affinity tags on the protein can interfere with the folding, complex formation, and functionality of the protein. It is sensible to employ, in parallel, different types of constructs with both C-terminal and N-terminal tags. If the functionality of one form of the tagged protein is reduced or lost, due to the position of the tag, the other form of tagged protein can potentially be used thereby reducing chances of a negative impact of the tag. Combined use of Yeast ORF and Yeast GST-Tagged Collections (Gelperin *et al.*, 2005; Sopko *et al.*, 2006) provides this complementary approach for the purification of approximately 80% of all *S. cerevisiae* proteins.

Acknowledgments

We acknowledge the support of the BBSRC/EPSRC grant BB/C008219/1 "The Manchester Centre for Integrative Systems Biology (MCISB)." We thank Douglas B. Kell for his early contribution to this work.

REFERENCES

Boone, C., Bussey, H., and Andrews, B. J. (2007). Exploring genetic interactions and networks with yeast. *Nat. Rev. Genet.* **8,** 437–449.

Brondyk, W. H. (2009). Selecting an appropriate method for expressing a recombinant protein. *Methods Enzymol.* **463,** 131–147.

Carroll, K. M., Lanucara, F., and Eyers, C. E. (2011). Quantification of proteins and their modifications using QconCAT technology. *Methods Enzymol.* **500,** 113–131.

Chua, G., Morris, Q. D., Sopko, R., Robinson, M. D., Ryan, O., Chan, E. T., Frey, B. J., Andrews, B. J., Boone, C., and Hughes, T. R. (2006). Identifying transcription factor functions and targets by phenotypic activation. *Proc. Natl. Acad. Sci. USA* **103,** 12045–12050.

Cregg, J. M., Tolstorukov, I., Kusari, A., Sunga, J., Madden, K., and Chappell, T. (2009). Expression in the yeast Pichia pastoris. *Methods Enzymol.* **463,** 169–189.

Fatica, A., Cronshaw, A. D., Dlakić, M., and Tollervey, D. (2002). Ssf1p prevents premature processing of an early pre-60S ribosomal particle. *Mol. Cell* **9,** 341–351.

Franklin, J. B., and Ullu, E. (2010). Biochemical analysis of PIFTC3, the Trypanosoma brucei orthologue of nematode DYF-13, reveals interactions with established and putative intraflagellar transport components. *Mol. Microbiol.* **78,** 173–186.

Gavin, A.-C., Aloy, P., Grandi, P., Krause, R., Boesche, M., Marzioch, M., Rau, C., Jensen, L. J., Bastuck, S., Dümpelfeld, B., Edelmann, A., Heurtier, M.-A., *et al.* (2006). Proteome survey reveals modularity of the yeast cell machinery. *Nature* **440,** 631–636.

Gelperin, D. M., White, M. A., Wilkinson, M. L., Kon, Y., Kung, L. A., Wise, K. J., Lopez-Hoyo, N., Jiang, L., Piccirillo, S., Yu, H., Gerstein, M., Dumont, M. E., *et al.* (2005). Biochemical and genetic analysis of the yeast proteome with a movable ORF collection. *Genes Dev.* **19,** 2816–2826.

George, R., Chan, H.-L., Ahmed, Z., Suen, K. M., Stevens, C. N., Levitt, J. A., Suhling, K., Timms, J., and Ladbury, J. E. (2009). A complex of Shc and Ran-GTPase localises to the cell nucleus. *Cell. Mol. Life Sci.* **66,** 711–720.

Ghaemmaghami, S., Huh, W. K., Bower, K., Howson, R. W., Belle, A., Dephoure, N., O'Shea, E. K., and Weissman, J. S. (2003). Global analysis of protein expression in yeast. *Nature* **425,** 737–741.

Gold, L. (1990). Expression of heterologous proteins in *Escherichia coli. Methods Enzymol.* **185,** 11–14.

Hernández, H., Dziembowski, A., Taverner, T., Séraphin, B., and Robinson, C. V. (2006). Subunit architecture of multimeric complexes isolated directly from cells. *EMBO Rep.* **7,** 605–610.

Jackman, J. E., Kotelawala, L., Grayhack, E. J., and Phizicky, E. M. (2007). Identification and characterization of modification enzymes by biochemical analysis of the proteome. *Methods Enzymol.* **425,** 139–152.

Jackman, J. E., Grayhack, E. J., and Phizicky, E. M. (2008). The use of *Saccharomyces cerevisiae* proteomic libraries to identify RNA-modifying proteins. *Methods Mol. Biol.* **488,** 383–393.

Jarvis, D. L. (2009). Baculovirus-insect cell expression. *Methods Enzymol.* **463,** 191–222.

Junge, F., Schneider, B., Reckel, S., Schwarz, D., Dötsch, V., and Bernhard, F. (2008). Large-scale production of functional membrane proteins. *Cell. Mol. Life Sci.* **65,** 1729–1755.

Katzen, F., Chang, G., and Kudlicki, W. (2005). The past, present and future of cell-free protein synthesis. *Trends Biotechnol.* **23,** 150–156.

Kell, D. B. (2004). Metabolomics and systems biology: Making sense of the soup. *Curr. Opin. Microbiol.* **7,** 296–307.

Kell, D. B. (2006). Metabolomics, modelling and machine learning in systems biology—Towards an understanding of the languages of cells. The 2005 Theodor Bücher Lecture. *FEBS J.* **273,** 873–894.

King, R. D., Rowland, J., Oliver, S. G., Young, M., Aubrey, W., Byrne, E., Liakata, M., Markham, M., Pir, P., Soldatova, L. N., Sparkes, A., Whelan, K. E., *et al.* (2009). The automation of science. *Science* **324,** 85–89.

Krogan, N. J., Cagney, G., Yu, H., Zhong, G., Guo, X., Ignatchenko, A., Li, J., Pu, S., Datta, N., Tikuisis, N. P., Punna, T., Peregrín-Alvarez, J. M., *et al.* (2006). Global landscape of protein complexes in the yeast *Saccharomyces cerevisiae*. *Nature* **440,** 637–643.

Kung, L. A., and Snyder, M. (2006). Proteome chips for whole-organism assays. *Nat. Rev. Mol. Cell Biol.* **7,** 617–622.

Laemmli, U. K. (1970). Cleavage of structural proteins during the assembly of the head of bacteriophage T4. *Nature* **227,** 680–685.

Li, Z., Barajas, D., Panavas, T., Herbst, D. A., and Nagy, P. D. (2008). Cdc34p ubiquitin-conjugating enzyme is a component of the tombusvirus replicase complex and ubiquitinates p33 replication protein. *J. Virol.* **82,** 6911–6926.

Lindbo, J. A. (2007). High-efficiency protein expression in plants from agroinfection-compatible tobacco mosaic virus expression vectors. *BMC Biotechnol.* **7,** 52.

Makrides, S. C. (1996). Strategies for achieving high-level expression of genes in Escherichia coli. *Microbiol. Rev.* **60,** 512–538.

Malhotra, A. (2009). Tagging for protein expression. *Methods Enzymol.* **463,** 239–258.

Malys, N., and McCarthy, J. E. G. (2006). Dcs2, a novel stress-induced modulator of m7G pppX pyrophosphatase activity that locates to P bodies. *J. Mol. Biol.* **363,** 370–382.

Malys, N., and McCarthy, J. E. G. (2011). Translation initiation: Variations in the mechanism can be anticipated. *Cell. Mol. Life Sci.* **68,** 991–1003.

Malys, N., Chang, D.-Y., Baumann, R. G., Xie, D., and Black, L. W. (2002). A bipartite bacteriophage T4 SOC and HOC randomized peptide display library: Detection and analysis of phage T4 terminase (gp17) and late σ factor (gp55) interaction. *J. Mol. Biol.* **319,** 289–304.

Malys, N., Carroll, K., Miyan, J., Tollervey, D., and McCarthy, J. E. G. (2004). The 'scavenger' m7G pppX pyrophosphatase activity of Dcs1 modulates nutrient-induced responses in yeast. *Nucleic Acids Res.* **32,** 3590–3600.

Mendes, P., Messiha, H., Malys, N., and Hoops, S. (2009). Enzyme kinetics and computational modeling for systems biology. *Methods Enzymol.* **467,** 583–599.

Messiha, H. L., Malys, N., and Carroll, K. M. (2011). Towards a full quantitative description of yeast metabolism: A systematic approach for estimating the kinetic parameters of isoenzymes under *in-vivo* like conditions. *Methods Enzymol.* **500,** 215–231.

Norris, M. G., and Malys, N. (2011). What is the true enzyme kinetics in the biological system? An investigation of macromolecular crowding effect upon enzyme kinetics of glucose-6-phosphate dehydrogenase. *Biochem. Biophys. Res. Commun.* **405,** 388–392.

Panigrahi, A. K., Schnaufer, A., and Stuart, K. D. (2007). Isolation and compositional analysis of trypanosomatid editosomes. *Methods Enzymol.* **424,** 3–24.

Puig, O., Caspary, F., Rigaut, G., Rutz, B., Bouveret, E., Bragado-Nilsson, E., Wilm, M., and Seraphin, B. (2001). The tandem affinity purification (TAP) method: A general procedure of protein complex purification. *Methods* **24,** 218–229.

Rigaut, G., Shevchenko, A., Rutz, B., Wilm, M., Mann, M., and Seraphin, B. (1999). A generic protein purification method for protein complex characterization and proteome exploration. *Nat. Biotechnol.* **17,** 1030–1032.

Sabate, R., De Groot, N. S., and Ventura, S. (2010). Protein folding and aggregation in bacteria. *Cell. Mol. Life Sci.* **67,** 2695–2715.

Smallbone, K., Malys, N., Messiha, H. L., Wishart, J. A., and Simeonidis, E. (2011). Building a kinetic model of trehalose biosynthesis in *Saccharomyces cerevisiae*. *Methods Enzymol.* **500,** 355–370.

Smith, P. K., Krohn, R. I., Hermanson, G. T., Mallia, A. K., Gartner, F. H., Provenzano, M. D., Fujimoto, E. K., Goeke, N. M., Olson, B. J., and Klenk, D. C. (1985). Measurement of protein using bicinchoninic acid. *Anal. Biochem.* **150,** 76–85.

Sopko, R., Huang, D., Preston, N., Chua, G., Papp, B., Kafadar, K., Snyder, M., Oliver, S. G., Cyert, M., Hughes, T. R., Boone, C., and Andrews, B. (2006). Mapping pathways and phenotypes by systematic gene overexpression. *Mol. Cell* **21,** 319–330.

Tabor, S., and Richardson, C. C. (1985). A bacteriophage T7 RNA polymerase/promoter system for controlled exclusive expression of specific genes. *Proc. Natl. Acad. Sci. USA* **82,** 1074–1078.

Yin, J., Li, G., Ren, X., and Herrler, G. (2007). Select what you need: A comparative evaluation of the advantages and limitations of frequently used expression systems for foreign genes. *J. Biotechnol.* **127,** 335–347.

Zerbs, S., Frank, A. M., and Collart, F. R. (2009). Bacterial systems for production of heterologous proteins. *Methods Enzymol.* **463,** 149–168.

Zhu, H., Bilgin, M., Bangham, R., Hall, D., Casamayor, A., Bertone, P., Lan, N., Jansen, R., Bidlingmaier, S., Houfek, T., Mitchell, T., Miller, P., *et al.* (2001). Global analysis of protein activities using proteome chips. *Science* **293,** 2101–2105.

SECTION FIVE

ENZYMATIC ASSAYS IN SYSTEMS BIOLOGY RESEARCH

CHAPTER TWELVE

Towards a Full Quantitative Description of Yeast Metabolism: A Systematic Approach for Estimating the Kinetic Parameters of Isoenzymes under *In vivo* like Conditions

Hanan L. Messiha,[*,†] Naglis Malys,[*,‡] *and* Kathleen M. Carroll[*,†]

Contents

* Manchester Centre for Integrative Systems Biology, Manchester Interdisciplinary Biocentre, The University of Manchester, Manchester, United Kingdom
† School of Chemistry, The University of Manchester, Manchester, United Kingdom
‡ The University of Manchester, Manchester, United Kingdom

Methods in Enzymology, Volume 500
ISSN 0076-6879, DOI: 10.1016/B978-0-12-385118-5.00012-8

Abstract

In order to produce a full quantitative description of yeast metabolism, a number of kinetic parameters of enzymes that are important for energy metabolism must be determined experimentally. We aim to determine the prospective *in vivo* kinetic properties of a range of yeast-purified isoenzymes that are important in energy metabolism, with respect to the concentration of their substrates and products. This endeavor forms part of our systems biology pipeline to facilitate the production of bottom-up models of metabolism. Within this workflow, we implement an infrastructure for medium- to high-throughput determination of the kinetic properties of purified isoenzymes in *in vivo* like conditions. This includes the use of the KineticsWizard software for data capture and analysis. The captured experimental data are analyzed by the software and subsequently stored in appropriate repositories (MeMo-RK and SABIO-RK). While we focus initially on glycolysis in *Saccharomyces cerevisiae*, our methodology is generic and can be widely applied to the study of other enzymes and pathways in yeast and other organisms.

1. Introduction

Systems biology aims at a system-level understanding of biological systems (Kitano, 2002a,b). Understanding how biological processes work and are regulated requires the biochemical mechanisms underlying these processes to be described in full. This can be achieved by creating comprehensive mathematical models that predict cellular behavior. Integration of experimental, computational, and theoretical approaches is therefore required (Kitano, 2005). A challenging aim of systems biology is to develop detailed kinetic models for simulating and predicting the dynamic responses of metabolic networks.

Modeling can be approached through "bottom-up" or "top-down" approaches (Bruggeman and Westerhoff, 2006), and both approaches are being used within the Manchester Centre for Integrative Systems Biology (MCISB).

In bottom-up approaches, the properties and interactions of molecules or of elemental subsystems are defined and characterized to construct models that try to explain how the interactions account for functional behaviors of the system that embeds them (e.g., kinetic model of reaction network). The model's predictions are evaluated by performing targeted experiments on the real system. Depending on the outcome, the model is adjusted and process is repeated, or additional experimental data that may be captured to improve the model are identified. The computer models allow for exploring the behavior of the model system under different environmental conditions, sometimes leading to discovery of new levels of organization in the system models.

Top-down approaches start with observations at the network level and successively add detail to initial rough representations (e.g., Ideker and Lauffenburger, 2003). The properties, functions, and behaviors of the whole system (system data) are regarded together. Models of component composition and interactions are then constructed to explain both the experimental data and the correlations between component properties and system behavior. The model itself may then be used to direct future experimental work in a targeted experimental design. As the model is refined, through iteration and adjustment as a result of targeted experimentation, the outcome may lead to a greater understanding of the system and further hypotheses. This potentially leads to a complete description of the component composition, and its organizational dynamics and functionality.

As a proof of principle for the bottom-up approach in systems biology, we apply this to study metabolism in the model organism *Saccharomyces cerevisiae*. This is an ideal organism in which to explore techniques for the development of integrated models of important cellular systems due to its well-characterized genome and it being highly amenable to genetic manipulations and to high-throughput technologies. Our aim is to develop an operational strategy enabling the understanding of more than 95% of the carbon flux in *S. cerevisiae* (Simeonidis *et al.*, 2010) and in the process to develop a set of tools (theoretical and experimental) for bottom-up systems biology that can be generalized to other organisms. In order to achieve this, we identify the reactions that carry the majority of carbon flux in and then identify the corresponding metabolic pathways. The targeted pathways are studied by full characterization of their components, and then fully annotated computational models are developed and validated. This study consequently encompasses a wide range of experimental and computational methods, including the production and purification of proteins and isoenzymes of interest, enzymatic assays, and determination of kinetic parameters, enzyme concentration measurements by liquid chromatography–mass spectroscopy (LC–MS), measurements of metabolites by gas chromatography–mass spectroscopy (GC–MS) and LC–MS, as well as the development of computational tools to handle, analyze, and store data.

The kinetic parameters of the enzymes in the pathways of interest are one of the most crucial components to measure accurately in order to construct reliable models. Here, we illustrate the approach used to measure the kinetic parameters of isoenzymes presenting our methodology in determining them for the glycolysis pathway, a key metabolic pathway in yeast metabolism. We have expressed and purified yeast *S. cerevisiae* isoenzymes and run enzymatic assays to determine their kinetic parameters.

We will also discuss the methodology we use to analyze, store, and submit data to the enzyme database SABIO-RK (System for the Analysis of Biochemical Pathways Reaction Kinetics; Rojas *et al.*, 2007; Wittig *et al.*, 2006). The approach we use is generic and can be widely applied to similar studies in other organisms as well as to study other enzymes of other pathways.

2. Enzyme Kinetics for Systems Biology

For our systems biology workflow, the fluxes and kinetics of substrates, metabolites, and products in biomolecular pathways/networks need to be determined. The biochemical networks are sets of linked reactions, and their dynamics can be described as sets of coupled deterministic or stochastic equations representing the rate of change of the concentrations of the chemical components in the network. Setting up this set of coupled differential equations requires the input of molecular concentrations or particle numbers and a detailed set of kinetic parameters of enzymes (Stein *et al.*, 2008). Enzyme kinetics can be obtained by studying enzymes in isolation, by using parameter estimation strategies, or by calculating parameters from protein structures (Stein *et al.*, 2008). The determination of kinetic parameters using parameter estimation strategies (e.g., Borger *et al.*, 2006; Kremling *et al.*, 2004; Moles *et al.*, 2003) can result in inaccuracies because the kinetic models are often simplified, and hence do not yield the "true" kinetic parameters. As soon as the model is enlarged and the same data are again used for parameter estimation, the parameters will assume different values. In addition, gaps in our knowledge can be missed by falsely fitting them to a model (fitted away). However, in some cases (e.g., for eukaryotic signaling and eukaryotic gene expression), it is extremely difficult—if not impossible—to measure kinetic parameters, which leaves parameter estimation as the only option.

When constructing models of biochemical networks, it is necessary to measure kinetic parameters of enzymes under specified conditions. These parameters may be available in the literature and publically accessible databases, but only measured under different experimental conditions, or using an enzyme from an organism other than the one of interest. The construction of models relying on parameters from these sources might not be precise enough to resemble the real biological processes of interest, and therefore, it is important if not crucial to use parameters measured under conditions resembling the *in vivo* environment of the process of interest. Accordingly, in the production of new models, there is generally a requirement to make kinetic parameters for all the isoenzymes of interest.

Many models have measured these parameters *in vivo* where the maximum velocity (V_{max}) of the enzymatic reaction and the K_m (substrate concentration at half the maximum velocity) value for substrates and products of the isoenzymes of interest were measured in cell lysates. Although this approach has the advantage of measuring the kinetic parameters under *in vivo* like conditions, it still has limitations. Measuring V_{max} in cell lysate for a particular enzyme will produce the sum of the activities of all the isoenzymes of that particular enzyme that catalyzes a specific reaction or multiple reactions. In most cases, each enzyme has more than one isoform that catalyze the same reaction. These isoenzymes are yet structurally different, displaying different kinetic and regulatory properties. Estimating the V_{max} by this approach will not distinguish between the detailed kinetics of each individual isoenzyme which may be of great importance in constructing accurate biological models and fine tuning the metabolism. Cell lysate is also rich in physiological metabolites and enzymes which can interfere with the enzyme of interest especially when they include coupling enzymes. The existence of such metabolites and enzymes is likely to make the assays' overlay complicated requiring many control experiments in order to obtain the correct rates for the targeted reaction. For all these reasons, it is arguably more advantageous if the kinetic parameters can be measured for isoenzymes in isolation. This can be achieved by targeted production and purification of the isoenzymes of interest than determining their kinetic parameters (k_{cat} [turnover number] and K_m for their substrates and products) under the specified conditions. Because our metabolic model requires the V_{max} for the isoenzymes, accurate measurements of their concentrations in cell lysates must be obtained ($V_{max} = k_{cat}$ multiplied by enzyme concentration [E]). The latter is achieved by LC–MS through application of the QconCAT strategy (Section 4.4 of this chapter), and this enables the V_{max} of each isoenzyme to be calculated.

3. Production and Purification of Isoenzymes

Enzymes were expressed in *S. cerevisiae* strains that contain either over-expression plasmid with the open reading frame of interest under the control of the GAL1 inducible promoter and in fusion to a multiple tag (MORF collection, Open Biosystems; Gelperin *et al.*, 2005) or chromosome-integrated gene and tag fusion (TAP collection, Open Biosystems, Ghaemmaghami *et al.*, 2003). Protein expressions and purifications were performed according to Rigaut *et al.* (1999) and Gelperin *et al.* (2005) with modifications described in this issue of *Methods in Enzymology* (Malys *et al.*, 2011).

The success of each step of purification for each protein was assessed by analyzing samples from the intermediate and final protein preparations using SDS-PAGE. Amount and concentration of the purified enzyme were determined using QuantiPro™ BCA Assay Kit (Sigma–Aldrich) according to the recommendation of the manufacturer. In general, quality of the enzyme preparation was further assessed using 2100 Bioanalyzer (Agilent Technologies).

4. General Protocol for Enzymatic Assays

Having hundreds of proteins to be assayed is a challenge to the systems biology research community. The challenge to the experimentalist is not confined in running assays; moreover, it is how to produce and analyze the large amount of data generated in an accurate and reliable way and how to make such data available in the appropriate format for modeling purposes. This can be achieved by finding reasonable strategies to standardize the assay conditions, to run assays in high-throughput settings, and to manage enzyme kinetics data from instrument to browser.

4.1. Standardization of assay conditions

In order to standardize the assay conditions, the kinetic parameters of the purified yeast *S. cerevisiae* isoenzymes are required to be estimated in conditions as close as possible to the intracellular conditions. Generally, for all assays, a reaction mixture that consists of 100 mM MES (2-[*N*-morpholino]-ethanesulfonic acid), pH 6.5, 100 mM KCl, and 5 mM free magnesium in the form of $MgCl_2$ is used unless otherwise specified. The purified isoenzymes are prepared in the reaction buffer immediately before assaying. The concentration of free magnesium is set at 5 mM and if any reaction includes other reagents requiring magnesium (e.g., ATP, ADP, EDTA, etc.), extra magnesium is added to maintain the concentration. All measurements are carried out at 30 °C to mimic the temperature under which the cells were grown.

4.2. Towards high-throughput measurements of enzymatic activities

To attain a high-throughput strategy in performing enzymatic assays, we have settled on running spectrophotometric assays for the glycolytic isoenzymes measuring the consumption or production of NADH or NADPH by using one or more coupling reactions when needed, and all assays are carried out in a medium- to high-throughput measurements with a BMG Labtech NOVOstar plate reader (automated fluorescence/FP/absorbance reader, Offenburg, Germany) in 384-well format plates with a 60 μl reaction volume. Assays are automated so that all reagents in the reaction buffer (including any coupling enzymes) are in 45 μl, the enzyme (to be assayed) in 5 μl, and the substrate in 10 μl volumes. The reaction is usually started by the addition of substrate in almost all cases. Using such small reaction volumes in assays allows the use of small quantities of enzymes (5 μl per reaction); therefore, high-throughput production of enzymes is feasible and consequently speeds the production of kinetics data. It is noted that in almost all cases, the final enzyme concentrations used in the reaction are in the nanomolar range; therefore, small volumes of purified enzymes (e.g., as small as 50 μl) with a concentration in the micromolar range would be sufficient enough to measure the full set of their kinetic parameters by this method.

By running assays in 384-well format plates and using the plate reader for recording the absorbance time course measurements, we can collate huge amounts of kinetic data in a relatively short time period compared to using classic spectrophotometers for making similar measurements. Full sets of experiments can be run simultaneously in the same plate for each isoenzyme with multiple substrates, for all isoenzymes of the same enzyme and even for multiple enzymes with multiple reactions and substrates.

For each individual enzyme, the forward and the reverse reactions are assayed whenever possible. This depends on the nature of the reaction, the ability to produce active enzyme, and the availability of physiological substrates either commercially or through chemical synthesis. Assays for each individual enzyme were either developed or modified from previously published methodology to be compatible with the conditions of the assay reactions (e.g., pH compatibility or unavailability of commercial substrates). For optimizing and validating the assays, purified enzymes available commercially are used to validate the assays before being applied to the purified isoenzymes. All assays are coupled with enzyme(s) where NAD(P) or NAD (P)H is a product or substrate so that its formation or consumption can be monitored spectrophotometrically at 340 nm using an extinction coefficient ($\Sigma_{340\ \mathrm{nm}}$) of 6.620 mM^{-1} cm^{-1}, unless the reaction of a particular enzyme consumes or produces NADH or NADPH in which case no coupling enzymes are required. Some assays are modified by altering the concentration of coupling enzymes or other reagents to ensure that the rate measured is the rate of the reaction of interest (rate limiting). This is a critical step that needs special care to be taken on using coupling enzymes in assays (Takagahara *et al.*, 1983).

All measurements are based on at least duplicate determination of the reaction rates at each substrate concentration. For all assays, control experiments are run in parallel to check and correct for any unwanted background activity that may arise from impurities in coupling enzymes or any unwanted side reactions. This is rarely observed, but when it is found, the control rate is subtracted from each individual rate measured.

For enzymes having more than one substrate, the kinetic parameters for each substrate are measured at a fixed (saturating) concentration of the others. This assumes that no substrate inhibition exists at that fixed concentration.

For each isoenzyme, the initial rates at various substrate concentrations are determined and the data obtained are analyzed and stored mainly by the KineticsWizard software (Swainston *et al.*, 2010). This software was specifically constructed and employed to manage enzyme kinetics data obtained by our methodology. The details of managing the data are illustrated in Section 4.3. In very few situations, the data must be analyzed by other software, for example, by Grafit (Leatherbarrow, 2009) and also by COPASI (Hoops *et al.*, 2006). This is because the KineticsWizard in its current status only employs the Michaelis–Menten function for analysis assuming that all reaction mechanisms follow irreversible Henri–Michaelis–Menten kinetics.

$$v = \frac{V\mathrm{max}[S]}{K_{\mathrm{m}} + [S]} \tag{12.1}$$

where v is the initial velocity, V_{max} is the maximum value of the initial velocity, [S] is the substrate concentration, K_{m} is the substrate concentration at half the maximal velocity.

An updated version of the KineticsWizard is currently under development to include more kinetic functions. Not all enzymes exhibit these kinetics, for example, the glycolytic enzyme triose phosphate isomerase (TPI1) isoenzyme from *S. cerevisiae* showed a substrate inhibition pattern in the reaction at the direction of dihydroxyacetone phosphate production and hence the data must be analyzed by the substrate inhibition function which is available in COPASI (Mendes *et al.*, 2009):

$$v = \frac{V\mathrm{max}[\mathrm{S}]}{K_{\mathrm{m}} + [\mathrm{S}]\left(1 + \left(\frac{[\mathrm{S}]}{K_{\mathrm{i}}}\right)^{4}\right)} \tag{12.2}$$

where K_{i} is the inhibition constant.

4.3. Management of enzyme kinetics data

Having large amount of kinetics data created for systems biology research requires not only quick and reliable analysis but also tools to make it accessible for modelers and make it publicly available. Therefore, an integrative automated system has been created to manage kinetics data (Swainston *et al.*, 2010). This web-based system was introduced to support the whole workflow for deriving kinetic data from the laboratory instrument and to make it easily accessible. Managing enzyme kinetics data from instrument to browser involves four steps: data capture, analysis, submission, and visualization. The KineticsWizard has been developed to integrate the first three tasks (Swainston *et al.*, 2010). Data querying and visualization are preformed by manual access and by using web services for automated access to both the MeMo-RK (a local experimental raw data repository) and the SABIO-RK databases. MeMo-RK is a derivation of the MeMo database that was developed for storing metabolomics data (Spasić *et al.*, 2006), then it has been further amended to store the kinetics data (Swainston *et al.*, 2010). The SABIO-RK is a web-based application based on the SABIO relational database that contains information about biochemical reactions, their kinetic equations with their parameters, and the experimental conditions under which these parameters were measured (Rojas *et al.*, 2007; Wittig *et al.*, 2006).

The KineticsWizard was developed to integrate with the BMG Labtech NOVOstar plate reader that was employed in this study. It is launched automatically from the instrument software. This way data are captured, analyzed, and submitted to databases immediately (Swainston *et al.*, 2010).

For data capture, the KineticsWizard extracts experimental data from a spreadsheet where time course measurements of series of assays are running with various reactant concentrations. Here, the followings are specified: organism, gene name, reaction, reactant concentrations, isoenzyme concentration, buffer reagents, and coupling enzymes. All are defined with other metadata. This includes reaction conditions, pH, and temperature under which the assays were carried out. For data analysis, the kinetic parameters (k_{cat} and K_m) are determined by applying an appropriate fitting algorithm to the experimental time series data (Swainston *et al.*, 2010). As discussed in the previous section, the current version of the KineticsWizard provides a fitting algorithm that assumes irreversible Michaelis–Menten kinetics. Initial reaction rates are estimated by a straight line fit; this line can be dragged to perform manual refit, a feature that can be used to correct the automated fit for any noisy data. The initial rate values are to be corrected by subtracting any background rates from control experiments and fed into the Eadie–Hofstee linearized version of the Michaelis–Menten equation to provide estimates of k_{cat} and K_m. For data submission, the raw experimental data is submitted to MeMo-RK and the kinetic parameters and corresponding metadata is submitted to SABIO-RK (Swainston *et al.*, 2010).

Figure 12.1 represents an example that illustrates the output of the data analysis for two sets of experiments run in parallel to measure the kinetic parameters of PDC1, one of the isoenzymes of pyruvate decarboxylase enzyme (EC 4.1.1.1) of *S. cerevisiae*, for two preparations: PDC1 (MORF strain) and PDC1 (TAP strain). Table 12.1 shows the kinetic parameters measured for the three isoenzymes of pyruvate decarboxylase enzyme (PDC1, PDC5, and PDC6).

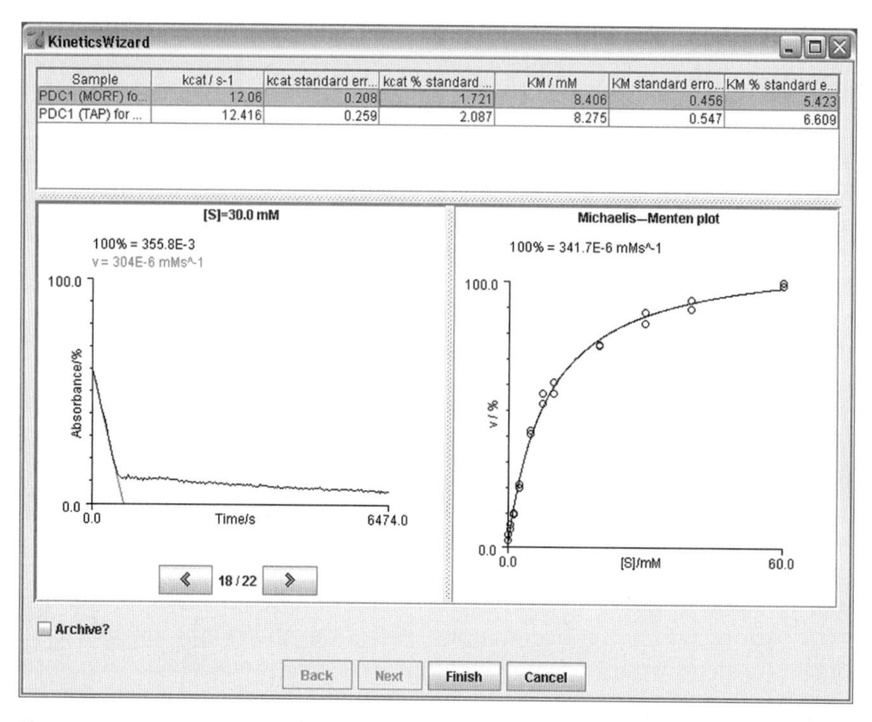

Figure 12.1 Displaying the results of the analysis of the kinetic data measured for two preparations of pyruvate decarboxylase isoenzyme 1 (PDC1) of *S. cerevisiae* (from MORF and TAP strains). Initial rates are plotted against substrate (pyruvate) concentration in the right-hand panel, which shows the Michaelis–Menten curve. The left-hand panel shows each assay in the data set and its automatically fitted initial rate. The top panel displays the calculated kinetic parameters k_{cat} and K_m together with their standard errors.

4.4. Determination of the absolute levels of isoenzymes

To calculate the V_{max} ($V_{max} = k_{cat} \times [E]$), the isoenzyme concentration [E] needs to be known. The number of molecules per cell is determined, which can be converted to a concentration through factoring in the cell volume (or an estimate of this). The absolute level of enzymes is determined experimentally in cell lysates (from cell cultures grown under defined turbidostat growth conditions) using specialized analytical instrumentation consisting of a liquid chromatograph coupled online to a mass spectrometer (LC–MS). We have applied the QconCAT technology to this analysis (Beynon *et al.*, 2005; Pratt *et al.*, 2006). The experimental details and methodologies are described elsewhere within this volume (Carroll *et al.*, 2011). In summary, the levels of analyte tryptic peptides are determined relative to labeled peptide analogues. The labeled peptides result from tryptic digestion of an affinity-purified recombinant protein obtained by heterologous gene expression in a host

Table 12.1 Summary of kinetic parameters measured for pyruvate decarboxylase isoenzymes of *S. cerevisiae*

Isoenzyme	k_{cat} (s^{-1})	K_m for pyruvate (mM)
PDC1 (from MORF strain)	12.06 ± 1.7%	8.41 ± 5.4%
PDC1 (from TAP strain)	12.42 ± 2.1%	8.28 ± 6.6%
PDC5 (from MORF strain)	2.2 ± 3.7%	4.99 ± 13.8%
PDC5 (from TAP strain)	10.32 ± 2.3%	7.08 ± 4.8%
PDC6 (from TAP strain)	9.21 ± 2.6%	2.92 ± 12.5%

Data were analyzed by the KineticsWizard.

organism, for example, *Escherichia coli*. Since trypsin cleaves C-terminal to lysine and arginine, the cell culture media is prepared with the corresponding labeled amino acids, $^{13}C_6$ arginine and $^{13}C_6$ lysine, in place of the usual C12 analogues of each of these. The labeled QconCAT protein is mixed with whole yeast lysates and analyzed by LC–MS following tryptic digestion as described (Carroll *et al.*, 2011). As the number of moles of labeled peptide is known, calculation of the ratios of heavy to light facilitates calculation of the number of moles of analyte relative to the standard. As the cell number is determined experimentally, we can deduce the number of moles per cell. This, in turn, can be converted to molecules per cell through multiplication by 6.0221415×10^{23}, that is, Avogadro's number[1]. The concentration of each isoenzyme [E] is then determined by multiplying the number of molecules per cell by the cell volume.

5. Assays for Measuring the Activities of the Glycolytic Isoenzymes of *S. cerevisiae*

All assays were performed in the reaction buffer which consists of 100 mM MES, pH 6.5, 100 mM KCl, and 5 mM free magnesium ($MgCl_2$) at 30 °C. The conditions used for each enzyme were as follows.

5.1. Hexokinase (EC 2.7.1.1) and glucokinase (2.7.1.2)

Hexokinase isoenzymes (HK1 and HK2) and glucokinase isoenzyme (GLK1) were assayed in the reaction buffer with 5 mM NADP, 50 U/ml glucose-6-P-dehydrogenase, at a fixed ATP concentration (2 mM), and various glucose concentrations (0–20 mM) to measure the kinetic parameters for glucose.

[1] Avogadro's number is defined as the number of molecules in 1 mol, that is, 1 mol = 6.0221415×10^{23} molecules.

The ATP dependence of the isoenzymes was assayed using the same reaction conditions, but at a fixed glucose concentration (20 mM) and various ATP concentrations (0–2.5 mM). Magnesium concentration was adjusted to ensure the availability of free 5 mM concentration by adding additional magnesium (equivalent to the ATP or ADP concentrations in the reaction).

5.2. Phosphoglucose isomerase (EC 5.3.1.9)

The activity of the PGI 1 isoenzyme was determined for both directions of the reaction. For the forward reaction, the activity was measured in the reaction buffer with 12.5 U/ml phosphofructokinase, 5 mM ATP, 1.5 U/ml aldolase, 50 U/ml triose phosphate isomerase, 4.3 U/ml glycerol-3-phosphate dehydrogenase, 1 mM AMP, and 0.2 mM NADH at various concentrations of glucose-6-phosphate (0–25 mM). The measured rate was divided by 2. For the reverse reaction, the activity was measured in the reaction mixture with 0.4 mM NADP and 10 U/ml of glucose-6-phosphate dehydrogenase at various concentrations of fructose-6-phosphate (0–2 mM).

5.3. Phosphofructokinase (EC 2.7.1.11)

The activity of the two subunits of phosphofructokinase(α-subunit, PFK1, and β-subunit, pFK2) in a complex prepared from TAP strain was measured in the reaction buffer with 1.5 U/ml aldolase, 50 U/ml triose phosphate isomerase, 4.3 U/ml glycerol-3-phosphate dehydrogenase, 1 mM AMP, and 0.2 mM NADH at fixed concentration of ATP (1 mM), and various concentrations of fructose-6-phosphate (0–3 mM) for measuring the kinetic parameters for fructose-6-phosphate and at fixed concentration of fructose-6-phosphate (3 mM), and various concentrations of ATP (0–1 mM) for measuring the kinetic parameters for ATP. The rate of the reaction was divided by 2 and magnesium concentration was adjusted to ensure the availability of free 5 mM concentration by adding additional magnesium (equivalent to the ATP or ADP concentrations in the reaction).

5.4. Fructose-1,6-biphosphate aldolase (EC 4.1.2.13)

Aldolase isoenzyme (FBA1) was assayed in the forward reaction in the reaction buffer with 0.15 mM NADH, 50 U/ml triose phosphate isomerase, 4.3 U/ml glycerol-3-phosphate dehydrogenase, and various concentrations of fructose-1,6-biphosphate (0–4 mM). The rate of the reaction was divided by 2.

5.5. Triose phosphate isomerase (EC 5.3.1.1)

The activity of the TPI1 isoenzyme was determined for both directions of the reaction. The activity of the forward reaction was measured in the reaction buffer with 2 mM NAD, 1 mM EDTA, 120 μM DTT, 4 mM sodium arsenate, and 2.5 U/ml glyceraldehyde-3-phosphate dehydrogenase at various concentrations of dihydroxyacetone phosphate (0–20 mM). Magnesium concentration was adjusted to ensure that free 5 mM magnesium is available. The activity in the reverse reaction was measured in the reaction buffer with 8.5 U/ml glycerol-3-phosphate dehydrogenase, and 0.15 mM NADH at various concentrations of glyceraldehyde-3-phosphate (0–49 mM).

5.6. Glyceraldehyde-3-phopsphate dehydrogenase (EC 1.2.1.12)

TDH1, TDH2, and TDH3 isoenzymes of glyceraldehyde-3-phopsphate dehydrogenase were assayed for both directions. The activity of the forward reaction was measured in the reaction buffer with 2 mM NAD, 1 mM EDTA, 120 μM DTT, and 100 mM sodium phosphate at various concentrations of glyceraldehyde-3-phosphate (0–30 mM). For the reverse reaction, the activity was measured in the reaction buffer with 1 mM ATP, 1 mM EDTA, 120 μM DTT, 0.15 mM NADH, and 50 U/ml phosphoglyceric phosphokinase with various concentration of 3-phosphoglycerate (0–8.3 mM). Magnesium concentration is adjusted to ensure that free 5 mM magnesium was available.

5.7. Phosphoglycerate kinase (EC 2.7.2.3)

The activity of PGK1 isoenzyme was measured for both directions of the reaction. For the forward reaction, only the kinetic parameters for ADP was measured in the reaction mixture with 50 mM K_2HPO_4, 5 mM NH_4Cl, 1 mM DTT, 80 g/ml glyceraldehyde-3-P-dehydrogenase, 5 mM NAD, and 10 mM glyceraldehye-3-phosphate at various concentrations of ADP (0–5 mM), the reaction was incubated to reach an equilibrium so that 1,3-bisphosphoglycerate was generated *in situ*. Then the PGK1 enzyme was added to start the reaction. The reaction was followed by measuring the reduction of NAD after the addition of the PGK1 to the reaction. The reverse direction activity was measured in the reaction buffer with 1 mM EDTA, 0.15 mM NADH, and 8.0 U/ml glyceraldehyde-3-P-dehydrogenase at a fixed concentration of ATP (1 mM) with various concentrations of 3-phosphoglycerate (0–16 mM) for measuring the kinetic parameters of 3-phosphoglycerate and at fixed concentration of 3-phosphoglycerate (16 mM) and various concentrations of ATP (0–1 mM) to measure the kinetic parameters of ATP. Magnesium concentration was adjusted to ensure that free 5 mM magnesium was available.

5.8. Phosphoglycerate mutase (EC 5.4.2.1)

The activity for the reverse reaction of PGM1 was measured in the reaction buffer with 0.15 mM NADH, 2 mM mercaptoethanol, 1 mM ATP, 6 U glyceraldehyde-3-phosphate dehydrogenase, 9 U triose phosphate isomerase, 2 U of phosphoglyceric phosphokinase, and various concentrations of 2-phosphoglycerate (0–10 mM). Magnesium concentration was adjusted to ensure that free 5 mM magnesium was available.

5.9. Enolase (EC 4.2.1.11)

The activity of the two isoenzymes ENO1 and ENO2 was measured in the direction of 2-phosphoglycerate to phosphoenolpyruvate formation. The reaction mixture contained 0.15 mM NADH, 10 mM ADP, 14 U/ml pyruvate kinase, 20 U/ml lactate dehydrogenase, and various concentrations of 2-phosphoglycerate (0–3.33 mM). Magnesium concentration was adjusted to ensure that free 5 mM magnesium was available.

5.10. Pyruvate kinase (EC 2.7.1.40)

The activity of the two isoenzymes of pyruvate kinase (CDC19, PYK2) was measured in the reaction buffer with 1 mM fructose-1,6-biphosphate, 20 U/ml lactate dehydrogenase, 0.15 mM NADH, ADP, and phosphoenolpyruvate. For the determination of the kinetic parameters for phosphoenolpyruvate, a fixed concentration of ADP (10 mM) was used and phosphoenolpyruvate concentration was varied (0–3.33 mM). On measuring the parameters for ADP, phosphoenolpyruvate concentration was fixed at 3.33 mM and ADP concentration was varied (0–10 mM). Magnesium concentration was adjusted to ensure that free 5 mM magnesium was available.

5.11. Pyruvate decarboxylase (EC 4.1.1.1)

The activity of PDC1, PDC5, and PDC6 isoenzymes was determined in the reaction mixture with 0.2 mM thiamine pyrophosphate, 88 U/ml alcohol dehydrogenase, 0.15 mM NADH, and various concentrations of pyruvate (0–50 mM).

5.12. Alcohol dehydrogenase (EC 1.1.1.1)

For the forward reaction (acetaldehyde to ethanol), the activity of the alcohol dehydrogenase isoenzyme ADH1 was determined in a reaction mixture with 83 mM potassium phosphate buffer, pH 6.5, 40 mM KCl and 5 mM $MgCl_2$, 0.18 mM NADH, and various concentrations of acetaldehyde (0–8 mM) in 60 μl reaction volumes in a microcuvette. The rate of NADH oxidation was followed by a temperature-controlled Jasco V-630 spectrophotometer. This is the only enzyme where the general assay reaction buffer was replaced due to an incompatibility arising from the reaction of MES buffer with the substrate (acetaldehyde).

6. Concluding Remarks

Systems biology research has developed rapidly in recent times and has thus instigated advances in the development of tools and methodologies to measure many of the cellular components that are essential for constructing mathematical models to understand and describe biological processes. Kinetic parameters of enzymes are a critical component of the biological models. Having hundreds of enzymes to be characterized, we have set out on a large scale effort to obtain enzyme kinetic data for the purpose of constructing models of yeast metabolism. The kinetic parameters are measured experimentally in medium- to high-throughput measurements for isoenzymes that are produced and purified for studying targeted metabolic pathways in yeast. The focus has been initially on the glycolysis pathway and has extended to include isoenzymes, beyond glycolysis, that catalyze other metabolic reactions with a significant flux in *S. cerevisiae* in turbidostat growth conditions.

The measurements are automated and assays are run under conditions similar to the *in vivo* conditions. We are also exploring the employment of a liquid handling robot with the aim of increasing the throughput of assay procedures. The KineticsWizard was developed to automate data analyses and submit data to the SABIO-RK database. Our approach is deliberately generic and can be applied widely to other systems.

ACKNOWLEDGMENTS

We are grateful to the Professors Douglas Kell, Pedro Mendes, and Hans Westerhoff and all colleagues in the MCISB. We thank Neil Swainston and Martin Golebiewski for developing the KineticsWizard for data analysis and storage, Farid Khan for setting up the equipments in the MCISB, and Daniel Jameson for reading and revising the chapter.
The authors would like to thank the BBSRC and EPSRC for funding the MCISB (BB/C008219/1), and the BBSRC funding through grant BB/F003501/1 for the MOSES project. This is a contribution from the Manchester Centre for Integrative Systems Biology.

REFERENCES

Beynon, R. J., Doherty, M. K., Pratt, J. M., and Gaskell, S. J. (2005). Multiplexed absolute quantification in proteomics using artificial QCAT proteins of concatenated signature peptides. *Nat. Methods* **2,** 587–589.

Borger, S., Libermeister, W., and Klipp, E. (2006). Prediction of enzyme kinetic parameters based on statistical learning. *Genome Inform.* **17,** 80–87.

Bruggeman, F. J., and Westerhoff, H. V. (2006). The nature of systems biology. *Trends Microbiol.* **15,** 45–50.

Carroll, K. M., Lanucara, F., and Eyers, C. E. (2011). Quantification of proteins and their modifications using QconCAT technology. *Methods Enzymol.* **500,** 113–131.

Gelperin, D. M., White, M. A., Wilkinson, M. L., Kon, Y., Kung, L. A., Wise, K. J., Lopez-Hoyo, N., Jiang, L., Piccirillo, S., Yu, H., Gerstein, M., Dumont, M. E., *et al.* (2005). Biochemical and genetic analysis of the yeast proteome with a movable ORF collection. *Genes Dev.* **19,** 2816–2826.

Ghaemmaghami, S., Huh, W., Bower, K., Howson, R. W., Belle, A., Dephoure, N., O'Shea, E. K., and Weissman, J. S. (2003). Global analysis of protein expression in yeast. *Nature* **425,** 737–741.

Hoops, S., Sahle, S., Gauges, R., Lee, C., Pahle, J., Simus, N., Singhal, M., Xu, L., Mendes, P., and Kummer, U. (2006). COPASI—A complex pathway simulator. *Bioinformatics* **22,** 3067–3074.

Ideker, T., and Lauffenburger, D. (2003). Building with a scaffold: Emerging strategies for high- to low-level cellular modelling. *Trends Biotechnol.* **21,** 255–262.

Kitano, H. (2002a). Computational systems biology: A brief overview. *Nature* **420,** 206–210.

Kitano, H. (2002b). Systems biology: A brief overview. *Science* **295,** 1662.

Kitano, H. (2005). International alliances for quantitative modeling in systems biology. *Mol. Syst. Biol.* **1,** 10.1038/msb4100011, News and Views.

Kremling, A., Fischer, S., Gadkar, K., Doyle, F. J., Sauter, T., Bullinger, E., Allgöwer, F., and Gilles, E. D. (2004). A benchmark for methods in reverse engineering and model discrimination: Problem formulation and solutions. *Genome Res.* **14,** 1773–1785.

Leatherbarrow, R. J. (2009). GraFit Version 7. Erithacus Software Ltd., Horley, UK.

Malys, N., *et al.* (2011). Protein production in *Saccharomyces cerevisiae* for systems biology studies. *Methods Enzymol.* **500,** 197–212.

Mendes, P., Messiha, H., Malys, N., and Hoops, S. (2009). Enzyme kinetics and computational modeling for systems biology. *Methods Enzymol.* **467,** 583–599.

Moles, C. G., Mendes, P., and Banga, J. R. (2003). Parameter estimation in biochemical pathways: A comparison of global optimization methods. *Genome Res.* **13,** 2467–2474.

Pratt, J. M., Simpson, D. M., Doherty, M. K., Rivers, J., Gaskell, S. J., and Beynon, R. J. (2006). Multiplexed absolute quantification for proteomics using concatenated signature peptides encoded by QconCAT genes. *Nat. Protoc.* **1,** 1029–1043.

Rigaut, G., Shevchenko, A., Rutz, B., Wilm, M., Mann, M., and Séraphin, B. (1999). A generic protein purification method for protein complex characterization and proteome exploration. *Nat. Biotechnol.* **17,** 1030–1032.

Rojas, I., Golebiewski, M., Kania, R., Krebs, O., Mir, S., Weidemann, A., and Wittig, U. (2007). SABIO-RK: A database for biochemical reactions and their kinetics. *BMC Syst. Biol.* **1**(Suppl. 1), S6.

Simeonidis, E., Murabito, E., Smallbone, K., and Westerhoff, H. V. (2010). Why does yeast ferment? A flux balance analysis study. *Biochem. Soc. Trans.* **38,** 1225–1229.

Spasić, I., Dunn, W. B., Velarde, G., Tseng, A., Jenkins, H., Hardy, N., Oliver, S. G., and Kell, D. B. (2006). MeMo: A hybrid SQL/XML approach to metabolomic data management for functional genomics. *BMC Bioinformatics* **7,** 281.

Stein, M., Gabdoulline, R. R., and Wade, R. C. (2008). Calculating enzyme kinetic parameters from protein structures. *Biochem. Soc. Trans.* **36,** 51–54.

Swainston, N., Golebiewski, M., Messiha, H. L., Malys, N., Kania, R., Kengne, S., Krebs, O., Mir, S., Sauer-Danzwith, H., Smallbone, K., Kell, D. B., Mendes, P., *et al.* (2010). Enzyme kinetics informatics: From instrument to browser. *FEBS J.* **277,** 3769–3779.

Takagahara, I., Yamauti, J., Fujii, K., Yamashita, J., and Horio, T. (1983). Theoretical and experimental analyses of coupled enzyme reactions. *J. Biochem.* **93,** 1145–1157.

Wittig, U., Golebiewski, M., Kania, R., Krebs, O., Mir, S., Weidemann, A., Anstein, S., Saric, J., and Rojas, I. (2006). SABIO-RK: Integration and curation of reaction kinetics data. Proceedings of the 3rd International workshop on Data Integration in the Life Sciences (DILS'06), Hinxton, UK, Lecture Notes in Bioinformatics, Vol. 4075, pp. 94–103.

CHAPTER THIRTEEN

Enzyme Kinetics for Systems Biology: When, Why and How

Malgorzata Adamczyk,* Karen van Eunen,[†,‡] Barbara M. Bakker,[†,‡] *and* Hans V. Westerhoff[‡,§]

Contents

* Warsaw University of Technology, Faculty of Chemistry, Institute of Biotechnology ul.Noakowskiego 3, Warsaw, Poland
† Department of Pediatrics, Center for Liver, Digestive and Metabolic Diseases, University Medical Center Groningen, University of Groningen, Groningen, The Netherlands
‡ Department of Molecular Cell Physiology, Faculty of Earth and Life Sciences, VU University, Amsterdam, The Netherlands
§ Manchester Centre for Integrative Systems Biology, Manchester Interdisciplinary BioCentre, The University of Manchester, Manchester, United Kingdom

Methods in Enzymology, Volume 500
ISSN 0076-6879, DOI: 10.1016/B978-0-12-385118-5.00013-X

Abstract

In vitro enzymatic assays of cell-free extracts offer an opportunity to assess *in vivo* enzyme concentrations. If performed under conditions that resemble the conditions *in vivo*, they may also reveal some of the capacities and properties of the same enzymes *in vivo*; we shall call this the *ex vivo* approach. The kinetic characterization of purified enzymes has yet a different utility for systems biology, as does the *in vivo* determination of enzyme activities. All these approaches are different, and it is becoming important that the appropriate approach be used for the intended purpose. Here, we therefore discuss five approaches to the measurement of enzyme activity in terms of the source of the enzyme activity, the identity of the assay medium, and the purpose of the assay.

1. Enzyme Kinetics for Systems Biology: Five Variations on the Theme

As shown in Table 13.1, there are at least five modalities in which the determination of enzyme activities may be important for systems biology. They differ in purpose. Depending on the purpose, it is better to assay the activity in extract or to determine it in a preparation of the purified enzyme. Likewise, the choice between assaying the enzyme in a medium that is best for the enzyme *per se* and assaying it in a medium that resembles the conditions *in vivo* depends on the purpose. We shall now discuss each of the five approaches a bit more in detail.

Table 13.1 Various approaches using enzyme activity assays with their names, conditions, and purpose

Approach name	Enzyme source	Conditions	Purpose
Enzymology	Purified	Diverse, enzyme optimized	Enzyme mechanism
In vitro lysate	Extract	Diverse, enzyme optimized	Estimation of *in vivo* protein concentration, for example, for hierarchical regulation analysis
Ex vivo	Extract	Homogeneous, *in vivo* like	Middle-out systems biology
Bottom-up	Purified	Homogeneous, *in vivo* like	Bottom-up systems biology
In vivo	*In vivo*	*In vivo*	*In situ* systems biology

1.1. Enzymology

Enzyme kinetics has a long history (Cornish-Bowden, 1995; Michaelis and Menten, 1913). In its beginnings, it was closely connected to the understanding of the biological basis of the chemical processes that occurred in living systems. Metabolic pathways explaining the conversion of food to products secreted by living organisms, and to chemical components of the organisms themselves, were identified by demonstrating that proteins purified from extracts from those organisms had specific chemical activities. Characteristic of the biochemical nature of these assays was that the rate exhibited a maximum when the substrate concentration was increased. This allowed for a characterization of the enzyme by the concentration of the substrate that drove half of this V_{max}, that is, the K_M. The maximum was usually proportional to the concentration of the enzyme and could thereby be used as a measure of the concentration of the latter. A branch of biochemistry called enzymology then developed. It determined how the protein responsible for the chemical reaction, and thereby called enzyme, carried out the often complex catalysis. Especially in the cases of Gibbs energy transduction between chemical reactions that are not interlinked chemically, between transport reactions, or between chemical reactions and transport reactions, this has been and for some cases remains (Boogerd *et al.*, 2011) challenging and highly interesting.

For some enzyme activities, the rates depended in complex ways on the concentrations of substrates, products, or other substances modifying the reaction rate allosterically. It was understood that this was important for the regulation of the processes in living organisms. The determination of the mechanisms of catalysis and this molecular regulation required accurate and quantitative determination of the concentration dependence of the reaction rate, which, in turn, required the reaction rate to be well above the experimental noise. This made it important to examine which reaction conditions, such as pH, temperature, and concentration of cofactors and coenzymes, produced the highest reaction rate that was most dependent on a factor of interest. Often, the conditions that were optimal for one enzyme were not optimal for a different one. We therefore describe such assay conditions as "enzyme-optimized conditions" (see Table 13.1).

In the early days of enzymology, enzymes would be detected in extracts of the living organism under study. However, such extracts also exhibited activities catalyzed by other enzymes. Therefore, it was much better to purify the enzyme from the extract, or, later, overexpress the genes encoding the enzyme and then purify the macromolecules from a suitable heterologous host organism.

Enzymology is important for systems biology. Systems biology investigates how biological functioning finds much of its origin in interactions between components of living organisms (Westerhoff and Palsson, 2004). The enzymes are among the more fundamental components (Westerhoff *et al.*, 2009). While the biochemical details of the reaction and regulation mechanisms are not essential for systems biology (although they are often highly interesting), the precise identity of the chemical reaction and the precise dependence of the reaction rate on the concentration of the metabolites are. The former is important for the establishment of what genome-wide networks of enzymes might do in terms of overall chemical processes (Herrgard *et al.*, 2008). The latter is important to determine whether and when and at what rate the network will actually carry out which of those possible processes.

Biophysical techniques been developed for *in vitro* activity measurement of purified enzymes including absorption and fluorescence spectroscopy, heat release in isothermal titration calorimetry (Olsen, 2006) as well as radiometry (Koto and Inoue, 1981), chromatography (Pietta *et al.*, 1997), NMR (Johnson *et al.*, 1977), and mass spectrometry (Rathore *et al.*, 2010). The absorption and fluorescence spectroscopy monitoring NADH or NADPH are popular detection methods because many biochemical reactions in living organisms involve these redox coenzymes and enable continuous measurement of NADH of NADPH oxidation or of NAD or NADP reduction by monitoring the change in absorbance at 340 nm. In the case of enzymes with activities that do not involve NAD(P)H/NAD(P) transformation, the enzyme reaction of interest can often be monitored indirectly by including an excess of an additional enzyme that couples the reaction to a different reaction involving coenzymes NADH or NADPH.

1.2. The *in vitro* lysate approach

Cell lysis followed by centrifugation and isolation of the supernatant should yield the cell's interior with all its metabolic enzymes hopefully still active. Usually, the extract is prepared by mechanical or enzymatic disruption of the plasma membrane and, sometimes, the cell wall at low temperature. In the assay medium, the extract is strongly diluted, preventing the enzymatic reaction measurement to be perturbed by metabolites present in the extract. In this "*in vitro* lysate approach," the assay conditions are often again (as in the enzymology approach) optimal for measurement of the enzyme activity of interest in the extract. The resulting conditions may well be identical to those developed for the kinetic assay conditions used in the enzymology of the same enzyme. As a consequence of these specific assay conditions, enzyme properties measured in this "*in vitro* lysate approach" may differ from those in the intact cell. Michaelis constants, catalytic rate constants, Hill coefficients, elasticity coefficients, and even apparent equilibrium constants tend to depend significantly on properties that often differ between enzyme-optimized conditions and the conditions in the living cell in which the enzyme originates. Modelers often use these parameters when more appropriate experimental data are not available. Then, they should at least annotate their model explicitly in terms of this limitation. It is not so bad perhaps that they may model their pathway of interest at a nonphysiological pH: much worse is that they may model the various enzymes of a single pathway as if at different pHs.

For other purposes, it may not be such a problem that the assay conditions are not physiological and differ between the enzymes in the same pathway. If, for instance, the amount of a certain enzyme needs to be compared between cells grown under two different conditions, then this may be achieved by comparing the V_{max} upon extraction measured in an assay medium that has nothing to do with the intracellular milieu but leads to high enzyme activities. For this, it is only required that under both assay conditions, that is *in vitro* and *in vivo*, the reaction rate is proportional to the enzyme concentration.

In cell extracts, a plethora of catalytic activities is present. Consequently, the detection method of a specific reaction requires the assay to be more specific than when the enzyme has been purified. Specificity may be achieved by the injection of a chemical substrate specific for the reaction of interest and a specific detection method. The two most popular detection methods applied to enzymes studies in cell extracts are spectrometric and fluorimetric detection, which are based on the continuous measurement of NADH of NADPH oxidation or of NAD or NADP reduction already discussed above.

1.3. The *ex vivo* approach

The approach that we call *ex vivo* here again measures enzyme activity in cell extracts, but now in an assay medium that is the same for all enzymes that derive from the same compartment in the same organism, and is similar to the actual intracellular environment. This should reduce the change in activity of enzymes as they are extracted. However, it is difficult to get the assay medium identical to the intracellular milieu if only because the latter tends to contain substances that intervene with an accurate determination of the enzyme's V_{max} (van Eunen *et al.*, 2010a,b) (Table 13.2). Recently, a medium was proposed to simulate the intracellular conditions of *Saccharomyces cerevisiae* (van Eunen *et al.*, 2010a,b). Especially for enzymes that act freely in the cell fluids, like the glycolytic enzymes, data derived from this medium may represent the *in vivo* situation better than those from the above-described approaches. Indeed, Van Eunen showed that a kinetic model based on such data predicted metabolite concentrations and fluxes—at different steady states and under dynamic conditions—much better than the same model parameterized with *in vitro* lysate data (Van Eunen, 2010) (Table 13.3). But for other metabolic routes, the differences between *in vivo* and *ex vivo* may well be more substantial than that they can be annihilated by this medium: physical interactions with structures in the cell, such as microtubules (Cortassa *et al.*, 1994), organelles (Laterveer *et al.*, 1995), or the plasma membrane are likely to be disturbed by breaking the cells open and by pelleting the nucleus and the plasma membrane away from the presumed (Srere, 1987) cytosol. Protein kinases and protein phosphatases, for instance, will be diluted, as will be their substrates. The ratio of the

Table 13.2 Physicochemical condition of *in vivo*-like media

Element	Measured amount (g per kg dry weight)	Calculated intracellular concentration (mM)	Concentration or ions in *in vivo*-like medium (mM)
Calcium	0.16 ± 0.07	1.9 ± 0.1	0.5
Potassium	28 ± 2	342 ± 30	300
Magnesium	2.6 ± 0.0	51 ± 1	2[a]
Sodium	1.3 ± 0.1	28 ± 3	20
Phosphate	20 ± 1	304 ± 14	50
Sulfate	3.0 ± 0.0	45 ± 0	2.5 – 10
Glutamate	–	75[b]	245

The biomass composition was determined by inductively coupled plasma atomic emission spectroscopy in samples from CEN.PK 113-7D (MATa) grown in glucose-limited (42 mM in the feeding medium) chemostat at 30 °C in CBS mineral media. Yeast cells were grown under respiratory conditions at a dilution rate 0.1 h^{-1}. The intracellular concentrations of the measured elements represent average total concentrations of chemical elements. Composition of *in vivo*-like media.
[a] Free magnesium.
[b] LC–MS measured (Canelas *et al.*, 2009).

Table 13.3 V_{max} values measured under the optimized and the *in vivo*-like conditions in the absence of phosphatase inhibitors

Enzyme	Optimized V_{max} (mmol min^{-1} g $protein^{-1}$)	*In vivo*-like V_{max} (mmol min^{-1}g $protein^{-1}$)
HXK	1.8 ± 0.1	0.80 ± 0.06
PGI reverse	4.0 ± 0.0	2.0 ± 0.1
PFK	0.69 ± 0.10	0.25 ± 0.0(0.80 ± 0.10[a])
ALD	0.76 ± 0.16	1.2 ± 0.1
TPI	97 ± 5	26 ± 0
GAPDH reverse	6.5 ± 0.2	3.2 ± 0.1
PGK reverse	10 ± 1	9.4 ± 0.3
ENO	0.99 ± 0.04	0.96 ± 0.06
PYK	3.6 ± 0.5	3.1 ± 0.1
PDC	0.65 ± 0.12	1.5 ± 0.1
ADH reverse	10 ± 0	3.5 ± 0.1

Errors represent standard errors of the mean of at least three independent cell-free extracts from steady-state sample from a single chemostat culture $D = 0.1\ h^{-1}$.
[a] V_{max} with saturation of Fru6P concentration (reproduced from van Eunen *et al.*, 2010a,b).

phosphorylation to the dephosphorylation rate may well change with a concomitant effect on the phosphorylation state of the enzymes. One would further have to ensure that during the extraction procedure, the protein complexes that exist *in vivo* remain intact, such that cooperativity through enzyme–enzyme interactions Srivastava and Bernhard (1987) is retained, possible metabolite channeling is unperturbed, and levels of covalent modifications of the enzymes are retained, also during the subsequent activity assay. In some cases, this may be achieved through quick and carefully employed extraction procedures. In other cases, the *in vivo* conditions may be mimicked by adding agents that cause macromolecular crowding (Minton, 1992; Rohwer *et al.*, 1998). In conclusion, the *ex vivo* approach may offer some direct insight into complex biological processes if carefully applied to suitable systems. When macromolecular structures play an important role, the *in vivo* approach, described in Section 1.5, is more appropriate.

1.4. The bottom-up approach

The approach to determine and understand enzyme kinetics in living cells that is most bottom-up may be to identify each enzyme on the basis of the genome sequence, clone and overexpress the gene in some host, purify the resulting enzyme, and assay that in a test tube. This approach is described in Chapter 12. Combined with proteomics and an assay medium that resembles the *in vivo* conditions (like in the *ex vivo* approach), repetition of this approach for all enzymes in a pathway should enable one to make a mathematical model of that pathway bottom-up and test whether one understands the functioning of the pathway on the basis of its components. This was pioneered by Heinrich and coworkers for human erythrocyte glycolysis. They started with models that assumed that metabolite concentrations were far below the corresponding Michaelis constants but subsequently added more and more kinetic realism (e.g., Werner and Heinrich, 1985). The aim has also been achieved to a significant extent by Bakker and Michels and coworkers for a human parasite causing sleeping sickness (Bakker *et al.*, 1997). It reproduced pathway flux and concentrations reasonably well (to the limited amount that this was testable). Further interrogation even produced understanding of an unsuspected mechanism for preventing a metabolic explosion and a possible *raison d'être* of the glycosomal compartment in this organism (Bakker *et al.*, 2000, Haanstra *et al.*, 2008). In these pioneering studies, the assay conditions were not yet representing the *in vivo* conditions. For the trypanosome model, however, most enzymes were assayed under mutually similar conditions.

This "bottom-up approach," also has limitations, however. First of all, it is laborious, not all enzymes can be purified into stable forms, and purification and then assay of purified transporters are fraught with problems. Moreover, in such an analysis, the proteins are taken out of their cellular context, cloned and expressed as single genes or in complexes, and purified by affinity chromatography, in an attempt to reconstitute *in vivo* features, *in vitro*. The growth conditions under which recombinant heterologous proteins are produced, or endogenous proteins are overproduced, typically satisfy a goal distinct from maintaining the physiological activity level, that is, optimization for maximum yield. These optimal growth conditions may be diverse for a set of proteins belonging to the same pathway and may cause partial deactivation of the recombinant enzyme, by proteolysis, aggregation, or covalent modification. In principle, measurement of the degree of posttranslational modifications within the physiological protein pool, followed by separate kinetic characterizations of the various enzyme forms, should get the methodology in check. This is laborious, however, and at or beyond the limits of current technology. Although in principle, the QconCAT technology is able to quantify the absolute concentrations of a protein and its modifications in a cell, labile phosphorylations constitute an exception, and the information on the complete spectrum of protein modifications is lacking (Johnson *et al.*, 2009). Posttranslational modifications include not only moderately stable labeling like serine-phosphorylation, acetylation, methylation, sumoylation, biotinylation but also labile redox modifications and histidine phosphorylations of proteins, which are hardly detectable in mass spectrometry (Slazano *et al.*, 2008).

1.5. *In vivo* enzyme activity

In vivo measurements of enzyme kinetics have been carried out by in-cell NMR and by in-cell enzyme assays in water-in-oil microdroplets involving fluorescence-activated cell sorting (Huebner *et al.*, 2008). In-cell NMR (rather than conventional *in vivo* NMR) as well as the microfluidic approach are still at a trial and development stage and have been tested with cells overexpressing proteins from a plasmid or cells injected with ^{15}N-labeled target proteins (Sakai *et al.*, 2006; Serber and Dotsch, 2001). Rapid perturbation of the metabolic state of living cells such as pioneered by Theobald *et al.* (1997) also gives *in situ* information about the kinetics of the enzymes, but this information is indirect and involves the kinetic properties of various enzymes at once. Although such experiments are highly valuable, we see them more as serving the purpose of validations of bottom-up systems biology models than as the providers of input in the form of the kinetic properties of individual enzymes. None of these techniques has been implemented into high-throughput workflows in systems biology. Although they will become highly important in the future, we shall here limit the discussion to *in vitro* approaches.

2. Three System Biology Approaches that Use Different Enzyme Kinetics

We will now discuss three different types of systems biology research in which enzyme kinetics plays a role but in ways that require different ones of the above-mentioned five approaches. One is the silicon cell approach, which aims at making precise mathematical replica of pathways in living cells. A second is hierarchical regulation analysis, which aims to determine how much intracellular processes are regulated by gene expression and how much by metabolic regulation. A third focuses on the metabolic part of the regulation and tries to establish whether *in situ* enzyme action varies because of changes in substrate concentration, changes in product concentration or allosteric effects.

2.1. The silicon cell and the testing of systems biology mechanisms

One type of systems biology asks whether we can understand the functioning of living cells and organisms on the basis of what we know about their components and about the interactions between the components. The approach starts from measurement of all molecular properties and nonlinear interactions (such as in Michaelis–Menten kinetics) and integrates these through precise mathematical models. This is called the silicon cell approach (Snoep and Westerhoff, 2004). The ultimate silicon cell approach would start from purified enzymes and would then assay these under *in vivo* conditions.

Where the interactions are linear, the properties of the system are just the sum of the properties of the components. Where the interactions are nonlinear, they can lead to qualitatively new properties, new in the sense that they not yet occur in the components. These are called "emergent properties." Mass is not an emergent property because the mass of a metabolic pathway is simply the sum of the masses of its components. The duration of the cell cycle *is* an emergent property, as none of its molecular components has a cycle time by itself (a single component would not cycle), hence they can never be added.

The ultimate test for the silicon cell approach is to validate all model predictions experimentally, a test that has not yet been achieved so far (Teusink *et al.*, 2000). Nevertheless, the route toward this goal has generated interesting results.

2.2. Hierarchical regulation analysis

Intracellular enzyme activities may change with time when the cell adjusts its biochemical functions to a change in circumstances. For quite a while, there have been effectively two schools of thought. One of these studied changes at the metabolic level that impinged on the enzymes and caused the changes in metabolic fluxes. The other school of thought exclusively looked at changes in transcription factor activity and consequent changes in transcription; after all, protein activities in the cells should be a projection of the genome, and consequently, changes effecting regulation should be dominated by those at the transcription/translation level. Both views could persist because flux changes are often accompanied both by changes in gene expression and by changes in the levels of metabolites. Of course, changes in metabolic fluxes might only be caused to 0.1% by transcriptional changes, which could themselves be due to changes in metabolism, and for 99.9% by changes in metabolism (or *vice versa*), but before such quantification, this could not be decided upon. Only if the changes were compared quantitatively, the jury could be out. Such a comparison was not endeavored, as it was not clear how metabolic and gene-expression regulation could be disentangled. After all, metabolic control analysis had shown that pathway fluxes are often controlled by multiple enzymes (Groen *et al.*, 1982), and that this flux control is, in turn, determined by the elasticities of response of all these enzymes to changes in metabolite concentrations (Kacser and Burns, 1973) and by network topology (Westerhoff and Arents, 1984; Westerhoff and Kell, 1987).

Ter Kuile and Westerhoff (2001) developed a perspective that does enable the dissection of gene-expression and metabolic regulation of reaction rates, which has been broadened to include the regulation of pathway fluxes (Westerhoff *et al.*, 2010). The approach is called hierarchical regulation analysis and is dealt with extensively in Chapter 27. Therefore, we here limit the discussion to the dual use of enzyme kinetic assays in this method and to the two different types of kinetic assays required.

One use is in the determination of the amount of enzyme. The hierarchical, or gene-expression (but see below), regulation coefficient for the rate of an intracellular process is defined in terms of the change in concentration of the enzyme as:

$$\rho_{\mathrm{h},i} \underline{\underline{\mathrm{def}}} \frac{\mathrm{dln}[\mathrm{enzyme}_i]}{\mathrm{dln}|v_i|}$$

Because changes in reaction rates are often only moderate, the changes in enzyme concentrations must be assayed accurately. Proteomics is not always accurate enough for this. Most, though not all (Daran-Lapujade *et al.*, 2007), applications of hierarchical regulation analysis have therefore measured enzyme activities in cell lysates as stand-ins for enzyme concentrations. The following relationship is then used:

$$[\mathrm{enzyme}_i] = \frac{V_{\mathrm{max},i}}{k'_{\mathrm{cat},i}}$$

Here, $k'_{\mathrm{cat,i}}$ refers to the k_{cat} under standard *in vitro* conditions, the same for all samples, that is, not covariant with the regulation of k_{cat} that may take place *in vivo* (which would be classified as metabolic regulation). As a consequence:

$$\frac{\mathrm{d}\ln\left[k'_{\mathrm{cat},i}\right]}{\mathrm{d}\ln|v_i|}=0$$

And therefore:

$$\rho_{\mathrm{h},i}\underline{\underline{\mathrm{def}}}\frac{\mathrm{d}\ln[\mathrm{enzyme}_i]}{\mathrm{d}\ln|v_i|}=\frac{\mathrm{d}\ln\left[V_{\mathrm{max},i}\right]}{\mathrm{d}\ln|v_i|}-\frac{\mathrm{d}\ln\left[k'_{\mathrm{cat},i}\right]}{\mathrm{d}\ln|v_i|}=\frac{\mathrm{d}\ln\left[V_{\mathrm{max},i}\right]}{\mathrm{d}\ln|v_i|}$$

The relationship shows that a relative change in enzyme concentration (dln [enzyme]) can be measured as a relative change in V_{max} in the extract, provided that the conditions are such that the *in vitro* k_{cat} is always the same.

These considerations reveal a subtlety in this approach to assaying enzyme concentration through *in vitro* V_{max}. If the regulation that occurs *in vivo* affects the k_{cat} of the enzyme in a way that is stable upon preparing the extract and subjecting the extract to the V_{max} assay, then that regulation will be classified, as hierarchical regulation. An example would be adenylylation of glutamine synthetase (Bruggeman *et al.*, 2005). It is fine to classify this regulation through signal transduction as hierarchical regulation (Kahn and Westerhoff, 1991), but it should not be classified as gene-expression regulation. To ensure that a V_{max} regulation is due to gene-expression regulation, one should establish that there is no covalent modification of the enzyme, or measure the protein concentration of the enzyme directly (Daran-Lapujade *et al.*, 2007).

2.3. Metabolic regulation analysis

In hierarchical regulation analysis, the metabolic regulation is often calculated, as 1 minus the hierarchical regulation. It should of course also be possible to determine metabolic regulation more directly, in terms of the elasticities of the enzyme with respect to the concentration changes of the various metabolites affecting it, and the changes of those concentrations (Sauro, 1990):

$$\rho_{\mathrm{m}}=\frac{\mathrm{d}\ln\left(f\left(S,P,X,k_{\mathrm{cat}},K_S,K_P,K_X,K_{\mathrm{eq}}\right)\right)}{\mathrm{d}\ln v}=\varepsilon_S\frac{\mathrm{d}\ln(S)}{\mathrm{d}\ln v}+\varepsilon_P\frac{\mathrm{d}\ln(S)}{\mathrm{d}\ln v}+\varepsilon_X\frac{\mathrm{d}\ln(X)}{\mathrm{d}\ln v}$$

The elasticities (ε's, partial log–log derivatives of the reaction rate with respect to the concentrations of the various metabolites influencing the reaction rate) should come from well-known rate equations, or from, or validated by, assays of the enzymes *in vitro*. The elasticities should be relevant for the *in vivo* conditions. Hence the assays should be of the *ex vivo* type, that is, under conditions that are close to those *in vivo*. It should be noted therefore that for hierarchical regulation analysis comprising both the hierarchical and the metabolic part, two different types of enzyme kinetic assays are needed.

3. Biological Material

Various factors concerning the preparation of the biological material have to be taken into consideration when aiming to obtain uniform yeast cultures for regulation or other analyses. Use of a chemically defined medium is necessary to precisely define growth conditions and help data reproducibility. It is particularly beneficial in studies on nutrient limitation. Traditionally, cells have been grown in batch culture and harvested in exponential growth phase. During that phase, cells divide at a maximal rate and their generation time remains constant. The cells would appear to be constant in time with regard to metabolite and enzyme concentrations. The latter constancy is not guaranteed, however. Some proteins may well be stabler than the duration of the cell cycle. Hence they may require multiple cell cycles for approaching their steady-state levels. Because control on growth rate of individual enzymes is small (Jensen *et al.*, 1993a; Kacser and Burns, 1973), remaining variations in enzyme levels may not affect growth rate, even though they are still being regulated. Technically it is difficult, though not impossible (Jensen *et al.*, 1993b), to maintain microorganisms in serial, low-density batch culture. Therefore, continuous cultures are important sources of well-defined intact cells.

Several continuous culture fermentation procedures have been developed over the years to study the metabolism of microorganisms as relevant for industrial fermentation or for the controlled production of commercially important secondary metabolites and enzymes. Chemostat cultures allow for steady-state growth over several generations (Van Hoek *et al.*, 1998). In chemostats, one of the nutrients, the carbon or nitrogen source, is limiting, and the specific growth rate is controlled by the experimenter via the supply of the limiting nutrient medium (Kleman *et al.*, 1991), and the concentration of the limiting substrate in the culture vessel is controlled by or through the microorganism (Snoep *et al.*, 1994). For instance, if the effect of reduced expression of a gene encoding a glycolytic enzyme is studied in glucose-limited chemostats at fixed dilution rate, then the mutant will be growing at higher glucose concentrations. In batch growth, this problem does not arise. In chemostats run at substantial cell density, the medium will contain appreciable levels of metabolic products. When the regulation by external glucose concentration is studied by comparing cells from chemostats run at different dilution rates, effects through the different product concentrations are likely. It should also be noted that chemostats do not necessarily solve the problem of cellular heterogeneity that occurs in batch cultures. Under respiratory growth conditions in chemostat at dilution rates below a dilution rate D of 0.25 h^{-1}, mixed populations of cells have been observed (Achilles *et al.*, 2006). Cells that consumed glucose at high rates had large cell size, while the cells that consumed glucose at low rate had lower content of RNA and a small cell size but high glucose carrier capacity. At higher dilution rates (from $D = 0.3$ h^{-1} upwards), the differences in glucose uptake diminished and the cell populations showed less distinct affinities to the substrate as measured by 2NBD-glucose uptake (Achilles *et al.*, 2006). Nutrient limitation in chemostats followed for generations may lead to adaptations at the regulation level triggered by the growth conditions, which is what would be studied as gene-expression regulation. The cells may also become subject to Baldwinian evolution or genetic assimilation (Waddington, 1952), or to plane selection of mutants (Groeneveld *et al.*, 2009, Jansen *et al.*, 2005).

Examples of flow-through systems alternative to chemostat are turbidostat and nutristat procedures. In a turbidostat, nutrient feed rate is feedback controlled based on the turbidity (optical density, cell mass) of the culture (Munson, 1970). The nutristat is based on feedback control of a nutrient concentration. An example of a nutristat is a glucose-stat, where glucose concentration remains unchanged during the time of cultivation. Both the turbidostat and the nutristat enable experiments at maximum growth rate, where the chemostat is unstable. The glucose-stat can be set up for glucose limitation as well, when the organism's Monod constant for glucose is not too low. As the threshold for carbon source limitation and starvation conditions is a characteristic feature of an individual microorganism, the glucose concentration setup has to be treated in a careful way.

The kinetic model of glycolysis in yeast published by Teusink *et al.* in 2001 was based on *in vitro* V_{max} values measured in cell extracts generated from lysed, compressed commercial yeast, that is, not at all from well-specified biological material. The reason was that this increased relevance for the baking industry. But concordantly, the systems biology value of the model is limited. The enzyme assays were carried out in a single medium (50 mM pipes, pH 7.0, 100 mM KCl, 5 mM Mg_2SO_4) that was not yet optimized for being similar to the *in vivo* milieu. Hence the method was a hybrid between the *ex vivo* and *in vitro* lysate methodologies. Since 2001, other research groups have implemented various cultivation methods.

In hierarchical regulation analysis, Rossell and coworkers (Rossell *et al.*, 2005, 2006) obtained cell extracts from pH-controlled, aerobic batch cultures, whereas van Eunen and colleagues (van Eunen *et al.*, 2009) implemented chemostat culturing under glucose-limited conditions. The differences need to be considered when interpreting and comparing the observed regulation. Regulation may well depend on what the cells have adapted to.

4. Protocols

4.1. Preparation of yeast cell-free extracts by homogenization

Downstream enzyme assays require extraction from cells of the enzymes in their active form. This is not as trivial as it may seem. For instance, pyruvate carboxylase is cold-labile, and incubation at low temperature (0–4 °C) has been shown to inactivate the enzyme (Scrutton and Utter, 1965). As many enzymes, however, are more stable on ice than at room temperature, the procedure outlined below is carried out at low temperature. This illustrates that any procedure, however, carefully designed must be reconsidered when applied to a new set of enzymes. Similarly, cell extracts that have not been supplemented with proteases inhibitors should be handled with great care to prevent enzyme degradation during extraction. Yet, van Eunen *et al.* (2010a,b) have shown that enzyme activities in their cell-free extracts did not decrease while treated with buffer without protease inhibitors. When preparing crude extract for *in vitro* protein production and synthetic biology, Jewett *et al.* (2008) did not add protease inhibitors either. However, if no measures are taken to inactivate proteolytic enzymes in the cell extracts, all the downstream manipulations have to be performed in ice-cold buffers with samples kept on ice. To avoid variation in cell-free extract composition due to centrifugation or degradation of protein complexes, the duration of each step of the protocol has to be strictly controlled. Further, to avoid a decrease in enzyme activities, it is advised to perform assays on freshly prepared cell-free extracts.

It would be ideal if the extraction buffer consisted of reagents that do not alter properties of the analyte such as its redox state. Addition of a reducing agent like dithiothreitol (DTT) may cause downstream effects on protein activity by interfering with cysteine thiols in catalytic sites of certain enzymes (Roche and Cate, 1976; Trotta *et al.*, 1974). However, supplementation of reaction buffer with DTT is necessary in the case of enzymes that perform forward and reverse redox action, for example, GAPDH.

4.1.1. Materials

Glass beads (425–600 μm, Sigma), 1.5-ml Eppendorf tubes, 2-ml screw-cap microvials, potassium dihydrogen phosphate (KH_2PO_4, Merck, 1.05108.0500), di-Potassium hydrogen phosphate (K_2HPO_4, Merck 1.05104.1000), magnesium chlorate (Fisher 232-094-6)

4.1.2. Preparation of solutions

1× Sonication buffer: 100 mM potassium buffer, pH 7.5, 2 mM $MgCl_2$, and 1 mM DTT. For 1 l of buffer, take 800 ml of 100 mM K_2HPO_4 and adjust the pH to 7.5 with 100 mM KH_2PO_4. Add 0.4 g l^{-1} $MgCl_2$ and the freshly prepared DTT.

4.1.3. Method

(Modified from de Jong-Gubbels *et al.*, 1995)

1. Thaw samples on ice until defrosted.
2. Centrifuge them at 17,000 RCF for 1 min at 4 °C. Resuspend the pellet in 1 ml of sonication buffer. Repeat the centrifugation step. Transfer the sample to a precooled screw lock tubes with 0.75 g of glass beads (425–600 μm, Sigma). Make sure to firmly finger tight the screw caps that will prevent opening the tubes in the cell processor.
3. Homogenize the samples in the Minibeadbeater in buckets filled with ice in the cold room (4 °C). We recommend shaking eight bursts of 10 s with shaking speed 3450 oscillations min^{-1}. Cool the samples between bursts on ice for 1 min.
4. The last step of the extraction procedure is to pellet cell debris and remaining intact cells by centrifugation for 10 min at 17,000 RCF at 4 °C. Transfer the homogenized samples into Eppendorf tubes. The supernatant contains enzymes and should be handled with care and kept on ice until used in an experiment.

4.2. Enzyme capacities (V_{max}) in submilliliter reaction volumes: Hexokinase as example

The following procedure has been designed for miniaturized enzyme assays in reaction volume of 300 μl in a 96-well microplate. Concentrated cell extracts (at a protein concentration 0.4–0.7 mg ml^{-1}) were used undiluted and at four dilutions (see below). The reaction rate was measured during incubation at 30 °C, as a reduction in NADH concentration over time monitored by change in absorbance at 340 nm using Zenyth3100 plate readers. Enzyme capacities were expressed as moles of substrate converted per minute per milligram of extracted protein. The automated method is adaptable to measurement of any enzyme that can be monitored in a NAD(P)H-linked assay.

Our example is that of a hexokinase assay with a coupling enzyme glucose-6-phosphate dehydrogenase. Using enzymes from different manufacturer might create a serious obstacle in generating reproducible data. Before using an enzyme from a new source, it is good laboratory practice to check activity units as well as the composition of the enzyme's storage solution. The protocol emphasizes practical aspects of an enzymatic assay, which can be performed in various ways with different handling of the reagents.

4.2.1. Materials

Microplates PS, flat bottom, 96 well (Greiner Bioone 655101)

4.2.2. Reagents

For the hexokinase assay: Imidazole (C3H4N2, Merck 1.04716.0050), nicotinamide adenine dinucleotide phosphate (NADP, Roche 10240354001), glucose-6-phosphodehydrogenase from yeast (G6PDH, Roche 10127671001), D-Glucose anhydrous (Merck 346351-250), magnesium chloride ($MgCl_2$, Fisher 232-094-6), adenosine-5′triphosphate disodium salt (ATP, Roche 10127523001), potassium dihydrogen phosphate (KH_2PO_4, Merck, 1.05108.0500), Di-potassium hydrogen phosphate (K_2HPO_4, Merck 1.05104.1000), magnesium chlorate (Fisher 232-094-6).

1× Sonication buffer: 100 mM potassium buffer, pH 7.5, 2 mM $MgCl_2$. For 1 l of buffer, take 800 ml of 100 mM K_2HPO_4 and adjust the pH to 7.5 with 100 mM KH_2PO_4. Add 0.4 g l^{-1} $MgCl_2$.

Master mix: 50 mM Imidazole/HCl buffer, pH 7.6, 1 mM NADP, 5 mM $MgCl_2$, 10 mM glucose, 1.8 U ml^{-1} G6PDH. Prepare the Master mix by diluting stock solutions: 0.5 M Imidazole, pH 7.6, 35 mM NADP, 1 M $MgCl_2$, 1 M glucose. Use glucose-6-phosphodehydrogenase without dilution. Stock solutions of NADP and $MgCl_2$ (may otherwise precipitate) must be prepared freshly or stored in small aliquots at −20 °C and used only once. While setting up the enzymatic assay, regardless of the given of hexokinase, redox cofactors like NADH/NAD, allosteric effectors, adenosine phosphonucleotides, and sugar phosphates have to be prepared freshly.

1. While working with a new batch of cell-free extract, always begin the assays with undiluted cell-free extract (Fig. 13.1).
2. Dilute the cell-free extracts in sonication buffer in Eppendorf tubes and keep on ice. Dilution factors should be optimized for each of enzymes separately. In our experience, enzyme activities determined in undiluted samples are not always proportional to activities after dilution (Fig. 13.2). In this case, some of the data points should be excluded, but a note mentioning the nonlinearity should be added to any reported results, as this could be relevant for extrapolation to the enzyme activity *in vivo*. Three independent assays performed on undiluted extract should be an

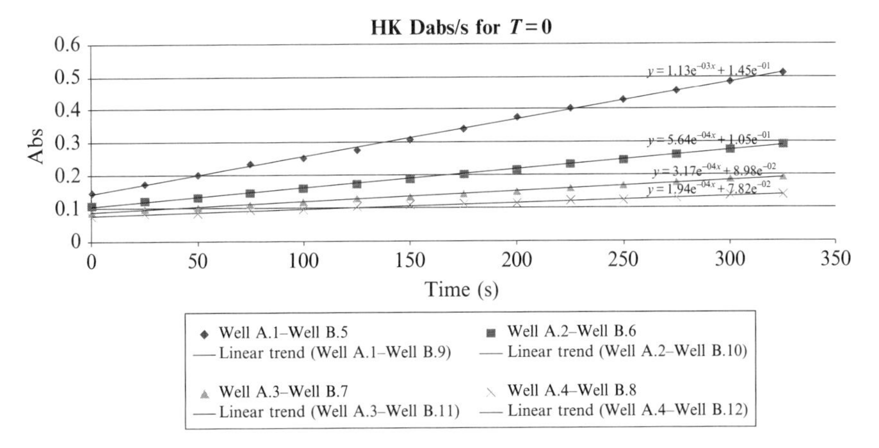

Figure 13.1 Initial rate of reaction catalyzed by hexokinase. Graph generated in Calculation Center software. The slope shows NADH accumulation linear with time in three out of four dilutions. Absorbance was measured at 340 nm at 30 °C in an automated spectrophotometer (Zenyth3100).

	1	2	3	4	5	6	7	8	9	10	11	12
A	B	B	$T=0$	$2x$	$T=30$	$2x$	$T=2$	$2x$	$T=4$	$2x$	$T=24$	$2x$
B	B	B	$4x$	$8x$	$4x$	$8x$	$4x$	$8x$	$4x$	$8x$	$4x$	$8x$
C												
D												
E												
F												
G												
H												

Figure 13.2 Proposed plate layout for multiple enzyme assays. B-blank contains sonication buffer, square A3–B4 contains undiluted extract $T = 0$ and $2\times$, $4\times$, $8\times$ dilutions of $T = 0$ unstarved sample, square A5-B6- $T = 30$ min starvation and dilutions, A7–B8- $T = 2$ h starvation and dilutions, A9–B10- $T = 4$h starvation and dilutions, $T = 24$ h starvation with its dilutions.

alternative for producing satisfactory data points that can further be used for initial rate analysis and V_{max} calculations.

3. Prepare Master mix in appropriate volume for several reactions and preincubate for 10 min at the temperature chosen for enzymatic assay. Transfer 5 μl of undiluted and diluted CFEs (cell-free extracts) into a microplate (Fig. 13.2). Add 265 μl of Master mix with a multichannel pipette and mix up and down. Mixing the components well is a very important step of the procedure. Read the plate at 340 nm for

2 min (20 cycles of 10 s each) to obtain a "blank" for background subtraction. It is necessary to read the background absorbance of a control reaction in the same wells to avoid a slight discrepancy in absorbance between wells.

4. Start the reaction by adding 30 μl of substrate, which is in this case, 10 mM ATP, to each well with a multichannel pipette. Add the substrate starting from the row with more diluted CFE (Fig. 13.2, row B). One will obtain a total reaction volume of 300 μl. Read the plate immediately at 340 nm for 15 min. The maximal reaction rate should be reached without the first 5 min. The slope during the initial rate period is the initial rate of the reaction.

   The above experiment should be repeated at four different extract concentrations and the slopes found should be plotted against the extract concentration. It should be checked that the rate observed is proportional to the extract concentration, which in practice may not be at low concentrations, because of a background rate of change in NADH, at high concentrations, because of a limiting factor in the assay. At intermediate concentrations, it may not be linear, for example, because the enzyme is engaged in a monomer–dimer equilibrium. In the former two cases, the straight part of the line should be populated by more points and then the slope should be used. In the latter case, a clearly visible flag should be shown that there is a difficulty extrapolating the V_{max} of the enzyme. For regulation analysis, one should then titrate every sample to the same observed rate and the sample dilution factor is then a proportional measure of the enzyme concentration in the extract. The assays should also be checked for independence of doubling of the substrate concentration and for constancy of pH when samples are added.

5. Calculate V_{max}:

$$V_{max}\left(\frac{\mu\text{mol}}{\text{min}\cdot\text{mg protein}}\right) = \frac{\text{Rate of absorbance change }(\text{s}^{-1})\cdot 60\cdot(\text{df})}{\text{Extracted protein concentration }(\text{mg}\cdot\text{ml}^{-1})\cdot \varepsilon_{\text{NADH}}(\text{mM}^{-1}\cdot\text{cm}^{-1})\cdot \text{L(cm)}}$$

The factor 60 is to convert seconds to minutes, df is the dilution factor of the sample into the reaction volume, ε_{NADH}*L was determined in an NADH calibration curve for a volume of 300 μl in a 96-wells plate and equals 4.354 the millimolar extinction coefficient of NADH at 340 nm (=6.222 $mM^{-1}cm^{-1}$), when *L*, the path length, is 0.7 cm (Greiner Bioone 655101 plate). For absolute measurements, the effective extinction coefficient of NADH should be calibrated by subjecting known aliquots of substrate to the assay. The V_{max} values obtained using this procedure represent the total activity of all isoenzymes in the cell at saturating concentration of the substrates and expressed relative to total cell protein. Alternatively to the estimation of initial rate of reaction, an enzyme progress curve analysis can be implemented for calculation of reaction rates.

The aforementioned approach has its own limitations, although it has been shown to provide detailed enzyme kinetics characterization, for example, when calculated with Copasi software (Mendes *et al.*, 2009).

4.3. Quantification of protein concentration

There are several colorimetric methods available for measurements of protein concentration. The method chosen has to be compatible with the analyte and with the various additions that have been made, to ensure high accuracy. If the sample (cell-free extract) contains reducing agents, like DTT or β-mercaptoethanol, the use of the Bradford or BCA protein assay-Reducing Agent Compatible kit is recommended. The BCA-RAC assay offers higher uniformity and reproducibility when compared to the Bradford assay. It can also be used in a format convenient for microplate assays. It should always be checked whether added compounds such as DTT interfere with the protein assay.

For determination of total protein content, the Lowry method is the most suitable, as the Coomassie dye ligands present in Bradford solution do not bind to free amino acids, peptides, and protein of low molecular weight (<3 kDa) and the BCA method is sensitive to protein composition.

4.4. Toward further protocols for *ex vivo* enzymatic assays

At the current stage of technological development, this can partly be achieved by reducing the dilution of cell extracts (mimicking crowding), by setting the reaction conditions to physiological pH, and by avoiding nonphysiological additives.

According to the literature, several factors including endogenous metabolite localization, metabolite channeling, and interactions with cytoskeleton and lipid rafts may modulate enzyme kinetics (Basu *et al.*, 1968; Keleti and Ovádi, 1988; Winkel, 2009). Several authors have proposed that enzyme complexes could be involved in metabolic regulation (Roche and Cate, 1976; Srere, 1987; Westerhoff, 1985). This type of organization into supramolecular complexes of sequential metabolic enzymes and structural components is referred to as "microcompartmentation." Specific interactions have been demonstrated *in vitro* between six of the eight sequential TriCarboxylicAcid cycle enzymes including citrate synthase (CS1) and malate dehydrogenase (MDH) (Haggie and Brindle, 1999, Srere, 1987). Glyceraldehyde-3-phosphate dehydrogenase, aldolase, phosphofructokinase, pyruvate kinase, and lactate dehydrogenase have been shown to assemble into a glycolytic enzyme complex on the inner surface of the human erythrocyte membrane. Associations of enzymes with membrane proteins, like in the case of phosphofructokinase in erythrocyte cells (Higashi *et al.*, 1979), or hexokinase with the mitochondrion, contractile proteins, or other enzymes to form multienzyme complexes, may constitute points of metabolic control. Definitive experimental evidence for these implications of these interactions *in vivo* is lacking, however.

Labile protein complexes are particularly prone to disassembly during exposure to nonphysiological ionic strength and pH. Ideally, therefore, the composition of solutions used in enzymatic assays should be as close as possible to the intracellular environment. The relatively weak interactions between the enzymes that are favored *in vivo* by the very high protein concentrations in the mitochondrial matrix (Aragón and Sols, 1991; Minton, 1992) are disrupted by the dilution that occurs during cell extraction. Media more accurately mimicking the crowded intracellular environment have been developed for *Escherichia coli* cell-free extracts, that is, the cytomin/glutamate–phosphate system for protein production, which substitutes for unnatural components (e.g., polyethylene glycol and pH buffers) and reduces the concentration of ion components (Jewett *et al.*, 2008). In the cytomin system, the central metabolism, catabolism, oxidative phosphorylation, transcription, and translation have been shown to be active, proving that crude extract systems can reproduce integrated metabolic function. This, however, does not mean that the rates of these processes are measured accurately. More work is needed on getting the *ex vivo* conditions right vis-à-vis the unavoidable dilution of the extract, and *in vivo* validations will be essential.

The fact that the volume of reaction in the enzymatic assays under *in vivo*-like conditions proposed in this work is 300 μl is a compromise with the requirements of using a semi-automated methodology. A robot-based platform would allow for further volume reduction with higher accuracy than conventional methods, and with reduction of the cost of the reagents utilized. With new technologies such as digital microfluidic devices, the proposed *in vivo*-like medium conditions may soon be implemented into high-throughput pipelines for enzyme kinetic characterization in reaction volumes mimicking that of the cell (i.e., the 30 fL estimated volume of the yeast cell; Jebrail and Wheeler, 2010; Miller and Wheeler, 2008).

5. Perspectives

Numerous reports on cell metabolism have demonstrated the feasibility of utilizing *in vitro* measurements of enzyme activities in cell-free extracts in small reaction volumes to generate data in medium-throughput systems biology workflows. This approach allows for reading approximately 96 reactions per 10 min. Data management can be automated, extracting the raw data, importing these into an excel spreadsheet and producing graphs, processing all calculation steps automatically. The perspective of generating such data under *ex vivo* conditions, resembling the *in vivo* milieu, is in urgent need for standardization of the procedures, in part because intracellular conditions are not always compatible with accurate assaying, because they are incompletely known and because they may vary. Yet, they may be essential for the realistic modeling efforts of metabolic pathways. Implementation of microfluidic platforms could be a next step in accelerating the enzyme assaying process when combined with new detection methods to cover the wide spectrum of enzymes. Pico to nanoliter scale droplets, each home to a single enzyme assay, can be easily handled by changing temperature. They should enable the titration of substrate concentration and the testing of new effectors/inhibitors in repetitive runs. Reagents cost can be reduced thanks to the small volume of reaction.

In summary, cell-free extracts in combination with well-designed *ex vivo* media that resemble the *in vivo* conditions may improve our ability to restore all the features of a living cell both *in vitro* and *in silico*. The features include the actual activity of the metabolic networks, sustained Gibbs energy transduction, and effective protein synthesis. Perhaps then, we shall achieve the complete predictive power of systems biology.

ACKNOWLEDGMENTS

The work of B. M. Bakker and H. V. Westerhoff is and has been supported by a Roslind Franklin Fellowship to B. M. Bakker, as well as by IOP Genomics, STW, the NGI-Kluyver Centre, NCSB, NWO-SysMo, BBSRC-MCISB, SysMO, ERASysBio and BRIC, EPSRC, AstraZeneca and by the EU grants BioSim, NucSys (and extensions), ECMOAN, and UniCellSys. The CEN.PK 113-7D strain was kindly donated by P Kotter, Euroscarf, Frankfurt. The software Calculation Centre was designed by Christopher Pfleger, Siemens, Germany.

REFERENCES

Achilles, J., Harms, H., and Müller, S. (2006). Analysis of living *S. cerevisiae* cell states—A three colour approach. *Cytometry A* **69A,** 173–177.

Aragón, J. J., and Sols, A. (1991). Regulation of enzyme activity in the cell: Effect of enzyme concentration. *FASEB J.* **5,** 2945–2950.

Bakker, B. M., Michels, P. A. M., Opperdoes, F. R., and Westerhoff, H. V. (1997). Glycolysis in bloodstream form *Trypanosoma brucei* can be understood in terms of the kinetics of the glycolytic enzymes. *J. Biol. Chem.* **272,** 3207–3215.

Bakker, B. M., Mensonides, F. I. C., Teusink, B., van Hoek, P., Michels, P. A. M., and Westerhoff, H. V. (2000). Compartmentation protects trypanosomes from the dangerous design of glycolysis. *Proc. Natl. Acad. Sci. USA* **97,** 2087–2092.

Basu, S., Kaufman, B., and Roseman, S. (1968). Enzymatic synthesis of ceramide-glucose and ceramide-lactose by glycosyltransferases from embryonic chicken brain. *J. Biol. Chem.* **243,** 5802–5804.

Boogerd, F. C., Ma, H., Bruggeman, F. J., van Heeswijk, W. C., Garcia-Contreras, R., Molenaar, D., Krab, K., and Westerhoff, H. V. (2011). AmtB-mediated NH_3 transport in prokaryotes must be active and as a consequence regulation of transport by GlnK is mandatory to limit futile cycling of NH_4^+/NH_3. *FEBS Lett.* **585,** 23–28.

Bruggeman, F. J., Boogerd, F. C., and Westerhoff, H. V. (2005). The multifarious short-term regulation of ammonium assimilation of Escherichia coli: Dissection using an in silico replica. *FEBS J.* **272,** 1965–1985.

Canelas, A. B., ten Pierick, A., Ras, C., Seifar, R. M., van Dam, J. C., van Gulik, W. M., and Heijnen, J. J. (2009). Quantitative evaluation of intracellular metabolite extraction techniques for yeast metabolomics. *Anal. Chem.* **81,** 7379–7389.

Cornish-Bowden, A. (1995). Fundamentals of Enzyme Kinetics. Portland Press, London.

Cortassa, S., Cáceres, A., and Aon, M. A. (1994). Microtubular protein in its polymerized or nonpolymerized states differentially modulates *in vitro* and intracellular fluxes catalyzed by enzymes of carbon metabolism. *J. Cell. Biochem.* **55,** 120–132.

Daran-Lapujade, P., Rossell, S., van Gulik, W. M., Luttik, M. A. H., de Groot, M. J. L., Slijper, M., Heck, A. J. R., Daran, J. M., de Winde, J. H., Westerhoff, H. V., Pronk, J. T., and Bakker, B. M. (2007). The fluxes through glycolytic enzymes in *Saccharomyces cerevisiae* are predominantly regulated at posttranscriptional levels. *Proc. Natl. Acad. Sci. USA* **104,** 15753–15758.

De Jong-Gubbels, P., Vanrolleghem, P., Heijnen, S., van Dijken, J. P., and Pronk, T. (1995). Regulation of carbon metabolism in chemostat cultures of *Saccharomyces cerevisiae* grown on mixtures of glucose and ethanol. *Yeast* **11,** 407–418.

Groen, A. K., Wanders, R. J., Westerhoff, H. V., van der Meer, R., and Tager, J. M. (1982). Quantification of the contribution of various steps to the control of mitochondrial respiration. *J. Biol. Chem.* **257,** 2754–2757.

Groeneveld, P., Stouthamer, A. H., and Westerhoff, H. V. (2009). Super life—How and why 'cell selection' leads to the fastest-growing eukaryote. *FEBS J.* **276,** 254–270.

Haanstra, J. R., van Tuijl, A., Kessler, P., Reijnders, W., Michels, P. A. M., Westerhoff, H. V., Parsons, M., and Bakker, B. M. (2008). Compartmentation prevents a lethal turbo-explosion of glycolysis in trypanosomes. *Proc. Natl. Acad. Sci. USA* **105,** 17718–17723.

Haggie, P. M., and Brindle, K. M. (1999). Mitochondrial citrate synthase is immobilized *in vivo*. *J. Biol. Chem.* **274,** 3941–3945.

Herrgard, M. J., Swainston, N., Dobson, P., Dunn, W. B., Arga, K. Y., Arvas, M., Bluthgen, N., Borger, S., Costenoble, R., Heinemann, M., Hucka, M., Le Novere, N., *et al.* (2008). A consensus yeast metabolic network reconstruction obtained from a community approach to systems biology. *Nat. Biotechnol.* **26,** 1155–1160.

Higashi, T., Richards, C. S., and Uyeda, K. (1979). The interaction of phosphofructokinase with erythrocyte membrane. *J. Biol. Chem.* **254,** 9542–9550.

Huebner, A., Olguin, F. L., Bratton, D., Whyte, G., Huck, S. T. W., de Mello, J. A., Edel, B. J., Abell, C., and Hollfelder, F. (2008). Development of quantitative cell-based enzyme assays in microdroplets. *Anal. Chem.* **80,** 3890–3896.

Jansen, M. L., Diderich, J. A., Mashego, M., Hassane, A., de Winde, J. H., Daran-Lapujade, P., and Pronk, J. T. (2005). Prolonged selection in aerobic, glucose-limited chemostat cultures of *Saccharomyces cerevisiae* causes a partial loss of glycolytic capacity. *Microbiology* **151**(5), 1657–1669.

Jebrail, M. J., and Wheeler, A. R. (2010). Let's get digital: Digitizing chemical biology with microfluidics. *Curr. Opin. Chem. Biol.* **14,** 1–8.

Jensen, P. R., Michelsen, O., and Westerhoff, H. V. (1993a). Control analysis of the dependence of Escherichia coli physiology on the H(+)-ATPase. *Proc. Natl. Acad. Sci. USA* **90,** 8068–8072.

Jensen, P. R., Westerhoff, H. V., and Michelsen, O. (1993b). The use of lac-type promoters in control analysis. *Eur. J. Biochem.* **211,** 181–191.

Jewett, M. C., Calhoun, K. A., Voloshin, A., Wuu, J. J., and Swartz, J. R. (2008). An integrated cell-free metabolic platform for protein production and synthetic biology. *Mol. Syst. Biol.* **4,** 220.

Johnson, H. L., Thomas, D. W., Ellis, M., Cary, L., and DeGraw, J. I. (1977). Application of 13C-NMR spectroscopy to *in vitro* analysis of enzyme kinetics. *J. Pharm. Sci.* **66**(11), 1660–1662.

Johnson, H., Eyers, C. E., Eyers, P. A., Beynon, R. J., and Gaskell, S. J. (2009). Rigorous determination of the stoichiometry of protein phosphorylation using mass spectrometry. *J. Am. Soc. Mass Spectrom.* **20**(12), 2211–2220.

Kacser, H., and Burns, J. A. (1973). The control of flux. *Symp. Soc. Exp. Biol.* **27,** 65–104.

Kahn, D., and Westerhoff, H. V. (1991). Control theory of regulatory cascades. *J. Theor. Biol.* **153,** 255–285.

Keleti, T., and Ovádi, J. (1988). Control of metabolism by dynamic macromolecular interactions. *Curr. Top. Cell. Regul.* **29,** 1–33.

Kleman, G. L., Chalmers, J. J., Luli, G. W., and Strohl, W. R. (1991). Glucose-stat, a glucose-controlled continuous culture. *Appl. Environ. Microbiol.* **57,** 918–923.

Koto, T., and Inoue, T. (1981). Estimation of kinetic parameters, amount of endogenous substrate and contaminating enzyme activity in a target enzyme reaction. *Biochim. Biophys. Acta* **15**(661), 1–11.

Laterveer, F. D., Gellerich, F. N., and Nicolay, K. (1995). Macromolecules increase the channeling of ADP from externally associated hexokinase to the matrix of mitochondria. *Eur. J. Biochem.* **232,** 569–577.

Mendes, P., Messiha, H., Malys, N., and Hoops, S. (2009). Enzyme kinetics and computational modeling for systems biology. *Methods Enzymol.* **467,** 583–599.

Michaelis, L., and Menten, M. L. (1913). The kinetics of the inversion effect. *Biochem. Z.* **49,** 333–369.

Miller, E. M., and Wheeler, A. R. (2008). A digital microfluidic approach to homogeneous enzyme assays. *Anal. Chem.* **80,** 1614–1619.

Minton, A. P. (1992). Confinement as a determinant of macromolecular structure and reactivity. *Biophys. J.* **63,** 1090–1100.

Munson, R. J. (1970). Turbidostats. *In* "Methods in Microbiology," (J. R. Norris and D. W. Ribbons, eds.), Vol. 2, pp. 349–376. Academic Press Inc., London.

Olsen, N. S. (2006). Applications of isothermal titration calorimetry to measure enzyme kinetics and activity in complex solutions. *Thermochim. Acta* **448,** 12–18.

Pietta, P. G., Mauri, P. L., Gardana, C., and Benazzi, L. (1997). Assay of soluble guanylate cyclase by isocratic high-performance liquid chromatographic. *J. Chromatogr. B Biomed. Sci. Appl.* **690,** 343–347.

Rathore, R., Pribil, P., Corr, J. J., Seibel, L. W., Evdokimov, A., and Greis, D. K. (2010). Multiplex enzyme assays and inhibitor screening by mass spectrometry. *J. Biomol. Screen.* **15**(8), 1001–1007.

Roche, T. E., and Cate, R. L. (1976). Evidence for lipoic acid mediated NADH and acetyl-CoA stimulation of liver and kidney pyruvate dehydrogenase kinase. *Biochem. Biophys. Res. Commun.* **72,** 1375–1383.

Rohwer, J. M., Postma, P. W., Kholodenko, B. N., and Westerhoff, H. V. (1998). Implications of macromolecular crowding for signal transduction and metabolite channeling. *Proc. Natl. Acad. Sci. USA* **95,** 10547–10552.

Rossell, S., van der Weijden, C. C., Kruckeberg, A. L., Bakker, B. M., and Westerhoff, H. V. (2005). Hierarchical and metabolic regulation of glucose influx in starved Saccharomyces cerevisiae. *FEMS Yeast Res.* **5,** 611–619.

Rossell, S., van der Weijden, C. C., Lindenbergh, A., van Tuijl, A., Francke, C., Bakker, B. M., and Westerhoff, H. V. (2006). Unraveling the complexity of flux regulation: A new method demonstrated for nutrient starvation in Saccharomyces cerevisiae. *Proc. Natl. Acad. Sci. USA* **103,** 2166–2171.

Sakai, T., Tochio, H., Tenno, T., Ito, Y., Kokubo, T., Hiroaki, H., and Shirakawa, M. (2006). In-cell NMR spectroscopy of proteins inside Xenopus laevis oocytes. *J. Biomol. NMR* **36**(3), 179–188.

Sauro, H. M. (1990). Quantification of metabolic regulation by effectors. *In* "Control of Metabolic Processes," (A. Cornish-Bowden and M. L. Cardenas, eds.), pp. 225–230. NATO ASI Series, Plenum Press, New York.

Scrutton, M. C., and Utter, M. F. (1965). Pyruvate carboxylase: Some physical and chemical properties of the highly purified enzyme. *J. Biol. Chem.* **240,** 1–9.

Serber, Z., and Dotsch, V. (2001). In-cell NMR spectroscopy. *Biochemistry* **40,** 14317–14323.

Slazano, M. A., D'Ambrosio, C., and Scaloni, A. (2008). Mass spectrometric characterization of proteins modified by nitric oxide-derived species. *Methods Enzymol.* **440,** 3–15.

Snoep, J. L., and Westerhoff, H. V. (2004). The silicon cell initiative. *Curr. Genomics* **5,** 687–697.

Snoep, J. L., Jensen, P. R., Groeneveld, P., Molenaar, D., Kholodenko, B. N., and Westerhoff, H. V. (1994). How to determine control of growth rate in a chemostat—Using metabolic control analysis to resolve the paradox. *Biochem. Mol. Biol. Int.* **33,** 1023–1032.

Srere, P. (1987). Complexes of sequential metabolic enzymes. *Annu. Rev. Biochem.* **56,** 89–124.

Srivastava, D. K., and Bernhard, S. A. (1987). Biophysical chemistry of metabolic reaction sequences in concentrated enzyme solution and in the cell. *Annu. Rev. Biophys. Biophys. Chem.* **16,** 175–204.

Ter Kuile, B. H., and Westerhoff, H. V. (2001). Transcriptome meets metabolome: Hierarchical and metabolic regulation of the glycolytic pathway. *FEBS Lett.* **500**(3), 169–171.

Teusink, B., Passarge, J., Reijenga, C. A., Esqalhado, E., van der Weijden, C. C., Schepper, M., Walsh, M. C., Bakker, B. M., van Dam, K., Westerhoff, H. V., and Snoep, J. L. (2000). Can yeast glycolysis be understood in terms of in vitro kinetics of the constituent enzymes? Testing biochemistry. *Eur. J. Biochem.* **267**(17), 5313–5329.

Theobald, U., Mailinger, W., Baltes, M., Rizzi, M., and Reuss, M. (1997). *In vivo* analysis of metabolic dynamics in *Saccharomyces cerevisiae*: I. Experimental observations. *Biotechnol. Bioeng.* **55,** 305–316.

Trotta, P. P., Pinkus, L. M., and Meister, A. (1974). Inhibition by dithiothreitol of the utilization of glutamine by carbamyl phosphate synthetase. Evidence for formation of hydrogen peroxide. *J. Biol. Chem.* **25,** 1915–1921.

van Eunen, K. (2010). The multifarious and dynamic regulation of the living cell. Ph D thesis, VU University Amsterdam.

van Eunen, K., Bouwman, J., Lindenbergh, A., Westerhoff, H. V., and Bakker, B. M. (2009). Time-dependent regulation analysis dissects shifts between metabolic and gene-expression regulation during nitrogen starvation in baker's yeast. *FEBS J.* **276,** 5521–5536.

van Eunen, K., Bouwman, J., Daran-Lapujade, P., Postmus, J., Canelas, A. B., Mensonides, F. I., Orij, R., Tuzun, I., van den Brink, J., Smits, G. J., van Gulik, W. M., Brul, S., *et al.* (2010a). Measuring enzyme activities under standardized *in vivo*-like conditions for systems biology. *FEBS J.* **277,** 749.

van Eunen, K., Dool, P., Canelas, A. B., Kiewiet, J., Bouwman, J., van Gulik, W. M., Westerhoff, H. V., and Bakker, B. M. (2010b). Time-dependent regulation of yeast glycolysis upon nitrogen starvation depends on cell history. *IET Syst. Biol.* **4,** 157–168.

Van Hoek, P., Van Dijken, J. P., and Pronk, J. T. (1998). Effect of specific growth rate on fermentative capacity of baker's yeast. *Appl. Environ. Microbiol.* **64,** 4226–4233.

Waddington, C. H. (1952). Selection of the genetic basis for an acquired character. *Nature* **169,** 278.

Werner, A., and Heinrich, R. (1985). A kinetic model for the interaction of energy metabolism and osmotic states of human erythrocytes. Analysis of the stationary "*in vivo*" state and of time dependent variations under blood preservation conditions. *Biomed. Biochim. Acta* **44,** 185–212.

Westerhoff, H. V. (1985). Organization in the cell soup. *Nature* **318,** 106.

Westerhoff, H. V., and Arents, J. C. (1984). 2 (completely) rate-limiting steps in one metabolic pathway—Resolution of a paradox using bacteriorhodopsin liposomes and control theory. *Biosci. Rep.* **4,** 23–31.

Westerhoff, H. V., and Kell, D. B. (1987). Matrix-method for determining steps most rate-limiting to metabolic fluxes in biotechnological processes. *Biotechnol. Bioeng.* **30,** 101–107.

Westerhoff, H. V., and Palsson, B. O. (2004). The evolution of molecular biology into systems biology. *Nat. Biotechnol.* **22,** 1249–1252.

Westerhoff, H. V., Kolodkin, A., Conradie, R., Wilkinson, S. J., Bruggeman, F. J., Krab, K., van Schuppen, J. H., Hardin, H., Bakker, B. M., Moné, M. J., Rybakova, K. N., Eijken, M., *et al.* (2009). Systems biology towards life in silico: Mathematics of the control of living cells. *J. Math. Biol.* **58,** 7–34.

Westerhoff, H. V., Verma, M., Nardelli, M., Adamczyk, M., van Eunen, K., Simeonidis, E., and Bakker, B. M. (2010). Systems biochemistry in practice: Experimenting with modelling and understanding, with regulation and control. *Biochem. Soc. Trans.* **38,** 1189–1196.

Winkel, B. S. (2009). Metabolite Channeling and Multi-enzyme Complexes. *Plant-Derived Natural Products* **2,** 195–208.

SECTION SIX

SAMPLE PREPARATION IN METABOLIC STUDIES

CHAPTER FOURTEEN

The Use of Continuous Culture in Systems Biology Investigations

Catherine L. Winder* *and* Karin Lanthaler†

Contents

Abstract

When acquiring data for systems biology studies, it is essential to perform the experiments in controlled and reproducible conditions. Advances in the fields of proteomics and metabolomics allow the quantitative analysis of the components of the biological cell. It is essential to include a method in the experimental pipeline to culture the biological system in controlled and reproducible

* Manchester Centre for Integrative Systems Biology and Department of Chemistry, Manchester Interdisciplinary Biocentre, University of Manchester, Manchester, United Kingdom
† Faculty of Life Sciences, University of Manchester, Manchester, United Kingdom

Methods in Enzymology, Volume 500
ISSN 0076-6879, DOI: 10.1016/B978-0-12-385118-5.00014-1

conditions to facilitate the acquisition of high-quality data. The employment of continuous culture methods for the growth of microorganisms is an ideal tool to achieve these objectives. This chapter will review the continuous culture approaches which may be applied in such studies, outline the experimental options which should be considered, and describe the approach applied in the production of steady-state cultures of *Saccharomyces cerevisiae*.

1. Introduction

Systems biology requires experiments to be performed in well-controlled and reproducible conditions. Recent advances in analytical technologies which allow the quantitative analysis of the proteome and metabolome (described in chapters 7 and 15) are pivotal in providing data of sufficient accuracy and precision. It is therefore important to include a method which allows the cells to be cultured in a defined, controlled, and reproducible manner to minimize inaccuracies at all stages of the experimental pipeline. The use of continuous culture approaches provides a means of rigorous and robust experimental design and is ideal for approaches in postgenomic studies and systems biology (Bull, 2010; Hoskisson and Hobbs, 2005). The ability to establish a state in which the concentrations of proteins and metabolites are constant for an extended period of time, in a controlled and reproducible environment, reduces the variability in the measured biological system. Once the steady state of the culture is established, there are minimal restrictions in terms of time-dependent sampling, such that representative samples may be taken over a period of time and retained for analysis by multiple experimental approaches, allowing measurement of the complement of molecules present in the biological system. This is obviously advantageous compared to batch culture methods in which the sampling and measurements taken would be restricted by time-dependent analyses due to the time-course of the culture. In addition, samples may be removed by the most appropriate method for each class of compounds to be investigated. For example, sampling the metabolome requires the quenching of cultures to halt the turnover of metabolites. This is achieved by spraying into an aqueous solution of methanol at -48 °C to reduce the culture temperature to below -20 °C. The addition of methanol is problematic in samples which are retained for proteomic analysis, and thus a separate sample is removed under different conditions to overcome this problem. This allows multiple samples to be collected at the same biological condition.

1.1. Chemostat

The chemostat was originally introduced in the 1950s as a method to culture a bacterial population at a reduced growth rate for an indefinite period (Monod, 1950; Novick and Szilard, 1950) and is the most widely used approach to establish steady-state culture for various applications (Daran-Lapujade *et al.*, 2009; Ferenci, 2008). The initial studies provided the basis for the first theoretical description of continuous culture systems, which allowed the prediction of the steady-state concentrations of biomass and the residual concentration of the limiting substrate. In liquid medium, microbial growth is usually exponential and can be described by

$$(1/x)(\mathrm{d}X/\mathrm{d}t) = \mathrm{d}(\ln X)/\mathrm{d}t = \mu = \ln 2/t_\mathrm{d} \tag{14.1}$$

where X is the biomass expressed as dry weight per volume (g L^{-1}), t is the time, μ is the specific growth rate (h^{-1}), and t_d is the doubling time (h). The μ and t_d can be assumed to be constant if all substrates necessary for growth are present in excess. Monod (1942) was the first to show that there is a correlation between the specific growth rate μ and the concentration of the limiting substrate according to

$$\mathrm{d}X/\mathrm{d}t = X\mu_{\max}[s/(K_\mathrm{s} + s)] \tag{14.2}$$

where X is the biomass concentration at a given time, $\mu_{\max}$ is the maximum specific growth rate (h^{-1}), s is the concentration of the limiting substrate, and K_s is the saturation constant and equals the substrate concentration at $0.5 \times \mu_{\max}$. By converting Eq. (14.2) to $(\mathrm{d}X/X)(1/\mathrm{d}t) = \mu_{\max}[s/(K_\mathrm{s} + s)]$ and $(\mathrm{d}X/X)(1/\mathrm{d}t)$, being the increase of biomass over time or specific growth rate μ. Eq. (14.2) can be written as

$$\mu = \mu_{\max}[s/(K_\mathrm{s} + s)] \tag{14.3}$$

Herbert *et al.* (1956) concluded that μ can be set at any value between 0 and $\mu_{\max}$ and requires the concentration of the limiting substrate to be held constant at the appropriate set point value. This cannot be achieved in batch fermentations but is one of the key features of chemostat cultivations (Daran-Lapujade *et al.*, 2009). In the chemostat, fresh nutrients are added to the culture at a fixed flow rate, the biomass and the products of metabolism are removed from the vessel at the same flow rate to maintain a fixed culture volume. The culture is maintained in a growth phase, where the doubling time and cell density are fixed and rigorous control of the environment is achieved. In the steady-state condition, the specific growth rate is equivalent to the dilution rate applied in the experiment and can only be operated at a growth rate below μ_{max}. If the dilution rate exceeds the maximum specific growth rate of the microorganism, the culture will become unstable and washout will occur. In a chemostat setup, the rate at which the growth-limiting substrate is supplied to the culture vessel is called the dilution rate (*D*) and thus equals the specific growth rate as described by Eqs. (14.4) and (14.5). In a chemostat operated system, the microorganisms are growing at a rate which is described by Eq. (14.1) and biomass is simultaneously washed out at a rate given by Eq. (14.4)

$$-\mathrm{d}X/\mathrm{d}t = DX \tag{14.4}$$

Hence, an increase in biomass is equal to the growth minus the output.

$$\mathrm{d}X/\mathrm{d}t = \mu X - DX \tag{14.5}$$

In the steady state, the biomass is constant and $\mathrm{d}X/\mathrm{d}t$ is therefore zero and $\mu = D$. The dilution rate (*D*) is determined by the flow rate (*F*) expressed in mL h^{-1} of the feed pump and the culture's working volume (*V*) in mL, such that the hourly dilution rate is calculated by

$$D = \frac{F}{V}. \tag{14.6}$$

The constant medium supply and working volume results in a constant growth rate of the culture. The chemostat may be operated at different growth rates and limited by different growth-limiting nutrients in the media, which is generally selected to be carbon, nitrogen, sulfur, or phosphorus limitation (Castrillo *et al.*, 2007). The disadvantages with using the chemostat approach to establish steady-state cultures is that the observed results will be specific to the particular nutrient limitation, and the experimental conditions will be strongly selective for advantageous mutations in the population. The limitation of specific nutrients and selected growth rate produces characteristic profiles of the gene expression, metabolism, and physiology of the culture (Ferenci, 2008). The influence of the limited nutrient in the chemostat approach is now considered to be more complex than originally described in Monod's studies (Ferenci, 1999; Snoep *et al.*, 2009) and thus will need to be considered in the modeling of biological systems. It should also be remembered that the objective of one of the original manuscripts describing the use of a chemostat (Novick and Szilard, 1950) was to investigate evolutionary studies in bacterial populations, and thus the genetic makeup of the culture will change by the occurrence of mutants with the capacity to outcompete the parental wild type (Delneri *et al.*, 2008).

1.2. Auxostats

There are a number of alternative approaches which may be applied to establish steady-state microbial cultures; these are generally classified as auxostats. In the auxostat, steady-state conditions are established in relation to the measurement of a growth-associated parameter and the steady state is held constant by slight adjustment of the dilution rate (Bull, 2010). Such examples include the turbidostat which measures the cell density by photoelectric or laser devices (Bryson and Szybalski, 1952), pH auxostats (Bijmans *et al.*, 2009; Groeneveld *et al.*, 2009; Martin and Hempfling, 1976), the permittistat which measures the electrical capacitance of the culture (Davey *et al.*, 1996; Markx *et al.*, 1991), the cytostats which perform both the measurement and cell sorting (Gilbert *et al.*, 2009), and the Evolugator which facilitates the evolution of microbial strains for characteristics of specific interest (de Crecy *et al.*, 2007, 2009).

1.2.1. Turbidostat

The first documentation of the turbidostat culture was by Bryson and Szybalski (1952); the turbidostat operates by maintaining the biomass at a constant value, which lies below the maximum biomass yield. All nutrients are still present in excess. The system works by a feedback loop to control the pump which supplies media into the vessel in response to the concentration of biomass detected in the culture. Media are added to the vessel when the biomass concentration is equal to or above the set point, and the media feed is stopped when the concentration of the culture decreases below the set point. In the turbidostat, the same basic growth equations as observed in the chemostat apply, but the setup does not allow the substrates to become limiting ($s_{\mathrm{residual}} \gg K_s$). This is achieved by keeping the total biomass concentration in the vessel constant but at about 70–75% of the level of the maximum achievable biomass yield (g biomass/g carbon source used) for the growth medium. This is achieved by controlling the addition of fresh growth medium and ensuring that the working volume is constant. According to Eq. (14.3), if the concentration of the limiting substrate is much higher than its K_s ($s \gg K_s$), it follows that the term $[s/(K_s + s)]$ is very close to 1, and hence $\mu = \mu_{\mathrm{max}}$. This means that a turbidostat can be operated without any nutrient limitations, but only at μ_{max} as opposed to a chemostat, where the growth rate can be set, but the culture is nutrient limited. The advantage of applying the turbidostat approach is therefore that the growth of the culture is performed at the maximum specific growth rate of the microorganism because the nutrients are not limited in the media. The measured profile will therefore not be indicative of a particular nutrient limitation. However, as with the chemostat approach, the method may still be selective toward mutants, albeit toward mutants with a growth rate higher than that observed by the parental strain. This is advantageous in some applications (Gilbert *et al.*, 2009; Groeneveld *et al.*, 2009); however, in investigations where adaptations or mutations are not desirable, it is important to minimize the number of generations of the culture (Paquin and Adams, 1983) and the growth rate should be constantly monitored to check for consistency.

1.2.2. Permittistat

The permittistat was originally described by Markx *et al.* (1991) in the anaerobic culture of yeast and is a similar approach to the turbidostat in that the feedback loop operates on the basis of the biomass measured. In the turbidostat, optical methods are employed to measure the turbidity of the culture, whereas in the permittistat, the electrical capacitance of the cell suspension is measured. The determination of biomass based on turbidity measurements is prone to inaccuracy because it includes the total cells and debris present in the culture and the linearity range is limited. In addition, particulate solids and gas bubbles (Harris and Kell, 1985), and the formation of biofilms on the optical sensor, will interfere with turbidity measurements (Northrop, 1954). The permittistat method provides a signal which is linear over a wide range of biomass measurements (Harris and Kell, 1983), and the presence of dead cells, gas bubbles, and nonbiomass solids do not interfere with the determination of biomass in the cell suspension (Davey *et al.*, 1996; Harris *et al.*, 1987). The development of dielectric spectroscopy into a commercial device to monitor biomass is documented in the literature (Davey *et al.*, 1993, 1996; Harris and Kell, 1985; Markx *et al.*, 1991). In brief, the approach applied to monitor the biomass in our studies is an online method based on the electrical capacitance of the culture at low radio frequencies (RFs). A RF electrical field is produced by applying a low current RF field to the culture around four electrodes present on the biomass probe inserted into the vessel. The RF electrical field polarizes the intact cells within the cell suspension such that the obtained results provide an accurate measurement of the viable cells within the culture. The biomass is controlled to a predefined level using the permittivity measurements of the culture thereby providing a direct measurement of the viable cells.

2. Experimental Considerations

There are a number of experimental considerations that should be undertaken when planning steady-state cultures. The first step is to decide the method to use in establishing the steady-state culture. We will describe the use of the permittistat (Davey *et al.*, 1996; Markx *et al.*, 1991) to establish steady-state cultures growing at μ_{max} and the practical considerations undertaken during the planning of the experiments.

2.1. Microorganism of choice and culture maintenance

In our studies, we have selected the baker's yeast *Saccharomyces cerevisiae* as the model organism. The selected auxotrophic strain BY4743 (*MATa/α his3Δ1/his3Δ1 leu2Δ0/leu2Δ0 lys2Δ0/LYS2 MET15/met15Δ0 ura3Δ0/ ura3Δ0* ydl227c:*kan*MX4/YDL227c) can be purchased from EURO-SCARF (Accession number Y23935). For long-term storage, the cultures are maintained in YPD with the addition of 15% glycerol at −80 °C and for short-term storage on YPD agar plates with the addition of 200 μg mL^{-1} G418. An aliquot of the culture is removed from long-term storage at the start of each experiment, and purity of the culture is established by plating on YPD agar plates prior to inoculation of the overnight culture.

2.2. Media

In the operation of the permittistat, the nutrients in the media are not limited to allow growth of the culture at μ_{max}. However, the objective of the experiment may influence the composition of the media employed in the investigation. The authors have used two types of media in their steady-state cultures, a minimal defined media (Table 14.1) and a synthetic metabolic footprinting medium with the addition of amino acids, nucleotide bases, and organic acids (Allen *et al.*, 2003). The rationale in the medium choice may be driven by either biological or analytical reasons. Examples of the justification include (a) culturing the microorganism in an environment which closely represents the natural habitat, (b) the exposure to a particular condition, (c) to probe a particular metabolic pathway(s), or (d) to facilitate particular analysis of the culture medium. The minimal medium was specifically employed in experiments to perform carbon balance measurements and carbon flux analysis, whereas the synthetic metabolic footprint medium is employed to interrogate different areas of intracellular metabolism.

In the preparation of the media, high-quality chemicals and water should be used to minimize the introduction of contaminants into the culture. This is particularly important when the samples are analyzed by mass spectrometry because low concentration contaminants would be detected and may complicate the analysis of the data. Trace elements and vitamins are present at micromolar and nanomolar concentrations in the final solutions, and concentrated stock solutions are used to provide accuracy. Errors in the addition of the trace elements and vitamins may affect the biomass yield and growth rate of the culture. Glassware should be cleaned to minimize cross contamination, and low-reactivity plastics are used to minimize contamination. If filter-sterilization is required for heat-liable components, the vessels and water should be previously autoclaved to minimize contamination. The media are prepared between 24 and 36 h before use to ensure that it is not

Table 14.1 F1 medium, preparation of stock solutions

	Component	Final conc. (gL^{-1})	Stock solution (g)
Solution 1 (Dissolve in 1800 mL deionized water)	NH_4SO_4	5	10
	KH_2PO_4	3	6
	$MgSO_4 \cdot 7H_2O$	0.55	1.1
	NaCl	0.1	0.2
	$CaCl_2 \cdot 2H_2O$	0.09	0.18
	Uracil	0.11	2.2
	Histidine	0.15	3
	Leucine	0.1	2
Solution 2 (Dissolve in 50 mL deionized water and filter sterilize)	$ZnSO_4 \cdot 7H_2O$	0.00007	0.014
	$CuSO_4 \cdot 5H_2O$	0.00001	0.002
	H_3BO_3	0.00001	0.002
	KI	0.00001	0.002
Solution 3 (Dissolve in 50 mL deionized water and filter sterilize)	$FeCl_3 \cdot 6H_2O$	0.00005	0.01
Solution 4 (Dissolve in 100 mL deionized water and filter sterilize)	Inositol	0.062	6.2
	Thiamine/HCL	0.014	1.4
	Pyridoxine	0.004	0.4
	Ca-pantothenate	0.004	0.4
	Biotin	0.0003	0.03
Solution 5 (Dissolve in 5 L deionized water and autoclave)	Glucose	20	400

contaminated prior to use. Finally, regular mixing of the media in the reservoir is required to ensure that nutrient gradients do not develop.

2.3. Selection of biomass set point

In the operation of a turbidostat or permittistat, the biomass is controlled below the maximum achievable yield for the selected growth medium to ensure that the growth is not limited by any of the components in the media. In general, a set point of between 70% and 75% is recommended although we have investigated the effect of different set points ranging between 50% and 100% of the maximum biomass yield. At 100%, the experimental conditions are comparable to those of a chemostat and the dilution rate was below that observed in unlimited growth. The dilution rate was equivalent to the maximum growth rate of the cultures when the biomass was set between 60% and 75%. Decreasing the biomass to 50% of biomass yield increased the dilution rate above the specific growth rate of the culture. This resulted in washout of the culture because the dilution rate exceeded the specific growth rate of the culture and clumping of biomass was observed around the vessel components. It is important in systems biology studies that the cultures are homogeneous; it is therefore not advisable to employ growth conditions in which clumping of the cells was observed such that both planktonic and biofilm-derived cells may be present in the samples. In our experiments, we employ a biomass set point of 75% in which we did not observe clumping of the cells.

2.4. Experimental time scale and replicates

In continuous cultures, adaptations of the microbial community to environment conditions increase with operation time and potentially selecting for mutations in the culture. If the mutants outcompete the parental strain, the genetic makeup of the population will alter potentially changing the phenotype of the culture. It is therefore essential to minimize the number of generations by which the culture is allowed to progress and monitor the growth rate to check that it does not change during the duration of the experiment.

It is important to include biological replicates in any experiment to assess both the biological and analytical (or technical) variability of the experiment. In microbial studies, biological replicates are those which originate from different cultures which have been grown in identical conditions and ideally from the same inoculum. Analytical (or technical) replicates result from either multiple samples from the same culture (at the same time point in time-dependent studies) or repeated measurements from the same biological sample. In our experiments, we perform a minimum of three biological replicates for measurements of the proteome and metabolome. In order to check the reproducibility of our cultures, we monitor a number of growth-associated parameters in the steady-state cultures before the collection of sample commences. These include (1) the growth rate, (2) the biomass yield (monitored in both the batch phase of growth and steady state), (3) the extracellular glucose concentration, (4) the purity of the culture, and (5) the off-gas analysis of CO_2 and O_2.

2.5. Off-line measurements

A number of measurements are routinely performed during the duration of the experiment. These include plating of the cultures and microscopic examination to check the purity of the culture, (at this point a viable plate count may be performed) and the dry cell weight of the culture is measured. A robust and accurate method may be required to count the cells when the quantification measurements of the proteins and metabolites are to be expressed per cell. The Cellometer Auto M10 (Nexcelom Bioscience) is employed for cell counting and has the advantage that the images can be stored for reevaluation if required. In addition, a sample of the culture is fixed in 5% formaldehyde for storage. The quantification of metabolites is performed by mass spectrometry, the concentration of glucose is routinely monitored in the exometabolome by Gas Chromatography Time of Flight Mass Spectrometry (GC–ToF–MS).

2.6. Off-gas analysis

Tandem gas analysis is performed during the steady-state cultures to (a) monitor the consumption of oxygen and production of carbon dioxide during the cultures and compare replicates and to (b) ensure consistent readings throughout the steady state of the culture and to monitor for any synchronization of the culture. *S. cerevisiae* cultures are known to display oscillatory behavior (Bull, 2010; Hoskisson and Hobbs, 2005; Silverman *et al.*, 2010). The authors have only observed this behavior in nutrient limited conditions; however, it is important in our studies that the culture does not synchronize and the monitoring of the off-gases analysis provides us with such evidence.

3. Operation of the Permittistat

A number of commercial bioreactor systems are available. The bioreactor system that the authors use is from Applikon Biotechnology (Holland) controlled with the operating software BioExpert and connected to a Biomass monitor 220 from Aber Instruments (UK) providing the permittistat function. The working volume of the vessels is between 470 mL and 2.7 L, and the control loops are present for temperature, pH, dissolved O_2, and antifoam.

3.1. Preparation of bioreactor

The design of the top plate of the bioreactor is outlined in Fig. 14.1. In the preparation of the vessel for sterilization, all of the tubing lines are secured with hose clamps. Approximately 500 mL of deionized water is added to the vessel and the top plate is secured with mill nuts. The pH probe is calibrated by external calibration with buffers at pH 4 and 7. The biomass and dissolved oxygen probes are added to the vessel, and protective covers are placed on the electrical components (a polarization module is placed on the dissolved oxygen probe) and covered with aluminum tin foil. The tubing gas inlet pipe is clamped off during the autoclave cycle, and filters, the foam level sensor, and push-fit valves are covered with aluminum tin foil and secured with autoclave tape. Addition bottles (500 mL) are added to the

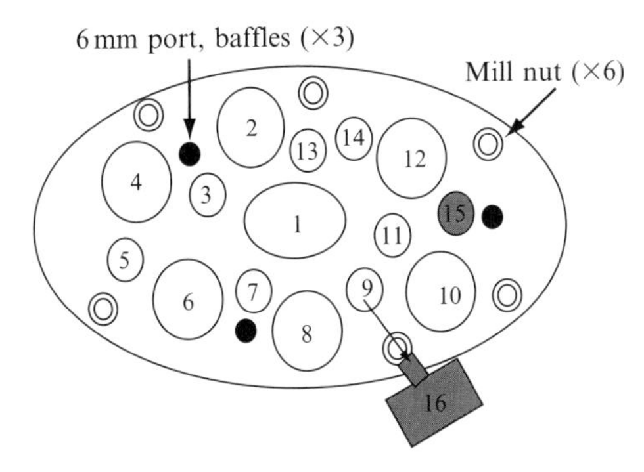

Figure 14.1 The setup of the bioreactor top plate, (1) Central port M30 × 1 for stirrer assembly, (2) M18 × 1.5 port, triple inlet assembly for acid, base and antifoam addition, (3)10 mm port, temperature pocket (temperature probe inserted), (4) G3/4″ port, biomass probe, (5) 10 mm port, addition tube for inoculum bottle, (6) M18 × 1.5 port for dO2 probe, (7) 10 mm port, foam level sensor, (8) M18 × 1.5 port, air condenser outlet (the condenser is cooled via cooling water, and a 0.22 μm filter is attached to the air-out line), (9) 12 mm port, fixed length sample tube (connected via tubing to manual sampler), (10) M18 × 1.5 port, pH probe, (11) 12 mm port, adjustable height sample tube (the culture is removed from the vessel via a variable speed pump), (12) M18 × 1.5 port, Heat exchange (connected to cooling water), (13) 10 mm port, addition tube to supply media via an antigrow bac tube (the sterilized media is added via a variable speed pump under the regulation of the feedback mechanism), (14) 10 mm port, air sparger (the air is passed through a 0.22 μm filter), (15) 10 mm port, blind stopper, and (16) manual sampler for universal bottles.

acid, base, and antifoam lines, and an inoculum bottle is attached. The vessels are sterilized by heat sterilization (121 °C, 15 min), the autoclave contains an internal thermometer which is placed in the temperature pocket to check that the sterilization temperature is achieved. If a larger volume is added to the vessel, the autoclave cycle should be checked and optimized. The empty reservoir bottles for the waste are heat sterilized (121 °C, 15 min), filters are covered with aluminum tin foil and the lids are released prior to entry into the autoclave.

3.2. Preparation of media

Vitamins and trace elements should be prepared as stock solutions and filter sterilized as outlined in Table 14.1. Glucose stocks are sterilized in the 10 L reservoir bottles (an overnight autoclave cycle is used). The components (solution 1, 1800 mL, solution 2, 2.5 mL, solution 3 2.5 mL, solution 4, 10 mL, solution 5, 5000 mL) are aseptically added together into a 10 L bottle and the volume is made up to 10 L with sterile water. The components are transferred to a 20 L reservoir using a variable speed pump and the process is repeated to provide a 20 L volume of media. The media is prepared 24–36 h prior to use to ensure that it is not contaminated.

3.3. Preparation of starter culture

An overnight culture is inoculated from a single colony into a flask containing 250 mL of sterile F1 medium (Table 14.1). The culture is incubated for 18 h at 30 °C, 200 rpm.

3.4. Setting up of the bioreactor

The bioreactors are assembled (connecting sensors to respective cables, temperature probe inserted, air supply connected, heating blanket added). *Note*: if the polarization module was not placed on the dO2 probe during the autoclave process, the probe should be connected and allowed to stand for 24 h before use. The off-gas analysis (Tandem gas analyzer, Magellan Instruments, UK) is connected and the tubes are checked to ensure that there is no moisture in the lines as this would cause damage to the analyzer. Media reservoir and waste lines are aseptically connected, and acid (1 M H_2SO_4), base (1 M NaOH), and antifoam (3% BDH Dow Corning 1500) are aseptically added to the bottles. The cooling water is attached to the condenser and heat exchange port, and the tubing is secured with hose clamps. All pumps are purged to remove the dead volume, the water is removed from the vessel, and media are added to the vessel at the working volume. The height of the sample tube on the waste line is adjusted to the desirable working volume (2 L). The control loops are started to control the growth parameters, the air supply is started, and the foam level sensor is adjusted to above the level of the media and the dissolved oxygen probe is calibrated. The growth parameters are set at, 650 rpm agitation control, 1.5 L min^{-1} air supply, 30 °C temperature, and pH 4.5.

3.5. Setting up of the biomass monitor

The biomass monitor should be switched on for a minimum of 30 min before use. The measurement settings on the biomass monitor were adjusted to the optimal values for yeast cells. These are a frequency of 500 kHz, the capacitance is measured at two frequencies with the polarization correction set to "ON" (mode 2f and Pol. C) and the low pass filter (LPF) was set to 10. The biomass probe is electrolytically cleaned by applying ±5 V d.c. across the electrodes for 4 s. The biomass probe was checked after 10 min to ensure that it is free from bubbles. Prior to inoculation of the culture, the capacitance of the biomass monitor is set to 0 pF cm^{-1}, the zero baseline should remain stable within ±0.2 pF cm^{-1}.

3.6. Calibration of tandem gas analyzer

The tandem gas analyzer is calibrated for both oxygen and carbon dioxide gases at a flow rate of 0.5 L min^{-1}. A nitrogen supply is used to produce the zero reading for both gasses, and oxygen is calibrated based on the content in air. A minimum of two concentrations of carbon dioxide in nitrogen are used to calibrate for carbon dioxide. The calibration of the gas analysis is performed within the BioExpert software.

3.7. Operation of the permittistat

A new experimental run is started and a data file is created with the relevant details. An access interval time is selected of 2 min and the logging of data is started. The inoculum is aseptically added to the vessel and the inoculum tube is clamped off to maintain the pressure in the system for the off-gas analysis. The batch growth of the culture is monitored (via the biomass measurements) until stationary phase is reached. Off-line analyses are performed to monitor the $OD_{600\ nm}$ and dry weight of the culture, culture purity, and extracellular glucose measurements. The capacitance of the culture at 75% of the maximum biomass yield is calculated, and a control statement is programmed into the software to control the media pump. The pump speed is set such that it would provide a dilution rate equivalent to the growth rate at μ_{max}. The pump is started on the waste feed line, note our system uses a variable height pipe to pump the culture from the vessel to maintain a fixed volume. The antifoam control loop is switched off after the batch growth of the culture. The culture is deemed to be in steady state when the biomass remained constant over three residence times (Castrillo *et al.*, 2007) and off-line measurements are carried out to measure the $OD_{600\ nm}$, dry weight, culture purity, extracellular glucose, and ethanol. The media intake rate is calculated (by both volume and weight change of the reservoir vessels). The dilution rate (D h^{-1}) of the vessel is calculated as described in Section 1.2.1, Eq. (14.6). The growth rate and parameters are checked throughout the experimental period to ensure that there are no observable changes in the culture.

4. Sampling of Biomass for Proteome and Metabolome Analyses

The preparation of samples for analysis of the proteome and metabolome is covered in other chapters in this volume. In sampling the culture for proteomic analyses, 50 mL of culture are removed from the vessel and harvested by centrifugation at 4 °C, before snap freezing in liquid nitrogen and storing at −80 °C. For metabolomic analyses, the culture requires quenching to stop metabolism. The culture (10 mL) is sprayed (directly from the bioreactor) into a quenching solution (40 mL of 60:40 methanol: water, with 0.85% ammonium hydrogen carbonate, at −48 °C) to reduce the temperature of the culture to below −20 °C. The biomass is harvested by centrifugation at −9 °C, and the biomass is snap frozen in liquid nitrogen and stored at −80 °C. An aliquot of the quenching solution is retained for analysis to assess for the leakage of metabolites. Samples of the exometabolome are collected by syringe filtering (0.2 μm pore size) the culture and snap freezing the filtrate. The filtrate is stored at −80 °C.

ACKNOWLEDGMENTS

The authors wish to thank BBSRC for financial support of The Manchester Centre for Integrative Systems Biology (BBC0082191).

REFERENCES

Allen, J., *et al.* (2003). High-throughput classification of yeast mutants for functional genomics using metabolic footprinting. *Nat. Biotechnol.* **21,** 692–696.

Bijmans, M. F. M., *et al.* (2009). Sulfate reduction at pH 4.0 for treatment of process and wastewaters. *Biotechnol. Prog.* **26,** 1029–1037.

Bryson, V., and Szybalski, W. (1952). Microbial selection. *Science* **116,** 45–46.

Bull, A. T. (2010). The renaissance of continuous culture in the post-genomics age. *J. Ind. Microbiol. Biotechnol.* **37,** 993–1021.

Castrillo, J. I., *et al.* (2007). Growth control of the eukaryote cell: a systems biology study in yeast. *J. Biol.* **6,** 4.

Daran-Lapujade, P., *et al.* (2009). Chemostat-based micro-array analysis in baker's yeast. *Adv. Microb. Physiol.* **54,** 257–311.

Davey, C. L., *et al.* (1993). Introduction to the dielectric estimation of cellular biomass in real-time, with special emphasis on measurements at high-volume fractions. *Anal. Chim. Acta* **279,** 155–161.

Davey, H. M., *et al.* (1996). Oscillatory, stochastic and chaotic growth rate fluctuations in permittistatically controlled yeast cultures. *Biosystems* **39,** 43–61.

de Crecy, E., *et al.* (2007). Development of a novel continuous culture device for experimental evolution of bacterial populations. *Appl. Microbiol. Biotechnol.* **77,** 489–496.

de Crecy, E., *et al.* (2009). Directed evolution of a filamentous fungus for thermotolerance. *BMC Biotechnol.* **9,** 74.

Delneri, D., *et al.* (2008). Identification and characterization of high-flux-control genes of yeast through competition analyses in continuous cultures. *Nat. Genet.* **40,** 113–117.

Ferenci, T. (1999). 'Growth of bacterial cultures' 50 years on: Towards an uncertainty principle instead of constants in bacterial growth kinetics. *Res. Microbiol.* **150,** 431–438.

Ferenci, T. (2008). Bacterial physiology, regulation and mutational adaptation in a chemostat environment. *Adv. Microb. Physiol.* **53,** 169–229.

Gilbert, A., *et al.* (2009). Rapid strain improvement through optimized evolution in the cytostat. *Biotechnol. Bioeng.* **103,** 500–512.

Groeneveld, P., *et al.* (2009). Super life—How and why 'cell selection' leads to the fastest-growing eukaryote. *FEBS J.* **276,** 254–270.
Harris, C. M., and Kell, D. B. (1983). The radio-frequency dielectric-properties of yeast-cells measured with a rapid, automated frequency-domain dielectric spectrometer. *Bioelectrochem. Bioenerg.* **11,** 15–28.
Harris, C. M., and Kell, D. B. (1985). The estimation of microbial biomass. *Biosensors* **1,** 17–84.
Harris, C. M., *et al.* (1987). Dielectric permittivity of microbial suspensions at radio frequencies—A novel method for the real-time estimation of microbial biomass. *Enzyme Microb. Technol.* **9,** 181–186.
Herbert, D., *et al.* (1956). The continuous culture of bacteria—A theoretical and experimental study. *J. Gen. Microbiol.* **14,** 601–622.
Hoskisson, P. A., and Hobbs, G. (2005). Continuous culture—Making a comeback? *Microbiology-Sgm* **151,** 3153–3159.
Markx, G. H., *et al.* (1991). The permittistat—A novel type of turbidostat. *J. Gen. Microbiol.* **137,** 735–743.
Martin, G. A., and Hempfling, W. P. (1976). Method for regulation of microbial-population density during continuous culture at high growth-rates. *Arch. Microbiol.* **107,** 41–47.
Monod, J. (1942). Recherces sur la croissance des cultures bacteriennes. Hermann & Cie, Paris.
Monod, J. (1950). La technique De culture continue theorie Et applications. *Ann I Pasteur Paris* **79,** 390–410.
Northrop, J. H. (1954). Apparatus for maintaining bacterial cultures in the steady state. *J. Gen. Physiol.* **38,** 105–115.
Novick, A., and Szilard, L. (1950). Experiments with the chemostat on spontaneous mutations of bacteria. *Proc. Natl. Acad. Sci. USA* **36,** 708–719.
Paquin, C., and Adams, J. (1983). Frequency of fixation of adaptive mutations is higher in evolving diploid than haploid yeast populations. *Nature* **302,** 495–500.
Silverman, S. J., *et al.* (2010). Metabolic cycling in single yeast cells from unsynchronized steady-state populations limited on glucose or phosphate. *Proc. Natl. Acad. Sci. USA* **107,** 6946–6951.
Snoep, J. L., *et al.* (2009). Control of specific growth rate in Saccharomyces cerevisiae. *Microbiology-Sgm* **155,** 1699–1707.

CHAPTER FIFTEEN

Sample Preparation Related to the Intracellular Metabolome of Yeast: Methods for Quenching, Extraction, and Metabolite Quantitation

Warwick B. Dunn*,†,‡ *and* Catherine L. Winder*,†

Contents

Abstract

The determination of intracellular metabolite concentrations in *Saccharomyces cerevisiae* cell systems requires appropriate experimental methods to (a) collect cells and rapidly inhibit metabolism (quenching), (b) fracture cell walls and extract

* Manchester Centre for Integrative Systems Biology, University of Manchester, Manchester, United Kingdom
† Department of Chemistry, Manchester Interdisciplinary Biocentre, University of Manchester, Manchester, United Kingdom
‡ Centre for Advanced Discovery and Experimental Therapeutics, School of Biomedicine, University of Manchester, Manchester, United Kingdom

Methods in Enzymology, Volume 500

ISSN 0076-6879, DOI: 10.1016/B978-0-12-385118-5.00015-3

metabolites from within the cellular envelope(s), and (c) detect and quantify metabolites. A range of methods are applied for each of these processes, and no single method is appropriate for all metabolites. For example, the physicochemical diversity of metabolites, including solubility in water or organic solvents, is large. No single extraction solvent is appropriate for all metabolites reported in *S. cerevisiae*, and multiple solvent systems for extraction employing water, methanol, and chloroform at different pH are recommended for targeted extraction of metabolites. In this chapter, methods for the targeted study of organic acids present in the tricarboxylic acid cycle will be described. These include (a) the quenching of metabolism in batch cell cultures, (b) a single extraction method which provides the extraction of a wide diversity of metabolites, and (c) an analytical method applying gas chromatography–mass spectrometry for targeted analysis of six organic acids present in the tricarboxylic acid cycle metabolic pathway.

1. Introduction

Initial studies to investigate the biology of yeast cultures were observed as early as the 1950s (Wiken and Richard, 1955). Today, *Saccharomyces cerevisiae* is widely applied as a model organism in system biology research (Castrillo *et al.*, 2007; Mustacchi *et al.*, 2006) and offers many advantages. Yeast has a short cycle time of typically less than 270 min, is cheap, is easy to grow (including in large volumes), and is easy to store. The genome of this eukaryote is relatively simple, containing approximately 6000 genes (http://www.yeastgenome.org/cache/genomeSnapshot.html) and has been fully sequenced. A wide range of yeast genes have orthologs in humans. This allows the study of human genes in a simple system, as biological process and structure are highly conserved to a high degree throughout eukaryotic life. Yeast is easy to manipulate biologically and genetically and without the requirement for ethical approval which is essential for human studies. Many applications have been observed which apply yeast in systems biology research from cell cycle regulation (Barik *et al.*, 2010) to the effect of calorific restriction on biological aging (Goldberg *et al.*, 2006).

One area of systems biology research in yeast and humans is that of metabolism (Mo *et al.*, 2007; Nicholson and Wilson, 2003; Teusink *et al.*, 2000). The synthesis and catabolism of small molecular weight chemicals in biological systems for biomass production (e.g., amino acids to proteins, lipids to cell wall), energy production and storage (e.g., adenosine triphosphate, ATP), and other functions required for survival, growth, and reproduction are encompassed in metabolic processes. Metabolites are low molecular weight chemicals (<1500 Da) present in biological systems. They have a diversity of physicochemical properties and concentrations in biological systems and are involved in multiple biochemical processes (metabolism, allosterism, and posttranslational modifications being a few examples). The complete complement of metabolites in a biological system is defined as the metabolome, and multiple metabolomes can be associated with a single organism. For example, growing yeast cultures may be characterized by both the intracellular and extracellular metabolome, defined as the endo- and exometabolome, respectively. Metabolites in the intracellular metabolome result from uptake of nutrients from the external environment and intracellular metabolism, whereas the extracellular metabolome is the result of metabolite uptake into the cell (e.g., glucose from the growth media) and secretion of metabolites from the cells (e.g., ethanol or carbon dioxide).

Metabolism can be viewed as a network. A yeast systems biology community consensus metabolic reconstruction is available (Herrgard *et al.*, 2008) which describes 1168 metabolites and 1761 metabolic reactions. The reconstruction is not complete and, for example, areas of lipid metabolism are not fully detailed (Dobson *et al.*, 2010; Nookaew *et al.*, 2008).

Metabolomics focuses on the study of metabolites and the interactions between metabolites and (a) other metabolites and proteins (metabolism), (b) metabolite–protein interactions (e.g., allosterism), and (c) metabolite–RNA interactions (e.g., riboswitches). In metabolomics, a typical workflow includes the design of biological experiments, followed by sample collection, preparation, and analysis, and then data preprocessing and data analysis (Brown *et al.*, 2005; Dunn *et al.*, 2011). Two specific types of metabolomic study are available (Dunn *et al.*, 2011). The first operates from a position of limited biological knowledge where biological experiments are designed to interrogate the metabolic network on a holistic level without *a priori* knowledge of the metabolites of specific interest. The objective is to acquire data on a wide range of metabolites followed by determination of those metabolites whose relative concentrations change following a perturbation (e.g., gene knockout, change in growth nutrients). These studies are defined as metabolomics if all metabolites in the system are detected. However, this is currently not feasible with any single analytical technology or analytical method, and combinations of complementary techniques are advisable for coverage of the metabolome. In reality, a wide range of metabolites are detected but not all in any single study, and these are defined as metabolite profiling, metabolic profiling, or untargeted analysis. These are discovery-phase studies, sometimes defined as inductive studies (Kell and Oliver, 2003).

At the opposite end of the spectrum are targeted methods which are interested in only a limited number of metabolites, defined as being of specific biological interest and usually related as defined by a specific pathway in metabolism. Here, absolute concentrations of metabolites are determined for each metabolite, applying sample preparation and analytical methodologies with high accuracy, precision, sensitivity, and specificity. This strategy is applied for hypothesis testing or systems biology *in silico* modelling approaches.

A number of issues have to be considered when studying the intracellular metabolism of yeast. These include (a) the rapid metabolic flux in central metabolism which requires the rapid halting or inhibition of metabolism (quenching) to achieve accurate and reproducible results, (b) the solubility of metabolites in aqueous or organic solvent systems, and (c) the requirement for relative or absolute concentration determination. The processes of sample collection, metabolite extraction, and targeted quantitation of metabolites present in the intracellular metabolome will be described, with specific consideration of the issues which should be addressed during the experimental design.

2. Sample Collection from Batch Cultures with Quenching of Intracellular Metabolism

The flux observed in intracellular metabolism operates at timescales of seconds depending on the area of metabolism being studied. This has been experimentally defined at specific environmental conditions (Mashego *et al.*, 2006).

The procedure of sample collection should provide a sample which is representative of the biological system at the time of sampling. For the study of intracellular metabolism, where sampling and metabolite detection are separated, this requires the rapid inhibition of metabolism so to provide a sample representative of the culture at the time of sampling. This process of metabolic inhibition is defined as quenching.

Quenching procedures applied for yeast cultures rapidly transfer a sample from the culture to a quenching solution operating at either low or high temperatures. Traditionally, the quenching solution was not applied. Instead, cells were separated from the growth medium and washed applying filtration (Saez and Lagunas, 1976). However today, the manual process of transferring a culture sample is performed using a pipette for batch cultures or applying a pressurized system for continuous cultures. De Koning and colleagues provided one of the first reported methods applying spraying of the culture into a quenching solution (De Koning and Van Dam, 1992). More recently, an automated system for sampling and quenching yeast continuous cultures has been developed and allowed multiple samples to be collected per second to determine the dynamics of yeast metabolism (Mashego *et al.*, 2006). However, most laboratories still apply a manual process, and it should be noted that changes in the concentrations of metabolites whose metabolic flux is high and operates on subsecond timescales may be expected to be observed during the sampling process.

A range of quenching solutions have been applied to sample yeast cultures (Canelas *et al.*, 2008; Castrillo *et al.*, 2003; Ewald *et al.*, 2009; VillasBoas *et al.*, 2007; Villas-Boas and Bruheim, 2007; Villas-Boas *et al.*, 2005). Methanol/water solutions at −48 °C or similar temperatures are commonly applied with the ratio of methanol:water being studied, and the quenching solution volume: culture volume of a minimum of 3:1 being recommended. This method provides an appropriate rapid reduction in temperature of the culture (typically to less than −20 °C) which stops enzymatic activity and consequently metabolic flux. This method also ensures a low ratio of methanol to water in the final sample solution. The ratio of methanol/water in the final sample solution can be important, as a large organic solvent content can result in partial loss of cell membrane integrity and leakage of intracellular metabolites into the quenching solution (Canelas *et al.*, 2008). The rapid decrease in temperature has also been implicated in the process of intracellular metabolite leakage though a cold-shock process, admittedly in *Escherichia coli* (Wittmann *et al.*, 2004) in which metabolite leakage is more pronounced than in *S. cerevisiae* cultures. For this reason, it is recommended to analyze the quenching solution and exometabolome to determine whether leakage is observed and whether leakage is small or large. This can be expected to be dependent on the metabolites studied (Villas-Boas *et al.*, 2005). To minimize metabolite leakage, it is recommended to minimize the contact time of cells and organic solvents. After sample collection, centrifugation is performed at a low *g* force and low temperature (less than 0 °C) to produce a cell pellet and supernatant. The supernatant can be removed followed by storage of the cell pellet at −80 °C prior to extraction. To ensure that bias is not introduced during the quenching process, it is advised to ensure that the sample collection time is identical for all samples.

Applying a glycerol-based quenching solution has also been applied in yeast metabolism studies, though operating at temperatures above 0 °C (Villas-Boas and Bruheim, 2007). The reasoning for this is that glycerol and the higher temperatures will eliminate leakage of intracellular metabolites through the action of organic solvents and cold-shock. However, it has been reported that the approach may be unsuitable due to the dominance of the glycerol peak (from the quenching solution) in the gas chromatography–mass spectrometry (GC–MS) chromatograms which masked other metabolites (Spura *et al.*, 2009). Methods applying solutions containing a nonsalt buffer (e.g., tricine) have also been developed to eliminate leakage of intracellular metabolites (Castrillo *et al.*, 2003). The use of the salt ammonium carbonate is also appropriate, as during lyophilization, the salt can be removed as gaseous ammonia and carbon dioxide, thereby preventing the introduction of artifacts in the data during the sample preparation stage. It is reported that reducing the temperature to 0 °C will reduce metabolic flux for enzymes which operate at an optimal efficiency at approximately 30 °C; however, reducing the temperatures to below −20 °C will eliminate the majority of metabolic flux.

Combined quenching and extraction methods have also been reported, the most common being the boiling ethanol method (Gonzalez *et al.*, 1997) where samples are plunged into a boiling aqueous ethanol solution which acts to rapidly create cell wall disruption and to inactivate intracellular enzymes through degradation of 3D structure at an elevated temperature in a single process.

The culture medium is sampled along with cells in the quenching process. Metabolites are present in the culture medium for two reasons: (a) metabolites present in the growth medium and not fully transported into the intracellular metabolome during growth or (b) metabolites secreted from the intracellular volume during growth (e.g., lactate or ethanol). Metabolites present in the culture medium are constrained within the sampled culture solution, either being present in the solution or absorbed to cell walls. After centrifugation to pellet cells and removal of the quenching solution, metabolites can be either absorbed to cell walls or present in small volumes of the quenching solution remaining. A second quick centrifugation step can reduce the volume of quenching solution remaining. In targeted studies, the presence of extracellular metabolites can be problematic as the detected concentration of a metabolite is a composite of the intracellular metabolite concentration and the concentration of metabolite absorbed to the cell wall or constrained in remaining quenching solution. The relative contribution of the metabolite concentration from the extracellular environment is dependent on the metabolite, the growth medium, the growth conditions, and the quenching method applied. A wash step, as described in this chapter, is recommended when the presence of extracellular metabolites is expected to contribute to the total metabolite concentration in targeted or nontargeted methods. However, it is recommended to test whether a wash step is required, as cell wall permeabilization during the initial quenching process, if low temperatures and organic solvents are applied, can result in further leakage of metabolites during the wash process and result in an under estimation of the intracellular metabolite concentration. Therefore, validation of the method by assaying the quenching and wash solutions without and with one or more wash steps is recommended.

A second source of metabolites or chemicals which can interfere with the chemical analysis is the consumables (e.g., centrifuge tubes) used during quenching and extraction processes. It is recommended that a "blank" test is performed, that is the quenching and extraction procedures are performed with the absence of a yeast culture (i.e. quenching in a pure water solution) and the final extraction solution assayed for the presence of metabolites and other interfering chemicals. If these are observed, other consumables should be tested until suitable consumables are found which do not result in contamination of the samples.

For many different areas of metabolism, the authors have observed that applying a methanol/water quenching solution containing a salt (e.g., ammonium bicarbonate) at low temperature is the most appropriate method to apply for both targeted and nontargeted applications. The protocol for quenching of batch cultures is described below and in Fig. 15.1.

2.1. Materials and instruments

- Methanol, water, and ammonium hydrogen carbonate, all of analytical grade quality or higher
- Dry ice, solid carbon dioxide

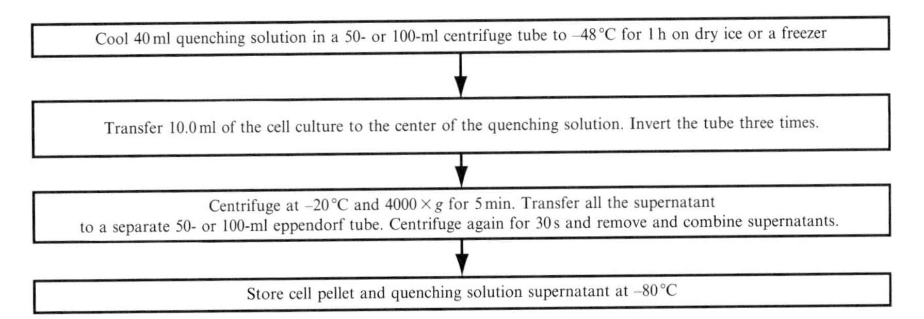

Figure 15.1 Schematic workflow for the quenching of yeast batch cultures.

- 50- and/or 100-ml centrifuge tubes, composed of plastics which are chemically resistant and of high quality
- Digital thermometer to measure solution temperatures in the range +40 to −50 °C
- Centrifuge capable of operating with 50-ml centrifuge tubes at temperatures of −20 °C and a centrifugal force of 4000×*g*
- Quenching solution composed of 60:40 (v/v) methanol:water containing ammonium hydrogen carbonate at a final concentration of 0.85% (w/v) and at a temperature of −48 °C. The solution should be adjusted, if required, to pH 5.5 with acid and base and cooled to −48 °C in a dry ice bath or −80 °C freezer
- Saline solution, 0.85% (w/v) sodium chloride dissolved in water
- Liquid nitrogen and appropriate safety equipment

2.2. Method for quenching of metabolism and sample collection from batch cultures

2.2.1. Cool 40-ml of quenching solution in a 50- or 100-ml centrifuge tube to −48 °C for a minimum of 1 h.

2.2.2. Using a pipette, accurately and rapidly transfer 10-ml of culture into the quenching solution. Expel in the center of the tube to ensure no cells stick to the sides of the centrifuge tube where they may freeze and rupture.

2.2.3. Rapidly invert the tube three times to ensure the culture does not stick to the inner surface of the centrifuge tube.

2.2.4. Pellet the cell fraction by centrifugation for 5 min at a temperature below 0 °C and preferably at −20 °C. The maximum force should be 4000×*g*.

2.2.5. Rapidly transfer all combined media and quenching solution supernatant to a separate 50-ml centrifuge tube.

2.2.6. Repeat step 2.2.4 for 30 s and transfer remaining supernatant to the 50-ml centrifuge tube used in step 2.2.5 using a pipette.

2.2.7. Snap-freeze the samples in liquid nitrogen.

2.2.8. Store the cell pellet and combined media and quenching solution supernatant at −80 °C.

Note, in continuous cultures (as described elsewhere in this volume), the samples are transferred directly from the bioreactor to quenching solution. The bioreactor is under slightly negative pressure, and the length of the sample tube is minimized to enable the rapid transfer of the sample from the vessel to the quenching solution. The quenching solution is cooled to −48 °C in a graduated bottle suitable for the sampling port of the bioreactor, and the same ratio of sample to quenching solution is used as described in step 2.2.2. The sample is inverted three times to ensure adequate mixing and transferred to a 50-ml centrifuge tube. The procedure is followed from step 2.2.4 to harvest the biomass from the culture.

Growth of *S. cerevisiae* can be performed in a range of growth media from minimal media containing a limited number of metabolites required for growth to complex media containing many metabolites (e.g., metabolic footprinting media; Allen *et al.*, 2003). In applications where the metabolites to be studied are present in the growth media, an additional step of washing the cell pellet is recommended to remove residual metabolites absorbed to the cell walls or present from the process of cell pelleting.

2.2.9. To the cell pellet remaining in step 2.2.6 (and before storage at −80 °C), add 10-ml of saline solution and vortex mix for 15 s to suspend the cell pellet.

2.2.10. Pellet the cell fraction by centrifugation for 5 min at a temperature of 4 °C. The maximum force should be 4000×*g*.

2.2.11. Transfer the saline solution supernatant to a 50-ml centrifuge tube.

2.2.12. Repeat step 2.2.9. for 30 s and transfer the supernatant to the 50-ml centrifuge tube.

2.2.13. Snap-freeze the samples in liquid nitrogen

2.2.14. Store the cell pellet and saline solution supernatant at −80 °C.

3. Extraction of Polar and Nonpolar Metabolites from the Intracellular Metabolome

The intracellular metabolome of *S. cerevisiae* contains an estimated 1168 metabolites (Herrgard *et al.*, 2008), although the true number of metabolites is expected to be higher than the estimate. For example, recent research has shown that many lipids have not been included in reconstructions of yeast metabolism (Dobson *et al.*, 2010; Nookaew *et al.*, 2008).

The intracellular metabolome is a composite of metabolic pathways incorporating both polar and nonpolar (or lipid) metabolites.

The physical and chemical characteristics of metabolites are diverse. These include molecular weight and solvent solubility among others. The solubility of metabolites in aqueous or organic solvents (e.g., methanol) is important so to optimize extraction procedures which provide the highest recovery of metabolites. Polar metabolites including sugars and organic acids will dissolve efficiently in water or water/organic solvent solutions, but these metabolites will not dissolve in nonpolar solvents such as chloroform. Nonpolar metabolites, including lipids, will dissolve efficiently in chloroform. Therefore, the extraction procedure applied is dependant on the metabolites of specific interest.

A range of extraction methods have been applied. The origins of many methods applied today originated in methods published in the twentieth century, and two papers have reviewed and assessed these methods (De Koning and Van Dam, 1992; Saez and Lagunas, 1976). These reported methods had the objective to extract specific metabolites or classes of metabolites, that is, provided the targeted study of specific metabolites or areas of metabolism. These included acid and alkaline extraction solutions to extract acid stable and reduced nucleotides, respectively (Saez and Lagunas, 1976), and the use of a chloroform-based extraction at -40 °C and at neutral pH for the determination of glycolytic metabolites (De Koning and Van Dam, 1992). Methods applied in research today employ either of water/organic solvent solutions, chloroform, or methanol/water/chloroform solvent solutions to include some specificity in the extraction method and reduce the complexity of the extraction solution. These methods have been reviewed and assessed (Castrillo *et al.*, 2003; Mashego *et al.*, 2007; VillasBoas *et al.*, 2007; Villas-Boas *et al.*, 2005, 2007; Winder *et al.*, 2008). However, it can be prudent to perform a composite extraction method with water, organic solvents, and chloroform to provide two separate extraction solutions: (1) a water-organic solvent solution containing polar metabolites and (2) a chloroform solution containing nonpolar metabolites including lipids. This allows multiple classes of metabolites to be extracted from a single sample and eliminates the necessity for different samples to be collected for the study of different metabolite classes. The application of organic solvents also disrupts enzyme structure and eliminates further enzymatic activity and metabolic flux.

However, there is no ideal method available to provide the efficient extraction of all metabolites from a yeast culture, whether for a targeted or nontargeted metabolomic study. The large variations in physical and chemical properties and concentrations (i.e., dynamic range of concentrations) of the complete complement of metabolites in the yeast metabolome ensure that no single method is applicable for all metabolites.

Yeast is composed of multiple internal compartments, and metabolites will be present in some or all of these compartments. Metabolites can be present in solution but are also bound to other biochemicals, including cell or compartment membranes, proteins, and nucleic acids. Any extraction method should ensure that metabolites are released into the extraction solution. Methods to determine whether release of bound metabolites is achieved have not yet been developed, though they are required in the future. The accuracy of metabolite quantification in targeted studies is dependent on the release of bound metabolites.

The extraction process is performed to disrupt cell walls of the organism, including intracellular organelles in eukaryotic organisms, and the release of the intracellular metabolites into the extraction solution. Cell wall disruption can be performed by applying a range of methods. These include multiple freeze–thaw cycles (VillasBoas *et al.*, 2007; Villas-Boas *et al.*, 2005, 2007), sonication (Dejonggubbels *et al.*, 1995), and bead beating (Sporty *et al.*, 2008). Each has advantages and limitations. The authors apply multiple freeze–thaw cycles which are simple to perform and allow multiple samples to be processed simultaneously for higher throughput. This process involves rapid freezing of the extraction solvent and sample in liquid nitrogen followed by slow thawing on dry ice. This repeated process of freeze/thaw cycles creates expansion of the water volume in the cell and cell wall rupture. The process is also performed at low temperatures ensuring metabolism is not reactivated during the extraction process. Recently, methods for automation of this process in multiwall plates have been reported (Ewald *et al.*, 2009).

The protocol here will describe the combined extraction of polar and nonpolar metabolites in a single extraction method. The method is applicable to a wide range of polar (including amino and organic acids) and nonpolar (lipids including fatty acids and phosphocholines) metabolite classes. The authors typically extract 15 mg dry weight of cell biomass to allow detection of many metabolites at physiological concentrations. To increase sensitivity, two separate cell pellets of 15 mg dry weight biomass can be extracted and the extraction solutions combined before lyophilization. This assumes that the two cell pellets are from steady state cultures or are sampled at the same time in time-dependent batch cultures. The protocol is described in Fig. 15.2.

3.1. Materials and instruments

- Methanol, water, and chloroform, all of analytical grade quality or higher
- Dry ice, solid carbon dioxide
- Liquid nitrogen in an appropriate container and with ladle or tweezers for sample removal.
- 2-ml centrifuge tubes, composed of plastics which are chemically resistant and of high quality

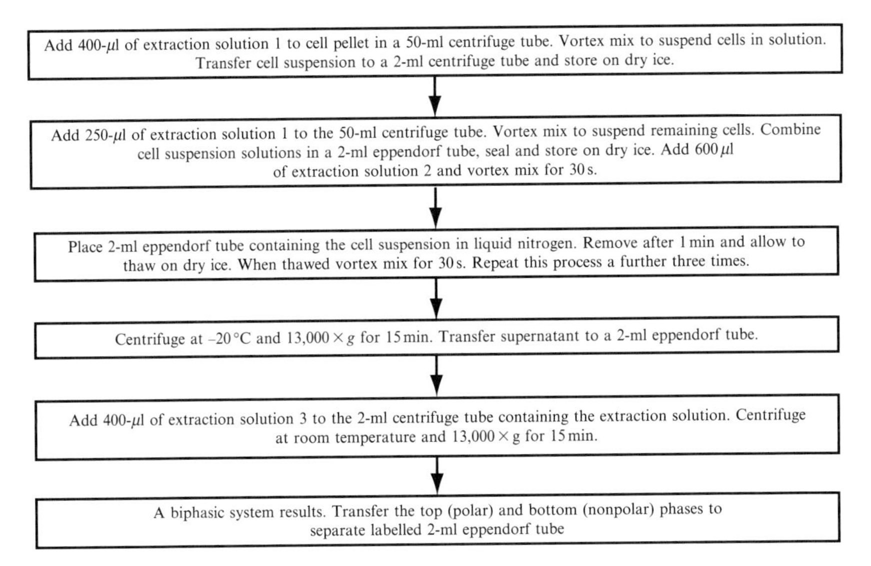

Figure 15.2 Schematic workflow for the extraction of polar and nonpolar metabolites from the intracellular metabolome of yeast.

- Centrifuge capable of operating with 2-ml centrifuge tubes at temperatures of −9 °C and a force of 13,000×*g*
- Vortex mixer
- Extraction solution 1, 80/20 methanol/water
- Extraction solution 2, chloroform
- Extraction solution 3, water

3.2. Method for combined extraction of polar and non polar metabolites

3.2.1. To the cell pellet in a 50-ml centrifuge tube, add 400-μl of extraction solution 1. Store on dry ice.

3.2.2. Suspend the cell pellet in the extraction solution through a process of vortex mixing.

3.2.3. Transfer the cell suspension to a 2-ml centrifuge tube using a 1000-μl pipette. Store on dry ice.

3.2.4. Add 250 μl of extraction solution 1 to the 50-ml centrifuge tube and vortex mix to suspend any remaining cell biomass. Transfer the extraction solution to the 2-ml centrifuge tube described in step 3.3.3. to combine with the 400-μl extraction solution aliquot. Add 600-μl of extraction solution 2 and close the 2-ml eppendorf lid. Vortex mix for 30 s.

3.2.5. Place the 2-ml eppendorf tube into liquid nitrogen for 1 min.

3.2.6. After 1 min, remove from liquid nitrogen and place on dry ice. Allow to thaw. Once thawed, vortex mix for 30 s.
3.2.7. Repeat steps 3.2.5 and 3.2.6 for a further three times.
3.2.8. To separate the cell biomass and extraction solution, centrifuge at -9 °C and 13,000×*g* for 15 min.
3.2.9. Transfer the extraction solution supernatant to a separate 2-ml eppendorf tube and store on dry ice.
3.2.10. Add 350-μl of extraction solution 3 to the extraction solution recovered in step 3.2.9. Vortex mix for 30 s.
3.2.11. Centrifuge the 2-ml centrifuge tube at 13,000×*g* for 15 min at room temperature.
3.2.12. A biphasic solution will result. The upper layer is composed of methanol and water and polar metabolites. The lower layer is composed of chloroform and nonpolar metabolites.
3.2.13. Transfer the upper layer into a separate 2-ml eppendorf tube and seal. Label as methanol/water extract.
3.2.14. Transfer the lower layer into a separate 2-ml eppendorf tube and seal. Label as chloroform extract.

The two extraction solutions are available for sample analysis. This can include direct analysis of the extraction solution or analysis after further sample processing including fractionation processes (e.g., solid phase extraction) or lyophilization. As many plastics are not fully chloroform resistant, it is recommended to perform a blank set of extractions which followed the Standard Operating Procedure (SOP) described but with no biomass used.

4. Targeted Quantification of Organic Acids Applying Gas Chromatography–Mass Spectrometry

Following sample preparation, targeted or untargeted analysis can be performed depending on the experimental objectives. The role and methods for untargeted analysis applying mass spectrometry and NMR platforms have been discussed (Dunn *et al.*, 2011).

Targeted analysis has the objective to determine the absolute concentration of metabolites. These data are applied in hypothesis testing or during *in silico* model construction and validation in systems biology. Chromatographic systems coupled to mass spectrometry detection systems are typically used, and further discussions are based on the assumption that these systems are applied. Targeted methods differ significantly from untargeted analysis methods. A higher level of accuracy and precision are generally required. Sample preparation and analysis methods can be developed to increase specificity by separating the metabolites of interest from other metabolites and sample matrix. Solid phase extraction is one method applied to fractionate sample extracts. Alternatively, tandem mass spectrometry can be applied to increase specificity as was discussed in chapter 2.

In targeted methods, appropriate authentic chemical standards are required and are applied to calibrate the analytical system to provide the required accuracy during quantification. These are absent in untargeted methods. Two strategies can be applied, external calibration or standard addition (Robards *et al.*, 1997). External calibration employs calibration solutions of different concentrations and which extend over the linear calibration range of the analytical system employed. All calibration solutions are analyzed, peak areas or heights are determined, and a linear calibration curve defining the relationship between concentration and chromatographic peak area or height is calculated. The unknown concentration of metabolites in samples can then be determined by sample analysis under identical analytical conditions and through comparison of the peak area determined for the sample and the calibration curve.

Reproducible analytical methods are required for high accuracy. To provide accurate results, an internal standard should be applied to compensate for process variations. These can include variation in injection volume or differences in the sample matrix between the samples and the often less complex chemical standard solutions. The internal standard is commonly a structural analogue of the metabolite and typically can be an isotopic analogue, for example, $^{13}C_6$ glucose can be applied as an internal standard for glucose. Chromatographic properties will be similar, but the mass of $^{12}C_6$ glucose and internal standard are different and can be applied to detect each separately when using mass spectrometry as a detection system. In studies where appropriate internal standards are not available or the price is restrictive, two options are available. The first is to apply a single internal standard for multiple metabolites of similar chemical structure. For example, $^{13}C_6$ glucose can be applied as an internal standard for all C_6 sugars. This is a compromise as the retention time may differ for different metabolites and the coeluting species will vary across the chromatographic run. Here, a separate internal standard for each metabolite is preferable. The second is to apply the standard addition method. Here, separate aliquots of the sample are spiked with solutions containing increasing concentrations of the metabolites of interest. Each solution is analyzed and a linear calibration curve is plotted. The intercept of the calibration curve on the abscissa defines the concentration of the metabolite in the original sample. This method is highly appropriate when internal standards are not available or when the sample matrix is significantly different to the chemical standard solution matrix.

To provide confidence in the results, it is recommended to analyze quality control (QC) samples whose concentration is known. If the quantitation method provides the expected concentrations (typically within a 10%

error range), then the results for biological samples can be defined as accurate. Two QC samples of different concentrations are applied: one low concentration and one high concentration. Also, to provide reproducibility at the start of an analytical run, it is recommended to inject 5 QC samples for GC–MS (Begley *et al.*, 2009) and 10 QC samples for LC–MS (Zelena *et al.*, 2009).

The concentration of each metabolite can be reported in a number of ways. Typically, the mass or moles of metabolite per mass of biomass (e.g., microgram of metabolite per gram dry weight) is reported. Alternatively, if the cell count is available, the number of molecules per cell can be reported.

An example will be provided for the targeted analysis of six organic acids (citric, fumaric, malic, malonic, succinic, and 2-oxoglutaric), five of which are involved in the tricarboxylic acid (TCA) cycle. GC–MS was employed and external calibration with internal standards was performed. The method described assumes the linear calibration range of the system employed for each metabolite is known. If not, a set of standard solutions should be prepared in the range 0.1–5000 $\mu mol\ l^{-1}$ and analyzed to determine the limit of detection and linear calibration range.

4.1. Materials and instruments

- An appropriate GC–MS system. In the example described here, an Agilent 6890N gas chromatograph with 7683 autosampler (Agilent Technologies, Cheadle, UK) was coupled with a Leco Pegasus III electron impact-time-of-flight mass spectrometer (Leco, Stockport, UK). The analytical method applied is employed for quantification of a range of metabolites including metabolites present in the glycolysis pathway.
- An internal standard solution composed of $^{13}C_6$ citric acid and $^{13}C_4$ fumaric acid, each at a concentration of 100 $\mu mol\ l^{-1}$. The two isotopic internal standards are available from Sigma-Aldrich (Gillingham, UK). The tribasic $^{13}C_6$ citric acid will be applied as an internal standard for citric acid, and the dibasic $^{13}C_4$ fumaric acid will be applied as an internal standard for other dibasic organic acids (fumaric, malic, malonic, succinic, and 2-oxoglutaric acid).
- A minimum of six standard solutions containing citrate, fumaric, malic, malonic, succinic, and 2-oxoglutaric acids at increasing concentrations dependent on the linear calibration range of the GC–MS employed. The chemical standards are available from Sigma-Aldrich.
- 20 mg ml^{-1} *O*-methoxylamine hydrochloride in pyridine, each of analytical grade quality or higher.
- *N*-Methyl-*N*-trimethylsilyl-trifluoroacetamide (MSTFA), of analytical grade quality or higher.
- Appropriate 2 ml vials, screw caps, vial inserts, and springs.

- Aluminum block heater system capable of operating at 60 °C and of holding 2-ml centrifuge tubes.
- 20, 100, and 1000-μl pipettes.

4.2. Methods for quantitation of organic acids

4.2.1. Prepare an internal standard solution containing $^{13}C_6$ citric acid and $^{13}C_4$ fumaric acid, each at a concentration of 500 μmol l^{-1}. The solution can be prepared by dissolution of 19.8 mg of $^{13}C_6$ citric acid and 12.0 mg of $^{13}C_4$ fumaric acid in 10.0 ml of 60/40 methanol/water to prepare internal standard solution IS1 (10,000 μmol l^{-1}). Dissolve 500-μl of IS1 in 9.50 ml of 60/40 methanol/water to prepare internal standard solution IS2. Vortex mixing may be required for full dissolution.

4.2.2. Prepare a minimum of six standard solutions which extend from the limit of detection across the linear calibration range of the instrument. Each standard solution should contain all six metabolites and be dissolved in 60/40 methanol/water. A stock solution of 10,000 μmol l^{-1} can be prepared by dissolution of 19.1 mg citrate, 11.6 mg fumaric, 13.4 mg malic, 10.4 mg malonic, 11.8 mg succinic, and 14.6 mg 2-oxoglutaric acid in 10 ml 60/40 methanol/water. Accurately measure the mass of each metabolite during the weighing process and apply these masses during the calculation of the calibration curve parameters. Dilution of this stock solution can be performed to prepare standard solutions of different concentrations. The example described here involved the preparation of standard solutions at 1, 10, 25, 50, 100, 250, and 500 μmol l^{-1} by dilution of a 10,000 μmol l^{-1} stock solution. Vortex mixing may be required for full dissolution. Aliquot 100 μl of the internal standard solution into 500-μl aliquots of each standard solution in 2-ml eppendorf tubes. Ensure all tubes are labeled appropriately. Lyophilize one replicate of each standard solution.

4.2.3. Repeat step 4.2.2. to prepare two independent and lyophilized sets of standard solution samples.

4.2.4. Prepare two QC sample solutions containing each metabolite at a concentration of 25 or 250 μmol l^{-1}. Apply the method described in step 4.2.2. Ensure all tubes are labeled appropriately.

4.2.5. Take a 500-μl aliquot of each sample and transfer to a 2-ml eppendorf tube. Add 100 μl of internal standard solution. Ensure all tubes are labeled appropriately and lyophilize.

4.2.6. Prepare a blank solution by transferring 500-μl aliquots of a 60/40 methanol/water solution and 100 μl of internal standard solution to a 2-ml eppendorf tube and lyophilize.

4.2.7. On the day of analysis, randomize all samples, standard solutions, and QC samples. Preparation of all lyophilized standards and samples is performed as described in steps 4.2.8–4.2.11. This employs a two-stage derivatization process.

4.2.8. Add 50-μl of the 20 mg ml^{-1} *O*-methoxylamine hydrochloride in pyridine solution to each eppendorf tube. Vortex mix for 30 s and then heat at 60 °C for 40 min.

4.2.9. Remove all eppendorf tubes from the heater block and add 50-μl MSTFA to each tube. Vortex mix for 30 s and then heat at 60 °C for 40 min.

4.2.10. After both chemical derivatization steps are completed, remove all eppendorf tubes from the heater block, allow to cool for 5 min, and then centrifuge (15 min, 13,000×*g*) to remove any particulate matter.

4.2.11. Transfer the supernatant to autosampler vials and seal with a screw cap. The samples and standards and blanks are ready for analysis.

4.2.12. Analyze samples applying the GC–MS method described in Table 15.1. Apply an analysis order similar to that described in Table 15.2.

4.2.13. Apply the data processing software appropriate for the GC–MS system used to calculate the peak areas (or heights) for each metabolite in each sample or standard analyzed. In this example, the authors used ChromaTof software v2.25 (Leco).

Table 15.1 Analytical method employed for the targeted quantification of organic acids applying gas chromatography–mass spectrometry

Parameter	Value
Sample volume injected (μl)	1.0
Inlet temperature (°C)	250
Split ratio	1:10
Helium flow rate (ml min^{-1})	0.8
Acquisition rate (Hz)	20
Initial GC temperature (°C)	70
Start temp hold time (min)	4.5
Ramp speed (°C min^{-1})	9
Final temperature (°C)	300
Hold final temperature (min)	5
Detector voltage (V)	1700
Run time (min)	46.2

Table 15.2 Typical analytical run order for targeted quantification of organic acids applying gas chromatography–mass spectrometry

Injection order	Sample/standard information
1	QC (250 μmol l^{-1})
2	QC (250 μmol l^{-1})
3	QC (250 μmol l^{-1})
4	QC (250 μmol l^{-1})
5	QC (250 μmol l^{-1})
6	Blank standard solution
7	Standard solution (1 μmol l^{-1})
8	Standard solution (10 μmol l^{-1})
9	Standard solution (25 μmol l^{-1})
10	Standard solution (50 μmol l^{-1})
11	Standard solution (100 μmol l^{-1})
12	Standard solution (250 μmol l^{-1})
13	Standard solution (500 μmol l^{-1})
14	Blank standard solution
15	QC (250 μmol l^{-1})
16	Sample A
17	Sample B
18	Sample C
19	Sample D
20	Sample E
21	Sample F
22	QC (25 μmol l^{-1})
23	QC (250 μmol l^{-1})
24	Sample G
25	Sample H
26	Sample I
27	Sample J
28	Sample K
29	Sample L
30	QC (25 μmol l^{-1})
31	QC (250 μmol l^{-1})
32	Blank standard solution
33	Standard solution (1 μmol l^{-1})
34	Standard solution (10 μmol l^{-1})
35	Standard solution (25 μmol l^{-1})
36	Standard solution (50 μmol l^{-1})
37	Standard solution (100 μmol l^{-1})
38	Standard solution (250 μmol l^{-1})
39	Standard solution (500 μmol l^{-1})

4.2.14. Construct a linear calibration curve for each set of seven standard solutions applying the peak area data for calibration standards and blanks and determine the correlation coefficient and limit of detection.

4.2.15. Using the calibration curve, peak area data for the two sets of calibration standards and the peak area data for each sample determine the concentration of each metabolite in each sample applying the calibration curve data for both sets of calibration solutions bracketing the sample. Calculate the average concentration for the two concentrations calculated. This relates to the concentration in the final derivatization solution volume of 100 μl.

4.2.16. Repeat step 4.2.15 for the QC samples. If the measured and expected (25 and 250 $\mu mol\ l^{-1}$) concentrations have an error of less than 10%, continue with step 4.2.17. If not, reassess steps 4.2.13.–4.2.15. If the errors are still larger than 10%, repeating the analytical experiment should be considered.

4.2.17. To calculate the concentration in the final extraction solution, volume of 1000 μl for the polar phase divide by ten the concentration calculated in step 4.2.15.

4.2.18. To calculate the number of moles in the final extraction solution, divide the concentration calculated in step 4.2.15 by 1000.

4.2.19. To calculate the number of moles per gram of dry weight biomass, multiple the number of moles calculated in step 4.2.18 by 1/the dry weight of biomass in 10 ml of culture.

4.2.20. To calculate the number of molecules in the final extraction solution, divide the number of moles calculated in step 4.2.18 by 6.022×10^{23}.

Typical results acquired from a continuous culture of *S. cerevisiae* are shown in Table 15.3.

Table 15.3 Metabolite concentrations determined for citrate, fumaric, malic, malonic, succinic, and 2-oxoglutaric acid in two continuous cultures of *S. cerevisiae*

		1	2
Metabolite	CHEBI: ID	Molecules per cell	Molecules per cell
2-Oxoglutaric acid	CHEBI: 30915	7,943,939	7,437,545
Citric acid	CHEBI: 30769	25,746,091	19,924,599
Fumaric acid	CHEBI: 18012	25,386,560	23,682,815
Malic acid	CHEBI: 6650	47,164,063	42,491,673
Malonic acid	CHEBI: 30794	233,537,548	204,520,158
Succinic acid	CHEBI: 15741	25,449,017	19,992,599

ACKNOWLEDGMENTS

W. B. D. and C. L. W. wish to thank BBSRC for financial support of The Manchester Centre for Integrative Systems Biology (BBC0082191). WBD wishes to thank NIHR and NMDA for financial support of CADET.

REFERENCES

Allen, J., Davey, H. M., Broadhurst, D., Heald, J. K., Rowland, J. J., Oliver, S. G., and Kell, D. B. (2003). High-throughput classification of yeast mutants for functional genomics using metabolic footprinting. *Nat. Biotechnol.* **21,** 692–696.

Barik, D., Baumann, W. T., Paul, M. R., Novak, B., and Tyson, J. J. (2010). A model of yeast cell-cycle regulation based on multisite phosphorylation. *Mol. Syst. Biol.* **6,** 405.

Begley, P., Francis-McIntyre, S., Dunn, W. B., Broadhurst, D. I., Halsall, A., Tseng, A., Knowles, J., Goodacre, R., Kell, D. B., and Consortium, H. (2009). Development and performance of a gas chromatography-time-of-flight mass spectrometry analysis for large-scale nontargeted metabolomic studies of human serum. *Anal. Chem.* **81,** 7038–7046.

Brown, M., Dunn, W. B., Ellis, D. I., Goodacre, R., Handl, J., Knowles, J. D., O'Hagan, S., Spasic, I., and Kell, D. B. (2005). A metabolome pipeline: From concept to data to knowledge. *Metabolomics* **1,** 39–51.

Canelas, A. B., Ras, C., ten Pierick, A., van Dam, J. C., Heijnen, J. J., and Van Gulik, W. M. (2008). Leakage-free rapid quenching technique for yeast metabolomics. *Metabolomics* **4,** 226–239.

Castrillo, J. I., Hayes, A., Mohammed, S., Gaskell, S. J., and Oliver, S. G. (2003). An optimized protocol for metabolome analysis in yeast using direct infusion electrospray mass spectrometry. *Phytochemistry* **62,** 929–937.

Castrillo, J. I., Zeef, L. A., Hoyle, D. C., Zhang, N., Hayes, A., Gardner, D. C. J., Cornell, M. J., Petty, J., Hakes, L., Wardleworth, L., Rash, B., Brown, M., *et al.* (2007). Growth control of the eukaryote cell: A systems biology study in yeast. *J. Biol.* **6**(4).

De Koning, W., and Van Dam, K. (1992). A method for the determination of changes of glycolytic metabolites in yeast on a subsecond time scale using extraction at neutral pH. *Anal. Biochem.* **204,** 118–123.

Dejonggubbels, P., Vanrolleghem, P., Heijnen, S., Vandijken, J. P., and Pronk, J. T. (1995). Regulation of carbon metabolism in chemostat cultures of Saccharomyces-cerevisiae grown on mixtures of glucose and ethanol. *Yeast* **11,** 407–418.

Dobson, P. D., Smallbone, K., Jameson, D., Simeonidis, E., Lanthaler, K., Pir, P., Lu, C. A., Swainston, N., Dunn, W. B., Fisher, P., Hull, D., Brown, M., *et al.* (2010). Further developments towards a genome-scale metabolic model of yeast. *BMC Syst. Biol.* **4**(145).

Dunn, W. B., Broadhurst, D. I., Atherton, H. J., Goodacre, R., and Griffin, J. L. (2011). Systems level studies of mammalian metabolomes: The roles of mass spectrometry and nuclear magnetic resonance spectroscopy. *Chem. Soc. Rev.* **40,** 387–426.

Ewald, J. C., Heux, S., and Zamboni, N. (2009). High-throughput quantitative metabolomics: Workflow for cultivation, quenching, and analysis of yeast in a multiwell format. *Anal. Chem.* **81,** 3623–3629.

Goldberg, A., Gregg, C., Boukh-Viner, T., Bourque, S., Kyryakov, P., Guo, T., Aziz, Z., Cyr, D., Naimi, A., Negoita, F., Oluoha, A., Salah, M. H., *et al.* (2006). Molecular systems biology of aging: Using calorie-restricted yeast as a model system for defining a modular network controlling chronological aging. *Mol. Cell. Proteomics* **5,** 550.

Gonzalez, B., Francois, J., and Renaud, M. (1997). A rapid and reliable method for metabolite extraction in yeast using boiling buffered ethanol. *Yeast* **13,** 1347–1355.

Herrgard, M. J., Swainston, N., Dobson, P., Dunn, W. B., Arga, K. Y., Arvas, M., Bluthgen, N., Borger, S., Costenoble, R., Heinemann, M., Hucka, M., Le Novere, N., *et al.* (2008). A consensus yeast metabolic network reconstruction obtained from a community approach to systems biology. *Nat. Biotechnol.* **26,** 1155–1160.

Kell, D. B., and Oliver, S. G. (2003). Here is the evidence, now what is the hypothesis? The complementary roles of inductive and hypothesis-driven science in the post-genomic era. *Bioessays* **26,** 99–105.

Mashego, M. R., van Gulik, W. M., Vinke, J. L., Visser, D., and Heijnen, J. J. (2006). In vivo kinetics with rapid perturbation experiments in Saccharomyces cerevisiae using a second-generation BioScope. *Metab. Eng.* **8,** 370–383.

Mashego, M. R., Rumbold, K., De Mey, M., Vandamme, E., Soetaert, W., and Heijnen, J. J. (2007). Microbial metabolomics: Past, present and future methodologies. *Biotechnol. Lett.* **29,** 1–16.

Mo, M. L., Jamshidi, N., and Palsson, B. O. (2007). A genome-scale, constraint-based approach to systems biology of human metabolism. *Mol. Biosyst.* **3,** 598–603.

Mustacchi, R., Hohmann, S., and Nielsen, J. (2006). Yeast systems biology to unravel the network of life. *Yeast* **23,** 227–238.

Nicholson, J. K., and Wilson, I. D. (2003). Understanding 'global' systems biology: Metabonomics and the continuum of metabolism. *Nat. Rev. Drug Discov.* **2,** 668–676.

Nookaew, I., Jewett, M. C., Meechai, A., Thammarongtham, C., Laoteng, K., Cheevadhanarak, S., Nielsen, J., and Bhumiratana, S. (2008). The genome-scale metabolic model iIN800 of Saccharomyces cerevisiae and its validation: A scaffold to query lipid metabolism. *BMC Syst. Biol.* **2**(71).

Robards, K., Haddad, P. R., and Jackson, P. E. (1997). Principles and Practice of Modern Chromatographic Methods. Academic Press, London.

Saez, M. J., and Lagunas, R. (1976). Determination of intermediary metabolites in yeast—Critical-examination of effect of sampling conditions and recommendations for obtaining true levels. *Mol. Cell. Biochem.* **13,** 73–78.

Sporty, J. L., Kabir, M. M., Turteltaub, K. W., Ognibene, T., Lin, S. J., and Bench, G. (2008). Single sample extraction protocol for the quantification of NAD and NADH redox states in Saccharomyces cerevisiae. *J. Sep. Sci.* **31,** 3202–3211.

Spura, J., Reimer, L. C., Wieloch, P., Schreiber, K., Buchinger, S., and Schomburg, D. (2009). A method for enzyme quenching in microbial metabolome analysis successfully applied to gram-positive and gram-negative bacteria and yeast. *Anal. Biochem.* **394,** 192–201.

Teusink, B., Passarge, J., Reijenga, C. A., Esgalhado, E., van der Weijden, C. C., Schepper, M., Walsh, M. C., Bakker, B. M., van Dam, K., Westerhoff, H. V., and Snoep, J. L. (2000). Can yeast glycolysis be understood in terms of in vitro kinetics of the constituent enzymes? Testing biochemistry. *Eur. J. Biochem.* **267,** 5313–5329.

Villas-Boas, S. G., and Bruheim, P. (2007). Cold glycerol-saline: The promising quenching solution for accurate intracellular metabolite analysis of microbial cells. *Anal. Biochem.* **370,** 87–97.

VillasBoas, S. G., Nielsen, J., Smedsgaard, J., Hansen, M. A. E., and RoessnerTunali, U. (2007). Metabolome Analysis: An Introduction. John Wiley and Sons, New Jersey.

Villas-Boas, S. G., Hojer-Pedersen, J., Akesson, M., Smedsgaard, J., and Nielsen, J. (2005). Global metabolite analysis of yeast: Evaluation of sample preparation methods. *Yeast* **22,** 1155–1169.

Wiken, T., and Richard, O. (1955). Further research on the importance of oxygen in the fermentation of glucose with wine yeast. *Schweiz. Z. Pathol. Bakteriol.* **18,** 970–980.

Winder, C. L., Dunn, W. B., Schuler, S., Broadhurst, D., Jarvis, R., Stephens, G. M., and Goodacre, R. (2008). Global metabolic profiling of Escherichia coli cultures: An evaluation of methods for quenching and extraction of intracellular metabolites. *Anal. Chem.* **80,** 2939–2948.

Wittmann, C., Kromer, J. O., Kiefer, P., Binz, T., and Heinzle, E. (2004). Impact of the cold shock phenomenon on quantification of intracellular metabolites in bacteria. *Anal. Biochem.* **327,** 135–139.

Zelena, E., Dunn, W. B., Broadhurst, D., Francis-McIntyre, S., Carroll, K. M., Begley, P., O'Hagan, S., Knowles, J. D., Halsall, A., Wilson, I. D., Kell, D. B., and Husermet Consortium (2009). Development of a robust and repeatable UPLC-MS method for the long-term metabolomic study of human serum. *Anal. Chem.* **81,** 1357–1364.

CHAPTER SIXTEEN

Plant Metabolomics and Its Potential for Systems Biology Research: Background Concepts, Technology, and Methodology

J. William Allwood,* Ric C. H. De Vos,†,|| Annick Moing,‡,# Catherine Deborde,‡,# Alexander Erban,§ Joachim Kopka,§ Royston Goodacre,*,¶ *and* Robert D. Hall†,||

Contents

* Manchester Interdisciplinary Biocentre, School of Chemistry, University of Manchester, Manchester, United Kingdom
† Plant Research International, Wageningen University and Research Centre (Wageningen UR), Wageningen, The Netherlands
‡ INRA, UMR1332 Biologie du Fruit & Pathologie Centre INRA de Bordeaux, INRA—Université de Bordeaux, IBVM, BP 81, Villenave d'Ornon, France
§ Max Plank Institute of Molecular Plant Physiology, Am Mühlenberg 1, Golm, Germany
¶ Manchester Centre for Integrative Systems Biology, Manchester Interdisciplinary Biocentre, University of Manchester, Manchester, United Kingdom
|| Centre for BioSystems Genomics, Wageningen, The Netherlands
Plateforme Métabolome-Fluxome Bordeaux, Génomique Fonctionnelle Bordeaux, IBVM, Centre INRA de Bordeaux, BP 81, Villenave d'Ornon, France

Methods in Enzymology, Volume 500
ISSN 0076-6879, DOI: 10.1016/B978-0-12-385118-5.00016-5

Abstract

The "metabolome" comprises the entire complement of small molecules in a plant or any other organism. It represents the ultimate phenotype of cells, deduced from the perturbation of gene expression and the modulation of protein function, as well as environmental cues. Extensive advances over the past decade, regarding the high-throughput (HTP) nature of "omics" research, have given birth to the expectation that a type of "systems level" overview may soon be possible. Having such a global overview of the molecular organization of a plant in the context of a particular set of genetic or environmental conditions, be it at cell, organ, or whole plant level, would clearly be very powerful. Currently, we are far from achieving this goal; however, within our hands, plant metabolomics is an HTP and informative "omics" approach to both sample generation and data generation, as well as raw data preprocessing, statistical analysis, and biological interpretation. Within this chapter, we aim to describe the great attention given to experimental design to ensure that the correct sample set and control are included and to, thereby, enable reliable statistical analysis of the data. For as comprehensive metabolite coverage as possible, we advocate the use of multiparallel approaches; thus, we describe a step-by-step standardized method for Nuclear magnetic resonance spectroscopy, as well as discussing with reference to standardized methodologies the techniques of gas chromatography-time of flight/mass spectrometry, and liquid chromatography–mass spectrometry.

1. An Introduction to Plant Metabolomics

1.1. The concept

> Science is facts. Just as houses are made of stones, so is science made of facts: but a pile of stones is not a house and a collection of facts is not necessarily science.
>
> Henri Poincare, French philosopher and mathematician

The "metabolome" comprises the entire complement of small molecules in a plant or any other organism. It represents the ultimate phenotype of cells, deduced from the perturbation of gene expression and the modulation of protein function. These, in turn, are under the influence of genome mutation and environmental cues (Saito and Matsuda, 2010). The metabolome of an individual plant is therefore a highly dynamic component which has both temporal (developmental, seasonal, etc.) and spatial (location, cultivation conditions, etc.) constraints (Fiehn, 2002). Furthermore, within the individual plant, the metabolome can be further spatially defined as organs, tissues, and even individual cells that have readily distinguishable metabolite profiles (Ebert *et al.*, 2010; Schad *et al.*, 2005; Sumner *et al.*, 2011). Plants have been considered as nature's most prolific biochemists (Hall, 2006) and therefore represent huge, natural, compound libraries which have to date been poorly mined and exploited (Fernie, 2007; Saito and Matsuda, 2010). The vast majority of plant species have never been fully characterized biochemically, and further, the great majority of plant compounds, which have been detected, have never been structurally identified. The vast richness of this natural resource represents huge potential for future efforts both in scientific research (allowing a more holistic understanding of the complexity of plant metabolism and its control) and in an applied context (e.g., novel drug, flavor, biocidal, compound discovery).

1.2. Fields of application

For all the above reasons, ever since the first scientific publication on plant metabolomic analysis (Roessner *et al.*, 2000), the field has expanded enormously and has opened up into a wide range of contrasting areas of application. Through metabolomics, plant biologists have gained the opportunity to delve deeply into cell biochemistry, and the technology has quickly gained a strong foot-hold in research, as shown, for example, by all the chapters in Harrigan and Goodacre (2003), Saito *et al.* (2006), Weckwerth (2007), and Hall (2011). Early efforts concentrated on a few "model" species (e.g., Arabidopsis, tomato, and potato); however, approaches have subsequently been applied to many species, both cultivated and wild, to answer considerably diverse biological questions of scientific and industrial relevant nature. Extensive use of metabolomics approaches has been made to enhance our understanding of the broad influences on plant metabolism due to, for example, mutations (Bino *et al.*, 2005; Yonekura-Sakakibara *et al.*, 2008), environmental perturbation (Ahuja *et al.*, 2010; Allwood *et al.*, 2006, 2008, 2010; Choi *et al.*, 2006; Jansen *et al.*, 2008; Kaplan *et al.*, 2004; Sanchez *et al.*, 2008, 2010; van Dam and van der Meijden, 2011; Ward *et al.*, 2010a), genetic introgression (Fernie and Schauer, 2009; Keurentjes *et al.*, 2006; Schauer *et al.*, 2006), and organ development (Tikunov *et al.*, 2010) to name but a few. Many previous investigations have focused on the primary metabolites where the pathways are generally well known, and the genetic background is also well documented (Fernie and Schauer, 2009; Giavalisco *et al.*, 2009; Schauer *et al.*, 2005). Further, authentic standards are generally available thus entailing both definitive identification and full quantification. However, extensive investigations have also been initiated on the highly diverse and much less well characterized secondary metabolite pathways. Such pathways are often biologically significant in that they are associated with, for example, abiotic or biotic stress resistance and other environment and organism interactions, food quality traits, such as taste and fragrance, or biocidal/pharmaceutical activity. Advances have been made particularly in the areas of polyphenolics (Bovy *et al.*, 2010; Tikunov *et al.*, 2010), alkaloids (Kim *et al.*, 2010), brassinosteroid and glucosinolate (Hall *et al.*, 2010) research. For an extensive overview of novel developments, the reader is referred to recent volumes such as Saito *et al.* (2006), Baginsky and Fernie (2007), and Hall (2011).

1.3. Sampling: From whole plant to single cell

Key to any metabolomics experiment is the experimental design in order to ensure that the correct sample set and control are included and thus to enable reliable statistical analysis of the data. As with all "omics"-type approaches, it is recommended to seek advice from an experienced statistician at the earliest possible stage. Full account must be taken of the key statistical challenge which is that the number of metabolite features generally exceeds the number of samples. This entails that correct experimental design and sampling regimes in terms of numbers of repetitions, treatments, time intervals, control samples, etc. are essential. Plant metabolomics experiments are generally performed at the organ level (developing fruit, whole leaf, root etc.), and it is often recommended to have pooled samples per replicate to reduce the level of biological variation. Strict control of the preharvest conditions (both cultivation conditions and harvesting regime—time of day, location on the plant, time from harvesting to metabolic quenching, etc.) is essential for sample uniformity (Beale and Ward, 2011;

Biais *et al.*, 2011; Gibon and Rolin, 2011). Taking materials from the field or wild, as required for environmental metabolomic studies, where little or no control of sample prehistory is possible, entails that even greater account must be taken of biological variation. Recently, improvements in instrument sensitivity and opportunities, for example, laser dissection microscopy and microinjection/microextraction, have generated opportunities to reduce sample size (Ebert *et al.*, 2010; Moco *et al.*, 2009; Schad *et al.*, 2005; Sumner *et al.*, 2011). Single cell metabolomic imaging has also become possible, for example, extracellular leaf trichomes (Sumner *et al.*, 2011). Whereas pools of identical cell types should become accessible to metabolomic studies, true single cell metabolomics may still be far off. Using whole organs entails that even known, significant differences in tissue biochemistry have to be ignored due to sampling limitations. Therefore, we typically obtain only a global overview of metabolism which takes little account of the intercellular differences which are often of great biological relevance (e.g., see Moco *et al.*, 2007a).

1.4. Metabolite profiling technologies

> Any sufficiently advanced technology is indistinguishable from magic.
>
> Arthur C. Clarke, physicist and science fiction author

The measurement of plant metabolites is difficult (Fernie, 2007), and the challenges faced in analysis of plant material are clear from the outset. The structural diversity and the associated contrasting physicochemical properties of many common plant metabolites (e.g., monoterpenoid volatiles, polar amino acids, or hydrophobic lipids; cf. Fernie, 2007; Saito and Matsuda, 2010) are a potential analytical nightmare. The large variation in relative abundance (up to 6 orders of magnitude concentration difference, i.e., "dynamic range") has the inevitable consequence that no single extraction/separation/detection methodology will even come close to suffice our needs. Common approaches have arisen for both separation (liquid chromatography and gas chromatography, as well as capillary electrophoresis; Kopka, 2006; Sumner, 2006; Timischl *et al.*, 2008) and detection (nuclear magnetic resonance (NMR) and mass spectrometry; Allwood and Goodacre, 2010; Verhoeven *et al.*, 2006; Ward and Beale, 2006) being the most routinely applied. Many efforts have focused on method standardization to produce comparable data between different instruments and laboratories (Allwood *et al.*, 2009; De Vos *et al.*, 2007; Lisec *et al.*, 2006; Ward *et al.*, 2010b). To gain as comprehensive coverage as possible requires multiparallel complementary approaches, combining appropriate extraction, separation, and detection technologies. The Holy Grail of truly holistic coverage will likely never be possible or indeed even, affordable, despite the increasing technological sophistication. These purely practical and economic reasons introduce a degree of bias into the generation of results. However, educated choices for suitable technology combinations and combined hyphenated approaches have already been widely reported (see Hall, 2011; Hardy and Hall, 2011; Saito *et al.*, 2006; Weckwerth, 2007). Complementary approaches require increasingly sophisticated procedures for the handling of multiplexed datasets, data merging, and efficient and robust integrative data mining strategies (Fiehn *et al.*, 2011; Goodacre *et al.*, 2004; Kopka *et al.*, 2011; Redestig *et al.*, 2011).

1.5. The "utopia" of a systems level understanding

Nothing shocks me – I'm a scientist.

Indiana Jones, fictitious Professor of Archaeology, and adventurer

The recent trends and extensive advances regarding the HTP nature of "omics" research have given birth to the expectation that a type of "systems level" overview may soon be possible (see Baginsky and Fernie, 2007). Having such a global overview of the molecular organization of a plant in the context of a particular set of genetic or environmental conditions, be it at cell, organ, or whole plant level would clearly be very powerful. However, we are far from achieving this goal, although taking steps toward it should be loudly applauded. Current "omics" methodologies are already providing the most detailed insights ever into how genomes are organized and how plants function at the molecular level. There are clear opportunities to exploit Arabidopsis as a model organism (Beale and Sussman, 2010), and simpler systems such as suspension cell cultures (Allwood *et al.*, 2010; Farag *et al.*, 2008) offer certain advantages regarding rapidity and uniformity of experimentation, isotopic labeling, and flux analyses (Huege *et al.*, 2007; Sulpice *et al.*, 2010). However, individual technological limitations, as detailed above, for example, for metabolomics, when piled up in a multi-omics manner, still represent a significant bottleneck to designing a truly holistic, systems level approach. Further, for whole plants, parallel platforms for HTP (nonmolecular) phenotyping are also still in their infancy (Keurentjes, 2009). These remaining challenges are by no means facile—only the future will reveal how far and fast we can proceed to overcome these hurdles. Nevertheless, large steps are already being made, and our ambition to develop appropriate methodologies and procedures to combine diverse datasets from multidisciplinary approaches will help enhance and finally establish systems biology fully within the field of plant science. In the following sections, an overview is given on considerations in plant growth, harvest, and preparation of material, and the current standardized methodologies being used across our laboratories for GC-TOF/MS and HPLC–PDA–MS will be discussed with reference to published standardized methodologies, a previously unpublished reference protocol for ^{1}H NMR will also be presented. Of course, robust statistical analysis with an array of multivariate and univariate statistical tools is essential for mining the key information from the vast amounts of data that are generated by metabolomics platforms. Further metabolite identification at later experimental stages is also of key importance and introduces yet further analytical platforms such as FT-ICR–MS (Allwood *et al.*, 2011), Orbitrap MS (Dunn *et al.*, 2008; van der Hooft *et al.*, 2011), and HPLC–NMR (Glauser *et al.*, 2008). To present methodologies for complex metabolite identification technologies and statistical analyses is beyond the scope of this chapter, and in the case of plant metabolomics, the selection of appropriate tools for both statistical analyses and in-depth metabolite identification is very much dependent upon the experiment and biochemistry (Kopka *et al.*, 2011).

2. Considerations in Plant Growth and Preparation of Plant Material

Plant science experiments generally involve the variation of at least one genetic or environmental factor in order to induce responses which will provide information for knowledge generation. In order to decrease the experimental error for all the steps preceding that of biochemical analysis, the experiment has to be designed and planned from plant growth to organ harvest, sample constitution, and distribution to the chemical analyst(s).

2.1. Experimental design and plant growth

A plant metabolomics study is usually characterized by several hundreds (even thousands) of metabolites or metabolite features that by far exceed the sample number, increasing false discovery risks. In order to decrease the experimental error, it is crucial to identify the major sources of error at the environmental, biological, and technical levels, for each step from plant growth to sample storage and distribution. This will help to optimize the "experimental design," that is, the procedure planning each step of the experiment in order to obtain data describing the relationships between the factor(s) and variables affecting the process of interest and its consequences (Anderson and Whitcomb, 2007). This can be facilitated by performing limited preliminary experiments before planning full-scale experiments. Then, for plant growth, a design has to be chosen (Anderson and Whitcomb, 2007; Hinkelmann and Kempthorne, 2008; Rocke, 2004) such as factorial design or optimal design, blocking, or randomization, with labeling of individual plants and possibly organs per plant. Beyond recording minimum information about plant growth conditions as advised in the metabolomics standard initiative (Fiehn *et al.*, 2007, 2008), management of the environment during culture and recording the changes of some major environmental variables (e.g., fertilization, temperature, and/or light) are crucial. This is the case even in controlled environments such as greenhouses or growth chambers, as small variations (e.g., location and seasonal variations in greenhouse or limited malfunctioning of growth chamber control) can cause changes in the biochemical status. In the field, recording of daily minimum and maximum temperatures, rainfall, insulation duration, and irrigation dates and recording of chemical compounds and treatment dates for pathogen/pest control may be valuable. Plant growth metadata documentation and storage in databases is crucial not only for the ongoing experiment but also for the data reanalysis and the possibility to perform data meta-analyses combining several experiments (Hannemann *et al.*, 2009). Therefore, the format of the metadata concerning plant growth should allow easy recording and sharing. Replication of biological samples is indispensable, and should be preferred to extraction and analytical replication, if all are not feasible due to high analytical cost (for further statistical analysis of the metabolic profiles using univariate or multivariate methods, see Broadhurst and Kell, 2006; Goodacre *et al.*, 2007; Kopka *et al.*, 2011; Trygg *et al.*, 2006). A compromise has to be made between sample number and throughput. The number and constitution of biological replicates has to be defined in the experimental design. Replication is crucial to incorporate and measure biological variations and also provides a sample set representative of the population of plants, organs, or tissues.

2.2. Harvest

For harvest, the plant or organ age or developmental stage, time with reference to the light period, and duration of harvest and sample processing constitute a potential source of uncontrolled variability if their procedures are not carefully standardized. This is especially important when the harvest sessions of a given experiment are numerous or when each session requests several operators to limit its duration. The age, or preferably the developmental stage, of the plants or their organs needs to be defined relative to standardized growth conditions and/or phenology descriptors, by using dedicated ontology's (e.g., Plant Ontology at http://www.plantontology.org/ for phenology) or reference articles (e.g., Boyes *et al.*, 2001, for Arabidopsis, or Brukhin *et al.*, 2003, for tomato) when available. For seedlings in controlled conditions, hours or days postimbibition may be used. For fruits, in cases of controlled pollination and growth under the same temperature conditions, harvest is usually expressed as days-post-anthesis but degree-days post-anthesis can also be used (Bonhomme, 2000). If the age of the organ is unknown, well-defined criteria, such as organ aspect, color (e.g., a color code with a standard scale for fruit maturation of a given species), and size, will improve sample homogeneity. Even for a given organ with clearly defined development stage(s), the time and method of sampling must be precise (Fiehn *et al.*, 2007). As the metabolite composition of the leaf (Gibon *et al.*, 2006; Kim and Verpoorte, 2010; Urbanczyk-Wochniak *et al.*, 2005) and also fruit (Klages *et al.*, 2001; Ma and Cheng, 2003) has been shown to vary during a day and night cycle, the time of harvest has to be precisely defined and its duration limited. The position of the organ has to be defined (e.g., fruit position in a cluster or sun vs. shade exposition), as it may also have an impact on its composition (Gautier *et al.*, 2005; Ma and Cheng, 2003; Nothmann and Rylski, 1983; Pereira *et al.*, 2005, 2006). For harvest in an open field, similar weather conditions for each harvest are recommended. The harvested organs should be handled with care and stored in adapted packaging whilst preventing shocks and oxygen limitation. Enzyme activities need to be quenched (AP Rees and Hill, 1994) which is ideally achieved by freeze clamping or immediately plunging small organs (e.g., leaves, flowers, or small fruits) or dissected tissues into liquid nitrogen. For larger organs, transportation has to be performed quickly with precautions (avoiding temperature fluctuation or hygrometry stresses), and their processing has to then be undertaken immediately after harvest. To ensure representativeness of the physiological variations of the organ of interest, it is recommended to harvest several organs of each variety in a given condition to constitute a sample set. For instance, for cherry tomato, 18 fruits from 9 plants were pooled to make a representative fruit sample in a study on volatiles (Tikunov *et al.*, 2005). For Arabidopsis, harvest of 15 plant rosettes to constitute 5 samples of 3 rosettes at each time point was carried out in a study of diurnal changes of primary metabolism (Gibon *et al.*, 2006). Several (at least three, or better, at least five) sample sets of each variety have to be independently prepared. Pooling of plants or organs that will average the variability arising from uncontrolled factors is especially interesting when the cost of sample preparation is low but that of biochemical analysis is high.

2.3. Sampling

Sampling should provide samples in a form and quantity suitable for metabolomic analyses and representative of the total population of plants or organs to be analyzed. Special care must be taken if the plant material is intrinsically heterogeneous (e.g., an organ such as a flower or a fruit made of several tissues). Contamination, loss, metabolism, and any other alteration of the sample have to be minimized (Markert, 1995; Wagner, 1995). Plant organs may have to be cleaned (Markert, 1995), but usually for plant metabolomics studies, sterile conditions are not required. When a specific tissue needs to be dissected before freezing, as the sample must be representative of the entire tissue in the organ, material must be taken from several parts of the organ, which should be defined. As the use of knives and scalpels causes wound stresses affecting the metabolism, working as fast as possible is recommended. Ideally, surfaces that are exposed after cutting should be minimized, while volume should allow rapid freezing of the organ pieces as indicated above. Microdissection requires specific devices and procedures (Balestrini and Bonfante, 2008; Ebert *et al.*, 2010). After complete freezing in liquid nitrogen, the pieces can be ground immediately, or transferred into tagged plastic bags and stored at −80 °C until grinding is possible, before distribution into tubes and storage at −80 °C. It is recommended that all samples for a given experiment follow exactly the same procedure before, during, and after grinding. Sample grinding is usually required to optimize solvent extraction and also helps to homogenize the sample material (Markert, 1995); however, possible contamination or volatilization of certain compounds of interest (Markert, 1995) has to be minimized. For metabolomics, samples are usually ground frozen despite the possible modification of the volatile composition upon freezing as shown for strawberry (Douillard and Guichard, 1990). For HTP metabolomics studies, robotized automation of grinding has to be considered. Labeling and traceability of samples is crucial in quality assurance and quality control (QC) throughout the sample preparation process, from the field to the sample storage location and through distribution to chemical analysts. At each step (harvest, constitution of biological replicates, storage, and distribution of samples), the samples should be clearly identified (possibly by barcoding) and referenced in a file, or better a database, with a unique identifier.

2.4. Sample storage

All samples of an experiment should follow the same storage conditions. The conditions and duration of sample storage have to be controlled and recorded. Studies on the effect of long-term storage of plant samples on their metabolites remain rare (Phillips *et al.*, 2005; Ryan and Robards, 2006). Therefore, when working on several families of compounds using a combination of extractions and analytical methods, it would be helpful to propose a few compounds as "markers" of good storage conditions and duration. Samples can be stored as organ, tissue pieces, or ground tissue(s) at −80 °C, although solid phase micro extraction trapping of volatile compounds (Tholl *et al.*, 2006) can be performed upon entire organs when small or tissue pieces. Depending on the intended analyses, tissue samples can be stored as fresh-frozen or lyophilized samples. Fresh-frozen samples are necessary for analytical determination of highly volatile compounds as freeze-drying will result in a loss of volatiles (Julkunen-Titto and Tahvanaiem, 1989; Keinänen and Julkunen-Titto, 1996). It must also be carefully considered that freeze-drying can also cause a loss of some metabolites through irreversible binding to cell walls or membranes, although positively it protects against enzyme activities and microbial decomposition during storage (Dunn *et al.*, 2005a). Storage conditions have to be controlled, as stability during sample storage is an important factor that is rarely measured. The time in frozen storage was shown to modify some aromatic components in fruit (Ma *et al.*, 2007). Usually, samples for metabolomics are stored at −80 °C. It is recommended that lyophilized samples are stored at −20 °C (Salminem, 2003) and in dry conditions (e.g., in tubes enclosed in sealed plastic bags containing a desiccant such as silica gel). For specific purposes, tissue samples and extracts may have to be stored in tubes flushed with nitrogen or argon gas to protect them against oxidation (Erban *et al.*, 2007). For sample transport to the analytical laboratory, rapid transport of samples using an ice-chest, a Dewar, and liquid nitrogen is recommended. If shipment is necessary, dry ice is recommended for fresh-frozen samples. If lyophilized samples are shipped to the analytical laboratory, dry conditions need to be ensured. During storage or after shipment, sample quality has to be verified using at least visual inspection or ideally physicochemical analyses or targeted biochemical analyses performed at different time points (Fish and Davis, 2003; Phillips *et al.*, 2005). More detailed information and guidelines for experimental design and sampling for plant metabolomics are provided in Biais *et al.* (2011), Gibon and Rolin (2011), and Kim and Verpoorte (2010).

3. GC-TOF/MS Metabolite Profiling, Recommended Experimental Reference Protocols, and Data Preprocessing Approaches

3.1. An introduction to GC-TOF/MS metabolite profiling

Since early studies on herbicide modes of action (Sauter *et al.*, 1988), GC–MS has experienced a renaissance in HTP fingerprinting and profiling of genetically modified (e.g., Fernie *et al.*, 2004; Roessner *et al.*, 2001a,b) or experimentally challenged plant samples (e.g., Cook *et al.*, 2004; Kaplan *et al.*, 2004; Urbanczyk-Wochniak and Fernie, 2005), and the spatial analysis of individual plant organs (Biais *et al.*, 2009, 2010), of tissues (Schad *et al.*, 2005), and even of individual cell types (Ebert *et al.*, 2010). Metabolic phenotyping via GC–MS profiling has become an integral part of plant functional genomics (Fernie *et al.*, 2004; Fiehn *et al.*, 2000a; Roessner *et al.*, 2002) and is on the verge of becoming a routine technology with the potential to generate comparable data between laboratories (Allwood *et al.*, 2009). This fact substantially contributes to the development of metabolomics as a fourth Rosetta stone for plant functional genomics and molecular physiology (Fiehn *et al.*, 2000a; Trethewey, 2004; Trethewey *et al.*, 1999). Metabolite profiling with GC–MS involves six general steps: (i) extraction of metabolites, which should be as comprehensive as possible, while avoiding degradation or modification of metabolites (e.g., Kopka *et al.*, 2004). (ii) Derivatization of metabolites making them amenable to GC. (iii) GC separation. High-resolution GC and also 2D GC × GC can be highly reproducible as it involves automated sample injection robotics, highly standardized conditions of gas flow, temperature programming, and standardized capillary column material. (iv) Ionization of compounds. Electron impact (EI) is most widely used, as it has low susceptibility to suppression effects and produces reproducible fragmentation. (v) Time resolved detection of molecular and fragment ions. Mass separation and detection preferably by TOF detectors that can be tuned to fast scanning rates. (VI) Acquisition and evaluation of GC–MS data files.

3.2. GC-TOF/MS plant metabolite profiling: Recommended experimental procedures

The essence of metabolite profiling is discovery of novel marker metabolites and determination of relative changes of metabolite pool sizes in comparison to reference samples (Bino *et al.*, 2004; Fernie *et al.*, 2004). Current systems biology approaches add further demands for multiparallel absolute quantification of pool sizes. Both absolute and relative quantification necessitate thorough control experiments, monitoring of GC–MS system performance and contaminations. QC samples are of key importance, and these are best prepared by pooling equal volumes of material from all of the biological samples to be analyzed. Alternatively, with long-standing experience, a chemically defined mixture of authenticated reference compounds (e.g., Strehmel *et al.*, 2008) that mimics the metabolic composition of the investigated biological material can be employed. Both the synthetic mixtures and biological QC samples are then subjected to the same sample extraction, instrumental analyses (ideally distributed across the analytical run), and data processing, thus providing quality checks for technical and analytical error, and quantitative calibration for the final processed data. The perhaps most widely applied extraction protocol for GC–MS plant profiling involves a fractionation of nonpolar (chloroform fraction) and polar metabolites (water–methanol fraction comprising largely primary metabolites, e.g., sugars, alcohols, organic acids, amino acids, amines) (e.g., Erban *et al.*, 2007; Lisec *et al.*, 2006; Roessner *et al.*, 2000). This method focused on primary metabolism is highly recommended by the authors and available as full reference protocols within Lisec *et al.* (2006) and Erban *et al.* (2007), an overview is also provided (Fig. 16.1). GC–MS detection of polar, nonvolatile metabolites demands a two-step derivatization by methoxyamination and silylation (Fiehn *et al.*, 2000a,b; Roessner *et al.*, 2000, 2001a). The imprecise dispensing of reagent volumes and the

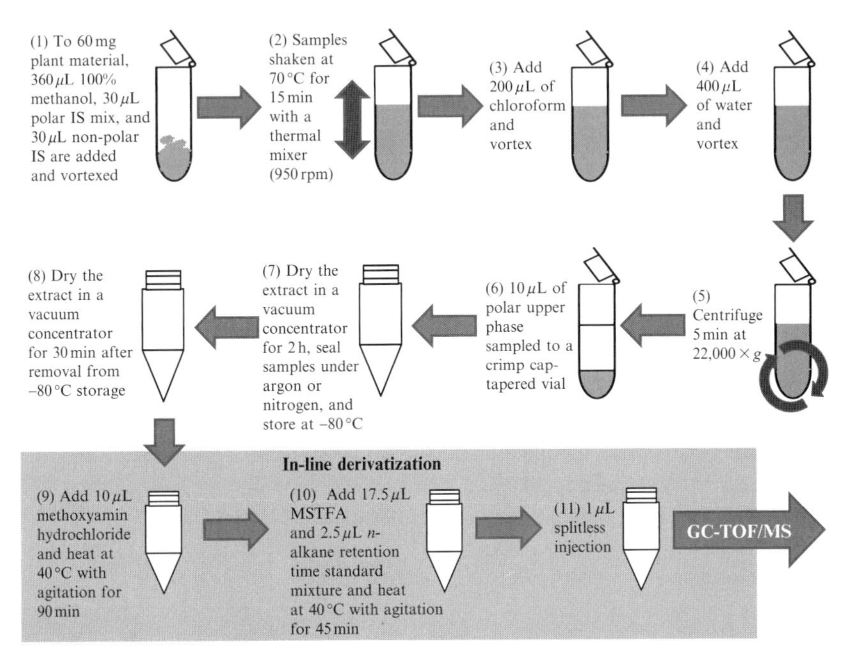

Figure 16.1 Preparation of polar plant extracts and chemical derivatization for GC-TOF/MS analysis. An illustration of the plant polar metabolite extraction scheme and in-line derivatization (gray box) as presented in the reference protocol of Erban *et al.* (2007).

variable timing of the silylation reaction are major sources of analytical variability. Typically, 50–100 samples are processed, and thus, the exposure time to the silylation reagent between the first and the last analyzed sample can differ considerably. We therefore recommend employing automated and timed in-line derivatization and sample injection as fully detailed in the reference protocol of Erban *et al.* (2007). Our recommended approach to GC-TOF/MS metabolite profiling demands that standardized consumables and parameters be applied to each of the autosampler and GC, as well as TOF/MS settings for mass ion detection, all of which are fully detailed in the reference protocol of Erban *et al.* (2007). Example chromatograms produced via this reference protocol are given (Fig. 16.2).

3.3. Data preprocessing and metabolite classification in plant GC-TOF/MS profiling

Automated data processing tools are required to (i) generate numerical sample/metabolite or sample/chemical feature matrices for statistical calculations and (ii) need to provide means of linking such chemical features

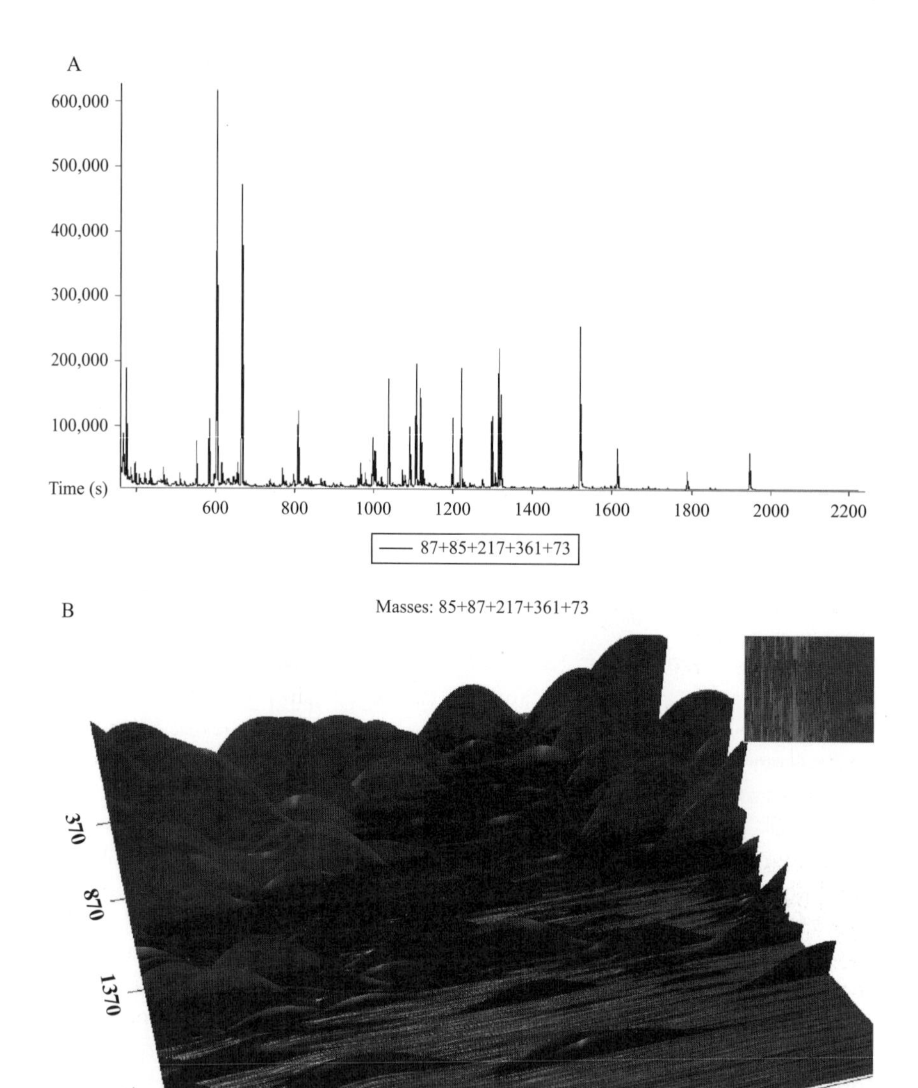

Figure 16.2 An example GC chromatogram for a polar extract of red wine. (A) Although a method is not presented for GC × GC-TOF/MS analyses, an example of a GC × GC chromatogram (B) is given for the same derivate sample; this illustrates the complexity of metabolites that can be resolved by GC x GC compared to conventional GC.

to a metabolite identity. In GC–MS profiling studies, retention indices and mass spectra are used for this purpose (Wagner *et al.*, 2003). Tools which enable mass spectral deconvolution and retention index calculation for compound recognition, and peak height or peak area retrieval for relative or absolute quantification, plus last but not least the analysis of mass isotopomer distributions for stable isotope tracing and flux studies have been developed; refer for example to the motivation of the TagFinder software development (Luedemann *et al.*, 2008). Different data preprocessing approaches have been implemented in academia. One general approach first applies mass spectral deconvolution and subsequently quantifies based on deconvoluted (mass fragment) abundances; care must be taken to avoid deconvolution errors (Dunn *et al.*, 2005b). An alternative approach first retrieves peak height or peak areas of all recorded mass traces for statistical analysis prior to metabolite identification. Standardized data preprocessing procedures have been described earlier, for example, MetAlign (Lommen, 2009; Lommen *et al.*, 2010) and TagFinder (Luedemann *et al.*, 2008, 2011). A recent study explored seven alternative modes of data preprocessing and established high agreement of all alternative preprocessing modes (Allwood *et al.*, 2009). Both of the above mentioned approaches to GC–MS data preprocessing are indeed valid, and the selection of the most appropriate is experimentally dependent. If the objective is a long-term metabolomic profiling study, then it is indeed best to attempt to identify all metabolite features requiring use of full data set deconvolution through either AMDIS (Lisec *et al.*, 2006) or LECO ChromaTOF (Begley *et al.*, 2009). However, if the experiment involves a metabolite fingerprinting approach where the investigator only wishes to identify statistically significant metabolite differences between experimental groups, for example, in gene function analyses, then a more rapid approach would involve use of generic software such as TagFinder (Luedemann *et al.*, 2008, 2011) or MetAlign (Lommen, 2009; Lommen *et al.*, 2010). The current more urgent challenge may be seen in solving the automated and exact identification of metabolites within complex mixtures. Even today, matching and classification of mass spectra or mass spectral tags (MSTs) requires human evaluation and judgment. Many software tools exist, which either support MST recognition in comparison to reference libraries such as NIST or freely available libraries such as the GMD (Golm Metabolome Database: Kopka *et al.*, 2005) or aid the interpretation of unknown mass spectral features that in the absence of authenticated reference compounds cannot be linked to a compound structure (e.g., Hummel *et al.*, 2010). However, for unambiguous identification, a laboratory performing routine metabolic profiling should establish an in-house library of pure authenticated reference standards analyzed after subjection to the same derivatization and analytical parameters as the profiled samples.

4. HPLC–PDA–QTOFMS Metabolite Profiling, Recommended Experimental Reference Protocols, and Data Preprocessing Approaches

4.1. An introduction to LC–MS metabolite profiling

HPLC coupled to MS (in short LC–MS) is the preferred method for metabolic profiling of semi-polar secondary metabolites such as phenolic acids, flavonoids, alkaloids, polyamines, saponins, and glucosinolates. These compounds can be effectively extracted with aqueous alcohol solutions and directly analyzed without prior derivatization. In contrast to EI applied in GC-TOF/MS, ionization in the liquid phase of LC–MS typically involves soft ionization techniques, such as electrospray ionization (ESI) or atmospheric pressure chemical ionization, resulting in protonated (in positive mode) or deprotonated (in negative mode) molecular ions. Modern high-resolution instruments with exact mass detection, such as TOF/MS, ion cyclotron FT-MS, or Orbitrap FT-MS, nowadays enable the profiling of hundreds to thousands of compounds in crude plant extracts, combined with elemental formulae calculations of the detected masses (Allwood and Goodacre, 2010). Using an essentially unbiased procedure that takes into account all metabolite mass signals from the LC–MS raw data files, detailed information on the relative abundance of hundreds of both known and, as yet, unknown semi-polar metabolites can be obtained (de Vos *et al.*, 2007; von Roepenack-Lahaye *et al.*, 2004). LC–MS-based metabolomics approaches frequently make use of C_{18}-based reversed phase columns to obtain optimal separation of the large variety of semi-polar compounds that can be present in crude plant extracts (de Rijke *et al.*, 2003; de Vos *et al.*, 2007; Hanhineva *et al.*, 2008; Huhman and Sumner, 2002; Iijima *et al.*, 2008; Moco *et al.*, 2006; Stobiecki *et al.*, 2006; von Roepenack-Lahaye *et al.*, 2004). However, by choosing dedicated columns, a range of primary metabolites including several polar organic acids and amino acids can also be reliably analyzed using LC–MS (Tolstikov and Fiehn, 2002). Many semi-polar metabolites present in plant materials, such as glucosinolates, (poly)phenols, and a range of other glycosylated compounds, can be readily detected by LC–MS in ESI-negative mode. However, plant metabolites that easily form proton adducts, for example, polyamines, alkaloids, and anthocyanins, can be better detected in positive mode. Thus, LC–MS analysis of samples in both positive and negative ionization mode will provide the most comprehensive insight into the plant metabolic composition (Fait *et al.*, 2008; Hanhineva *et al.*, 2008; von Roepenack-Lahaye *et al.*, 2004). This approach has proven valuable for the profiling of secondary metabolites in many plant species including Arabidopsis, cabbage, tomato, potato, strawberry, apple, lettuce, etc. (de Vos *et al.*, 2007; Keurentjes *et al.*, 2006; Moco *et al.*, 2006; Vorst *et al.*, 2005) as well as different plant tissues (Adato *et al.*, 2009; Moco *et al.*, 2007a) and plant products (Capanoglu *et al.*, 2008).

4.2. HPLC-PDA-QTOFMS plant metabolite profiling: Recommended experimental procedures

As for GC-TOF/MS metabolic profiling, for LC–MS analysis, QC samples should be employed to monitor technical and analytical performance, and quality assurance checks upon the preprocessed data during statistical analysis. For LC–MS extraction and analysis, we recommend the reference protocols fully detailed within de Vos *et al.* (2007). The extraction typically uses 500 mg fresh weight and 1.5 mL of extraction solvent (99.875% MeOH + 0.125% formic acid, FA), that is, a volume:fresh weight ratio of 3:1, which will result in a final concentration of 75% MeOH + 0.1% FA, assuming 95% water content of plant material. The mixture is sonicated for 15 min prior to centrifugation, and the supernatant is filtered through 0.45 or 0.22 μm PTFE filters and directly analyzed (De Vos *et al.*, 2007). Detection sensitivity and mass spectra obtained by soft ionization LC–MS are highly dependent on the type of mass spectrometer, ionization source and mode, and chromatographic system used. The recommended LC–MS protocol is based on a Waters Alliance 2795 HT HPLC with a 2996 PDA detector and a QTOF MS Ultima detector with an ESI lock spray (0.1 mg/mL leucine enkephaline) for online mass correction (De Vos *et al.*, 2007; Moco *et al.*, 2006). The LC–MS instrumentation is operated precisely as described within the protocol of De Vos *et al.* (2007). Although it is possible to attain a global profile in a single ionization mode, it is highly recommended to perform analysis across both polarities due to preferential ionization of different compound classes in the two ionization modes. The protocol applies relatively long chromatographic runs of 1 h per a sample. A typical chromatogram obtained by this LC–QTOF MS protocol is given in Fig. 16.3. After adequate modification of the protocol according to instrument-specific settings and analysis conditions, it is also applicable to Ultra (U)HPLC (permitting shorter sample run times) and alternative MS platforms, such as LC-Orbitrap FT-MS and LC–QTOF MS Premier.

4.3. Data preprocessing of plant LC–MS profiling experiments

MetAlign is one of the free software packages available for untargeted processing of LC–MS, and GC–MS, raw data files from various instrument vendors (download at www.metalign.wur.nl). It performs local baseline correction, unbiased peak picking, and mass peak alignment for up to hundreds of samples, resulting in a matrix of mass signal intensities × samples (Lommen, 2009). It is specifically suitable for MS instruments

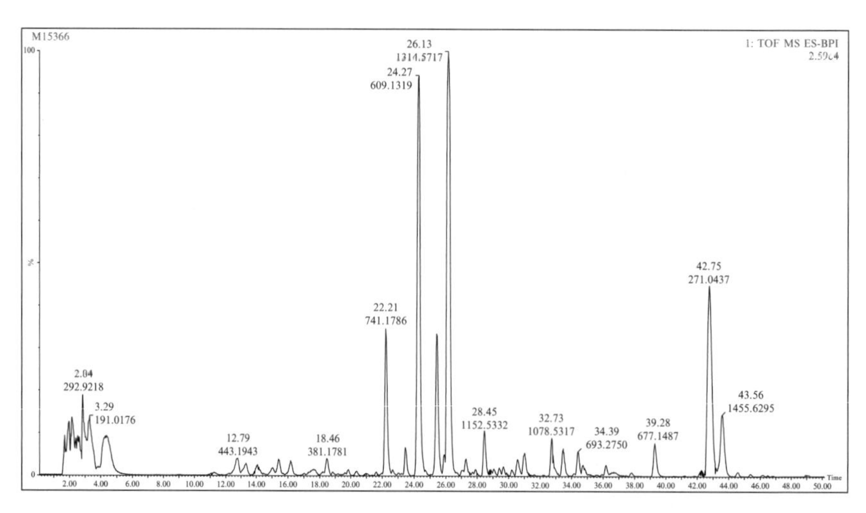

Figure 16.3 LC-QTOF MS chromatogram (base-peak intensity) of an aqueous-methanol extract from peel of ripe tomato fruit numbers at peaks indicate retention time and accurate mass on top of peak. Typically, thousands of mass signals representing hundreds of metabolites, both known and unknown compounds, are detected in such a tomato sample, depending upon variety, tissue, and developmental stage (Iijima *et al.*, 2008; Moco *et al.*, 2006).

with a limited dynamic range in mass accuracy, such as the described QTOF Ultima, as it provides the option to select an intensity window for accurate mass calculation. Postprocessing, the data of the QC samples can be used to check for consistent peak picking and alignment. Typically, about 70–80% of all peaks should be present in all QC samples with an overall intensity variation of less than 20% and a mass deviation of less than 3 ppm for known compounds. The generated peak table can be further filtered for inconsistent and low (noisy) or saturated signals. Mass signals derived from the same metabolite can be grouped according to their corresponding retention time and relative intensities across the samples, for example, by using a mass spectra reconstruction approach (Tikunov *et al.*, 2005, 2010). In this approach, the LC–MS signals originating from the same metabolite (molecular ion and its fragments, adducts, and isotopes) are all clustered and replaced by a single representative metabolite signal. Assignment of metabolites can be performed by matching the exact mass of the observed molecular ion with plant accurate mass databases, for example, MotoDB (http://appliedbioinformatics.wur.nl/moto/), KNapSAck (http://kanaya.naist.jp/knapsack_jsp/top.html), and Massbank (http://www.massbank.jp/index.html). Data-dependent tandem-MS/MS and MS^n fragmentation experiments may be performed to further identify the significant secondary metabolites (de Vos *et al.*, 2007; Moco *et al.*, 2007b; Muth *et al.*, 2008; van der Hooft *et al.*, 2011).

5. ^{1}H NMR: Experimental Reference Protocols and Data Preprocessing for Plant Metabolite Profiling

5.1. An introduction to NMR spectroscopy

Despite NMR spectroscopy having been one of the leading technologies in metabolomics since 2000 (Ludwig and Viant, 2010), this technique is underused in plant compared to mammalian metabolomics. NMR can be mono- or multidimensional. Until recently, largely, 1D ^{1}H NMR has been applied in metabolomics, but use of HSQC (heteronuclear single-quantum coherence spectroscopy) and 2D JRES (2D ^{1}H ^{1}J-resolved) spectroscopy are powerful and often essential tools for metabolite identification. The success of 1D ^{1}H NMR in metabolomics stems from rapid spectral acquisition and the possibility to achieve relative or absolute quantification. In addition, NMR provides detailed molecular structural information, and the sample is not destroyed during the NMR measurement. Detailed technical issues involved in the NMR methods are beyond the scope of this chapter, and hence it is recommended to consult previous publications for comprehensive descriptions of NMR technology (Ross *et al.*, 2007) and applications in plant biochemistry (Eisenreich and Bacher, 2007).

5.2. Extraction of polar metabolites

Plant sample preparation for NMR typically consists of multiple steps. Samples are always analyzed immediately after extraction, and the method of extraction has a crucial bearing on the proceeding detection of metabolites. In contrast to tissues, sap or fruit juice sample preparation is rapid, but nevertheless, enzymes in sap or juice should be deactivated, either with deuterated methanol and/or with a heating step (Biais *et al.*, 2009; Moing *et al.*, 2004). In comparison to chromatographic methods coupled or not with mass spectrometry, the amount of sample material required for NMR analysis is much larger: 20–50 mg dry weight (~200–500 mg fresh weight) instead of 60 mg fresh weight for GC–MS, for instance. As water is the main component of plant tissue and is composed of two hydrogen atoms which are NMR detectable, it is important to remove it by lyophilization or cryodessication. To be more efficient, this step should be performed after cryogrinding, as fine plant powder helps facilitate the freeze-drying process as well as that of extraction (Kim and Verpoorte, 2010). Sap or fruit juice samples are not lyophilized, but application of a presaturation technique to suppress the signal of water is required (see Section 5.2.2). Several extraction protocols using methanol, ethanol, water (for polar/semi-polar metabolites), and/or chloroform (for a polar/nonpolar metabolites) with variations in the relative amount of each solvent, one- or multistep processes, at different temperatures (i.e., extraction at 4 °C to preserve secondary metabolites) are described in the literature. A comparison of the results obtained by NMR for the metabolome of dried Arabidopsis cells with four different commonly used protocols lead to similar polar metabolite profiles (Gromova and Roby, 2010). The impacts of extraction protocols (solvent composition, pH, one- vs. two-step processes, lyophilization) on the stability of extracts and on the quality of metabolomic profiles of Arabidopsis mature leaves were described by Kaiser *et al.* (2009). There are two common methods for extraction of plant constituents in NMR profiling: an expensive, yet simple, method using deuterated NMR solvents and a cheap but more time-consuming approach with nondeuterated solvents (Kim and Verpoorte, 2010; Schripsema, 2010). For the protocols detailed below, we will focus on polar metabolite extracts (refer to Table 16.1 for a list of all chemicals and laboratory apparatus).

5.2.1. Extraction with deuterated solvents

The main advantage of extraction with deuterated solvents is HTP capacity, as after extraction and centrifugation, the supernatant is directly transferred to an NMR tube for data acquisition. One drawback is the difficulty to control pH (in fact pD) and ionic force for the samples within a series. Metabolomic studies using this approach are mainly based on methanol–water extraction in proportions ranging from 20% to 70% (v/v) methanol. The presence of two deuterated solvents in the NMR tube may disturb the locking process of the NMR spectrometer as a deuterated solvent is used as a magnetic field lock signal for the spectrometer. One way to cope with this problem is to disable the autolocking procedure of the spectrometer, for instance, by choosing the parameter "lock no_auto" available in Bruker spectrometers.

The following protocol was established by Ward *et al.* (2003, 2010b).

1. 15–30 mg of dry plant powder is weighed into 1.5-mL Eppendorf tubes. $D_2O—CD_3OD$ (1 mL, 80/20, v/v) containing 0.05% (w/v) TSP-d_4 (sodium salt of trimethylsilylpropionic acid) is added to each sample. The contents of the tube are mixed thoroughly and heated at 50 °C in a water bath for 10 min.
2. After cooling (5 min), the samples are next spun down in a microcentrifuge for 5 min; 800 μL of the supernatant is transferred to a clean Eppendorf tube and heated at 90 °C for 2 min. This heat step is compulsory to ensure denaturation of enzymes.

Table 16.1 Chemicals, materials, equipment, and instrumentation required for extraction with nondeuterated solvents

Chemicals	1. D_2O (Euriso-Top™ 99.9%) 2. Dipotassium phosphate K_2HPO_4 (MM = 174.2 g/mol) and monopotassium phosphate KH_2PO_4 (MM = 136.1 g/mol) (VWR™ Normapur) 3. Ethanol (VWR™ Normapur 95%) 4. EDTA disodium salt (MM = 374.28 g/mol) (Sigma™) 5. Liquid nitrogen (Air Liquide™) 6. MilliQ™ water 7. Potassium hydroxide KOH (MM = 56.1 g/mol) (VWR™ Normapur) 8. 3-Trimethylsilyl[2,2,3,3-^{2}H] propionic acid sodium salt (TSP) (Sigma-Aldrich™ 26,991-3)
Materials	1. 13-mL polypropylene tubes (Sarstedt™ ref 55.515 & ref 60.541.003) 2. Gilson tips for PipetMan® P5000, P1000, P100, and P10 3. NMR tube Wilmad 507-PP-7
Equipment	1. Safety glasses 2. Cryogenic gloves (Tempshield™, blue) 3. Lab coat
Instrumentation	1. Centrifuge (Jouan™ KR25i) 2. Cryogrinder (Freezer/Mill 6750 SPEX or UMC5 STEPHAN) 3. Dessicator/Silica gel (VWR™ Chameleon C 2.5–6 mm) 4. Desktop mini centrifuge (Qualitron™ DW41) 5. Freeze-drier (Bioblock™ DURA-DRY) 6. Heating bath (Heidolph™) 7. pH-Meter (MeterLab™ PHM210) 8. PipetMan® P5000, P1000, P100, and P10 9. Sonicator (Bioblock™ 88155) 10. Speed-vac (Thermo™ SC210A) 11. Vortex (Heidolph™ 94323) 12. Weighing scale (Sartorius™, 1602MP, $d = 0.1$ mg)

3. The samples are then cooled at 4 °C for 45 min prior to further centrifugation for 5 min (4 °C); 700 μL of the supernatant is transferred to a 5-mm NMR tube (5-mm economy NMR tubes WG-1226).
4. Deuterated water can be replaced by potassium phosphate buffer in D_2O (pH 6) to avoid shift of resonances due to different pH values of extracts. However, the concentration of the buffer is limited by its stability in the presence of methanol; precipitation should be avoided.

5.2.2. Extraction with nondeuterated solvents

The following describes the three steps of a typical analysis of broccoli florets adapted from the method of Moing *et al.* (2004).

1. For the first extraction, approximately 50 mg ($\pm$2 mg) of dried powder is put in a 15-mL polypropylene tube with screw cap. Next, 2 mL of ethanol/water (80/20, v/v) is added and mixed with a vortex for 5 s. The tube is heated for 15 min at 80 °C. Every 5 min, the sample is mixed with a vortex for 5 s. The tube is then centrifugated for 10 min at 30,000$\times g$ (4 °C). The first supernatant (S1) is transferred into a 15-mL polypropylene tube. The second and third extractions are carried out as the first on the remaining pellet only with 2 mL of ethanol/water (50/50, v/v) and 3 mL of ultrapure water, respectively. The third supernatant (S3) is transferred into the tube containing S1 and S2 and mixed with a vortex for 5 s. The combined supernatants are centrifuged for 10 min at 30,000$\times g$ (4 °C). The resulting supernatant is transferred into a 15-mL polypropylene tube, frozen and dried under vacuum (Speed-vacuum concentrator) over night in order to remove three-quarters of the initial volume and is then frozen in liquid nitrogen and freeze-dried for 24 h to remove nonbound water.
2. The dried extracts are solubilized in 500 μL of 400 mM phosphate buffer in D_2O supplemented with 5 mM Na_2EDTA (ethylene diamine tetraacetic acid disodium salt) and titrated to pH 6.0 $\pm$ 0.05 with 1 M KOH in D_2O. The phosphate buffer and KOD solutions have to be prepared the day of the titration to avoid D/H exchange. It is recommended to use EDTA addition to buffer solution in order to improve spectral resolution of organic acids (like malic and citric acids). It is noteworthy that addition of EDTA will in some cases add supplementary signals in the ^{1}H NMR spectrum (Han *et al.*, 2007). The titrated extracts are frozen in liquid nitrogen and lyophilized again for 24 h to decrease the signal of residual water. The dried, purified, titrated extracts are stored in a dry atmosphere (dessicator) or at -20 °C (if storage is longer than a week) before ^{1}H NMR analysis.
3. The samples should be reconstituted on the day of NMR measurement with 500 μL of D_2O from a freshly opened bottle and 5 μL of a 2% (v/v) TSP/D_2O solution, corresponding to 0.02% of TSP. The sample is mixed with a vortex for 5 min and then centrifuged for 5 min at 20,000$\times g$ (room temperature). The supernatant is transferred into a clean 5-mm NMR tube and closed with a cap, and NMR acquisition is performed within 12 h.

5.3. Considerations and recommendations for spectral acquisition

5.3.1. Instrumentation

Metabolomic profiling by NMR demands the use of high-field instruments with superconducting magnet coils operating at liquid helium temperature. At present, the highest commercially available field is 23.5 T, equivalent to a ^{1}H frequency of 1 GHz. For plant metabolite profiling, the routine fields used have a ^{1}H frequency of 400–600 MHz. An NMR spectrometer is also composed of a console (high-frequency channels for, excitation and acquisition, gradient amplifier unit), a probe head, and an autosampler. Automatically, tuneable broadband inverse gradient probe heads (i.e., ATMA-BBi by Bruker and ProTune Auto-X ID by Varian) flushed with nitrogen gas (coming from a nitrogen separator which enriches N_2 to about 98% of the compressed air) and temperature controlled are recommended. The chemical shift of water is temperature dependant. Inverse or indirect detection probe heads provide optimal signal-to-noise ratio for ^{1}H detection. The diameter of such standard probe heads is usually 5 mm and requires a sample volume of 500–600 μL (Ross *et al.*, 2007). Autosampler or robotic sample changers permit the mechanical change of NMR tubes by a preset program, thus allowing HTP sample handling. Concerning tubes, the use of the same quality of tube and cap from the same vendor for the whole series of samples is recommended. Tubes with outer diameter of 5 mm or less are convenient. The acquisition method described within the following section is designed for a capillary or coaxial NMR system, rather than performing quantification via locking to the TSP signal, and it is performed using an internal standard solubilized within deuterated solvent.

5.3.2. Acquisition

Most commonly, one-pulse sequence methods are preferred for quantitative NMR profiling of polar metabolites in samples depleted of water. For samples that contain water, the water signal must be suppressed and a pulse sequence based on the start of the nuclear Overhauser effect spectroscopy, known as noesy-presat, is used. The noesy-presat technique, with its highest chemical shift selectivity over the other presaturation methods, is the technique of choice for the acquisition of high-quality and reproducible spectra from aqueous samples (Ross *et al.*, 2007). Nevertheless, this pulse sequence is very sensitive to lock phase, and this parameter should be checked at the beginning of the workflow. Tables 16.2 and 16.3 describe the spectral acquisition and processing parameters used for a typical quantitative 1D ^{1}H NMR analysis and for a semi-quantitative 1D ^{1}H NMR analysis with presaturation (noesy-presat), respectively. Some care should be taken with sample temperature equilibration before NMR measurement. When the tube is placed in the probe head, a delay of 2–3 min is

Table 16.2 Recommended spectral acquisition parameters for quantitative 1D ^{1}H NMR spectrum and for semi-quantitative 1D ^{1}H NMR with presaturation (noesy-presat) of polar metabolites using a 500-MHz NMR spectrometer with an automatically tuneable inverse probe

Parameter	Setting 1D ^{1}H qNMR	Setting noesy-presat
Lock solvent	D_2O	D_2O
Solvent	100–400 mM potassium or sodium phosphate buffer in D_2O (pH 6 or 7)[a]	100–400 mM potassium or sodium phosphate buffer in D_2O/H_2O (10%/90%, v/v) (pH 6 or 7)
Chelating agent	1–5 mM Na_2EDTA in D_2O	1–5 mM Na_2EDTA in D_2O
Calibrating standard for chemical shift scale	0.5 mM TSP	0.5 mM TSP
Sample temperature (K)	300 or 303[b]	300 or 303
Pulse sequence in Bruker library	Normal one-pulse sequence (zg)	Noesyzgpr1d
Water suppression	—	Mixing time 100 ms
Quantification standard	ERETIC[c]	ERETIC
Excitation pulse angle (°)	90	90
Relaxation delay (s)	20–35[d]	2–4
Number of data points	32,768 or 65,536	65,536
Sweep width (ppm)	12	20
Number of transients or scans	64	64
Typical acquisition time (min)	Typically 15–20	Typically 5

[a] The choice of pH for a given extract preparation may be governed by the need not only to resolve signals of certain metabolites (Kaiser *et al.*, 2009) but also to avoid the precipitation of certain metabolites (oxalate buffer in D_2O, pH 4, to protect tartaric acid in berry extracts; Pereira *et al.*, 2005).

[b] The choice of the temperature for a given extract NMR acquisition may be governed by the need to resolve signals of certain metabolites. Gromova and Roby (2010) used 293 K in order to shift the residual water peak and detect oxidized glutathione peak.

[c] ERETIC (Electronic REference To access *In vivo* Concentrations; Akoka *et al.*, 1999). This electronic reference is generated by the spectrometer with an adapted intensity and chemical shift position chosen by the spectroscopist. The main advantage is no chemical interaction between the internal standard and the constituents of the extract leading to erroneous quantification and no spectral overlapping with the resonances of the constituents of the extract. Attention should be paid when TSP is used as internal standard, especially with sap, or "juice" or samples where proteins were not removed. TSP can then interact with proteins and became less NMR visible thus impairing quantification (Kriat *et al.*, 1992). In addition, to be an accurate internal quantification standard, the intensity of the TSP signal should be as high as the median intensities of the spectrum, not too small but also not too high. It should be kept in mind that the receptor gain of the spectrometer is calculated on the biggest signal of the spectrum. The ERETIC method is an alternative to the use of internal standards. Recent Bruker spectrometers include ERETIC (indirect mode) in their basic configuration. Gromova and Roby (2010) used maleic acid (singlet at 6.023 ppm at pH 7) as internal standard for quantification.

[d] In order to take in account long T_1 relaxation times of compounds like organic acids (malate, citrate, etc.). Quantitative conditions require a relaxation delay $d_1 \geq 5 \times T_{1\ longest}$. In beer, Rodrigues *et al.* (2010) reported 3.8 s for acetic, 1.1 s for citric, 1.3 s for lactic, 1.5 s for malic, 2.9 s for pyruvic, and 1.7 s for succinic acids.

Table 16.3 Recommended spectral processing parameters for quantitative 1D ^{1}H NMR spectrum and for semi-quantitative 1D ^{1}H NMR with presaturation (noesy-presat) of polar metabolites using a 500-MHZ NMR spectrometer with inverse ambient probe

Parameters	Setting 1D ^{1}H qNMR	Setting noesy-presat
Fourier transformation	Exponential with line broadening of 0.3 or 0.5 Hz	Exponential with line broadening of 0.3 or 0.5 Hz
No of (real and imaginary) data points	32,768 or 65,536	65,536 to 4 × 65,536
Phasing	Required Manual	Required Manual
Baseline corrected	Manual (polynomial)	Manual or automated (polynomial)
Chemical shift calibration	Chemical shift of the TSP singlet signal set to 0 ppm	Chemical shift of the TSP singlet signal set to 0 ppm

required for this equilibration to take place, especially with "salted" samples. Special care should be also taken with spectrometer adjustments for each sample like shimming and determination of the 90° pulse angle. With recent NMR spectrometers, the determination and optimization of these parameters are routinely achieved by automation. NMR spectra of the technological replicate samples have to be acquired in a randomized order (Defernez and Colquhoun, 2003). In addition to biological samples, it is recommended to run a solvent blank and an extraction blank to identify impurities originating from the solvents (Kaiser *et al.*, 2009) or the extraction procedure (i.e., phthalates from plastic ware). For quantification purposes, reference samples with known concentrations of metabolites of interest should be measured in the same analytical run. For example, the calibration curve for glucose can be performed with five concentrations ranging from 0 to 50 mM using the same solvent as the biological samples. The mid-concentration of the calibration range should be equivalent to the expected concentration within the sample. The calibration curve of glucose is used for scaling the integral of the ERETIC signal. The absolute quantity of a metabolite in the NMR sample is determined from the measurement of the integral of one of its specific resonances scaled with ERETIC intensity. The integration limits are chosen at the valleys on each side of the peak or pattern base, and in the cases where these points are not at the baseline level, manual local baseline correction is applied. In addition, specific calibration curves for fructose, glutamate, and glutamine have to be performed separately, and this is governed by the requirement of having a response factor of a part of their signal to accurately quantify (Moing *et al.*, 2004). Automation of phasing and baseline correction can be used, but only with limited reliability and human visual inspection of processed spectra is compulsory to detect artifacts of automated processing. Figure 16.4 provides examples of 1D ^{1}H qNMR spectra of polar extracts from melon flesh and broccoli floret obtained after hydroethanolic extraction described in Section 5.2.2. Table 16.4 demonstrates a spectral processing method that is applied to the raw data in order to permit latter statistical analysis. The presented postacquisition data processing steps differ if the entire spectral signature is analyzed with data reduction into about 300–1200 spectral regions (fingerprinting), or if absolute quantification of identified compounds is chosen. For binning or bucketing, commercially available softwares include Amix (Bruker) or KnowItAll Metabolomics Editions (BioRad). The file containing the intensity values of each bin or bucket could be either transferred to generic statistical softwares like SIMCA-P (Umetrics) or processed in statistics modules of the above softwares.

5.3.3. Metabolite identification or spectral assignment

It is worth noting that the chemical shift range of ^{1}H is small and rarely beyond 10 ppm, which means that ^{1}H NMR spectra of complex mixtures like plant extracts are crowded and resonance overlap occurs. In addition, most of the ^{1}H signals have a multiplet structure. Most metabolites have multiple interdependent resonances across the spectrum. So checking for the presence of all the chemical shifts and patterns of the suspected metabolite and crosschecking for the expected proportions of these resonances between patterns are the traditional way of performing metabolite structural elucidation with 1D ^{1}H NMR. The following describes a typical pipeline of NMR spectral assignment. The first step for the assignments of metabolites in the NMR spectra is made by comparing the observed proton chemical shifts.

1. With literature, for example, using Fan (1996) as a good start.
2. With in-house or public/commercial database values of authentic compounds acquired under the same solvent conditions (see Saito and Matsuda, 2010), for example, Prime (http://prime.psc.riken.jp/), HMDB (http://www.hmdb.ca/), and commercial databases (Chenomx Metabolite Database and BBIOREFCODE, Bruker NMR Metabolic Profiling Database).
3. With in-house or public knowledge bases dedicated to NMR metabolomic profiles, for example, MeRy-B (http://www.cbib.u-bordeaux2.fr/MERYB/).
4. By spiking the samples with authentic or commercial compounds if available, by acquiring further spectra, and demonstrating coresonance of the peaks from the extract sample and the added compound.

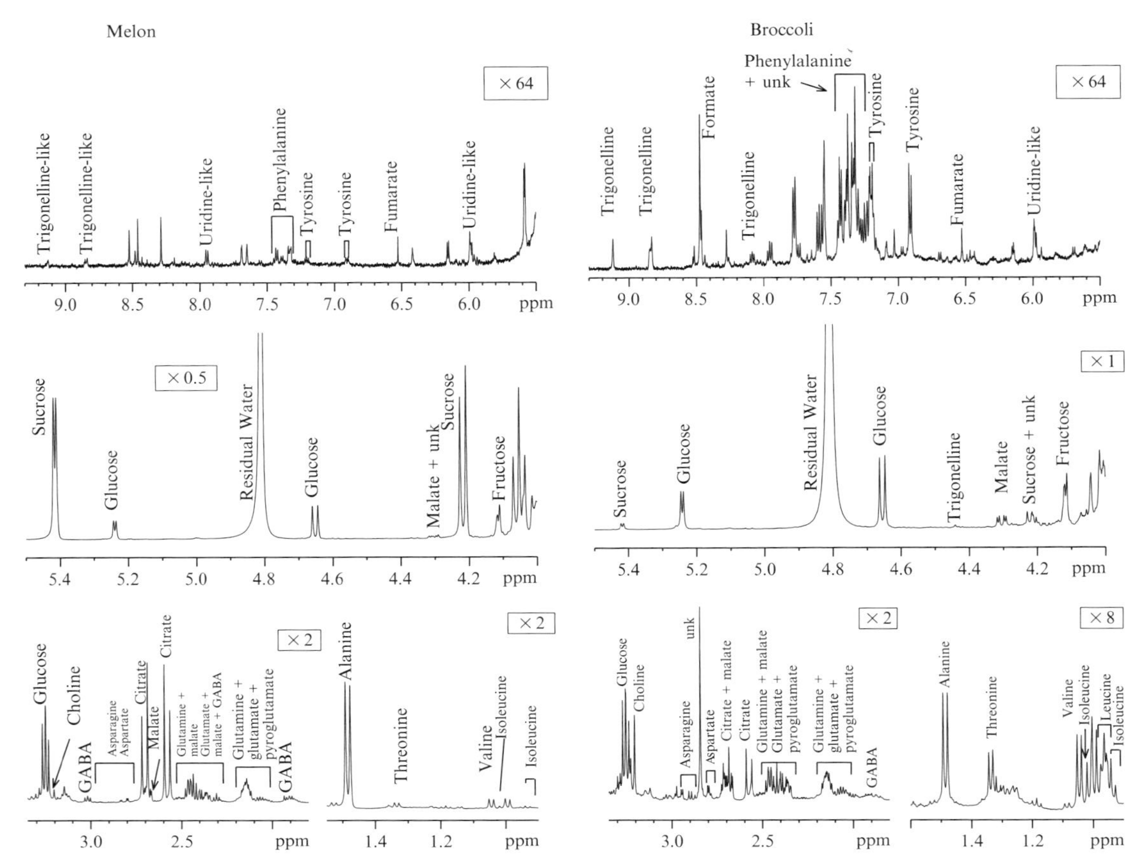

Figure 16.4 Examples of 1D ^{1}H qNMR spectra for polar extract from melon flesh and from broccoli floret tissue obtained after hydroethanolic extraction described in Section 5.2.2. Unk, unknown compound; GABA, gamma-aminobutyric acid.

Table 16.4 Recommended postprocessing steps for metabolic fingerprinting study of a batch of 1D spectra

Step	Setting 1D ^{1}H qNMR or noesy-presat
Identification and removal of unwanted spectral regions	Residual water signal (5.0–4.5 ppm) Residual extraction solvent Impurities of extraction process[a]
Normalization	Each area of bin is divided by the total spectra area (without the excluded spectra regions) or divided by the area of the bin containing the ERETIC signal
Binning or bucketing	Regular bin (or bucket) with size of 0.01, 0.02, or 0.04 ppm or variable-sized bin[b]
Further scaling	Mean centering
Multivariate data analysis	PCA, PLS, etc.

[a] Determined by running a solvent blank and an extraction blank (all the process without the biological sample).

[b] In some cases, even if attention was paid to pH titration, the ionic force of samples in the same batch can slightly differ and variable sized bin can be used to take into account shift of chemical shifts (Ross *et al.*, 2007).

The next stage for structural assignment is to record 2D NMR spectra, such as HSQC, and to compare the ^{1}H and ^{13}C chemical shifts measured following the same approach as described above from points 1 to 4 (Fan, 1996; Gromova and Roby, 2010). For 2D JRES NMR (see Ludwig and Viant, 2010), few examples in plant metabolomics are described in literature (e.g., Tobacco: Choi *et al.*, 2006; Brassica: Liang *et al.*, 2006). Combining 1D NMR for quantification and 2D NMR for unambiguous identification of the observed metabolites justifies why NMR is a useful method for plant metabolite profiling, especially of bulk metabolite compounds. For specific identification purposes, LC–NMR, LC–SPE–NMR, and LC–NMR/MS can also be used (Exarchou *et al.*, 2005; Glauser *et al.*, 2008); LC–SPE is especially useful as a technique to isolate and bulk a pure metabolite prior to NMR characterization.

6. Concluding Remarks

Although the application of systems biology level research to the study of plants is very much still in its infancy, it is hoped that this chapter has highlighted the potential of plant metabolomics for future applications of multi-"omic"/systems level research. In fact, there are already many good examples where metabolomics and microarray transcriptomics have been combined to provide maximum information (Hirai *et al.*, 2004; Nikiforova *et al.*, 2004; Yonekura-Sakakibara *et al.*, 2008). Although this chapter has focused purely upon introducing the reader to the field of plant metabolomics and the production of high-quality and reproducible metabolite profiles, it is hoped that the reader can see the future benefit of the use of such methods for the production of informative and comparable data between research laboratories (Allwood *et al.*, 2009; de Vos *et al.*, 2007; Erban *et al.*, 2007; Lisec *et al.*, 2006; Ward *et al.*, 2010b). In conclusion, many aspects of metabolomics are now firmly embedded in biological and systems biology sciences. Further, we have entered a phase focused on dealing with the final elimination of bottle necks/automation and exploration of the limits such as miniaturization of samples, enhancement from relative to absolute quantification for systems biology/modeling applications, and enhanced coverage using higher dimensional hyphenated technologies such as GC $\times$ GC-TOF/MS. With further ongoing technological developments making "omics" data more reproducible, as well as improvements in data modeling, and perhaps, applying multi-omic level analysis to model plant species such as Arabidopsis using both artificial cell cultures and their respective soil grown plants, we feel that the plant science field will greatly benefit from systems biology and that such approaches will eventually benefit many areas of research within plant physiology and food nutrition.

ACKNOWLEDGMENTS

J. W. A. and all authors would like to thank the EU for experimental funding as part of the Framework VI initiative within the META-PHOR project (FOOD-CT-2006-036220). J. W. A. would also like to acknowledge CR-UK for current research funding. R. G. would like to thank the UK BBSRC and EPSRC (BBC0082191) for financial support of the MCISB (Manchester Centre for Integrative Systems Biology). R. D. H and R. C. H. D. V. acknowledge the Centre for Biosystems Genomics, which is part of the Netherlands Genomics Initiative, for additional funding. A. M. and C. D. would like to thank Dr. S. Bernillon for his critical reading of Section 5 and the Metabolome Facility of Bordeaux Functional Genomics Centre for support.

REFERENCES

Adato, A., Mandel, T., Mintz-Oron, S., Venger, I., Levy, D., Yativ, M., Dominguez, E., Wang, Z., De Vos, C. H. R., Jetter, R., Schreiber, L., Heredia, A., *et al.* (2009). Fruit-surface flavonoid accumulation in tomato is controlled by a SlMYB12-regulated transcriptional network. *PLoS Genet.* **5**(12), e1000777, 10.1371/journal.pgen.1000777.

Ahuja, I., De Vos, C. H. R., Bones, A. M., and Hall, R. D. (2010). Plant molecular stress programmes face climate change. *Trends Plant Sci.* **15,** 664–674.

Akoka, S., Barantin, L., and Trierweiler, M. (1999). Concentration measurement by proton NMR using the ERETIC method. *Anal. Chem.* **71,** 2554–2557.

Allwood, J. W., and Goodacre, R. (2010). An introduction to liquid chromatography–mass spectrometry instrumentation applied in plant metabolomic analyses. *Phytochem. Anal.* **21,** 33–47.

Allwood, J. W., Ellis, D. I., Heald, J. K., Goodacre, R., and Mur, L. A. J. (2006). Metabolomic approaches reveal that phosphatidic and phosphatidyl glycerol phospholipids are major discriminatory metabolites in responses by *Brachypodium distachyon* to challenge by *Magnaporthe grisea*. *Plant J.* **46,** 351–368.

Allwood, J. W., Ellis, D. I., and Goodacre, R. (2008). Metabolomic technologies and their application to the study of plants and plant-host interactions. *Physiol. Plant.* **132,** 117–135.

Allwood, J. W., Erban, A., de Koning, S., Dunn, W. B., Luedemann, A., Lommen, A., Kay, L., Löscher, R., Kopka, J., and Goodacre, R. (2009). Inter-laboratory reproducibility of fast gas chromatography–electron impact–time of flight mass spectrometry (GC-EI-TOF/MS) based plant metabolomics. *Metabolomics* **5,** 479–496.

Allwood, J. W., Clarke, A., Goodacre, R., and Mur, L. A. J. (2010). Dual metabolomics: A novel approach to understanding plant-pathogen interactions. *Phytochemistry* **71,** 590–597.

Allwood, J. W., Parker, D. P., Beckmann, M., Draper, J., and Goodacre, R. (2011). Fourier transform ion cyclotron resonance mass spectrometry (FT-ICR-MS) for plant metabolite profiling and metabolite identification. *In* "Plant Metabolomics Methods," (N. W. Hardy and R. D. Hall, eds.), Humana Press, Otawa, USA (in press).

Anderson, M. J., and Whitcomb, P. J. (2007). *DOE Simplified: Practical Tools for Effective Experimentation.* 2nd edn. Productivity Press, New York.

AP Rees, T., and Hill, S. A. (1994). Metabolic control analysis of plant metabolism. *Plant Cell Environ.* **17,** 587–599.

Baginsky, S., and Fernie, A. R. (eds.), (2007). Plant Systems Biology, Birkhauser, Basel, Switzerland.

Balestrini, R., and Bonfante, P. (2008). Laser microdissection (LM): Applications to plant materials. *Plant Biosyst.* **142,** 331–336.

Beale, M. H., and Sussman, M. R. (2010). Metabolomics of Arabidopsis thaliana. *In* "The Biology of Plant Metabolomics," (R. D. Hall, ed.). Blackwell-Wiley, London (Chapter 6, 157–180).

Beale, M. H., and Ward, J. L. (2011). Sample preparation using Arabidopsis. *In* "Plant Metabolomics Methods," (N. W. Hardy and R. D. Hall, eds.), Humana Press, Otawa, USA (in press).

Begley, P., Francis-McIntyre, S., Dunn, W. B., Broadhurst, D. I., Halsall, A., Tseng, A., Knowles, J., HUSERMET ConsortiumGoodacre, R., and Kell, D. B. (2009). Development and performance of a gas chromatography-time-of-flight mass spectrometry analysis for large-scale nontargeted metabolomic studies of human serum. *Anal. Chem.* **81,** 7038–7046.

Biais, B., Allwood, J. W., Deborde, C., Xu, Y., Maucourt, M., Beauvoit, B., Dunn, W. B., Jacob, D., Goodacre, R., Rolin, D., and Moing, A. (2009). 1H-NMR, GC-EI-TOF/MS, and dataset correlation for fruit metabolomics: Application to spatial metabolite analysis in melon. *Anal. Chem.* **81,** 2884–2894.

Biais, B., Beauvoit, B., Allwood, J. W., Deborde, C., Maucourt, M., Goodacre, R., Rolin, D., and Moing, A. (2010). Metabolic acclimation to hypoxia revealed by metabolite gradients in melon fruit. *J. Plant Physiol.* **167,** 242–245.

Biais, B., Bernillon, S., Deborde, C., Cabasson, C., Rolin, D., Tadmor, Y., Burger, J., Schaffer, A. A., and Moing, A. (2011). Precautions for harvest, sampling, storage and transport of crop plant metabolomics samples. *In* "Plant Metabolomics Methods," (N. W. Hardy and R. D. Hall, eds.), Humana Press, Otawa, USA (in press).

Bino, R. J., Hall, R. D., Fiehn, O., Kopka, J., Saito, K., Draper, J., Nikolau, B. J., Mendes, P., Roessner-Tunali, U., Beale, M. H., Trethewey, R. N., Lange, B. M., *et al.* (2004). Potential of metabolomics as a functional genomics tool. *Trends Plant Sci.* **9,** 418–425.

Bino, R. J., de Vos, C. H. R., Lieberman, M., Hall, R. D., Bovy, A., Jonker, H. H., Tikunov, Y., Lommen, A., Moco, S., and Levin, I. (2005). The light-hyperresponsive high pigment-2(dg) mutation of tomato: Alterations in the fruit metabolome. *New Phytol.* **166,** 427–438.

Bonhomme, R. (2000). Bases and limits to using 'degree.day' units. *Eur. J. Agronomy.* **13,** 1–10.

Bovy, A. G., Gomez-Roldan, V., and Hall, R. D. (2010). Strategies to optimize the flavonoid content of tomato fruit. *In* "Recent Advances in Polyphenol Research," (C. Santos-Buelga, M.-T. Escribano-Bailon, and V. Lattanzio, eds.), pp. 138–162. Blackwell-Wiley, London.

Boyes, D. C., Zayed, A. M., Ascenzi, R., McCaskill, A. J., Hoffmann, N. E., Davis, K. R., and Görlach, J. (2001). Growth stage-based phenotypic analysis of Arabidopsis. A model for high throughput functional genomics in plants. *Plant Cell* **13,** 1499–1510.

Broadhurst, D. I., and Kell, D. B. (2006). Statistical strategies for avoiding false discoveries in metabolomics and related experiments. *Metabolomics* **2,** 171–196.

Brukhin, V., Hernould, M., Gonzalez, N., Chevalier, C., and Mouras, A. (2003). Flower development schedule in tomato *Lycopersicon esculentum* cv. sweet cherry. *Sex. Plant Reprod.* **15,** 311–320.

Capanoglu, E., Beekwilder, J., Boyacioglu, D., de Vos, R., and Hall, R. (2008). Changes in antioxidant and metabolite profiles during production of tomato paste. *J. Agric. Food Chem.* **56,** 964–973.

Choi, Y. H., Kim, H. K., Linthorst, H. J. M., Hollander, J. G., Lefeber, A. W. M., Erkelens, C., Nuzillard, J. M., and Verpoorte, R. (2006). NMR metabolomics to revisit the tobacco mosaic virus infection in *Nicotiana tabacum* leaves. *J. Nat. Prod.* **69,** 742–748.

Cook, D., Fowler, S., Fiehn, O., and Thomashow, M. F. (2004). A prominent role for the CBF cold response pathway in configuring the low-temperature metabolome of Arabidopsis. *Proc. Natl. Acad. Sci. USA* **101,** 15243–15248.

de Rijke, E., Zappey, H., Ariese, F., Gooijer, C., and Brinkmen, U. A. Th. (2003). Liquid chromatography with atmospheric pressure chemical ionization and electrospray ionization mass spectrometry of flavonoids with triple-quadrupole and ion-trap instruments. *J. Chromatogr. A* **984,** 45–58.

de Vos, C. H. R., Moco, S., Lommen, A., Keurentjes, J. J. B., Bino, R. J., and Hall, R. D. (2007). Untargeted large-scale plant metabolomics using liquid chromatography coupled to mass spectrometry. *Nat. Protoc.* **2,** 778–791.

Defernez, M., and Colquhoun, I. J. (2003). Factors affecting the robustness of metabolite fingerprinting using 1H NMR spectra. *Phytochemistry* **62,** 1009–1017.

Douillard, C., and Guichard, E. (1990). The aroma of strawberry (*Fragaria ananassa*): Characterisation of some cultivars and influence of freezing. *J. Sci. Food Agric.* **50,** 517–531.

Dunn, W. B., Bailey, N. J. C., and Johnson, H. E. (2005a). Measuring the metabolome: Current analytical technologies. *Analyst* **130,** 606–625.

Dunn, W. B., Overy, S., and Quick, W. P. (2005b). Evaluation of automated electrospray-TOF mass spectrometry for metabolic fingerprinting of the plant metabolome. *Metabolomics* **1,** 137–148.

Dunn, W. B., Broadhurst, D., Brown, M., Baker, P. N., Redman, C. W. G., Kenny, L. C., and Kell, D. B. (2008). Metabolite profiling of serum using ultra performance liquid chromatography and the LTQ-Orbitrap mass spectrometry system. *J. Chromatogr. B* **871,** 288–298.

Ebert, B., Zoeller, D., Erban, A., Fehrle, I., Hartmann, J., Niehl, A., Kopka, J., and Fisahn, J. (2010). Metabolic profiling of *Arabidopsis thaliana* epidermal cells. *J. Exp. Bot.* **61,** 1321–1335.

Eisenreich, W., and Bacher, A. (2007). Advances of high-resolution NMR techniques in the structural and metabolic analysis of plant biochemistry. *Phytochemistry* **68,** 2799–2815.

Erban, A., Schauer, N., Fernie, A. R., and Kopka, J. (2007). Nonsupervised construction and application of mass spectral and retention time index libraries from time-of-flight gas chromatography-mass spectrometry metabolite profiles. *In* "Methods in Molecular Biology," (W. Weckwerth, ed.), Vol. 358, (Part I), pp. 19–38. Springer, New York, USA.

Exarchou, V., Krucker, M., van Beek, T. A., Vervoort, J., Gerothanassis, I. P., and Albert, K. (2005). LC–NMR coupling technology: Recent advancements and applications in natural products analysis. *Magn. Reson. Chem.* **43,** 681–687.

Fait, A., Hanhineva, K., Beleggia, R., Dai, N., Rogachex, I., Nikiforova, V. J., Fernie, A. R., and Aharoni, A. (2008). Reconfiguration of the achene and receptacle metabolic networks during strawberry fruit development. *Plant Physiol.* **148,** 730–750.

Fan, T. W. M. (1996). Metabolite profiling by one- and two-dimensional NMR analysis of complex mixtures. *Prog. Nucl. Magn. Reson. Spectrosc.* **28,** 161–219.

Farag, M. A., Huhman, D. V., Dixon, R. A., and Sumner, L. W. (2008). Metabolomics reveals novel pathways and differential mechanistic and elicitor-specific responses in phenylpropanoid and isoflavonoid biosynthesis in *Medicago truncatula* cell cultures. *Plant Physiol.* **146,** 387–402.

Fernie, A. R. (2007). The future of metabolic phytochemistry: Larger numbers of metabolites, higher resolution, greater understanding. *Phytochemistry* **68,** 2861–2880.

Fernie, A. R., and Schauer, N. (2009). Metabolomics-assisted breeding: A viable option for crop improvement? *Trends Genet.* **25,** 39–48.

Fernie, A. R., Trethewey, R. N., Krotzky, A. J., and Willmitzer, L. (2004). Metabolite profiling: From diagnostics to systems biology. *Nat. Rev. Mol. Cell Biol.* **5,** 763–769.

Fiehn, O. (2002). Metabolomics—The link between genotypes and phenotypes. *Plant Mol. Biol.* **48,** 155–171.

Fiehn, O., Kopka, J., Dörmann, P., Altmann, T., Trethewey, R. N., and Willmitzer, L. (2000a). Metabolite profiling for plant functional genomics. *Nat. Biotechnol.* **18,** 1157–1161.

Fiehn, O., Kopka, J., Trethewey, R. N., and Willmitzer, L. (2000b). Identification of uncommon plant metabolites based on calculation of elemental compositions using gas chromatography and quadrupole mass spectrometry. *Anal. Chem.* **72,** 3573–3580.

Fiehn, O., Sumner, L. W., Rhee, S. Y., Ward, J., Dickerson, J., Lange, B. M., Lane, G., Roessner, U., Last, R., and Nikolau, B. (2007). Minimum reporting standards for plant biology context information in metabolomic studies. *Metabolomics* **3,** 195–201.

Fiehn, O., Wohlgemuth, G., Scholz, M., Kind, T., Lee, D. Y., Lu, Y., Moon, S., and Nikolau, B. (2008). Quality control for plant metabolomics: Reporting MSI-compliant studies. *Plant J.* **53,** 691–704.

Fiehn, O., Kind, T., and Kumar Barupal, D. (2011). Data processing, metabolomic databases and pathway analysis. *In* "The Biology of Plant Metabolomics," (R. D. Hall, ed.). Blackwell-Wiley, London (in press).

Fish, W., and Davis, A. (2003). The effects of frozen storage conditions on lycopene stability in watermelon tissue. *J. Agric. Food Chem.* **51,** 3582–3585.

Gautier, H., Rocci, A., Buret, M., Grasselly, D., and Causse, M. (2005). Fruit load or fruit position alters response to temperature and subsequently cherry tomato quality. *J. Sci. Food Agric.* **85,** 1009–1016.

Giavalisco, P., Kohl, K., Hummel, J., Seiwert, B., and Willmitzer, L. (2009). C-13 isotope-labeled metabolomes allowing for improved compound annotation and relative quantification in liquid chromatography-mass spectrometry-based metabolomic research. *Anal. Chem.* **81,** 6546–6551.

Gibon, Y., and Rolin, D. (2011). Precautions for harvest, sampling, storage and transport of crop plant metabolomics samples. *In* "Plant Metabolomics Methods," (N. W. Hardy and R. D. Hall, eds.), Humana Press, Otawa, USA (in press).

Gibon, Y., Usadel, B., Blaesing, O., Kamlage, B., Hoehne, M., Trethewey, R., and Stitt, M. (2006). Integration of metabolite with transcript and enzyme activity profiling during diurnal cycles in Arabidopsis rosettes. *Genome Biol.* **7,** R76.

Glauser, G., Guillarme, D., Grata, E., Boccard, J., Thiocone, A., Carrupt, P.-A., Veuthey, J.-L., Rudaz, S., and Wolfender, J.-L. (2008). Optimized liquid chromatography-mass spectrometry approach for the isolation of minor stress biomarkers in plant extracts and their identification by capillary nuclear magnetic resonance. *J. Chromatogr. A* **1180,** 90–98.

Goodacre, R., Vaidyanathan, S., Dunn, W. B., Harrigan, G. G., and Kell, D. B. (2004). Metabolomics by numbers—Acquiring and understanding global metabolomics data. *Trends Biotechnol.* **22,** 245–252.

Goodacre, R., Broadhurst, D., Smilde, A. K., Kristal, B. S., Baker, J. D., Beger, R., Bessant, C., Connor, S., Calmani, G., Craig, A., Ebbels, T., Kell, D. B., *et al.* (2007). Proposed minimum reporting standards for data analysis in metabolomics. *Metabolomics* **3,** 231–241.

Gromova, M., and Roby, C. (2010). Toward *Arabidopsis thaliana* hydrophilic metabolome: Assessment of extraction methods and quantitative 1H NMR. *Physiol. Plant.* **140,** 111–127.

Hall, R. D. (2006). Plant metabolomics: From holistic hope, to hype, to hot topic. *New Phytol.* **169,** 453–468.

Hall, R. D. (ed.) (2011). Biology of Plant Metabolomics, Wiley-Blackwell, Oxford.

Hall, R. D., De Vos, C. H. R., and Ward, J. (2010). Plant metabolomics applications in the Brassicaceae: Added value for science and industry. *Acta Horticulturae* **867,** 191–206.

Han, S., Mathias, E. V., and Ba, Y. (2007). Proton NMR determination of $Mg2^+$ and $Ca2^+$ concentrations using tetrasodium EDTA complexes. *J. Chem.* **1,** 1–5.

Hanhineva, K., Rogachev, I., Kokko, H., Mintz-Oron, S., Venger, I., Kärenlampi, S., and Aharoni, A. (2008). Non-targeted analysis of spatial metabolite composition in strawberry (Fragaria × ananassa) flowers. *Phytochemistry* **69,** 2463–2481.

Hannemann, J., Poorter, H., Usadel, B., Blasing, O. E., Finck, A., Tardieu, F., Atkin, O. K., Pons, T., Stitt, M., and Gibon, Y. (2009). Xeml Lab: A tool that supports the design of experiments at a graphical interface and generates computer-readable metadata files, which capture information about genotypes, growth conditions, environmental perturbations and sampling strategy. *Plant Cell Environ.* **32,** 1185–1200.

Hardy, N. W., and Hall, R. D. (eds.), (2011). Plant Metabolomics Methods, Humana Press, Otawa, USA (in press).

Harrigan, G. G., and Goodacre, R. (eds.), (2003). Metabolic Profiling: Its Role in Biomarker Discovery and Gene Function Analysis, Kluwer Academic Publishers, Boston/Dordrecht/London.

Hinkelmann, K., and Kempthorne, O. (2008). *Design and Analysis of Experiments. Vol. 1. Introduction to Experimental Design.* 2nd edn. Wiley, New Jersey.

Hirai, M. Y., Yano, M., Goodenowe, D. B., Kanaya, S., Kimura, T., Awazuhara, M., Arita, M., Fujiwara, T., and Saito, K. (2004). Integration of transcriptomics and metabolomics for understanding of global responses to nutritional stresses in *Arabidopsis thaliana. Proc. Natl. Acad. Sci. USA* **101,** 10205–10210.

Huege, J., Sulpice, R., Gibon, Y., Lisec, J., Koehl, K., and Kopka, J. (2007). GC-EI-TOF-MS analysis of in vivo-carbon-partitioning into soluble metabolite pools of higher plants by monitoring isotope dilution after (13CO2)-labelling. *Phytochemistry* **68,** 2258–2272.

Huhman, D. V., and Sumner, L. W. (2002). Metabolic profiling of saponins in *Medicago sativa* and *Medicago truncatula* using HPLC coupled to an electrospray ion-trap mass spectrometer. *Phytochemistry* **59,** 347–360.

Hummel, J., Strehmel, N., Selbig, J., Walther, D., and Kopka, J. (2010). Decision tree supported substructure prediction of metabolites from GC-MS profiles. *Metabolomics* **6,** 322–333.
Iijima, Y., Nakamura, Y., Ogata, Y., Tanaka, K., Sakurai, N., Suzuki, H., Okazaki, K., Kitayama, M., Kanaya, S., Aoki, K., and Shibata, D. (2008). Metabolite annotations based on the integration of mass spectral information. *Plant J.* **54,** 949–962.
Jansen, J. J., Allwood, J. W., Marsden-Edwards, E., van der Putten, W. H., Goodacre, R., and van Dam, N. M. (2008). Metabolomic analysis of the interaction between plants and herbivores. *Metabolomics* **5,** 150–161.
Julkunen-Titto, R., and Tahvanaiem, J. (1989). The effect of the sample preparation method of extractable phenolics of *Salicaceae* species. *Planta Med.* **55,** 55–58.
Kaiser, K. A., Barding, G. A., Jr., and Larive, C. K. (2009). A comparison of metabolite extraction strategies for 1H-NMR-based metabolic profiling using mature leaf tissue from the model plant *Arabidopsis thaliana. Magn. Reson. Chem.* **47,** S147–S156.
Kaplan, F., Kopka, J., Haskell, D. W., Zhao, W., Schiller, K. C., Gatzke, N., Sung, D. Y., and Guy, C. L. (2004). Exploring the temperature-stress metabolome of Arabidopsis. *Plant Physiol.* **136,** 4159–4168.
Keinänen, K., and Julkunen-Titto, R. (1996). Effect of sample preparation method on birch (*Betula pendula* Roth) leaf phenolics. *J. Agric. Food Chem.* **44,** 2724–2727.
Keurentjes, J. J. B. (2009). Genetical metabolomics: Closing in on phenotypes. *Curr. Opin. Plant Biol.* **12,** 223–230.
Keurentjes, J. J. B., Fu, J., De Vos, C. H. R., Lommen, A., Hall, R. D., Bino, R. J., Van der Plas, L. H. W., Jansen, R. C., Vreugdenhil, D., and Koornneef, M. (2006). The genetics of plant metabolism. *Nat. Genet.* **38,** 842–849.
Kim, H. K., and Verpoorte, R. (2010). Sample preparation for plant metabolomics. *Phytochem. Anal.* **21,** 4–13.
Kim, H. K., Wilson, E. G., Choi, Y. H., and Verpoorte, R. (2010). Metabolomics: A tool for anti-cancer lead-finding from natural products. *Planta Med.* **76,** 1094–1102.
Klages, K., Donnison, H., Wunsche, J., and Boldingh, H. (2001). Diurnal changes in non-structural carbohydrates in leaves, phloem exudate and fruit in 'Braeburn' apple. *Aust. J. Plant Physiol.* **28,** 131–139.
Kopka, J. (2006). Gas chromatography mass spectrometry. *In* "Plant Metabolomics," (K. Saito, R. Dixon, and L. Willmitzer, eds.), pp. 3–20. Springer-Verlag, Berlin.
Kopka, J., Fernie, A. F., Weckwerth, W., Gibon, Y., and Stitt, M. (2004). Metabolite profiling in plant biology: Platforms and destinations. *Genome Biol.* **5,** 109–117.
Kopka, J., Schauer, N., Krueger, S., Birkemeyer, C., Usadel, B., Bergmueller, E., Doermann, P., Weckwerth, W., Gibon, Y., Stitt, M., Willmitzer, L., Fernie, A. R., *et al.* (2005). GMD@CSB.DB: The Golm metabolome database. *Bioinformatics* **21,** 1635–1638.
Kopka, J., Walther, D., Allwood, J. W., and Goodacre, R. (2011). Progress in chemometrics and biostatistics for plant applications: Or a good red wine is a bad white wine. *In* "The Biology of Plant Metabolomics," (R. D. Hall, ed.). Blackwell-Wiley, London (Chapter 10, 317–342).
Kriat, M., Confort-Gouny, S., Vion-Dury, J., Sciaky, M., Viout, P., and Cozzone, P. J. (1992). Quantitation of metabolites in human blood serum by proton magnetic resonance spectroscopy. A comparative study of the use of formate and TSP as concentration standards. *NMR Biomed.* **5,** 179–184.
Liang, Y. S., Choi, Y. H., Kim, H. K., Linthorst, H. J. M., and Verpoorte, R. (2006). Metabolomic analysis of methyl jasmonate treated *Brassica rapa* leaves by 2-dimensional NMR spectroscopy. *Phytochemistry* **67,** 2503–2511.
Lisec, J., Schauer, N., Kopka, J., Willmitzer, L., and Fernie, A. R. (2006). Gas chromatography mass spectrometry-based metabolite profiling in plants. *Nat. Protoc.* **1,** 387–396.
Lommen, A. (2009). MetAlign: Interface-driven, versatile metabolomics tool for hyphenated full-scan mass spectrometry data preprocessing. *Anal. Chem.* **81,** 3079–3086.

Lommen, A., Gerssen, A., Oosterink, J. E., Kools, H. J., Ruiz-Aracama, A., Peters, R. J. B., and Mol, H. G. J. (2010). Ultrafast searching assists in evaluating sub-ppm mass accuracy enhancement in U-HPLC/Orbitrap MS data. *Metabolomics* **7**(1), 15–24.

Ludwig, C., and Viant, M. R. (2010). Two-dimensional J-resolved NMR spectroscopy: Review of a key methodology in the metabolomics toolbox. *Phytochem. Anal.* **21,** 22–32.

Luedemann, A., Strassburg, K., Erban, A., and Kopka, J. (2008). TagFinder for the quantitative analysis of gas chromatography–mass spectrometry (GC-MS)-based metabolite profiling experiments. *Bioinformatics* **24,** 732–737.

Luedemann, A., Erban, A., von Malotky, L., and Kopka, J. (2011). Processing of comprehensive fingerprint data matrices from nominal mass GCMS data using TagFinder. *In* "Plant Metabolomics Methods," (N. W. Hardy and R. D. Hall, eds.), Humana Press, Otawa, USA (in press).

Ma, F., and Cheng, L. (2003). The sun-exposed peel of apple fruit has higher xanthophyll cycle-dependent thermal dissipation and antioxidants of the ascorbate/glutathione pathway than the shaded peel. *Plant Sci.* **165,** 819–827.

Ma, Y. K., Hu, X. S., Chen, J., Chen, F., Wu, J. H., Zhao, G. H., Liao, X. J., and Wang, Z. F. (2007). The effect of freezing modes and frozen storage on aroma, enzyme and micro-organism in Hami melon. *Food Sci. Technol. Int.* **13,** 259–267.

Markert, B. (1995). Sample preparation (cleaning, drying, homogenization) for trace element analysis in plant matrices. *Sci. Total Environ.* **176,** 45–61.

Moco, S., Bino, R. J., Vorst, O., Verhoeven, H. A., de Groot, J., van Beek, T. A., Vervoort, J., and de Vos, C. H. R. (2006). A liquid chromatography-mass spectrometry-based metabolome database for tomato. *Plant Physiol.* **141,** 1205–1218.

Moco, S., Capanoglu, E., Tikunov, Y., Bino, R. J., Boyacioglu, D., Hall, R. D., Vervoort, J., and de Vos, C. H. R. (2007a). Tissue specialization at the metabolite level is perceived during the development of tomato fruit. *J. Exp. Bot.* **58,** 4131–4146.

Moco, S., Vervoort, J., de Vos, C. H. R., and Bino, R. J. (2007b). Metabolomics technologies and metabolite identification. *Trends Anal. Chem.* **26,** 855–866.

Moco, S., Schneider, B., and Vervoort, J. (2009). Plant micrometabolomics: The analysis of endogenous metabolites present in a plant cell or tissue. *J. Proteome Res.* **8,** 1694–1703.

Moing, A., Maucourt, M., Renaud, C., Gaudillère, M., Brouquisse, R., Lebouteiller, B., Gousset-Dupont, A., Vidal, J., Granot, D., Denoyes-Rothan, B., Lerceteau-Kohler, E., and Rolin, D. (2004). Quantitative metabolic profiling by 1-dimensional H-1-NMR analyses: Application to plant genetics and functional genomics. *Funct. Plant Biol.* **31,** 889–902.

Muth, D., Marsden-Edwards, E., Kachlicki, P., and Stobiecki, M. (2008). Differentiation of isomeric malonylated flavonoid glyconjugates in plant extracts with UPLC-ESI/MS/MS. *Phytochem. Anal.* **19,** 444–452.

Nikiforova, V. J., Gakière, B., Kempa, S., Adamik, M., Willmitzer, L., Hesse, H., and Hoefgen, R. (2004). Towards dissecting nutrient metabolism in plants: A systems biology case study on sulphur metabolism. *J. Exp. Bot.* **55,** 1861–1870.

Nothmann, J., and Rylski, I. (1983). Effects of floral position and cluster size on fruit-development in eggplant. *Scientia Hortic.* **19,** 19–24.

Pereira, G. E., Gaudillère, J. P., van Leeuwen, C., Hilbert, G., Lavialle, O., Maucourt, M., Deborde, C., Moing, A., and Rolin, D. (2005). 1H NMR and chemometrics to characterize mature grape berries in four wine-growing areas in Bordeaux, France. *J. Agric. Food Chem.* **53,** 6382–6389.

Pereira, G. E., Gaudillère, J. P., Pieri, P., Hilbert, G., Maucourt, M., Deborde, C., Moing, A., and Rolin, D. (2006). Microclimate influence on mineral and metabolic profiles of grape berries. *J. Agric. Food Chem.* **54,** 6765–6775.

Phillips, K. M., Wunderlich, K. M., Holden, J. M., Exler, J., Gebhardt, S. E., Haytowitz, D. B., Beecher, G. R., and Doherty, R. F. (2005). Stability of 5-methyltetrahydrofolate in frozen fresh fruits and vegetables. *Food Chem.* **92,** 587–595.

Redestig, H., Szymanski, J., Hirai, M. Y., Selbig, J., Willmitzer, L., Nikoloski, Z., and Saito, K. (2011). Data integration, metabolic networks and systems biology. *In* "The Biology of Plant Metabolomics," (R. D. Hall, eds.), Blackwell-Wiley, London (Chapter 9, 261–316).

Rocke, D. M. (2004). Design and analysis of experiments with high throughput biological assay data. *Semin. Cell Dev. Biol.* **15,** 703–713.

Rodrigues, J. E. A., Erny, G. L., Barros, A. S., Esteves, V. I., Brandao, T., Ferreira, A. A., Cabrita, E., and Gil, A. M. (2010). Quantification of organic acids in beer by nuclear magnetic resonance (NMR)-based methods. *Anal. Chim. Acta* **674,** 166–175.

Roessner, U., Wagner, C., Kopka, J., Trethewey, R. N., and Willmitzer, L. (2000). Simultaneous analysis of metabolites in potato tuber by gas chromatography-mass spectrometry. *Plant J.* **23,** 131–142.

Roessner, U., Luedemann, A., Brust, D., Fiehn, O., Linke, T., Willmitzer, L., and Fernie, A. R. (2001a). Metabolic profiling allows comprehensive phenotyping of genetically or environmentally modified plant systems. *Plant Cell* **13,** 11–29.

Roessner, U., Willmitzer, L., and Fernie, A. R. (2001b). High-resolution metabolic phenotyping of genetically and environmentally diverse plant systems—Identification of phenocopies. *Plant Physiol.* **127,** 749–764.

Roessner, U., Willmitzer, L., and Fernie, A. R. (2002). Metabolic profiling and biochemical phenotyping of plant systems. *Plant Cell Rep.* **21,** 189–196.

Ross, A., Schlotterbeck, G., Dieterle, F., and Senn, H. (2007). NMR spectroscopy techniques for application to metabonomics. *In* "The Handbook of Metabonomics and Metabolomics," (J. C. Lindon, J. K. Nicholson, and E. Holmes, eds.), pp. 55–112. Elsevier, Oxford, UK.

Ryan, D., and Robards, K. (2006). Analytical chemistry considerations in plant metabolomics. *Sep. Purif. Rev.* **35,** 319–356.

Saito, K., and Matsuda, F. (2010). Metabolomics for functional genomics, systems biology and biotechnology. *Annu. Rev. Plant Biol.* **61,** 463–489.

Saito, K., Dixon, R., and Willmitzer, L. (eds.), (2006). Plant Metabolomics, Springer-Verlag, Berlin.

Salminem, J. P. (2003). Effects of sample drying and storage, and choice of extraction solvent and analysis method on the yield of birch leaf hydrolyzable tannins. *J. Chem. Ecol.* **29,** 1289–1305.

Sanchez, D. H., Siahpoosh, M. R., Roessner, U., Udvardi, M. K., and Kopka, J. (2008). Plant metabolomics reveals conserved and divergent metabolic responses to salinity. *Physiol. Plant.* **132,** 209–219.

Sanchez, D. H., Szymanski, J., Erban, A., Udvardi, M. K., and Kopka, J. (2010). Mining for robust transcriptional and metabolic responses to long-term salt stress: A case study on the model legume *Lotus japonicus*. *Plant Cell Environ.* **33,** 468–480.

Sauter, H., Lauer, M., and Fritsch, H. (1988). Metabolite profiling of plants—A new diagnostic technique. *Abstr. Pap. Am. Chem. Soc.* **195,** 129.

Schad, M., Mungur, R., Fiehn, O., and Kehr, J. (2005). Metabolic profiling of laser microdissected vascular bundles of *Arabidopsis thaliana*. *Plant Methods* **1,** 2.

Schauer, N., Zamir, D., and Fernie, A. R. (2005). Metabolic profiling of leaves and fruit of wild species tomato: A survey of the *Solanum lycopersicum* complex. *J. Exp. Bot.* **56,** 297–307.

Schauer, N., Semel, Y., Roessner, U., Gur, A., Balbo, I., Carrari, F., Pleban, T., Perez-Melis, A., Bruedigam, C., Kopka, J., Willmitzer, L., Zamir, D., *et al.* (2006). Comprehensive metabolic profiling and phenotyping of interspecific introgression lines for tomato improvement. *Nat. Biotechnol.* **24,** 447–454.

Schripsema, J. (2010). Application of NMR in plant metabolomics: Techniques, problems and prospects. *Phytochem. Anal.* **21,** 14–21.

Stobiecki, M., Skirycz, A., Kerhoas, L., Kachliki, P., Muth, D., Einhorn, J., and Mueller-Roeber, B. (2006). Profiling of phenolic glycosidic conjugates in leaves of *Arabidopsis thaliana* using LC/MS. *Metabolomics* **2,** 197–219.

Strehmel, N., Hummel, J., Erban, A., Strassburg, K., and Kopka, J. (2008). Estimation of retention index thresholds for compound matching using routine gas chromatography-mass spectrometry based metabolite profiling experiments. *J. Chromatogr. B* **871,** 182–190.

Sulpice, R., Sienkiewicz-Porzucek, A., Osorio, S., Krahnert, I., Stitt, M., Fernie, A. R., and Nunes-Nesi, A. (2010). Mild reductions in cytosolic NADP-dependent isocitrate dehydrogenase activity result in lower amino acid contents and pigmentation without impacting growth. *Amino Acids* **39,** 1055–1066.

Sumner, L. W. (2006). Current status and forward-looking thoughts on LC/MS metabolomics. *In* "Plant Metabolomics," (K. Saito, R. Dixon, and L. Willmitzer, eds.), Springer-Verlag, Berlin.

Sumner, L. W., Yang, D. S., Bench, B. J., Watson, B. S., Li, C., and Jones, A. D. (2011). Spatially—resolved metabolomics—challenges for the future. *In* "The Biology of Plant Metabolomics," (R. D. Hall, ed.). Blackwell-Wiley, London. (Chapter 11, 343-366).

Tholl, D., Boland, W., Hansel, A., Loreto, F., Rose, U. S. R., and Schnitzler, J. P. (2006). Practical approaches to plant volatile analysis. *Plant J.* **45,** 540–560.

Tikunov, Y., Lommen, A., de Vos, C. H. R., Verhoeven, H. A., Bino, R. J., Hall, R. D., and Bovy, A. G. (2005). A novel approach for nontargeted data analysis for metabolomics. Large-scale profiling of tomato fruit volatiles. *Plant Physiol.* **139,** 1125–1137.

Tikunov, Y., de Vos, C. H. R., Gonzalez-Paramas, A. M., Hall, R. D., and Bovy, A. G. (2010). A role for differential glycoconjugation in the emission of phenylpropanoid volatiles from tomato fruit discovered using a metabolic data fusion approach. *Plant Physiol.* **152,** 55–70.

Timischl, B., Dettmer, K., Kaspar, H., Thieme, M., and Oefner, P. J. (2008). Development of a quantitative, validated capillary electrophoresis-time of flight–mass spectrometry method with integrated high-confidence analyte identification for metabolomics. *Electrophoresis* **29,** 2203–2214.

Tolstikov, V. V., and Fiehn, O. (2002). Analysis of highly polar compounds of plant origin: Combination of hydrophilic interaction chromatography and electrospray ion mass trap spectrometry. *Anal. Biochem.* **301,** 298–307.

Trethewey, R. N. (2004). Metabolite profiling as an aid to metabolic engineering in plants. *Curr. Opin. Plant Biol.* **7,** 196–201.

Trethewey, R. N., Krotzky, A. J., and Willmitzer, L. (1999). Metabolic profiling: A Rosetta stone for genomics? *Curr. Opin. Plant Biol.* **2,** 83–85.

Trygg, J., Gullberg, J., Johansson, A. I., Jonsson, P., and Moritz, T. (2006). Chemometrics in metabolomics—An introduction. *In* "Plant Metabolomics," (K. Saito, R. Dixon, and L. Willmitzer, eds.), pp. 117–128. Springer-Verlag, Berlin.

Urbanczyk-Wochniak, E., and Fernie, A. R. (2005). Metabolic profiling reveals altered nitrogen nutrient regimes have diverse effects on the metabolism of hydroponically-grown tomato (*Solanum lycopersicum*) plants. *J. Exp. Bot.* **56,** 309–321.

Urbanczyk-Wochniak, E., Baxter, C., Kolbe, A., Kopka, J., Sweetlove, L. J., and Fernie, A. R. (2005). Profiling of diurnal patterns of metabolite and transcript abundance in potato (*Solanum tuberosum*) leaves. *Planta* **221,** 891–903.

van Dam, N. M., and van der Meijden, E. (2011). A role for metabolomics in plant ecology. *In* "The Biology of Plant Metabolomics," (R. D. Hall, ed.). Blackwell-Wiley, London. (Chapter 4, 87–108).

van der Hooft, J. J., Vervoort, J., Bino, R. J., Beekwilder, J., and de Vos, R. C. (2011). Polyphenol identification based on systematic and robust high-resolution accurate mass spectrometry fragmentation. *Anal. Chem.* **83,** 409–416.

verhoeven, H. A., de Vos, C. H. R., Bino, R. J., and Hall, R. D. (2006). Plant metabolomics strategies based upon quadrupole time of flight mass spectrometry. *In* "Plant Metabolomics," (K. Saito, R. Dixon, and L. Willmitzer, eds.), pp. 33–48. Springer-Verlag, Berlin.

von Roepenack-Lahaye, E., Degenkolb, T., Zerjeski, M., Franz, M., Roth, U., Wessjohann, L., Schmidt, J., Scheel, D., and Clemens, S. (2004). Profiling of Arabidopsis secondary metabolites by capillary liquid chromatography coupled to electrospray ionization quadrupole time-of-flight mass spectrometry. *Plant Physiol.* **134,** 548–559.

Vorst, O., de Vos, C. H. R., Lommen, A., Staps, R. V., Visser, R. G. F., Bino, R. J., and Hall, R. D. (2005). A non-directed approach to the differential analysis of multiple LC-MS-derived metabolic profiles. *Metabolomics* **1,** 169–180.

Wagner, G. (1995). Basic approaches and methods for quality assurance and quality control in sample collection and storage for environmental monitoring. *Sci. Total Environ.* **176,** 63–71.

Wagner, C., Sefkow, M., and Kopka, J. (2003). Construction and application of a mass spectral and retention time index database generated from plant GC/EI-TOFMS metabolite profiles. *Phytochemistry* **62,** 887–900.

Ward, J. L., and Beale, M. H. (2006). NMR spectroscopy in plant metabolomics. *In* "Plant Metabolomics," (K. Saito, R. Dixon, and L. Willmitzer, eds.), pp. 81–92. Springer-Verlag, Berlin.

Ward, J. L., Harris, C., Lewis, J., and Beale, M. H. (2003). Assessment of 1H-NMR spectroscopy and multivariate analysis as a technique for metabolite fingerprinting of *Arabidopsis thaliana. Phytochemistry* **62,** 949–957.

Ward, J. L., Forcat, S., Beckmann, M., Bennett, M., Miller, S. J., Baker, J. M., Hawkins, N. D., Vermeer, C. P., Lu, C. A., Lin, W. C., Truman, W. M., Beale, M. H., *et al.* (2010a). The metabolic transition during disease following infection of *Arabidopsis thaliana* by Pseudomonas syringae pv. tomato. *Plant J.* **63,** 443–457.

Ward, J. L., Baker, J. M., Miller, S. J., Deborde, C., Maucourt, M., Biais, B., Rolin, D., Moing, A., Moco, S., Vervoort, J., Lommen, A., Schäfer, H., *et al.* (2010b). An inter-laboratory comparison demonstrates that [1H]-NMR metabolite fingerprinting is a robust technique for collaborative plant metabolomic data collection. *Metabolomics* **6,** 263–273.

Weckwerth, W. (ed.), (2007). Metabolomics: Methods and Protocols. Methods in Molecular Biology, Humana Press, Totowa, USAVol. 358.

Yonekura-Sakakibara, K., Tohge, T., Matsuda, F., Nakabayashi, R., Takayama, H., Niida, R., Watanabe-Takahashi, A., Inoue, E., and Saito, K. (2008). Comprehensive flavonol profiling and transcriptome coexpression analysis leading to decoding gene-metabolite correlations in Arabidopsis. *Plant Cell* **20,** 2160–2176.

CHAPTER SEVENTEEN

The Study of Mammalian Metabolism through NMR-based Metabolomics

Reza Salek,* Kian-Kai Cheng,*,† *and* Julian Griffin*,‡

Contents

Abstract

High-resolution NMR spectroscopy has been widely used to monitor metabolism almost since the technique's development. It is now one of the principle technologies used in metabolomics, to profile the metabolite compliment of a cell, tissue, organism, or biofluid. This chapter describes how tissue extracts are prepared for NMR spectroscopy and, in particular, focuses on two approaches based on perchloric acid and methanol/chloroform extractions. This is followed

* Department of Biochemistry, University of Cambridge, Cambridge, United Kingdom
† Department of Bioprocess Engineering, Faculty of Chemical and Natural Resources, Engineering, Universiti Teknologi Malaysia, 81310 UTM Skudai, Johor, Malaysia
‡ The Cambridge Systems Biology Centre, University of Cambridge, Cambridge, United Kingdom

Methods in Enzymology, Volume 500
ISSN 0076-6879, DOI: 10.1016/B978-0-12-385118-5.00017-7

by a description of key NMR experiments that can be used to profile tissue extracts, biofluids, or intact tissues. While these NMR techniques should be optimized for a particular sample set, we provide some tried and tested starting parameters for these experiments which should allow the user to acquire good quality spectra.

1. Introduction

Since the advent of functional genomics, approaches have been developed that aim to profile all the entities contained within a tier of organization within a cell, tissue, or organism. In much the same way that transcriptomics aims to profile all the mRNA found within a biological matrix, and proteomics aims to profile all the proteins within a biological matrix, the aim of metabolomics is to profile all the metabolites that are found in a cell, tissue, organism, or biofluid. The term metabolome was first coined by Oliver *et al.* (1998) and Tweeddale *et al.* (1998) and is closely related to the term metabonome coined by Nicholson *et al.* (1999).

Although no analytical tool can completely profile a metabolome of even the simplest mammalian cell, the approaches are uncovering an ever complicated and expanding set of metabolites that characterize mammalian metabolism. The approaches are finding wide applicability to functional genomics, biomarker discovery, toxicology, and modeling of metabolic networks in all mammalian species that are commonly used in the laboratory as well as in man.

The aim of metabolomics is to measure the relative or absolute concentrations of all the metabolites present. In many metabolomic studies, changes in these concentrations are then used to infer changes in the metabolic flux of specific pathways and biochemical changes related to a genetic or environmental perturbation. However, a danger of this approach is that metabolic fluxes may change with no observable change in the concentration of many of the metabolites present in the cell. To measure changes in metabolic fluxes, stable isotope techniques are commonly used. In these approaches, specific substrates are synthesized replacing ^{12}C with ^{13}C or ^{1}H with ^{2}D, and then either followed across time or the resultant labeling patterns are modeled. These stable isotope techniques are now often referred to as fluxomics (Massou *et al.*, 2007).

Given the relative youth of the field, there are no widely accepted methods for performing metabolomics, and many laboratories develop approaches specific to the problems under investigation. It would be impractical to try to survey all these techniques, and so in this chapter, the intent is to describe commonly used approaches for creating tissue extracts and consider what special precautions should be considered for biofluid analysis, particularly with a focus on the subsequent use of NMR spectroscopy. Two brief sections discuss the general considerations required for performing metabolomic studies by NMR spectroscopy. While the experimental parameters described should not be treated as definitive values, they should provide reasonable starting points for NMR-based metabolomic experiments.

2. Tissue Extraction

The tissue extraction process is one of the most important steps in any tissue-based metabolomic study. If the extraction fails to partition the metabolite of interest into the solvent, fails to quench metabolism sufficiently quickly, or modifies the metabolites to be extracted, then the overall experiment will fail at the first steps. There are a wide range of extraction procedures used in mammalian metabolomics, but we have chosen two of the more commonly used extraction procedures to describe.

Prior to extraction, one must consider how the tissue is to be collected. In animal studies, tissue will most commonly be collected at postmortem. To reduce the effects of postmortem degradation, tissue should be excised and frozen rapidly. The most commonly used approach is to freeze clamp tissues, whereby tongs are chilled in liquid nitrogen and then used to clamp the freshly excised tissue (McIlwain and Bachelard, 1985). However, for small tissue samples (< 100 mg), we have found that placing the tissue in a plastic eppendorf tube and then placing this directly in a dewar of liquid nitrogen also quenches metabolism rapidly. However, this approach is not always rapid enough to quench metabolism. In studies of tumor metabolism, in order to reliably estimate the total pool size of lactate funnel freezing of the animal's brain, liquid nitrogen is poured directly on to the skull of an anesthetized animal (Griffin *et al.*, 2003). One problem with this approach is that it can be difficult to subsequently dissect out specific regions of tissue, and this may require the careful use of hacksaws! Alternatively, death by microwave can be used to denature proteins in the brain but maintains the compliment of metabolites (Melo *et al.*, 2005).

2.1. Perchloric acid extraction

From our previous experience, this method is ideal for aqueous metabolites where metabolism must be quenched rapidly (Griffin *et al.*, 1998). In this respect, it has been widely used in highly oxidative tissues that are prone to postmortem degradation, and, in particular, it has found wide use in neuroscience. The extraction is particularly useful for maintaining the concentration of Adenosine-5′-triphosphate (ATP). However, perchloric acid (PCA) is an oxidizing agent and will oxidize metabolites such as glutathione—all the glutathione in perchloric tissue extracts is found as GSSG rather than a ratio of Glutathione (GSH) to Glutathione disulfide (GSSG). Further, PCA can oxidize proteins and so should not be used for combined metabolomic and proteomic studies.

2.1.1. Materials and equipments

All chemicals are available from a variety of standard chemical suppliers (e.g., from Sigma–Aldrich).

Aqueous 6% PCA, aqueous 1 M KOH, dry ice, or liquid nitrogen.

2.1.2. Tissue extraction procedure

1. During the processing steps described below, the samples are chilled on ice while waiting for further processing.
2. Homogenize frozen samples (20–100 mg wet weight or 10^6–10^7 cells) in 1.0 ml of ice-cold 6% PCA. For this, we favor either a pestle and mortar which has been chilled on dry ice or a tissue lyser (Qiagen supplies a suitable tissue lyser: TissueLyser II) where the sample tray has been chilled on dry ice.
3. To aid tissue extraction, vortex mixing and/or sonication can be used. The final product should look like a fine suspension to ensure the contact of the extraction media with the tissue is optimal. The sample-solvent system is further mixed by vortex mixing (three 20-s pulses) and/or sonication (3–5-min period). At this stage, it is not possible to chill the samples while in the bath of the sonicator, but sufficient mixing should have taken place to denature the proteins present to eliminate enzymatic activity.
4. Centrifugation of the sample-solvent system is performed for 10 min at $16{,}100\times g$.
5. Neutralize each sample to pH 7 with 1 M KOH. While pH paper can be used during the neutralization process, it is advised to use a pH meter with a suitably sized anode for improved accuracy.
6. Centrifugation of the sample-solvent system is performed for 5 min at $16{,}100\times g$.
7. Transfer the supernatant to a new centrifuge tube.
8. Wash the precipitate with 400 μl of distilled water and transfer the wash solution to the supernatant collected in step 7.
9. Repeat step 8.
10. Lyophilize the final supernatant and store at a temperature of less than -20 °C.

2.2. Chloroform/methanol extraction

The PCA approach described above can be quite time consuming as the tissue extracts have to be neutralized individually and the exact amount of KOH solution is determined manually. It also focuses on polar metabolites present in the aqueous fraction. To address these problems, we have favored a modification of the Bligh and Dyer method (1959; and for the modification used Le Belle *et al.*, 2002). No neutralization step is required, and both aqueous and lipophilic metabolites can be extracted simultaneously. Another important advantage is that it avoids oxidation of metabolites and the proteins. Care must be taken not to contaminant the aqueous fraction with the organic fraction during the separation step.

2.2.1. Materials and equipments

All chemicals are available from a variety of standard chemical suppliers (e.g., Sigma–Aldrich).

Analytical grade chloroform, analytical grade methanol, double distilled water, dry ice, or liquid nitrogen.

2.2.2. Tissue extraction procedure

1. During the processing steps described below, the samples are chilled on ice while waiting for further processing.
2. Frozen samples (20–100 mg wet weight or 10^6–10^7 cells) are homogenized in 600 μl of ice-cold chloroform/methanol (1:2 proportion). For this, we favor either a pestle and mortar which has been chilled on dry ice or a Qiagen tissue lyser (Qiagen supplies a suitable tissue lyser: TissueLyser II) where the sample tray has been chilled on dry ice.
3. To aid tissue extraction, vortex mixing and/or sonication can be used. The final product should look like a fine suspension to ensure the contact of the extraction media with the tissue is optimal. The sample-solvent system is further mixed by vortex mixing (three 20-s pulses) and/or sonication (3–5-min period). At this stage, it is not possible to chill the samples while in the bath of the sonicator but sufficient mixing should have taken place to denature the proteins present to eliminate enzymatic activity.
4. Centrifugation of the sample-solvent system is performed for 10 min at $16100 \times g$.
5. Add 200 μl of chloroform and 200 μl of water to the sample-solvent system, vortex mix and then perform centrifugation ($16{,}100 \times g$, 20 min).
6. After centrifugation, the chloroform (lower layer) and aqueous methanol phases (upper layer) are physically separated with a protein pellet at the center. Transfer each phase to separate eppendorfs' labeled chloroform or water–methanol.

7. Reextract the protein pellet using 300 μl of ice-cold chloroform/methanol (1:2 proportion). Repeat steps 5–6 and combine the second step of aqueous methanol and chloroform extracts with the first fractions (this step may be repeated to improve recovery).
8. The final aqueous methanol fraction is lyophilized and is stored at a temperature of less than −20 °C.
9. The final chloroform fraction is dried under a steady stream of nitrogen gas and is stored at a temperature of less than −20 °C.

2.3. NMR spectroscopy of tissue extracts

Tissue extracts are relatively easy to study by ^{1}H NMR spectroscopy. During the extraction process, proteins are often precipitated out of solution and the lipid fraction separated from the aqueous fraction by application of a modified Bligh and Dyer method (see above). For these reasons, the aqueous fraction will produce spectra containing sharp well-defined ^{1}H NMR resonances, whereby metabolites can be identified using information from a combination of chemical shifts and spin-coupling patterns (Bothwell and Griffin, 2011). Depending on the tissue, the number of metabolites that can be quantified range from 20 to 50 metabolites. While numerous one-dimensional pulse sequences are used for collecting data, we favor the NOESYPR1D pulse sequence. As a starting set of parameters, we have employed the pulse sequence on a Bruker Avance III spectrometer interfaced with an 11.7-Tesla superconducting magnet, equipped with a 5-mm TXI ATM probe (Bruker BioSpin GmbH, Rheinstetten, Germany) using 5-mm tubes. Spectra are acquired at a proton frequency of 500.3 MHz, temperature of 27 °C, and with the following pulse sequence parameters: relaxation delay = 2.0 s, t_m = 50 ms, with t_1 fixed at 4 μs. Typically, spectra are acquired with 128 scans collected into 64 K data points with an acquisition time of 4.09 s and spectral width of 16.00 ppm (Salek *et al.*, 2008).

While one-dimensional spectra can have significant congestion, a number of two-dimensional techniques exist which can be used either to confirm assignments or to detect low-concentration metabolites that are coresonant with high-concentration metabolites in one dimension. Two-dimensional techniques include J-resolved spectroscopy (JRES), correlation spectroscopy (COSY), and total correlation spectroscopy (TOCSY) for ^{1}H–^{1}H couplings and heteronuclear single quantum coherence (HSQC) spectroscopy and heteronuclear multiple bond coherence (HMBC) spectroscopy. It is beyond the scope of this chapter to fully describe these forms of spectroscopy, but the interested reader is referred to a review by Braun *et al.* (1998).

In addition to analyzing the aqueous fraction, the lipid fraction can also be analyzed in a similar manner. For this, samples are dissolved in either deuterated chloroform ($CDCl_3$) or a combination of deuterated chloroform and methanol (CD_3OD). If the latter is used, a dual solvent suppression pulse sequence is required to remove the effects of residual protonated methanol and chloroform.

3. Analysis of Biofluids by NMR Spectroscopy

Although the majority of mammalian biofluids have been analyzed by NMR spectroscopy, the most frequently studied biofluids are urine and blood plasma/serum. Urine is rich in a number of high-concentration metabolites and can be analyzed almost immediately with very little sample preparation—some centrifugation may be necessary to remove particulate matter, but as the NMR spectrometer does not come into contact with the sample, the high salt concentrations do not interfere with data acquisition as for mass spectrometry instruments. Both blood plasma and serum can be analyzed. As a word of caution, analyses should not use mixes of these (i.e., some samples as plasma and some samples as serum) as their metabolome differs markedly, particularly in the lipoprotein proportion. One problem with the analysis of serum and plasma by NMR spectroscopy is that a number of chemical shift references, and in particular, TSP, bind to lipoproteins and so may interfere with the resultant spectra. To address this, a two-tube system can be adopted which has worked well in our hands (see below).

3.1. NMR spectroscopy of urine samples

3.1.1. Materials and equipment

All chemicals are available from a variety of standard chemical suppliers (e.g., Sigma–Aldrich). Deuterated water (D_2O), 3-(trimethylsilyl)propionic-2,2,3,3-d_4 acid, sodium salt, 98% D (TSP), Na_2HPO_4, NaH_2PO_4.

NMR tubes—volumes below—are given for Wilmad 5-mm NMR tubes used in a standard 5-mm broad band inverse probe interfaced with the NMR spectrometer. For example, we currently use a 5-mm TXI ATM probe interfaced with a Bruker Advance III spectrometer and an 11.7 T superconducting magnet (Bruker BioSpin GmbH, Rheinstetten, Germany).

3.1.2. Protocol

1. Dilute 200 μl of urine in 400 μl of 0.24 M sodium phosphate buffered D_2O, pH 7.0, containing 1 mM TSP.
2. Remove particulate matter by centrifugation (10 min at 16,100×*g*) and transfer the supernatant to 5-mm NMR tubes. A minimum of 500 μl must be transferred to ensure adequate shimming during the acquisition

process. For samples with significant particulate matter, it may be necessary to add 100 μl of D_2O to ensure an adequate volume.

3. Load NMR tubes into the NMR magnet and acquire data. For ^1H NMR spectroscopy, spectra are acquired using a solvent suppression pulse sequence. We commonly use a method referred to as NOESYPR on Bruker spectrometers based on the start of the NOESY pulse sequence. This pulse sequence has become favored as it ensures a flat baseline even for samples with significant protein content which can affect the magnetic properties of the sample. Typically, values for this pulse sequence are relaxation delay = 2 s, t_1 = 3 μs, mixing time = 150 ms, and solvent presaturation is applied during the relaxation time and the mixing time. Typically, spectra are collected using 128 transients which are collected into 16 K data points over a spectral width of 12 ppm at 37 °C. Other favored solvent suppression pulse sequences are WET and WATERGATE (Braun *et al.*, 1998). However, one must not mix these water suppression pulse sequences as they can affect the resonances around the water peak.

3.2. NMR spectroscopy of blood plasma or serum with one NMR tube

3.2.1. Materials and equipment

All chemicals are available from a variety of standard chemical suppliers (e.g., Sigma–Aldrich). Deuterated water (D_2O), 3-(trimethylsilyl)propionic-2,2,3,3-d_4 acid, sodium salt, 98% D (TSP), NaCl.

NMR tubes—volumes below—are given for Wilmad 5-mm NMR tubes used in a standard 5-mm broad band inverse probe interfaced with the NMR spectrometer. For NMR spectroscopy, we currently use a 5-mm TXI ATM probe interfaced with a Bruker Advance III spectrometer and an 11.7 T superconducting magnet (Bruker BioSpin GmbH, Rheinstetten, Germany). However, a range of NMR spectrometers can be used.

3.2.2. Protocol

1. Dilute 200 μl of blood plasma in 400 μl of 0.9% saline buffered D_2O, pH 7.0. This solution may contain TSP, though some laboratories prefer to leave the TSP out or replace with dimethyl-4-silapentane-1-ammonium trifluoroacetate because of the binding properties of TSP (Alum *et al.*, 2008).
2. Load NMR tubes into the NMR magnet and acquire data. For ^1H NMR spectroscopy, spectra are acquired using a solvent suppression pulse sequence. We commonly use a method referred to as NOESYPR on Bruker spectrometers based on the start of the NOESY pulse sequence. Typically, values for this pulse sequence are relaxation delay = 2 s, t_1 = 3 μs, mixing time = 150 ms, solvent presaturation is applied during the relaxation time and the mixing time. Typically, spectra are collected

using 128 transients which are collected into 16 K data points over a spectral width of 12 ppm at 37 °C (McCombie *et al.*, 2009).

3. To improve the detection of low molecular aqueous metabolites, the solvent suppression pulse sequence can be complimented with a Carr Purcell Meiboom and Gill (CPMG) pulse sequence where a series of spin echoes are used to edit the NMR signal, reducing contributions from lipids (Nicholson *et al.*, 1995). For this, spin echo times are 20–40 ms, consisting of a series of 180° pulses interspersed with delays of 500 μs–1 ms for the spin echo. Solvent suppression can be provided by presaturation or combining the CPMG and NOESYPR1D pulse sequences.
4. To improve the detection of lipids, diffusion-ordered spectroscopy (DOSY) is commonly used to selectively remove fast diffusing (low molecular weight) metabolites. To perform this type of spectroscopy, we have used a probe with a gradient of a 5 G/cm magnetic field gradient applied along the magic angle axis to attenuate metabolite resonances with a standard diffusion Longitudinal Eddy-current Delay (LED) bipolar pulse program (Bruker GmbH, Karlsruhe, Germany). Spectra were acquired for 32 fractional increments of the gradient field, 32 scans per increment with a 50 ms gradient delay (*D*), gradient pulse length of 1 ms (*d*), and a 100-ms gradient recovery time (*t*) (Griffin *et al.*, 2001).

3.3. NMR spectroscopy of blood plasma or serum with a two-tube NMR system

3.3.1. Materials and equipment

All chemicals are available from a variety of standard chemical suppliers (e.g., Sigma–Aldrich). Deuterated water (D_2O), 3-(trimethylsilyl)propionic-2,2,3,3-d_4 acid, sodium salt, 98% D (TSP), NaCl.

NMR tubes—volumes below—are given for Wilmad 5-mm NMR tubes used in a standard 5-mm broad band inverse probe interfaced with the NMR spectrometer. For NMR spectroscopy, we currently use a 5-mm TXI ATM probe interfaced with a Bruker Advance III spectrometer and an 11.7-T superconducting magnet (Bruker BioSpin GmbH, Rheinstetten, Germany). However, a range of NMR spectrometers can be used. 1.7-mm OD capillary tubes (New Era, Vineland, NJ, USA) and ceramic spacers (New Era) are needed.

3.3.2. Protocol

1. In this two-tube system, intact blood serum/plasma (60–80 μl) is loaded into a capillary tube (1.7 mm OD) which is then inserted into an outer 5-mm NMR tube that contains 500 μl D_2O with 0.9%, w/v sodium chloride and 0.1 mM of TSP. A ceramic spacer is used to hold the 1.7-mm tube in place. Data are acquired as defined in Section 3.2.

Figure 17.1 shows a comparison between NMR spectra of a human blood plasma sample, using a two-tube system and conventional single-tube system with and without TSP. Then spectra are acquired using the parameters described in Section 3.2.

4. A Brief Overview of Directly Measuring Metabolites in Mammalian Tissues by High-Resolution Magic Angle Spinning ^{1}H NMR Spectroscopy

Being noninvasive, NMR spectroscopy can be performed on intact tissues and even *in vivo* (here, it is often referred to as magnetic resonance spectroscopy or MRS; Bothwell and Griffin, 2011). A useful half-way house technique between the solution state and *in vivo* MRS spectroscopy is high-resolution magic angle spinning (HRMAS) NMR spectroscopy. In this approach, intact tissue samples are spun at an angle to the magnetic field (54.7°, termed the magic angle). This process of spinning removes the

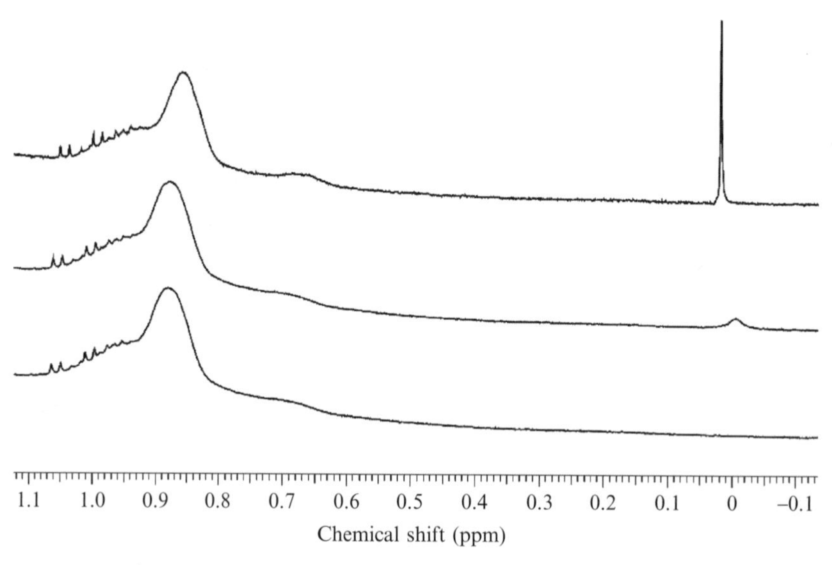

Figure 17.1 ^{1}H NMR spectra (δ-0.10–1.10) of a human blood plasma sample using: (top) a two-tube system, 80 μl blood plasma in capillary, and 0.1 mM TSP in D_2O with 0.9% sodium chloride in outer tube. (Middle) a single-tube system, 100 μl blood plasma was added to 500 μl D_2O (containing 0.9% sodium chloride and 0.1 mM TSP). TSP binds with protein and results in a low and broad peak which is difficult to quantify. (Bottom) Single-tube system, 100 μl blood plasma was added to 500 μl D_2O (containing 0.9% sodium chloride and no TSP). Notice the TSP peak at 0.0 ppm.

effects of dipole coupling and chemical shift anisotropy, both of which would normally cause line broadening for *in vivo* spectroscopy. In this manner, spectra can be obtained with line widths approaching those found in solution state, typically 1–3 Hz.

This approach has been very useful for following metabolism in intact tumors and allows the monitoring of both lipophilic and aqueous metabolites together. Further, spectral editing techniques like DOSY, T_1-, or T_2-weighted spectroscopy can be performed to probe the physical environment of the tumors (Griffin *et al.*, 2003). Because of the high spinning speeds, one must consider the heating effect of the process. However, this can be circumvented by cooling the gas used to drive the spinning of the sample to 4 °C, and spectra can be acquired at just above the freezing point of the tissue.

4.1. Materials and equipment

In order to carry out HRMAS ^{1}H NMR spectroscopy, a HRMAS probe is required along with a pneumatic unit required to spin the sample. In addition, samples are loaded into rotors (e.g., 3 mm zirconium oxide rotors for the Bruker HRMAS probes). Improved spectral quality can be achieved using a Teflon plug to hold the sample in place. For NMR spectroscopy, we currently use a HRMAS ^{1}H NMR probe interfaced with a Bruker Advance III spectrometer and an 11.7-T superconducting magnet (Bruker BioSpin GmbH, Rheinstetten, Germany). However, a range of NMR spectrometers can be used.

Deuterium oxide (D_2O), 3-(trimethylsilyl)propionic-2,2,3,3-d_4 acid, sodium salt, 98% D (TSP) can all be obtained from chemical suppliers (e.g., Sigma–Aldrich).

4.2. Protocol

1. Tissue samples (weighing 5–10 mg) are washed with D_2O to remove excess blood and are then placed into a zirconium oxide MAS rotor alongside 10 μl of D_2O (deuterium lock reference) containing 10 mM 3-trimethylsilyl propionic acid (TSP, chemical shift reference and useful for shimming purposes).
2. HRMAS ^{1}H NMR spectra are acquired at 277 K by chilling the air used for the drive inlet of the HRMAS probe. A conventional solvent suppressed pulse/acquire sequence based on the start of the 2D NOESY pulse sequence is applied. Parameters are TR = 2 s, SW = 10 kHz, 32k data points, 5 kHz spinning rate; water suppression during the mixing time of 150 ms and relaxation delay. The spinning rate is chosen to ensure that the residual spinning side bands of the water resonance are outside the spectral range where metabolites are to be detected.

3. To further characterize the metabolites within the tissue T_1, T_2- and diffusion-weighted spectra can be acquired (Griffin *et al.*, 2001). To observe low molecular weight metabolites, the CPMG pulse sequence is commonly used (Fig. 17.2—see Section 3.2 for further details). For diffusion spectroscopy, we have used a stimulated echo pulse sequence incorporating bipolar gradients with 32 increments of a

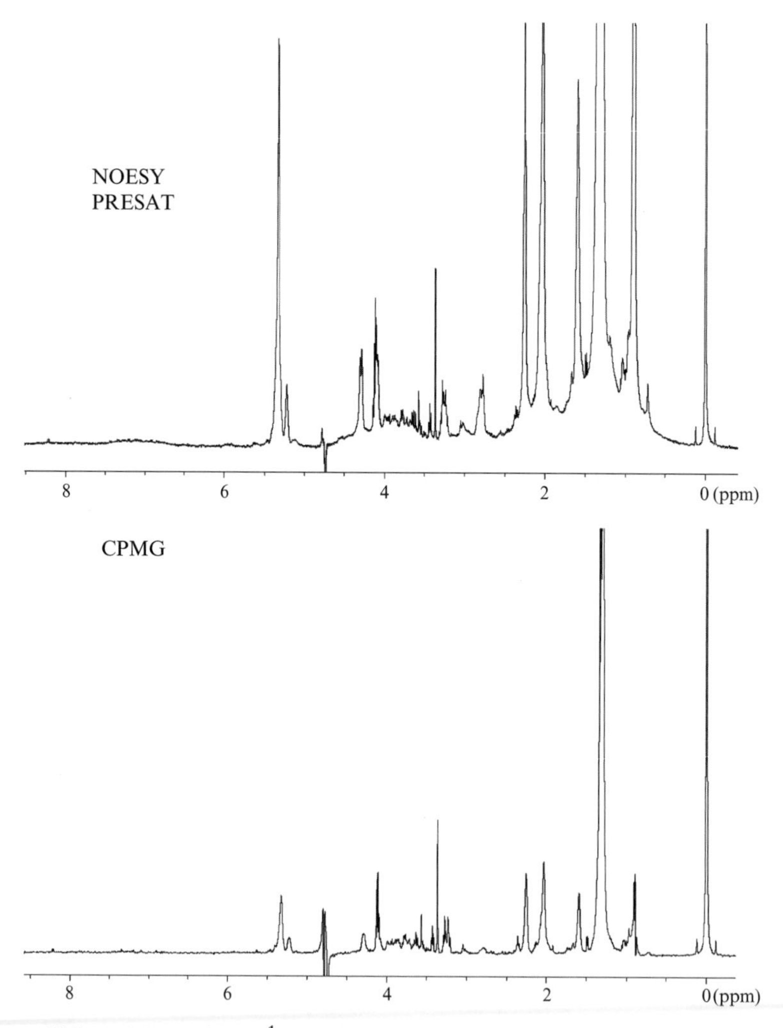

Figure 17.2 High-resolution ^{1}H NMR spectra of a piece of breast tissue (10 mg) using two different pulse sequences. The NOESY PRESAT pulse sequence produces a spectrum with good solvent suppression. The CPMG pulse sequence reduces the effect of lipids contributing to the spectra, allowing the easier detection of aqueous metabolites.

53.1 G/cm field gradient placed along the magic angle axis. Sine-shaped gradients of 2.5 ms ($\delta/2$ for a bipolar sequence) are used with 100 ms intergradient delay yielding Δ of 98.3 ms. The 100-ms delay was repeated three times for each sample. Thirty-two transients were acquired for each increment using a 16k time domain over a spectral width of 8.4 kHz. Diffusion coefficients for metabolites can be calculated by fitting the NMR signal as a function of *b*-value into a single exponential, and all the manufacturers of NMR spectrometers have routines within their software to perform this.

5. Data Processing

In order to carry out further analysis on NMR spectra, it is necessary to convert the spectra into a numerical format. One of the simplest and widely used approaches involves integration between fixed width regions of the NMR spectrum, often referred to as buckets. This approach, while simple, circumvents problems with minor shifts in chemical shift across the dataset and also allows the rapid conversion of spectra into a data matrix suitable for further univariate and multivariate statistics. Most commonly, buckets are set at 0.02–0.04 ppm regions, excluding the water containing region of the spectrum due to its high variability.

One problem with this approach is that resonances can straddle two buckets. To circumvent this, Advanced Chemistry Developments Labs (www.acdlabs.com) have developed an intelligent bucketing routine to ensure peaks fall within buckets. However, if enough NMR spectra are acquired, one can describe the variation in chemical shifts readily with statistics rather than resorting to bucketing. Thus, some use the raw data directly to process spectra at natural resolution (e.g., the STOCSY approach; Cloarec *et al.*, 2005). There are also software products like that produced by Chenomx that use peak fitting to estimate the concentration of metabolites, relying on a large library of standards (Weljie *et al.*, 2006). More recently, methods have been developed for the automatic peak fitting of spectra (Rubtsov and Griffin, 2007).

Finally, as well as needing to process the data, users also have to be able to identify what metabolites are present. There are now a number of NMR spectral databases, including the Human Metabolome Database (HMDB) (Wishart *et al.*, 2007), BioMagResBank (BMRB; http://www.bmrb.wisc.edu/metabolomics/), and Madison Metabolomics Consortium Database (MMCD; Cui *et al.*, 2008). A number of two-dimensional techniques exist which can be used to confirm assignments. Two-dimensional techniques include JRES, COSY, and TOCSY for ^{1}H–^{1}H couplings and HSQC spectroscopy and HMBC spectroscopy. It is beyond the scope of this chapter to fully describe these forms of spectroscopy, but the interested reader is referred to a review by Braun *et al.* (1998).

6. Conclusions

While there are no standardized methods for ^{1}H NMR spectroscopy, we have described a series of extraction procedures and NMR spectroscopic experiments that should provide a good starting place for performing the analysis of total pool sizes by ^{1}H NMR spectroscopy. However, these should be merely seen as starting points, and the user is encouraged to develop and validate these parameters to optimize signal-to-noise and hence the results obtained.

Acknowledgments

Work in the Griffin group is supported by grants from the Medical Research Council (MRC) UK (G0801841), the Biotechnology and Biological Sciences Research Council (BBSRC) UK (BB/H013539/1), European Union Framework 7 (METACANCER) and (INHERITANCE), and the Wellcome Trust (093148/Z/10/Z).

References

Alum, M. F., Shaw, P. A., Sweatman, B. C., Ubhi, B. K., Haselden, J. N., and Connor, S. C. (2008). 4, 4-Dimethyl-4-silapentane-1-ammonium trifluoroacetate (DSA), a promising universal internal standard for NMR-based metabolic profiling studies of biofluids, including blood plasma and serum. *Metabolomics* **4,** 122–127.

Bligh, E. G., and Dyer, W. J. (1959). A rapid method of total lipid extraction and purification. *Can. J. Biochem. Physiol.* **37,** 911–917.

Bothwell, J. H., and Griffin, J. L. (2011). An introduction to biological nuclear magnetic resonance spectroscopy. *Biol. Rev. Camb. Philos. Soc.* **86**(2), 493–510.

Braun, S., Kalinowski, H.-O., and Berger, S. (1998). 150 and More Basic NMR Experiments. A Practical Course. 2nd edn. Wiley-VCH, Weinheim.

Cloarec, O., Dumas, M. E., Craig, A., Barton, R. H., Trygg, J., Hudson, J., Blancher, C., Gauguier, D., Lindon, J. C., Holmes, E., and Nicholson, J. (2005). Statistical total correlation spectroscopy: An exploratory approach for latent biomarker identification from metabolic 1H NMR data sets. *Anal. Chem.* **77**(5), 1282–1289.

Cui, Q., Lewis, I. A., Hegeman, A. D., Anderson, M. E., Li, J., Schulte, C. F., Westler, W. M., Eghbalnia, H. R., Sussman, M. R., and Markley, J. L. (2008). Metabolite identification via the Madison Metabolomics Consortium Database. *Nat. Biotechnol.* **26,** 162.

Griffin, J. L., Rae, C., Dixon, R. M., Radda, G. K., and Matthews, P. M. (1998). Excitatory amino acid synthesis in hypoxic brain slices: Does alanine act as a substrate for glutamate production in hypoxia? *J. Neurochem.* **71**(6), 2477–2486.

Griffin, J. L., Williams, H. J., Sang, E., and Nicholson, J. K. (2001). Abnormal lipid profile of dystrophic cardiac tissue as demonstrated by one- and two-dimensional magic-angle spinning (1)H NMR spectroscopy. *Magn. Reson. Med.* **46**(2), 249–255.

Griffin, J. L., Lehtimäki, K. K., Valonen, P. K., Gröhn, O. H., Kettunen, M. I., Ylä-Herttuala, S., Pitkänen, A., Nicholson, J. K., and Kauppinen, R. A. (2003). Assignment of 1H nuclear magnetic resonance visible polyunsaturated fatty acids in BT4C gliomas undergoing ganciclovir-thymidine kinase gene therapy-induced programmed cell death. *Cancer Res.* **63**(12), 3195–3201.

Le Belle, J. E., Harris, N. G., Williams, S. R., and Bhakoo, K. K. (2002). A comparison of cell and tissue extraction techniques using high-resolution 1H-NMR spectroscopy. *NMR Biomed.* **15**(1), 37–44.

Massou, S., Nicolas, C., Letisse, F., and Portais, J. C. (2007). NMR-based fluxomics: Quantitative 2D NMR methods for isotopomers analysis. *Phytochemistry* **68**(16–18), 2330–2340.

McCombie, G., Browning, L. M., Titman, C. M., Song, M., Shockcor, J., Jebb, S. A., and Griffin, J. L. (2009). Omega-3 oil intake during weight loss in obese women results in remodeling of plasma triglyceride and fatty acids. *Metabolomics* **5**(3), 363–374.

McIlwain, H., and Bachelard, H. S. (1985). Biochemistry and the Central Nervous System. 5th edn. Churchill Livingstone, Edinburgh.

Melo, T. M., Nehlig, A., and Sonnewald, U. (2005). Metabolism is normal in astrocytes in chronically epileptic rats: a 13C NMR study of neuronal-glial interactions in a model of temporal lobe epilepsy. *J. Cereb. Blood Flow Metab.* **25,** 1254–1264.

Nicholson, J. K., Foxall, P. J., Spraul, M., Farrant, R. D., and Lindon, J. C. (1995). 750 MHz 1H and 1H-13C NMR spectroscopy of human blood plasma. *Anal. Chem.* **67**(5), 793–811.

Nicholson, J. K., Lindon, J. C., and Holmes, E. (1999). 'Metabonomics': Understanding the metabolic responses of living systems to pathophysiological stimuli via multivariate statistical analysis of biological NMR spectroscopic data. *Xenobiotica* **29**(11), 1181–1189.

Oliver, S. G., Winson, M. K., Kell, D. B., and Baganz, F. (1998). Systematic functional analysis of the yeast genome. *Trends Biotechnol.* **16**(9), 373–378.

Rubtsov, D. V., and Griffin, J. L. (2007). Time-domain Bayesian detection and estimation of noisy damped sinusoidal signals applied to NMR spectroscopy. *J. Magn. Reson.* **188**(2), 367–379.

Salek, R. M., Colebrooke, R. E., Macintosh, R., Lynch, P. J., Sweatman, B. C., Emson, P. C., and Griffin, J. L. (2008). A metabolomic study of brain tissues from aged mice with low expression of the vesicular monoamine transporter 2 (VMAT2) gene. *Neurochem. Res.* **33**(2), 292–300.

Tweeddale, H., Notley-McRobb, L., and Ferenci, T. (1998). Effect of slow growth on metabolism of Escherichia coli, as revealed by global metabolite pool ("metabolome") analysis. *J. Bacteriol.* **180**(19), 5109–5116.

Weljie, A. M., Newton, J., Mercier, P., Carlson, E., and Slupsky, C. M. (2006). Targeted profiling: Quantitative analysis of 1H NMR metabolomics data. *Anal. Chem.* **78**(13), 4430–4442.

Wishart, D. S., Tzur, D., Knox, C., Eisner, R., Guo, A. C., Young, N., Cheng, D., Jewell, K., Arndt, D., Sawhney, S., Fung, C., Nikolai, L., *et al.* (2007). HMDB: The Human Metabolome Database. *Nucleic Acids Res.* **35**(Database issue), D521–D526.

SECTION SEVEN

MATHEMATICAL MODELLING IN SYSTEMS BIOLOGY

CHAPTER EIGHTEEN

Building a Kinetic Model of Trehalose Biosynthesis in *Saccharomyces cerevisiae*

Kieran Smallbone, Naglis Malys, Hanan L. Messiha, Jill A. Wishart, *and* Evangelos Simeonidis

Contents

Manchester Centre for Integrative Systems Biology, The University of Manchester, Manchester, United Kingdom

Methods in Enzymology, Volume 500
ISSN 0076-6879, DOI: 10.1016/B978-0-12-385118-5.00018-9

Abstract

In this chapter, we describe the steps needed to create a kinetic model of a metabolic pathway based on kinetic data from experimental measurements and literature review. Our methodology is presented by utilizing the example of trehalose metabolism in yeast. The biology of the trehalose cycle is briefly reviewed and discussed.

1. Introduction

The emergent field of systems biology involves the study of the interactions between the components of a biological system and how these interactions give rise to the function and behavior of this system (e.g., the enzymes and metabolites in a metabolic pathway). Nonlinear processes dominate the dynamic behavior of such biological networks, and hence intuitive verbal reasoning approaches are insufficient to describe the resulting complex system dynamics (Lazebnik, 2002; Mendes and Kell, 1998; Szallasi *et al.*, 2006). Nor can such approaches keep pace with the large increases in -omics data (such as metabolomics and proteomics) and the accompanying advances in high-throughput experiments and bioinformatics. Rather, experience from other areas of science has taught us that quantitative methods are needed to develop comprehensive theoretical models for interpretation, organization, and integration of this data. Once viewed with scepticism, we now realize that mathematical models, continuously revised to incorporate new information, must be used to guide experimental design and interpretation.

We focus here on mathematical models of cellular metabolism (Klipp *et al.*, 2005; Palsson, 2006; Wiechert, 2002). In recent years, two major (and divergent) modeling methodologies have been adopted to increase our understanding of metabolism and its regulation. The first is constraint-based modeling (Covert *et al.*, 2003; Price *et al.*, 2004), which uses physicochemical constraints such as mass balance, energy balance, and flux limitations to describe the potential behavior of an organism. The biochemical structure of (at least the central) metabolic pathways is more or less well known, and hence the stoichiometry of such a network may be deduced. In addition, the flux of each reaction through the system may be constrained through, for example, knowledge of its V_{max}, or irreversibility considerations. From the steady-state solution space of all possible fluxes, a number of techniques have been proposed to deduce network behavior, including flux balance and extreme pathway or elementary mode analysis. In particular, flux balance analysis (FBA) (Kauffman *et al.*, 2003) highlights the most effective and efficient paths through the network in order to achieve a particular objective function, such as the maximization of biomass or ATP production.

The key benefit of FBA lies in the minimal amount of biological knowledge and data required to make quantitative inferences about network behavior. This apparent "free lunch" comes at a price, as constraint-based modeling is concerned only with fluxes through the system and makes neither inferences nor predictions about cellular metabolite concentrations. By contrast, kinetic modeling aims to characterize the mechanics of each enzymatic reaction, in terms of how changes in metabolite concentrations affect local reaction rates. However, a considerable amount of data is required to parameterize a mechanistic model; if complex reactions like phosphofructokinase are involved, an enzyme kinetic formula may have 10 or more kinetic parameters (Wiechert, 2002). The determination of such parameters is costly and time consuming, and moreover, many may be difficult or impossible to determine experimentally. The *in vivo* molecular kinetics of some important processes like oxidative phosphorylation and many transport mechanisms are almost completely unknown, so that modeling assumptions about these metabolic processes are necessarily highly speculative.

Because precise kinetic formulas are missing for many enzymes, simplified or phenomenological approaches are used frequently to facilitate modeling. One well-known approach is the power law formalism which uses an exponential expression for each reaction step (Voit, 1991). Alternatives include linlog kinetics, which draws ideas from thermodynamics and metabolic control analysis (Hatzimanikatis and Bailey, 1997; Smallbone *et al.*, 2007) and convenience kinetics, which provides a more realistic approximation to the underlying enzymatic mechanisms (Liebermeister and Klipp, 2006). Finally, thermodynamic flow–force relationships are a way to relate thermodynamics to kinetics (Westerhoff and van Dam, 1987).

In this chapter, we describe the steps needed to create a kinetic model of a metabolic pathway. We use trehalose metabolism in yeast as an example to apply the methodology required for this purpose. But before addressing the mathematics, we need to understand the biology underlying the system.

2. Biological Background

The view of the role of trehalose, extensively studied in baker's yeast, has changed over the years. Its apparent role is to function as a carbohydrate reservoir, but it has now gained new importance as a crucial part of a stabilizing mechanism for proteins and cellular membranes under stress conditions such as heat shock (Crowe *et al.*, 1984; Singer and Lindquist, 1998). Under a stress environment, *Saccharomyces cerevisiae* is able to increase the concentration of trehalose up to 15% of cell dry mass (François and Parrou, 2001). The metabolic pathway that produces trehalose is believed to regulate glucose uptake, particularly when the cell exists in an adverse environment. For the above reasons, and also because of its numerous applications in the cosmetics, food, and pharmaceutical industries, trehalose has become an important biotechnological product. It has also been found that trehalose 6-phosphate (T6P), an intermediate of trehalose biosynthesis, plays a key role in the control of glycolytic flux (Blázquez *et al.*, 1993).

Trehalose is a disaccharide synthesized from two glucose subunits through a pathway elucidated in yeast over 50 years ago (Cabib and Leloir, 1958). The trehalose pathway in yeast consists of a small number of reactions, but these are "arranged" in a metabolic cycle and are governed by a highly complicated regulatory system of inhibitions and/or activations. Due to this complexity, the operation of the pathway is difficult to study experimentally. Glucose is converted into glucose 6-phosphate (G6P), which, together with uridine diphosphate (UDP) glucose, leads to the formation of T6P and, subsequently, trehalose. Trehalose can, in turn, be hydrolyzed into two glucose molecules, thereby closing the *trehalose cycle* (see Fig. 18.1).

2.1. T6P synthase complex

In *S. cerevisiae*, the enzymes that catalyze the reactions of trehalose biosynthesis, T6P synthase (TPS1) and T6P phosphatase (TPS2), form a complex with two other stabilizing, noncatalytic proteins (TSL1 and TPS3). While this complex provides the primary mechanism for trehalose biosynthesis, TPS1 can function independently of the other units of the complex and T6P can be dephosphorylated by other (unspecific) phosphatases (Bell *et al.*, 1998). An important property of this protein complex is its strong temperature activation, with an optimum at around 45 °C.

T6P synthase activity for the complex has affinities for G6P and UDP glucose higher than their concentration in the cells (see Tables 18.1 and 18.2). The enzyme is also strongly noncompetitively inhibited by phosphate. Accordingly, the rate of trehalose synthesis is heavily influenced by changes in both substrate concentration and temperature.

2.2. Response to stress

Cells subjected to stress respond with an interplay of transcriptional and posttranslational changes that lead to a stress resistant state. Trehalose belongs to the early metabolic response, as exposure to various stresses leads to a rapid increase in its concentration. Its primary role is to protect membranes from desiccation, though it can also protect proteins from denaturation in hydrated cells.

In yeast, the transcription factors Msn2p and Msn4p control an environmental stress response (Berry and Gasch, 2008). Upon heat, osmotic shock, oxidative stress, and nutrient starvation, Msn2p/Msn4p are phosphorylated

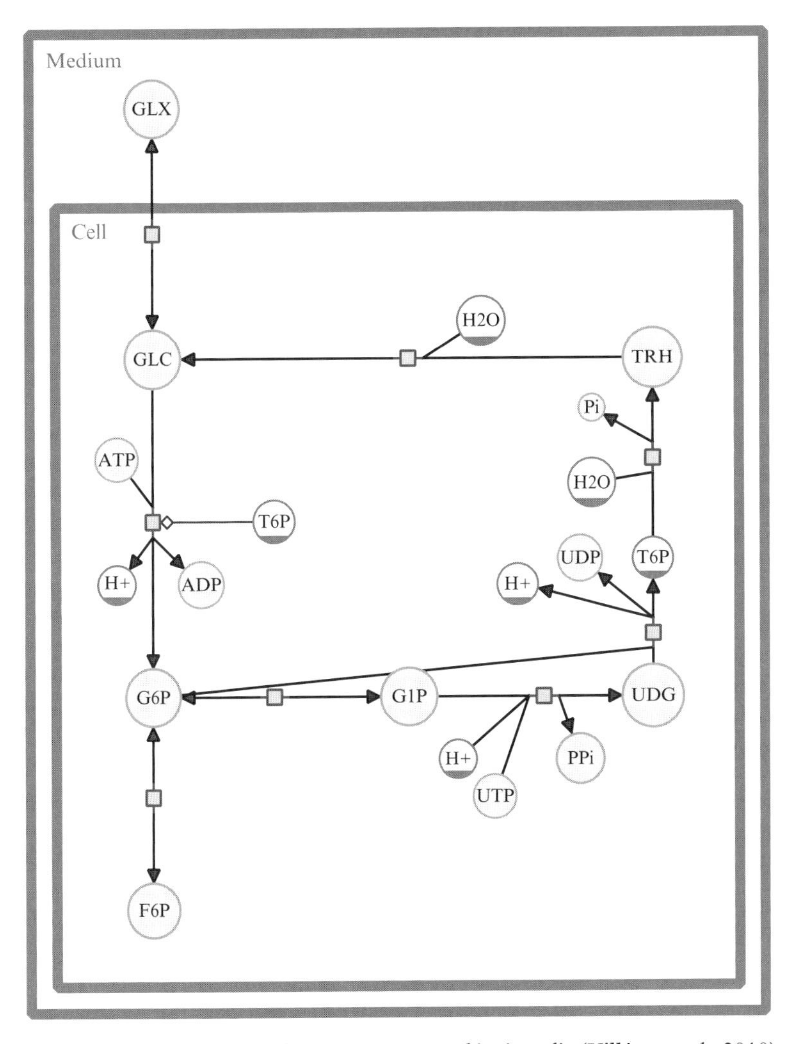

Figure 18.1 The trehalose cycle. Diagram created in Arcadia (Villéger *et al.*, 2010) and made available in Systems Biology Graphical Notation (SBGN, Le Novère *et al.*, 2009) format. Reactive metabolites are marked with a blue circle, ubiquitous metabolites with a yellow circle, reactions with a square, and regulators with a diamond. GLX, external glucose; GLC, glucose; G6P, glucose 6-phosphate; F6P, fructose 6-phosphate; G1P, glucose 1-phosphate; UTP, uridine triphosphate; UDG, UDP–glucose; PPi, diphosphate; T6P, trehalose 6-phosphate; UDP, uridine diphosphate; TRH, trehalose. (For interpretation of the references to color in this figure legend, the reader is referred to the Web version of this chapter.)

by protein kinase A and translocated to the nucleus. Here, these factors bind to the stress response element (STRE) in gene promoters stimulating the expression of a large group of stress response genes.

Table 18.1 Metabolite concentrations used in the model

Metabolite	Concentration (mM)	Reference
ADP	1.282	Pritchard and Kell (2002)
ATP	2.525	Pritchard and Kell (2002)
Fructose 6-phosphate	0.625	Pritchard and Kell (2002)
Glucose (extracellular)	100	Pritchard and Kell (2002)
Glucose (intracellular)	0.09675	Pritchard and Kell (2002)
Glucose 1-phosphate	0.1	Voit (2003)
Glucose 6-phosphate	2.675	Pritchard and Kell (2002)
Trehalose	0.05	Voit (2003)
Trehalose 6-phosphate	0.02	Voit (2003)
UDP	0.2815	–
UDP glucose	0.7	Voit (2003)
UTP	0.6491	–

[UDP] and [UTP] are calculated from [ATP] using ratios ATP:UDP and ATP:UTP of 8.97 and 3.89, respectively (Canelas *et al.*, 2009).

The two best-characterized cases of response to stress—temperature upshift and osmotic shock—show the induction of a wide class of genes including key players of carbohydrate and trehalose metabolism, in particular (Table 18.3). In the case of heat shock, cells subjected to a temperature of 33–38 °C experience a complicated set of changes; the slight accumulation of trehalose is primarily due to transcriptional activation of T6P1. However, if cells are subjected to acute heat shock (over 40 °C), the induction of STRE-containing genes is totally abolished. Rather, the efficacy to accumulate trehalose at high temperature is a consequence of both direct stimulation of T6P synthase complex activity and inhibition of trehalase activity (François and Parrou, 2001).

In addition to environmental stress responses, genes involved in trehalose metabolism can be induced if other gene functions are deficient, for example, scavenger decapping pyrophosphatase (De Mesquita *et al.*, 2003; Malys *et al.*, 2004).

2.3. Interaction with glycolysis

An unexpected link between the trehalose and glycolytic pathways is that mutations in TPS1 prevent cellular growth on glucose (Thevelein and Hohmann, 1995). Within a few seconds of sugar addition, TPS1 mutants accumulate large amounts of hexose monophosphates and fructose 1,6-bisphosphate and show a depletion of ATP and a reduction of intracellular phosphate. This reveals an imbalance between the upper (ATP consuming) and lower (ATP regenerating) parts of glycolysis. The imbalance must be

Table 18.2 Parameter values used in the model

Parameter	Value	Reference
Hexokinase		
K_{GLC}	0.08 mM	Pritchard and Kell (2002)
K_{ATP}	0.15 mM	Pritchard and Kell (2002)
K_{G6P}	30 mM	Pritchard and Kell (2002)
K_{ADP}	0.23 mM	Pritchard and Kell (2002)
K_{iTRH}	0.04 mM	Blázquez *et al.* (1993)
K_{eq}	2000	Pritchard and Kell (2002)
Phosphoglucomutase		
K_{G6P}	0.05 mM	Ray and Roscelli (1964)
K_{G1P}	0.023 mM	Daugherty *et al.* (1975)
K_{eq}	1/6	Wright *et al.* (1977)
UDP–glucose phosphorylase		
K_{UTP}	0.11 mM	Wright *et al.* (1977)
K_{iUTP}	0.11 mM	Wright *et al.* (1977)
K_{G1P}	0.32 mM	Wright *et al.* (1977)
K_{iUDG}	0.035 mM	Guranowski *et al.* (2004)
T6P synthase		
K_{G6P}	3.8 ($\pm$0.57) mM	MCISB
K_{UDPG}	0.886 ($\pm$0.16) mM	MCISB
T6P phosphatase		
K_{T6P}	0.5 mM	Vandercammen *et al.* (1989)
Trehalase		
K_{TRE}	2.99 ($\pm$0.45) mM	MCISB

MCISB enzymes were expressed in *S. cerevisiae* overexpression strains (Gelperin *et al.*, 2005), purified, and quantified as described in Malys *et al.* (2010).

rectified in wild-type yeast either through a decrease in the rate of the initial glycolytic steps or through an increase in substrates (such as free phosphate) to activate the second part of glycolysis.

The primary explanation for the observed phenotype of TPS1 mutants is that control is provided by T6P, which competitively inhibits hexokinase 2 at submillimolar concentrations (Blázquez *et al.*, 1993). But this is rather implausible; one would expect T6P to be channeled within the protein complex synthesizing trehalose and "it appears unlikely that the massive flux through yeast glycolysis would be entirely dependent on fortuitous seeping of T6P from the complex" (Thevelein and Hohmann, 1995). Moreover, a 50-fold overexpression of hexokinase 2 does not lead to a TPS1 null phenotype.

T6P inhibition of hexokinase is not the complete picture of glycolysis regulation by the trehalose cycle. A number of alternatives have been proposed to explain the TPS1 phenotype (Gancedo and Flores, 2004).

Table 18.3 Changes in enzymatic activities and transport steps induced by heat and osmotic shock

Reaction	Fold change in activity upon	
	Heat shock	Osmotic shock
Glucose transport	8	9
Hexokinase	8	15
G6P isomerase	1	1
Phosphoglucomutase	16	5
UDP–glucose phosphorylase	16	11
T6P synthase	12	6
T6P phosphatase	18	20
Trehalase	6	8

Heat shock data from Voit (2003) (processed from the transcriptomics data of Gasch *et al.*, 2000) and osmotic shock data from Rep *et al.* (2000).

It has been suggested that trehalose synthase could recycle inorganic phosphate to avoid a blockage of glycolysis at the glyceraldehyde 3-phosphate dehydrogenase step; but this seems unlikely as glycolysis flux is a 100-fold faster than trehalose formation, making it inconceivable that the pathway could recycle sufficient phosphate. A second proposal suggested that TPS1, hexokinase, and the glucose carrier could form a glucose-sensing complex, but this is very much hypothetical.

It has been suggested that trehalose turnover may function as a glycolytic safety valve (Blomberg, 2000). Comparing glycolysis to the turbo design of a motor, it is predicted that lack of a control mechanism results in a steady rise of sugar phosphates until ATP and phosphates are depleted (Teusink *et al.*, 1998). The recycling of trehalose via a futile ATP cycle could avoid substrate-accelerated death under stress.

Recent evidence suggests that the *TPS1* gene product modulates mitochondrial respiratory content through cAMP and interaction with hexokinase 2 (Noubhani *et al.*, 2009). Assuming this is correct, it makes sense physiologically that both glycolysis and mitochondrial activity are controlled by a common regulatory mechanism, in which when the glycolysis flux is inhibited, the mitochondrial respiration can be activated allowing the balancing of ATP between glycolysis and oxidative respiration.

One experimental feature not explained by any of the above hypotheses is the lack of fermentation in TPS1 mutants, despite their accumulation of hexose phosphates. An interesting suggestion is that the second half of glycolysis is activated by T6P or TPS1. But, again, this is hypothetical, and the only experimentally demonstrated connection found between trehalose biosynthesis and glycolysis is the inhibition of hexokinase by T6P.

3. Model Development

Despite its small size, the trehalose cycle is governed by a surprisingly complex control mechanism, and its interaction with glycolysis is poorly understood. Intuitive, verbal reasoning approaches are insufficient to describe the resulting complex system dynamics. Rather, quantitative methods are needed to develop comprehensive theoretical models for interpretation, organization, and integration of available data. We present here a methodology for building a kinetic model of the trehalose cycle, based on characterizations of the mechanism of each enzymatic reaction in the pathway. The model's kinetic constants were collected through experimentation, literature review, and with the help of a text-mining toolbox, KiPar, developed to retrieve kinetic parameters of interest from publicly available scientific literature (Spasić *et al.*, 2009).

A mathematical description of a kinetic metabolic model may be given in differential equation form as

$$x' = Nv(x, y, p), \quad x(0) = x_0,$$

and this may be used as a guide as to the data required to create and parameterize a kinetic model.

First, N is the stoichiometric matrix, which may be derived easily from the topology of the model (Fig. 18.1)—here derived from a recent genome-scale reconstruction of yeast metabolism (Dobson *et al.*, 2010). Symbol x denotes metabolite concentrations, which for our pathway are the concentrations of glucose, G6P, glucose 1-phosphate, UDP glucose, T6P, and trehalose. Symbol y denotes boundary metabolites, whose concentrations are not allowed to vary but do affect the reaction rates; in our case, these are ADP, ATP, fructose 6-phosphate, extracellular glucose, UDP, and UTP. Concentrations for both x and y must be defined (see Table 18.1), though note that only concentrations x will change over time. Finally, v denotes reaction rates; these are dependent on kinetic mechanisms, concentrations (x and y), and parameters (p)—typically V_{max} and K_m. Further information on the individual reactions is given below. Parameter values may be found in Table 18.2.

3.1. Hexokinase

$$\text{Glucose} + \text{ATP} \rightarrow \text{glucose 6-phosphate} + \text{ADP} + \text{H}^+$$

Following Pritchard and Kell (2002), hexokinase is modeled using bi–bi kinetics; in addition, we allow for the competitive inhibition of glucose by T6P:

$$\frac{v}{V_{\max}} = \frac{\frac{1}{K_{\mathrm{GLC}}K_{\mathrm{ATP}}}\left([\mathrm{GLC}][\mathrm{ATP}] - \frac{[\mathrm{G6P}][\mathrm{ADP}]}{K_{\mathrm{eq}}}\right)}{\left(1 + \frac{[\mathrm{GLC}]}{K_{\mathrm{GLC}}} + \frac{[\mathrm{G6P}]}{K_{\mathrm{G6P}}} + \frac{[\mathrm{T6P}]}{K_{\mathrm{T6P}}}\right)\left(1 + \frac{[\mathrm{ATP}]}{K_{\mathrm{ATP}}} + \frac{[\mathrm{ADP}]}{K_{\mathrm{ADP}}}\right)}.$$

3.2. Phosphoglucomutase

Glucose 6-phosphate $\leftrightarrow$ glucose 1-phosphate

Phosphoglucomutase is modeled using uni–uni kinetics:

$$\frac{v}{V\mathrm{max}} = \frac{\frac{1}{K_{\mathrm{G6P}}}\left([\mathrm{G6P}] - \frac{[\mathrm{G1P}]}{K_{\mathrm{eq}}}\right)}{1 + \frac{[\mathrm{G6P}]}{K_{\mathrm{G6P}}} + \frac{[\mathrm{G1P}]}{K_{\mathrm{G1P}}}}.$$

3.3. UDP–glucose phosphorylase

Glucose 1-phosphate + UTP + H^+ $\rightarrow$ UDP glucose + diphosphate

Following Wright *et al.* (1977), this is modeled using an ordered bi–bi mechanism:

$$\frac{v}{V\mathrm{max}} = \frac{\frac{[\mathrm{UTP}]}{K_{\mathrm{UTP}}}\frac{[\mathrm{G1P}]}{K_{\mathrm{G1P}}}}{\frac{K_{\mathrm{iUTP}}}{K_{\mathrm{UTP}}} + \frac{[\mathrm{UTP}]}{K_{\mathrm{UTP}}} + \frac{[\mathrm{G1P}]}{K_{\mathrm{G1P}}} + \frac{[\mathrm{UTP}]}{K_{\mathrm{UTP}}}\frac{[\mathrm{G1P}]}{K_{\mathrm{G1P}}} + \frac{K_{\mathrm{iUTP}}}{K_{\mathrm{UTP}}}\frac{[\mathrm{UDPG}]}{K_{\mathrm{iUDPG}}} + \frac{[\mathrm{G1P}]}{K_{\mathrm{G1P}}}\frac{[\mathrm{UDPG}]}{K_{\mathrm{iUDPG}}}}.$$

The product inhibition expression used does not include inhibition constants for diphosphate, as this product is assumed to be effectively hydrolyzed *in vivo*.

3.4. T6P synthase

Glucose 6-phosphate + UDP glucose
$\rightarrow$ trehalose 6-phosphate + UDP + H^+

This reaction is irreversible (Cabib and Leloir, 1958). Given this irreversibility and no information regarding the enzymatic mechanism, or product inhibition levels, we assume the reaction to be uninhibited by its products:

$$\frac{v}{V\mathrm{max}} = \frac{\frac{[\mathrm{G6P}]}{K_{\mathrm{G6P}}}\frac{[\mathrm{UDPG}]}{K_{\mathrm{UDPG}}}}{\left(1 + \left[\frac{\mathrm{G6P}]}{K_{\mathrm{G6P}}}\right)\left(1 + \frac{[\mathrm{UDPG}]}{K_{\mathrm{UDPG}}}\right)}.$$

It is clear that such a form will not capture the complexities of this step discussed above. Nonetheless, use of simple kinetics is most appropriate in the absence of clear experimental guidance.

3.5. T6P phosphatase

Trehalose 6-phosphate + water → trehalose + phosphate

T6P phosphatase is not inhibited by trehalose (Vandercammen *et al.*, 1989), so we model the reaction as

$$\frac{v}{V\text{max}} = \frac{\frac{[\text{T6P}]}{K_{\text{T6P}}}}{1 + \frac{[\text{T6P}]}{K_{\text{T6P}}}}.$$

3.6. Trehalase

Trehalose + water → 2 glucose

Trehalase is not inhibited by glucose (Wright *et al.*, 1977), so we model the reaction as

$$\frac{v}{V\text{max}} = \frac{\frac{[\text{TRE}]}{K_{\text{TRE}}}}{1 + \frac{[\text{TRE}]}{K_{\text{TRE}}}}.$$

Two reactions were modeled as per Pritchard and Kell (2002):

3.7. Glucose transport

Glucose[medium] ↔ glucose[cell]

3.8. G6P isomerase

Glucose 6-phosphate ↔ fructose 6-phosphate

3.9. V_{max}

V_{max} parameters have not been defined in Table 18.2. Such parameters typically vary wildly, even within a single species, and are highly dependent upon growth conditions. However, reaction fluxes through glycolysis (89.33 mM/min; Pritchard and Kell, 2002) and trehalose (0.5 mM/min;

Voit, 2003) are known. Using these fluxes and the other known parameter values, V_{max} values may be inferred.

3.10. Systems biology standards

Describing mathematical models as above is unwieldy and error-prone and naturally leads to difficulties in reproduction of results. Thus researchers have developed SBML (the Systems Biology Markup Language, Hucka *et al.*, 2003), a computer-readable format, for representing models of biological processes that is supported by many software packages. SBML can be combined with MIRIAM (the Minimum Information Requested In the Annotation of biochemical Models, Le Novère *et al.*, 2005) to annotate the entities of those models, for example, by marking-up the molecule "GLC" as CHEBI:17925 (http://www.ebi.ac.uk/chebi/searchId.do?chebiId=CHEBI:17925) allows its unambiguous identification and automatically links to many additional sources of information. For more information on these standards, see Krause *et al.* (2010).

The SBML model is available from BioModels.net (Li *et al.*, 2010), a modeling repository (see http://www.ebi.ac.uk/biomodels-main/MODEL1010010000).

4. Results

4.1. Heat shock

Response of the cell to heat shock is simulated by modifying the enzymatic activities and transport steps by the factors defined in Table 18.3. Expression of genes involved in trehalose turnover is stimulated under many types of environmental change. However, increase in production of trehalose is only measurable under specific conditions (e.g., heat shock). This suggests that the mRNA is not always translated into protein (and active enzyme). Moreover, the level of protein synthesis does not always correlate with transcription. Therefore, the increase in mRNA level and the increase in protein activity are incorporated into these values.

Results are presented in Table 18.4: we see a large increase in the levels of G6P and trehalose. This is in agreement with the previously measured 10- to 17-fold increase in trehalose when cells were grown under heat shock conditions at 36 °C (Hottiger *et al.*, 1987; Ribeiro *et al.*, 1994).

4.2. TPS1 mutant

The effects of knocking out TPS1 are simulated by reducing TPS1 activity to 1% of its normal level. Results are presented in Table 18.4: we see a build-up of some intermediates in the cycle (G6P and UDP glucose) and a

Table 18.4 Response of model concentrations to heat shock and TPS1 mutation

Metabolite	Heat shock	TPS1 mutant
Glucose	0.121	−0.146
Glucose 6-phosphate	0.728	0.00213
Glucose 1-phosphate	0.0787	0.642
UDP glucose	−0.0636	2.05
Trehalose 6-phosphate	0.0718	−1.67
Trehalose	0.565	−1.66

Results are presented as $\log([X]_{ss}/[X]_0)$, so a value of 2 refers to a $10^2 = 100$-fold increase in concentration.

drop in concentration of others (T6P and trehalose). A minor increase in G6P is observed, suggesting that the 50-fold decrease in T6P concentration (from 20 to 0.4 μM) may only have a minor affect on the hexokinase activity (the observed K_i of T6P on hexokinase 1 and hexokinase 2 are 200 and 40 μM, respectively, Blázquez *et al.*, 1993). This further questions the possibility of the inhibition of hexokinase by T6P.

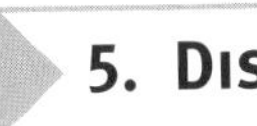

5. Discussion

Trehalose synthesis and hydrolysis have been established as an underlying molecular process for stress survival and adaptation to the environmental changes in *S. cerevisiae* and other microorganisms (Crowe *et al.*, 1984; Hottiger *et al.*, 1987; Singer and Lindquist, 1998). Discovered decades ago, its relationship with key biochemical processes involving carbohydrate and energy metabolism has not been clearly elucidated yet (Blomberg, 2000; François and Parrou, 2001; Gancedo and Flores, 2004; Noubhani *et al.*, 2009; Thevelein and Hohmann, 1995).

There is a need to further develop models of cellular metabolic processes and to analyze those with new tools and approaches derived from the systems biology perspective. By admitting that metabolism is truly a systemic process, one may begin to understand its emergent behavior as more than the sum of its constituent parts. Moreover, through providing a theoretical framework to which the vast array of available metabolic data may be fused, one may begin to uncover the nonlinear interactions that govern its complexities.

In the case of trehalose metabolism, a theoretical approach has got us some way to understanding the interactive dynamics of the pathway. In this chapter, we demonstrate how to generate a kinetic model and use it to investigate the effects of the heat shock and TPS1 mutation on trehalose metabolism. In the case of heat shock, the calculated increase in concentrations of intermediate metabolites reflects the previously measured elevation of glucose uptake and enzyme activities of trehalose metabolism (Ribeiro *et al.*, 1994). Moreover, the large response of trehalose concentration supports experimental observations (Hottiger *et al.*, 1987; Ribeiro *et al.*, 1994). Intriguingly, the key intermediate at a crossroad of glycolysis, pentose phosphate, glycogen, and trehalose pathways—G6P—is also greatly elevated, suggesting its critical role in stress response.

The use of our model to investigate the effect of TPS1 mutation showed an expected increase in the concentration of metabolites upstream to T6P, that is, G6P and UDP glucose and a decrease in the concentration of downstream derivatives, that is, T6P, trehalose, and glucose. In addition, our model suggests that the intracellular concentration of T6P is insufficient to reduce significantly the hexokinase activity, as it has been postulated previously (Blázquez *et al.*, 1993).

Due to the scope of this chapter, we limit ourselves to the kinetic description of a single pathway. Nevertheless, it would be an intriguing topic of future research to combine the model constructed here with a kinetic model of glycolysis, as there is a direct link through hexokinase and carbohydrate intermediates such as G6P and/or to extend the model to include glycogen (interesting for the TPS1 mutant). Another interesting extension of the model would be the addition of phosphate dynamics, which could influence the concentration of inorganic phosphate (inhibitor of T6P synthase) and the consumption and regeneration of ATP in the cell.

In this chapter, we have presented the necessary steps for constructing a kinetic model of a metabolic pathway, and examples of the kinds of analyses that can be performed with such a model. Application of such methodologies can help us decipher metabolic behavior, especially when combined with laboratory experiments that will test and verify the simulation results, and suggest new directions and hypotheses for the study of the pathway of interest.

ACKNOWLEDGMENTS

We are grateful for the financial support of the BBSRC and EPSRC through grant BB/C008219/1 "The Manchester Centre for Integrative Systems Biology (MCISB)." We also thank Michael Howard for his invaluable oversight, and our MCISB colleagues.

REFERENCES

Bell, W., *et al.* (1998). Composition and functional analysis of the *Saccharomyces cerevisiae* trehalose synthase complex. *J. Biol. Chem.* **273,** 33311–33319.

Berry, D. B., and Gasch, A. P. (2008). Stress-activated genomic expression changes serve a preparative role for impending stress in yeast. *Mol. Biol. Cell* **19,** 4580–4587.

Blázquez, M. A., *et al.* (1993). Trehalose-6-phosphate, a new regulator of yeast glycolysis that inhibits hexokinases. *FEBS Lett.* **329,** 51–54.
Blomberg, A. (2000). Metabolic surprises in *Saccharomyces cerevisiae* during adaptation to saline conditions: Questions, some answers and a model. *FEMS Microbiol. Lett.* **182,** 1–8.
Cabib, E., and Leloir, L. F. (1958). Biosynthesis of trehalose phosphate. *J. Biol. Chem.* **231,** 259–275.
Canelas, A. B., *et al.* (2009). Quantitative evaluation of intracellular metabolite extraction techniques for yeast metabolomics. *Anal. Chem.* **81,** 7379–7389.
Covert, M. W., Famili, I., and Palsson, B.Ø. (2003). Identifying constraints that govern cell behavior: A key to converting conceptual to computational models in biology? *Biotechnol. Bioeng.* **84,** 763–772.
Crowe, J. H., Crowe, L. M., and Chapman, D. (1984). Preservation of membranes in anhydrobiotic organisms—The role of trehalose. *Science* **223,** 701–703.
Daugherty, J. P., Kraemer, W. F., and Joshi, J. G. (1975). Purification and properties of phosphoglucomutase from Fleischmann's yeast. *Eur. J. Biochem.* **57,** 115–126.
De Mesquita, J. F., Panek, A. D., and de Araujo, P. S. (2003). *In silico* and *in vivo* analysis reveal a novel gene in *Saccharomyces cerevisiae* trehalose metabolism. *BMC Genomics* **4,** 45.
Dobson, P. D., *et al.* (2010). Further developments towards a genome-scale metabolic model of yeast. *BMC Syst. Biol.* **4,** 145.
François, J., and Parrou, J. L. (2001). Reserve carbohydrates metabolism in the yeast *Saccharomyces cerevisiae*. *FEMS Microbiol. Rev.* **25,** 125–145.
Gancedo, C., and Flores, C. L. (2004). The importance of a functional trehalose biosynthetic pathway for the life of yeasts and fungi. *FEMS Yeast Res.* **4,** 351–359.
Gasch, A. P., *et al.* (2000). Genomic expression programs in the response of yeast cells to environmental changes. *Mol. Biol. Cell* **11,** 4241–4257.
Gelperin, D. M., *et al.* (2005). Biochemical and genetic analysis of the yeast proteome with a movable ORF collection. *Genes Dev.* **19,** 2816–2826.
Guranowski, A., *et al.* (2004). Uridine 5'-polyphosphates (p(4)U and p(5)U) and uridine(5') polyphospho(5')nucleosides (Up(n)Ns) can be synthesized by UTP: Glucose-1-phosphate uridylyltransferase from Saccharomyces cerevisiae. *FEBS Lett.* **561,** 83–88.
Hatzimanikatis, V., and Bailey, J. E. (1997). Effects of spatiotemporal variations on metabolic control: Approximate analysis using (log)linear kinetic models. *Biotechnol. Bioeng.* **54,** 91–104.
Hottiger, T., Schmutz, P., and Wiemken, A. (1987). Heat-induced accumulation and futile cycling of trehalose in Saccharomyces cerevisiae. *J. Bacteriol.* **169,** 5518–5522.
Hucka, M., *et al.* (2003). The Systems Biology Markup Language (SBML): A medium for representation and exchange of biochemical network models. *Bioinformatics* **19,** 524–531.
Kauffman, K. J., Prakash, P., and Edwards, J. S. (2003). Advances in flux balance analysis. *Curr. Opin. Biotechnol.* **14,** 491–496.
Klipp, E., *et al.* (2005). Systems Biology in Practice: Concepts, Implementation and Application. Wiley-VCH, Weinheim.
Krause, F., *et al.* (2011). Sustainable modelling in systems biology: The roles of standards and semantic annotations. *Methods Enzymol.* **500,** (this issue).
Lazebnik, Y. (2002). Can a biologist fix a radio? Or, what I learned while studying apoptosis. *Cancer Cell* **2,** 179–182.
Le Novère, N., *et al.* (2005). Minimum Information Required In the Annotation of Models (MIRIAM). *Nat. Biotechnol.* **23,** 1509–1515.
Le Novère, N., *et al.* (2009). The systems biology graphical notation. *Nat. Biotechnol.* **27,** 735–741.
Li, C., *et al.* (2010). BioModels Database: An enhanced, curated and annotated resource for published quantitative kinetic models. *BMC Syst. Biol.* **4,** 92.

Liebermeister, W., and Klipp, E. (2006). Bringing metabolic networks to life: Convenience rate law and thermodynamic constraints. *Theor. Biol. Med. Model.* **3,** 41.
Malys, N., *et al.* (2004). The 'scavenger' m7G pppX pyrophosphatase activity of Dcs1 modulates nutrient-induced responses in yeast. *Nucleic Acids Res.* **32,** 3590–3600.
Malys, N., *et al.* (2011). Protein production in *S. cerevisiae* for systems biology studies. *Methods Enzymol.* **500,** (in this volume).
Mendes, P., and Kell, D. (1998). Non-linear optimization of biochemical pathways: Applications to metabolic engineering and parameter estimation. *Bioinformatics* **14,** 869–883.
Noubhani, A., *et al.* (2009). The trehalose pathway regulates mitochondrial respiratory chain content through hexokinase 2 and cAMP in Saccharomyces cerevisiae. *J. Biol. Chem.* **284,** 27229–27234.
Palsson, B.Ø. (2006). Systems Biology: Properties of Reconstructed Networks. Cambridge University Press, Cambridge.
Price, N. D., Reed, J. L., and Palsson, B.Ø. (2004). Genome-scale models of microbial cells: Evaluating the consequences of constraints. *Nat. Rev. Microbiol.* **2,** 886–897.
Pritchard, L., and Kell, D. B. (2002). Schemes of flux control in a model of *Saccharomyces cerevisiae* glycolysis. *Eur. J. Biochem.* **269,** 3894–3904.
Ray, W. J., and Roscelli, G. A. (1964). Phosphoglucomutase pathway—Investigation of phospho-enzyme isomerization. *J. Biol. Chem.* **239,** 3935.
Rep, M., *et al.* (2000). The transcriptional response of *Saccharomyces cerevisiae* to osmotic shock. Hot1p and Msn2p/Msn4p are required for the induction of subsets of high osmolarity glycerol pathway-dependent genes. *J. Biol. Chem.* **275,** 8290–8300.
Ribeiro, M. J. S., Silva, J. T., and Panek, A. D. (1994). Trehalose metabolism in *Saccharomyces cerevisiae* during heat-shock. *Biochim. Biophys. Acta* **1200,** 139–147.
Singer, M. A., and Lindquist, S. (1998). Multiple effects of trehalose on protein folding in vitro and in vivo. *Mol. Cell* **1,** 639–648.
Smallbone, K., *et al.* (2007). Something from nothing—Bridging the gap between constraint-based and kinetic modelling. *FEBS J.* **274,** 5576–5585.
Spasić, I., *et al.* (2009). KiPar, a tool for systematic information retrieval regarding parameters for kinetic modelling of yeast metabolic pathways. *Bioinformatics* **25,** 1404–1411.
Szallasi, Z., Stelling, J., and Periwal, V. (2006). System Modeling in Cellular Biology: From Concepts to Nuts and Bolts. MIT Press, Boston.
Teusink, B., *et al.* (1998). The danger of metabolic pathways with turbo design. *Trends Biochem. Sci.* **23,** 162–169.
Thevelein, J. M., and Hohmann, S. (1995). Trehalose synthase—Guard to the gate of glycolysis in yeast. *Trends Biochem. Sci.* **20,** 3–10.
Vandercammen, A., François, J., and Hers, H. G. (1989). Characterization of trehalose-6-phosphate synthase and trehalose-6-phosphate phosphatase of Saccharomyces cerevisiae. *Eur. J. Biochem.* **182,** 613–620.
Villéger, A. C., Pettifer, S. R., and Kell, D. B. (2010). Arcadia: A visualization tool for metabolic pathways. *Bioinformatics* **26,** 1470–1471.
Voit, E. O. (1991). Canonical Nonlinear Modeling: S-system Approach to Understanding Complexity. Van Nostrand Reinhold, New York.
Voit, E. O. (2003). Biochemical and genomic regulation of the trehalose cycle in yeast: Review of observations and canonical model analysis. *J. Theor. Biol.* **223,** 55–78.
Westerhoff, H. V., and van Dam, J. C. (1987). Mosaic Nonequilibrium Thermodynamics and Control of Biological Free-Energy Transduction. Elsevier, Amsterdam.
Wiechert, W. (2002). Modeling and simulation: Tools for metabolic engineering. *J. Biotechnol.* **94,** 37–63.
Wright, B. E., Tai, A., and Killick, K. A. (1977). 4th expansion and glucose perturbation of dictyostelium kinetic-model. *Eur. J. Biochem.* **74,** 217–225.

CHAPTER NINETEEN

Sustainable Model Building: The Role of Standards and Biological Semantics

Falko Krause,* Marvin Schulz,* Neil Swainston,† *and* Wolfram Liebermeister‡

Contents

* Max Planck Institute for Molecular Genetics, Dep. Computational Molecular Biology, Berlin, Germany
† Manchester Centre for Integrative Systems Biology, University of Manchester, United Kingdom
‡ Weizmann Institute of Science, Department of Plant Sciences, Rehovot, Israel

Methods in Enzymology, Volume 500
ISSN 0076-6879, DOI: 10.1016/B978-0-12-385118-5.00019-0

Abstract

Systems biology models can be reused within new simulation scenarios, as parts of more complex models or as sources of biochemical knowledge. Reusability does not come by itself but has to be ensured while creating a model. Most important, models should be designed to remain valid in different contexts—for example, for different experimental conditions—and be published in a standardized and well-documented form. Creating reusable models is worthwhile, but it requires some efforts when a model is developed, implemented, documented, and published. Minimum requirements for published systems biology models have been formulated by the MIRIAM initiative. Main criteria are completeness of information and documentation, availability of machine-readable models in standard formats, and semantic annotations connecting the model elements with entries in biological Web resources. In this chapter, we discuss the assumptions behind bottom-up modeling; present important standards like MIRIAM, the Systems Biology Markup Language (SBML), and the Systems Biology Graphical Notation (SBGN); and describe software tools and services for handling semantic annotations. Finally, we show how standards can facilitate the construction of large metabolic network models.

1. Sustainable Model Building

Computational models are indispensable for analyzing complex biological data, simulating and predicting the behavior of cells, and rationally engineering organisms in biotechnology. Models need to be not only complex enough to provide a detailed and quantitative picture of biochemical systems but also simple enough to be understandable, numerically tractable, and supported by available data. The resulting compromise will depend on the available data and on the purpose of the model. Although complex models covering all main constituents of a cell are a long-term aim of systems biology, it is practical to start with simple models. There are two complementary approaches: top-down modeling and bottom-up modeling. In top-down modeling, one first considers a coarse-grained overall model which is then gradually refined. Bottom-up modeling, on the contrary, starts with smaller models of biochemical pathways or even individual reactions, which are later combined to create more comprehensive models. A prototypical example for the latter, the glycolysis model built from enzymatic rate laws measured *in vitro*, has been presented in Teusink *et al.* (2000).

We can therefore assume that a pathway model does not only have a value in itself but could potentially be reused for bottom-up modeling. Moreover, models may be modified for simulating new experiments or may serve as a collection of biochemical knowledge. To enable a later reuse, modelers have to foresee and prepare this while creating and publishing their models. Such "sustainable model building" will be the topic of this chapter.

There are two main requirements for model reuse: on the one hand, the model elements (e.g., reaction kinetics) should remain valid in other contexts. In general, *realistic* models that faithfully describe all relevant processes are better reusable than *pragmatic* black-box models that are just meant to fit available data irrespective of biochemical mechanisms. In the latter case, the model elements will not have a clear meaning and can hardly be used for any other purposes. On the other hand, model reuse requires that the models are available and easy to use. This involves a clear and complete documentation and the use of standardized, preferably computer-readable formats, allowing other scientists to easily understand the models and to process them with their computer tools. Important requirements for published models have been formulated by the MIRIAM (Minimum Information Required in the Annotation of Models) initiative (Le Novère *et al.*, 2005).

2. How to Create Reusable Models

2.1. The bottom-up modeling paradigm

Bottom-up modeling, the approach of building complex models from smaller preexisting parts, requires that model elements will remain valid even if they are placed in a new context. This may in fact be expected if a model is *realistic*, describing all relevant processes faithfully and to sufficient detail, if all interactions between coupled systems are covered, and if the biochemical mechanisms themselves will not be drastically affected by the different context. In fact, without any such assumptions, modeling would be of limited use. However, the way models are built in practice—including fitting to limited data, pruning of elements that are not supported by the present data—may not always lead to reliable realistic models.

Typical examples are the values of fitted parameters which usually depend on the model in which they were estimated. If we modify a part of a model and reestimate all parameters, this will usually change all parameter values, not just those in the modified part. Thus, the parameter values are not a faithful picture of reality but pragmatically chosen to yield the best results in the context of the model. In general, some of the estimated parameter values will be trustworthy, others will be not. Some parameters may be nonidentifiable, because different parameter values (or combinations) fit the data equally well. This may not play a role for some forward simulations, but it can become problematic if we are interested in the numerical parameter values or if the model is reused in a different context.

A similar problem occurs if parameters, although mathematically identifiable, do not represent the quantity the modeler had in mind. This can happen quite easily: for instance, imagine a cell in which two enzymes, A and B, are acting on the same substrate. If we fit a model containing only enzyme A to the experimental data, the enzyme activity will be overestimated because it has to account for both enzymes in the real cell. Again, such parameters cannot be used for bottom-up modeling and have to be refitted after the model has been modified or merged with other models.

2.2. Guidelines for creating reusable models

Model reusability needs to be considered during all stages of model building. The following guidelines can help:

- *Study autonomous biological systems.* In bottom-up modeling, we assume autonomous modules that interact (communicate) via a few known connections. Of course, this approach works best if the biological systems studied *do* behave like modules. Sometimes, this can be supported by carefully choosing the system under study. A prominent example is synthetic biology. When building artificial genetic circuits, researchers transfect cells with plasmids coding for proteins from a different organism. If this transfer of modules between cells works experimentally, such systems should also be good candidates for bottom-up modeling. In contrast, when modeling biochemical pathways *in vivo*, we are facing the full complexity of the cell. Therefore, we should make sure that connections to other pathways play at least a minor role—for example, currency metabolites like ATP should show relatively constant levels in the experiment—which then justifies our model of an isolated pathway.
- *Standard operation procedures for experiments.* Standard experimental conditions (e.g., fixed growth rates and temperatures, defined growth media, and microbial strains) should be chosen early in a project. On the one hand, this will help to cover different experiments by the same model. On the other hand, models developed from different experiments will be more likely to be compatible.
- *Choose a widely used mathematical formalism.* Most systems biology models are formulated in one of the major paradigms: Boolean or qualitative models, kinetic models (ordinary differential equations), constraint-based flux models, spatial models (partial differential equations), or stochastic simulations of particle numbers. Since coupling models from different formalisms is usually difficult, the choice of a well-accepted formalism is helpful for reuse.
- *Adopt conventions from existing models.* Network reconstructions define conventions about names, identifiers, and the level of details to which pathways are modeled. Adopting such existing conventions in other models can improve reusability.
- *Check model parameters for identifiability.* After fitting a model, confidence ranges for the parameters and problems with parameter identifiability should be reported. This may remind other modelers to mistrust and reestimate problematic parameters if necessary.
- *Document which circumstances are irrelevant for the model.* A computational model does not only describe biochemical processes, but it also implies—

usually implicitly—that other processes, which are not part of the model, are less relevant. Such negative statements can become important when the model is used for bottom-up modeling. Therefore, if certain factors are known to be irrelevant for the process, it is good to state this explicitly in a publication.

3. How to Implement Models in Standard Formats

3.1. Why respect standards?

The wish that models should be clearly described and easily exchangeable has led to standardization efforts like MIRIAM (Le Novère *et al.*, 2005; www.ebi.ac.uk/miriam/), Systems Biology Markup Language (SBML; Hucka *et al.*, 2003; www.sbml.org), and Systems Biology Graphical Notation (SBGN; Le Novère *et al.*, 2009; www.sbgn.org), and to model repositories like the BioModels Database (Le Novère *et al.*, 2006; www.ebi.ac.uk/biomodels-main/), which collects and curates computer-encoded versions of published models. These tools and resources are presented below. Every modeler can participate in the "sustainable modeling" movement by using these resources and respecting some rules of good practice. The extra effort for constructing and annotating a model—and for restricting yourself to common formats and conventions—will pay off in the long run because models that are well accessible will also have a higher impact. Several scientific journals already request the use of standard formats like SBML, compliance to the MIRIAM rules, and sometimes submission to the BioModels Database as preconditions for publication. Finally, encoding and annotating a model in standard formats may help to spot or avoid inconsistencies in the model that would otherwise remain unnoticed.

Despite all advantages, it is not always possible to adhere to standards. By definition, standards restrict the room of individuals and tend to be a least common denominator. The SBML format, for instance, has been developed as an exchange format and implements features that are shared by many software tools. As the field of systems biology progresses, new modeling approaches will typically go beyond the scope of existing standards and result in novel features that are not yet supported by SBML.

3.2. Systems Biology Markup Language

The SBML (Hucka *et al.*, 2003) is an XML-based exchange format for structural and dynamic biological models. As SBML is meant to be readable by many different modeling tools, this language represents the minimum information needed to understand the model's structure and to simulate its

Table 19.1 Description of SBML core elements

Element	Meaning	Important attributes
Compartments	Physical regions	Size
Species	Molecules in a certain compartment	InitialAmount InitialConcentration Compartment
Reactions	Processes converting species	Reactants Products Modifiers KineticLaw
Parameters	Numerical values	Value
Rules	Definitions for parameters	Variable
Events	Conditional rules	Trigger EventAssignments

dynamic behavior. Individual tools may support additional features, which depend on the tools' native data formats.

SBML represents the elements of a biochemical model by nested XML elements. We describe some of the key elements in Table 19.1, while for the others, the reader is referred to the documentation at http://sbml.org.

3.3. Systems Biology Graphical Notation

The SBGN (Le Novère *et al.*, 2009) is a standard for the visualization of biological network models. It is intended to describe a system in detail and without any ambiguities to a human reader. Further, SBGN diagrams can be written by computer programs, for example, for converting the structure of the reactions in a model and its semantic annotations into graphics. SBGN consists of the three independent diagrammatic languages shown in Fig. 19.2. They allow to represent the information in a model in different ways:

- The *Process Description Language* (PDL) describes biochemical reactions responsible for the system's temporal behavior. The species and reactions appear as nodes connected by arcs. The symbols (called "glyphs") carry semantic information; different glyphs stand for certain types of species, for example, small chemical entities, genes, or proteins. It is also possible to visualize protein complexes as a set of its constituents. Also the arcs can carry semantic information about a reaction, for instance, whether a reaction is reversible, or describe the specific role of a modifier.
- The *Entity Relationship Language* (ERL) neglects the temporal behavior of the model and focuses on the relations, for example, interactions of proteins that are shared between species. In contrast to the reactions

considered in the PDL, relations are supposed to represent independent information and, since they do not affect each other, they can be extracted from different sources (experiments/articles). This independence avoids combinatorial problems, which can, for example, arise in the PDL if a protein can be modified at many different sites and all potential modification state are modeled as separate species.

- The *Activity Flow Language* (AFL) visualizes how information is propagated through a biological system, while the conversion of any kind of matter is neglected. Therefore, nodes do not describe entities, but their activities, while edges define the effects these activities have on each other. Again, semantic information on the activity is given by the shape of the node, for example, the activity of a gene and the corresponding protein carry identical labels, but differ in their shapes (Fig. 19.1).

Although the three languages cover completely different aspects of a model, they also share common elements. One example is the visualization of physical compartments, which is identical and has the same restrictions in all languages.

3.4. Modeling tools and model repositories

SBML is currently supported by more than 200 tools (http://sbml.org; November 2010). These tools fall under various different categories, and some of them can also be used for multiple purposes.

- *Model development tools.* These are used to create new SBML models from scratch, for example, COPASI (Hoops *et al.*, 2006).

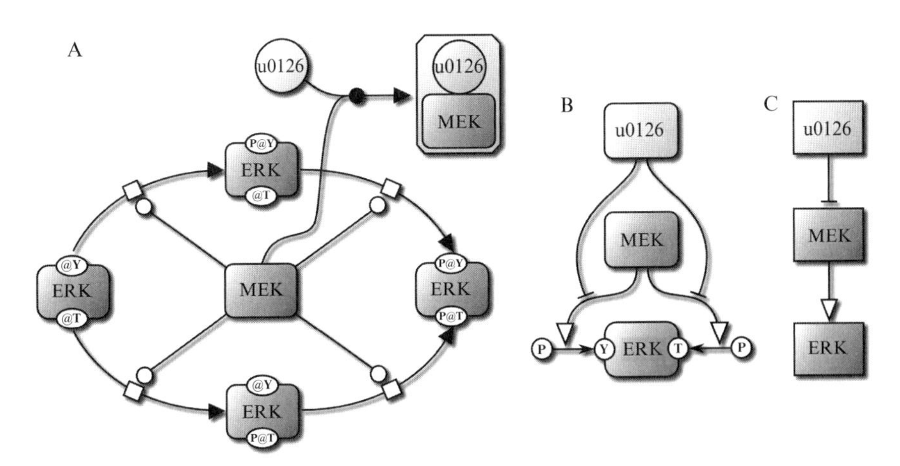

Figure 19.1 SBGN diagrams. (A) Process description diagram. (B) Entity relationship diagram. (C) Activity flow diagram. All three diagrams show the same process (activation of ERK by the kinase MEK) but emphasize different aspects.

- *Visualization tools.* These produce a human-readable representation of the reaction network defined by the SBML model, for example, CellDesigner (Funahashi *et al.*, 2008).
- *Simulation tools.* Simulation tools can be used to calculate the dynamic behavior of a model over time, for example, SBML ODE solver (Lu *et al.*, 2008) for continuous models, Dizzy (Ramsey *et al.*, 2005) for stochastic models, or MesoRD (Hattne *et al.*, 2005) for spatial models.
- *Metabolic control/flux balance analysis tools.* These investigate simplified dynamical aspects of a model, for example, TinkerCell (Chandran *et al.*, 2009).
- *Management tools.* These tools create models from pathway structures, kinetic information, and numerical values stored in different Web resources, for example, Pathway Tools (Karp *et al.*, 2009).
- *Conversion tools.* Conversion tools link SBML to other model description languages, for example, CellML2SBML (Schilstra *et al.*, 2006), a converter from the CellML format (Lloyd *et al.*, 2004), or SyBiL (Ruebenacker *et al.*, 2009), which converts BioPAX (Stromback and Lambrix, 2005) files to SBML.
- *Tools supporting SBGN graphics.* SBGN diagrams can be drawn by 18 software tools (August, 2010; http://sbgn.org). Examples are CellDesigner, a visual model creation and simulation software, and the BioModels Database model repository. To automatically convert SBML models into SBGN diagrams, semantic information about the model elements needs to be provided in the model's SBO terms (Le Novère *et al.*, 2007), for example, describing the roles of modifiers (enzyme, inhibitor, or activator), and MIRIAM annotations (Le Novère *et al.*, 2005), which specify the biological meaning of SBML elements. Apart from the tools offering full SBGN support, many tools can display the network structure of a model as a bipartite graph of species and reactions, for instance, using GraphViz (Ellson *et al.*, 2002) for rendering.

The reuse of kinetic models is further supported by Web repositories for published models. These repositories store models from classic articles as well as models submitted prior to publication.

- *BioModels Database* (Le Novère *et al.*, 2006). This is the largest model repository of models in SBML format. In its 17th release, it contains 249 fully curated, MIRIAM-compliant models, and 224 others. Further, it implements the most recent standards of the systems biology community, for example, MIRIAM-compliant annotations and model visualization in the SBGN. The models in its curated section have been edited by professional curators to ensure a high quality in the annotations and to guarantee that the simulation results of encoded models agree with published results in the original article.
- *JWS Online* (Olivier and Snoep, 2004). The JWS Online features 90 models (August 2010). Its main advantage is its online simulation

environment, enabling users to directly run the models without having to install an own simulation software.

- *PathGuide.* A large number of Web repositories for models of various pathways are listed in the repository PathGuide (Bader *et al.*, 2006; http://www.pathguide.org). It describes the scope of these resources and lists the formats they support for model export.

4. How to Document and Annotate Models

4.1. The MIRIAM rules for model publishing

The MIRIAM (Le Novère *et al.*, 2005) is a set of rules ensuring that a model can be unambiguously understood and its simulation results can be reproduced. To be MIRIAM compliant, a model needs to be encoded in a public, machine-readable format and comply with the standard in which it is encoded. Further, it must clearly refer to a single reference description (usually, a scientific paper), must show the same structure as in the reference description, and its simulation must reproduce all relevant results from the original publication.

In addition, certain pieces of information have to be provided in the encoded model. For *model attribution*, it has to be annotated with the preferred model name, a citation of the reference description, the name and contact information of model creators—that is, the persons who encoded the model—the date and time of creation and last modification, and the terms of distribution. Second, the MIRIAM proposal requires that the individual model elements are annotated with references to external data resources, which specify their biological meaning. These semantic annotations and the technical infrastructure to handle them will be discussed in the following sections.

4.2. Controlled vocabularies, taxonomies, and ontologies

Systems biology has evolved through the fusion of molecular and computational biology, biophysics, control theory, and other fields of science. Each of these fields brought along their concepts and their scientific language. The mix of these languages can lead to ambiguities in their terms and hinders communication. In addition, biologists have for a long time used various *ad hoc* notations to describe different aspects of processes, objects, and observations. A striking example is the multitude of naming schemes for genes and proteins, which are often incompatible between and even within organisms.

A way to avoid ambiguity is to define a list of mandatory terms representing defined concepts. Such lists are known as *controlled vocabularies* (CV). CVs can range from small lists of physical units to large systematic naming schemes, for example, for the open reading frames in yeast. If the terms of a CV can be classified under more general concepts, this leads to a hierarchy of concepts and the controlled vocabulary becomes a *taxonomy*. While taxonomies are limited to a small number of relations arising from the hierarchy (e.g., "is a," "is parent," "is child," "is sibling," etc.), an *ontology* as defined in information science can contain an arbitrary number of relationships (e.g., "has part," "is version of"). The terms "ontology," "taxonomy," and "controlled vocabulary" are often used synonymously, and to some extent, this is correct. A taxonomy can be regarded as a simple kind of ontology, while both ontologies and taxonomies comprise CVs (Fig. 19.2).

The most well-known ontology in systems biology, and historically one of the first, is the gene ontology (GO). It is mainly used as a taxonomy and consists of three parts: biological processes (e.g., "GO:0023045: signal transmission via conformational transition"), molecular functions (e.g., "GO:0043028: caspase regulator activity"), and cellular components (e.g., "GO:0005829: cytosol"). Another important example is the systems biology ontology (SBO), which is used in the SBML format to declare the biochemical roles of SBML elements. The SBO contains terms like "SBO:0000012—mass action rate law" that can be attached to a kinetic law or "SBO:0000027—Michaelis constant" which can be used to describe the role of a parameter.

As GO and other ontologies turned out to be useful, more and more ontologies were developed. To join these efforts and to avoid duplication, the Open Biomedical Ontologies (OBO) initiative was formed. The OBO foundry, a core group of ontology developers, is trying to establish common principles with which nonoverlapping and compatible ontologies can be developed. OBO ontologies are distributed in the *Web Ontology Language* (OWL) format and in their own "obo" format. On its Web page, OBO currently hosts almost 100 ontologies, which can be freely downloaded and used. The Ontology Lookup Service (http://www.ebi.ac.uk/ontology-lookup/) provides a Web-based search engine for all OBO ontologies.

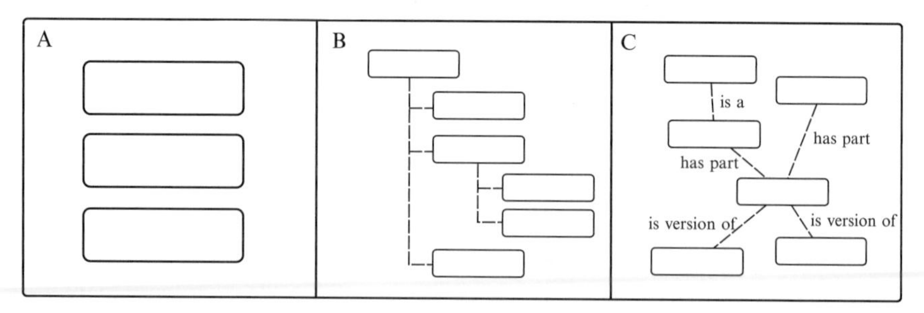

Figure 19.2 Each box with rounded corners represents a term that has a defined concept. A collection of such terms is called a controlled vocabulary (A). If this collection is organized hierarchically, it is referred to as taxonomy (B). In an ontology (C), the relation between the terms can be defined freely.

Several language formats were developed to store ontologies. The most widespread among these language formats are so-called *semantic Web languages*. The term "semantic" originates from the idea to store information in a way that makes the semantics of the information computationally accessible. The semantic Web language formats *Resource Description Framework* (RDF) and OWL can be processed automatically by several software systems. Possible applications are automated reasoning (e.g., if A *has part* B and B *has part* C, then A *has part* C) or the detection of logical contradictions (e.g., if A *is a* B and B *is not a* C then the statement A *is a* C is wrong).

4.3. MIRIAM-compliant annotations in SBML models

The SBML format has adopted the MIRIAM rules for semantic annotations which link model components to data items in biological Web resources. In contrast to the SBO terms, which define the role of a component in a model or specify *how* something is described, the annotation defines the biological meaning of a component or *what* is described by the model. In the following, the latter will be referred to as MIRIAM annotations.

A MIRIAM annotation in SBML is realized with the help of the RDF format, which is XML based as SBML itself. RDF was developed as a semantic Web language, allowing to express information in a way that its meaning—that is, the semantics—can be processed by software. The main concept of RDF is taken from human language: triplets consisting of a subject, a predicate, and an object. In a MIRIAM annotation, the model component is the subject, the data item of a Web resource is the object, and the relationship between the model component and the Web resource item is the predicate. Here is an example:

> The model component `compartment1` (*subject*) is (*predicate*) cytosol—as defined by web resource Gene Ontology Object Identifier GO:0005829 (*object*).

More formally, MIRIAM annotations in SBML are defined by

- the *subject* as the value of the `meta_id` attribute of a component,
- the *predicate* as a term from the BioModels qualifiers list,
- the *object* as a pair of a MIRIAM URNs (as defined by the MIRIAM Resources; www.ebi.ac.uk/miriam/) and the resource identifier encoded as a URN, for example, `urn:miriam:obo.go:GO\%3A0005829`.

MIRIAM Resources is a Web resource that defines a list of URN names, for example, `urn:miriam:kegg.compound` or `urn:miriam:ec-code` that represent Web resources in a nonambiguous way. MIRIAM

URNs were developed to provide unique and stable identifiers for Web resources that have multiple incarnations in the World Wide Web, for example, Enzyme Nomenclature: http://us.expasy.org/enzyme/, http://www.genome.jp/dbget-bin/www_bfind?enzyme, or http://www.ebi.ac.uk/intenz/, or Web resources that consist of several subresources, for example, KEGG Compound, KEGG Drug, or KEGG Reaction, or resources that change their name.

The MIRIAM Resources also provide Web services that can translate a MIRIAM URN into the name of a resource or into a hyperlink that can be used to access the data object on the World Wide Web, for example, http://www.ebi.ac.uk/ego/GTerm?id=GO:0005829.

Also the predicate or *qualifier* is chosen from a controlled list of terms, in this case, provided by the Web resource BioModels qualifiers. BioModels qualifiers define two sets of terms: the first set—the biological qualifiers—can be used for relationships that are relevant for components, while the second set—the model qualifiers—contains terms related to the entire model.

Biology qualifiers	Model qualifiers
encodes, hasPart, hasProperty, hasVersion, is, isDescribedBy, isEncodedBy, isHomologTo, isPartOf, isPropertyOf, isVersionOf, occursIn	is, isDerivedFrom, isDescribedBy

MIRIAM Resources and the BioModels qualifiers are both actively curated and updated frequently. The maintainers encourage the submission of new resources and terms and include them after an internal review process. While the usage of MIRIAM URNs has become a quasi-standard in SBML models, it is not enforced by the SBML format and, for example, Life Science Identifiers (LSIDs) could be used as an alternative.

5. Tools for Model Building and Annotation

5.1. Web services

The idea of distributing computing operations over several physically separated systems has existed since the beginning of the computer era. One way to achieve this distribution is the use of remote procedure calls (RPC). Given two software systems that need to communicate with each other, a RPC can be realized by finding a standardized way of transmitting information between the systems. The most widespread RPC protocol today is the Simple Object Access Protocol (SOAP). It was originally developed for business to business communication and has also become popular for bioinformatics applications.

SOAP is an XML-based protocol that, among other things, standardizes the way how data objects are encoded into XML (serialization) and how the serialized data objects and possible error messages are transmitted. Since SOAP is usually used on top of the HTTP protocol, the SOAP messages can easily be sent through the internet. Programming libraries that help to generate and parse SOAP messages are available for all major programming languages and are usually accompanied by a server software that can send and receive these messages. A SOAP-based interface is also called a *Web service*.

To automatically generate interfaces to SOAP Web services, the Web Service Description Language (WSDL) provides a standardized method of describing the location, input, and outputs of such services. In a recent effort, the Web resource BioCatalogue (http://www.biocatalogue.org/) has started to collect and list available bioinformatics Web services.

Today, the word "Web service" also comprises another popular concept called Representational State Transfer (REST) services (or short RESTful services). REST assumes that the HTTP protocol already contains all facilities needed for creating RPCs. In its original form, REST services assume that a Web service provides resources (in this case, data objects or information) that can be accessed with a unique identifier. To retrieve and manipulate the resources, the following HTTP methods can be used:

Example URI	Resource type	HTTP GET	HTTP POST	HTTP PUT	HTTP DELETE
http://www.protein.org/resources/	Collection of elements	List resources	Replace resources	Create resources	Delete resources
http://www.protein.org/resources/1234	Element	Retrieve resource	Update or replace resource	Create resource	Delete resource

Today's RESTful Web services do not strictly follow this scheme. Most bioinformatics RESTful Web services only use the GET and POST methods as an interface for the functions they provide. To serialize data objects, RESTful Web services usually rely on custom XML formats, JavaScript Object Notation (JSON), or simple plain text formats.

The advantage of REST over SOAP is that REST is far more lightweight. REST messages do not bear the overhead of the XML encoding of messages used by SOAP. HTTP clients as well as XML and JSON parser libraries are available for virtually every programming language and arguably simpler in their integration. Until recently, RESTful Web services lacked a standardized method of description. This has now changed by the development of the Web Application Description Language (WADL) and WSDL 2.0.

This example shows how to create a SOAP client in the Python programming language.

```
1. from SOAPpy import WSDL #import the SOAPpy library
2. #get a server object from a WSDL file
3. server=WSDL.Proxy('http://www.xmethods.net/sd/2001/
   TemperatureService.wsdl')
4. result = server.getTemp('90210')
```

This example shows how to create a REST client in the Python programming language.

```
1. import urllib, simplejson
2. # send HTTP GET request
3. u=urllib.urlopen('http://www.semanticsbml.org/
   semanticSBML/annotate/search.json?q=atp')
4. result = simplejson.reads(u.read() ) #deserialise the returned string
   data into a Python object
```

5.2. Workflow engines

The standardization of remote programming interfaces allowed the construction of workflow management systems. Such systems allow users to visually connect different Web services to create complex workflows. One of the most successful workflow management systems in bioinformatics is the Taverna Workbench (Hull *et al.* (2006)) shown in Fig. 19.3. Here is a common usage example:

1. Import a WSDL file of a Web service.
2. Drag and drop the imported Web service to the workflow canvas.
3. A box appears in the canvas, representing a Web service with input and output connectors on its sides.
4. Steps 1–3 are repeated to create a second Web service on the canvas.
5. The input and outputs are connected by clicking on an output connector and dragging it to an input connector.
6. Input and output nodes from the workflow system (e.g., file reader and file writer) are created on the canvas and connected to the inputs like in step 4.
7. The workflow is run in workflow system by pressing "run" button.

While workflow management systems often allow users to construct complicated workflows without any programming knowledge, this is

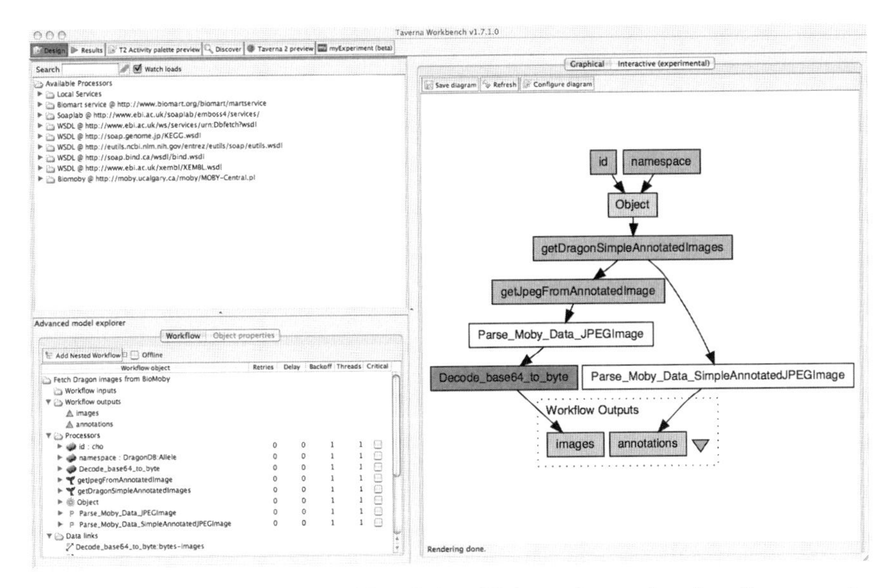

Figure 19.3 The Taverna Workbench is a Web service engine that allows you to visually create complex workflows based on local services and remote procedure calls.

sometimes hindered by the incompatibility of service outputs and inputs. Workflow systems solve this obstacle by

- providing a variety of local programs that can be integrated into the workflow
- allowing the creation of local converter programs (glue code)
- restricting the user to selected services, whose output can be converted to internal data objects that are compatible with each other.

The standardization of biochemical models helps to overcome these incompatibilities. Standardized element annotations further help to extend or combine models and to extract information from them.

The Taverna Workbench and other popular bioinformatics workflow management systems are focused on SOAP-based Web services. In the past few years, however, a series of REST-focused general purpose workflow management systems emerged: Yahoo Pipes, Google Mashups (now Google App Engine), Intel MashMaker, Microsoft Popfly, and Lotus Mashups. Their adoption in bioinformatics is currently low but their potential is promising.

5.3. Tools for managing biochemical names and identifiers

Semantic annotations can connect model elements with the entries from various biological Web resources and thereby describe their biological meaning. However, finding appropriate annotations and comparing entries from different resources, as is sometimes required during model merging, remain a challenge. Collection of this information is simplified by a number of online services, Web services, or programming libraries. Table 19.2 lists a number of interfaces for different application domains. We shall now discuss some tools focusing on systems biology.

- *libAnnotationSBML.* The libAnnotationSBML (Swainston and Mendes, 2009) is a Java library that extends libSBML, a library for the handling of SBML files, and links annotations in SBML models to the Web services of KEGG, ChEBI, UniProt, and SBO by a unified API. Among other things, it allows the extraction of names and cross-references for Web resource entries and the ability to check whether reactions appearing in a model are physically balanced (by counting atom numbers stored in ChEBI). The Web service queries of libAnnotationSBML are built dynamically from MIRIAM Web services and are therefore always up to date.
- *Saint.* The capacity to directly integrate annotations into SBML models is provided by the Web site and library Saint (Lister *et al.*, 2009). It is based on the Web resources UniProtKB (Apweiler *et al.*, 2009), STRING (Jensen *et al.*, 2009), Pathway Commons, SBO, and GO (Harris *et al.*, 2004). In addition to the conversion of names and resource elements, Saint enables the user to automatically extend an SBML model at a selected species element by reactions from STRING and Pathway Commons. The Web site is available at http://biodev.ncl.ac.uk:8081/saint/uk.ac.cisban.saint.Annotator/Annotator.html.
- *libSBAnnotation.* The Python library libSBAnnotation integrates information from the resources NCBI Taxonomy (Sayers *et al.*, 2009), GO, KEGG, Reactome (Matthews *et al.*, 2009), Interpro (Hunter *et al.*, 2009), Uniprot, SGD (Engel *et al.*, 2010), and SBO. While the libraries and tools mentioned before provide unified online interfaces, libSBAnnotation extracts information from the Web resources and stores it locally in an ontology. While this information might not always be up to date, libSBAnnotation is fast, enables the inference of incorrect cross-references or intra-resource reference, and features a fuzzy name search. Functions of the libSBAnnotation can be called via RESTful Web services at www.semanticsbml.org.
- *BridgeDb.* The BridgeDb framework (Van Iersel *et al.*, 2010) can act as a link between a tool and Web resources and provides access to information stored in flat files and relational databases. It is built on top of the Web services EnsMart (Kasprzyk *et al.*, 2004), PICR (Côté *et al.*, 2007), Synergizer (Berriz and Roth, 2008), CRONOS (Waegele *et al.*, 2009), and BridgeWebservice. The stand-alone BridgeWebservice is available at http://www.bridgedb.org/.

Table 19.2 Additional name and ID mapping tools combining information from multiple Web resources

	Application domain: metabolites, genes, and proteins	Resources	Web service	URL
Ontology Lookup Service	−/+/−	>40 OBO ontologies	+	www.ebi.ac.uk/ontology-lookup/
BioMart	+/+/+	Ensemble, UniProt, ...	−	www.biomart.org
Pathway Commons	−/−/+	HPRD, HumanCyc, ...	+	http://pathwaycommons.org
DICT	−/+/+	PIR, UniProtKB, ...	+	www.bridgedb.org/wiki/BridgeWebservice
CRONOS	−/+/+	RefSeq, UniProt, Ensemble	+	http://mips.helmholtz-muenchen.de/genre/proj/cronos/
MatchMiner	−/+/−	UCSC, LocusLink, ...	−	http://discover.nci.nih.gov/matchminer/index.jsp
PICR	−/−/+	UniProt	+	www.ebi.ac.uk/Tools/picr/
Synergizer	−/+/+	Ensemble, EntrezGene	+	http://llama.med.harvard.edu/synergizer/doc/

6. How to Create and Draw a Simple Model

6.1. Model creation

Existing tools make it easy to translate biochemical reactions into an SBML model and draw the model as an SBGN diagram. Let us illustrate this with a simple example, the phosphofructokinase reaction. First, we look up the reaction formula in KEGG (Kanehisa *et al.*, 2010): we search for "6-phosphofructokinase," select the first enzyme (2.7.1.11), and choose this enzyme's first detailed reaction (KEGG reaction identifier R04779):

```
ATP + beta-D-Fructose 6-phosphate <=> ADP + beta-D-Fructose 1,6-bisphosphate
```

The sum formula leads to the following differential equation system:

$$\frac{\mathrm{d}[\mathrm{ATP}]}{\mathrm{d}t} = -v, \quad \frac{\mathrm{d}[\mathrm{F6P}]}{\mathrm{d}t} = -v, \quad \frac{\mathrm{d}[\mathrm{ADP}]}{\mathrm{d}t} = v, \quad \frac{\mathrm{d}[\mathrm{F16BP}]}{\mathrm{d}t} = v. \tag{19.1}$$

which may, for instance, describe the isolated enzyme in a test tube. Since we do not know the kinetic mechanism of this reaction, we substitute it by a standard rate law. We choose the common modular rate law (Liebermeister *et al.*, 2010)

$$v = \frac{v_{+}^{\mathrm{M}} \frac{[\mathrm{ATP}]}{k_{\mathrm{ATP}}^{\mathrm{M}}} \frac{[\mathrm{F6P}]}{k_{\mathrm{F6P}}^{\mathrm{M}}} - v_{-}^{\mathrm{M}} \frac{[\mathrm{ADP}]}{k_{\mathrm{ADP}}^{\mathrm{M}}} \frac{[\mathrm{F16BP}]}{k_{\mathrm{F16BP}}^{\mathrm{M}}}}{\left(1 + \frac{[\mathrm{ATP}]}{k_{\mathrm{ATP}}^{\mathrm{M}}}\right)\left(1 + \frac{[\mathrm{F6P}]}{k_{\mathrm{F6P}}^{\mathrm{M}}}\right) + \left(1 + \frac{[\mathrm{ADP}]}{k_{\mathrm{ADP}}^{\mathrm{M}}}\right)\left(1 + \frac{[\mathrm{F16BP}]}{k_{\mathrm{F16BP}}^{\mathrm{M}}}\right) - 1}, \tag{19.2}$$

a generalized form of the reversible Michaelis–Menten kinetics for bimolecular reactions.

In the next step, we use the modeling tool COPASI (Hoops *et al.*, 2006) to create an SBML model of our equation system. When COPASI starts, a new model is created by default. We edit the biochemical part of the model by adding a reaction with the chemical equation "ATP + F6P = ADP + F16BP; PFK" and committing this reaction. By default, a new compartment and the species ATP, F6P, ADP, and F16BP are created and all numerical values are set to 0.1 (except for the compartment volume which is set to 1). We then add a rate equation to the reaction by clicking "new rate law" and insert the formula (19.2), written in the syntax

```
PFK * (vmf*ATP/katp*F6P/kf6p-vmr*ADP/kadp*F16BP/kf16bp)/( (1+ATP/katp+F6P/kf6p)+(1+ADP/kadp+F16BP/kf16bp)-1).
```

As a next step, we change the description of ATP and F6P to "substrate," of ADP and F16BP to "product," and of PFK to "modifier."

After committing this rate law, we can select it at the reaction screen and set all parameters in the symbol definition to 1. Finally, we export this reaction to SBML via the "File" menu.

If an SBML model has been created manually, it should be checked by the online validation service at http://sbml.org/Facilities/Validator.

6.2. Model annotation

Next, we add SBO terms to the model to specify the meaning of individual model elements using the "annotate" function of semanticSBML (Krause *et al.*, 2010; www.semanticsbml.org). We assign "SBO:0000014" (enzyme) to the PFK and "SBO:0000013" (catalyst) to its role in the reaction. The other species are annotated with "SBO:0000247" (simple chemical).

Further, it is possible to add annotations describing the chemical species. For this purpose, semanticSBML provides a fuzzy string search for finding appropriate annotations and can also suggest possible annotations based on the element names and identifiers appearing in the SBML model. The principle of MIRIAM annotations can also be transferred to other language formats. While it has been officially adopted in the CellML format, there is also an online service (Annotate Your Model at http://semanticsbml.org/aym) that allows users to annotate data tables and models written in programming languages like Matlab, C++, Python, or Mathematica.

6.3. Model drawing

Finally, to visualize our reaction in an SBGN drawing (see Fig. 19.4), we import the model to Arcadia (Villeger *et al.*, 2010). The square in the middle depicts the reaction, which is connected with its substrates and products. The special connection between the reaction and PFK shows that PFK acts as a catalyst.

7. Annotations in Genome-Scale Network Reconstructions

The development of metabolic network reconstructions has increased in recent years. It now covers a range of organisms and has been applied to a number of research topics including metabolic engineering, genome annotation, evolutionary studies, network property analysis, and interpretation of "omics" data sets (Oberhardt *et al.*, 2009).

The utility of such reconstructions for third parties is greatly enhanced by the utility of semantic annotations that unambiguously describe each compartment, metabolite, and enzyme within the reconstruction (Herrgård

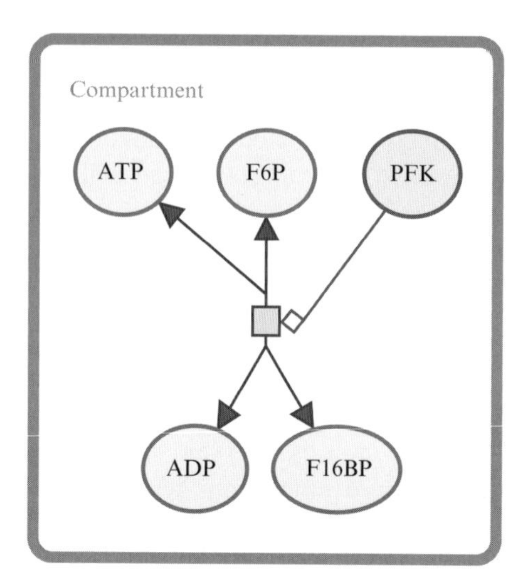

Figure 19.4 SBGN diagram of the phosphofructokinase reaction. The graphic was produced with the software Arcadia (Villeger *et al.*, 2010).

et al., 2008). Further, the use of semantic annotations and the exploitation of Web services which provide access to the database entries that they describe can greatly facilitate the development process of generating genome-scale models.

Over the past decade, a large number of metabolic reconstructions have been developed covering a range of organisms across a number of taxonomic branches. A recent development, in which the process of rapidly generating metabolic reconstructions from genome sequences has been automated (Henry *et al.*, 2010), promises to increase the coverage of such reconstructions yet further. Such a body of metabolic reconstructions will prove to be a valuable resource, providing the systems biology with a detailed description of the metabolic capabilities of a large range of organisms. Such reconstructions can be exploited computationally with such strategies as flux balance analysis (Orth *et al.*, 2010), which hold great promise for applications such as industrial bioproduct optimization through metabolic engineering and drug target identification through prediction of differential metabolic flux in both healthy and diseased conditions.

Most of these reconstructions, however, are reliant on proprietary naming of metabolites and enzymes. This poses less of a problem for smaller scale models of known biochemical pathways, in which the identity of participants can generally be inferred from such proprietary naming, but can cause problems when describing novel or less well-known metabolic pathways. Also, reliance of proprietary naming greatly impedes the use of automated computational analysis of large-scale networks.

To mitigate these issues, a recent metabolic reconstruction for *Saccharomyces cerevisiae* and a *Homo sapiens* reconstruction in development have followed MIRIAM recommendations, in which most components of the network are annotated with unambiguous semantic terms (Herrgård *et al.*, 2008). In these cases, the metabolic reconstruction was generated as a consensus between existing models.

The first stage of this merging process—highlighting the issues described above—was to provide unified identifiers for terms in each model. This was performed computationally with an automated annotation tool generated with libAnnotationSBML. This tool takes an existing model in SBML format and iterates through each compartment, metabolite, and enzyme, taking the supplied name and searching this against the Web service interface to GO, ChEBI, and UniProt, respectively. In cases where a component name matches a number of entries, each of the entries, along with their synonyms, is returned to the user, providing the facility for manual selection. In the case of the human metabolic reconstruction, involving the merging of two existing models, 63% of metabolites from the first model (Ma *et al.*, 2007) and 57% from the second model (Duarte *et al.*, 2007) could be identified in this way.

This left a number of components that needed to be identified by hand, and this process was undertaken during focused meetings ("jamborees"; Thiele and Palsson, 2010b), in which participants hand-curate reconstructions before publication. The determination of such missing identifiers was performed through a process of searching both existing databases and literature. Where "new" metabolites were identified, which were not present in existing databases, these were submitted to the ChEBI database (de Matos *et al.*, 2010), a process that provides both a unique identifier and also contributes to the development of data resources upon which the reconstruction process depends. In addition to determining identifiers, the jamboree process includes the assignment of a confidence degree to each of the metabolic reactions in the reconstruction. These take the form of literature references in terms of PubMed identifiers and GO Evidence Code terms describing confidence with terms such as "Inferred by Curator" or "Inferred from Experiment." Such annotations increase the usability of reconstructions, and can form the basis of subsequent curation work, in which effort can be concentrated upon validating reactions containing no literature evidence or annotated with low confidence terms.

At the end of this process, the original models should share common identifiers, allowing them to be automatically merged by tools such as semanticSBML (Krause *et al.*, 2010) or libAnnotationSBML. The reconstruction process then continues with a number of steps including determination of metabolite protonation state under an assumed intracellular pH, and mass and charge balancing of metabolic reactions. Both of these steps are requirements for generating models over which flux balance analysis may be performed (Thiele and Palsson, 2010a), and both steps can be automated on semantically annotated models.

The cheminformatics package, MARVIN (ChemAxon Kft., Budapest, Hungary), can be exploited to determine metabolite protonation state. MARVIN provides a pK_a prediction algorithm that requires a SMILES string (a textual representation of a chemical structure) to predict the dominant protonation state of a molecule at a given pH. These SMILES strings can be harvested automatically from the ChEBI Web services for those metabolites annotated with ChEBI terms. The chemical formula and charge can then be used in mass and charge balancing algorithms. This provides the twin benefits of ensuring consistency within the reconstruction, such that each reaction is accurately defined with the correct stoichiometry of participants and also acts as a validation step for the metabolic annotations, as incorrectly annotated species will provide incorrect formulae and charge that will manifest themselves in the reaction balancing process.

The development of such reconstructions, richly annotated with semantic terms, provides a valuable resource that can be further exploited with the development of computational techniques as yet undeveloped. Annotating genome-scale models with semantic terms provides a live link to the bioinformatics resources from which the terms are taken. As these resources mature, the range of data linked to the model increases. For example, annotating enzymes with UniProt terms allows the automated extraction of protein sequences for each of these enzymes, which may subsequently be used in binding or posttranslational modification site predictions. Recent developments of the ChEBI database provide links to the enzyme kinetic parameter database SABIO-RK, opening the possibility of automating the process of parametrizing annotated models with kinetic equations and parameters describing their metabolic reactions. It is hoped that the richness of information stored in such resources will be matched by the development of software to exploit such information, allowing genome-scale reconstructions to fulfill their potential as comprehensive compendiums of the biochemical capabilities of the cell.

ACKNOWLEDGMENTS

This work was supported by the BMBF SysMO Project Translucent [contract number 0313982A], the International Max Planck Research School for Computational Biology and Scientific Computing, the German Research Foundation [CRC 618], and the European Commission [BaSysBio, Grant number LSHG-CT-2006-037469]. The authors thank the EPSRC and BBSRC for their funding of the Manchester Centre for Integrative Systems Biology (http://www.mcisb.org), BBSRC/EPSRC Grant BB/C008219/1.

REFERENCES

Apweiler, R., Martin, M., O Donovan, C., Magrane, M., Alam-Faruque, Y., Antunes, R., Barrell, D., Bely, B., Bingley, M., Binns, D., *et al.* (2009). The universal protein resource (UniProt) in 2010. *Nucleic Acids Res.* **38,** D142–D148.

Bader, G., Cary, M., and Sander, C. (2006). Pathguide: A pathway resource list. *Nucleic Acids Res.* **34,** D504.

Berriz, G., and Roth, F. (2008). The Synergizer service for translating gene, protein and other biological identifiers. *Bioinformatics* **24,** 2272.

Chandran, D., Bergmann, F., and Sauro, H. (2009). TinkerCell: Modular CAD tool for synthetic biology. *J. Biol. Eng.* **3,** 19.

Côté, R., Jones, P., Martens, L., Kerrien, S., Reisinger, F., Lin, Q., Leinonen, R., Apweiler, R., and Hermjakob, H. (2007). The Protein Identifier Cross-Referencing (PICR) service: Reconciling protein identifiers across multiple source databases. *BMC Bioinformatics* **8,** 401.

de Matos, P., Alcántara, R., Dekker, A., Ennis, M., Hastings, J., Haug, K., Spiteri, I., Turner, S., and Steinbeck, C. (2010). Chemical entities of biological interest: An update. *Nucleic Acids Res.* **38,** D249–D254.

Duarte, N., Becker, S., Jamshidi, N., Thiele, I., Mo, M., Vo, I., Srivas, R., and Palsson, B. (2007). Global reconstruction of the human metabolic network based on genomic and bibliomic data. *Proc. Natl. Acad. Sci. USA* **104,** 1777–1782.

Ellson, J., Gansner, E., Koutsofios, L., North, S., and Woodhull, G. (2002). Graphviz—Open source graph drawing tools. *In* Graph Drawing, pp. 594–597. Springer.

Engel, S., Balakrishnan, R., Binkley, G., Christie, K., Costanzo, M., Dwight, S., Fisk, D., Hirschman, J., Hitz, B., Hong, E., *et al.* (2010). Saccharomyces Genome Database provides mutant phenotype data. *Nucleic Acids Res.* **38,** D433.

Funahashi, A., Matsuoka, Y., Jouraku, A., Morohashi, M., Kikuchi, N., and Kitano, H. (2008). Cell Designer 3.5: A versatile modeling tool for biochemical networks. *Proc. IEEE* **96,** 1254–1265.

Harris, M., Clark, J., Ireland, A., Lomax, J., Ashburner, M., Foulger, R., Eilbeck, K., Lewis, S., Marshall, B., Mungall, C., *et al.* (2004). The Gene Ontology (GO) database and informatics resource. *Nucleic Acids Res.* **32,** D258.

Hattne, J., Fange, D., and Elf, J. (2005). Stochastic reaction-diffusion simulation with MesoRD. *Bioinformatics* **21,** 2923.

Henry, C., DeJongh, M., Best, A., Frybarger, P., Linsay, B., and Stevens, R. (2010). High-throughput generation, optimization and analysis of genome-scale metabolic models. *Nat. Biotechnol.* **28,** 977–982.

Herrgård, M., Swainston, N., Dobson, P., Dunn, W., Arga, K., Arvas, M., Büthgen, N., Borger, S., Costenoble, R., Heinemann, M., Hucka, M., Le Novère, N., *et al.* (2008). A consensus yeast metabolic network reconstruction obtained from a community approach to systems biology. *Nat. Biotechnol.* **26,** 1155–1160.

Hoops, S., Sahle, S., Gauges, R., Lee, C., Pahle, J., Simus, N., Singhal, M., Xu, L., Mendes, P., and Kummer, U. (2006). COPASI—A COmplex PAthway SImulator. *Bioinformatics* **22,** 3067.

Hucka, M., Finney, A., Sauro, H., Bolouri, H., Doyle, J., Kitano, H., Arkin, A., Bornstein, B., Bray, D., Cornish-Bowden, A., Cuellar, A. A., Dronov, S., *et al.* (2003). The systems biology markup language (SBML): A medium for representation and exchange of biochemical network models. *Bioinformatics* **19,** 524–531.

Hull, D., Wolstencroft, K., Stevens, R., *et al.* (2006). Taverna: A tool for building and running workflows of services. *Nucleic Acids Res.* **34,** W729.

Hunter, S., Apweiler, R., Attwood, T., Bairoch, A., Bateman, A., Binns, D., Bork, P., Das, U., Daugherty, L., Duquenne, L., *et al.* (2009). InterPro: The integrative protein signature database. *Nucleic Acids Res.* **37,** D211.

Jensen, L., Kuhn, M., Stark, M., Chaffron, S., Creevey, C., Muller, J., Doerks, T., Julien, P., Roth, A., Simonovic, M., *et al.* (2009). STRING 8—A global view on proteins and their functional interactions in 630 organisms. *Nucleic Acids Res.* **37,** D412.

Kanehisa, M., Goto, S., Furumichi, M., Tanabe, M., and Hirakawa, M. (2010). KEGG for representation and analysis of molecular networks involving diseases and drugs. *Nucleic Acids Res.* **38,** D355.

Karp, P., Paley, S., Krummenacker, M., Latendresse, M., Dale, J., Lee, T., Kaipa, P., Gilham, F., Spaulding, A., Popescu, L., *et al.* (2009). Pathway Tools version 13.0: Integrated software for pathway/genome informatics and systems biology. *Brief. Bioinform.*

Kasprzyk, A., Keefe, D., Smedley, D., London, D., Spooner, W., Melsopp, C., Hammond, M., Rocca-Serra, P., Cox, T., and Birney, E. (2004). EnsMart: A generic system for fast and flexible access to biological data. *Genome Res.* **14,** 160.

Krause, F., Uhlendorf, J., Lubitz, T., Schulz, M., Klipp, E., and Liebermeister, W. (2010). Annotation and merging of SBML models with semanticSBML. *Bioinformatics* **26,** 421–422.

Le Novère, N., Finney, A., Hucka, M., Bhalla, U., Campagne, F., Collado-Vides, J., Crampin, E., Halstead, M., Klipp, E., Mendes, P., *et al.* (2005). Minimum information requested in the annotation of biochemical models (MIRIAM). *Nat. Biotechnol.* **23,** 1509–1515.

Le Novère, N., Bornstein, B., Broicher, A., Courtot, M., Donizelli, M., Dharuri, H., Li, L., Sauro, H., Schilstra, M., Shapiro, B., Snoep, J., and Hucka, M. (2006). BioModels Database: A free, centralized database of curated, published, quantitative kinetic models of biochemical and cellular systems. *Nucleic Acids Res.* **34,** D689–D691.

Le Novère, N., Courtot, M., and Laibe, C. (2007). Adding semantics in kinetics models of biochemical pathways. Proceedings of the Second International Symposium on experimental standard conditions of enzyme characterizations, pp. 137–153.

Le Novère, N., Hucka, M., Mi, H., Moodie, S., Schreiber, F., Sorokin, A., Demir, E., Wegner, K., Aladjem, M., Wimalaratne, S., Bergmann, F., Gauges, R., *et al.* (2009). The systems biology graphical notation. *Nat. Biotechnol.* **27,** 735–741.

Liebermeister, W., Uhlendorf, J., and Klipp, E. (2010). Modular rate laws for enzymatic reactions: Thermodynamics, elasticities, and implementation. *Bioinformatics* **26,** 1528–1534.

Lister, A., Pocock, M., Taschuk, M., and Wipat, A. (2009). Saint: A lightweight integration environment for model annotation. *Bioinformatics* **25,** 3026.

Lloyd, C., Halstead, M., and Nielsen, P. (2004). CellML: Its future, present and past. *Prog. Biophys. Mol. Biol.* **85,** 433–450.

Lu, J., Muller, S., Machne, R., and Flamm, C. (2008). SBML Ode Solver library: Extensions for inverse analysis. Proceedings of the Fifth International Workshop on Computational Systems Biology, WCSB, .

Ma, H., Sorokin, A., Mazein, A., Selkov, A., Selkov, E., Demin, O., and Goryanin, I. (2007). The Edinburgh human metabolic network reconstruction and its functional analysis. *Mol. Syst. Biol.* **3,** 135.

Matthews, L., Gopinath, G., Gillespie, M., Caudy, M., Croft, D., De Bono, B., Garapati, P., Hemish, J., Hermjakob, H., Jassal, B., *et al.* (2009). Reactome knowledgebase of human biological pathways and processes. *Nucleic Acids Res.* **37,** D619.

Oberhardt, M., Palsson, B., and Papin, J. (2009). Applications of genome-scale metabolic reconstructions. *Mol. Syst. Biol.* **5,** 320.

Olivier, B., and Snoep, J. (2004). Web-based kinetic modelling using jws online. *Bioinformatics* **20,** 2143–2144.

Orth, J., Thiele, I., and Palsson, B. (2010). What is flux balance analysis? *Nat. Biotechnol.* **28,** 245–248.

Ramsey, S., Orrell, D., and Bolouri, H. (2005). Dizzy: Stochastic simulation of large-scale genetic regulatory networks. *J. Bioinform. Comput. Biol.* **3,** 415–436.

Ruebenacker, O., Moraru, I., Schaff, J., and Blinov, M. (2009). Integrating BioPAX knowledge with SBML models. *IET Syst. Biol.* **3,** 317–328.

Sayers, E., Barrett, T., Benson, D., Bryant, S., Canese, K., Chetvernin, V., Church, D., DiCuccio, M., Edgar, R., Federhen, S., *et al.* (2009). Database resources of the National Center for Biotechnology Information. *Nucleic Acids Res.* **37,** 3124.

Schilstra, M., Li, L., Matthews, J., Finney, A., Hucka, M., and Le Novère, N. (2006). CellML2SBML: Conversion of CellML into SBML. *Bioinformatics* **22,** 1018.

Stromback, L., and Lambrix, P. (2005). Representations of molecular pathways: An evaluation of SBML, PSI MI and BioPAX. *Bioinformatics* **21,** 4401.

Swainston, N., and Mendes, P. (2009). libAnnotationSBML: A library for exploiting SBML annotations. *Bioinformatics* **25,** 2292.

Teusink, B., Passarge, J., Reijenga, C., Esgalhado, E., van der Weijden, C., Schepper, M., Walsh, M., Bakker, B., van Dam, K., Westerhoff, H., and Snoep, J. (2000). Can yeast glycolysis be understood in terms of in vitro kinetics of the constituent enzymes? Testing biochemistry. *Eur. J. Biochem.* **267,** 5313–5329.

Thiele, I., and Palsson, B. (2010a). A protocol for generating a high-quality genome-scale metabolic reconstruction. *Nat. Protoc.* **5,** 93–121.

Thiele, I., and Palsson, B. (2010b). Reconstruction annotation jamborees: A community approach to systems biology. *Mol. Syst. Biol.* **6,** 361.

Van Iersel, M., Pico, A., Kelder, T., Gao, J., Ho, I., Hanspers, K., Conklin, B., and Evelo, C. (2010). The BridgeDb framework: Standardized access to gene, protein and metabolite identifier mapping services. *BMC Bioinformatics* **11,** 5.

Villeger, A., Pettifer, S., and Kell, D. (2010). Arcadia: A visualization tool for metabolic pathways. *Bioinformatics* **26,** 1470.

Waegele, B., Dunger-Kaltenbach, I., Fobo, G., Montrone, C., Mewes, H., and Ruepp, A. (2009). CRONOS: The cross-reference navigation server. *Bioinformatics* **25,** 141.

CHAPTER TWENTY

From Reaction Networks to Information Flow—Using Modular Response Analysis to Track Information in Signaling Networks

Pascal Schulthess[*,†] *and* Nils Blüthgen[*,†]

Contents

Abstract

Even if the biochemical details of signaling networks are known, it is often hard to track how information flows through the network. In combination with experimental techniques, modular response analysis has proven useful in analyzing the quantitative information transfer in signal transduction networks. The sensitivity of a target (e.g., transcription factor, protein) to an upstream stimulus (e.g., growth factor) can be determined by a so-called response coefficient. We have used this methodology to analyze how information flows in networks where the details of the mechanisms in the networks are known, but parameters are lacking. Using a Monte Carlo approach, we apply this method to track the routes of information flow. More specifically, we determine whether a given species has no, positive or negative influence on any other species in the network. Surprisingly, one can uniquely determine whether a molecule activates or inhibits another one in more than 99% of the interactions solely from the topology of the reaction network. To exemplify the methodology, we briefly discuss three signaling networks of different complexity: (i) a Wnt

[*] Institute of Pathology, Charité-Universitätsmedizin Berlin, Charitéplatz 1, Berlin, Germany

[†] Institute for Theoretical Biology, Humboldt University of Berlin, Invalidenstraße 43, Berlin, Germany

Methods in Enzymology, Volume 500

ISSN 0076-6879, DOI: 10.1016/B978-0-12-385118-5.00020-7

signaling pathway model with 15 species, (ii) a MAPK signaling pathway model with 200 species, and (iii) a large-scale signaling network of the entire cell with over 6000 species.

1. Introduction

Even for very small and seemingly simple motifs that we often find in signal transduction, it is hard to systematically deduce whether one molecule activates or deactivates the other. Take, for example, a phosphorylation event, where the kinase K converts the protein A into an active, phosphorylated form A_p (see Fig. 20.1A). Our intuition tells us that K activates A_p. And indeed, if one analyzes the direct effects among the species, one would say that an increase in K leads to a decrease in A and an increase in A_p (see Fig. 20.1B), thereby K activates A_p. However, a decrease in A would also lead to a decrease in A_p, and therefore K indirectly inhibits A_p (see Fig. 20.1C). Thus, K has both a positive and an indirect negative effect on A_p. Of course, for such simple motifs, one could deduce simple rules on how to determine whether a molecule activates or deactivates the other. In our example, a rule could be to investigate the effect on the product of a reaction only. However, this simple example illustrates that it may become very difficult to extract such rules in larger, complex networks, especially if molecules contain multiple modification sites or form complexes.

Therefore, we decided to use the framework of modular response analysis (MRA, see Kholodenko *et al.*, 1997, 2002) to systematically analyze whether the influence of one molecule on another molecule in the network is positive or negative. Section 2 serves as an introduction to the mathematical theory behind MRA. After giving a short overview of conservation analysis in Section 3, we utilize MRA and conservation analysis in Section 4 to extract information flow from reaction networks with the help of a Monte Carlo approach.

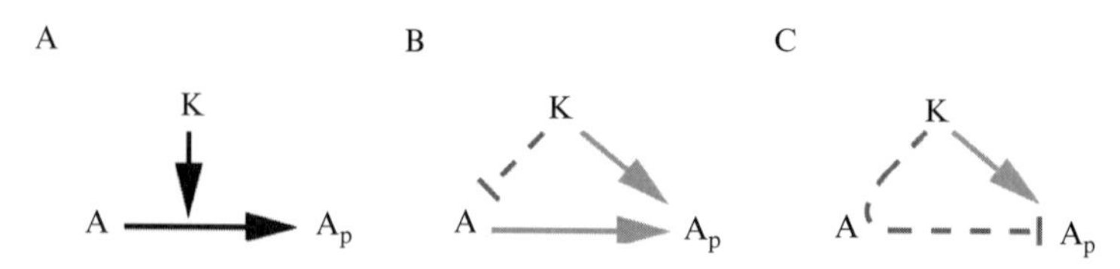

Figure 20.1 Simple reaction network. (A) A kinase K drives the phosphorylation of A to A_p. (B) Direct interactions in the network. (C) Global interactions between K and A_p which are a combination of activation (green solid arrow) and inhibition (red dashed arrow). (For interpretation of the references to color in this figure legend, the reader is referred to the Web version of this chapter.)

2. Modular Response Analysis

MRA has been developed as a framework to analyze interactions between species in hierarchical, information processing networks (Bruggeman *et al.*, 2002; Kholodenko *et al.*, 1997, 2002). It has been proven useful in reverse-engineering of network topologies (Santos *et al.*, 2007), and it has been used to analyze noise in regulatory networks (Bruggeman *et al.*, 2009). MRA is derived from and thus its notations are closely related to metabolic control theory (Heinrich and Rapoport, 1974a,b; Kacser and Burns, 1973). However, as classical metabolic control theory aims at investigating metabolic fluxes, MRA is applicable to systems where there is no flux between different nodes in the network but information transfer.

In the following, we will introduce the mathematical concepts behind MRA exemplified with a small example system, which is described in Fig. 20.2. A substrate S transforms to a product P catalyzed by enzyme E, and P decays back to S. We define the concentration vector $\mathbf{c}(t)$, which collects the concentrations of all species in the network: $\mathbf{c}(t) = [E(t), P(t), S(t)]^{\mathrm{T}}$. There are two fluxes in the network:

$$1: S \overset{E}{\rightarrow} P; \quad 2: P \rightarrow S; \ \mathrm{E} \leftrightarrow \emptyset \tag{20.1}$$

Assuming mass-action and Michaelis–Menten kinetics, the reaction velocities are then given by:

$$\mathbf{v}(\mathbf{c}(t)) = \begin{bmatrix} v_1(\mathbf{c}(t)) \\ v_2(\mathbf{c}(t)) \\ v_3(\mathbf{c}(t)) \end{bmatrix} = \begin{bmatrix} \dfrac{k_1 E(t) S(t)}{K_{\mathrm{M}} + S(t)} \\ k_2 P(t) \\ k_{+3} - k_{-3} E(t) \end{bmatrix}. \tag{20.2}$$

From the reactions in Eq. (20.1), the stoichiometric matrix $\mathbf{N}$ can be constructed as

$$\mathbf{N} = \begin{bmatrix} 0 & 0 & 1 \\ 1 & -1 & 0 \\ -1 & 1 & 0 \end{bmatrix}, \tag{20.3}$$

where each row refers to the corresponding species in $\mathbf{c}(t)$, and each column describes the corresponding reaction rate in $\mathbf{v}(\mathbf{c}(t))$. One can then describe the dynamics of the concentrations in the system with the following differential equation:

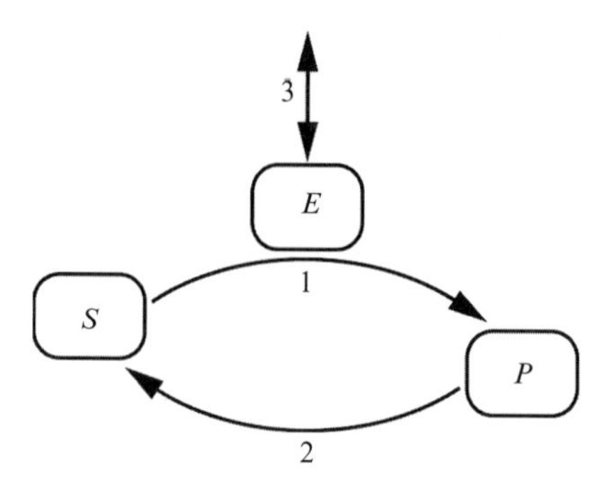

Figure 20.2 Example network. The transformation of substrate S to product P is catalyzed by enzyme E. P then decays to S. Enzyme E is also produced and can decay. Reactions 1 and 2 are irreversible, while reaction 3 is reversible.

$$\frac{\mathrm{d}}{\mathrm{d}t}\mathbf{c}(t) = \mathbf{N}\mathbf{v}(\mathbf{c}(t)) = \mathbf{f}(\mathbf{c}(t)), \tag{20.4}$$

with $\mathbf{c} \in \mathbb{R}^m$, $\mathbf{N} \in \mathbb{R}^{m\times n}$, and $\mathbf{v} \in \mathbb{R}^n$. One property, the so-called unscaled elasticity coefficient, is of particular importance for MRA. The elasticity coefficient characterizes the sensitivity of a reaction rate to perturbations in a species' concentration. Therefore, it provides a measure of the change in a reaction rate in response to a change in a species' concentration. Negative elasticity values represent inhibitory effects of a species on the reaction rate, while positive elasticity values denote activating effects.

For a whole reaction network, this can be expressed by the elasticity coefficient matrix $\boldsymbol{\epsilon} \in \mathbb{R}^{n\times m}$ at steady state $\bar{\mathbf{c}}$. It is formally defined as the partial derivative of $\mathbf{v}(\mathbf{c}(t))$ with respect to $\mathbf{c}(t)$ such that

$$\boldsymbol{\epsilon} = \left.\frac{\partial \mathbf{v}(\mathbf{c}(t))}{\partial \mathbf{c}(t)}\right|_{\bar{\mathbf{c}}} = \frac{\partial \bar{\mathbf{v}}}{\partial \bar{\mathbf{c}}}. \tag{20.5}$$

For the example network of Eqs. (20.1)–(20.3), the unscaled elasticity coefficient matrix evaluated at steady state reads:

$$\boldsymbol{\epsilon} = \begin{bmatrix} \dfrac{k_1 \bar{S}}{K_\mathrm{M} + \bar{S}} & 0 & \dfrac{k_1 K_\mathrm{M} \bar{E}}{(K_\mathrm{M} + \bar{S})^2} \\ 0 & k_2 & 0 \\ -k_{-3} & 0 & 0 \end{bmatrix}. \tag{20.6}$$

An effect of a species on a reaction has, in turn, effects on the rate by which species change. These effects are captured by the Jacobian matrix $\mathbf{J} \in \mathbb{R}^{m\times m}$, which is the linearization of the underlying dynamical system. It can be calculated from the stoichiometric matrix and the elasticity matrix:

$$\mathbf{J} = \frac{\partial \mathbf{f}(\mathbf{c}(t))}{\partial \mathbf{c}(t)} = \frac{\partial(\mathbf{N}\mathbf{v}(\mathbf{c}(t)))}{\partial \mathbf{c}(t)} = \mathbf{N}\frac{\partial \mathbf{v}(\mathbf{c}(t))}{\partial \mathbf{c}(t)} = \mathbf{N}\boldsymbol{\epsilon}. \tag{20.7}$$

For our example system, the Jacobian matrix reads:

$$\mathbf{J} = \begin{bmatrix} -k_{-3} & 0 & 0 \\ \dfrac{k_1\bar{S}}{K_M+\bar{S}} & -k_2 & \dfrac{k_1 K_M \bar{E}}{(K_M+\bar{S})^2} \\ -\dfrac{k_1\bar{S}}{K_M+\bar{S}} & k_2 & -\dfrac{k_1 K_M \bar{E}}{(K_M+\bar{S})^2} \end{bmatrix}. \tag{20.8}$$

The entries of the Jacobian can be used to determine whether one node activates or inhibits another node. If an entry $\mathbf{J}_{(k,l)}$ of the matrix is positive, then an increase in node l increases the rate by which node k changes. Consequently, node k influences node l positively.

In order to quantify the strength of the interaction, MRA then defines a so-called local response matrix $\mathbf{r} \in \mathbb{R}^{m\times m}$, which can be interpreted as a normalization of the Jacobian matrix. This matrix can be calculated by dividing the rows of the Jacobian matrix by its diagonal elements:

$$\mathbf{r} = -(\,\mathrm{diag}(\,\mathbf{J}))^{-1}\mathbf{J}. \tag{20.9}$$

An entry of this matrix quantifies how a concentration in steady state changes when one perturbs the value of one concentration while others remain constant. As for the Jacobian matrix, a nonzero entry implies that there is a direct link between the corresponding species. Positive and negative entries imply activation and inhibition, respectively. Therefore, the structure of the local response matrix describes the direct interactions between species.

However, if all variables are allowed to change, not only direct interactions but also indirect interactions, that is, interactions over several intermediates, come into play. These influences are given by the global response matrix $\mathbf{R} \in \mathbb{R}^{m\times m}$. Interestingly, this global response matrix can be calculated simply by inversion of the local response matrix:

$$\mathbf{R} = -\mathbf{r}^{-1}. \tag{20.10}$$

The influences between species can be categorized into three different cases. First, species k is inactivated by species l if the (k,l)-th entry is less than zero.

Second, species k activates species l if the (k,l)-th entry is greater than zero. And third, there is no influence from species l to species k if the (k, l)th entry is zero.

For our small example system, the local response matrix reads:

$$\mathbf{r} = \begin{bmatrix} -1 & 0 & 0 \\ \dfrac{k_1 \bar{S}}{k_2(K_\mathrm{M} + \bar{S})} & -1 & \dfrac{k_1 K_\mathrm{M} \bar{E}}{k_2(K_\mathrm{M} + \bar{S})^2} \\ -\dfrac{(K_\mathrm{M} + \bar{S})\bar{S}}{K_\mathrm{M}\bar{E}} & \dfrac{k_2(K_\mathrm{M} + \bar{S})^2}{k_1 K_\mathrm{M} \bar{E}} & -1 \end{bmatrix}. \tag{20.11}$$

In this example, we immediately see one problem of the approach: The matrix $\mathbf{r}$ has linearly dependent columns (2nd and 3rd column), that is, its determinant is 0. Therefore, an inverse of $\mathbf{r}$ does not exist, and one cannot calculate the global response matrix. The reason for the linear dependence between columns is that there exist conserved moieties, that is, that linear combinations of concentrations remain constant in the system. In our example, the sum of the concentrations of S and P will remain constant.

3. Conservation Analysis

In order to calculate the global response matrix, one needs to reduce the system to variables that are independent. There exists a manifold of algorithms to systematically reduce the system; readers are referred, for example, to Reder (1988) and Vallabhajosyula *et al.* (2006). Once we identified the conserved moieties, we can reorder the species such that the first species (e.g., E and P in the case of the example system) are linearly independent species and the remaining ones are at the end of the vector (e.g., S). Note that there are typically many choices of which species one can use as independent variables. (One could, e.g., also use E and S as independent species.) After reordering the species, one needs to reorder the stoichiometric matrix $\mathbf{N}$ accordingly, which will then decompose into:

$$\mathbf{N} = \begin{bmatrix} \mathbf{N_R} \\ \mathbf{N_0} \end{bmatrix}, \tag{20.12}$$

where $\mathbf{N_R} \in \mathbb{R}^{m_0 \times n}$ has linearly independent rows. Note that the number of rows m_0 of $\mathbf{N_R}$ equals the rank of the stoichiometric matrix $\mathbf{N}$. Further, from conservation analysis, one obtains a link matrix $\mathbf{\Lambda} \in \mathbb{R}^{m \times m_0}$ that relates the stoichiometric matrix of the full system and the one of the reduced system by

$$\mathbf{N} = \mathbf{\Lambda N_R}. \tag{20.13}$$

Along with this relationship, we can now reformulate Eq. (20.7) such that

$$\mathbf{J} = \mathbf{\Lambda N_R \epsilon}, \tag{20.14}$$

from which the Jacobian matrix for the reduced system results:

$$\mathbf{J_R} = \mathbf{N_R \epsilon \Lambda}. \tag{20.15}$$

For our example system, the matrices $\mathbf{\Lambda}$, $\mathbf{N_R}$, and $\mathbf{J_R}$ would read:

$$\mathbf{\Lambda} = \begin{bmatrix} 1 & 0 \\ 0 & 1 \\ 0 & -1 \end{bmatrix}, \quad \mathbf{N_R} = \begin{bmatrix} 0 & 0 & 1 \\ 1 & -1 & 0 \end{bmatrix}, \tag{20.16}$$

and

$$\mathbf{J_R} = \begin{bmatrix} -k_{-3} & 0 \\ \dfrac{k_1 \bar{S}}{K_M + \bar{S}} & -\dfrac{k_1 K_M \bar{E}}{(K_M + \bar{S})^2} - k_2 \end{bmatrix}. \tag{20.17}$$

The local response matrix of the reduced system $\mathbf{r_R}$ can now be determined by normalization with the help of Eq. (20.9):

$$\mathbf{r_R} = \begin{bmatrix} -1 & 0 \\ \dfrac{k_1 (K_M + \bar{S}) \bar{S}}{k_1 K_M \bar{E} + k_2 (K_M + \bar{S})^2} & -1 \end{bmatrix}. \tag{20.18}$$

This matrix can be inverted to gain the global response matrix of the reduced system

$$\mathbf{R_R} = \begin{bmatrix} 1 & 0 \\ \dfrac{k_1 (K_M + \bar{S}) \bar{S}}{k_1 K_M \bar{E} + k_2 (K_M + \bar{S})^2} & 1 \end{bmatrix}. \tag{20.19}$$

From this, the interaction diagram in Fig. 20.3 can be deduced. Note that for our small example system, the local and global interactions between the linear independent species E and P are the same when we neglect self-loops, as there are no intermediate species present.

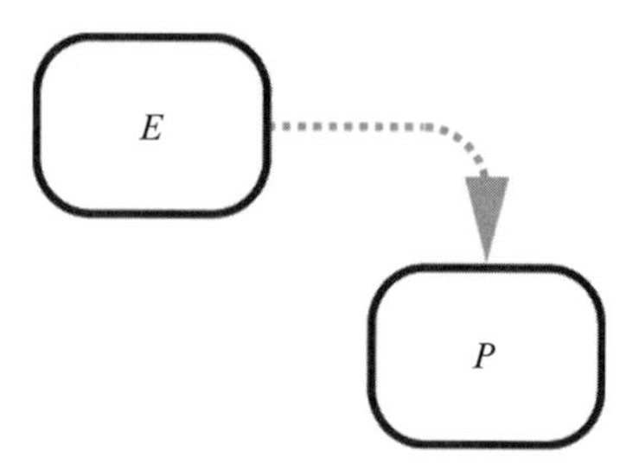

Figure 20.3 Response diagram. Interactions were deduced from the local and the global response matrices of Eqs. (20.18) and (20.19). Activations are marked by a green dashed arrow. (For interpretation of the references to color in this figure legend, the reader is referred to the Web version of this chapter.)

In Section 4, we present a recipe of how to deal with large reaction networks. Further, we perform Monte Carlo computations on the local response matrix and analyze how and to what extend the interactions between species in a system can be determined only from the systems topology.

4. From Reaction Schemes to Influence Networks Using a Monte Carlo Approach

In the following, we exemplify how we can use MRA to extract information flow from reaction networks. Typically, parameters within large-scale networks are unknown or display high uncertainty. Here, we address the question whether we can still deduce qualitatively how information is passed through the network. To do so, we will employ a Monte Carlo approach, where we sample the parameters from a distribution. It was shown that such an approach is very useful when incomplete knowledge about parameters is present (Murabito *et al.*, 2011; Steuer *et al.*, 2006). First, we will reformulate MRA Eq. (20.5) in terms of normalized elasticity coefficients $\widetilde{\epsilon}$:

$$\widetilde{\epsilon} = \frac{\partial \ln \bar{\mathbf{v}}}{\partial \ln \bar{\mathbf{c}}} = \frac{\bar{\mathbf{c}}}{\bar{\mathbf{v}}} \frac{\partial \bar{\mathbf{v}}}{\partial \bar{\mathbf{c}}} = \frac{\bar{\mathbf{c}}}{\bar{\mathbf{v}}} \epsilon. \tag{20.20}$$

In matrix notation and solving for ϵ, Eq. (20.20) yields

$$\epsilon = \bar{\mathbf{V}} \widetilde{\epsilon} \bar{\mathbf{C}}^{-1}, \tag{20.21}$$

where $\bar{\mathbf{V}} = \mathrm{diag}(\bar{\mathbf{v}})$ and $\bar{\mathbf{C}} = \mathrm{diag}(\bar{\mathbf{c}})$. In Michaelis–Menten-type enzymatic reactions, these normalized elasticity coefficients range typically between zero and one for substrates, zero and minus one for products,

and equal one for enzymes (see also Heinrich and Rapoport, 1974a,b, or any review on metabolic control analysis).
Therefore, the reduced Jacobian is provided by

$$\mathbf{J_R} = \mathbf{N_R}\bar{\mathbf{V}}\tilde{\boldsymbol{\epsilon}}\bar{\mathbf{C}}^{-1}\boldsymbol{\Lambda}, \tag{20.22}$$

while the local response matrix to investigate reads:

$$\mathbf{r_R} = -(\,\mathrm{diag}(\mathbf{J_R}))^{-1}\mathbf{J_R}. \tag{20.23}$$

These equations are then used to perform Monte Carlo sampling. $\mathbf{N_R}$ and $\boldsymbol{\Lambda}$ are known. Further, the signs of the entries in $\tilde{\boldsymbol{\epsilon}}$ are known and we decided to sample $\tilde{\boldsymbol{\epsilon}}$ from a uniform distribution between 0 and 1, and set the sign accordingly. In contrast, the values in the matrices $\bar{\mathbf{V}}$ and $\bar{\mathbf{C}}$ are unknown. Therefore, we chose to sample $\bar{\mathbf{V}}$ and $\bar{\mathbf{C}}$ according to a lognormal distribution with parameters $\mu = 1$ μM and $\sigma = 0.8$ μM (which corresponds to a mean of 3.74 μM and standard deviation of 3.54 μM). For s samples, this results in s different local response matrices $\mathbf{r}^s_{\mathbf{R}}$.

In a next step, we compare the individual entries $r^s_{R_{k,l}}$ of each sampled system to conclude that (i) an interaction is an inhibition for all s samples if all $r^s_{R_{k,l}} < 0$, (ii) an interaction is an activation for all s samples if all $r^s_{R_{k,l}} > 0$, and (iii) an interaction varies between inhibition and activation when $r^s_{R_{k,l}} \neq 0$ for all s samples and the sign of $r^s_{R_{k,l}}$ changes at least ones in all samples s.

We will now demonstrate the Monte Carlo simulations on the intermediate-scale model of the Wnt pathway. The Wnt signaling plays an important role in carcinogenesis. We use the kinetic model of the canonical Wnt pathway (Fig. 20.4). The model was first derived by Lee *et al.* (2003). If Wnt is not present, the so-called destruction complex consisting of APC, Axin, and GSK3 forms and phosphorylates β-catenin. Phosphorylated β-catenin is a substrate for ubiquitination and thus enters proteolysis. When Wnt is present, it binds to cell surface receptor Frizzled which, in turn, activates disheveled (Dsh). The active form of Dsh inhibits the destruction complex. Then, less destruction complex is present to phosphorylate β-catenin. Subsequently, access β-catenin accumulates and translocates to the nucleus where it regulates genes together with the TCF.

From the conservation analysis, we find that there are four conserved moieties in the model. Therefore, the Wnt model can be reduced from 15 overall species to 11 independent species. The unscaled elasticity coefficient matrix $\boldsymbol{\epsilon}$ can be calculated according to Eq. (20.5). In order to test our approach, we removed the exact kinetic laws, initial conditions, and parameter values defined in the model by Lee *et al.* (2003). We only sustained the signs in $\boldsymbol{\epsilon}$ to preserve the original model structure. We then generated many parameter sets sampled from the aforementioned distributions and

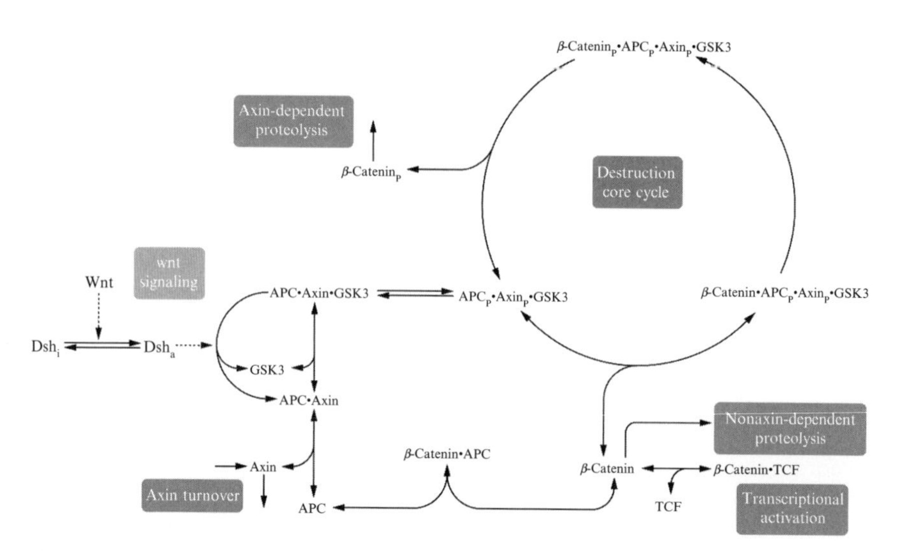

Figure 20.4 Reaction scheme of the Wnt model. Protein complexes are denoted by the names of their components separated by bullets, while phosphorylated, inactive, and active components are marked with a lowercase *p*, a lowercase *i*, or a lowercase *a*, respectively. Single- and double-headed arrows denote irreversible and reversible reactions, respectively. The individual reactions and their role in the Wnt pathway are outlined in the text.

calculated the local response matrix for every sample for each of the parameter sets. When analyzing the effects of the number of Monte Carlo samples, we find that 1000 is a typical number, where even for large-scale networks, larger sample sizes do not provide further information (data not shown). For the reduced Wnt model, we find that the signs of only three local response coefficients change over the different sample sets. All other 26 interactions could be uniquely determined using the Monte Carlo approach to be either activation (12 interactions) or inhibition (14 interactions). We visualized the elements that are positive, negative, or changing between positive and negative over all samples in the interaction diagram of Fig. 20.5. As the diagonal of the local response matrix is minus one by definition, we ignored to visualize the self-inhibitions in the diagram.

Within the sampling process, we can further calculate the global response matrix for every sample in the same way we calculated the local response matrix. The biological most meaningful global interaction is the effect from input (Dsh_i) to output (TCF). The analysis of the corresponding element in $\mathbf{R}^s_\mathbf{R}$ yields a changing sign for different samples. When we now repeat the whole sampling process a thousand times (i.e., 1000 times 1000 samples), the corresponding element in the global response matrix is positive in 99.3 ± 0.26% of the cases. This result strongly suggests that Dsh_i has an activating effect on TCF.

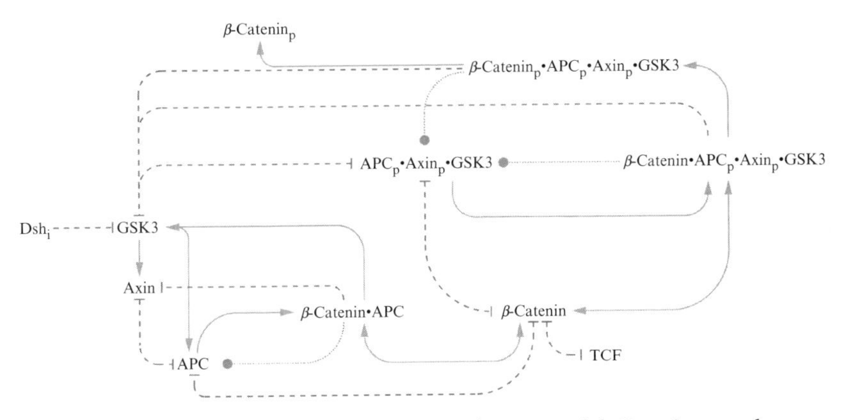

Figure 20.5 Direct interactions in the reduced Wnt model. Protein complexes are denoted by the names of their components separated by bullets, while phosphorylated, inactive, and active components are marked with a lowercase *p*, a lowercase *i*, or a lowercase *a*, respectively. Single-headed arrows describe a directed influence, while double-headed arrows depict a mutual interaction between the connected species. Inhibition and activation are shown with red dashed and green solid arrows, respectively, while blue dotted arrows mask a nondeterminable interaction type. (For interpretation of the references to color in this figure legend, the reader is referred to the Web version of this chapter.)

The high percentage of uniquely determinable local interactions leads to the hypothesis that the knowledge of interaction type and the stoichiometric matrix is sufficient to unveil the directionality of information flow also in larger systems. To test this hypothesis, we applied our approach to larger and very large signal transduction models. As one example, we used an SBML (Bornstein *et al.*, 2008) implementation of the MAPK model by Schoeberl *et al.* (2002) that has 97 species and 148 reactions. Conservation analysis yields five conserved moieties. As another example, we translated the whole signaling part of the Reactome database (Matthews *et al.*, 2009; Vastrik *et al.*, 2007) into a stoichiometric and an elasticity coefficient matrix with 6232 species and 3652 reactions. Conservation analysis shows that about half of the present species are linearly dependent in this large network.

In Fig. 20.6, we compare the four models (e.g., Example, Wnt, MAPK, and Reactome) with respect to the distributions of the signs in the local response matrices over all 1000 samples. What becomes apparent is that even for larger models, the fraction of interactions that change their signs over the different sampling sets are very low. For the MAPK model, there were only about 1.3% of all possible interactions, while the Reactome model yielded a fraction of 0.6‰ of the nonunique interactions. Especially in the Reactome network, the percentage of species that interact directly is obviously very low. Here, also the fraction of unambiguous positive and negative interactions is very low as well (1.5‰ and 1.8‰).

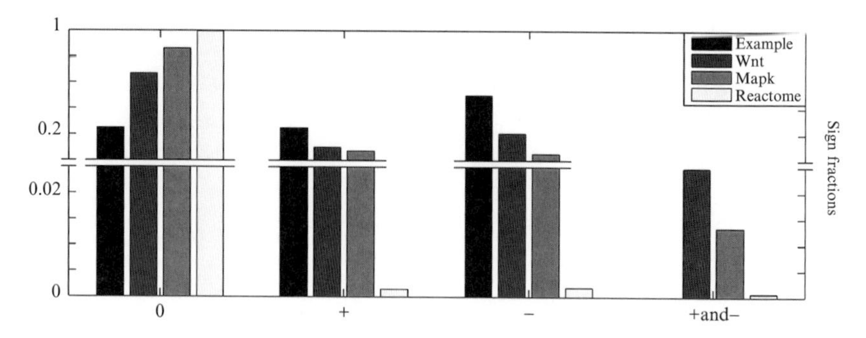

Figure 20.6 Sign distribution. Fractions of signs in the local response matrix with respect to the number of elements in **r**. 0, +, and − represent the fraction of elements that are always zero, always positive, and always negative, respectively. The fourth column of bars represents elements that change their sign over different samples.

In conclusion, our Monte Carlo simulations using modular response and conservation analysis show that the sign of interaction (activation or inhibition) is in most cases already defined through the kinetic scheme of the reaction network and does not require knowledge of the parameters. More than 99% of the signs of the local response matrix could be identified uniquely from the structure of the reaction network. That means, even without knowing the reaction network in detail, the direction and sign of information flow can often be determined.

5. Conclusion

Information flow is the important property of biochemical signaling networks. It determines the effect of a stimulus on the readout of a network. In this chapter, we presented a recipe to tackle this question on a computational and qualitative level. MRA serves as a handy instrument to gain an insight to otherwise hidden network properties. In combination with conservation analysis, we could show that a very high percentage of species interactions could be determined solely from the topology of the network. This method is applicable to molecular reaction systems where not all parameters of kinetic laws are known, especially for large and very large signaling networks.

Acknowledgments

We thank Bente Kofahl for providing a Mathematica notebook of the Wnt model. We also thank Ralf Steuer for very helpful discussions, and Franziska Witzel for critically reading the chapter. This work was supported by CancerSys (EU FP7) and FORSYS (BMBF).

REFERENCES

Bornstein, B. J., Keating, S. M., Jouraku, A., and Hucka, M. (2008). Libsbml: An Api library for sbml. *Bioinformatics* **24,** 880–881.

Bruggeman, F. J., Westerhoff, H. V., Hoek, J. B., and Kholodenko, B. N. (2002). Modular response analysis of cellular regulatory networks. *J. Theor. Biol.* **218,** 507–520.

Bruggeman, F. J., Bluethgen, N., and Westerhoff, H. V. (2009). Noise management by molecular networks. *PLoS Comput. Biol.* **5,** e1000506.

Heinrich, R., and Rapoport, T. A. (1974a). A linear steady-state treatment of enzymatic chains. General properties, control and effector strength. *Eur. J. Biochem.* **42,** 89–95.

Heinrich, R., and Rapoport, T. A. (1974b). A linear steady-state treatment of enzymatic chains. Critique of the crossover theorem and a general procedure to identify interaction sites with an effector. *Eur. J. Biochem.* **42**(1), 97–105.

Kacser, H., and Burns, J. A. (1973). The control of flux. *Symp. Soc. Exp. Biol.* **27,** 65–104.

Kholodenko, B. N., Hoek, J. B., Westerhoff, H. V., and Brown, G. C. (1997). Quantification of information transfer via cellular signal transduction pathways. *FEBS Lett.* **414,** 430–434.

Kholodenko, B. N., Kiyatkin, A., Bruggeman, F. J., Sontag, E., Westerhoff, H. V., and Hoek, J. B. (2002). Untangling the wires: A strategy to trace functional interactions in signaling and gene networks. *Proc. Natl. Acad. Sci. USA* **99,** 12841–12846.

Lee, E., Salic, A., Krueger, R., Heinrich, R., and Kirschner, M. W. (2003). The roles of apc and axin derived from experimental and theoretical analysis of the Wnt pathway. *PLoS Biol.* **1,** 116–132.

Matthews, L., Gopinath, G., Gillespie, M., Caudy, M., Croft, D., de Bono, B., Garapati, P., Hemish, J., Hermjakob, H., Jassal, B., Kanapin, A., Lewis, S., *et al.* (2009). Reactome knowledgebase of human biological pathways and processes. *Nucleic Acids Res.* **37,** D619–D622.

Murabito, E., Smallbone, K., Swinton, J., Westerhoff, H. V., and Steuer, R. (2011). A probabilistic approach to identify putative drug targets in biochemical networks. *J. R. Soc. Interface* **8,** 880–895.

Reder, C. (1988). Metabolic control theory: A structural approach. *J. Theor. Biol.* **135,** 175–201.

Santos, S. D. M., Verveer, P. J., and Bastiaens, P. I. H. (2007). Growth factor-induced mapk network topology shapes erk response determining. pc-12 cell fate. *Nat. Cell Biol.* **9,** 324–330.

Schoeberl, B., Eichler-Jonsson, E., Gilles, E. D., and Mueller, G. (2002). Computational modeling of the dynamics of the map kinase cascade activated by surface and internalized egf receptors. *Nat. Biotechnol.* **20,** 370–375.

Steuer, R., Gross, T., Selbig, J., and Blasius, B. (2006). Structural kinetic modeling of metabolic networks. *Proc. Natl. Acad. Sci. USA* **103,** 11868–11873.

Vallabhajosyula, R. R., Chickarmane, V., and Sauro, H. M. (2006). Conservation analysis of large biochemical networks. *Bioinformatics* **22,** 346–353.

Vastrik, I., D'Eustachio, P., Schmidt, E., Joshi-Tope, G., Gopinath, G., Croft, D., de Bono, B., Gillespie, M., Jassal, B., Lewis, S., Matthews, L., Wu, G., *et al.* (2007). Reactome: A knowledge base of biologic pathways and processes. *Genome Biol.* **8,** R39.

CHAPTER TWENTY-ONE

Whole-Genome Metabolic Network Reconstruction and Constraint-Based Modeling[☆]

Charles R. Haggart,[*,†] Jennifer A. Bartell,[*] Jeffrey J. Saucerman,[*,†] *and* Jason A. Papin[*]

Contents

Abstract

With the advent of modern high-throughput genomics, there is a significant need for genome-scale analysis techniques that can assist in complex systems analysis. Metabolic genome-scale network reconstructions (GENREs) paired

[☆] This chapter and Chapter 3 in Section 8 are focused on large-scale metabolic model reconstruction. The approaches and viewpoints of the authors differ, and the editors would advise readers to consider both chapters together for a broader understanding of the methodologies and issues within this field.

[*] Department of Biomedical Engineering, University of Virginia, Charlottesville, Virginia, USA
[†] Robert M. Berne Cardiovascular Research Center, University of Virginia, Charlottesville, Virginia, USA

Methods in Enzymology, Volume 500
ISSN 0076-6879, DOI: 10.1016/B978-0-12-385118-5.00021-9

with constraint-based modeling are an efficient method to integrate genomics, transcriptomics, and proteomics to conduct organism-specific analysis. This text explains key steps in the GENRE construction process and several methods of constraint-based modeling that can help elucidate basic life processes and development of disease treatment, bioenergy solutions, and industrial bioproduction applications.

1. Introduction

The rapid expansion of methods to utilize organism-specific whole-genome sequences provided by high-throughput sequencing technology has provided clarity in areas scientists have long puzzled over. We have greatly improved our ability to probe for answers on a system-wide level in addition to our classic reductionist investigative strategy. The substantial amounts of data now being harvested on phenotypic, genetic, protein, and molecular scales are driving the development of computational systems analysis ever faster in a search for ways to organize and contextualize raw data into a coherent picture. One promising area within the field of systems biology is reconstruction of organism-specific genome-scale metabolic networks accompanied by the development of a wide range of constraint-based modeling approaches. The metabolic genome-scale network reconstruction (GENRE) provides a framework to organize all available information about an organism's metabolism through careful construction and curation of a computational network that links the cell's genome and gene expression to metabolic reaction fluxes and biomass and energy production and consumption. Once completed, the power of each GENRE can be realized as a model for probing a cell's genotype–phenotype relationship via constraint-based modeling.

Building an organism-specific GENRE is a promising opportunity to improve metabolic understanding. If the reconstructed network model can be validated through experimental investigations of growth rates, phenotypes, reaction fluxes, and gene expression, it may be immediately useful as a method for drug discovery or the development of strategies for optimizing byproduct metabolite yields. If the model predictions vary from experimental validation, the discrepancies offer the researcher a roadmap for iterative *in vivo* experimentation, *in silico* modeling, and annotation refinement that further the understanding of the organism's metabolism. Also, the GENRE itself, if well constructed and curated, is a directed method of collecting and organizing all currently available knowledge of the organism's metabolism in a functional manner.

A cell's phenotype is subject to the constraints imposed by its genome and extracellular environment. By quantifying these constraints in an appropriate manner, one can predict the range of possible cellular phenotypes *in silico* without acquiring large quantities of kinetic metabolic data. While constraint-based modeling of metabolic networks is far more quantitative than interaction-based (i.e., graph-based) network models (Papin *et al.*, 2003), it does not allow for characterization of the dynamics of metabolism with the same level of confidence as a mechanistic, kinetic model (Raman and Chandra, 2009). However, given the lack of experimentally measured metabolic reaction parameters for most organisms, kinetic modeling is not a widely applicable approach to modeling cellular biochemical networks. Constraint-based modeling presents an alternative approach to better understanding and predicting the behavior of a biochemical network under various environmental and genetic perturbations.

As excellent reviews of the reconstruction process are available (Durot *et al.*, 2009; Feist *et al.*, 2009; Rocha *et al.*, 2008), including highly detailed protocols for reconstructing genome-scale metabolic networks (Thiele and Palsson, 2010), the present work focuses on providing an introduction to the reconstruction process, including important methods, tools, and validation techniques, as well as the constraint-based modeling approaches that utilize the curated information within each GENRE.

2. Metabolic Network Reconstruction

Metabolic network reconstructions comprise the stoichiometry of reactions necessary for nutrient usage in synthesis and degradation of basic metabolites and more complex compounds, specific genes whose protein products are associated with these biochemical reactions, and supporting annotation and literature references. This list can then be converted to mathematical form and combined with constraint-based modeling approaches detailed in the second half of this chapter to predict metabolic phenotype. The reconstruction of a metabolic network of an organism with a completed genome sequence consists of the following steps: (1) genome annotation, (2) automated network reconstruction, (3) network refinement, (4) *in vitro* experimentation, and (5) gap analysis. These steps are often completed concurrently or iteratively to improve the accuracy of the model as shown in Fig. 21.1.

2.1. Genome annotation

Since the first genome was sequenced via high-throughput technology in 1998, technological advancement has decreased the required cost and time to obtain a full genome sequence (Kircher and Kelso, 2010). However, for completed genome sequences to be useful, they need to be annotated using standardized gene ontologies that provide consistency while identifying genes and cataloging their function and regulation in relation to biological processes (Giglio *et al.*, 2009). Automated annotation programs (ERGO, J. Craig Venter

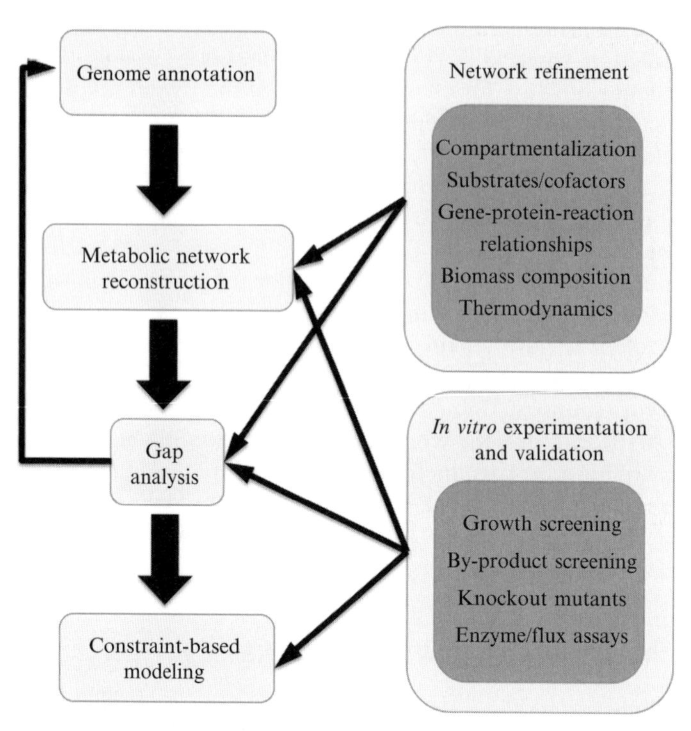

Figure 21.1 Flow diagram illustrates the iterative multistep methodology for reconstructing and simulating the behavior of an organism-specific genome-scale metabolic network model.

Institute, Integrated Microbial Genomes) provide annotations that usually require manual curation to supplement organism-specific depth and detail as well as verify the automated annotation assignments (Yang *et al.*, 2010). Some effort has been made to design methods for evaluating annotation quality dependent on the degree of sequence match and phylogenetic relatedness to other annotated genomes (Yang *et al.*, 2010). Alignment tools such as BLAST are commonly used to annotate gene function based on orthology with previously annotated genomes provided in online databases, followed by other more intensive methods relying on phylogenetic grouping (Kuzniar *et al.*, 2008). Both general and organism-specific databases of genome annotations are available as listed in the appropriate section of Table 21.1, which also details other useful tools in the reconstruction process.

2.2. Automated network reconstruction

Most metabolic reconstruction efforts begin with an automated draft reconstruction from a publicly annotated genome available online. Much effort is being directed to development of comprehensive automation suites. One

Table 21.1 Valuable resources for GENRE curation and analysis efforts

Genome annotation resources		
ERGO	Automated and manual annotation with pathway mapping and bioinformatics tools	http://ergo.integratedgenomics.com
Comprehensive microbial resource	Database of all prokaryotic genomes sequenced to date with automated and/or manual annotation	http://cmr.jcvi.org/tigr-scripts/CMR/CmrHomePage.cgi
Integrated microbial genomes	Automated genome annotation and comparative genomics analysis	http://img.jgi.doe.gov
ENSEMBL	Sequence access portal with annotation, visualization, and analysis tools	http://www.ensemblgenomes.org/
NCBI Entrez Genome	Currently over 12,000 sequenced and/or annotated records covering all major organism groups	http://www.ncbi.nlm.nih.gov/sites/genome
KEGG	Databases on sequenced genomes, orthology, hierarchy, reactions, proteins, and pathway maps	http://www.genome.jp/kegg/
UCSC genome bioinformatics	Browser and database useful for comparing genomes and annotations	http://genome.ucsc.edu/
GO	Database of standardized annotation terms	http://www.geneontology.org/
Gene sequence alignment tools		
BLAST	Local alignment search of protein/nucleotide sequences, does not allow gaps when matching sequences	http://blast.ncbi.nlm.nih.gov/
FASTA	Local alignment search of protein/nucleotide sequences. allows gaps when matching sequences	http://www.ebi.ac.uk/Tools/fasta/
BLAT	Local alignment search matching sequence greater than 25 bp long	http://genome.ucsc.edu/cgi-bin/hgBlat?org=human

(*Continued*)

Table 21.1 (*Continued*)

Enzyme/protein databases		
UniProt/Swiss-Prot	Manual annotation of protein sequence function	http://www.uniprot.org
UniProt/TrEMBL	Automatic annotation of protein sequence function	http://www.uniprot.org
NCBI-RefSeq	Curated DNA, RNA, and protein sequence annotations	http://www.ncbi.nlm.nih.gov/RefSeq/
BRENDA	Enzymes with associated reactions, pathways, substrates, and cofactors	http://www.brenda-enzymes.org/
Automated reconstruction		
SEED	Automated annotation and draft reconstruction of user-input genomes	http://www.theSEED.org/
PathwayTools	Draft reconstruction with visualization tools and pathway gap prediction	http://bioinformatics.ai.sri.com/ptools/
metaSHARK	Draft reconstruction with ability to overlay information on generic metabolic networks	http://bmbpcu36.leeds.ac.uk/shark/
KOBAS	Automated annotation and reconstruction using KEGG orthology	http://kobas.cbi.pku.edu.cn/
ASGARD	Automated annotation and reconstruction	http://sourceforge.net/projects/asgard-bio/
Analysis		
CellNetAnalyzer	Matlab-based environment for topological network analysis of metabolism and cell signaling	http://www.mpi-magdeburg.mpg.de/projects/cna/cna.html
Metatool	Analysis suite that finds a set of elementary modes (nondecomposable-constrained steady-state networks) for a reconstruction	http://pinguin.biologie.uni-jena.de/bioinformatik/networks/

such effort, the Model SEED project, moves from an unannotated, sequenced genome to a draft metabolic network with gap filling and verification features. Programs such as Pathway Tools and others listed in Table 21.1 require a previously annotated genome to build a draft reconstruction and offer various visualization and analysis tools. These programs produce at minimum lists of genes, their associated reactions, and corresponding Enzyme Commission (EC) numbers. It is also possible to use appropriate ontology keywords to retrieve core metabolic genes and reactions from an annotation. However, regardless of the automated tool used, manual reconstruction efforts are always necessary to ensure organism-specific metabolic characteristics are included (DeJongh *et al.*, 2007; Karp *et al.*, 2010).

2.3. Network refinement

Manual reconstruction begins with an initial evaluation of the completeness of the automated draft reconstruction. Each reaction must be evaluated for its necessity to the model; accurate stoichiometry; reaction direction and reversibility; and role in metabolite production, usage, and recycling. Other concerns include thermodynamic feasibility and energy constraints. Integral to a successful reconstruction, this manual evaluation can require a significant amount of dedicated work to complete.

2.3.1. Key organizational tools

Research groups commonly use the few techniques listed below to speed the reconstruction process and provide consistency between different reconstruction projects.

(i) *Spreadsheet organization*: A thorough and well-organized annotation record, stored in a spreadsheet or as a file in systems biology markup language (SBML) format with embedded annotations (Herrgard *et al.*, 2008), should include the gene name, all pertinent gene abbreviations, the reaction equation with consistent reactant, substrate, and product symbols, balanced stoichiometry, an indicator of reversibility, the metabolic subsystem that utilizes the reaction, the associated protein and its EC number, any literature references used to adjust or add the reaction, a rating of confidence in the annotation entry, and comments about any adjustments or questions about the entry.

(ii) *Reaction confidence level*: In curating the reconstruction, it is helpful to keep track of the quality of the various annotations and changes. A standard method is to apply a classification system to each entry, ranging from class 1 (high confidence in the data) to class 4 (low confidence). This judgment is still qualitative, so consistency in class assignment within the reconstruction is important. When adding a reaction based solely on sequence matches with another organism, the confidence level of the added reaction should be based on the phylogenetic relatedness of the organism. It is also vital to record the addition of enzymes that have not been associated with a particular gene in the organism's sequence data but are known to be necessary in a certain pathway. Examples of metabolic reconstructions that have utilized reaction confidence ratings include iMO1056 (Oberhardt *et al.*, 2008) and human Recon 1 (Duarte *et al.*, 2007).

2.3.2. Organism-specific curation

After canonical metabolic reactions have been collected via the automated draft reconstruction, a review of literature pertaining to the metabolism and function of the organism is necessary to identify organism-specific characteristics that should be integrated. Reconstructions of related organisms can be searched for specific functionality such as substrate usage preference and unique metabolic products that may be replicated in the organism of interest. For example, the reconstruction of *Aspergillus niger* included pathways related to steroid synthesis based on the biology of other aspergilli fungi, while a gap in lipopolysaccharide synthesis in the reconstruction of *Pseudomonas aeruginosa* was filled using gene similarity in the *Pseudomonas fluorescens* annotation (Andersen *et al.*, 2008; Oberhardt *et al.*, 2008).

(i) *Compartmentalization and exchange reactions*: When building the reconstruction, care must be taken to check the transport and exchange potential of a particular metabolite so that appropriate limits can be implemented as constraints on model behavior. Localization and transport of a metabolite between the cytosol and organelles or within the cytosol may be complicated, particularly in eukaryotic organisms, and the order of transport may be critical to metabolic processes (Hao *et al.*, 2010). Exchange reactions can be implemented that account for the movement of metabolites across intra- and intercellular membranes, creating compartments within the *in silico* network to represent these physical boundaries.

(ii) *Gene–protein–reaction relationships (GPRs)*: While data for the reconstruction are based on gene annotations, constraint-based modeling can proceed using solely the enzymes and associated reaction stoichiometries. However, to simulate the phenotypic effects of gene knockout (KO) experiments and enable integration of gene expression microarray data, genes must be connected to their associated enzymes. Boolean logic is often used to define this relationship. When multiple genes encode enzyme subunits that are each required for catalysis of a reaction, an "AND" statement is used to convey necessity of each gene for that reaction. Alternately, an "OR" statement is used to reflect the requirement of genes that encode isozymes that catalyze the same reaction independently of one another. Some enzymes may perform similar functions with unclear substrate specificity where it is difficult to correlate a particular gene with a particular pathway (Andersen *et al.*, 2008) and may require approximation during modeling. The incorporation of GPRs has been incredibly valuable to researchers interested in studying the impact of different growth states, environment, or degree of infection measured with microarrays and connecting the gene expression results to a modeled reconstruction to investigate changes in metabolism (Oberhardt *et al.*, 2010).

(iii) *Biomass composition*: Biomass, the key components from which an organism is built and sustains itself, may be determined from experimental measures, described here, or through more computational methods described in Section 3. DNA, RNA, lipids, fatty acids, cell wall components, cytosolic solutes and ions, and cofactors must be present in sufficient amounts to enable cell operation (Senger, 2010). Experimentally, biomass composition can be determined from carbon and amino acid content measured at different population levels and growth states (Gonzalez *et al.*, 2010). The number of factors that should be included varies with the organism's metabolic characteristics, but even biomass composition of the minimal cell model *Mycoplasma genitalium* includes 61 different factors (Suthers *et al.*, 2009).

(iv) *Thermodynamics*: Assessing reaction thermodynamics can alleviate excessive constraining of the model's directionality (Henry *et al.*, 2009). Directionality of reactions (reversible vs. irreversible) in the model and feasible rates of reaction are based on thermodynamic favorability determined from Gibbs free energy changes. Experimentally determined values can be found in literature, but available data are rarely sufficient to address a genome-scale reconstruction and more often have been obtained for a subset of metabolism or major canonical pathways. Methods for computationally estimating the free energy values of reactions include group contribution methods that can be performed online at Web GCM using reaction information from KEGG and BRENDA (Jankowski *et al.*, 2008).

2.4. *In vitro* experimentation and validation

In conjunction with *in silico* reconstruction efforts that rely on annotations, databases, and literature review, valuable information can also be collected through *in vivo* and *in vitro* experimentation. Growing an organism on specific carbon sources via high-throughput Biolog microplates or more traditional techniques is an easy method for initial model validation because carbon source can be limited *in silico*. The organism's carbon source usage and production of byproducts during growth can be investigated by measuring media metabolite concentration using high-performance chromatography. Many organisms have also been examined to determine survival genes through mutagenesis studies that create single gene mutations

(Cameron *et al.*, 2008; Gallagher *et al.*, 2007; Jacobs *et al.*, 2003). Phenotypic characteristics such as substrate usage and byproduct synthesis by a particular mutant can be predicted *in silico* and then compared to *in vitro* experiments to validate and improve GENRE accuracy. Investigating reaction fluxes and static and dynamic localization of specific enzymes in a living cell for comparison with or incorporation into a GENRE is possible via metabolic mapping of fluorescent substrates and cofactors within live cells (Van Noorden, 2010) and pulse labeling, using ^{13}C as a radiotracer to quantify distribution of labeled molecules (Niittylae *et al.*, 2009).

3. Constraint-Based Modeling Methods

Thus far, we have introduced the necessary steps to assemble an *in silico* metabolic GENRE whose components include stoichiometric coefficients for the reactants and products of each metabolic reaction, a Boolean GPR rule-set which defines the genes associated with each reaction, and notation for reversible and irreversible reactions. Each of these constrains the behavior of the metabolic model they comprise and limits the attainable metabolic phenotype of the cell. An organism's metabolic phenotype is often quantified by growth rate, the amount of biomass added per unit time. Because the reactions that produce individual biomass components are integrated into the GENRE (e.g., DNA, RNA, amino acids, lipids, carbohydrates), we can also characterize metabolic phenotype by the amount of flux (metabolite mass per dry organism mass per unit time) that is carried through these reactions in the GENRE. Constraint-based modeling aims to reduce the number of possible flux profiles and identify one that best predicts the metabolic phenotype of the organism under specified genetic and environmental conditions.

Constraints limit the number of possible flux profiles for a given organism and allow us to make more accurate *in silico* predictions of metabolic phenotype. These constraints are generally assigned to one of four categories: physicochemical constraints (e.g., mass and energy balance), environmental constraints (e.g., temperature, pH, substrate availability), spatial/topological constraints (e.g., organelle compartmentalization), and self-regulatory constraints (Price *et al.*, 2004). The rapid growth of complete genome sequencing and curation of metabolic GENREs and the utility of constraint-based modeling have driven the improvement and innovation of methods for constraint-based modeling. Here, we will introduce a range of these methods developed for constraint-based modeling of genome-scale metabolism using GENREs constructed with the methods detailed in Section 2. In particular, we will focus on the quantitative constraints and motivation for method development.

3.1. GENRE-to-model implementation

To test model predictions against experiments, the constraining elements of the GENRE are given mathematical structure, primarily within the sparse $\boldsymbol{S}$ matrix, which quantifies the stoichiometric relationship between reactant and product metabolites for each reaction in the GENRE (Fig. 21.2). The nonzero elements of $\boldsymbol{S}$ are the stoichiometric coefficients collected in the reconstruction process that correspond to the ith metabolite within the jth metabolic reaction, such that each of the m rows represents a metabolite ($M_{1,2,3}$) and each of n columns represents a reaction, including intracellular biochemical conversions ($R_{1,2}$), metabolite exchange (uptake, $X_{1,2}$; secretion, X_3), metabolite demand (cell maintenance, D_1), and intercompartment metabolite transport (extracellular to cytosolic space, $T_{1,2}$). To account for a metabolite's presence in separate compartments, one row should be added to the $\boldsymbol{S}$ matrix for each compartment within which the metabolite exists. By convention, reactants have a negative coefficient and products have a positive one, while flux, $\boldsymbol{v}_i$, is positive for forward biochemical conversions, transport reactions into the cytosol, and exchange reactions out of the system (i.e., secretion), and negative flux indicates a reverse intracellular reaction, an uptake exchange reaction, and transport from the cytosol to another compartment. The GPR relationships between genes and reactions should be coded as a Boolean vector with an element for each reaction, while flux bounds and an objective function should be implemented in vector form, as described in Section 3.2. One can carry out most of the constraint-based methods that follow with these few mathematical and Boolean structures.

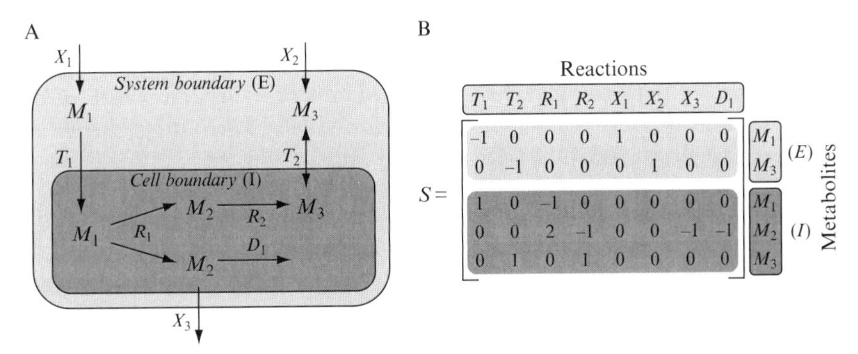

Figure 21.2 Conversion of GENRE reaction stoichiometries to S matrix. (A) Schematic of a toy cell/organism with defined boundaries (E/I), metabolites (M), and reactions (X, T, R, D). (B) Five metabolites and seven reactions comprise the S matrix converted from the cell/organism of (A).

3.2. Flux balance analysis

The most widely used constraint-based method in the field of metabolic engineering and systems analysis of metabolism is flux balance analysis (FBA), and most of the methods described herein are either explicit alternatives or add-on improvements to FBA. The number of review papers dedicated solely to FBA in recent years (Kauffman *et al.*, 2003; Lee *et al.*, 2006; Raman and Chandra, 2009) gives an idea of the wide applicability of this approach. Recent articles provide greater detail to support carrying out an FBA (Oberhardt *et al.*, 2009) as well as a concise, visually aided description of what is achieved by performing an FBA (Orth *et al.*, 2010). FBA allows for the *in silico* prediction of a flux profile that optimizes some predefined cellular objective without making any experimental measurements. It has become a standard tool for simulating the effects of genetic perturbations or environmental conditions on an organism's growth rate or rate of byproduct synthesis (Raman and Chandra, 2009). FBA has also provided a means to simulate the metabolic effects of global gene expression and other large "omics" datasets for an organism or a specific human cell or tissue type (Becker and Palsson, 2008; Jerby *et al.*, 2010; Shlomi *et al.*, 2008).

The underlying theory for FBA assumes that cellular metabolism is at a steady state (i.e., constant growth rate) (Savinell and Palsson, 1992; Varma and Palsson, 1994), and all reaction fluxes ($\boldsymbol{v}_i$) and metabolite concentrations (C_i) have reached a steady state, such that the net amount of each metabolite added to and removed from the cell must equal zero (Eq. (21.1)). Given this, FBA uses linear programming (LP) to find a steady-state flux profile ($\boldsymbol{v}$) that optimizes the organism's objective function (e.g., maximal growth rate) while satisfying a set of physicochemical and environmental constraints. While the objective function can be unique for each FBA problem, a common choice for microorganisms is maximization of biomass (e.g., proteins, nucleic acids, carbohydrates, and lipids necessary for growth), motivated by the idea that evolutionary pressures select for higher growth rates of these organisms. In some organisms, however, alternative objective functions such as maximization (minimization) of ATP production (utilization) are more appropriate, as will be further discussed below. The primary physical constraint that an FBA problem considers is mass balance, requiring that the amount of each metabolite that is (1) transported into or (2) produced within the cell be balanced by the amount of that metabolite that is (3) consumed within or (4) secreted from the cell. The mass balance constraint (Eq. (21.1)) provides m metabolite balance equations to solve for n unknown reaction fluxes, and because there are usually more reactions than metabolites, the system of equations is underdetermined. As such, the collection of flux vectors that satisfy the mass balance constraints span a multidimensional space, often referred to as the *null space* of the $\boldsymbol{S}$ matrix.

$$\frac{dC_i}{dt} = \sum_{j=1}^{n} S_{ij}v_j = 0 \quad \forall i \in M \tag{21.1}$$

$$\alpha_j \leq v_j \leq \beta_j \quad \forall j \in N \tag{21.2}$$

It is important to note that many of the flux profile vectors, $\boldsymbol{v}$, in the null space are physically impossible to achieve (Orth *et al.*, 2010). In order to refine and reduce the size of the predicted flux space and exclude irrelevant flux profiles, the standard FBA implementation constrains individual reaction fluxes, $\boldsymbol{v}_j$, by defining lower and upper bounds (α_j and β_j, respectively; Eq. (21.2)). Some of these bounds are based on literature and experimentation and are collected in the GENRE process. These bounds can be applied to enforce environmental constraints, such as the availability of a substrate in growth media, by setting the upper and lower uptake exchange bounds to the same negative value (recall that exchange flux is negative for uptake and positive for secretion). Similarly, setting the lower and upper bounds on an exchange reaction to zero and some arbitrarily high value, respectively, will inhibit metabolite uptake, while leaving the rate of secretion of the metabolite unconstrained. Bounds can also be set on internal biochemical reactions by constraining the lower bound to zero to enforce thermodynamic irreversibility of a reaction, by setting the bounds to reflect the kinetic limitations of a particular enzyme or membrane transporter, or by constraining the flux to zero to simulate a gene KO. This approach is often used to identify genes essential for growth, both for identification of drug targets and maximizing the production of targeted byproducts (Feist and Palsson, 2008; Oberhardt *et al.*, 2009). Several of the methodologies described below impose more sophisticated or problem-specific constraints on the flux bounds to minimize the possible flux space and provide more accurate predictions of metabolic network behavior, though a standard FBA implementation typically enforces constraints on substrate uptake, reaction irreversibility, and flux capacity.

Once the mass balance constraint and flux bounds are set, an objective function ($\boldsymbol{v}_{\text{obj}}$, Eq. (21.3)) must be defined for the LP optimization problem. The optimization problem will find a flux profile, $\boldsymbol{v}$, which maximizes or minimizes the objective function and satisfies the mass balance constraints and flux bounds described above. In those GENREs for which maximization of biomass production is the organism's objective, a "biomass reaction" is added as a column to the $\boldsymbol{S}$ matrix with stoichiometric coefficients in proportion to each metabolite's *in vivo* contribution to one unit of biomass determined as explained in Section 2.3.2 (Oberhardt *et al.*, 2009). To implement the chosen objective function, a "cost" vector ($\boldsymbol{c}$) is defined, whose elements correspond to each reaction (i.e., columns of $\boldsymbol{S}$, elements of $\boldsymbol{v}$). To maximize the desired reaction(s), all elements of the cost vector are set equal to zero except those corresponding to the chosen objective reaction(s) (Eq. (21.4)). FBA identifies an optimal flux vector whose elements corresponding to the nonzero elements of the cost vector are maximized, subject to the defined constraints. For this example, the first and the second-to-last reactions comprise the objective function and would thus be maximized.

$$max\left(v_{\text{obj}}\right) = c \cdot v \tag{21.3}$$

$$c = [1\ 0\ 0 \cdots 0\ 1\ 0] \tag{21.4}$$

3.3. Dynamic flux balance analysis

The steady-state assumption required of FBA is not appropriate for *in silico* modeling of organisms in all situations, most notably during a "diauxic shift" from one carbon source to another. To address this, dynamic flux balance analysis was developed, based on an earlier approach (Varma and Palsson, 1994), implementing both dynamic (nonlinear programming) and static (LP) optimization of an objective function and applying constraints to the rates of change of flux in addition to the typical FBA constraints (Mahadevan *et al.*, 2002). The static optimization—a more tractable computational approach for large biochemical networks—performs a series of FBA problems at the beginning of each discrete time interval. Using both approaches, Mahadevan and colleagues correctly predicted the timing of acetate production and the sequence of substrate utilization (glucose before acetate) in *Escherichia coli*, validating the use of this constraint-based method to simulate the dynamic behavior of genome-scale metabolism.

3.4. Flux variability analysis

The redundancy inherent in most biochemical networks corresponds to the possibility for numerous alternative flux distributions that yield the same maximal growth rate or ATP synthesis rate (Price *et al.*, 2002). Put in graphical terms, the multidimensional flux space of possible flux profiles often includes multiple points with an identical optimal objective function value. In metabolic pathway terms, different combinations of intermediate reaction fluxes give rise to the same flux through the objective reaction (e.g., biomass), and each of these alternative pathways may be biologically meaningful. FBA is often used to identify a reference *in silico* flux profile with which to compare constraint-based behavior of a genetically or environmentally perturbed counterpart (see Section 3.5). As such, it is critical to understand flux variability, as two alternate FBA reference flux profiles could lead to a significantly different assessment of the perturbed state.

A recent report by Smallbone and Simeonidis introduced a computational method to identify a unique flux profile by analyzing the geometry of the multidimensional flux space defined by the standard FBA constraints (Smallbone and Simeonidis, 2009). This method identifies a minimal flux solution without any thermodynamically infeasible fluxes. An earlier approach by Mahadevan and coworkers called flux variability analysis (FVA) uses constraints to assess the variability of each reaction flux toward an optimized objective flux value (Mahadevan and Schilling, 2003). FVA first uses FBA to determine the optimal objective flux (Eqs. (21.3)–(21.5)), which is used with mass balance and flux bounds (Eqs. (21.1) and (21.2)) as constraints for the subsequent one-by-one maximization and minimization of each reaction flux within the GENRE (Eqs. (21.6) and (21.7)). This method quantifies the maximum and minimum fluxes for each reaction that are consistent with the optimal objective function value.

$$v_{\mathrm{obj}} = \max\left(c_j \cdot v_j\right) \tag{21.5}$$

$$\max v_j, \quad \forall j \in N, j \neq \mathrm{obj} \tag{21.6}$$

$$\min v_j, \quad \forall j \in N, j \neq \mathrm{obj} \tag{21.7}$$

3.5. Minimization of metabolic adjustment

FBA assumes that each organism's metabolic network has been tuned through evolution for some objective function, be it maximal growth rate or energy efficiency (e.g., minimal ATP utilization). While this assumption may be valid for wild-type (WT) organisms that have evolved over many hundreds or thousands of generations, it may be less appropriate for engineered mutants, genetically modified in a controlled laboratory environment and unexposed to the same evolutionary forces or number of reproduction cycles. As such, Segre and colleagues developed minimization of metabolic adjustment (MOMA), hypothesizing that mutant organisms are unable to immediately adapt their metabolic network to achieve the WT objective function but instead display some suboptimal flux profile intermediate to the FBA-determined optima of both the WT (WT-FBA) and genetically perturbed (KO-FBA) organisms (Segre *et al.*, 2002). This approach finds a suboptimal flux profile ($\boldsymbol{v}$) that is a minimal Euclidean distance from the WT-FBA flux profile ($\boldsymbol{v}^{\mathrm{FBA}}$; Eq. (21.8)) and is mathematically formalized as a quadratic programming problem (Eq. (21.9)), subject to the standard FBA mass balance and flux capacity constraints (Eqs. (21.1) and (21.2)), including constraining all KO gene-associated reaction fluxes to zero (Eq. (21.10)). Ultimately, FBA combined with MOMA provides a more accurate prediction of the immediate metabolic response to KO than FBA does on its own (Segre *et al.*, 2002).

$$D\left(v^{\mathrm{FBA}}, v\right) = \sqrt{\sum_{j=1}^{N} \left(v_j^{\mathrm{FBA}} - v_j\right)^2} \quad \forall j \in N \tag{21.8}$$

$$\min\left(v^{\mathrm{FBA}} - v\right)^{\mathrm{T}}\left(v^{\mathrm{FBA}} - v\right) \tag{21.9}$$

$$v_k = 0, \quad \forall k \in A \tag{21.10}$$

3.6. Regulatory on/off minimization

The Euclidean distance metric introduced in MOMA finds a suboptimal KO-associated flux distribution with many component fluxes slightly altered from their WT-FBA state. Shlomi and colleagues developed an alternative method, regulatory on/off minimization (ROOM), which seeks a flux profile for the mutant organism with a minimal number of significant flux changes from the WT-FBA flux profile (Shlomi *et al.*, 2005). This method is based on the assumptions that a KO organism will minimize the costs associated with adapting its gene regulatory network and that these costs are quantifiable—using a Boolean on/off framework—and independent of the magnitude of change in gene expression and associated reaction flux.

ROOM does not explicitly incorporate any gene regulatory constraints into the model but accounts for them via a binary cost variable for each reaction (y_j set to one for fluxes that are significantly different from the WT-FBA; Eq. (21.12)). This quantification of an altered flux is assumed to reflect a concordant change in expression of the necessary enzyme-encoding gene, and the associated cost to the organism of regulating that gene's expression. The objective of this mixed-integer linear programming problem is to minimize the number of fluxes in the mutant GENRE that are different from their WT-FBA counterparts (Eq. (21.11)), while satisfying the FBA mass balance, irreversibility, and flux capacity constraints (Eqs. (21.1) and (21.2)), and constraining the set of fluxes (A) associated with the given KOs to zero (Eq. (21.10)). The upper and lower thresholds by which flux changes are deemed significant ($\boldsymbol{v}^{\mathrm{FBA}+}$ and $\boldsymbol{v}^{\mathrm{FBA}-}$, respectively) are user defined via relative (δ) and absolute (ε) flux tolerance terms (Eqs. (21.13) and (21.14)). Accordingly, if y_j is set to zero, $\boldsymbol{v}_j$ is constrained by these bounds, while setting y_j to one adds to the cost sum (Eq. (21.11)) and leaves $\boldsymbol{v}_j$ unbounded (Eqs. (21.15) and (21.16)). This method was shown to improve predictions of flux and steady-state growth rate over MOMA, though it performed very similarly to FBA (Shlomi *et al.*, 2005).

$$\min \sum_{j=1}^{n} y_j \tag{21.11}$$

$$y_j \in \{0, 1\} \tag{21.12}$$

$$v_j^{\mathrm{FBA}+} = v_j^{\mathrm{FBA}} + \delta \left| v_j^{\mathrm{FBA}} \right| + \varepsilon \tag{21.13}$$

$$v_j^{\mathrm{FBA}-} = v_j^{\mathrm{FBA}} - \delta \left| v_j^{\mathrm{FBA}} \right| - \varepsilon \tag{21.14}$$

$$v_j - y_j \left(v_{\max,j} - v_j^{\mathrm{FBA}+} \right) \leq v_j^{\mathrm{FBA}+} \tag{21.15}$$

$$v_j - y_j \left(v_{\max,j} - v_j^{\mathrm{FBA}-} \right) \leq v_j^{\mathrm{FBA}-} \tag{21.16}$$

3.7. Objective function search methods

Both MOMA and ROOM proposed an alternative to the FBA assumption that organisms are evolved for a specific objective function. Though MOMA and ROOM were developed to address this assumption in the context of a gene KO, the question of a universal WT objective function for an organism remains. To address this, Schuetz and colleagues tested the ability of 11 unique FBA objective functions (or combinations thereof) to accurately predict fluxes in an *E. coli* model of central metabolism (Schuetz *et al.*, 2007) and reported that no single *in silico* objective function was the most accurate predictor of ^{13}C-determined *in vivo* fluxes across many different growth conditions. Several constraint-based approaches have been developed to identify the best objective function for an organism in a specified environment, including ObjFind, BOSS, and a Bayesian probability-based selection method.

3.7.1. ObjFind

ObjFind (Burgard and Maranas, 2003) uses bilevel programming to find a set of positive weights, $\boldsymbol{c}_j$, that maximize the sum of optimal fluxes (Eqs. (21.18) and (21.19)), while minimizing the sum-squared difference between the optimal flux profile, $\boldsymbol{v}_j$, and the experimentally measured fluxes, $v_j^\star$ (Eq. (21.17), where E is the set of all experimentally measured fluxes and P is the set of all reactions that could potentially be cellular objectives). By comparing these optimal weights to those given by a hypothesized objective function, such as biomass, a researcher is able to assess the accuracy of the hypothesized objective function. However, this approach requires that all components of the true objective function be included within the $\boldsymbol{S}$ matrix, *a priori*, and will return the wrong objective if a component reaction is not included in the GENRE.

$$\min_{c_j} \sum_{j \in E} \left(v_j - v_j^* \right)^2 \tag{21.17}$$

$$\max_{v_j} \sum_{j \in P} c_j v_j \tag{21.18}$$

$$\sum_{j \in P} c_j = 1, \quad c_j \geq 0, \quad \forall j \in P \tag{21.19}$$

3.7.2. Biological objective solution search

Gianchandani and coworkers addressed the *a priori* requirement of ObjFind by adding a generic "objective reaction," $\boldsymbol{v}_{\text{obj}}$, to the stoichiometric $\boldsymbol{S}$ matrix (Gianchandani *et al.*, 2008). This Biological Objective Solution Search (BOSS) requires only the $\boldsymbol{S}$ matrix and experimental isotopomer flux data. BOSS adds the generic objective reaction with unknown stoichiometry as an additional column in the $\boldsymbol{S}$ matrix (Eq. (21.20), where m is the total number of metabolites) and performs an FBA on this updated set of mass balance constraints and flux bounds (Eqs. (21.1) and (21.2)), defining $\boldsymbol{v}_{\text{obj}}$ as the objective function to maximize (Eq. (21.22)).

$$S_{i,\text{obj}} = \left[S_{1,\text{obj}} \; S_{2,\text{obj}} \cdots S_{m,\text{obj}} \right] \tag{21.20}$$

$$\min \sum_{j \in N} \left(v_j^{\text{BOSS}} - v_j^* \right)^2 \tag{21.21}$$

$$\max v_{\text{obj}} \tag{21.22}$$

Similar to ObjFind, this inner optimization serves as a constraint on the outer optimization, which minimizes the sum-squared difference between the *in silico* flux profile ($\boldsymbol{v}^{\text{BOSS}}$) and the measured *in vivo* fluxes ($\boldsymbol{v}^{\star}$). This approach will identify the optimal objective reaction not only if it already exists in the network (i.e., with identical stoichiometry as an existing column in $\boldsymbol{S}$) but also if it is a combination of existing reactions or a reaction that was omitted from the GENRE altogether. To validate this method, BOSS was used to identify the objective function in the central metabolic network of *Saccharomyces cerevisiae* and found the best two reactions to be nearly identical to the precursor biomass synthesis reaction and the ATP maintenance reaction, both commonly used FBA objective functions.

3.7.3. Bayesian discrimination

Knorr and colleagues introduced a method which uses FBA and *in vivo* flux measurements to compare any number of candidate objective functions (Knorr *et al.*, 2007). This method determines the most probable objective function by calculating posterior probabilities. Specifically, it calculates the probability of each objective function (F_x), given the product of differences between predicted and measured data (Y), normalized by the sum of all posterior probabilities (Eq. (21.23)). While this approach again requires the definition of candidate objective functions *a priori*, the validation used only a few *in vivo* measurements (growth rate, oxygen uptake rate, succinate uptake rate, and acetate production rate) to compare to *in silico* predictions.

$$\pi(F_x|Y) = \frac{p(F_x|Y)}{\sum_z p(F_z|Y)} \tag{21.23}$$

3.8. Multiple metabolic objectives

Selecting the most biologically relevant objective function is critical to accurately predicting cellular metabolism by FBA or other constraint-based methods, though more than one objective function is occasionally desired. For example, metabolic engineering often aims to identify the genetic manipulations for an organism that will optimize its synthesis rate of a desired byproduct (e.g., ethanol in *E. coli*). An appropriate approach must optimize the organism's intrinsic objective function (e.g., max growth) in parallel with this secondary engineered objective. Burgard and colleagues developed OptKnock, a constraint-based method, to tackle this problem of multiple objective functions (Burgard *et al.*, 2003). Similar to the bilevel optimizations of ObjFind and BOSS, OptKnock uses an FBA framework to maximize the cellular objective, $\boldsymbol{v}_{\text{obj}}$, subject to mass balance and flux bounds constraints (Eqs. (21.1) and (21.2)), as well as gene KO constraints (Eqs. (21.25)–(21.28), where K is the maximum number of allowable KOs across the set of all reactions, N). In parallel, this approach varies the number and identity of KO genes to maximize byproduct secretion, $\boldsymbol{v}_{\text{byproduct}}$ (Eq. (21.24)). More recently, numerous constraint-based metabolic engineering methods have been developed to identify optimal gene deletions with more computational efficiency than OptKnock (OptGene and GDLS) and to identify optimal gene knock-ins (OptStrain) or manipulations to increase or decrease gene expression (OptReg) (Lun *et al.*, 2009; Patil *et al.*, 2005; Pharkya and Maranas, 2006; Pharkya *et al.*, 2004).

$$\max_{y_j} \ v_{\text{byproduct}} \tag{21.24}$$

$$\max_{y_j} \ v_{\text{obj}} \tag{21.25}$$

$$v_j^{\min} \cdot y_j \leq v_j \leq v_j^{\max} \cdot y_j, \quad \forall j \in N \tag{21.26}$$

$$y_j = \{0, 1\} \tag{21.27}$$

$$\sum_{j \in N} (1 - y_j) \leq K \tag{21.28}$$

3.9. Constraint-based modeling software

The bulk of the aforementioned constraint-based modeling approaches are LP problems, which can be solved using a standard solver of which there are many open source (LP_Solve, glpk) and proprietary versions (Gurobi, CPLEX, LINDO). The constraint-based reconstruction and analysis (COBRA) toolbox has become a popular means to apply many of these constraint-based methods (e.g., FBA, MOMA, and FVA) to GENRE-derived models (Becker *et al.*, 2007). This toolbox is open source and has been written to operate in the MATLAB programming environment using the SBML. A recently developed open source software platform, OptFlux, combines strain optimization methods (e.g., OptKnock) as well as standard constraint-based approaches FBA, MOMA, and ROOM (Rocha *et al.*, 2010).

4. Summary

In this chapter, we have attempted to detail the key steps in the metabolic reconstruction process as well as introduce several constraint-based modeling techniques useful for various applications and desired outcomes. Again, it is critical to use the highest quality annotation available and dedicate significant time to refinement of the network reconstructions with well-curated databases and quality literature references in addition to the results of automated reconstruction techniques. Some common methods of collecting experimental data for both model improvement and validation are detailed, but many more are available that may be more appropriate for specific applications. The constraint-based modeling techniques presented are widely used by systems metabolism researchers and are accompanied by additional information helpful during difficult portions of the modeling process such as objective function definition and variability analysis. The field is a fast growing one, with an exponential rise in new reconstructions and further investigations into comparative model building techniques and reconstruction consistency, and we hope this text will be a helpful reference in these efforts.

ACKNOWLEDGMENTS

This work was supported by funding from the National Science Foundation (CAREER Grant #0643548 to JAP), the National Institutes of Health and National Institute of General Medical Sciences (Grant #GM088244 to JAP), the Cystic Fibrosis Research Foundation (Grant #1060 to JAP), and the Robert M. Berne Cardiovascular Research Center (USPHS Grant #HL007284-33 Basic Cardiovascular Research Training Grant providing postdoctoral support to CRH).

REFERENCES

Andersen, M. R., Nielsen, M. L., and Nielsen, J. (2008). Metabolic model integration of the bibliome, genome, metabolome and reactome of Aspergillus niger. *Mol. Syst. Biol.* **4,** 178.

Becker, S. A., and Palsson, B. O. (2008). Context-specific metabolic networks are consistent with experiments. *PLoS Comput. Biol.* **4**(5), e1000082.

Becker, S. A., Feist, A. M., Mo, M. L., Hannum, G., Palsson, B. O., *et al.* (2007). Quantitative prediction of cellular metabolism with constraint-based models: The COBRA toolbox. *Nat. Protoc.* **2**(3), 727–738.

Burgard, A. P., and Maranas, C. D. (2003). Optimization-based framework for inferring and testing hypothesized metabolic objective functions. *Biotechnol. Bioeng.* **82**(6), 670–677.

Burgard, A. P., Pharkya, P., and Maranas, C. D. (2003). Optknock: A bilevel programming framework for identifying gene knockout strategies for microbial strain optimization. *Biotechnol. Bioeng.* **84**(6), 647–657.

Cameron, D. E., Urbach, J. M., and Mekalanos, J. J. (2008). A defined transposon mutant library and its use in identifying motility genes in Vibrio cholerae. *Proc. Natl. Acad. Sci. USA* **105**(25), 8736–8741.

DeJongh, M., Formsma, K., Boillot, P., Gould, J., Rycenga, M., *et al.* (2007). Toward the automated generation of genome-scale metabolic networks in the SEED. *BMC Bioinformatics* **8,** 139.

Duarte, N. C., Becker, S. A., Jamshidi, N., Thiele, I., Mo, M. L., *et al.* (2007). Global reconstruction of the human metabolic network based on genomic and bibliomic data. *Proc. Natl. Acad. Sci. USA* **104**(6), 1777–1782.

Durot, M., Bourguignon, P. Y., and Schachter, V. (2009). Genome-scale models of bacterial metabolism: Reconstruction and applications. *FEMS Microbiol. Rev.* **33**(1), 164–190.

Feist, A. M., and Palsson, B. O. (2008). The growing scope of applications of genome-scale metabolic reconstructions using Escherichia coli. *Nat. Biotechnol.* **26**(6), 659–667.

Feist, A. M., Herrgard, M. J., Thiele, I., Reed, J. L., and Palsson, B. O. (2009). Reconstruction of biochemical networks in microorganisms. *Nat. Rev. Microbiol.* **7**(2), 129–143.

Gallagher, L. A., Ramage, E., Jacobs, M. A., Kaul, R., Brittnacher, M., *et al.* (2007). A comprehensive transposon mutant library of Francisella novicida, a bioweapon surrogate. *Proc. Natl. Acad. Sci. USA* **104**(3), 1009–1014.

Gianchandani, E. P., Oberhardt, M. A., Burgard, A. P., Maranas, C. D., and Papin, J. A. (2008). Predicting biological system objectives de novo from internal state measurements. *BMC Bioinformatics* **9,** 43.

Giglio, M. G., Collmer, C. W., Lomax, J., and Ireland, A. (2009). Applying the Gene Ontology in microbial annotation. *Trends Microbiol.* **17**(7), 262–268.

Gonzalez, O., Oberwinkler, T., Mansueto, L., Pfeiffer, F., Mendoza, E., *et al.* (2010). Characterization of growth and metabolism of the haloalkaliphile Natronomonas pharaonis. *PLoS Comput. Biol.* **6**(6), e1000799.

Hao, T., Ma, H. W., Zhao, X. M., and Goryanin, I. (2010). Compartmentalization of the Edinburgh Human Metabolic Network. *BMC Bioinformatics* **11,** 393.

Henry, C. S., Zinner, J. F., Cohoon, M. P., and Stevens, R. L. (2009). iBsu1103: A new genome-scale metabolic model of Bacillus subtilis based on SEED annotations. *Genome Biol.* **10**(6), R69.

Herrgard, M. J., Swainston, N., Dobson, P., Dunn, W. B., Arga, K. Y., *et al.* (2008). A consensus yeast metabolic network reconstruction obtained from a community approach to systems biology. *Nat. Biotechnol.* **26**(10), 1155–1160.

Jacobs, M. A., Alwood, A., Thaipisuttikul, I., Spencer, D., Haugen, E., *et al.* (2003). Comprehensive transposon mutant library of Pseudomonas aeruginosa. *Proc. Natl. Acad. Sci. USA* **100**(24), 14339–14344.

Jankowski, M. D., Henry, C. S., Broadbelt, L. J., and Hatzimanikatis, V. (2008). Group contribution method for thermodynamic analysis of complex metabolic networks. *Biophys. J.* **95**(3), 1487–1499.

Jerby, L., Shlomi, T., and Ruppin, E. (2010). Computational reconstruction of tissue-specific metabolic models: Application to human liver metabolism. *Mol. Syst. Biol.* **6,** 401.

Karp, P. D., Paley, S. M., Krummenacker, M., Latendresse, M., Dale, J. M., *et al.* (2010). Pathway Tools version 13.0: Integrated software for pathway/genome informatics and systems biology. *Brief. Bioinform.* **11**(1), 40–79.

Kauffman, K. J., Prakash, P., and Edwards, J. S. (2003). Advances in flux balance analysis. *Curr. Opin. Biotechnol.* **14**(5), 491–496.

Kircher, M., and Kelso, J. (2010). High-throughput DNA sequencing—Concepts and limitations. *Bioessays* **32**(6), 524–536.

Knorr, A. L., Jain, R., and Srivastava, R. (2007). Bayesian-based selection of metabolic objective functions. *Bioinformatics* **23**(3), 351–357.

Kuzniar, A., van Ham, R. C., Pongor, S., and Leunissen, J. A. (2008). The quest for orthologs: Finding the corresponding gene across genomes. *Trends Genet.* **24**(11), 539–551.

Lee, J. M., Gianchandani, E. P., and Papin, J. A. (2006). Flux balance analysis in the era of metabolomics. *Brief. Bioinform.* **7**(2), 140–150.

Lun, D. S., Rockwell, G., Guido, N. J., Baym, M., Kelner, J. A., *et al.* (2009). Large-scale identification of genetic design strategies using local search. *Mol. Syst. Biol.* **5,** 296.

Mahadevan, R., and Schilling, C. H. (2003). The effects of alternate optimal solutions in constraint-based genome-scale metabolic models. *Metab. Eng.* **5**(4), 264–276.

Mahadevan, R., Edwards, J. S., and Doyle, F. J., 3rd (2002). Dynamic flux balance analysis of diauxic growth in Escherichia coli. *Biophys. J.* **83**(3), 1331–1340.

Niittylae, T., Chaudhuri, B., Sauer, U., and Frommer, W. B. (2009). Comparison of quantitative metabolite imaging tools and carbon-13 techniques for fluxomics. *Methods Mol. Biol.* **553,** 355–372.

Oberhardt, M. A., Puchalka, J., Fryer, K. E., Martins dos Santos, V. A., and Papin, J. A. (2008). Genome-scale metabolic network analysis of the opportunistic pathogen Pseudomonas aeruginosa PAO1. *J. Bacteriol.* **190**(8), 2790–2803.

Oberhardt, M. A., Chavali, A. K., and Papin, J. A. (2009). Flux balance analysis: Interrogating genome-scale metabolic networks. *Methods Mol. Biol.* **500,** 61–80.

Oberhardt, M. A., Goldberg, J. B., Hogardt, M., and Papin, J. A. (2010). Metabolic network analysis of Pseudomonas aeruginosa during chronic cystic fibrosis lung infection. *J. Bacteriol.* **192**(20), 5534–5548.

Orth, J. D., Thiele, I., and Palsson, B. O. (2010). What is flux balance analysis? *Nat. Biotechnol.* **28**(3), 245–248.

Papin, J. A., Price, N. D., Wiback, S. J., Fell, D. A., and Palsson, B. O. (2003). Metabolic pathways in the post-genome era. *Trends Biochem. Sci.* **28**(5), 250–258.

Patil, K. R., Rocha, I., Forster, J., and Nielsen, J. (2005). Evolutionary programming as a platform for in silico metabolic engineering. *BMC Bioinformatics* **6,** 308.

Pharkya, P., and Maranas, C. D. (2006). An optimization framework for identifying reaction activation/inhibition or elimination candidates for overproduction in microbial systems. *Metab. Eng.* **8**(1), 1–13.

Pharkya, P., Burgard, A. P., and Maranas, C. D. (2004). OptStrain: A computational framework for redesign of microbial production systems. *Genome Res.* **14**(11), 2367–2376.

Price, N. D., Papin, J. A., and Palsson, B. O. (2002). Determination of redundancy and systems properties of the metabolic network of Helicobacter pylori using genome-scale extreme pathway analysis. *Genome Res.* **12**(5), 760–769.

Price, N. D., Reed, J. L., and Palsson, B. O. (2004). Genome-scale models of microbial cells: Evaluating the consequences of constraints. *Nat. Rev. Microbiol.* **2**(11), 886–897.

Raman, K., and Chandra, N. (2009). Flux balance analysis of biological systems: Applications and challenges. *Brief. Bioinform.* **10**(4), 435–449.

Rocha, I., Forster, J., and Nielsen, J. (2008). Design and application of genome-scale reconstructed metabolic models. *Methods Mol. Biol.* **416,** 409–431.

Rocha, I., Maia, P., Evangelista, P., Vilaca, P., Soares, S., *et al.* (2010). OptFlux: An open-source software platform for in silico metabolic engineering. *BMC Syst. Biol.* **4,** 45.

Savinell, J. M., and Palsson, B. O. (1992). Optimal selection of metabolic fluxes for in vivo measurement. I. Development of mathematical methods. *J. Theor. Biol.* **155**(2), 201–214.

Schuetz, R., Kuepfer, L., and Sauer, U. (2007). Systematic evaluation of objective functions for predicting intracellular fluxes in Escherichia coli. *Mol. Syst. Biol.* **3,** 119.

Segre, D., Vitkup, D., and Church, G. M. (2002). Analysis of optimality in natural and perturbed metabolic networks. *Proc. Natl. Acad. Sci. USA* **99**(23), 15112–15117.

Senger, R. S. (2010). Biofuel production improvement with genome-scale models: The role of cell composition. *Biotechnol. J.* **5**(7), 671–685.

Shlomi, T., Berkman, O., and Ruppin, E. (2005). Regulatory on/off minimization of metabolic flux changes after genetic perturbations. *Proc. Natl. Acad. Sci. USA* **102**(21), 7695–7700.

Shlomi, T., Cabili, M. N., Herrgard, M. J., Palsson, B. O., and Ruppin, E. (2008). Network-based prediction of human tissue-specific metabolism. *Nat. Biotechnol.* **26**(9), 1003–1010.

Smallbone, K., and Simeonidis, E. (2009). Flux balance analysis: A geometric perspective. *J. Theor. Biol.* **258**(2), 311–315.

Suthers, P. F., Dasika, M. S., Kumar, V. S., Denisov, G., Glass, J. I., *et al.* (2009). A genome-scale metabolic reconstruction of Mycoplasma genitalium, iPS189. *PLoS Comput. Biol.* **5**(2), e1000285.

Thiele, I., and Palsson, B. O. (2010). A protocol for generating a high-quality genome-scale metabolic reconstruction. *Nat. Protoc.* **5**(1), 93–121.

Van Noorden, C. J. (2010). Imaging enzymes at work: Metabolic mapping by enzyme histochemistry. *J. Histochem. Cytochem.* **58**(6), 481–497.

Varma, A., and Palsson, B. O. (1994). Stoichiometric flux balance models quantitatively predict growth and metabolic by-product secretion in wild-type Escherichia coli W3110. *Appl. Environ. Microbiol.* **60**(10), 3724–3731.

Yang, Y., Gilbert, D., and Kim, S. (2010). Annotation confidence score for genome annotation: A genome comparison approach. *Bioinformatics* **26**(1), 22–29.

SECTION EIGHT

UNDERSTANDING SYSTEMS BIOLOGY

CHAPTER TWENTY-TWO

Hands-on Metabolism: Analysis of Complex Biochemical Networks Using Elementary Flux Modes

Sascha Schäuble, Stefan Schuster, *and* Christoph Kaleta

Contents

Abstract

The aim of this chapter is to discuss the basic principles and reasoning behind elementary flux mode analysis (EFM analysis)—an important tool for the analysis of metabolic networks. We begin with a short introduction into metabolic pathway analysis and subsequently outline in detail fundamentals of EFM analysis by way of a small example network. We discuss issues arising in the reconstruction of metabolic networks required for EFM analysis and how they can be circumvented. Subsequently, we analyze a more elaborate example network representing photosynthate metabolism. Finally, we give an overview of applications of EFM analysis in biotechnology and other fields and discuss issues arising when applying methods from metabolic pathway analysis to genome-scale metabolic networks.

Department of Bioinformatics, Friedrich Schiller University Jena, Jena, Germany

Methods in Enzymology, Volume 500
ISSN 0076-6879, DOI: 10.1016/B978-0-12-385118-5.00022-0

1. Introduction

Exploring the structure of metabolic networks is a key step in order to understand the fundamental properties of living systems (Price *et al.*, 2004; Ruppin *et al.*, 2010). Such networks summarize the capabilities of a subsystem or the entire metabolism of an organism and show, for instance, how a set of source compounds the organism can find in its environment can be converted into the constituent metabolites and macromolecules of which it consists.

However, it is often not possible to identify important routes in metabolic networks from their reactions alone (Schuster *et al.*, 1999). Due to this problem a large array of methods that allow one to analyze such networks based on network stoichiometry and constraints upon fluxes and concentrations have been developed (Price *et al.*, 2004). These methods are often referred to as constraint-based methods. One important method for identifying pathways in metabolic networks is the concept of elementary flux modes (EFMs; Schuster *et al.*, 1999) and the related extreme pathways (Schilling *et al.*, 2000). EFMs correspond to minimal sets of reactions that can work together at steady state while obeying thermodynamical constraints on the direction of reaction fluxes that make some reactions practically irreversible at physiological conditions. The fundamentals of that approach are based on earlier work by Clarke (1981). An EFM is minimal in the sense that removing one reaction will preclude any steady-state flux through the remaining set of reactions of which it consists.

In this work, we will give a detailed introduction into EFM analysis and the concepts upon which this method is based. We will outline this method by way of several example networks that also allow us to demonstrate principal avenues of its application. Further, we will give a brief overview on recent works applying the concept of EFMs. Finally, we will address challenges that arise from the advent of genome-scale metabolic networks that summarize the metabolic capabilities of entire organisms and how they can be met in the context of EFM analysis.

2. Elementary Flux Modes

The concept of EFMs (Schuster and Hilgetag, 1994; Schuster *et al.*, 1999) allows one to decompose a reaction network into well-defined metabolic pathways. As stated above, they are defined as minimal sets of reactions that allow for a steady-state flux that uses irreversible reactions only in the thermodynamically feasible direction. To apply EFM analysis, only information on the reaction stoichiometry of the network and information on reaction directionality is required, information that is often much easier to obtain than details on the precise kinetics governing reactions. The reaction stoichiometry indicates for each reaction the number of metabolite molecules consumed or produced: the stoichiometric coefficient is negative if a metabolite is a substrate of a reaction and positive if it is a product.

The information of reaction stoichiometry is gathered in the stoichiometric matrix **N**, where m_{ij} of **N** is the coefficient of the *i*th metabolite taking part in the *j*th reaction:

$$\mathbf{N} = \begin{pmatrix} m_{11} & m_{12} & \cdots & m_{1m} \\ m_{21} & m_{22} & \cdots & m_{2m} \\ \vdots & & \ddots & \\ m_{n1} & m_{n2} & \cdots & m_{nm} \end{pmatrix}$$

Note that external metabolites are often not included in **N**, as they are not considered in the context of the steady-state condition that will be discussed more thoroughly below. Instead they are often part of a parameter vector **p** as their concentrations are assumed to be constant.

At this point it is important to note that while most applications of EFM analysis focus on small-molecule metabolism an application to more complex compounds such as DNA and proteins is not precluded. Thus, EFM and extreme pathway analysis have also been used to investigate signaling and regulatory networks (Behre and Schuster, 2009; Gianchandani *et al.*, 2006, 2009).

In order to define a systems boundary, the set of metabolites of a reaction network is decomposed into internal and external metabolites. While internal metabolites are required to be balanced within the network and thus subjected to the steady-state condition, external metabolites are assumed to be buffered. Mechanisms that buffer these metabolites can be, for instance, supply from a growth medium, drain through dilution or participation in a large number of reactions beyond the scope of the metabolic network. Metabolites that are, thus, often considered as external are metabolites of the growth medium, biomass precursors such as amino acids and nucleotides as well as energy currency cofactors such as ATP, NADH, or NADPH.

Under these premises, an EFM can be understood as a path through the network that transforms a set of external (substrate) metabolites into another set of external (product) metabolites over the intermediate of a set of balanced internal metabolites. Figure 22.1 illustrates the principle of EFMs in a biologically relevant pathway—the TCA cycle, the powerplant that provides many organisms with sufficient energy.

Note that for two reasons it is of utter importance to determine the state of a metabolite, that is, which metabolite can be defined as external or internal. First, the simulated uptake or production of particular metabolites, or the maintenance of energy greatly depends on this definition. Second, especially in

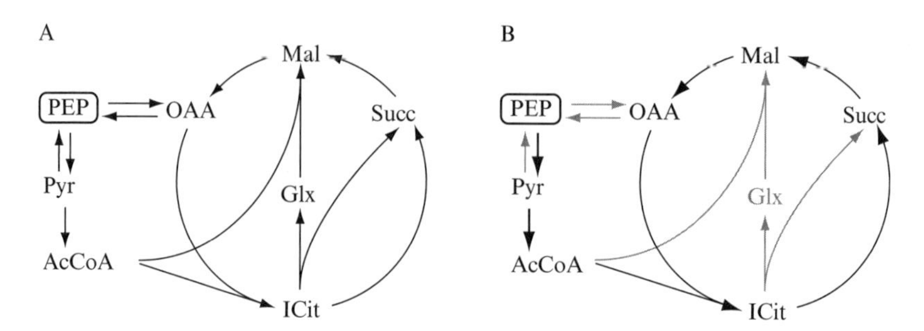

Figure 22.1 EFMs in a simplified model of the TCA cycle including adjacent reactions of glycolysis. The model comprises the metabolites phosphoenolpyruvate (PEP), pyruvate (Pyr), acetyl-Coenzyme-A (AcCoA), isocitrate (ICit), succinate (Succ), malate (Mal), oxaloacetate (OAA), and glyoxylate (Glx) where only the metabolite PEP is set to external status (boxed metabolite). (A) Shows all reactions considered, while (B) displays one valid EFM that completely oxidises PEP. Release and fixation of CO_2 has been omitted for clarity.

dense networks, where the number of EFMs grows exponentially with the size of the network (Klamt and Stelling, 2002), determining the state of metabolites has a very strong influence on the number of resulting EFMs (Dandekar *et al.*, 2003; Gagneur and Klamt, 2004; Schuster *et al.*, 2002b).

In fact, examining particularly compact networks is rather common, since metabolic networks often contain bi- or even multimolecular reactions. By transforming a multitude of metabolites into another, these reactions significantly increase the complexity of the network and hence the number of EFMs. However, as computer performance increases and more efficient algorithms to compute EFMs are developed, networks of increasing size can be investigated using EFM analysis (Gagneur and Klamt, 2004; Pfeiffer *et al.*, 1999; Terzer and Stelling, 2008; Urbanczik and Wagner, 2005).

2.1. Mathematical background

Although specific kinetic data is not required, some general assumptions about the kinetics of the system under consideration have to be made for EFM analysis.

For simplicity, it is reasonable to assume that the metabolites are homogeneously distributed and that the cell does not exhibit a time-dependent inflow or outflow behavior. Thus, the time-course of the concentrations of metabolites can be described formally by the differential equation:

$$\frac{d\mathbf{S}}{dt} = \mathbf{NV}(\mathbf{S}, \mathbf{p}) - \mu\mathbf{S} \tag{22.1}$$

where $\mathbf{S} = (S_1, S_2, \ldots, S_n)^{\mathrm{T}}$, $\mathbf{V}$, and $\mathbf{p}$ are the vectors of the concentrations of internal metabolites, net reaction rates, and parameters, respectively.

The growth rate μ can be taken into account using two approaches. The first approach is to consider dilution by growth to be so small for the time frame of the analysis that it can be neglected. Alternatively, growth can be modeled within the stoichiometric matrix $\mathbf{N}$ through addition of an artificial biomass reaction that drains components of the cell, or their precursors, in their relative amounts.

Under the time frame considered, it is reasonable to assume that the concentrations of internal metabolites stay constant (Clarke, 1981; Pfeiffer *et al.*, 1999; Schuster and Hilgetag, 1994). Hence, their production and consumption need to be balanced. Thus, Eq. (22.1) can be simplified by setting it equal to zero:

$$\mathbf{NV}(\mathbf{S}, \mathbf{p}) = 0 \tag{22.2}$$

Obviously, as organisms undergo dynamic processes, no concentration will be completely constant over time. Nevertheless, this assumption also holds in the approximate context as long as no intermediates accumulate or are depleted to a considerable extent over time, for instance, in the case of oscillations.

A first simple approach to analyze the steady-state condition (Eq. (22.2)) is the computation of the null space of $\mathbf{N}$. The null space or kernel refers to the Euclidean subspace of all vectors $\mathbf{V}$ fulfilling Eq. (22.2). It already features some simple pathways in the network and can be computed by utilizing standard methods of linear algebra (Strang, 2009) such as Gaussian elimination. Note that basis vectors can be understood as steady-state flux distributions across the system.

However, this set of basis vectors is not unique, that is, there can be several sets of basis vectors. Further, they do not take into account the irreversibility of some reactions and might use them in a thermodynamically infeasible direction. To overcome these drawbacks, it is required that a subvector of $\mathbf{V}^{\mathrm{irr}}$ of $\mathbf{V}$, in which the coefficients correspond to irreversible reactions, satisfies

$$\mathbf{V}^{\mathrm{irr}} \geq 0 \tag{22.3}$$

Hence, a linear inequality system is formed by the Eqs. (22.2) and (22.3).

Now, a *flux mode* $\mathbf{V}^* \in \mathbb{R}^r$ with r being the number of reactions within a network is defined as follows:

(i) *steady-state condition*: $\mathbf{V}^\star$ satisfies Eq. (22.2)
(ii) *sign restriction*: $\mathbf{V}^\star$ satisfies Eq. (22.3)

A first important observation is that if a flux vector $\mathbf{V}$ fulfils Eqs. (22.2) and (22.3), also $\lambda\mathbf{V}$ with $\lambda > 0$ fulfils both equations. Hence, the analysis should be restricted to a set of flux modes in which none can be derived as a simple scalar/multiple of another.

Another important observation from Eqs. (22.2) and (22.3) is that two flux modes $\mathbf{V}_1$ and $\mathbf{V}_2$ can be linearly combined by $\lambda_1\mathbf{V}_1 + \lambda_2\mathbf{V}_2$ with λ_1, $\lambda_2 > 0$ and we again obtain a flux mode. Thus, the analysis can be restricted to *elementary* flux modes. A flux mode is called elementary if suppressing the flux through any reaction used by it implies that there is no flux through the remaining reactions that satisfies Eqs. (22.2) and (22.3) (Schuster and Hilgetag, 1994). This is equivalent to stating that a flux mode $\mathbf{V}'$ is called elementary if there is no other flux mode $\mathbf{V}''$ that uses a proper subset of the reactions of $\mathbf{V}'$ (Schuster *et al.*, 2002a). In mathematical terms, this statement can be formulated by defining the support "supp" of a flux mode $\mathbf{V}$, which includes all nonzero elements of $\mathbf{V}$ as

$$\text{supp}(\mathbf{V}) = \{i | V_i \neq 0\} \tag{22.4}$$

Then, a flux mode $\mathbf{V}'$ is called elementary if there exists no other flux mode $\mathbf{V}''$ such that

$$\text{supp}\left(\mathbf{V}''\right) \subseteq \text{supp}(\mathbf{V}'). \tag{22.5}$$

Thus, EFMs can be defined as suggested by Schuster and Hilgetag (1994):

Definition 1 An EFM is a minimal set of enzymes that can operate at steady state with all irreversible reactions used in the correct direction.

An interesting property of the set of EFMs of a reaction network is that they are unique up to scaling by a factor $\lambda > 0$. Moreover, every flux mode $\mathbf{V}$ (which hence satisfies Eqs. (22.2) and (22.3)) can be written as positive linear combination of EFMs:

$$V = \sum_k \eta_k \mathrm{e}^{(k)}, \quad \eta > 0 \tag{22.6}$$

where $\mathrm{e}^{(k)}$ refers to the set of EFMs and η_k represents a scaling factor.

In Fig. 22.2, all of the EFMs of the example network in Fig. 22.1 are displayed. Again, the metabolite PEP is the only external species, whereas all other metabolites are considered to be intermediates and thus set to internal status (balanced at steady state). The model gives rise to four EFMs, of which two correspond to the interconversion of PEP and OAA and the interconversion of PEP and Pyr, respectively (Fig. 22.2B and C). Additionally to these two trivial EFMs, we find two EFMs that constitute all possible pathways on which PEP can be oxidized by this network. The first EFM

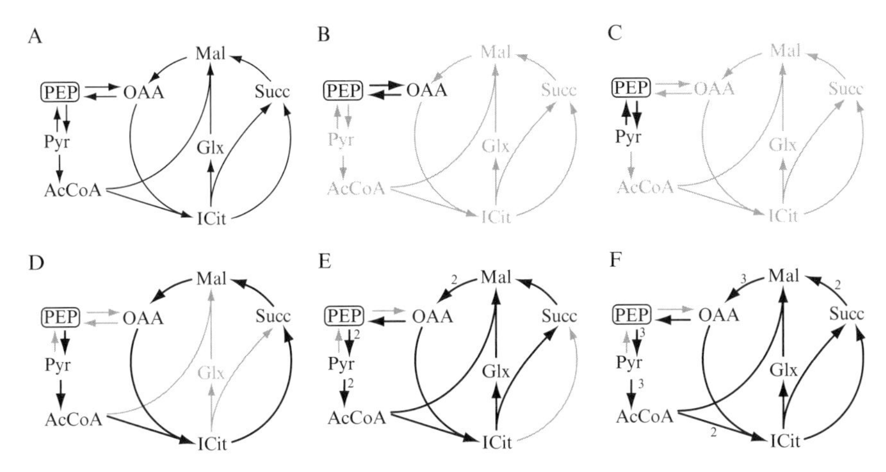

Figure 22.2 The simplified TCA cycle in (A), as depicted Fig. 22.1 gives rise to four EFMs, shown in (B)–(E). The flux in (F) would utilize all reactions, which is a violation of condition (iii) and thus not a regular EFM, since it uses the combined reactions of the EFMs in (D) and (E). Abbreviations are explained in Fig. 22.1.

(Fig. 22.2D) corresponds to the classical scheme of PEP oxidation along the TCA cycle. The other pathway, in contrast, utilizes reactions from the glyoxylate shunt to oxidize PEP. Until some years ago, the enzymes of the glyoxylate shunt have only been considered to play a role in gluconeogenesis/anaplerosis rather than catabolism of glucose (Fig. 22.2E). However, they have been found to be used for PEP oxidation during growth of *Escherichia coli* (Fischer and Sauer, 2003) and *Mycobacterium tuberculosis* (Beste and McFadden, 2010) on low glucose concentrations. The corresponding pathway has been called PEP-glyoxylate cycle. Why are there no more EFMs? One might argue that in principle there should exist at least one more EFM that uses the complete TCA cycle and the PEP-glyoxylate cycle simultaneously (Fig. 22.2F). While this can be observed in *E. coli* (Fischer and Sauer, 2003), the resulting set of reactions does not constitute an EFM since it is not minimal, that is, it is just a superposition of two pathways. In consequence, any steady-state flux that uses the TCA cycle and the PEP-glyoxylate cycle simultaneously corresponds to a linear combination of two EFMs.

3. Application

3.1. Network reconstruction

In order to apply EFM analysis, an accurate model is required. In the most simple case, the particular metabolic network of interest is already published and ready for an EFM analysis, as is the case for classic model organisms, such as *Escherichia coli* or *Bacillus subtilis*. Although a continuously rising number of models are published, it is often necessary to reconstruct a model from scratch if data concerning an organism of interest is not publicly available or insufficient. One should not underestimate the notable efforts and complications that come along with a model reconstruction effort. This becomes even more crucial, as more and more sequenced genomes are publicly available and allow for a genome-scale reconstruction and analysis of metabolic networks. Even in these cases, the reconstruction task can consume months until a high-quality model is developed (Feist *et al.*, 2009; Ruppin *et al.*, 2010; Thiele and Palsson, 2010).

For instance, one might encounter the obstacle that databases like KEGG (Kanehisa *et al.*, 2010) or MetaCyc (Caspi *et al.*, 2010) that comprise metabolic pathway data contain contradicting information. This can be due to different conventions concerning the protonation of metabolites or differences in the reversibility assignment of reactions. The latter can be thermodynamically feasible in forward and backward direction or only in one of them. This issue is of critical importance, as those databases commonly form the starting point for the reconstruction process. Since data from different databases often do not agree with each other, including data from multiple sources, such as genomic, proteomic, or bibliomic sources, is most recommended if not mandatory, in order to reconstruct a high-quality model.

Note that genome-scale metabolic models are nevertheless abstractions from real world biological pathways and do not cover the complete set of reactions featured by the underlying organism *per se*, for instance, if some reactions have not yet been identified. Moreover, if no gene ID can be found in public databases, it is rather common that corresponding reactions, for instance, transporters are artificially introduced into the system in order to fill open gaps. This occurs very often in eukaryotes that feature different compartments and cell types. Although reasonable, this should naturally be regarded as a danger to the correctness of a model and hence be considered when reconstructing and ultimately analyzing a metabolic network. Therefore, subsequent analysis can only be considered accurate if the underlying model is as complete as sources will allow. If information is missing or inconsistent, sometimes biochemical expertise can be used to improve on this situation.

One good starting point for a reconstruction are central metabolic pathways as they are described in biochemistry textbooks. Although these networks represent only a small fraction of a complete biological system, they are commonly found in most organisms and have already been investigated for decades.

3.2. Application to photosynthate metabolism

In order to demonstrate how EFM analysis can be used in detail we will analyze a metabolic model of photosynthesis. This model (Fig. 22.3; Appendix) includes the photosynthate metabolism of the chloroplast stroma and

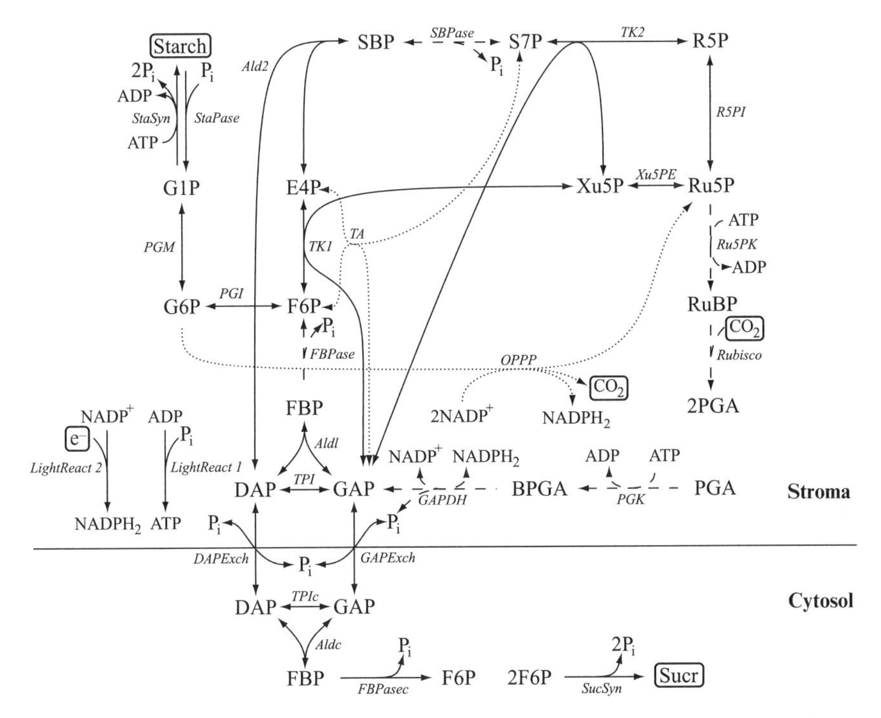

Figure 22.3 Reactions of the Calvin cycle, including parts of glycolysis. Unidirectional (bidirectional) arrows indicate irreversible (reversible) reactions. Dashed reactions are activated whereas dotted reactions are downregulated during the day. Boxed metabolites are set to external and all others to internal status for EFM analysis. Abbreviations of metabolites: PGA, 3-phosphoglycerate; BPGA, glycerate-1, 3-bisphosphate; GAP, glyceraldehyde-3-phosphate; DAP, dihydroxyacetone phosphate; FBP, fructose-1, 6-bisphosphate; F6P, fructose-6-phosphate; E4P, erythrose-4-phosphate; SBP, sedoheptulose-1,7-bisphosphate; S7P, sedoheptulose-7-phosphate; R5P, ribose-5-phosphate; Ru5P, ribulose-5-phosphate; RuBP, ribulose-1, 5-bisphosphate; X5P, xylulose-5-phosphate; G6P, glucose-6-phosphate; G1P, glucose-1-phosphate.

comprises reactions from the Calvin cycle, which is primarily regulated by the thioredoxin system (Schürmann and Jacquot, 2000). In contrast to the work of Poolman *et al.* (2003), who analyzed a similar model with respect to triose phosphate export in detail, we will investigate the production of sucrose and starch. Sucrose and starch are the major carbohydrate storage compounds in plants. Moreover, sucrose is the carbohydrate transported in the phloem and also corresponds to the sugar commonly used in everyone's kitchen.

If all reactions are taken into account, the system gives rise to 42 EFMs in total. As we will limit the analysis on EFMs during daytime (where the two dotted reactions in Fig. 22.3 are inactive), only 12 EFMs are relevant. These EFMs are listed in Table 22.1 and characterize three groups of EFMs of which some are shown in Fig. 22.4. These groups can be differentiated

Table 22.1 Twelve EFMs active during the day

Group	No.	Overall reaction	Enzymes
I	(1)	No net transformation	-TPI GAPExch -DAPExch TPIc
I	(2)	No net transformation	StaSyn StaPase Lightreact1
II	(3)	$6CO_2 + 12e^- \rightarrow$ Starch	6 Rubisco 12 PGK 12 GAPDH 5 TPI 3 Ald1 3 FBPase PGI PGM StaSyn 2 TK1 2 Ald2 2 SBPase 2 TK2 2 R5PI 4 Xu5PE 6 Ru5PK 19 Lightreact1 12 Lightreact2
II	(4)	$6CO_2 + 12e^- \rightarrow$ Starch	6 Rubisco 12 PGK 12 GAPDH 3 Ald1 3 FBPase PGI PGM StaSyn 2 TK1 2 Ald2 2 SBPase 2 TK2 2 R5PI 4 Xu5PE 6 Ru5PK 5 GAPExch -5 DAPExch 19 Lightreact1 5 TPIc 12 Lightreact2
III	(5)	$12CO_2 + 24e^- \rightarrow$ Sucr	12 Rubisco 24 PGK 24 GAPDH 8 TPI 4 Ald1 4 FBPase 4 TK1 4 Ald2 4 SBPase 4 TK2 4 R5PI 8 Xu5PE 12 Ru5PK 4 GAPExch 36 Lightreact1 2 TPIc 2 Aldc 2 FBPasec SucSyn 24 Lightreact2
III	(6)	$12CO_2 + 24e^- \rightarrow$ Sucr	12 Rubisco 24 PGK 24 GAPDH 10 TPI 4 Ald1 4 FBPase 4 TK1 4 Ald2 4 SBPase 4 TK2 4 R5PI 8 Xu5PE 12 Ru5PK 2 GAPExch 2 DAPExch 36 Lightreact1 2 Aldc 2 FBPasec SucSyn 24 Lightreact2
III	(7)	$12CO_2 + 24e^- \rightarrow$ Sucr	12 Rubisco 24 PGK 24 GAPDH 12 TPI 4 Ald1 4 FBPase 4 TK1 4 Ald2 4 SBPase 4 TK2 4 R5PI 8 Xu5PE 12 Ru5PK 4 DAPExch 36 Lightreact1 -2 TPIc 2 Aldc 2 FBPasec Suc Syn 24 Lightreact2

III	(8)	$12CO_2 + 24e^- \rightarrow Sucr$	12 Rubisco 24 PGK 24 GAPDH 4 Ald1 4 FBPase 4 TK1 4 Ald2 4 SBPase 4 TK2 4 R5PI 8 Xu5PE 12 Ru5PK 12 GAPExch -8 DAPExch 36 Lightreact1 10 TPIc 2 Aldc 2 FBPasec SucSyn 24 Lightreact2
III	(9)	$12CO_2 + 4 \times Starch + 24e^- \rightarrow 3 \times Sucr$	12 Rubisco 24 PGK 24 GAPDH 10 TPI -4 PGI -4 PGM 4 TK1 4 Ald2 4 SBPase 4 TK2 4 R5PI 8 Xu5PE 12 Ru5PK 6 GAPExch 6 DAPExch 4 StaPase 36 Lightreact1 6 Aldc 6 FBPasec 3 SucSyn 24 Lightreact2
III	(10)	$12CO_2 + 4 \times Starch + 24e^- \rightarrow 3 \times Sucr$	12 Rubisco 24 PGK 24 GAPDH 16 TPI -4 PGI -4 PGM 4 TK1 4 Ald2 4 SBPase 4 TK2 4 R5PI 8 Xu5PE 12 Ru5PK 12 DAPExch 4 StaPase 36 Lightreact1 -6 TPIc 6 Aldc 6 FBPasec 3 SucSyn 24 Lightreact2
III	(11)	$12CO_2 + 4 \times Starch + 24e^- \rightarrow 3 \times Sucr$	12 Rubisco 24 PGK 24 GAPDH 4 TPI -4 PGI -4 PGM 4 TK1 4 Ald2 4 SBPase 4 TK2 4 R5PI 8 Xu5PE 12 Ru5PK 12 GAPExch 4 StaPase 36 Lightreact1 6 TPIc 6 Aldc 6 FBPasec 3 SucSyn 24 Lightreact2
III	(12)	$12CO_2 + 4 \times Starch + 24e^- \rightarrow 3 \times Sucr$	12 Rubisco 24 PGK 24 GAPDH -4 PGI -4 PGM 4 TK1 4 Ald2 4 SBPase 4 TK2 4 R5PI 8 Xu5PE 12 Ru5PK 16 GAPExch -4 DAPExch 4 StaPase 36 Lightreact1 10 TPIc 6 Aldc 6 FBPasec 3 SucSyn 24 Lightreact2

The group classification is given in the first column, while the net transformation is given in the third and the fluxes of the reactions in the fourth column. An overview of the reaction scheme is given in Fig. 22.3.

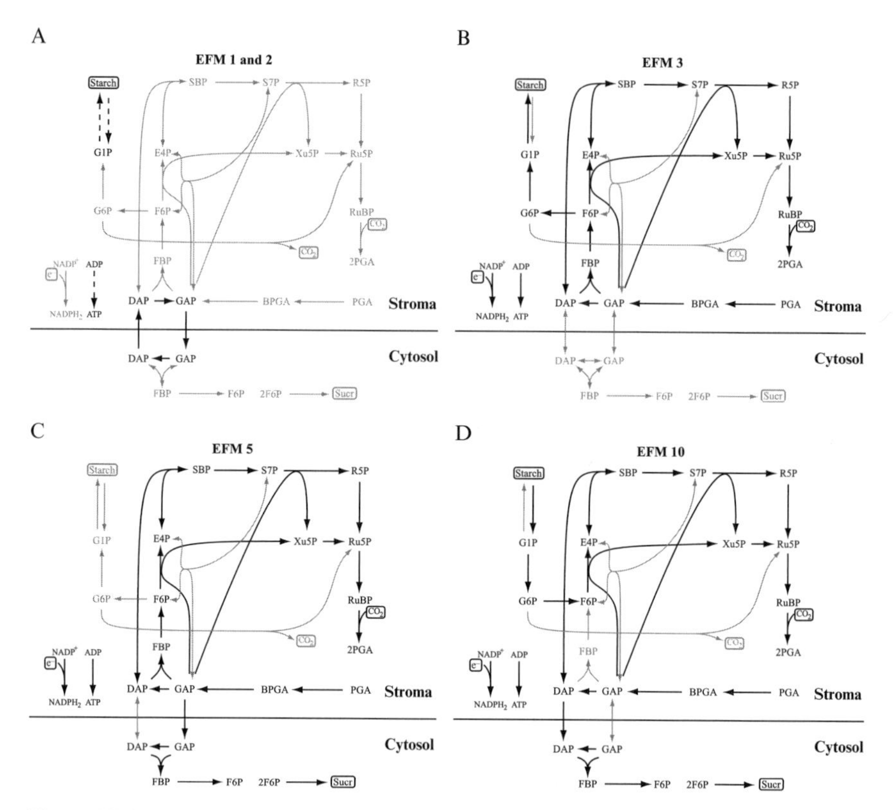

Figure 22.4 Selected EFMs of the model of Fig. 22.3. (A) Displays two cycling EFMs that show no net transformation (EFM 1—solid arrows, EFM 2—dashed arrows). The EFM in (B) describes starch production, via Rubisco. Sucrose synthesis from CO_2 alone or from CO_2 and starch is displayed in (C) and (D), respectively. The selected EFMs as well as further EFMs are described in Table 22.1 and in the text. Names of enzymes are given in Fig. 22.3.

according to their net transformation of external metabolites: in the first group, no net transformation occurs while EFMs of the second and third group produce starch and sucrose, respectively.

The first group comprises two cycles that are present during the day. The first reflects reversible DAP and GAP exchange across the cytosolic membrane with concurrent antiport of P_i (EFM 1, Fig. 22.4A). As this cycle is not driven thermodynamically by external metabolites, it is detailed balanced and can, thus, not carry any steady-state flux due to thermodynamical reasons. The second EFM is a futile cycle, as it consumes ATP from the light reactions, while interconverting starch and G1P back and forth (EFM 2, Fig. 22.4A). As energy is available in excess through the light reactions, both starch and sucrose can be produced at daytime.

The EFMs synthesizing starch (EFMs 3 and 4, see also Fig. 22.4) can be differentiated by their location. While EFM 3 (Fig. 22.4B) occurs solely in the stroma, EFM 4 (not shown in Fig. 22.4) comprises reactions of the stroma and the cytosol. Nevertheless, both EFMs require 19 mol of ATP and 12 mol of NADPH, provided by the light reactions in order to produce 1 mol of starch (i.e., the metabolite that represents one hexose unit in starch) and are thus equally "expensive" in the context of this model.

Sucrose can be produced solely from carbon dioxide (EFMs 5–8) or both, from carbon dioxide and starch (EFMs 9–12). An interesting question is which of the pathways is energetically more efficient, since both use the same amount of ATP and NADPH provided by the light reactions, but produce different quantities of sucrose from different sets of precursors. Producing 3 mol of sucrose from carbon dioxide alone requires an investment of three times 36 mol of ATP and 24 mol of NADPH provided by the light reactions and hence an overall consumption of 108 mol of ATP and 72 mol of NADPH. Producing 3 mol of sucrose from starch and carbon dioxide requires an investment of 36 mol of ATP, 24 mol of NADPH, and 4 mol of starch. Producing 4 mol of starch through EFMs of group II requires an additional investment of four times 19 mol of ATP and 12 mol of NADPH. In consequence, an overall $36 + 4 \times 19 = 112$ mol of ATP and $24 + 4 \times 12 = 72$ mol of NADPH are consumed to produce 3 mol of sucrose through EFMs 9–12. Thus, producing sucrose through EFMs 5–8 is energetically more favorable. Hence, while sucrose biosynthesis from light and starch is in principle possible, it is slightly more expensive for plants to rely also on starch deposits, if the light source is not sufficient. For further selected EFMs, see Fig. 22.4.

Since Table 22.1 comprises 12 EFMs, there are 30 EFMs left. All these 30 EFMs have in common that they either use reactions of the oxidative pentose phosphate pathway (OPPP in Fig. 22.3) or the transaldolase. These reactions are downregulated during the day and, thus, not available (Schürmann and Jacquot, 2000). All of these EFMs additionally use Rubisco or sedoheptulose-bisphosphatase (SBPase), reactions which are downregulated during the night (Schürmann and Jacquot, 2000). In consequence, these 30 EFMs are also downregulated during the night.

Hence, neither starch production nor sucrose synthesis is possible during night in the context of this model. Even if an ATP source were available during night, additional reactions that are downregulated during this phase of the day would be required. A detailed description and analysis of triose phosphate biosynthesis in plants is given in Poolman *et al.* (2003).

3.3. Overview of further applications of EFM analysis

EFM analysis has been used intensively to analyze the metabolic capabilities of organisms. One important example is the analysis of metabolic pathways in the central metabolism of *Escherichia coli* that provided evidence for the existence of an alternative pathway for the complete oxidation of glucose apart from the TCA cycle (Liao *et al.*, 1996; Schuster *et al.*, 1999). This pathway, now called PEP-glyoxylate cycle, after it had been experimentally confirmed (Fischer and Sauer, 2003), makes use of the glyoxylate shunt and skips several reactions of the TCA cycle (Fig. 22.2E). Recently, this pathway has also been reported in *Mycobacterium tuberculosis* (Beste and McFadden, 2010). Another example is the analysis of a long-standing issue in biochemistry: the question whether even-chain fatty acids can be converted into carbohydrates in humans at steady state. Even though there exists a connected route between acetyl-CoA—the product of β-oxidation of even-chain fatty acids—and glucose along the TCA cycle, there can be no steady-state flux along this route at steady state. Analyzing a metabolic network comprising relevant reactions, de Figueiredo *et al.* (2009b) could indeed show that there exists no elementary mode within this network that converts fatty acids into glucose thus confirming earlier results (Weinman *et al.*, 1957).

Another important application of EFM analysis is the investigation of the susceptibility of metabolic networks to perturbations. A useful property of EFMs is that the knockout of an enzyme can be simulated by removing all EFMs that contain a reaction that is catalyzed by that enzyme. In consequence, enzyme deficiencies can be easily analyzed and EFM analysis has been used to understand metabolic pathways within a mutant of *E. coli* that lacks an outer membrane and the cell wall (Kenanov *et al.*, 2010). Moreover, it has been used to analyze medical implications of enzyme deficiencies in human erythrocytes (Çakir *et al.*, 2004; Schuster and Kenanov, 2005). On a larger scale, EFM analysis has been used to analyze and compare different networks with regard to their susceptibility to random perturbations (Behre *et al.*, 2008; Stelling *et al.*, 2002; Wilhelm *et al.*, 2004).

EFM analysis can also be used to facilitate the interpretation and integration of large-scale experimental data sets. The analysis of such data sets is and has been an important focus in Systems Biology since data analysis is currently lagging behind data generation (Palsson and Zengler, 2010). EFM analysis is a suitable tool in this endeavor as has been shown in a study which investigates regulatory adaptations of *E. coli* to different carbon sources (Stelling *et al.*, 2002) and in another work in which transcriptomic changes during the response to stresses in yeast have been analyzed (Schwartz *et al.*, 2007).

One particular successful area of application of EFM analysis is biotechnology. It has been used in the design and implementation of several strains of *E. coli* overproducing various biotechnological products of interest (Carlson *et al.*, 2002; Trinh and Srienc, 2009; Trinh *et al.*, 2006; Unrean *et al.*, 2010). The motivation behind these applications is to reduce the space of admissible fluxes toward a smaller space in which the production of the desired product is presumably coupled to a cellular objective such as maximizing the growth rate. This is achieved by knocking out genes that remove EFMs that have low yields in the desired product. Such knockout strategies can be identified by using the concept of minimal cut sets (Klamt, 2006)-minimal sets of reactions that allow to block a given set of EFMs. Also other theoretical tools based on EFM analysis that allow one to identify genetic modifications that increase the flux to a desired product have been proposed recently (Boghigian *et al.*, 2010; Bohl *et al.*, 2010; Hädicke and Klamt, 2010).

3.4. Pathway analysis in genome-scale metabolic networks

EFMs can only be enumerated in small to medium-scale metabolic networks. These restrictions come from the exponential increase in the number of EFMs with network size (Klamt and Stelling, 2002). In recent years, the algorithms to compute EFMs have been considerably improved (Gagneur and Klamt, 2004; Terzer and Stelling, 2008; Urbanczik and Wagner, 2005) such that networks with up to 25 million EFMs could be analyzed (Terzer and Stelling, 2008). However, the estimated number of EFMs in genome-scale metabolic networks is much larger. For instance, it has been estimated that the number of extreme pathways, a subset of the EFMs, in a genome-scale metabolic network of humans is approximately 10^{29} (Yeung *et al.*, 2007). Thus, even with drastic improvements in algorithms and hardware the enumeration of all EFMs in most genome-scale metabolic networks appears to be infeasible, apart from the difficulty to analyze such a large set of EFMs.

To counter this problem, recently, several approaches that allow for a pathway-based analysis even in genome-scale metabolic networks have been developed. They can be divided into two types. The first type aims to enumerate a subset of the EFMs or extreme pathways that fulfill certain biotechnological criteria (de Figueiredo *et al.*, 2009a; Kaleta *et al.*, 2009a; Xi *et al.*, 2009) and the second analyses sets of reactions that are part of a pathway of the entire system within a specific subsystem, so-called elementary flux patterns (Kaleta *et al.*, 2009b).

The advantage of the first methodology is that a subset of EFMs can be enumerated without requiring to compute the entire set as was necessary using classical algorithms for EFM computation. If one is interested in the shortest EFMs using a specific reaction, a mixed-integer linear programming formulation can be used to compute EFMs (de Figueiredo *et al.*, 2009a). Alternatively, large sets of EFMs using a specific reaction can be obtained using a genetic algorithm that allows one to sample EFMs randomly (Kaleta *et al.*, 2009a). Also random sampling methods have been used to identify subsets of extreme pathways in large-scale networks (Xi *et al.*, 2009).

The second methodology, elementary flux pattern analysis (Kaleta *et al.*, 2009b), allows one to identify pathways within a subsystem of metabolism that are compatible with a steady-state flux of the entire system. Due to this strong link of elementary flux patterns with the entire system it is further possible to analyze how a subsystem integrates into the remaining system. Using this approach, many methods building on EFM analysis can also be applied to subsystems of genome-scale metabolic networks. Another interesting property of elementary flux patterns is that they put no constraints on the connectivity between the reactions of the subsystem that are considered. In consequence, it is even possible to analyze subsystems that comprise two sets of reactions that do not interface each other through common substrates or products. This can be of importance when analyzing, for instance, the dependencies between a set of reactions within a parasite and a set of reactions within the host in order to understand how both organisms interact on a metabolic level. Thus, critical metabolic dependencies between the host and the parasite can be identified in order to develop new drugs that affect the parasite but have only a small deleterious effect on the host. The analysis of such consortium pathways going across two or more organisms is a topic of high current interest (Bordbar *et al.*, 2010; Raghunathan *et al.*, 2009).

4. Conclusion

EFM analysis is a useful tool that allows the decomposition of biochemical networks into minimal constituent pathways. Thus, EFM analysis has already seen a wide array of applications ranging from theoretical works to biotechnology and has made significant contributions to research in these fields. However, EFM analysis in its classical form cannot be applied to genome-scale metabolic networks which has been considered as a downturn of this method. But with the development of new tools and concepts that port EFM analysis to such networks some hurdles have already been vanquished. In consequence, even in the era of genome-scale metabolic networks, EFM analysis is still a central tool for the analysis of metabolic pathways.

ACKNOWLEDGMENTS

Financial support from the German Ministry for Research and Education (BMBF) to C. Kaleta within the framework of the Forsys Partner initiative (grant FKZ 0315285E) and to S. Schäuble within the framework of the GerontoSys initiative (grant FKZ 0315581D) is gratefully acknowledged. Further, we thank Luís Filipe de Figueiredo and Ines Heiland for useful discussions on the topic of metabolic pathway analysis.

Appendix

Reaction scheme of the photosynthate metabolism in metatool format (discussed in Section 3.2).

```
-METEXT
CO2 Starch e Sucr
-CAT
#reactions of the Calvin cycle (down at night)
Rubisco : CO2 + RuBP = > 2 PGA
PGK : PGA + ATP = BPGA + ADP
GAPDH : BPGA + NADPH = NADP + GAPs + Pis
FBPase : FBPs = > F6Ps + Pis
SBPase : SBP = > S7P + Pis
Ru5PK : Ru5P + ATP = > RuBP + ADP
StaSyn : G1P + ATP = > ADP + 2 Pis + Starch
Lightreact2 : ADP + Pis = > ATP
Lightreact1 : NADPred : NADP + e = > NADPH
#stroma reactions
TPI : GAPs = DAPs
Ald1 : DAPs + GAPs = FBPs
TK1 : F6Ps + GAPs = E4P + X5P
Ald2 : E4P + DAPs = SBP
TK2 : GAPs + S7P = X5P + R5P
R5PI : R5P = Ru5P
Xu5PE : X5P = Ru5P
PGI : F6Ps = G6Ps
PGM : G6Ps = G1P
StaPase : Starch + Pis = > G1P
#cytoplasmatic reactions
TPIc : GAPc = DAPc
Aldc : GAPc + DAPc = FBPc
FBPasec : FBPc = > F6Pc + Pic
SucSyn : 2 F6Pc = > Sucr + 2 Pic
#transport reactions
GAPExch : GAPs + Pic = Pis + GAPc
DAPExch : DAPs + Pic = Pis + DAPc
#unique in oxPPP #down at day
TA : F6Ps + GAPs = E4P + S7P
OPPP : G6Ps + 2 NADP = > 2 NADPH + CO2 + Ru5P
```

REFERENCES

Behre, J., and Schuster, S. (2009). Modeling signal transduction in enzyme cascades with the concept of elementary flux modes. *J. Comput. Biol.* **16,** 829–844.

Behre, J., Wilhelm, T., von Kamp, A., Ruppin, E., and Schuster, S. (2008). Structural robustness of metabolic networks with respect to multiple knockouts. *J. Theor. Biol.* **252,** 433–441.

Beste, D. J. V., and McFadden, J. (2010). Systems biology of the metabolism of Mycobacterium tuberculosis. *Biochem. Soc. Trans.* **38,** 1286–1289.

Boghigian, B. A., Shi, H., Lee, K., and Pfeifer, B. A. (2010). Utilizing elementary mode analysis, pathway thermodynamics, and a genetic algorithm for metabolic flux determination and optimal metabolic network design. *BMC Syst. Biol.* **4,** 49, doi: 10.1186/1752-0509-4-49.

Bohl, K., de Figueiredo, L. F., Hädicke, O., Klamt, S., Kost, C., Schuster, S., and Kaleta, C. (2010). CASOP GS: Computing intervention strategies targeted at production improvement in genome-scale metabolic networks. *In* "Lecture Notes in Informatics," (D. Schomburg and A. Grote, eds.), Vol. P-173, pp. 71–80. Gesellschaft für Informatik, Bonn.

Bordbar, A., Lewis, N. E., Schellenberger, J., Palsson, B. O., and Jamshidi, N. (2010). Insight into human alveolar macrophage and *M. tuberculosis* interactions via metabolic reconstructions. *Mol. Syst. Biol.* **6,** 422, doi: 10.1038/msb.2010.68.

Çakir, T., Tacer, C. S., and Ülgen, K. O. (2004). Metabolic pathway analysis of enzyme-deficient human red blood cells. *Biosystems* **78,** 49–67.

Carlson, R., Fell, D., and Srienc, F. (2002). Metabolic pathway analysis of a recombinant yeast for rational strain development. *Biotechnol. Bioeng.* **79,** 121–134.

Caspi, R., Altman, T., Dale, J. M., Dreher, K., Fulcher, C. A., Gilham, F., Kaipa, P., Karthikeyan, A. S., Kothari, A., Krummenacker, M., Latendresse, M., Mueller, L. A., *et al.* (2010). The MetaCyc database of metabolic pathways and enzymes and the BioCyc collection of pathway/genome databases. *Nucleic Acids Res.* **38,** D473–D479, doi: 10.1093/nar/gkp875.

Clarke, B. L. (1981). Complete set of steady states for the general stoichiometric dynamical system. *J. Chem. Phys.* **75,** 4970–4979.

Dandekar, T., Moldenhauer, F., Bulik, S., Bertram, H., and Schuster, S. (2003). A method for classifying metabolites in topological pathway analyses based on minimization of pathway number. *Biosystems* **70,** 255–270.

de Figueiredo, L. F., Podhorski, A., Rubio, A., Kaleta, C., Beasley, J. E., Schuster, S., and Planes, F. J. (2009a). Computing the shortest elementary flux modes in genome-scale metabolic networks. *Bioinformatics* **25,** 3158–3165.

de Figueiredo, L. F., Schuster, S., Kaleta, C., and Fell, D. A. (2009b). Can sugars be produced from fatty acids? A test case for pathway analysis tools. *Bioinformatics* **25,** 152–158.

Feist, A. M., Herrgard, M. J., Thiele, I., Reed, J. L., and Palsson, B.Ø. (2009). Reconstruction of biochemical networks in microorganisms. *Nat. Rev. Microbiol.* **7,** 129–143.

Fischer, E., and Sauer, U. (2003). A novel metabolic cycle catalyzes glucose oxidation and anaplerosis in hungry *Escherichia coli*. *J. Biol. Chem.* **278,** 46446–46451.

Gagneur, J., and Klamt, S. (2004). Computation of elementary modes: a unifying framework and the new binary approach. *BMC Bioinformatics* **5,** 175, doi: 10.1186/1471-2105-5-175.

Gianchandani, E. P., Papin, J. A., Price, N. D., Joyce, A. R., and Palsson, B.Ø. (2006). Matrix formalism to describe functional states of transcriptional regulatory systems. *PLoS Comput. Biol.* **2,** e101, doi: 10.1371/journal.pcbi.0020101.

Gianchandani, E. P., Joyce, A. R., Palsson, B.Ø., and Papin, J. A. (2009). Functional states of the genome-scale *Escherichia coli* transcriptional regulatory system. *PLoS Comput. Biol.* **5,** e1000403, doi: 10.1371/journal.pcbi.1000403.

Hädicke, O., and Klamt, S. (2010). CASOP: A computational approach for strain optimization aiming at high productivity. *J. Biotechnol.* **147,** 88–101.

Kaleta, C., de Figueiredo, L. F., Behre, J., and Schuster, S. (2009a). EFMEvolver: Computing elementary flux modes in genome-scale metabolic networks. *In* "Lecture Notes in Informatics—Proceedings," (I. Grosse, S. Neumann, S. Posch, F. Schreiber, and P. Stadler, eds.), Vol. P-157, pp. 179–189. Gesellschaft für Informatik, Bonn.

Kaleta, C., de Figueiredo, L. F., and Schuster, S. (2009b). Can the whole be less than the sum of its parts? Pathway analysis in genome-scale metabolic networks using elementary flux patterns. *Genome Res.* **19,** 1872–1883.

Kanehisa, M., Goto, S., Furumichi, M., Tanabe, M., and Hirakawa, M. (2010). KEGG for representation and analysis of molecular networks involving diseases and drugs. *Nucleic Acids Res.* **38,** D355–D360, doi: 10.1093/nar/gkp896.

Kenanov, D., Kaleta, C., Petzold, A., Hoischen, C., Diekmann, S., Siddiqui, R. A., and Schuster, S. (2010). Theoretical study of lipid biosynthesis in wild-type *Escherichia coli* and in a protoplast-type L-form using elementary flux mode analysis. *FEBS J.* **277,** 1023–1034.

Klamt, S. (2006). Generalized concept of minimal cut sets in biochemical networks. *Biosystems* **83,** 233–247.

Klamt, S., and Stelling, J. (2002). Combinatorial complexity of pathway analysis in metabolic networks. *Mol. Biol. Rep.* **29,** 233–236.

Liao, J. C., Hou, S. Y., and Chao, Y. P. (1996). Pathway analysis, engineering, and physiological considerations for redirecting central metabolism. *Biotechnol. Bioeng.* **52,** 129–140.

Palsson, B.Ø., and Zengler, K. (2010). The challenges of integrating multiomic data sets. *Nat. Chem. Biol.* **6,** 787–789.

Pfeiffer, T., Sánchez-Valdenebro, I., Nuño, J. C., Montero, F., and Schuster, S. (1999). METATOOL: For studying metabolic networks. *Bioinformatics* **15,** 251–257.

Poolman, M. G., Fell, D. A., and Raines, C. A. (2003). Elementary modes analysis of photosynthate metabolism in the chloroplast stroma. *Eur. J. Biochem.* **270,** 430–439.

Price, N. D., Reed, J. L., and Palsson, B.Ø. (2004). Genome-scale models of microbial cells: Evaluating the consequences of constraints. *Nat. Rev. Microbiol.* **2,** 886–897.

Raghunathan, A., Reed, J., Shin, S., Palsson, B.Ø., and Daefler, S. (2009). Constraint-based analysis of metabolic capacity of *Salmonella typhimurium* during host-pathogen interaction. *BMC Syst. Biol.* **3,** 38, doi: 10.1186/1752-0509-3-38.

Ruppin, E., Papin, J. A., de Figueiredo, L. F., and Schuster, S. (2010). Metabolic reconstruction, constraint-based analysis and game theory to probe genome-scale metabolic networks. *Curr. Opin. Biotechnol.* **21,** 502–510.

Schilling, C. H., Letscher, D., and Palsson, B.Ø. (2000). Theory for the systemic definition of metabolic pathways and their use in interpreting metabolic function from a pathway-oriented perspective. *J. Theor. Biol.* **203,** 229–248.

Schürmann, P., and Jacquot, J. P. (2000). Plant thioredoxin systems revisited. *Annu. Rev. Plant Physiol. Plant Mol. Biol.* **51,** 371–400.

Schuster, S., and Hilgetag, C. (1994). On Elementary Flux Modes in biochemical reaction systems at steady state. *J. Biol. Syst.* **2,** 165–182.

Schuster, S., and Kenanov, D. (2005). Adenine and adenosine salvage pathways in erythrocytes and the role of S-adenosylhomocysteine hydrolase. A theoretical study using elementary flux modes. *FEBS J.* **272,** 5278–5290.

Schuster, S., Dandekar, T., and Fell, D. A. (1999). Detection of elementary flux modes in biochemical networks: A promising tool for pathway analysis and metabolic engineering. *Trends Biotechnol.* **17,** 53–60.

Schuster, S., Hilgetag, C., Woods, J. H., and Fell, D. A. (2002a). Reaction routes in biochemical reaction systems: Algebraic properties, validated calculation procedure and example from nucleotide metabolism. *J. Math. Biol.* **45,** 153–181.

Schuster, S., Pfeiffer, T., Moldenhauer, F., Koch, I., and Dandekar, T. (2002b). Exploring the pathway structure of metabolism: Decomposition into subnetworks and application to *Mycoplasma pneumoniae. Bioinformatics* **18,** 351–361.

Schwartz, J. M., Gaugain, C., Nacher, J. C., de Daruvar, A., and Kanehisa, M. (2007). Observing metabolic functions at the genome scale. *Genome Biol.* **8,** R123, doi: 10.1186/gb-2007-8-6-r123.

Stelling, J., Klamt, S., Bettenbrock, K., Schuster, S., and Gilles, E. D. (2002). Metabolic network structure determines key aspects of functionality and regulation. *Nature* **420,** 190–193.

Strang, G. (2009). Introduction to Linear Algebra. 4 edn. Wellesley-Cambridge, Wellesley.

Terzer, M., and Stelling, J. (2008). Large-scale computation of elementary flux modes with bit pattern trees. *Bioinformatics* **24,** 2229–2235, doi: 10.1093/bioinformatics/btn401.

Thiele, I., and Palsson, B.Ø. (2010). A protocol for generating a high-quality genome-scale metabolic reconstruction. *Nat. Protoc.* **5,** 93–121, doi: 10.1038/nprot.2009.203.

Trinh, C. T., and Srienc, F. (2009). Metabolic engineering of *Escherichia coli* for efficient conversion of glycerol to ethanol. *Appl. Environ. Microbiol.* **75,** 6696–6705.

Trinh, C. T., Carlson, R., Wlaschin, A., and Srienc, F. (2006). Design, construction and performance of the most efficient biomass producing *E. coli* bacterium. *Metab. Eng.* **8,** 628–638.

Unrean, P., Trinh, C. T., and Srienc, F. (2010). Rational design and construction of an efficient *E. coli* for production of diapolycopendioic acid. *Metab. Eng.* **12,** 112–122.

Urbanczik, R., and Wagner, C. (2005). An improved algorithm for stoichiometric network analysis: Theory and applications. *Bioinformatics* **21,** 1203–1210.

Weinman, E. O., Srisower, E. H., and Chaikoff, I. L. (1957). Conversion of fatty acids to carbohydrate: Application of isotopes to this problem and role of the Krebs cycle as a synthetic pathway. *Physiol. Rev.* **37,** 252–272.

Wilhelm, T., Behre, J., and Schuster, S. (2004). Analysis of structural robustness of metabolic networks. *Syst. Biol. (Stevenage)* **1,** 114–120.

Xi, Y., Chen, Y. P. P., Cao, M., Wang, W., and Wang, F. (2009). Analysis on relationship between extreme pathways and correlated reaction sets. *BMC Bioinformatics* **10**(Suppl 1), S58, doi: 10.1186/1471-2105-10-S1-S58.

Yeung, M., Thiele, I., and Palsson, B.Ø. (2007). Estimation of the number of extreme pathways for metabolic networks. *BMC Bioinformatics* **8,** 363, doi: 10.1186/1471-2105-8-363.

CHAPTER TWENTY-THREE

How to Obtain True and Accurate Rate-Values

Sef J. Heijnen *and* Peter J. T. Verheijen

Contents

Biotechnology Department, Delft University of Technology, Julianalaan 67, 2628 BC Delft, The Netherlands

Methods in Enzymology, Volume 500
ISSN 0076-6879, DOI: 10.1016/B978-0-12-385118-5.00023-2

Abstract

In metabolic flux calculations, the uptake and secretion rates (for substrate, O_2, CO_2, growth, (by)-products) are essential to arrive at correct calculated fluxes. Surprisingly, a lot of research has been published on the methods of flux calculations, but much less attention has been spent on the methods to obtain accurate and true uptake and secretion rates which are used as input. Therefore, this contribution focuses on

- the experimental issues to obtain these rates;
- forgotten processes as pitfalls in these calculations;
- methods to detect these based on element conservation principles;
- experimental design which leads to most accurate rates; and
- the statistical methods to obtain accurate uptake and secretion rates.

1. Introduction

A cell can be considered as a small chemical factory that converts substrates into products and biomass. The net conversion rate of a compound is called here the q-rate. This is expressed as an amount (converted) per unit of time and per unit of cell amount.

The q-value for a compound i must be calculated from its compound balance

$$\frac{\mathrm{d}(V(t)c_i(t))}{\mathrm{d}t} = q_i(t)V(t)c_x(t) + F_{\mathrm{in}}c_{\mathrm{in},i} - F_{\mathrm{out}}c_i(t), \tag{23.1}$$

which represents the equation accumulation = conversion + transport. Here, V is volume; F_{in} and F_{out} are the in- and outgoing flows; and $c_{\mathrm{in},i}$, c_i, and c_x are the concentrations of compound i and the biomass.

The calculated q-values are used

- to build a Black Box kinetic model of the organism;
- to perform flux calculations of the intracellular reactions;
- to calculate yields;
- to redesign a new organism by metabolic engineering;
- to design production processes using a given organism; and
- to perform economic calculations.

This shows that it is important that the calculated q-values are true and accurate.

A q-value is not true when there is a systematic error because of systematic measurement errors or because the used balance is not complete due to forgotten processes. This aspect will be discussed in Section 3.

The accuracy of the calculated q-value is expressed by its error. As the q-rate is derived from the compound balance of Eq. (23.1) into which the measured values of flows (e.g., F_{in} and F_{out}), volume (V), and concentrations ($c_{in,i}$, c_i, and c_x) are entered. The errors in q are then obtained from error propagation calculation (discussed in Section 5).

In addition, it is recommended to reflect on the experimental design to obtain q-values which allow

- to minimize error propagation into q_i
- to discriminate the presence or absence of forgotten processes which require redundancy in measurements (Section 5)

Finally, using the redundant information and having ascertained that the calculated q-value is true (absence of forgotten processes), it is possible to improve the accuracy of the q-values (reconciliation, Section 6).

However, the starting point is that we need to obtain accurate measurements of the flow, concentration, and volume variables, which are present in each compound balance. Concentration measurements do not present problems, nor do flow rate measurements. However, quantification of the broth amount, represented as its volume V is a major point of concern, which is therefore discussed in Section 2.

2. Quantification of Broth Amount in the Cultivation Vessel

2.1. Problems in measuring the broth total volume V or broth mass M

The biomass specific uptake and secretion rate $q_i(t)$ of compound i is obtained from the balance for compound i as shown above. The volumetric liquid flow rates F_{in}, F_{out} and the volumetric concentrations c_x, $c_{in,i}$, c_i are measured using flow sensors and standard sampling and analysis protocols.

The broth volume $V(t)$ must also be obtained from measurement which is difficult as appears from the following discussion:

- One can measure the height of the broth surface in the fermentor (which is not that easy). This height does not provide the broth liquid volume $V(t)$, but the volume of liquid + all gas bubbles present in the broth. This volume of all gas bubbles can vary rapidly and is difficult to monitor separately.
- A second possibility is to weigh the whole fermentor to obtain $M(t)$. Problems in accurate weighing arise because the fermentor is usually not free standing, especially for large ($>10\ m^3$) fermentors. Small lab fermentors are also not free standing because many fixed tube connections exist to feed vessels. Often flexible tubes are used to solve this problem, but this is expensive and risky (e.g., infection).
- An alternative to obtain broth mass is to measure the pressure difference between the fermentor bottom and the headspace. The pressure drop is proportional to the broth mass present per m^2 of fermentor, and the total mass is calculated using the free cross-sectional area of the fermentor. A first problem with measuring the pressure difference is that the rotating stirrer and the sparged gas bubbles at the bottom of the fermentor create a turbulent liquid flow which creates dynamic pressure fluctuations, which complicates the pressure measurement and hence the calculated broth mass. For example, a change in stirrer rotational speed or airflow will change the pressure drop while the broth mass does not change. Apart from measurement problems in broth mass $M(t)$ another problem is that the broth density, needed to convert $M(t)$ into $V(t)$, depends on broth composition and therefore is time dependent and often unknown. Only in fermentations with diluted aqueous solutions (low c_x, c_p) one can assume $\rho \approx 1000\ kg/m^3$ and constant. For fermentations with high product concentration, this is not allowed.

Clearly experimental quantification of $V(t)$ or $M(t)$ is a major problem.

2.2. Calculation of *V* using a volume balance is a wrong approach

We have seen that accurate measurement of broth liquid volume V is nearly impossible in lab scale as well as in large scale fermentors.

When measurement is not possible then one can apply calculation. In many textbooks, it is suggested that the broth volume V can be calculated using the volume balance.

$$\frac{dV(t)}{dt} = \sum \text{volumetric liquid flows entering broth} - \sum \text{volumetric flows leaving broth.} \tag{23.2}$$

This balance is absolutely wrong because it is assumed that density is constant and also the volumes of molecules (e.g., O_2, CO_2) exchanged between gas and broth are neglected. The density depends on the composition of the broth as the batch culture undergoes changes. A consequence is that the broth volume changes even without inflow and outflow. Further, in a stationary situation, the amount of ingoing and outgoing mass flows are exactly the same, the volumetric flows can differ because of the density difference between inflow and outflow.

Example Liquid volume is not conserved

Consider 1 kg pure water (density at 25 °C is 998.2 g/l) which has $V = 1001.8$ ml. Also consider 1 kg pure ethanol (density is 789.3 g/l at 25 °C) which has a volume of 1266.9 ml. When we mix these amounts of water and ethanol

- one obtains 1 kg (water) + 1 kg (ethanol) = 2 kg aqueous solution of ethanol. This solution (50%, w/w), according to thermodynamic tables, has a density of 914.7 g/l leading to a volume of 2186.5 ml.
- One does not obtain a total volume of 1001.8 ml (water) + 1266.9 ml (ethanol) = 2268.7 ml.

In reality, the measured volume is 2186.5, hence 80 ml volume has disappeared. The volume change is due to molecular interaction between water and ethanol.

2.3. The broth total mass balance in (fed) batch provides a calculated *M*

The only reliable approach to quantify the broth amount present in a fermentor is to abandon $V(t)$ and instead use broth mass $M(t)$. The broth mass is difficult to be measured (Section 2.1), but fortunately it can be calculated very accurately from the broth total mass balance on the fermentor. This broth total mass balance is defined in kg/h,

$$\frac{\mathrm{d}M(t)}{\mathrm{d}t} = \sum \text{mass flows } \textit{entering} \text{ broth} - \sum \text{mass flows } \textit{leaving} \text{ broth}. \tag{23.3}$$

This balance is extremely simple and gives highly accurate values for $M(t)$, if one takes all broth related mass flows into account.

Examples of mass flows (kg/h) entering the broth are

- All liquid feeds, such as feeds of alkali or acid for pH control, antifoam for foam control, feed solution containing the C-source, feed of N-source solution (e.g., an aqueous NH_3-solution).

- The mass of all molecules transported into the broth liquid. For example, O_2 is transferred from gas to broth liquid, where it is consumed. This is a very substantial mass, for example, an O_2 transfer rate of 200 mol O_2/m^3 h, makes a total mass transfer of 640 kg O_2/m^3 broth in 100 h.

Examples of mass flows (kg/h) leaving the broth are

- Mass associated with samples taken from the fermentor. This is especially relevant for lab scale fermentations where the total sample mass can easily add up to 10% of the total broth mass.
- Withdrawals of broth as occurs in continuous fermentation processes, where the broth amount in the fermentor is maintained constant by a continuous broth outflow or due to sampling.
- The mass of molecules transferred from the broth liquid into the gas phase.

A first example is CO_2, produced by heterotrophic organism, which is transferred from broth to gas phase. The mass amount of CO_2 loss follows from analysis of the CO_2 in the off-gas and the total molar gas outflow. For example, a CO_2-production of 100 mol/h during 100 h leads to a mass loss of 440 kg/m^3 broth.

A second example is water evaporation. Water equilibrates between broth and the sparged gas bubbles. Because the gas entering the fermentor (e.g., air) is dry (to enable proper sterilization), there is net loss of water to the gas phase. The amount of water present in the gas phase follows from the equilibrium water pressure. This pressure depends on temperature and the broth liquid composition and can be found in thermodynamic tables or calculated using equilibrium thermodynamics. For example, the vapor pressure p_{H_2O} of pure water is 0.031 and 0.20 bar at respectively 25 and 60 °C. Assume a pressure $p = 1.5$ bar at 25 °C, then the $x_{H_2O} = p_{H_2O}/p = 0.0207$. A typical airflow is 1 vvm (*v*olume of gas per *v*olume of broth per *m*inute), which is equivalent to 2400 mol/h gas for 1 m^3 of broth. This gas then contains 50 mol H_2O l/h, which represents 0.9 kg/h. Hence, air sparging over a period of 100 h leads to a water loss of 90 kg/m^3 broth.

A third example is the loss of a volatile product (or substrate), for example, ethanol due to gas sparging. Consider a fermentor, broth volume V with an ethanol concentration c_{eth} mol/m^3. A gas is sparged (e.g., to supply O_2 or remove CO_2) at a molar flow rate (mol/h). It can be assumed that, due to its high volatility, ethanol in gas and liquid are in equilibrium. This equilibrium can be described as a Henry type relation

$$c_{eth} = \alpha_{eth} p x_{eth}. \tag{23.4}$$

Here, c_{eth} is the ethanol concentration in the broth (mol/m^3), p is the gas pressure (bar), x_{eth} is the equilibrium mole fraction ethanol in the gas. α_{eth} reflects the ethanol volatility and has units mol/m^3/bar.

Usually, lab fermentors have a condenser in the outgas channel and the condensed ethanol (and water) will flow back to the fermentor. The condenser will decrease the ethanol mole fraction in the outlet gas with a factor β to βx_{eth} ($0 < \beta < 1$). The value of β depends on numerous condenser related factors (gas flow rate, condenser cooling surface area, condenser temperature compared to fermentor temperature), which are constant in a given fermentor setup.

We can now write for the ethanol loss rate (mol/h) from the broth to the gas,

$$\text{Ethanol loss rate} = R_{eth} = F_{G,out}\beta x_{eth}. \tag{23.5}$$

We can now eliminate x_{eth} using the volatility relation (23.4) and eliminate c_{eth} to introduce the total amount of ethanol present in broth, called M_{eth} ($=Mc_{eth}/\rho$).

$$R_{eth} = F_{G,out}\beta \frac{c_{eth}}{\alpha_{eth} p} = \left[\frac{F_{G,out}\beta\rho}{M\alpha_{eth} p}\right] M_{eth} = k_{eth} M_{eth}. \tag{23.6}$$

The term in brackets represents a first-order rate constant k_{eth} (h^{-1}) for ethanol evaporation, which depends on

- β, it reflects the condenser function;
- $F_{G,out}/M$, more gas sparged per broth mass, the higher the ethanol loss;
- p, a higher pressure decreases loss; and
- T, because it influences α_{eth}. High temperatures cause more loss.

The last three can be directly influenced during the operation. It is easy to obtain k_{eth} for a given fermentor from a simple evaporation experiment, where M_{eth} is followed in time by analyzing the ethanol concentration c_{eth} and the broth mass M. Clearly, k_{eth} changes when one changes the above variables in the fermentor setup.

Setting up the broth total mass balance is easy, once all mass flows are identified. The broth total mass balance allows an accurate calculation of the broth mass $M(t)$ present in a (fed)batch fermentor over time, as shown in the examples below for batch and fed batch. Its application to a chemostat fermentation is discussed in the next section.

Example Total mass balance calculations in a fed batch

Consider a fed batch with a constant amount of biomass which produces citric acid ($C_6H_8O_7$) from glucose ($C_6H_{12}O_6$) according to the following stoichiometry

$$-1C_6H_{12}O_6 - 3(1/2)O_2 + 1C_6H_8O_7 + 2H_2O$$

There is a constant feed flow of a glucose solution (0.20 m^3/h, 771.2 kg glucose/m^3, density 1287.7 kg/m^3). All glucose is converted. In the beginning, the broth volume is $V(0) = 100\ m^3$. The broth density is 1000 kg/m^3. There is no loss of water during the fermentation (due to a condenser, which is present in the off-gas).

Question

Calculate the broth volume and citrate concentration in g/l after 100 h of feeding.

Answer

The total mass balance must be applied to obtain the answer.

Mass added is

- feed $0.20 \times 1287.7 = 257.5$ kg/h
- Glucose added is converted to citric acid and equals $0.20 \times 771.2/180 = 0.857$ kmol/h. This requires $0.857 \times 1.50 \times 32 = 41.1$ kg/h O_2 according to the above O_2-stoichiometry which is transferred from gas to broth

Mass removed

There is no H_2O loss by evaporation and no mass removal because CO_2-production.

The broth total mass balance gives now the mass after 100 h:

$$M(100\,\mathrm{h}) = 100,000 + 100 \times (257.5 + 41.1) = 129,860\,\mathrm{kg}.$$

To obtain the end volume, we need the density. The converted glucose feed has led to 0.857 kmol/h citric acid, leading after 100 h to

$$0.857 \times 100 \times 192 = 16,454\,\mathrm{kg}\ \text{citric acid}.$$

This leads to a mass concentration of citric acid of 16,454/129,860 = 0.127 kg/kg broth. Thermodynamic tables show that this solution has $\rho = 1054$ kg/m^3. Hence,

$$V(100\,\mathrm{h}) = \frac{129,860}{1054} = 123.2\,\mathrm{m}^3.$$

Note that, using the "volume balance" one obtains 120 m^3 of end volume.

In a chemostat, the total mass balance provides the total mass outflow but not the broth mass. In continuous fermentation processes, the broth total mass balance can be used to calculate the total mass flow of broth F_{out} (kg/h) which leaves the fermentor. The steady state broth total mass balance follows as:

$$\sum \text{mass flows } \textit{entering} \text{ broth} = \sum \text{mass flows } \textit{leaving} \text{ broth}. \quad (23.7)$$

The mass flows entering the broth in the fermentor are

- the nutrient solution measured by a mass flow sensor or from a balance upon which the storage vessel with nutrient solution is placed;
- solutions for pH control, and other solutions added (e.g., antifoam or precursor), whose mass rate is measured similarly; and
- molecules transferred from gas phase to broth, for example, O_2, a gaseous electron donor or gaseous NH_3 which is used as nitrogen source.

The mass flows leaving the broth in the fermentor are

- molecules transferred from broth to gas phase, for example, CO_2, evaporating H_2O and volatile compounds (e.g., ethanol), as described in Section 2.2; and
- the broth mass leaving the fermentor F_{out} (kg/h).

Because usually all, except F_{out}, contributions to the broth total mass balance are known, this balance allows to calculate F_{out} (kg/h).

In contrast to expectation, the total mass balance does not allow to calculate the broth mass M in a chemostat, because, M is absent from the steady state total mass balance. M must therefore be measured.

Experimental quantification of the broth mass M present in a continuous fermentation process is not trivial. In a lab fermentor, usually one keeps a constant total working volume, using an outflow device (overflow pipe or surface suction pipe). This volume consists of broth and the gas bubbles present in the broth. Especially the total gas bubble volume can change significantly (up or down) due to surface active compounds such as added antifoam or surface active metabolites produced and consumed by the organism. Therefore, the broth mass M can, despite a constant working volume, change significantly. To obtain the broth mass M present in a chemostat there are two possibilities.

- Usually, at the end of a chemostat experiment one empties completely the fermentor broth content and weighs this broth, to obtain M.
- Alternatively, one can put the fermentor on a balance and control the broth mass M at a fixed value by broth-withdrawal. This is the preferred method but is only accurate for fermentor volumes of several tons. For small fermentors, the weighing is complicated by the tubing's attached to the fermentor.

In full continuous scale fermentors, one usually applies pressure sensors to obtain M, which is not without problems (see Section 2.2). A more accurate method is to use weight sensors under the whole vessel.

2.4. Reformulating individual compound balances on total mass and not volume basis

The traditional (see Section 1) compound balance is volume based using volumetric concentrations $c_i(t)$ for compound i, volume based flow rates (F_{in}, F_{out}), and the total broth volume $V(t)$.

In dilute fermentation processes with low biomass and product concentrations due to the use of dilute (<10 g/l) nutrient solutions, the broth density can be assumed nearly independent of the dilute broth composition ($\rho = 1000$ kg/m^3 for dilute aqueous broth). In such a dilute situation, which typically occurs at lab scale, the traditional "volume based compound balance" approach does introduce only a very minor error.

However, in high cell density fermentations and in fermentations with pure feed compounds or with high product concentrations this "volume" approach introduces large errors. It then becomes necessary to use the correct approach, which is to use mass based compound balances.

The compound balance—in mol/h—for compound i present in broth based on broth total mass M and mass based concentration $c_{\text{M},i}$ can be written

$$\frac{\mathrm{d}\left(M(t)c_{\text{M},i}(t)\right)}{\mathrm{d}t} = q_i(t)c_{\text{M},x}(t)M(t) + F_{\text{M,in}}c_{\text{M,in},i} + F_{\text{M,out}}c_{\text{M},i}(t). \quad (23.8)$$

In this compound balance, the following definitions apply.

- The subscript M indicates the use of mass instead of volume basis.
- $c_{\text{M},i}$ is the " concentration" in mol of compound per kg broth.
- $c_{\text{M,in},i}$ is in mol of compound per kg of feed solution.
- $F_{\text{M,in}}$ is the mass flow (kg/h) of feed solution into the vessel.
- $F_{\text{M,out}}$ is the mass flow (kg/h) of broth liquid leaving the vessel.

The mass flows F_{M} are easily quantified using weighing balances under the feed storage waste vessel and broth outflow vessels (at lab scale) or using mass flow meters at full scale. Another significant advantage of using strictly mass based in- and outflow rates (kg/h) is that these quantities are also needed in the total mass balance, which is needed to determine the broth mass $M(t)$ in (fed)batch or the mass outflow rate $F_{\text{M,out}}$ in a continuous fermentation process.

Apart from mass based flow rate measurement, one needs to specify concentrations on kg broth basis. Conventionally, one analyzes compounds in the supernatant, after removal of biomass (to avoid concentration changes in the sample). We can write for a specific concentration that

$$\underbrace{\frac{\text{mol}}{\text{kg broth}}}_{\text{needed in mass balance}} = \underbrace{\frac{\text{mol}}{\text{kg supernatant}}}_{\text{analyzed}} \times \alpha, \quad (23.9)$$

where α is the mass fraction of the supernatant in the broth, when the analyzed compound is negligibly present in the biomass (otherwise, e.g., due to active export, α is just a ratio, which could be >1). The difference between broth and supernatant is the wet biomass. Therefore, α is related to the biomass concentration $c_{M,x}$ (Cmol/kg broth).

Assuming that wet biomass has a water mass fraction f_{H_2O} (≈ 0.75) and a dry cell weight of 25 g, it follows for

$$\alpha = 1 - \frac{25/1000\, c_{M,x}}{1 - f_{H_2O}} \frac{\text{kg supernatant}}{\text{kg broth}}. \tag{23.10}$$

For example, $c_{M,x} = 100$ g dry cell weight/kg broth $= 4$ Cmol/kg broth and $f_{H_2O} = 0.75$ one obtains $\alpha = 0.60$ which is far below 1, showing that we need to take this broth-supernatant issue into account.

A final concentration aspect is related to a proper analysis of the biomass concentration in chemostats. Here, it is always assumed that c_x in the fermentor and in the outflow are the same. This is not guaranteed, because the broth outflow may not be ideal (Noorman *et al.*, 1991). It is, therefore, recommended to measure always both $c_{x,\text{fermentor}}$ and $c_{x,\text{out}}$.

Finally, kinetics (such as the hyperbolic rate of substrate uptake) needs to be expressed in *volume*-based concentrations c_i (mol/m^3); hence $c_i = \rho c_{M,i}$ and broth density ρ is still needed, which is obtained from thermodynamic tables and broth composition.

3. Forgotten Processes

3.1. Introduction

Usually our idea of cellular metabolism revolves around the ideal organism where a single limiting substrate is converted into biomass and one product, and we obtain values for q_s, μ, q_p. These are used to construct Black Box kinetic models and to calculate the "other rates" (O_2, CO_2, heat, H^+, etc.). This stoichiometric and kinetic information is then the basis for

- comparing organism performance,
- construction of a kinetic model, and
- designing bioprocesses.

However, organisms are quite flexible in producing "by-products" in response to changing extracellular conditions, leading to forgotten conversion processes. Well-known examples are

- "Overflow" of metabolites present in a product pathway. An important example is the "overflow" of metabolites from central metabolism.

The demand of the products of central metabolism often becomes imbalanced with their supply rate at high substrate supply. To avoid the intracellular accumulation of these compounds, the cell secretes products which are easily made from these compounds. For example, pyruvate surplus is converted into lactate, ethanol, or acetate; 6-phosphogluconate surplus is converted into gluconic acid; and acetyl CoA surplus is converted into acetate.

- Unknown metabolic activity occurs. For example, in the penicillin fermentation the precursor PAA (phenyl acetic acid) is used to provide the "side chain" of the produced penicillin. Ideally, one wishes that this expensive compound is converted for 100% into the produced penicillin. However, often it is found that the mole ratio of consumed PAA to produced penicillin is higher than 1. This points to other fates for PAA. It is well-known that PAA is partially catabolized by oxidation of PAA into *p*-OH-PAA.
- Unknown chemical degradation occurs to by-products. Here it is important to realize that many products (antibiotics, proteins, etc.) are chemically unstable in aqueous environment; hence degradation of the main product occurs. For example, penicillin G hydrolyses to penicilloic acid, following first-order kinetics, with the degradation rate constant being dependent on pH, T, composition. Unnoticed degradation of product leads to an underestimation of q_p (Heijnen *et al.*, 1979).
- An important environmental condition is mechanical shear on cells due to the use of stirrers and aeration. This leads to dead cells which can lyse where the intracellular compounds (proteins, cell-wall compounds) accumulate dissolved in the broth. When lysis is unnoticed μ is underestimated (Taymaz-Nikerel *et al.*, 2010).

Not only forgotten conversion processes occur, but there are also forgotten transport processes. Examples are evaporation of volatile substrate or a volatile product or absorption of a product in, for example, plastic tubing.

In the next paragraph, we will discuss the impact of these forgotten processes on compound balance based q_s, q_p, μ-values (Sections 3.2–3.4).

We should realize that identification of these "forgotten" processes offers immediate possibilities to improve product formation and product yield. Therefore, methods to detect these forgotten processes (Section 4) are not only relevant for true q-values but also contribute to improve immediately the process economy.

The most basic set of q-rates consists of q_s, q_p, μ, which are used to construct the basic kinetic model (Herbert-Pirt, $q_p(\mu)$-relation). Therefore, it is relevant to discuss how their compound balance based calculation is effected due to forgotten degradation (Sections 3.2 and 3.3) and forgotten transport processes (Section 3.4).

3.2. Cell death influences the calculation of μ

The occurrence of cell death needs to be quantified, which is not easy. We identify different methods. Its quantification depends on the assumptions made about the dead cells.

- Dead cells remain intact. Then, in the fermentor, we have both living and dead cells, which can be distinguished using viability staining and microscopy.
- Dead cells do completely lyse (dissolve) leading to nonconsumable lysis products in the supernatant. These lysis products are most easily measured using the accumulation of total organic carbon (TOC) in the supernatant. Under substrate limited condition, the organic carbon in the supernatant is due to presence of residual organic substrate, products, and known by-products. Because these are analyzed by specific methods, one expects that the result of TOC matches the carbon present in the analyzed compounds. Very often one finds that TOC gives much higher values, which is a serious indication of cell death or secretion of unknown by-products.

Because cellular components not only contain carbon but also nitrogen (protein, RNA), a confirmation of cell death can also be obtained by Total Kjeldahl-N analysis (K_j-N) of the supernatant. K_j-N represents the sum of NH_4^+ and organic N present. In the ideal case K_j-N $=$ NH_4^+–N, then there is no organic nitrogen secreted. Usually K_j-N $\gg$ NH_4^+, indicating presence of proteins and RNA released from cells or N-containing metabolites (e.g., amino acids). Off course one can also analyze protein and RNA in the supernatant.

For the two different mechanisms, cell dead can be quantified as follows.

3.2.1. Viability staining, leads to k_d

Cell death leads to c_x^{dead} and c_x^{live} in the cultivation system, where the concentration of dead and living cells can be obtained from viability staining. If we take a chemostat, then the compound balance of dead cells leads to

$$0 = R_x^{\text{dead}} - F_{\text{out}} c_x^{\text{dead}}. \tag{23.11}$$

Cell death is characterized by the specific biomass death rate k_d (h^{-1}),

$$k_d = \frac{\text{Rate of formation of dead cells}}{\text{Total amount of living cells}} = \frac{R_x^{\text{dead}}}{V c_x^{\text{live}}}. \tag{23.12}$$

Equations (23.11) and (23.12) give for k_d,

$$k_d = \frac{F_{\text{out}} c_x^{\text{dead}}}{V c_x^{\text{live}}} = \frac{D c_x^{\text{dead}}}{c_x^{\text{live}}}, \tag{23.13}$$

introducing the dilution rate D ($=F_{out}/V$). This relation shows that the fraction of living cells present in the cultivation vessel follows from

$$\frac{c_x^{live}}{c_x^{live}+c_x^{dead}}=\frac{D}{D+k_d}. \tag{23.14}$$

This shows, assuming a constant cell death rate k_d, that at high D ($D \gg k_d$), the percentage of living cells approaches 100%, but at low D, for example, at $D = k_d$, 50% dead cells are present. This is indeed experimentally observed.

3.2.2. Supernatant—TOC, leads to μ_{lysis}

The TOC_{sup} value in supernatant ($Cmol/m^3$) can be corrected for carbon in residual substrate and for all known secreted organic products and by-products $TOC_{(b)p}$, leading to TOC^{dead} in $Cmol/m^3$ supernatant. One assumes that the dead cells nearly completely converted into soluble molecules, a process called cell lysis. The remaining dry matter in the cultivation vessel is then active biomass. Usually, one analyzes the organic carbon of the total broth TOC_{broth}. The living biomass concentration TOC^{live} (in $Cmol/m^3$) follows now from

$$\begin{aligned} c_x^{live} &= TOC^{live} = TOC_{broth} - TOC_{sup} \quad \text{and} \\ c_x^{dead} &= TOC^{dead} = TOC_{sup} - TOC_s - TOC_{(b)p}. \end{aligned} \tag{23.15}$$

The compound balance on dead cells Eq. (23.11) together with the equations above give a relation for the total production rate, R_X^{dead} in Cmol/h, of dead cells in a chemostat as

$$R_x^{dead} = F_{out}\left(TOC_{sup} - TOC_s - TOC_{(b)p}\right). \tag{23.16}$$

The biomass specific lysis rate μ_{lysis} follows then:

$$\mu_{lysis} = \frac{R_x^{dead}}{Vc_x^{live}} = D\frac{TOC_{sup} - TOC_s - TOC_{(b)p}}{TOC_{broth} - TOC_{sup}}. \tag{23.17}$$

3.2.3. Due to cell death μ changes, $\mu \neq D$

Due to the presence of cell death, the calculation of μ from the compound balance on living cells also changes. μ (h^{-1}) is defined as

$$\mu = \frac{\text{Rate of production of living cells}}{\text{Amount of living cells}} = \frac{R_x^{live}}{Vc_x^{live}}. \tag{23.18}$$

The rate of production of living cells follows from the compound balance on living cells in a chemostat,

$$0 = R_x^{\text{live}} - R_x^{\text{dead}} - F_{\text{out}} c_x^{\text{live}}. \tag{23.19}$$

Then, substitute Eqs. (23.19) into (23.18) for the rate R_x^{live}. Then together with Eq. (23.12) leads to an expression for the growth rate,

$$\mu = \frac{F_{\text{out}} c_x^{\text{live}} + k_d V c_x^{\text{live}}}{V c_x^{\text{live}}} = D + k_d. \tag{23.20}$$

Note that k_d is replaced by μ_{lysis} when the cell death is quantified using lysis and TOC as in Eq. (23.17).

This shows that the presence of the process of cell death,

- changes the relation $\mu = D$ (no cell death) into $\mu = D + k_d$
- might lead to the presence of two cell types, living and dead depending on the assumed cell death mechanism.

Accurate information on c_x^{live} (living cells), μ_{lysis} (or k_d), and μ is important because true q-values should be based on living cells.

3.3. Product degradation influences the calculation of q_p

Cellular products are preferably secreted into the aqueous environment. Simple compounds such as lactic acid are highly stable. However others, for example, antibiotics, are unstable and usually they degrade with a first-order rate constant k_p into a by-product, where 1 mol product leads to 1 mol by-product. This has two consequences:

First, the calculation of R_p and q_p needs to take product degradation into account. We can write the product compound balance in a chemostat

$$0 = R_p - k_p V c_p - F_{\text{out}} c_p. \tag{23.21}$$

With the dilution rate D $(=F_{\text{out}}/V)$, this results in an

$$q_p = \frac{R_p}{V c_x} = \frac{k_p V c_p + F_{\text{out}} c_p}{V c_x} = \left(k_p + D\right)\frac{c_p}{c_x}. \tag{23.22}$$

This shows that, when degradation is not noticed, q_p is underestimated. In, for example, penicillin fermentation, product degradation is an important aspect. Typically, $k_p = 0.0015\ \text{h}^{-1}$ and $D = 0.006\ \text{h}^{-1}$, such that q_p is off by 25%.Second, the accumulation of by-product is affected. This is derived from the by-product compound balance,

$$0 = k_p V c_p - F_{out} c_{bp}. \tag{23.23}$$

This gives directly the ratio of concentration by-product over product

$$\frac{c_{bp}}{c_p} = \frac{k_p}{D}. \tag{23.24}$$

This ratio increases for low D. For the typical case above, this ratio is 25%.

3.4. Evaporation of volatile compounds influences the calculation of q_p, q_s

Many products are volatile. Ethanol is a well-known example. Also esters and larger chain alcohols (such as flavors and fragrances) are highly volatile. Usually, cultivation vessels are aerated (bubble sparging) for O_2-supply. This leads to transport of a volatile compound (substrate or product or, e.g., NH_3) from the broth to the gas phase. The product compound balance now reads

$$0 = R_p - F_{out} c_p - F_G x_p. \tag{23.25}$$

The last term is the rate of evaporation, which follows from analyzing the volume fraction of volatile product (x_p = mole fraction) in the gas outflow. F_{out} is the gas outflow (mol/h). In case of gas–liquid equilibrium, which easily happens for very volatile compounds, the liquid phase product concentration c_p (mol/m^3) and x_p are connected through a Henry type of relation,

$$c_p = \alpha x_p p, \tag{23.26}$$

where α_p is the Henry related constant (mol/(m^3 bar)) and p is the gas pressure (bar). Equations (23.25) and (23.26) then give

$$R_p = \left[F_{out} + \frac{F_G}{\alpha_p p}\right] c_p. \tag{23.27}$$

This shows that the evaporation contribution increases at higher gas flow, lower pressure, and higher volatility (lower α_p). For very low volatile compounds (α_p high), the second (evaporation transport) term is negligible to the transport in liquid (F_{out}), and the evaporation based transport can be neglected. When the substrate is volatile, the R_s and hence q_s needs to be calculated by taking this loss into account in a similar way.

4. Detecting Forgotten Processes and Systematic Errors

4.1. Principles of element conservation

Uptake and secretion rates are subject to element and charge conservation, which means that conservation relations exist between these rates. Consider aerobic growth of an organism in glucose. In Table 23.1, we identify the element composition (and from here we use the term "element" to encompass also charge and degree of reduction) of the involved compounds.

Here, we used an experimental determined biomass element composition (for 1 Cmol) which represents a fairly average for many organisms (Roels, 1983). Each row represents one conservation relation (for carbon, hydrogen, oxygen, nitrogen, and electric charge). Using this element constraint matrix, we can write the conservation relations, for example, for carbon (first row)

$$6q_s + q_{CO_2} + \mu = 0, \tag{23.28}$$

where the q-rates have negative and positive signs depending on whether it is respectively taken up or secreted. There are seven different q_i: glucose (s), O_2, CO_2, proton (H^+), N-source (NH_4^+), biomass (x), and water (H_2O). There are five constraints, so there are only two free rates.

A frequently used other constraint is the "degree of reduction balance." The degree of reduction (symbol γ) is a suitable stoichiometric quantity for each compound, which represents the electrons present in that compound as can be calculated using redox half reaction. The definition is based on element γ's and in the case of nitrogen the molecular form is distinguished.

The degree of reduction, γ, now follows from the element composition and Table 23.2. For example, for glucose, $C_6H_{12}O_6$, $\gamma = 6 \times 4 + 12 + 1$

Table 23.1 Element constraint matrix for aerobic growth with glucose

	Glucose	O_2	CO_2	H^+	NH_4^+	Biomass	H_2O
C	6	0	1	0	0	1	0
H	12	0	0	1	4	1.8	2
O	6	2	2	0	0	0.50	1
N	0	0	0	0	1	0.20	0
Charge	0	0	0	+1	+1	–	0

Table 23.2 The degree of reduction for selected atoms

C	4
H	1
O	−2
N in biomass (NH_4^+)	−3
N in product (NO_3^-)	+5
Depending on N-source (N_2)	0
+Charge	−1
−Charge	+1

$+ 6 \times (-2) = 24$; and for O_2, $\gamma = 2 \times (-2) = -4$. These values are indeed found in the redox half reactions for glucose and O_2.

The suitability of γ is due to the property that $\gamma = 0$ for H_2O, CO_2, HCO_3^-, N-source, H^+, OH^-. This means that the γ-balance only contains q_i or substrate, electron acceptor, biomass, and product.

In principle, one could also add an additional row containing molar enthalpies of formation of each compound. This would represent, the first law of thermodynamics, implying the energy balance.

4.2. O_2 and CO_2 balances to calculate R_{O_2} and R_{CO_2}

In order to make C- or γ-balances, one needs O_2 and CO_2 rates. The rate of O_2-consumption by organisms present in broth, R_{O_2} in mol/h, by definition is obtained from the broth O_2-balance. This balance is however highly problematic due to the unknown kinetics of the O_2-transfer term to the gas: $K_L A(c^*_{L,O_2} - c_{L,O_2})$, where $K_L A$ easily changes because $K_L A$ is very sensitive to gas bubble hold up and bubble size, which are very susceptible to changes, for example, surface active agents.

However, an accurate value of this transfer term can be obtained from the gas phase O_2-balance. O_2 and CO_2 concentrations in gases are practically measured in mole fractions (x is volume fraction)

$$\frac{d(N_{tot} x_{O_2})}{dt} = F_{G,in} x_{in,O_2} - F_{G,out} x_{O_2} - K_L A\left(c^*_{L,O_2} - c_{L,O_2}\right). \qquad (23.29)$$

The O_2-gas phase balance is in mol/h and there is obviously *no* O_2-consumption term. There is only the accumulation term and *three* different transport terms.

- The *accumulation term* contains the measured mole fraction of O_2, x_{O_2}, and the total amount, N_{tot} in mol, of *all* gas molecules present in the gas phase of the vessel (such as O_2, CO_2, N_2, H_2O). N_{tot} can easily be obtained with the ideal gas law from measured gas quantities: pressure p, V_G (total volume of gas bubbles and headspace), and temperature (in K),

$$N_{tot} = \frac{pV_G}{RT}. \quad (23.30)$$

Suppose that in the fermentor, $p = 1$ bar (10^5 Pa), $T = 310$ K, and $V_G = 1$ m^3. $R = 8.314$ J/mol K. Then, the fermentor contains 38.8 mol of gas.

The other three terms on the gas phase O_2-balance are transport terms:

- $F_{G,in}x_{in,O_2}$ is the convective inflow term, where $F_{G,in}$ (mol/h) is the flow rate of the gas and x_{in,O_2} is its O_2-fraction.
- $F_{G,out}x_{O_2}$ is the convective outflow term, where $F_{G,out}$ (mol/h) is the flow rate of the gas.
- x_{O_2} is the O_2-fraction of the gas in the vessel. Because we assume ideal mixing of the gas phase, the oxygen fraction in all gas bubbles is assumed the same.
- $K_LA(c^*_{L,O_2} - c_{L,O_2})$ is the well-known O_2-transfer term from gas to broth.

Note that this term has a negative value, because O_2 leaves the gas phase. K_LA is difficult because its value is hard to predict.

For the calculation of the steady state R_{O_2}, we combine the gas phase O_2-balance, Eq. (23.29) with accumulation term set to 0, and the broth O_2-balance,

$$0 = R_{O_2} + K_LA\left(c^*_{L,O_2} - c_{L,O_2}\right). \quad (23.31)$$

Addition of the two oxygen balances eliminates the difficult transfer term, leading to the following *combined* broth and gas phase O_2-compound balance to calculate R_{O_2}. In fact, it is the compound balance of the whole vessel, broth, and gas phase together,

$$0 = R_{O_2} + F_{G,in}x_{in,O_2} - F_{G,out}x_{O_2}. \quad (23.32)$$

This combined O_2-balance immediately gives R_{O_2} when measurements are available on x_{in,O_2}, x_{O_2}, $F_{G,in}$, and $F_{G,out}$. This equation simply states that the difference between the rate of O_2-inflow and O_2-outflow of the cultivation vessel equals the rate of O_2-consumption. Note that $R_{O_2} < 1$.

A completely analogous equation can be derived for CO_2, leading to the following combined CO_2-balance to calculate R_{CO_2}:

$$0 = R_{CO_2} + F_{G,in}x_{in,CO_2} - F_{G,out}x_{CO_2}. \quad (23.33)$$

A point of major importance is to recognize that $F_{G,out}$ is often assumed to be equal to $F_{G,in}$. This assumption is generally wrong and leads to severe errors in calculated R_{O_2} and R_{CO_2}. One of the reasons for a difference in the total gas flow rates is that often the molar consumption rate of O_2 is not equal to the molar production rate of CO_2. This means that we need to quantify $F_{G,out}$. This is easily done by using the "gas phase N_2-balance" (N_2 is present in the sparged air), because N_2 is not consumed or produced in most bioprocesses. Hence, in steady state

$$0 = F_{G,in}x_{in,N_2} - F_{G,out}x_{N_2}. \tag{23.34}$$

In cultivation vessels, the analyzed gases usually contain *only* O_2, N_2, CO_2, and hence $x_{N_2} + x_{O_2} + x_{CO_2} = 1$. Therefore, the equation above gives the outflow,

$$F_{G,out} = F_{G,in}\frac{\left(1 - x_{in,O_2} - x_{in,CO_2}\right)}{\left(1 - x_{O_2} - x_{CO_2}\right)}, \tag{23.35}$$

as function of measured quantities: $F_{G,in}$, x_{in,O_2}, x_{in,CO_2}, x_{O_2}, and x_{CO_2}. The example below shows a calculation and a pitfall.

Example O_2-consumption and CO_2-production rate in a chemostat

Question

In a chemostat, one sparges 1000 mol/h of air (21% O_2 and 79% N_2) to supply O_2 for microbial growth. One measures the fraction of O_2 and CO_2 in the gas leaving the fermentor and obtains 20.0% O_2 and 2.0% CO_2, the remaining is N_2-gas. Calculate the consumed O_2 and produced CO_2 in mol/h.

Answer

The usual, *but wrong* calculation is to multiply the airflow rate with the concentration difference. This gives an O_2-consumption of $1000 \times (0.21 - 0.20) = 10$ mol/h and a CO_2-production of $1000 \times 0.02 = 20$ mol/h. This calculation assumes that the gas outflow in mol/h equals the gas inflow. This is not true and leads to wrong values.

The gas outflow is obtained from the N_2-balance. The N_2-inflow equals 790 mol/h. The N_2-outflow equals $F_{G,out} \times (1 - 0.2 - 0.02)$. They are equal so $F_{G,out} = 1013$ mol/h. This is only 1.3% higher than the inflow.

The correct O_2-consumption now follows from the O_2-gas phase balance with the real values of gas flow rates as: $1000 \times 0.21 - 1013 \times 0.20 = 7$ mol/h. This is *30% lower* than the previous incorrectly calculated value of 10 mol/h. Hence, the small increase in gas flow has a highly significant effect on the calculated O_2 consumption. The explanation is that the O_2-consumption follows from subtracting two large numbers, 210–200 or

210–203, where the second number is only 1.5% wrong. The correct CO_2-production follows as $1013 \times 0.02 = 20.26$ mol/h, which shows that the wrong calculation (20 mol/h) only differs little here.

Fermentors are usually sparged with dry air, because of air sterilization. The sparged gas flow then relates to mol/h of dry air. Inside the fermentor the gas becomes nearly immediately saturated with water vapor. The mole fraction of water x_{H_2O} therefore rises from 0 (dry air) to p_{sat}/p. p_{sat} is the water saturation partial pressure (dependent on temperature and broth liquid composition), p is the pressure. Therefore, the gas which leaves the fermentor contains water. In a similar way, a volatile product (e.g., ethanol) leads to a mole fraction of product in the gas phase. The presence of water and product vapor has several consequences:

- Loss of mass from the fermentor due to water and product evaporation.
- The molar gas outflow increases due to the presence of water and volatile products. This would affect the calculation of gas phase O_2, CO_2 balance. In practice, the analysis of O_2 and CO_2 is done in dried off-gas (to avoid condensation) and x_{O_2}, x_{CO_2} relate to the dry gas. Volatiles then play no role in O_2 and CO_2 rates calculations.

4.3. Using conservation of C and degree of reduction to detect forgotten processes

Usually one obtains R_s, R_p, and R_x from concentration, volume, and flow rate measurements and the proper compound balances for substrate, product, and biomass.

If these rates are correct and there are no forgotten processes, the measured CO_2-production, R_{CO_2}—obtained from gas phase CO_2 compound balances—should agree with R_{CO_2} calculated from the C-balance using R_s, R_p, and R_x. Often the measured R_{CO_2} is smaller and the C-balance shows a gap, quantified by

$$\text{GAP} = \frac{\left|\sum \text{positive terms} - \sum \text{negative terms}\right|}{\text{Largest sum}}. \quad (23.36)$$

There are two possible explanations for this gap:

- The missing carbon is present in the broth as unknown secreted aqueous products. This is easily checked by analyzing TOC in the broth supernatant. Usually, TOC in supernatant is much larger than the carbon present in the analyzed compounds in supernatant; due to, for example, cell lysis products and secreted unknown metabolites. These can be distinguished because cell lysis products are polymers (such as protein, RNA) which can be analyzed separately.

- When TOC in supernatant equals the carbon present in known product and residual substrate, then the missing carbon is in the gas phase (e.g., volatile product or substrate). This volatility property severely limits the number of potential candidates.

If the C-balance indicates missing carbon, then also the balance of degree of reduction, which involves R_{O_2}, should show a deficit gap. One can then analyze the broth supernatant, with the chemical oxygen demand (COD) analysis or total oxygen demand (TOD) analysis. In this analysis, all organic compounds in the supernatant are combusted and the O_2-consumption during combustion is measured (in mol/h). The mole "degree of reduction of these compounds in the broth supernatant" equals $4 \times N_{COD}$. If the amount of degree of reduction present in broth supernatant closes the γ-balance, then the missing compounds are present in broth. If not, then they are in the gas phase.

This shows that the C- and γ-balance (using R_{O_2} and R_{CO_2}) are highly informative for detecting forgotten conversion or transport processes. In addition, one can use the N and charge balance to shed light on the nature of the missing compound as shown below.

4.4. Examples on the use of conservation principles to detect forgotten processes

When conservation of C, γ, N, and electric charge based on the available R_i is violated, then there is serious evidence of so-called systematic errors. There are several possible causes

- *Systematic measurement errors*: The available R_i has been compromised due to systematic deviations in concentrations, volumes, and flow measurements. This can be checked by analytical procedures alone.
- *Forgotten (transport and/or conversion) processes (Section 3)*: Below two examples are shown how to use conservation principles to track down forgotten processes.

Example Use of conservation principles to find forgotten conversion process

Consider an aerobic growth experiment of *Escherichia coli* on glucose in a chemostat. The scientist's mind is on growth only and in his idea, the growth system involves only glucose, NH_4^+ (N-source), H_2O, H^+, CO_2, O_2, and biomass.

From his experiment he obtains, using proper measurements and compound balance calculations, the following six rates: $R_s = -0.65$ mol/h glucose, $R_x = +1.50$ Cmol/h biomass, $R_{O_2} = -1.30$ mol/h O_2, $R_{CO_2} = +1.33$ mol/h CO_2, $R_H = +0.82$ mol/h H^+, and $R_{NH_4} = -0.28$ mol/h NH_4^+.

The only unknown rate is R_{H_2O}, hence the five conservation constraints will lead to four linear relations between the above measured six rates. These four relations are obtained by eliminating R_{H_2O} from the five conservation constraints by combining the H and O conservation relation.

The four relations are

N-balance: $R_{NH_4} + 0.20R_x = 0$

Filling in the measured rates gives $-0.28 + 0.20 \times 1.50 = -0.02 \neq 0$. The N-GAP (Eq. (23.36)) is 0.02/0.30 or 6.6%, which is acceptable.

Charge balance: $R_{NH_4} + R_H = 0$

Filling in the rates gives $-0.28 + 0.82 = -0.54 \neq 0$. The charge GAP is then 0.54/0.82 or 66%, which is very big.

Here is a *tremendous discrepancy; we miss more than 66% of charge.* There must be production of a negatively charged compound or consumption of a positively charged compound. The last possibility is ruled out, because the prepared medium contains glucose only. Metabolic knowledge suggests acetate ($C_2H_3O_2^-$) as a product candidate. The charge balance then suggests an acetate production of 0.54 mol/h $C_2H_3O_2^-$.

Carbon balance: $6R_s + R_x + R_{CO_2} = 0$

Here, acetate as a possible product has been neglected. Filling in the R_i obtained gives $6 \times (-0.65) + 1.50 + 1.33 = -1.07 \neq 0$. This gives GAP = 1.07/3.90 or 27%.

3.9 Cmol/h is consumed, yet only 2.83 Cmol/h is accounted for. We miss 1.07 mol/h which points at a possible by-product. If this by-product would be acetate, the carbon balance predicts $R_{ace} = 1.07/2 = +0.535$ mol/h. This matches with the rate obtained from the charge balance. Of course, the C-balance could be wrong due to, for example, an error in biomass production. However, the N-balance confirms the correctness of R_X, and this rules out this possibility.

Degree of reduction balance: $24R_s + 4.2R_x + (-4)R_{O_2} = 0$

Here the possible by-product was neglected. Filling in the experimental rates: $24 \times (-0.65) + 4.2 \times (1.50) + (-4) \times (-1.30) = -4.1 \neq 0$. The GAP = 4.1/15.6 or 26%.

Clearly, there is a gap in the electron balance of 4.1/15.6 = 26%. This points again to an unknown by-product. Assuming acetate, $\gamma = 8$, the GAP is explained by $R_{ace} = 4.1/8 = +0.51$ mol/h acetate. The conclusion is that conservation constraints for charge, C and γ point to a production of about 0.50 mol/h acetate. This is easily checked by analyzing acetate in the broth.

Example Use of conservation relations to find a forgotten transport processes

A thermophilic organism produces ethanol from glucose under anaerobic conditions at pH 5.0 and at a temperature of 60 °C. In a chemostat ($V = 0.010\ m^3$), one feeds a growth medium with a rate of $1.00 \times 10^{-3}\ m^3/h$ containing 1000 mol glucose/m^3. To keep the chemostat anaerobic and to remove the produced CO_2, one sparges N_2-gas into the chemostat at a rate of 10.0 mol/h N_2 (≈250 l/h).

The broth outflow rate is measured and is a little lower—due to water vapor—as the inflow rate, being $0.97 \times 10^{-3}\ m^3/h$. The broth is analyzed and contains biomass (310 Cmol/m^3), residual glucose (10 mol/m^3), glycerol ($C_3H_8O_3$ 110 mol/m^3), and ethanol (1550 mol/m^3). The gas leaving the fermentor is analyzed for CO_2 showing a volume fraction of 15.1% CO_2.

Question 1

Calculate R_i for biomass, glycerol, glucose, ethanol, and CO_2.

Answer

Using the broth compound balances and the measured concentrations and flows, one obtains $R_X = +0.3007$ Cmol/h, $R_{glyc} = +0.1067$ mol/h, $R_s = -0.9903$ mol/h, and $R_{eth} = +1.503$ mol/h.

Using the gas phase N_2-balance, one calculates a total gas outflow rate of $10/0.849 = 11.78$ mol/h (15.1% CO_2 and 84.9% N_2) leading to a CO_2-production rate of $R_{CO_2} = 1.779$ mol/h.

Question 2

Check the carbon and degree of reduction balance.

Answer

Using the C-balance, one expects to find $6 \times 0.9903 = 1 \times 0.3007 + 3 \times 0.1067 + 2 \times 1.503 + 1 \times 1.779$ or $5.942 \neq 5.406$. We are missing *10%* of the carbon.

Using the balance of degree of reduction, one expects to find $24 \times 0.9903 = 4.2 \times 0.3007 + 14 \times 0.1067 + 12 \times 1.503$ or $23.77 \neq 20.80$. We are missing *13%* of the electrons.

Question 3

The analytical procedures are checked for errors but all measured flow rates and concentrations are accurate. Hence, one suspects that a secretion process is performed by the organism. It is suggested to perform a

TOC analysis in the broth supernatant and one finds TOC = 3472 Cmol/m^3. Does this show in the broth the presence of an unknown compound?

Answer

The TOC of the analyzed compounds (glycerol, ethanol, glucose) amounts to $3 \times 110 + 2 \times 1550 + 6 \times 10 = 3490$ Cmol/m^3. This compares well to the TOC = 3472 mol/m^3. Hence, the unknown product is *not* present in the liquid.

Question 4

What can you tell about the unknown product?

Answer

Let us assume it is a single unknown product. This represents $5.942 - 5.406 = 0.536$ Cmol/h (from the C-balance) and $23.77 - 20.80 = 2.97$ mol/h from the γ-balance. This tells us that its degree of reduction of product carbon is $\gamma = 2.97/0.536 = 5.54$. Hence the product is highly reduced and is not present in the liquid (Question 3), and therefore must have escaped into the gas phase and it is probably highly volatile. This suggests that a significant amount of ethanol is transported out of the broth into the gas phase.

Question 5

How much ethanol do you expect in the gas phase?

Answer

The missing carbon represents $0.536/2 = 0.27$ mol/h ethanol, which leads to a mole fraction of $0.27/11.78 = 0.023$ ethanol in the off-gas.

Question 6

The off-gas analysis shows the presence of ethanol at a mole fraction level of 0.0207. Calculate now the true R_{CO_2} and R_{eth}.

Answer

In the calculation of the gas outflow rate, the presence of ethanol was not taken into account. The gas outflow rate now follows as $10/(1 - 0.151 - 0.0207) = 12.07$ mol/h. $R_{CO_2} = 0.151 \times 12.07 = 1.82$ mol/h.

The ethanol transport through the gas phase is $0.0207 \times 12.07 = 0.25$ mol/h. The transport of ethanol in the liquid phase out of the fermentor was 1.503 mol/h; hence the total ethanol production equals $R_{eth} = 1.503 + 0.25 = 1.78$ mol/h.

5. Propagation of Statistical Errors: Accuracy of Calculated Rates

5.1. Introduction

Rates of consumption or production R_i (mol/h) of individual compounds in biological processes are obtained from

a. properly defined compound balances for each compound i, and
b. measurements of concentrations of compound i, volumes, flow rates as function of time, which are used in compound balances.

Examples are R_x and R_s obtained from biomass- and substrate-compound balances, or R_{O_2} and R_{CO_2} obtained from gas phase O_2 and CO_2 balances. The calculated "compound balance derived" R has an error due to propagation of the errors (noise) in experimental measurements into R_i as obtained from the compound balance equations.

For biological systems, one needs minimally two (without noncatabolic product) or three (with noncatabolic product) different experimentally obtained "compound balance based" R (e.g., R_x, R_s, R_p) in order to calculate the other R_i using the element balances (R_o, R_c, R_n, R_h, R_{H_2O}). These notations are called "element balance derived" R_i. These derived R_i also have an error due to propagation of the errors present in the used compound balance based R_i through the conservation relations. In this chapter, the propagation of errors in compound balance based R_i and element balance derived R_i are treated separately, respectively, in Sections 5.2 and 5.3.

5.2. Statistical error propagation in compound balance based R_i

Suppose that one needs to obtain R_x from a steady-state biomass compound balance from a chemostat experiment. When there is no biomass present in the inflow and ideal broth outflow, the steady-state biomass compound balance gives R_x

$$0 = R_x - F_{out}c_x \Rightarrow R_x = F_{out}c_x. \tag{23.37}$$

In this example, we take the measurements as $F_x = 55.0 \pm 1.0$ ml/h and $c_x = 6.4 \pm 0.3$ mg/ml. This gives $R_x = 55 \times 6.4 = 352$ mg/h.

Repeated measurements of F_x and c_x give each time slightly different values due to sampling and analysis errors resulting in the indicated error. The error σ_R in R_x is now calculated using the error propagation rule for multiplications and divisions, which entails that the square of the relative errors are added to get the square of the relative error in the product or division,

$$\left(\frac{\sigma_R}{R_x}\right)^2 = \left(\frac{\sigma_F}{F_x}\right)^2 + \left(\frac{\sigma_c}{c_x}\right)^2. \qquad (23.38)$$

This gives the relative error in the rate as $\sqrt{(1/55)^2 + (0.3/6.4)^2} = 0.0503$ and then $\sigma_R = 0.0503 \times 352 = 18$, so $R_x = 352 \pm 18$ mg/h.

In a similar way, one can obtain R_s for a chemostat substrate balance,

$$0 = R_s + F_{in}c_{s,in} - F_{out}c_s \Rightarrow R_s = -F_{in}c_{s,in} + F_{out}c_s. \qquad (23.39)$$

The errors in each of the two terms on the right hand side can be calculated from Eq. (23.38) and the final error in the rate is obtained using the error propagation rule for additions and subtractions, which entails that the square of the absolute errors are added to obtain the square of the absolute error of R_i,

$$\sigma_R^2 = \sigma_{in}^2 + \sigma_{out}^2. \qquad (23.40)$$

A similar calculation is illustrated in the example. However, an interesting point to note is that covariance of the errors in R_x and in R_s exists. The biomass and substrate-compound balance both contain F_{out} as measurement. This means that the errors calculated for R_x and R_s are covariant, which is important when this is the dominating error for further calculations, for example, with the elemental balances.

Experimental design can minimize the error associated with R_i. For example, consider R_s which follows from the substrate balance, Eq. (23.39). The flows F_{in} and F_{out} are usually not very different. The minimal error in R_s is obtained when $c_{s,in}$ and c_s are maximally different. For substrate limited conditions, c_s is usually very small, which leads to the smallest error in R_s, as $R_s \approx -F_{in}c_{in,s}$. This is not always so. For example, $R_{NH_4^+}$ and R_{O_2} are calculated from compound balances in which both ingoing and outgoing transport terms have significant contributions. Experimental design should aim at an as large as possible concentration difference between in- and outgoing concentrations.

Example Calculation of $R_{NH_4^+}$ from the broth NH_4^+-compound balance

Consider, for example, $R_{NH_4^+}$ obtained from the broth NH_4^+-compound balance. Here, it is a matter of experimental design to choose c_{in,NH_4^+} such that there is a large difference in the ammonium concentration in broth, $c_{NH_4^+}$.

Consider two different feed media cases—same concentration of limiting substrate and different NH_4^+ concentration, $c_{NH_4^+}$, with the same relative error of 2%, and the same flows, $F_{in} = 200.0 \pm 2.0$ ml/h and $F_{out} = 210.0 \pm 2.1$ ml/h with a measurement error of 1%.

Case 1. $c_{in,NH_4^+} = 200.0 \pm 4.0$ mmol/L and $c_{NH_4^+} = 180.0 \pm 3.6$ mmol/L
Case 2. $c_{in,NH_4^+} = 30.00 \pm 0.60$ mmol/L and $c_{NH_4^+} = 19.00 \pm 0.38$ mmol/L

In both cases, the same $R_{NH_4^+} = -2.2$ mmol/h is found. For example, consider Case 1. Using the NH_4^+-compound balance expressed in mmol/ml, gives $0 = R_{NH_4^+} + 0.200 \times 200 - 0.18 \times 210 = -40.0 + 37.8 = -2.2$ mmol/h.

The error in the NH_4^+-inflow term follows from Eq. (23.38) as 0.89 mmol/h; the NH_4^+-outflow has an error of 0.85 mmol/h. The difference, $R_{NH_4^+}$, has then a huge error, $\sqrt{(0.89^2 + 0.85^2)} = 1.2$ mmol/h, from Eq. (23.40). The result for the two cases is

Case 1. $R_{NH_4^+} = -2.2 \pm 1.2$ mmol/h and
Case 2. $R_{NH_4^+} = -2.20 \pm 0.16$ mmol/h.

The experimental design of Case 2 leads to a significantly more accurate $R_{NH_4^+}$.

5.3. Statistical error propagation in element balances derived R_i

The problem of error propagation from, for example, R_s, R_x, and R_p through element balances to other calculated rates, is more easily explained with an example.

Example Error propagation in R_i calculated from conservation equations

Consider aerobic growth of *Saccharomyces cerevisiae* on glucose. From the balance calculations of O_2 and CO_2 in the gas phase together with gas phase measurements of O_2- and CO_2-fractions and air flow, one can obtain the following values for R_{O_2} and R_{CO_2}.

Case 1a R_{O_2} and R_{CO_2} are available

Given are $R_{O_2} = -1.000$ mol/h and $R_{CO_2} = +1.050$ mol/h

The unknown rates are for biomass, glucose, NH_4^+, H^+, H_2O, which can be calculated from the five conservation constraints. Solution gives the following overall growth reaction with R_i as coefficients (in bold the given compound balance derived rates).

$$-0.3417C_6H_{12}O_6 - \mathbf{1.000}O_2 - 0.200NH_4^+ + 1.000C_1H_{1.8}O_{0.5}N_{0.2} + \mathbf{1.050}CO_2 + 0.200H^+ + 1.450H_2O$$

Case 1b R_{O_2} and R_{CO_2} are available + noise effect

Given are $R_{O_2} = -1.010$ mol/h and $R_{CO_2} = +1.040$ mol/h

We assume that a second measurement of the off-gas, due to noise, gives slightly different values for the O_2- and CO_2-fractions and subsequently the associated rates are less than 4% off. This second set of values, using the element balances, gives us the overall growth reaction:

$$-0.2733C_6H_{12}O_6 - \mathbf{1.010}O_2 - 0.12NH_4^+ + 0.60C_1H_{1.8}O_{0.5}N_{0.2} + \mathbf{1.040}CO_2 + 0.12H^+ + 1.28H_2O$$

This overall reaction is very different and shows that there is 40% less biomass production and 20% less substrate consumption. The resulting yields are 2.93 Cmol/mol glucose (Case 1a) and 2.19 Cmol/mol glucose (Case 1b), showing that the calculated biomass yield is also very different. Assuming that the first overall reaction is correct, process and economic calculations based on the second overall reaction would be disastrous.

Case 2. R_s and R_{O_2} are available

It is now interesting to see what happens when we choose a different pair of experimentally obtained compound balance based rates. We analyze the glucose and flow rates of nutrient solutions (in) and broth (out). This allows using the broth glucose compound balance to calculate R_s. In addition, we still have the O_2-off-gas measurement. This leads to the compound balance based values for R_s and R_{O_2}:

The base case is $R_{O_2} = -1.000$ mol/h and $R_s = -0.3417$ mol/h, which has exactly the same growth equation as in Case 1a. The alternative case, which the base case with a small deviation, we consider here $R_{O_2} = -1.010$ mol/h and $R_s = -0.3300$ mol/h, with the same or comparable deviation as in Case 1b. The overall growth reaction now becomes

$$-0.3300C_6H_{12}O_6 - \mathbf{1.010}O_2 - 0.1848NH_4^+ + 0.9238C_1H_{1.8}O_{0.5}N_{0.2} + 1.0562CO_2 + 0.1848H^+ + 1.4257H_2O$$

This result is not very different from the result in Case 1a. Clearly, the measurement noise in the second pair (R_{O_2}, R_s) of "compound balance calculated" rates does propagate less than in case of the first pair (R_{O_2}, R_{CO_2}).

5.4. Conclusion

The examples show that error propagation in compound balance and conservation relation derived R_i is potentially a very serious problem, because very unreliable values of R_i can be obtained. Therefore, one can and must do a careful experimental design in choosing the best (Section 6.2) "compound balance based" R_i, with least error propagation in conservation equations and minimize the statistical measurement errors in compound balances by choosing the measurements to be done in the experiments including the choice of feed concentrations and gas flow rate (Section 6.3).

6. Experimental Design to Obtain True and Accurate *R*-Values

To obtain true and accurate R_i (and q_i), one can perform *a priori* experimental design to minimize error propagation. First we need to set up the compound balances (Section 6.1). Using the compound balances and a global idea on consumed and produced amounts, one can choose independent variables (Section 6.2, Stage 1) which minimize the errors in compound balance based R_i. Thereafter, we can minimize the measurement effort using the minimal set of compound balances in combination with element balances (Section 6.3) to obtain all R_i. It is very important that one designs the experimental measurements to create redundancy to check for forgotten processes (Section 6.4) followed by reconciliation, when forgotten processes are absent.

6.1. Setting up the equations to calculate R_i

Consider a chemostat where an organism grows aerobically on a limiting substrate. In this chemostat, the organism grows substrate limited. There is no evaporation of the substrate, no product formation, and negligible loss of H_2O due to evaporation because the off-gas passes a condenser which leads the condensed water flow back in the fermentor.

The O_2 supply and CO_2-removal are performed by air-sparging. The N-source is NH_4^+, which leads to alkali need for pH control. The pH is 5, meaning that the amount of CO_2/HCO_3^- in the broth is low enough to neglect their contribution in the combined (liquid and gas phase) CO_2-compound balance. Also there is hardly any (at pH 5) HCO_3^- present. Finally, there is biomass lysis.

In this growth process, the following molecules are interconverted: substrate, O_2, CO_2, H^+, NH_4^+, H_2O, active biomass, and dead biomass.

To obtain the R_i for each compound, we can formulate the respective compound balances of each compound in mmol/h. In these balances, R_i are in mmol/h, c_i in the liquid is in mmol/kg, F (liquid flow) is in kg/h, x_i (in gas phase) are in mole fraction, F_G (gas flow) is in mmol/h.

The lysed biomass is quantified as the difference between TOC in supernatant and the TOC present as products and residual substrate $TOC_{sup} - TOC_s - TOC_p$.

This leads to the following steady-state balances in broth and gas phase given in Table 23.3. Note that the water compound balance (which would lead to R_{H_2O}) is absent. The H_2O balance is replaced by the total broth mass balance. Here is a choice that one could either use a complete set of compound balances or replace one of those with an overall balance. There is basically no difference with either approach, but the water content is often not measured, while the total mass can be determined. In this total mass balance for fermentor broth, one needs (apart from F_{in} and F_{out}) to take into account the addition of mass to the broth through the O_2-mass flow and removal of mass through CO_2-mass flow (from bubbles) and the alkali flow. If there are volatile compounds (Section 3.4), then also their mass flow must be taken into account.

Into these nine balances one can provide measurements

- *liquid concentrations*: $c_{in,s}$, c_{in,NH_4^+}, c_s, $c_{NH_4^+}$, c_x^{live}, TOC_{sup}, TOC_p, and TOC_s either in mol/m^3 or in the more practical mol/kg.
- *gas phase concentrations*: x_{in,O_2}, x_{in,CO_2}, x_{O_2}, and x_{CO_2} as mole fractions.
- *liquid flow rates*: F_{in}, F_{out}, and F_{alk} in either m^3/h or kg/h (such that concentration × flow rate gives mol/h).
- *gas flow rates*: $F_{G,in}$ and $F_{G,out}$ in mol/h.

Table 23.3 Compound balances for an aerobic fermentation. M_w is the molecular weight

Balance	Equation
Broth substrate	$0 = R_s + F_{in}c_{in,s} - F_{out}c_s$
Broth NH_4^+	$0 = F_{in}c_{in,NH_4^+} - F_{out}c_{NH_4^+} + R_{NH_4^+}$
Broth dead (lysed)	$0 = 0 - F_{out}TOC_{sup} - TOC_s - TOC_p + R_x^{dead}$
Broth living biomass	$0 = 0 - F_{out}c_x^{live} + R_x^{live} - R_x^{dead}$
Broth charge balance	$F_{alk}c_{in,OH} + R_{H^+} = 0$
Gas + broth O_2	$0 = F_{G,in}x_{in,O_2} - F_{G,out}x_{O_2} + R_{O_2}$
Gas + broth CO_2	$0 = F_{G,in}x_{in,CO_2} - F_{G,out}x_{CO_2} + R_{CO_2}$
Gas N_2	$0 = F_{G,in}(1 - x_{in,O_2} - x_{in,CO_2}) - F_{G,out}(1 - x_{O_2} - x_{CO_2})$
Total mass balance	$0 = F_{in} + F_{alk} - F_{out} + (- R_{O_2}) M_w(O_2) - R_{CO_2}M_w(CO_2)$

These nine balances contain seven R_i as unknowns (R_s, $R_{NH_4^+}$, R_x^{dead}, R_x^{live}, R_{O_2}, R_{CO_2}, and R_{H^+}).

Hence we have a redundant system if we measure everything. In practice, there are two flow measurements which are usually not directly determined from basic measurements.

- gas flow out ($F_{G,out}$) because one assumes usually that this equals the measured gas inflow (which is absolutely not correct, Section 4.2)
- the liquid inflow or outflow because one usually measures only one liquid flow. This can be the outflow (which then sets the dilution rate) and the liquid inflow is controlled but not measured to maintain a constant broth mass in the fermentor. Alternatively, one can measure the nutrient solution mass inflow rate, and the outflow is not measured.

When indeed only one gas and one liquid flow are measured (e.g., $F_{G,in}$ and F_{out}), then the unknown gas outflow rate $F_{G,out}$ and nutrient feed rate F_{in} add to the seven unknown R_i, leading to nine unknowns and nine balance equations. The nine balances with the measurements can then be solved to provide the values for the seven unknown R_i, $F_{G,out}$, and F_{in}.

The final equation to obtain q_i is the definition equation

$$q_i = \frac{R_i}{Mc_x^{live}}. \tag{23.41}$$

Clearly, to obtain q_i from R_i one does need the measurement value for the total broth mass M in the fermentor in kg and c_x^{live}, which is the living biomass dry matter in Cmol/kg broth using appropriate measurements methods. This results in q_i-values in units of mol/(h Cmol).

6.2. Stage 1: Select measurement effort with minimal error propagation for element balance derived R_i

In practice, the measurements needed to obtain all rates R_i from the compound balances (Section 5.1) require a lot of time and resources. In principle, all R_i can be obtained using much less measurements by using the additional requirement of element–charge conservation. These requirements generate extra relations between the various rates. For a simple aerobic growth system, there are five conversation constraints between the seven R_i. These constraints suggest that we only need to elaborate two R_i from two compound balances, requiring a lot less measurement effort. These constraints are most easily written in matrix notation. Assuming glucose as substrate and NH_4^+ as N-source, and using the standard biomass composition the following matrix relation gives the conservation constraints. We will develop the approach here using this example as guide. Then the conservation equations are

$$\begin{array}{l} C: \\ H: \\ O: \\ N: \\ \text{charge}: \end{array} \overbrace{\begin{bmatrix} 6 & 0 & 0 & 1 & 0 & 0 & 1 \\ 12 & 0 & 4 & 0 & 2 & 1 & 1.8 \\ 6 & 2 & 0 & 2 & 1 & 0 & 0.5 \\ 0 & 0 & 1 & 0 & 0 & 0 & 0.2 \\ 0 & 0 & 1 & 0 & 0 & 1 & 0 \end{bmatrix}}^{\mathbf{E}} \overbrace{\begin{pmatrix} R_s \\ R_{O_2} \\ R_{NH_4^+} \\ R_{CO_2} \\ R_{H_2O} \\ R_{H^+} \\ R_x \end{pmatrix}}^{\mathbf{R}} = 0. \quad (23.42)$$

$\mathbf{R}$ is the column vector containing all R_i. The constraint matrix $\mathbf{E}$ contains the coefficients for the elements (C, H, O, N) and electric charge as five rows and the components (glucose S, O_2, NH_4^+, CO_2, H_2O, H^+, X) as seven columns. The element composition for each component can be read in its column (e.g., biomass in the seventh column). The five conservation relations in matrix vector notation follow as:

$$\mathbf{ER} = \mathbf{0}. \quad (23.43)$$

From compound balance calculations, there is error propagation from measurement errors (concentrations, mass, flows) into R_i. The constraints allow to choose 2 R_i to be obtained from their compound balance and appropriate measurements. These chosen measured R_i are collected in column vector $\mathbf{R}^m$. The remaining five rates, collected in a column vector $\mathbf{R}^c$, are calculated using the five conservation constraints,

$$\mathbf{E}^m\mathbf{R}^m + \mathbf{E}^c\mathbf{R}^c = \mathbf{0} \Rightarrow \mathbf{R}^c = -(\mathbf{E}^c)^{-1}\mathbf{E}^m\mathbf{R}^m. \quad (23.44)$$

This is an explicit expression of the unknown rates as they are related to the rates obtained from measurement, and it assumes that the inverse in the equation exists. In this calculation, the errors in the "measured" $\mathbf{R}^m$ propagate into $\mathbf{R}^c$ through the 5×2 matrix $[(\mathbf{E}^c)^{-1}\mathbf{E}^m]$.

To calculate the error propagation, we need to elaborate the above conservation relations. This requires some computational approach. The easiest is to follow statistics whereby the covariance matrix of the "measured" rates, $\mathbf{R}^m$, is transformed to the covariance matrix of the calculated rates, $\mathbf{R}^c$, making use of the matrices defined in Eq. (23.44). A brute force method is to perturb each of the "measured" rates, $\mathbf{R}^m$, by its known error and then sum quadratically all these effects for each of the calculated rates, $\mathbf{R}^c$, separately.

The result is an estimate for the errors of all of the rates, and how we get to these is less important. These errors will vary depending on the choice of the "measured" rates, $\mathbf{R}^m$. As we have seen in Section 5.3, it made quite a difference whether one chooses to measure the oxygen uptake and carbon dioxide production rate, or to measure the oxygen uptake together with the substrate uptake rate. We can evaluate each of these designs by their effect on the errors in the calculated rates. Of course, one chooses that design with the smallest errors, or with the smallest error for a key rate.

Example Aerobic growth from two different measurement setups

We will examine the case of Section 5.3, where we considered aerobic growth of *S. cerevisiae* on glucose with two possible choices of "measured" rates, the set R_{O_2} and R_{CO_2} and the set R_{O_2} and R_s. (for data, see Section 5.3). In addition, it is assumed that any rate, obtained from a compound balance, has a relative error of 1%.

Equation (23.44) is the basis for the calculation for the first set, it is as follows:

$$\underset{\mathbf{E}^m}{\begin{bmatrix} 0 & 1 \\ 0 & 0 \\ 2 & 2 \\ 0 & 0 \\ 0 & 0 \end{bmatrix}} \underset{\mathbf{R}^m}{\begin{pmatrix} R_{O_2} \\ R_{CO_2} \end{pmatrix}} + \underset{\mathbf{E}^c}{\begin{bmatrix} 6 & 0 & 0 & 0 & 1 \\ 12 & 4 & 2 & 1 & 1.8 \\ 6 & 0 & 1 & 0 & 0.5 \\ 0 & 1 & 0 & 0 & 0.2 \\ 0 & 1 & 1 & 1 & 0 \end{bmatrix}} \underset{\mathbf{R}^c}{\begin{pmatrix} R_s \\ R_{NH_4^+} \\ R_{H_2O} \\ R_{H^+} \\ R_x \end{pmatrix}} = 0 \tag{23.45}$$

and then

$$\mathbf{R}^c = \begin{pmatrix} R_s \\ R_{NH_4^+} \\ R_{H_2O} \\ R_{H^+} \\ R_x \end{pmatrix} = -(\mathbf{E}^c)^{-1}\mathbf{E}^m\mathbf{R}^m = \begin{bmatrix} -3\frac{1}{3} & -3\frac{1}{2} \\ -4 & -4 \\ 8 & 9 \\ 4 & 4 \\ 20 & 20 \end{bmatrix} \begin{pmatrix} R_{O_2} \\ R_{CO_2} \end{pmatrix} \tag{23.46}$$

Substitution of $R_{O_2} = -1.000$ mol/h and $R_{CO_2} = +1.050$ mol/h gives the same $\mathbf{R}^c$-values as in Section 5.3 case. However, this is not interesting here. We are interested in the errors in the calculated rates. We will only quote the result, also for the other set of rates (R_{O_2}, R_s) in Table 23.4.

It is obvious that the second set is the preferred set, as all errors are significantly smaller than that of the first set.

The message is that one should choose $\mathbf{R}^m$ such that the resulting errors in the calculated R_i are as low as possible. This analysis can be done *a priori*, because one knows the constraint matrix $\mathbf{E}$.

Table 23.4 Comparison of errors in rates obtained from two measurement setups

Rate	Value	Error: R_{O_2} and R_{CO_2} measured (%)	Error: R_{O_2} and R_s measured (%)
R_s	−0.3417	15	(1)
R_{O_2}	−1.00	(1)	(1)
R_{NH_4}	−0.20	29	2.2
R_{CO_2}	1.05	(1)	0.91
R_{H_2O}	1.45	8.5	0.72
R_H	0.20	29	2.2
R_x	1.00	29	2.2

6.3. Stage 2: Minimize error propagation in compound balance based R_i

The above selected few compound balances to be used to obtain the minimal set of R_i requires to reflect on how to set up the experiment, called experimental design, where one chooses the independent variables, such that the error propagation into compound balanced based R_i is minimal. The independent variables are F_{out}, $F_{G,in}$, $c_{in,s}$, c_{in,NH_4^+}, x_{in,O_2}, x_{in,CO_2}, $c_{in,OH}$ (in alkali), and total broth mass M. These independent variables should be chosen such that the difference between the ingoing and outgoing terms in the compound balance is as large as possible. For example, one should choose c_n^{in} such that, for examlple, not only 10% of the added NH_4^+ is converted into biomass but 70% (see Section 4.2). Also one should choose the gas flow such that there is significant depletion of O_2 and large enrichment of CO_2 in the gas phase (e.g., order 1% and not 0.1%).

6.4. Stage 3: Create redundancy to check for forgotten processes to eliminate systematic errors

The minimal measurements approach (Sections 5.2 and 5.3) leads to a best choice for a minimum set of R_i (Section 5.2) to be elaborated from measurements using proper choices of independent feed concentrations and flows (Section 5.3). A serious problem which cannot be tackled by this minimal approach is the possibility of forgotten processes. To detect such a situation, we need to measure more than minimally needed, we need redundancy. This measurement redundancy allows to formulate equations (such as C- or γ-balance) between R_i obtained from compound balances. In Sections 3.3 and 3.4, one finds examples illustrating the search for forgotten processes using redundant measurement information.

We need to emphasize here the importance of designing the experiments and its measurements in such a way that redundancy is created.

Especially in biological processes, there are endless possibilities that the organism does produce by-products, or there are volatile substrates, etc. (forgotten processes).

It is obvious that this information is highly valuable, because it has immediate economic impact (allowing to eliminate previous unknown losses and modifying the organism to avoid by-product formation).

Consider the example in Section 5.2 for a simple growth system and the experimentally obtained compound balance based rates are R_s, R_{CO_2}, R_x, and $R_{NH_4^+}$. There are now two more R_i than minimally required (Section 2). Elimination of the three nonmeasured rates (R_{O_2}, R_{H_2O}, and R_{H+}) from the five element balances leads to two relations which only contain the four measured rates. These relations are

$$\left.\begin{array}{l} 0.2R_x + R_{NH_4^+} \approx 0 \\ R_x + 6R_s + R_{CO_2} \approx 0 \end{array}\right\} \Rightarrow \begin{bmatrix} 0.2 & 1 & 0 & 0 \\ 1 & 0 & 6 & 1 \end{bmatrix} \begin{pmatrix} R_x \\ R_{NH_4^+} \\ R_s \\ R_{CO_2} \end{pmatrix} \approx 0 \Rightarrow \mathbf{E}^{\mathrm{red}}\mathbf{R}^{\mathrm{m}} \approx \mathbf{0}, \tag{23.47}$$

where a matrix notation is introduced, with $\mathbf{E}^{\mathrm{red}}$ called the redundancy matrix. We expect that the measured values are such that the equations above are close to 0 on the right hand side. The redundancy matrix $\mathbf{E}^{\mathrm{red}}$ can be derived as in the example above, but it can also be calculated from $\mathbf{E}^{\mathrm{c}}$ and $\mathbf{E}^{\mathrm{m}}$:

$$\mathbf{E}^{\mathrm{red}} = \mathbf{E}^{\mathrm{m}} - (\mathbf{E}^{\mathrm{c}})^{\dagger}\mathbf{E}^{\mathrm{m}}, \tag{23.48}$$

where $\mathbf{E}^{\mathrm{c}}$ and $\mathbf{E}^{\mathrm{m}}$ are the element matrices related to, respectively, the calculated and the measured rates as in Eq. (23.44). The † indicates the Moore–Penrose pseudo-inverse (Van Der Heijden *et al.*, 1994a,b).

Example Use of redundant experimental rate information: more accuracy

From experimental data, the following rates were obtained using measurements and proper compound balances for *S. cerevisiae* growing aerobically on glucose in a chemostat using NH_4^+ as N-source, with $R_x = +1.23$ Cmol/h, $R_s = -0.41$ mol/h, and $R_{O_2} = -1.07$ mol/h; and four unknown rates $R_{NH_4^+}$, R_{CO_2}, R_{H_2O}, and R_{H^+}.

We have three measured rates, and we can now write the conversion equation (using *a*, *b*, etc. for the yet unknown rates) as follows:

$$-0.41C_6H_{12}O_6 - 1.07O_2 + aNH_4^+ + 1.23C_1H_{1.8}O_{0.5}N_{0.2} + bCO_2 + cH^+ + dH_2O$$

We can now write the five element balances for each of the elements, C, H, O, N, and charge. Four those can be used to eliminate a, b, c, and d. For example, from the N-balance, $a + 1.23 \times 0.2 = 0$, it follows directly that $a = -0.246$. At the end there remains *one* equation, which *only* contains rates determined by direct measurements:

$$24R_s + 4.2R_x - 4R_{O_2} \approx 0$$

The redundancy matrix has now one row (24, 4.2, −4). The gap (Eq. (23.36)) is about 4%, which compares well with typical experimental accuracies of a few percent. Considering experimental noise one performs reconciliation to improve the accuracy of the estimates (see Section 6.5).

6.5. Reconciliation: An example

We consider the aerobic chemostat cultivation for *E. coli* (Taymaz-Nikerel *et al.*, 2009) under glucose limited conditions. It is known that there occurs cell lysis. In the figure below, an overview of all relevant flows are given with the compounds entering and leaving the fermentor.

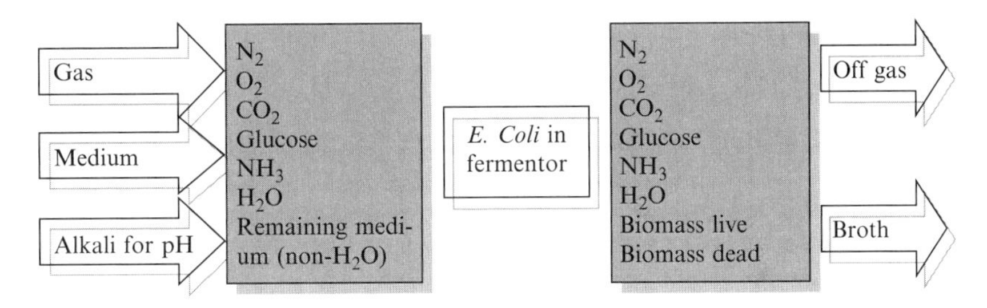

The nitrogen, oxygen, carbon dioxide, glucose, ammonia, live biomass, and dead biomass can be balanced individually such that for these uptake and secretion rates can be determined. For nitrogen, the uptake is given as zero. The biomass compounds are secreted only. Finally, the remaining water flow and other medium cannot be determined, other than by the overall mass balance.

6.5.1. Calculation of "measured" rate, R_i

The experimental design aimed at obtaining measurements (Table 23.5) which allow to calculate from the respective compound balances the following rates (in mmol/h) R_{O_2}, R_{CO_2}, R_s, R_{NH_3}, R_x^{live}, and R_x^{dead}. Table 23.5 shows the available measurements, with the symbols, values, and errors.

In order to get the oxygen and carbon dioxide rates, the gas outflow rate $F_{G,out}$ needs to be known. This is explicitly obtained from the N_2-gas phase balance. It is found that $F_{G,out} = 4470 \pm 91$ mmol/h, making use of error propagation.

Table 23.5 Values and errors of measured concentrations and flow rates

Flow	Measured quantity	Symbol	Value ± error	Units
Gas	Sparged air flow rate	$F_{G,in}$	4460 ± 89	mmol/h
	O_2-fraction in sparged air	x_{in,O_2}	0.2095 ± 0.0030	–
	CO_2-fraction in sparged air	x_{in,CO_2}	0.0004176 ± 0.0000059	–
Medium	Medium inflow rate	F_{in}	0.4029 ± 0.0040	kg/h
	Medium glucose concentration	$c_{in,s}$	151.5 ± 3.0	mmol/kg
	Medium NH_4^+ concentration	c_{in,NH_3}	145.0 ± 3.1	mmol/kg
Alkali for pH	Alkali flow rate for pH control	F_{alk}	0.0220 ± 0.0010	kg/h
Off-gas	Off-gas outflow rate	$F_{G,out}$	Unknown	mmol/h
	O_2-fraction in off-gas	x_{O_2}	0.1762 ± 0.0012	–
	CO_2-fraction in off-gas	x_{CO_2}	0.03550 ± 0.00025	–
Broth	Broth outflow rate	F_{out}	0.4179 ± 0.0042	kg/h
	Broth glucose concentration	c_s	0.200 ± 0.050	mmol/kg
	Broth NH_4^+ concentration	c_{NH_3}	35.0 ± 1.0	mmol/kg
	Organic carbon concentration in supernatant	TOC_{sup}	72.3 ± 2.0	mCmol/kg
	Biomass carbon concentration in broth	c_x	397 ± 12	mCmol/kg

Using the available compound balances (see Section 5.1) for O_2, CO_2, substrate, NH_3, living and dead biomass, one then obtains the values for the six different rates, R_i, with their errors (using the error propagation rules). The results are shown in Table 23.6.

6.5.2. Testing for errors and forgotten processes using residues of the redundancy equations (node analysis)

The above obtained rates could be used to set up Black Box kinetic models, to calculate economic performance or be used in flux balancing. Before we do this, it is very much recommended to perform an analysis on errors and/or forgotten processes. This analysis is based on the requirement that the rates shown in Table 23.7 should agree with fundamental constraints (see Section 4.1) based on elements, charge, and total mass.

We can now formulate the following five constraint relations using the constraint matrix and a vector, composed of its rates—R_{O_2}, R_{CO_2}, R_s, R_{NH_3}, R_x^{live}, R_x^{dead}, and R_{H_2O}—and the mass flows F_{in}, F_{alk}, and F_{out}.

The first four rows of Table 23.7 represent the element/charge atom conservation in mass-atom/h. The last row represents the total mass conservation relation in kg/h, whereby the nitrogen is left out as it is not taken up nor secreted from the fermentor. The charge or redox balance is not applied here.

The first seven columns refer to conversion rates (mmol/h) of all the compounds identified in this process, except for the extra medium (outside glucose, ammonia, and converted water) and of course the nitrogen.

The last three columns refer to liquid mass flows, entering broth (in kg/h for nutrient feed F_{in} and alkali flow F_{alk}) and broth leaving the fermentor (F_{out}).

Table 23.6 Compound rates, R_i (in mmol/h), from compound balances and their calculated errors

Compound balance	Derived rate	Value ± error (mmol/h)
Oxygen	R_{O_2}	-147 ± 18
Carbon dioxide	R_{CO_2}	156.8 ± 3.4
Glucose	R_s	-61.0 ± 1.4
Ammonia	R_{NH_3}	-43.8 ± 1.4
Live biomass	R_x^{live}	165.9 ± 5.3
Dead biomass	R_x^{dead}	30.21 ± 0.89
Water	R_{H_2O}	Unknown
Nitrogen	Given as zero	Used to obtain R_{O_2} and R_{CO_2}

Note the use of the nitrogen balance.

Table 23.7 Constraint matrix

	R_{O_2}	R_{CO_2}	R_s	R_{NH_3}	R_x^{live}	R_x^{dead}	R_{H_2O}	F_{in}	F_{alk}	F_{out}
C	0	1	6	0	1	1	0	0	0	0
H	0	0	12	3	2.0045	2.0045	2	0	0	0
N	0	0	0	1	0.2366	0.2366	0	0	0	0
O	2	2	6	0	0.4477	0.4477	1	0	0	0
Mass	-32×10^{-6}	-44×10^{-6}	–	–	–	–	–	+1	+1	−1

Further, mass contributions to the broth compound balance are not only due to F_{in}, F_{alk}, F_{out} but also refer to O_2 transfer from gas to broth (broth mass increase) and CO_2-stripping (mass removal from broth). Evaporation of water could also be included, when appropriate (e.g., absence of an off-gas condenser).

Given the available rates, R_i, Table 23.5, and measured mass flow rates the five constraint relations contain one unknown rate (R_{H_2O}). This one is eliminated using linear algebra, leading to $5 - 1 = 4$ linear constraint relations, which can be recognized as

- Carbon balance (no water involved)
- N-balance (no water involved)
- Degree of reduction balance[1] (obtained from H-balance—2× O-balance + 4× C-balance; this eliminates one equation with water).
- Total mass balance

These four constraint relations only contain nine known rates for which we have values (either from direct measurements such as F_{in}, F_{out}, and F_{alk}) or from compound balances for compounds with their measurements (which give R_{O_2}, R_{CO_2}, R_s, R_{NH_3}, R_x^{live}, R_x^{dead}, and R_{H_2O}).

This set of linear relations, containing only known rates, is called the redundant set of equations represented by the redundancy matrix, $\mathbf{E}^{red}$ (Table 23.8).

This matrix specifies four relations between the nine known rates and we can calculate the residue of each balance. This residue can be compared

- with its error which can be calculated for the residue (see Section 7.3). In the ideal case, the residue is within twice the expected residue error.
- with the sum of all positive (or negative) terms. In the ideal situation, the residue is only a few percent of the positive sum.

Of course, a residue value of exact 0 is not expected due to errors in the known rates.

Table 23.8 Redundancy matrix with only known rates

	R_{O_2}	R_{CO_2}	R_s	R_{NH_3}	R_x^{live}	R_x^{dead}	F_{in}	F_{alk}	F_{out}
C	0	1	6	0	1	1	0	0	0
N	0	0	0	1	0.2366	0.2366	0	0	0
γ	−4	0	24	0	4.3993	4.3993	0	0	0
Mass	-32×10^{-6}	-44×10^{-6}	–	–	–	–	+1	+1	−1

[1] Up to a point, the choice of combinations is unrestricted given that one of the two equations is eliminated. Here, the choice was given such that it could be given as a conserved quantity.

This residue or *node analysis* leads to two possible outcomes:

- If these expectations concerning the residues are fulfilled, the set of rates is consistent and there is no reason to contemplate forgotten processes or other errors (Sections 3 and 4).
- If the residues (Table 23.9) are much larger than expected, then it is necessary to search for possible errors or forgotten processes (Sections 3 and 4) before we proceed further.

6.5.3. Reconciliation

In many cases, the residue analysis shows that the linear constraints are satisfied given the statistical information. Subsequently, the constraints can then be used to reconcile the measurements such that the constraints are exactly satisfied. Mathematically, we can then formulate an optimization problem where for each measurement one obtains the best estimate of its value and a best estimate of its error.

The optimization problem is composed of equations which are

- all compound balances + N_2-gas phase balance which are linear in the rates and nonlinear in the measurements
- all element balances which are linear in the rates

This represents for our example

- seven compound balances (for substrate, live/dead biomass, NH_4^+, O_2, CO_2, N_2)
- five element balances (for C, H, O, N, charge, mass)

These 13 equations contain

- 14 known measured variables ($F_{G,in}$, x_{in,O_2}, x_{in,CO_2}, F_{in}, $c_{in,s}$, c_{in,NH_3}, F_{alk}, x_{O_2}, x_{CO_2}, F_{out}, c_s, c_{NH_3}, TOC_{sup}, c_x)
- nine unknown variables ($F_{G,out}$, R_{N_2}, R_{O_2}, R_{CO_2}, R_s, R_{NH_3}, R_x^{live}, R_x^{dead}, and R_{H_2O})

The 13 equations are 4 more than needed to calculate the nine unknown variables in these 13 equations.

Table 23.9 Residue of nodes

	Unit	Sum positive or negative terms	Residue ± error	Residue/error (to be <2)	Residue (%) (to be few %)
C	mmol/h	352.9	13 ± 10	+1.24	4
N	mmol/h	45.8	−2.6 ± 2.0	−1.32	−6
γ	mmol/h	1549	−20 ± 33	−0.62	−1
Mass	kg/h	0.4265	0.0048 ± 0.0059	+0.81	1%

This number 4 (difference between available number of equations and number of unknown variables to be calculated) is the number of redundancy equations and called degree of redundancy.

Because the 13 equations are nonlinear in the measurements (containing, e.g., products of flows and concentrations), we need nonlinear optimization to obtain the best estimates for the measurements, including their errors.

Table 23.10 shows the obtained best estimates (value and errors), along side to the original measurements.

It should be noted that the original measurement errors were independent and the estimated errors are dependent, as expressed in a covariance matrix (not given here).

One observes (Table 23.10) that the best estimates of the measurements are close to the measured values. Also, the errors in the estimates are smaller than the measurement errors, which is due to the redundant measurements which give more information, leading to smaller errors.

Having the best estimates of the measurements and errors and their variance–covariance matrix, one can use the compound balances to obtain the best estimates of values and errors of all rates (Table 23.11).

Again we observe that the best estimates are slightly different from the original values and that the errors are smaller. The errors given in Table 23.11 are only derived from the diagonal values of the variance–covariance matrix. Table 23.12 shows the complete correlation matrix of the errors in the calculated rates (so this is the variance–covariance matrix, scaled such that the diagonal is unity).

A strong correlation implies that the element balances impose a strong relation between two rates. An obvious one is the high correlation between the oxygen uptake and the carbon dioxide production rate. When the first increases, the second one decreases, that is, becomes more negative. So, larger oxygen uptake concurs with more CO_2 production.

6.5.4. q-Values: Best estimates and errors

Having the best information on all the $\hat{R}_i$ (Table 23.11) and on the living biomass concentration $\hat{c}_x$ (Table 23.10) allows now to obtain the best information on the q-rates, using the measured broth mass M and the best estimate biomass concentration:

$$\hat{q}_i = \frac{\hat{R}_i}{M\hat{c}_x}, \tag{23.49}$$

where $\hat{q}_i$, $\hat{R}_i$, M, and $\hat{c}_x$ are expressed, respectively, in mol/(h Cmol), mol/h, kg, and Cmol/kg. Here, M is 4.00 ± 0.02 kg and $\hat{c}_x$ is 389.0 ± 8.2 Cmmol/kg.

Table 23.10 Optimal estimates of measurements

Flow	Measured variable	Units	Measured value ± error	Estimated value ± error
Gas	$F_{G,in}$	mmol/h	4460 ± 89	4517 ± 83
	x_{in,O_2}	–	0.2095 ± 0.0030	0.2082 ± 0.0011
	x_{in,CO_2}	–	0.0004176 ± 0.0000059	0.0004176 ± 0.0000059
Medium	F_{in}	kg/h	0.4029 ± 0.0040	0.4002 ± 0.0029
	$c_{in,s}$	mmol/kg	151.5 ± 3.0	147.2 ± 1.7
	c_{in,NH_3}	mmol/kg	145.0 ± 3.1	150.4 ± 2.1
Alkali for pH	F_{alk}	kg/h	0.0220 ± 0.0010	0.0218 ± 0.0010
Off-gas	$F_{G,out}$	mmol/h	Unknown	4537 ± 83
	x_{O_2}	–	0.1762 ± 0.0012	0.1764 ± 0.0011
	x_{CO_2}	–	0.03550 ± 0.00025	0.03557 ± 0.00025
Broth	F_{out}	kg/h	0.4179 ± 0.0042	0.4195 ± 0.0029
	c_s	mmol/kg	0.200 ± 0.050	0.201 ± 0.050
	c_{NH_3}	mmol/kg	35.0 ± 1.0	34.4 ± 1.0
	TOC_{sup}	C mmol/kg	72.3 ± 2.0	72.1 ± 2.0
	c_x	C mmol/kg	397 ± 12	389.0 ± 8.2

Table 23.11 Optimal estimates of values and errors for R_i; $R_{N_2}=0$

Rate	Units	Derived value ± error	Estimated value ± error
R_{O_2}	mmol/h	−147 ± 18	−140.2 ± 3.2
R_{CO_2}	mmol/h	156.8 ± 3.4	159.5 ± 3.1
R_s	mmol/h	−61.0 ± 1.4	−58.81 ± 0.75
R_{NH_3}	mmol/h	−43.8 ± 1.4	−45.76 ± 0.88
R_x^{live}	C mmol/h	165.9 ± 5.3	163.2 ± 3.7
R_x^{dead}	C mmol/h	30.21 ± 0.89	30.24 ± 0.86
R_{H_2O}	mmol/h	Unknown	227.7 ± 3.2

Table 23.12 Correlation matrix of all calculated rates

	R_{O_2}	R_{CO_2}	R_s	R_{NH_3}	R_x^{live}	R_x^{dead}	R_{H_2O}
R_{O_2}	1	−0.99	0.46	−0.26	0.26	0.03	−0.86
R_{CO_2}	−0.99	1	−0.57	0.15	−0.15	−0.01	0.91
R_s	0.46	−0.57	1	0.73	−0.71	−0.16	−0.85
R_{NH_3}	−0.26	0.15	0.73	1	−0.97	−0.20	−0.27
R_x^{live}	0.26	−0.15	−0.71	−0.97	1	−0.03	0.26
R_x^{dead}	0.03	−0.01	−0.16	−0.20	−0.03	1	0.08
R_{H_2O}	−0.86	0.91	−0.85	−0.27	0.26	0.08	1

Table 23.13 q-Values: Original and best estimated values and errors

Rate	Units	Derived value ± error	Estimated value ± error
q_{O_2}	mmol/(h Cmol)	−92 ± 12	−90.1 ± 3.2
q_{CO_2}	mmol/(h Cmol)	98.8 ± 3.7	102.5 ± 3.2
q_s	mmol/(h Cmol)	−38.4 ± 1.5	−37.80 ± 0.64
q_{NH_3}	mmol/(h Cmol)	−27.6 ± 1.2	−29.41 ± 0.30
q_x^{live}	mmol/(h Cmol)	104.5 ± 1.2	104.88 ± 0.90
q_x^{dead}	mmol/(h Cmol)	19.03 ± 0.81	19.43 ± 0.73
q_{H_2O}	mmol/(h Cmol)	Unknown	146.3 ± 3.4

Note that in chemostat experiments, it is not possible to reconcile the total broth mass (M).

Table 23.13 shows the obtained q-values and their errors. Table 23.14 shows the correlation matrix.

These final obtained and validated $\hat{q}$-values (and the covariance matrix) are used with the stoichiometric matrix to obtain the most accurate values of intracellular fluxes and their errors (flux calculation) or are used for kinetic model building (different c_s conditions) or economical calculations.

Table 23.14 Correlation matrix of the errors in estimate q-values

	q_{O_2}	q_{CO_2}	q_s	q_{NH_3}	q_x^{live}	q_x^{dead}	q_{H_2O}
q_{O_2}	1	−1.00	0.94	0.32	−0.06	−0.49	−0.99
q_{CO_2}	−1.00	1	−0.95	−0.36	0.09	0.51	0.99
q_s	0.94	−0.95	1	0.63	−0.35	−0.66	−0.98
q_{NH_3}	0.32	−0.36	0.63	1	−0.83	−0.72	−0.47
q_x^{live}	−0.06	0.09	−0.35	−0.83	1	0.21	0.19
q_x^{dead}	−0.49	0.51	−0.66	−0.72	0.21	1	0.58
q_{H_2O}	−0.99	0.99	−0.98	−0.47	0.19	0.58	1

7. Mathematics of Reconciliation

In Section 6.5, the results of an optimization problem were presented involving nonlinear compound balances, linear constraints, and error propagation. The mathematical formulation of the problem, the error propagation, and the various statistical tests are the subject of this paragraph. Considerable literature (Crowe, 1996; Crowe *et al.*, 1983; Heyen and Kalitvenzeff, 2007) has been published in this area with applications in biotechnology (Van Der Heijden *et al.*, 1994a,b; Wang and Stephanopoulos, 1983).

Because compound balances are bilinear in the measurements, typically containing mathematical products of mass-(liquid phase) or molar-(gas phase) flow rates and concentrations (liquid phase) or mole fractions (gas phase), the reconciliation problem is nonlinear and is therefore best formulated as a nonlinear optimization problem.

7.1. Formulating the equations

The equations always consist of

- compound balances, providing rates
- element balances on the rates

In order to identify the compounds, one makes an inventory of all in- and outgoing streams. These streams are made up of different molecules or compounds. This is illustrated in the example of Section 6.5 (the figure at the beginning of that section). There are five in- and outgoing streams and there were nine different compounds identified.

The next step is the most difficult one, namely to set up compound balances from the available measurements. Each input and output flow of that compound needs to be described. It can happen, that a variable is not measured, but essential for the calculation of rates. In the data reconciliation approach that variable is taken along as unknown, with the expectation that the total set of equations also allow this variable to be estimated.

The last step is to identify the element balances. It is a matter of recognizing all the different elements of the compounds. Basically, they are the element (C, H, N, O, P, S, etc.) balances and the charge or redox balance. Also conserved moieties can be used rather than or in addition to the individual elements. Only those balances apply where all uptake and production rates for the specific element are present.

These equations contain variables which can be distinguished in

- measured variables in the compound balances (concentration, flow rates), to be put in a vector $\mathbf{x}_{\text{meas}}$
- variables in the compound balances without measurement, $\mathbf{x}_{\text{unmeas}}$
- the uptake and production rates, to be put in the vector $\mathbf{R}$.

The compound balances express the rates as function of measured and unmeasured variables, $\mathbf{R} = \mathbf{R}(\mathbf{x}_{\text{meas}}, \mathbf{x}_{\text{unmeas}})$. The element balances are summarized by the relation $\mathbf{E} \times \mathbf{R} = \mathbf{0}$, where the constraint matrix $\mathbf{E}$ contains the coefficients of the element balances.

7.2. Formulation of the nonlinear optimization problem

The solution of the above problem is based on the least squares method, whereby estimates for the measurements are introduced and we will require that these estimates are close to the actual measurements in a least squares sense. Let us designate the measurements with the vector $\mathbf{x}_{\text{meas}}$ and the estimate with a "hat" on it, $\hat{\mathbf{x}}_{\text{meas}}$. Similarly, the vectors $\mathbf{x}_{\text{unmeas}}$ and $\mathbf{R}$ have as estimates $\hat{\mathbf{x}}_{\text{unmeas}}$ and $\hat{\mathbf{R}}$.

The objective follows from statistics. If the measurement errors, σ_{meas}, are all normally distributed and independent of each other, the *weighted* least squares estimate requires that the differences between the measured and estimated variable divided by its measurement error should be minimized. However, the minimization should be such that the two sets of equations above are fulfilled. This optimization problem is posed as follows:

$$\frac{\min}{\hat{x}_{\text{meas}}, \hat{x}_{\text{unmeas}}, \hat{\mathbf{R}}} \sum_i \left(\frac{x_{\text{meas},i} - \hat{x}_{\text{meas},i}}{\sigma_{\text{meas},i}} \right)^2 \tag{23.50}$$
$$\text{subject to}\, \hat{\mathbf{R}} = \hat{\mathbf{R}}(\hat{\mathbf{x}}_{\text{meas}}, \hat{\mathbf{x}}_{\text{unmeas}}) \text{ and } \quad \mathbf{E} \times \hat{\mathbf{R}} = \mathbf{0}.$$

So the objective is a statistical measure that needs to be minimized by varying the estimates of the measurements and rates. (Those with the "hat.") It is these estimates that have to fulfill the relations between measured rates and measurements, and the linear relations imposed by the added balances. If these extra equations would not be there the estimates for the measurements would be equal to the measurement itself.

This approach is only possible, when there is indeed redundancy. The simple measure,

$$n_{dr} = \text{no. of measurements} + \text{no. of relations} - \text{no. of estimated variables} \quad (23.51)$$

is called degree of redundancy, which should be >0, for the data reconciliation to make sense.

Example In the case of *E. coli* of Section 5, there are

- 14 measurements
- six compound balances and six element balances (this includes the gaseous nitrogen balance, which was—in the example—explicitly used to calculate the off-gas flow), that is, 12 relations
- 14 variables measured, 1 unknown variable, and 7 rates, that is, 22 variables to estimate.

That means that the degree of redundancy is $14 + 12 - 22 = 4$.

The numerical implementation depends on the available software such as Matlab, Mathematica, and Maple. The Excel spreadsheet can also perform constraint optimization, and the problem formulated above is easily solved.

7.3. Statistical aspects

Data reconciliation leads to new estimates for the variables involved. As it is based on measurement data with known errors, it is possible to propagate the errors from the measurement data to the new estimates. This is slightly more involved than the simple error propagation discussed in Section 4, which can be used to propagate the error from the measurements, $\mathbf{x}_{meas} \pm \sigma_{meas}$, to rates directly derived from it, $\mathbf{R}_{meas}$ (note the one without "hat") such as those in Table 23.11. In data reconciliation, the error can be propagated from the measurements, $\mathbf{x}_{meas} \pm \sigma_{meas}$, to all estimated quantities, $\hat{\mathbf{x}}_{meas}$, $\hat{\mathbf{R}}$ (note the presence of the "hat"). We will leave out here the mathematics involved and simply state that dedicated software will indeed produce the standard errors of all estimates.

Example In the case of *E. coli* of Section 6.5, Table 23.10 shows the estimated values for the measured quantities. The standard errors have been reduced by 7% characterized by the median value of the changes involved.

The biggest relative change was from 0.0030 to 0.0011 for the fraction oxygen in the gas.

The results for the rates expressed in mmol/h are in Table 23.11 and expressed in mmol/h per unit biomass are in Table 23.13. A typical improvement of, respectively, 30% and 40% is obtained.

In fact, a complete covariance matrix can be deduced of all estimated variables, which might indicate underlying correlations in these estimates.

Besides the confidence intervals of each estimate, represented by the standard errors, the statistics also allows to test the internal consistency of all assumptions underlying the data reconciliation. These include the presence of gross errors and missing processes. There are three tests.

First of all, there is the global test, which considers the sum-of-squares result of the optimization. Without proof it can be shown that this should have a so-called Chi-square distribution where the number of degrees of freedom equal to the degree of redundancy of equation. In formula form,

$$SS = \sum_i \left(\frac{X_{\text{meas},\,i} - \hat{X}_{\text{meas},i}}{\sigma_{\text{meas},i}} \right)^2 \sim \chi^2(n_{\text{dr}} - 1). \tag{23.52}$$

Roughly said, the sum-of-squares should be on the same order as the degree of redundancy, which is well demonstrated in the case study. A more precise expression is to give the P-value. This is the probability that the experimental outcome is worse given that all the underlying equations and the statistical nature of the measurement error is expected.

Example In the case of *E. coli* of Section 6.5, SS = 7.227 with 4 as degree of reduction. It is indeed of the same order. The P-value is 01219.

A small value of P (typically <0.05) is considered as highly improbable. There are possible causes:

1. Gross error in the data (a data problem).
2. Forgotten processes (Section 3, a model problem).
3. Invalid approximations (a model problem).
4. The measurement errors were wrongly determined (an error problem).
5. New phenomena discovered (a model—opportunity).

The actual cause can only be found by carefully scrutinizing the data, and evaluation of all assumptions that have been made. At the end, some creative thinking will lead to the new phenomena and the new insight.

Second, there is the measurement test, where one compares the original measurements with the new estimates. A standard error for this difference can be derived, and the results are then presented as a residual being equal to the difference divided by its standard error, or as a *P*-value, giving the probability that the outcome could be worse. If only one or two measurements have a low *P*-value, then there is an indication that at that location in the experiment something interesting takes place.

Example In the case of *E. coli* of Section 6.5, the measurement test is summarized in Table 23.15, where the measured values and their estimates are shown together.

There is virtually not much difference. The residuals are mostly within the expected $\pm 2\sigma$ range. The *P*-values should not be compared with the usual 0.05 because one tests a group of variables together. In this case, a *P*-value of less than 0.002 is significant, which is not the case.

A third test evaluates the relations that have to be fulfilled, especially the balances. This is called the nodal test. In its simplest form, it is nothing else as identifying the redundant equations as done in Section 5.3. However, a more precise quantification can be given. It is mathematically possible to identify the few relations that can be evaluated with the measured data only. The estimation for the measurements will exactly fulfill the relations, but not the measurement themselves as they are subject to experimental error. The difference divided by the standard error derived for this difference is the residual, as above. Also the associated *P*-value can be calculated. If such *P*-value is very small, it indicates that the balance relation involved, has a distinct error, normally associated with a forgotten process.

Table 23.15 Measurement test applied to the *E. coli* case

Flow	Measured variable	Residual	*P*-value
Gas	$F_{G,in}$	−1.834	0.033
	x_{in,O_2}	0.472	0.319
	x_{in,CO_2}	1.799	0.036
Medium	F_{in}	0.969	0.166
	$c_{in,s}$	1.780	0.038
	c_{in,NH_3}	−2.333	0.0098
Alkali for pH	F_{alk}	1.042	0.145
Off-gas	$F_{G,out}$	Unknown	
	x_{O_2}	−0.472	0.319
	x_{CO_2}	−1.799	0.036
Broth	F_{out}	−0.531	0.298
	c_s	−1.78	0.038
	c_{NH_3}	−2.333	0.0098
	TOC_{sup}	0.912	0.181
	c_x	0.912	0.181

Table 23.16 Nodal test applied to the *E. coli* case

	Residue ± error	Residue/error (to be <2)	*P*-value
C	13 ± 10	+1.24	0.108
N	−2.6 ± 2.0	−1.32	0.093
γ	−20 ± 33	−0.62	0.267
Mass	0.0048 ± 0.0059	+0.81	0.209

Example In the case of *E. coli* of Section 6.5, the nodal test gives the results of Table 23.16 essentially confirming the consistency of the data.

A limit of 0.25% on the *P*-value is required. Here all observable balances fulfill this requirement.

Any of the three tests gives a pointer to possible experimental and model shortcomings. The global test is an overall test. The measurement test points toward potential experimental situations. The nodal test points at incompleteness in the compound and element balances.

7.4. Conclusion

The formulation of the data reconciliation problem as an optimization problem given by is a very general approach not limited by any constraints on the choice of variables and relations to take along. This complements the approach of the previous sections. There carefully the measurements were considered, from there rates (the so-called mass or compound balance rates) were deduced and finally the unknown rates were deduced from various balances (the so-called element balance rates). The results of the two approaches deviate normally not much.

The advantage of the latter approach is that it makes it more comprehensible to follow, and its disadvantage is that it in fact can give incorrect answers. The advantage of the full data reconciliation approach is its flexibility. Any relation and any measurement can be considered in this formulation without having to bother about the details of the derivation of the final estimates.

8. Conclusion

Biomass specific uptake and secretion rates (so-called *q*-rates) are essential

- to characterize the performance of an organism
- to perform cost prize calculation
- to calculate fluxes, which lead to targets for genetic improvement

Such q-rates can in principle not be measured using a sensor but need to be calculated using compound balances in combination with proper measurements (flow, total mass, concentration) and properly designed experiments. It is very relevant to perform redundant measurements to be able to detect, using conservation principles, forgotten processes and to improve accuracy of the calculated rates using statistical reconciliation methods.

REFERENCES

Crowe, C. M. (1996). Data reconciliation—Progress and challenges. *J. Process Cont.* **6**(2–3 spec.), 89–98.

Crowe, C. M., Campos, Y. A. G., *et al.* (1983). Reconciliation of process flow rates by matrix projection. Part I: Linear case. *AIChE. J.* **29**(6), 881–888.

Heijnen, J. J., Roels, J. A., *et al.* (1979). Application of balancing methods in modeling the penicillin fermentation. *Biotechnol. Bioeng.* **21**(12), 2175–2201.

Heyen, G., and Kalitvenzeff, B. (2007). Process monitoring and data reconciliation. *In* "Computer Aided Process and Product Engineering," (L. Puigjaner and G. Heyen, eds.), pp. 517–539. Wiley-VCH, Weinheim.

Noorman, H. J., Baksteen, J., *et al.* (1991). The bioreactor overflow device: An undesired selective separator in continuous cultures? *J. Gen. Microbiol.* **137**(9), 2171–2177.

Roels, J. A. (1983). Energetics and Kinetics in Biotechnology. Elsevier Science & Technology, Amsterdam.

Taymaz-Nikerel, H., de Mey, M., *et al.* (2009). Development and application of a differential method for reliable metabolome analysis in *Escherichia coli*. *Anal. Biochem.* **386**(1), 9–19.

Taymaz-Nikerel, H., Borujeni, A. E., *et al.* (2010). Genome-derived minimal metabolic models for *Escherichia coli* MG1655 with estimated *in vivo* respiratory ATP stoichiometry. *Biotechnol. Bioeng.* **107**(2), 369–381.

Van Der Heijden, R. T. J. M., Heijnen, J. J., *et al.* (1994a). Linear constraint relations in biochemical reaction systems: I. Classification of the calculability and the balanceability of conversion rates. *Biotechnol. Bioeng.* **43**(1), 3–10.

Van Der Heijden, R. T. J. M., Romein, B., *et al.* (1994b). Linear constraint relations in biochemical reaction systems: II. Diagnosis and estimation of gross errors. *Biotechnol. Bioeng.* **43**(1), 11–20.

Wang, N. S., and Stephanopoulos, G. (1983). Application of macroscopic balances to the identification of gross measurement errors. *Biotechnol. Bioeng.* **25**(9), 2177–2208.

CHAPTER TWENTY-FOUR

A Practical Guide to Genome-Scale Metabolic Models and Their Analysis☆

Filipe Santos,*,† Joost Boele,* *and* Bas Teusink*,†

Contents

☆ This chapter and Chapter 4 in Section 7 are focused on large-scale metabolic model reconstruction. The approaches and viewpoints of the authors differ, and the editors would advise readers to consider both chapters together for a broader understanding of the methodologies and issues within this field.

* Amsterdam Institute for Molecules, Medicines and Systems/NISB, VU University Amsterdam, De Boelelaan 1085, Amsterdam, The Netherlands
† Kluyver Centre for Genomics of Industrial Fermentations/Netherlands Consortium Systems Biology, VU University Amsterdam, De Boelelaan 1085, Amsterdam, The Netherlands

Methods in Enzymology, Volume 500
ISSN 0076-6879, DOI: 10.1016/B978-0-12-385118-5.00024-4

Abstract

Genome-scale metabolic reconstructions and their analysis with constraint-based modeling techniques have gained enormous momentum. It is a natural next step after sequencing of a genome, as a technique that links top-down systems biology analyses at genome scale with bottom-up systems biology modeling scrutiny. This chapter aims at (systems) biologists that have an interest in, but no extensive knowledge of, applying genome-scale metabolic reconstruction and modeling to their organism. Rather than being comprehensive—excellent and extensive reviews exist on every aspect of this field—we give a rather personal account on our experience with the process of reconstruction and modeling. First, we place genome-scale metabolic models in the spectrum of modeling approaches, and rather extensively discuss, for nonexperts, the central concept in constraint-based modeling: the solution space that is bounded through constraints on fluxes. We subsequently provide an overview of the different steps involved in metabolic reconstruction and modeling, pointing to aspects that we found difficult, important, not well enough addressed in the current reviews, or any combination thereof. In this way, we hope that this chapter serves as a practical guide through the field.

1. Introduction

Today's interest in systems biology is largely fuelled by high-throughput techniques that generate large amounts of data. There is a general consensus that functional genomics has enormous potential in the life sciences, in particular, in biotechnology and medicine. How to use these technologies most efficiently, for fundamental understanding, biomarker discovery, or concrete biotech applications, is an area of active research. It is clear that the volume and complexity of the data are becoming too large to cope with by biologists alone, especially when the latter are poorly trained in advanced mathematics and computation (which is still largely the case). So there is an understandable need from the biologist's perspective for help in mining, interpreting, and using the datasets that they collect. Such activities require modeling of one form or the other (Ideker and Lauffenburger, 2003).

Biostatistics and bioinformatics offer help in the analysis of genome-scale data sets, but they often rely on purely mathematical and statistical analysis. Although extremely useful, it ignores what is often referred to as "legacy data," that is, the large body of biological knowledge that is often scattered in literature and therefore poorly accessible. Moreover, many of the techniques were not designed to incorporate *a priori* knowledge, even if it is available (Liao *et al.*, 2003). "Bottom-up" systems biologists, however, construct detailed mechanistic models that aim at a fundamental understanding of systems behavior (Bruggeman and Westerhoff, 2007). To achieve this, they start from the molecular properties of biological components and their interactions in

BOX 24.1 "Beginner's Kit" for Genome-Scale Metabolic Model Reconstruction and Their Analysis

The first step of model development is generally obtaining a draft model. Multiple tools can be used for this with different input demands. These can be from only an assembled genome (The SEED) to the combination of multiple genome sequences and respective curated models (AUTOGRAPH). Again depending on the algorithms employed in the initial reconstruction, the list of gene-reaction associations needs to be checked for consistency, gaps, and errors, and also attached to the list of external metabolites and the biomass equation. The latter two generally are obtained either from literature or from experimental data. This first (almost) "growing" model then enters the refinement stage in which the outputs of a succession of constraint-based analysis techniques available within various software packages (see Table 24.1) are compared with both literature and experimental data. For more detail, please see the main text and/or references to external resources (Fig. 24.1).

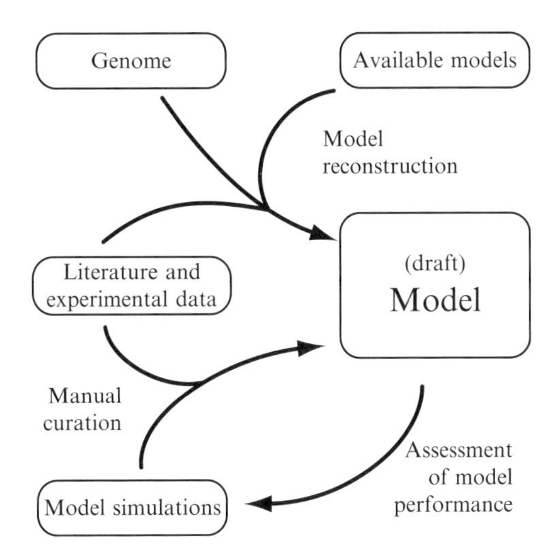

Figure 24.1 Overview of draft genome-scale metabolic model generation and iterative refinement cycle.

biological networks, be it metabolic, signaling, or gene regulatory, amongst others. As will be briefly discussed below, both approaches have their limitations. Using genome-scale reconstructions, and their corresponding models, may be considered as a "middle-out" approach, as they combine -omics data with more traditional modeling strategies. This approach is the focus of this chapter, in particular, its application to metabolic networks.

All aspects of genome-scale metabolic models have been extensively reviewed in recent years (Feist and Palsson, 2010; Feist *et al.*, 2009; Hyduke

Table 24.1 Selected list of external resources useful for the generation of genome-scale metabolic model and their analysis

Tool/data source	URL/reference	Description
AUTOGRAPH	www.biomedcentral.com/1471-2105/7/296	AUTOGRAPH is a semi-automatic approach to accelerate the process of genome-scale metabolic network reconstruction by taking full advantage of already manually curated networks
BiGG	bigg.ucsd.edu	BiGG is a knowledge base of biochemically, genetically, and genomically structured genome-scale metabolic network reconstructions
BioCyc	biocyc.org	BioCyc is a collection of >1000 pathway/genome databases. Each database in the BioCyc collection describes the genome and metabolic pathways of a single organism
BioMet Toolbox	129.16.106.142	The BioMet ToolBox is a web-based resource for analysis of high-throughput data, together with methods for flux analysis (fluxomics) and integration of transcriptome data exploiting the capabilities of metabolic networks described in genome scale models
BRENDA	www.brenda-enzymes.org	BRENDA is the main collection of enzyme functional data available to the scientific community.
COBRA	gcrg.ucsd.edu/Downloads/Cobra_Toolbox	The COBRA Toolbox for Matlab includes implementations of many of the commonly used forms of constraint-based analysis such as FBA, gene deletions, flux variability analysis, sampling, and batch simulations together with tools to read in and manipulate constraint-based models
KEGG	www.genome.jp/kegg/	KEGG (Kyoto Encyclopedia of Genes and Genomes) is a bioinformatics resource, popular for its visualization capabilities, that links genomes to pathways

Table 24.1 (*Continued*)

Tool/data source	URL/reference	Description
OptFlux	www.optflux.org	OptFlux incorporates strain optimization tasks and also allows the use of stoichiometric metabolic models for (i) phenotype simulation of both wild-type and mutant organisms, (ii) metabolic flux analysis, and (iii) pathway analysis through the calculation of elementary flux modes
Pathway Tools	bioinformatics.ai.sri.com/ptools	Pathway Tools is a comprehensive symbolic systems biology software system that supports several use cases in bioinformatics and systems biology
PubMed	www.ncbi.nlm.nih.gov/pubmed	PubMed is a service of the U.S. National Library of Medicine that includes over 19 million citations from MEDLINE and other life science journals for biomedical articles back to the 1950s
The SEED	www.theseed.org	The Model SEED automates as much as possible the development of genome-scale metabolic models requiring from the user only an assembled genome sequence
YANAsquare	yana.bioapps.biozentrum.uni-wuerzburg.de	YANA is a user friendly software package for the analysis of metabolic networks
INSD	http://www.insdc.org/	The International Nucleotide Sequence Databases (INSD) consists of an international collaboration between GenBank, DDBJ, and ENA resulting in the largest sequence repository

and Palsson, 2010; Liu *et al.*, 2010; Oberhardt *et al.*, 2009; Orth *et al.*, 2010; Teusink and Smid, 2006). In this chapter, therefore, we will try to guide the reader, whom we envisage is a biologist relatively new to systems biology in general, and to genome-scale metabolic models, in particular, through the various steps of the modeling process. The account on the many steps involved is necessarily not comprehensive, and rather personal, based on our experience of the process. It will hopefully help to understand the basic concepts and rationales, while providing many references to other resources where more detailed information can be pursued. We also provide a sort of "beginner's kit" of software that we found useful at different stages in the reconstruction and modeling process (see Box 24.1).

2. Genome-Scale Metabolic Models: Their Place in the Spectrum of Modeling Options

As seen above, two opposing strategies have been generally employed to model biological systems. The so-called bottom-up approach focuses mostly on the very detailed description of the different components of the system, while the "top-down" approach looks for (uneducated) correlations between different variables of the system. Genome-scale metabolic models can be seen as a "middle-out" strategy because they first delimit the genomic pool of the system, mainly through bioinformatic approaches, and from there model the potential interactions of these components.

It should be clear that all modeling strategies are good for specific tasks but have limited capabilities for others, and so it is important to understand the possibilities, limitations, and underlying assumptions before embarking on any of them. The direct comparison of genome-scale metabolic models with other modeling approaches places them in the spectrum of modeling options and helps understand their value.

2.1. Bottom-up (kinetic) models versus genome-scale metabolic models

The structure of a kinetic model is actually quite simple. Suppose, we have a simple system with two branches (Fig. 24.2).

We often assume an environment in which S is some infinite source with a constant concentration (often referred to as an external metabolite; Heinrich and Schuster, 1996; Schuster *et al.*, 2000), for example, glucose in the bloodstream or a nutrient in a chemostat. Similarly, $P1$ and $P2$, the products of this pathway, have constant concentrations and are also external metabolites. X is an internal metabolite, the concentration of which depends on the rates of production and consumption. In mathematical terms, this can be described as:

$$\frac{\mathrm{d}X}{\mathrm{d}t} = v_1 - v_2 - v_3 \qquad (24.1)$$

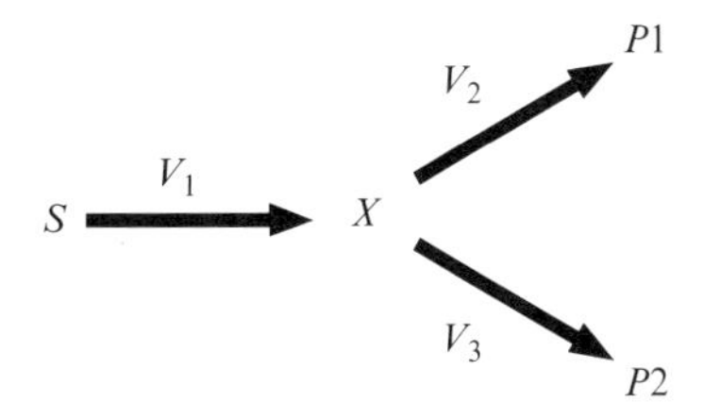

Figure 24.2 Simple metabolic network consisting of three reactions, three external metabolites (S, $P1$, and $P2$) that act as source or sink, and one internal metabolite (X).

The rates of the enzymes, v_i, are a function (f) of the kinetic parameters **p** (e.g., Michaelis–Menten constants V_{max} and K_m values) and concentrations of the metabolites, in this case:

$$v_i = f(\mathbf{p}, X, S, P1, P2) \tag{24.2}$$

Thus, Eq. (24.1) is a differential equation, the solution of which is a function that describes the time course of the concentration of X, that is,

$$X(t) = f(X_0, \mathbf{p}, S, P1, P2) \tag{24.3}$$

The X_0 indicates that the actual behavior is dependent on the initial concentration of X. As the rates depend on **p**, S, $P1$, and $P2$, so does X. If X is computed at each time point, Eq. (24.2) can be solved to find the rates of all the enzymes in time, that is, $v_i(t)$. Thus, such kinetic models can be constructed and validated through time-series of metabolites and flux measurements.

Detailed kinetic models can be used to explore the dynamics of a system, as well as the control structure of a pathway (Fell, 1997). Once a model is available, parameters and conditions (such as the concentration of S in our example) can be altered to study biologically relevant properties, such as bistabilities (Veening *et al.*, 2008), oscillations (Goldbeter, 1997), and robustness or homeostasis (Kitano, 2002, 2007), just to mention a few. In a practical sense, this understanding makes it possible to predict which steps to enhance, or which feedbacks to remove, for a particular modification of the pathway's behavior (e.g., enhanced production of a valuable product or improving a diseased state). Even though there are examples of successful applications of these types of kinetic models (Bakker *et al.*, 2010; Hoefnagel *et al.*, 2002), they face some serious limitations.

First, kinetic parameters **p** of all enzymes in a pathway are very rarely available, although databases such as BRENDA (Barthelmes *et al.*, 2007) and Sabio-RK (Rojas *et al.*, 2007) are a tremendous help. However, these kinetic parameters are often measured at different or nonphysiological conditions, such as at the optimal pH of the enzymes, not the physiological pH. Also, it remains quite controversial to what extent the *in vitro* kinetics reflects the *in vivo* kinetics (Teusink *et al.*, 2000). An alternative to the *in vitro* kinetic parameter determinations is to estimate "*in vivo* kinetic parameters" from time courses of metabolites after short timescale perturbations of the metabolic network (Mashego *et al.*, 2007; Theobald *et al.*, 1993; Visser *et al.*, 2004). Parameter estimation is a field by itself and not an easy task. We refer to a recent review on the topic (Banga and Balsa-Canto, 2008).

Second, these models necessarily represent relatively small, isolated metabolic pathways. As these pathways are embedded in a large metabolic network, the boundary conditions, that is, the exchanges of information with the rest of the system, become critically important in the success of the model predictions. It was shown that including the boundaries explicitly, even with uncertain parameter values, can improve significantly the predictive power of the model outcome (Liebermeister *et al.*, 2005). Several approaches, in fact, have been described that take the uncertainty in kinetic parameters explicitly into account, generating an ensemble of model outcomes that can be inspected for robust and more uncertain model predictions (Liebermeister and Klipp, 2005, 2006; Steuer *et al.*, 2006; Wang and Hatzimanikatis, 2006). Eventually such new approaches, together with further developments in metabolomics research and computing power, may substantially increase the size of systems for which kinetic models can be constructed by reverse engineering (Jamshidi and Palsson, 2010; Resendis-Antonio, 2009). However, at the moment these approaches scale poorly, and genome sizes are still a long way ahead.

Genome-scale metabolic models, however, do cover the total metabolic potential that is encoded in the genome of an organism, but this comes at a cost. They share with kinetic models the structure, that is, the stoichiometry of all reactions, but they leave out (almost) all of the kinetic details. This is possible because many pathways, such as the one depicted in Fig. 24.2, when analyzed over a sufficiently long time will reach a steady state, that is, a state in which all internal metabolites will be balanced by the producing and consuming reactions. In our example, this means that, in steady state:

$$\frac{dX}{dt} = v_1 - v_2 - v_3 = 0 \quad (24.4)$$

Equation (24.4) basically states that the concentration of X is constant in time, as it is balanced by its production and consumption rate. When the kinetics of the reactions are known, we can compute the steady state concentration of X, and the steady-state rates through the enzymes, which we then, in steady state, call *fluxes*:

$$X(t \to \infty) = X_{ss} = f(\mathbf{p}, S, P1, P2) \tag{24.5a}$$

$$V_{i,ss} = f(X_{ss}, \mathbf{p}, S, P1, P2) \tag{24.5b}$$

Equation (24.4) leads to a set of linear equations that link the rates v_i with each other, which does *not* require the parameters **p**, S, $P1$, and $P2$. However, to truly *predict* the rates v_i or the steady-state metabolite concentration from the conditions defined by S, $P1$, and $P2$, one *does* need this kinetic information, as is specified in Eqs. (24.5a) and (24.5b). Via Eq. (24.4), we only look at steady-state stoichiometric interrelationships between (steady-state) fluxes, nothing more. This difference with kinetic modeling is absolutely crucial for understanding the limitations of the analyses of genome-scale metabolic models discussed later on. Imposing steady-state relationships between rates (referred to as mass-balance constraints (Price *et al.*, 2004), for obvious reasons) constrains the possible states the system can be in only with respect to fluxes through the network, also referred to as the solution space (illustrated in Fig. 24.3).

The other type of constraints used in genome-scale metabolic modeling is called the capacity constraint. Such constraints also have a link with kinetics, as they represent any limitation that can be imposed on individual rates. Such can be measured V_{max} values that form an upper limit through

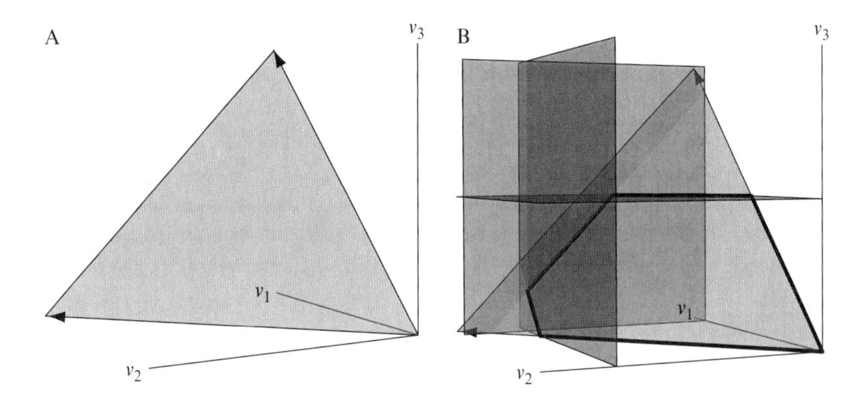

Figure 24.3 Illustration of the concept of solution space, and how mass-balance and capacity constraints result in a bounded space of possible flux states. (A) The steady-state solution space is two-dimensional (a plane), because there are three unknowns (three fluxes) and one equation: the mass balance of X, that is, Eq. (24.4). This plane stretches out to all directions but is depicted only in the positive quadrant for comparison with Fig. 24.3B. The arrows represent "extreme pathways," basis vectors that span the solution space, see Papin *et al.* (2004) for a review on the topic. (B) Specific capacity constraints on the fluxes v_1, v_2, and v_3 turn the plane into a bounded space within which all feasible flux states should lie. For each flux, we have set capacity constraints, $0 \leq v_i \leq v_{i,max}$.

the reaction catalyzed by the enzyme. More often, it contains a constraint on the directionality of a reaction, that is, the assessment whether a reaction can work in both directions under the physiological range of metabolite concentrations that is assumed (or sometimes measured) in the organism under study (Hoppe *et al.*, 2007; Kummel *et al.*, 2006a,b). Mass balance and capacity constraints together limit the solution space of all possible flux distributions (Fig. 24.3).

The analysis of genome-scale metabolic models, applying mass-balance and capacity constraints, is collectively named constraint-based modeling. Note that a detailed kinetic model will predict a specific steady state with fluxes that lie inside the solution space of the corresponding stoichiometric network (Teusink and Smid, 2006). Leaving out the kinetic details in constraint-based modeling comes at the cost of being able to predict the space of feasible states but not the specific state (unless specific assumptions are made that will be outlined later). Also information about the sensitivity of the steady state to parameters changes, as well as the trajectory how this state is reached, is not available through stoichiometry alone.

2.2. Top-down (biostatistical) models versus genome-scale metabolic models

Despite the -omics data explosion, it is clear that for true systems-level understanding, the abundance of relevant data, unfortunately, is matched by an abundance of less relevant data. This has to do with the inductive and open-ended nature of the typical -omics experiment. There is nothing wrong with this (well, a little perhaps; Lipton, 2005), and especially when done in a systematic way with good experimental design, there are good methods to turn data into knowledge (Kell, 2004; van der Werf, 2005). Statistics-based data analysis often gives qualitative interaction networks, or candidate components (genes, transcription factors, metabolites) that are scored to be related to the phenomenon under study. For many applications, this may be sufficient because these leads can be followed up by validation experiments. Also, the -omics data can simply be used for diagnostic purposes both in red and in white biotechnology, for example, to indicate a particular limitation or stress during a fermentation process (Knijnenburg *et al.*, 2007; Tai *et al.*, 2005), or to predict likely outcomes of cancer treatment (Kim and Paik, 2010), just to mention a few of the many, many applications of recent years.

Attempts to enhance the purely statistical analyses by integrating different data sets and *a priori* knowledge lie within the realm of integrative bioinformatics. We will not make an attempt to give a full account on that field but focus on where metabolic information was, or can be, used for integration. The construction of genome-scale models starts from the genome: using bioinformatic approaches, the sequenced genome is scanned for regions that encode proteins with enzymatic activity (Francke *et al.*, 2005; Reed *et al.*, 2006; Thiele and Palsson, 2010). Based on the presence of such genes, and supplemented with known biochemical and physiological information, putative metabolic pathways can be inferred. Thus, genome-scale metabolic models do not only contain reaction information (as was used above in comparison with kinetic models) but also the often many-to-man relationships between genes, proteins, and reactions (Reed *et al.*, 2006). The mapping of genes to proteins and to reactions allows for integration of these levels. Transcriptome data (gene level) can be mapped on metabolic maps in this way, providing a visual, metabolic context for the interpretation of the transcriptome data (Gehlenborg *et al.*, 2010; Kono *et al.*, 2009). By visual inspection, this may give leads to parts of the metabolic network that are regulated. For instance, the need for CO_2 in fermentations of *Lactobacillus plantarum* was identified in this way (Stevens *et al.*, 2008).

A computational counterpart of visual inspection was developed as reporter metabolites for proteome or transcriptome data (Patil and Nielsen, 2005). In these analyses, the differential expression of reactions that produce or consume a particular metabolite is scored for each metabolite and compared to an expected score (based on chance alone). Those metabolites, whose surrounding reactions change significantly, are then reporters for metabolic effects. CO_2 would be such a reporter metabolite in the *L. plantarum* example above. Reporter reactions are computed in a similar vein from metabolome data (Cakir *et al.*, 2006). Recently, a toolbox was developed that allows these types of analyses, called the BioMet Toolbox (Cvijovic *et al.*, 2010).

Also, the metabolic association of genes, via their mapping to the reaction network, has been used for functional association tests. Thus, genes close to each other on the metabolic map were found to have a stronger correlation in gene expression. Using constraint-based modeling, via the so-called flux coupling analysis (Burgard *et al.*, 2004), this relationship was resolved with a much higher functional resolution (Notebaart *et al.*, 2008, 2009). The key to this higher resolution is the appreciation that two reactions in series in a linear pathway are stoichiometrically fully coupled, whereas two reactions diverging from the same metabolite (such as reactions 2 and 3 in Fig. 24.2) are uncoupled, that is, they can carry flux in steady state independently from each other (Notebaart *et al.*, 2008).

Thus, genome-scale metabolic models provide a "context for content" (Palsson, 2004) and are very useful in -omics data integration. Reconstructions of other networks, such as transcription regulation networks (Herrgard *et al.*, 2004), signal transduction networks (Hyduke and Palsson, 2010), or the translation machinery (Thiele *et al.*, 2009), allow for further extension of top-down analyses in which biological data are integrated in a systematic way.

3. The Art of Making Genome-Scale Metabolic Models

The step-by-step instructions for making a genome-scale reconstruction have been published recently (Feist *et al.*, 2009; Thiele and Palsson, 2010), and so we will forego a detailed description of these steps in favor of placing emphasis on specific issues that we have experienced to be critical, difficult, easily overlooked, or a combination of these. The steps toward a reconstruction are (see Feist *et al.*, 2009; Thiele and Palsson, 2010, with many illustrations or a flow chart in Francke *et al.*, 2005; Henry *et al.*, 2010):

1. generating a draft reconstruction based on the genome sequence
2. identify and resolve errors, gaps, and inconsistencies in the network
3. define external metabolites and the biomass equation
4. validate the model with additional experiments

3.1. Generating a draft reconstruction based on the genome sequence

Tools and methods exist that take either a sequence or an annotated genome and produce a first draft of a metabolic reconstruction automatically. One such system is Pathway Tools (Karp *et al.*, 2010), which underlies the BioCyc collections including the famous EcoCyc database (Keseler *et al.*, 2009). It takes annotated genomes and identifies the metabolic genes based on EC code and name matching. Pathway Tools also includes algorithms for automatic gap filling. In our experience, the information in this resource is wonderful, but the automatically generated output of Pathway Tools results in a rather fragmented set of pathways that need extensive curation (Teusink *et al.*, 2005b). Also, the system is not explicitly designed for modeling, but rather, for making an inventory ("encyclopedia") of metabolic pathways, and their regulation.

The SEED, a pipeline system for metabolic reconstruction, has been recently made publicly available (Henry *et al.*, 2010). This system uses as input a sequenced genome and produces an SBML (Hucka *et al.*, 2003) model that is able to produce biomass (Feist and Palsson, 2010), one of the hallmarks of a completed reconstruction. However, we agree with Feist (Feist *et al.*, 2009), and have stressed ourselves (Francke *et al.*, 2005), that extensive manual curation cannot be fully replaced by automated methods. This is simply because there are too many important organism-specific choices that confer their own idiosyncrasies and set them apart from others. Expert domain knowledge about the organism under study is required to make these choices.

With this in mind, we ourselves have developed an automated method that uses orthology prediction between genes of a query genome and genes from already existing genome-scale metabolic models, to copy the gene–protein–reaction associations (Notebaart *et al.*, 2006). The philosophy of this AUTOGRAPH method is that information of the manual curation process of the existing genome-scale metabolic models is being reused as much as possible. Issues resolved during manual curation relate to (i) reaction stoichiometries of specific reactions not found in the databases (or with wrong stoichiometries, still not uncommon); (ii) choices in cofactor usage; and (iii) annotations of genes on the basis of gap filling, extensive bioinformatic analyses, or experimental results that have not found their way into the databases (yet). It does not remove the need for manual curation, but it clearly accelerates the curation process, in our experience. We expect the same to be true for the SEED pipeline, and the two methods are to a large extent complementary.

3.2. Identify and resolve errors, gaps, and inconsistencies in the network

After the initial draft of the model, manual curation involves going through the draft gene–protein–reaction associations to check for omissions, wrong assignments and gaps, and inconsistencies (for a review on gap filling, see Breitling *et al.*, 2008). One insightful inconsistency that we have found (Francke *et al.*, 2005) is the annotation of metA in *L. plantarum* to homoserine succinyl CoA transferase, while this organism lacks a TCA cycle and hence cannot make succinyl CoA. All databases nevertheless agree on this function for metA, presumably because it is a member of a distinct orthologous group that includes the *Escherichia coli* enzyme, which does take succinyl CoA. However, the protein from *B. subtilis*, also in this orthologous cluster, has been experimentally shown to take acetyl CoA, but somehow this result did not find its way into BRENDA or any other database. So, sequence similarity, even one-to-one orthology, is not a guarantee for function transfer. Interestingly, the model SEED has kept the succinyl CoA-dependent annotation and hence, the gap, but allows biomass production through methionine transport, even though domain experts will know that *L. plantarum* grows without methionine (Teusink *et al.*, 2005a). This teaches us that it is both wise and necessary to not always trust the databases (when tested on students, it was striking how much they trust information from the internet).

3.3. Define external metabolites and the biomass equation

For a reconstruction to become a model, there are a number of additional steps that involve a minimum of experimental data without which the model is of poor predictive value. They come in three flavors:

(i) *Biomass composition:* biomass composition must be defined, ideally for the different experimental conditions under study. The Supplementary Material in Oliveira *et al.* (2005), on the biomass of *L. lactis*, gives an extensive and useful account of the data and computation that is required. Protein, RNA, and lipid content change with growth rate and may affect the model simulations.

(ii) *Data on the bioenergetics*: this pertains to information on stoichiometry of pumps and the respiratory chain (P/O ratio), and on maintenance energy and ATP requirement for growth, often referred to as Y_{ATP} (Tempest and Neijssel, 1984). These data can be estimated by carefully monitoring product formation and biomass yield under a sufficient number of substrate/growth rate regimes; for excellent reviews on the techniques involved, see vanGulik and Heijnen (1995), Vanrolleghem *et al.* (1996), and Vanrolleghem and Heijnen (1998). Surprisingly, only very recently these parameters were carefully estimated for *E. coli* (Taymaz-Nikerel *et al.*, 2010). But there is an important assumption here: measurement and inclusion of energetic parameters are based on full coupling between ATP production by catabolism and ATP utilization through anabolic processes. Without this assumption, we cannot do the calculations, and the predictions will be imprecise (Teusink *et al.*, 2006).

(iii) *Essential and possible nutrients*: growth requirement data (e.g., on specific auxotrophies for vitamins and amino acids, or possible carbon, nitrogen and sulfur sources) are used to decide on the biological relevance of a gap, for example, if a vitamin is not essential, then a gap in the vitamin biosynthesis pathways may be benign (Teusink *et al.*, 2005b). These data can also be used to determine the potential external metabolites, that is, sources for the network, and the necessary transport systems to take them up.

Although it is not very often pointed out, sinks are equally important as sources, as they also require transport reactions, even for simple diffusion over the membrane. Easily overlooked sinks include by-products of biosynthesis pathways (e.g., folate biosynthesis produces glycolaldehyde), or diffusible substrates such as water and CO_2.

Because genome-scale models are not exclusively bottom-up but largely based on genome content, they tend to contain more (futile) cycles, parallel routes, and dead-end pathways which a biochemist or biochemical engineer would never put into his or her bottom-up model. These cycles are nevertheless potentially there in the network (Pinchuk *et al.*, 2010; Teusink *et al.*, 2006) and are either thermodynamically infeasible (substrate cycles; Beard *et al.*, 2002) or probably, and hopefully, regulated in such a way that they tend not to run in circles too much (in the case of futile cycles; if they do, however, we have uncoupling of ATP production and consumption by growth, and our bioenergetic parameters go down the drain). The dead-end pathways point to missing biochemical knowledge (or to reductive evolution, see e.g., Hols *et al.*, 2005) and can therefore be valuable leads to new discoveries, especially in combination with the data on substrate use and product formation. Metabolic reconstructions and their modeling therefore not only rely on gene annotation, but it can also help in functional annotation by identifying functions that have to be there.

3.4. Validate the model with additional experiments

Once the model can form biomass by providing all the necessary components that form the biomass, validation of the model requires comparison with additional experimental data. One such set of data are phenotypes of deletion strains; comprehensive sets are unfortunately usually not available but for model organisms such as *E. coli* and yeast. Their metabolic reconstructions have been extensively tested against the deletion strain phenotypes, with relatively high success rates in the order of 80–90% (Covert *et al.*, 2004; Kuepfer *et al.*, 2005). A more accessible set of data that can be used is physiological data on growth rates, yields, substrate utilization, and product formation rates. Especially chemostat data are useful, as they are the best experimental setup for establishing growth in steady state. The model should be compatible with the measured fluxes, or, at a higher level of validation, should be able to predict some of the fluxes as a function of some input fluxes (note that we cannot predict all fluxes *de novo*, only the steady-state dependencies between fluxes, Eq. (24.4)). Oberhardt (Oberhardt *et al.*, 2009) gives a good overview of the different models that exist and the extent to which they have been validated through experimental data.

4. Applications of Genome-Scale Metabolic Models

For an increasing number of microorganisms, a genome-scale metabolic model is available at the quality level that we have discussed above (Oberhardt *et al.*, 2009; Reed *et al.*, 2006). The group of Palsson developed a resource for such metabolic models (see Schellenberger *et al.*, 2010). These models can be used for a number of purposes, see a recent account on the applications of genome-scale metabolic models (Oberhardt *et al.*, 2009). A large set of different methods, constraint-based modeling techniques, have been developed in the past years to accommodate these goals. An excellent overview of these techniques is given in Price *et al.* (2004). In brief, successful use of genome-scale metabolic models have ranged from:

(i) exploration of gene lethality (Blank *et al.*, 2005; Covert *et al.*, 2004) and/or synthetic lethality (Costanzo *et al.*, 2010)
(ii) definition of metabolic context for integrative bioinformatics (Kharchenko *et al.*, 2004; Notebaart *et al.*, 2008; Patil and Nielsen, 2005)
(iii) the study of pathway evolution (Pal *et al.*, 2006; Papp *et al.*, 2004)
(iv) prediction of metabolic engineering strategies (Burgard *et al.*, 2003; Bro and Nielsen, 2004, Park *et al.*, 2010)
(v) prediction of adaptive evolution outcomes (Fong *et al.*, 2003, 2005; Ibarra *et al.*, 2002; Teusink *et al.*, 2009)
(vi) interpretation of fermentation data (Goffin *et al.*, 2010; Teusink *et al.*, 2006)

4.1. Flux balance analysis: The work horse of constraint based modeling

In many of the applications of genome-scale models, flux balance analysis (FBA) is used, and this is by far the most popular constraint-based modeling technique (Gianchandani *et al.*, 2010). For a more extensive primer on FBA, see Orth *et al.* (2010). FBA uses optimization of a certain objective function to find a subset of optimal states in the large solution space of possible states that is shaped by the mass balance and capacity constraints (see above; Orth *et al.*, 2010; Price *et al.*, 2004). We purposely state subset of optimal states, because the flux value of the objective function is unique, but the pathway distributions that give this optimal flux value, may not. There may be many different ways to Rome. If one is interested in the flux distributions, not only in the optimal value of the objective function, one needs to do a flux variability analysis (FVA) to check the uniqueness of the flux distributions (Mahadevan and Schilling, 2003). FVA maximizes and minimizes each reaction rate in the network at the optimal flux value: the resultant range in fluxes gives an indication how flexible the network is in reaching the optimum.

FBA basically tries to find a state (or set of states) that maximizes (or minimizes) some desired flux or linear combination of fluxes. The approach is an optimization technique, and hence it implies that the modeled system (e.g., organism) behaves optimally (by evolution). This may limit the understanding of how microorganisms behave in experiments, as they may not have had the time to adapt to the environment.

There are three ways to deal with the optimality issue. First, we can allow cells to adapt to their (constant) environment by laboratory evolution experiments (see application (v)). On the positive side, FBA offers an opportunity to predict evolutionary engineering outcomes, which may be guided by *in silico* analysis of specific deletions to enhance productivity

(Burgard *et al.*, 2003; Cvijovic *et al.*, 2010). It also provides optimal yields with which experimentally obtained yields can be benchmarked.

Second, alternatives to FBA have been proposed to circumvent the need for evolution, notably MOMA (minimization of metabolic adjustments; Segre *et al.*, 2002). MOMA requires a well-defined reference state and then tries to predict the response of the network to a perturbation (change in external conditions, deletion of a reaction) by assuming that the network is robust. Technically, it tries to minimize the (Euclidian) distance between the reference flux distribution and the new solution space (which has changed by the perturbation). In our experience, the major problem with MOMA is that it weighs all fluxes equally with respect to biological relevance, which is not very likely. Moreover, the response is dominated by high fluxes. Thus, MOMA predictions in our hands do not produce proper specific predictions of fluxes, but it has been useful to see if a certain perturbation is likely to cause severe growth problems even though the optimal FBA solution seems fine (e.g., see Teusink *et al.*, 2009).

Third, the optimality issue can just be ignored, simply because one does not care about a specific, quantitative prediction. For example, one may want to know if the model can produce biomass, or some compound, not necessarily how much. This goes for (synthetic) lethality predictions, and predictions of evolution of metabolic networks.

4.2. FBA predicts rates only through yield maximization

If one does want to optimize to obtain specific flux distributions, however, one requires a sensible objective function, and here some confusion and controversy have arisen in the literature. In a recent study, different objective functions were tested for the extent to which they could predict actual flux states under different conditions (Schuetz *et al.*, 2007). This study demonstrated that different objective functions were needed to describe the flux states under different conditions. Notably, under energy (or carbon) limitation, optimization of biomass yield appeared to be the best objective function. This is in line with earlier studies in which the biomass formation function was taken as objective to predict functional states (Edwards *et al.*, 2001).

However, many microorganisms display overflow metabolism, which is a wasteful lifestyle in terms of ATP generation and consequently, biomass yield. Yet it is observed even in glucose-limited chemostats above a certain critical dilution rate, such as ethanol fermentation in *Saccharomyces cerevisiae* (van Dijken *et al.*, 1993), acetate formation in *E. coli* (Vemuri *et al.*, 2006), or lactate formation in lactic acid bacteria (Thomas *et al.*, 1979). This behavior cannot be predicted by FBA, because it predicts rates through optimal yields. It could only be described by including additional capacity constraints on the oxidative phosphorylation pathways in the corresponding metabolic networks (Famili *et al.*, 2003; Varma and Palsson, 1994).

It is important to fully appreciate the point that FBA predicts rates through optimal yields, and not rates directly. If we take our example from Fig. 24.2, and suppose we want to maximize the production rate of $P1$ from S, we may formulate the FBA problem as:

$$\begin{aligned}&\max v_2\\&\text{given}:\\&\begin{cases}\text{mass balance constraint}:\\ \quad v_1 - v_2 - v_3 = 0\\ \text{capacity constraints}:\\ \quad 0 \leq v_1 \leq 10\\ \quad -\infty \leq v_2 \leq \infty\\ \quad 0 \leq v_3 \leq \infty\end{cases}\end{aligned} \tag{24.6}$$

The solution is easy in this case: v_1 should be 10, v_2 should be 10, and v_3 should be 0. But note that the rate of v_2 is fully dictated by the constraint on v_1. We could write the rate of v_2 as:

$$v_2 = v_1 \cdots Y_{P1,S} \tag{24.7}$$

where $Y_{P1,S}$ is the yield of $P1$ on S (1 in this case). As v_1 is fixed by the capacity constraint, the only way to maximize v_2 is to maximize the yield; this is also fixed in this simple case, but this is not so in larger systems while the capacity constraint is always required to bound the problem. For larger, real genome-scale models, this argument therefore remains (Schuster *et al.*, 2007): as the solution space is bounded by an input flux, FBA finds an optimal objective *flux* by optimizing the *yield* on the incoming substrate. Hence, if in a model there are two options to make ATP from glucose, fermentation (low ATP yield) and respiration (high ATP yield), optimization of ATP production rate will necessarily be achieved by respiration. Therefore, when applying biomass optimization in FBA, the underlying biological assumption is that *biomass yield maximization* was the strategy through which the organism has reached its fitness. It is clear that there are also other strategies that lead to fitness, and hence, the validity of FBA, using biomass yield as objective function, is organism and condition specific (Fong *et al.*, 2003; Ibarra *et al.*, 2002; Schuetz *et al.*, 2007).

Understanding the basic assumptions of FBA, we can turn the problem around and change the conditions in such a way that organisms are likely to grow efficiently, thereby increasing the predictive power of FBA. Growth yield maximization on poor substrates must have been the growth strategy under the conditions where FBA predictions were successful (Fong *et al.*, 2003; Teusink *et al.*, 2009). Growth yield maximization is not the best strategy under high glucose concentrations and explains why adaptation on glucose led the cells away from the line of optimality predicted by FBA (Ibarra *et al.*, 2002).

4.3. Using constraint-based modeling for discovery and interpretation: Sensitivity analysis

Even under conditions where FBA is not predictive, such as high glucose concentrations, FBA can be very useful when combined with experimental data on input *and output* fluxes (which could not be predicted by FBA). By using those measured fluxes as constraints, there are a number of interesting things one can do. We will discuss two of them.

First, we can set the measured flux data and then perform FVA to assess which parts of the network flux distribution is resolved by the measured fluxes and which are not. One basically maximizes and minimizes each reaction in the network, given the measured values as constraints (note we used FVA before to test the uniqueness of an FBA solution: in that case, fluxes were minimized and maximized with the optimal objective value as constraint). It thus gives a range of reaction flux values for each reaction: the larger the range, the poorer do the measured fluxes predict this flux. This analysis could be considered as a genome-scale analogue to the well-known metabolic flux analysis (Maertens and Vanrolleghem, 2010), but applicable to large, underdetermined, systems. Alternatively, the ranges of fluxes obtained this way may include products of fermentation that were not measured but may be required to obtain the proper mass (or redox) balances. In this way, we were pointed at degradation products of amino acids in a recent retentostat study (Goffin *et al.*, 2010).

Second, growth rate optimization under measured flux constraints will indicate which medium compounds could further contribute to growth. This can be done through sensitivity analysis, which comes for free during the optimization procedure. These sensitivity coefficients are called reduced costs, and they quantify to what extent the objective flux could be improved by changing a capacity constraint. For example, the reduced cost of reaction v_1 in the FBA problem of Eq. (24.6) is 1. We demonstrated the use of reduced costs by analysis of growth limitations of *L. plantarum* on a medium containing 3 carbon sources and 18 amino acids (Teusink *et al.*, 2006). It turned out that some amino acids contributed to growth via ATP production, even though the mechanism for this ATP production was not immediately obvious. In this way, we discovered a previously unknown transhydrogenase activity in the degradation of branched chain amino acids. The latter is a good example of how genome-scale metabolic models can lead to the discovery of new metabolic capacities. In a later study, this approach was used to understand the puzzling observation that some amino acids were being produced under conditions of extremely slow growth

(Goffin *et al.*, 2010). We can safely say that without the model, we would not have been able to solve these puzzles.

4.4. Final remarks

There is a general consensus that functional genomics has enormous potential in metabolic engineering and biotechnology. Systems biology, as a new interdisciplinary branch of science, is rapidly gaining momentum. The functional genomics toolbox has allowed a global view and thereby forced many (molecular) biologists to focus more on the system's behavior than on the behavior of one of its single components. Systems biology aims at understanding and ultimately predicting such system level behavior in terms of the underlying molecular components and their interactions. It has model building at its centre: models to integrate data, models to interpret the data, and models to make predictions. It is our view that such models will play a central role in the advancement of biology in this century, simply because we cannot grasp the complexity of biological systems by intuition alone.

Genome-scale metabolic models are a first approach to combine bottom-up modeling with top-down, genomics data-driven modeling. However, this can be used for exploration, for testing scenarios, for scanning conditions, and for eliminating impossibilities. However, these models can be used for interpretation and integration of high-throughput data. But these models are a middle-out approach, and probably only a temporary acceptance of our limitations. The challenge for the future will be to bring these models to life, that is, to make them dynamic and put them under physicochemical constraints and ultimately make them subject to our manipulation by understanding (exactly) how the control over flux and metabolite levels is distributed. Only then can we truly claim to be able to rationally engineer strains via computer-aided design.

REFERENCES

Bakker, B. M., *et al.* (2010). Systems biology from micro-organisms to human metabolic diseases: The role of detailed kinetic models. *Biochem. Soc. Trans.* **38,** 1294–1301.

Banga, J. R., and Balsa-Canto, E. (2008). Parameter estimation and optimal experimental design. *Essays Biochem.* **45,** 195–209.

Barthelmes, J., *et al.* (2007). BRENDA, AMENDA and FRENDA: The enzyme information system in 2007. *Nucleic Acids Res.* **35,** D511–D514.

Beard, D. A., *et al.* (2002). Energy balance for analysis of complex metabolic networks. *Biophys. J.* **83,** 79–86.

Blank, L. M., *et al.* (2005). Large-scale 13C-flux analysis reveals mechanistic principles of metabolic network robustness to null mutations in yeast. *Genome Biol.* **6,** R49.

Breitling, R., *et al.* (2008). New surveyor tools for charting microbial metabolic maps. *Nat. Rev. Microbiol.* **6,** 156–161.

Bro, C., and Nielsen, J. (2004). Impact of 'ome' analyses on inverse metabolic engineering. *Metab. Eng.* **6,** 204–211.
Bruggeman, F. J., and Westerhoff, H. V. (2007). The nature of systems biology. *Trends Microbiol.* **15,** 45–50.
Burgard, A. P., *et al.* (2003). Optknock: A bilevel programming framework for identifying gene knockout strategies for microbial strain optimization. *Biotechnol. Bioeng.* **84,** 647–657.
Burgard, A. P., *et al.* (2004). Flux coupling analysis of genome-scale metabolic network reconstructions. *Genome Res.* **14,** 301–312.
Cakir, T., *et al.* (2006). Integration of metabolome data with metabolic networks reveals reporter reactions. *Mol. Syst. Biol.* **2,** 50.
Costanzo, M., *et al.* (2010). The genetic landscape of a cell. *Science* **327,** 425–431.
Covert, M. W., *et al.* (2004). Integrating high-throughput and computational data elucidates bacterial networks. *Nature* **429,** 92–96.
Cvijovic, M., *et al.* (2010). BioMet Toolbox: Genome-wide analysis of metabolism. *Nucleic Acids Res.* **38**(Suppl.), W144–W149.
Edwards, J. S., *et al.* (2001). In silico predictions of Escherichia coli metabolic capabilities are consistent with experimental data. *Nat. Biotechnol.* **19,** 125–130.
Famili, I., *et al.* (2003). Saccharomyces cerevisiae phenotypes can be predicted by using constraint-based analysis of a genome-scale reconstructed metabolic network. *Proc. Natl. Acad. Sci. USA* **100,** 13134–13139.
Feist, A. M., and Palsson, B. O. (2010). The biomass objective function. *Curr. Opin. Microbiol.* **13,** 344–349.
Feist, A. M., *et al.* (2009). Reconstruction of biochemical networks in microorganisms. *Nat. Rev. Microbiol.* **7,** 129–143.
Fell, D. (1997). Understanding the Control of Metabolism. Portland Press, London.
Fong, S. S., *et al.* (2003). Description and interpretation of adaptive evolution of Escherichia coli K-12 MG1655 by using a genome-scale in silico metabolic model. *J. Bacteriol.* **185,** 6400–6408.
Fong, S. S., *et al.* (2005). Parallel adaptive evolution cultures of Escherichia coli lead to convergent growth phenotypes with different gene expression states. *Genome Res.* **15,** 1365–1372.
Francke, C., *et al.* (2005). Reconstructing the metabolic network of a bacterium from its genome. *Trends Microbiol.* **13,** 550–558.
Gehlenborg, N., *et al.* (2010). Visualization of omics data for systems biology. *Nat. Methods* **7,** S56–S68.
Gianchandani, E. P., *et al.* (2010). The application of flux balance analysis in systems biology. *Wiley Interdiscip. Rev. Syst. Biol. Med.* **2,** 372–382.
Goffin, P., *et al.* (2010). Understanding the physiology of *Lactobacillus plantarum* at zero growth. *Mol. Syst. Biol.* **6,** 413.
Goldbeter, A. (1997). Biochemical Oscillations and Cellular Rhythms—The Molecular Bases of Periodic and Chaotic Behaviour. Cambridge University Press, Cambridge, United Kingdom.
Heinrich, R., and Schuster, S. (1996). The Regulation of Cellular Systems. Chapman & Hall, New York.
Henry, C. S., *et al.* (2010). High-throughput generation, optimization and analysis of genome-scale metabolic models. *Nat. Biotechnol.* **28,** 977–982.
Herrgard, M. J., *et al.* (2004). Reconstruction of microbial transcriptional regulatory networks. *Curr. Opin. Biotechnol.* **15,** 70–77.
Hoefnagel, M. H., *et al.* (2002). Metabolic engineering of lactic acid bacteria, the combined approach: Kinetic modelling, metabolic control and experimental analysis. *Microbiology* **148,** 1003–1013.
Hols, P., *et al.* (2005). New insights in the molecular biology and physiology of Streptococcus thermophilus revealed by comparative genomics. *FEMS Microbiol. Rev.* **29,** 435–463.

Hoppe, A., *et al.* (2007). Including metabolite concentrations into flux balance analysis: Thermodynamic realizability as a constraint on flux distributions in metabolic networks. *BMC Syst. Biol.* **1,** 23.

Hucka, M., *et al.* (2003). The systems biology markup language (SBML): A medium for representation and exchange of biochemical network models. *Bioinformatics* **19,** 524–531.

Hyduke, D. R., and Palsson, B. O. (2010). Towards genome-scale signalling-network reconstructions. *Nat. Rev. Genet.* **11,** 297–307.

Ibarra, R. U., *et al.* (2002). *Escherichia coli* K-12 undergoes adaptive evolution to achieve in silico predicted optimal growth. *Nature* **420,** 186–189.

Ideker, T., and Lauffenburger, D. (2003). Building with a scaffold: Emerging strategies for high- to low-level cellular modelling. *Trends Biotechnol.* **21,** 255–262.

Jamshidi, N., and Palsson, B. O. (2010). Mass action stoichiometric simulation models: Incorporating kinetics and regulation into stoichiometric models. *Biophys. J.* **98,** 175–185.

Karp, P. D., *et al.* (2010). Pathway Tools version 13.0: Integrated software for pathway/genome informatics and systems biology. *Brief. Bioinform.* **11,** 40–79.

Kell, D. B. (2004). Metabolomics and systems biology: Making sense of the soup. *Curr. Opin. Microbiol.* **7,** 296–307.

Keseler, I. M., *et al.* (2009). EcoCyc: A comprehensive view of Escherichia coli biology. *Nucleic Acids Res.* **37,** D464–D470.

Kharchenko, P., *et al.* (2004). Filling gaps in a metabolic network using expression information. *Bioinformatics* **20**(Suppl. 1), I178–I185.

Kim, C., and Paik, S. (2010). Gene-expression-based prognostic assays for breast cancer. *Nat. Rev. Clin. Oncol.* **7,** 340–347.

Kitano, H. (2002). Systems biology: A brief overview. *Science* **295,** 1662–1664.

Kitano, H. (2007). Towards a theory of biological robustness. *Mol. Syst. Biol.* **3,** 137.

Knijnenburg, T. A., *et al.* (2007). Exploiting combinatorial cultivation conditions to infer transcriptional regulation. *BMC Genomics* **8,** 25.

Kono, N., *et al.* (2009). Pathway projector: Web-based zoomable pathway browser using KEGG atlas and Google Maps API. *PLoS One* **4,** e7710.

Kuepfer, L., *et al.* (2005). Metabolic functions of duplicate genes in Saccharomyces cerevisiae. *Genome Res.* **15,** 1421–1430.

Kummel, A., *et al.* (2006a). Putative regulatory sites unraveled by network-embedded thermodynamic analysis of metabolome data. *Mol. Syst. Biol.* **2,** 0034.

Kummel, A., *et al.* (2006b). Systematic assignment of thermodynamic constraints in metabolic network models. *BMC Bioinformatics* **7,** 512.

Liao, J. C., *et al.* (2003). Network component analysis: Reconstruction of regulatory signals in biological systems. *Proc. Natl. Acad. Sci. USA* **100,** 15522–15527.

Liebermeister, W., and Klipp, E. (2005). Biochemical networks with uncertain parameters. *IEE Proc. Syst. Biol.* **152,** 97–107.

Liebermeister, W., and Klipp, E. (2006). Bringing metabolic networks to life: Convenience rate law and thermodynamic constraints. *Theor. Biol. Med. Model.* **3,** 41.

Liebermeister, W., *et al.* (2005). Biochemical network models simplified by balanced truncation. *FEBS J.* **272,** 4034–4043.

Lipton, P. (2005). Testing hypotheses: Prediction and prejudice. *Science* **307,** 219–221.

Liu, L., *et al.* (2010). Use of genome-scale metabolic models for understanding microbial physiology. *FEBS Lett.* **584,** 2556–2564.

Maertens, J., and Vanrolleghem, P. A. (2010). Modelling with a view to target identification in metabolic engineering: A critical evaluation of the available tools. *Biotechnol. Prog.* **26,** 313–331.

Mahadevan, R., and Schilling, C. H. (2003). The effects of alternate optimal solutions in constraint-based genome-scale metabolic models. *Metab. Eng.* **5,** 264–276.

Mashego, M. R., *et al.* (2007). Metabolome dynamic responses of Saccharomyces cerevisiae to simultaneous rapid perturbations in external electron acceptor and electron donor. *FEMS Yeast Res.* **7,** 48–66.

Notebaart, R. A., *et al.* (2006). Accelerating the reconstruction of genome-scale metabolic networks. *BMC Bioinformatics* **7,** 296.

Notebaart, R. A., *et al.* (2008). Co-regulation of metabolic genes is better explained by flux coupling than network distance. *PLoS Comput. Biol.* **4,** e26.

Notebaart, R. A., *et al.* (2009). Asymmetric relationships between proteins shape genome evolution. *Genome Biol.* **10,** R19.

Oberhardt, M. A., *et al.* (2009). Applications of genome-scale metabolic reconstructions. *Mol. Syst. Biol.* **5,** 320.

Oliveira, A. P., *et al.* (2005). Modelling *Lactococcus lactis* using a genome-scale flux model. *BMC Microbiol.* **5,** 39.

Orth, J. D., *et al.* (2010). What is flux balance analysis? *Nat. Biotechnol.* **28,** 245–248.

Pal, C., *et al.* (2006). Chance and necessity in the evolution of minimal metabolic networks. *Nature* **440,** 667–670.

Palsson, B. (2004). Two-dimensional annotation of genomes. *Nat. Biotechnol.* **22,** 1218–1219.

Papin, J. A., *et al.* (2004). Comparison of network-based pathway analysis methods. *Trends Biotechnol.* **22,** 400–405.

Papp, B., *et al.* (2004). Metabolic network analysis of the causes and evolution of enzyme dispensability in yeast. *Nature* **429,** 661–664.

Park, J. H., *et al.* (2010). Fed-batch culture of Escherichia coli for L-valine production based on in silico flux response analysis. *Biotechnol Bioeng.* **108,** 934–946.

Patil, K. R., and Nielsen, J. (2005). Uncovering transcriptional regulation of metabolism by using metabolic network topology. *Proc. Natl. Acad. Sci. USA* **102,** 2685–2689.

Pinchuk, G. E., *et al.* (2010). Constraint-based model of Shewanella oneidensis MR-1 metabolism: A tool for data analysis and hypothesis generation. *PLoS Comput. Biol.* **6,** e1000822.

Price, N. D., *et al.* (2004). Genome-scale models of microbial cells: Evaluating the consequences of constraints. *Nat. Rev. Microbiol.* **2,** 886–897.

Reed, J. L., *et al.* (2006). Towards multidimensional genome annotation. *Nat. Rev. Genet.* **7,** 130–141.

Resendis-Antonio, O. (2009). Filling kinetic gaps: Dynamic modelling of metabolism where detailed kinetic information is lacking. *PLoS One* **4,** e4967.

Rojas, I., *et al.* (2007). Storing and annotating of kinetic data. *In Silico Biol.* **7,** S37–S44.

Schellenberger, J., *et al.* (2010). BiGG: A Biochemical Genetic and Genomic knowledgebase of large scale metabolic reconstructions. *BMC Bioinformatics* **11,** 213.

Schuetz, R., *et al.* (2007). Systematic evaluation of objective functions for predicting intracellular fluxes in Escherichia coli. *Mol. Syst. Biol.* **3,** 119.

Schuster, S., *et al.* (2000). A general definition of metabolic pathways useful for systematic organization and analysis of complex metabolic networks. *Nat. Biotechnol.* **18,** 326–332.

Schuster, S., *et al.* (2007). Is maximization of molar yield in metabolic networks favoured by evolution? *J. Theor. Biol.* **252,** 497–504.

Segre, D., *et al.* (2002). Analysis of optimality in natural and perturbed metabolic networks. *Proc. Natl. Acad. Sci. USA* **99,** 15112–15117.

Steuer, R., *et al.* (2006). Structural kinetic modelling of metabolic networks. *Proc. Natl. Acad. Sci. USA* **103,** 11868–11873.

Stevens, M. J., *et al.* (2008). Improvement of Lactobacillus plantarum aerobic growth as directed by comprehensive transcriptome analysis. *Appl. Environ. Microbiol.* **74,** 4776–4778.

Tai, S. L., *et al.* (2005). Two-dimensional transcriptome analysis in chemostat cultures. Combinatorial effects of oxygen availability and macronutrient limitation in Saccharomyces cerevisiae. *J. Biol. Chem.* **280,** 437–447.

Taymaz-Nikerel, H., *et al.* (2010). Genome-derived minimal metabolic models for Escherichia coli MG1655 with estimated in vivo respiratory ATP stoichiometry. *Biotechnol. Bioeng.* **107,** 369–381.
Tempest, D. W., and Neijssel, O. M. (1984). The status of YATP and maintenance energy as biologically interpretable phenomena. *Annu. Rev. Microbiol.* **38,** 459–486.
Teusink, B., and Smid, E. J. (2006). Modelling strategies for the industrial exploitation of lactic acid bacteria. *Nat. Rev. Microbiol.* **4,** 46–56.
Teusink, B., *et al.* (2000). Can yeast glycolysis be understood in terms of in vitro kinetics of the constituent enzymes? Testing biochemistry. *Eur. J. Biochem.* **267,** 5313–5329.
Teusink, B., *et al.* (2005). In silico reconstruction of the metabolic pathways of Lactobacillus plantarum: Comparing predictions of nutrient requirements with those from growth experiments. *Appl. Environ. Microbiol.* **71,** 7253–7262.
Teusink, B., *et al.* (2006). Analysis of growth of *Lactobacillus plantarum* WCFS1 on a complex medium using a genome-scale metabolic model. *J. Biol. Chem.* **281,** 40041–40048.
Teusink, B., *et al.* (2009). Understanding the adaptive growth strategy of *L. plantarum* by *in silico* optimisation. *PLoS Comput. Biol.* **5,** e1000410.
Theobald, U., *et al.* (1993). In vivo analysis of glucose-induced fast changes in yeast adenine nucleotide pool applying a rapid sampling technique. *Anal. Biochem.* **214,** 31–37.
Thiele, I., and Palsson, B. O. (2010). A protocol for generating a high-quality genome-scale metabolic reconstruction. *Nat. Protoc.* **5,** 93–121.
Thiele, I., *et al.* (2009). Genome-scale reconstruction of Escherichia coli's transcriptional and translational machinery: A knowledge base, its mathematical formulation, and its functional characterization. *PLoS Comput. Biol.* **5,** e1000312.
Thomas, T. D., *et al.* (1979). Change from homo- to heterolactic fermentation by Streptococcus lactis resulting from glucose limitation in anaerobic chemostat cultures. *J. Bacteriol.* **138,** 109–117.
van der Werf, M. J. (2005). Towards replacing closed with open target selection strategies. *Trends Biotechnol.* **23,** 11–16.
van Dijken, J. P., *et al.* (1993). Kinetics of growth and sugar consumption in yeasts. *Antonie Van Leeuwenhoek* **63,** 343–352.
vanGulik, W. M., and Heijnen, J. J. (1995). A metabolic network stoichiometry analysis of microbial growth and product formation. *Biotechnol. Bioeng.* **48,** 681–698.
Vanrolleghem, P. A., and Heijnen, J. J. (1998). A structured approach for selection among candidate metabolic network models and estimation of unknown stoichiometric coefficients. *Biotechnol. Bioeng.* **58,** 133–138.
Vanrolleghem, P. A., *et al.* (1996). Validation of a metabolic network for Saccharomyces cerevisiae using mixed substrate studies. *Biotechnol. Prog.* **12,** 434–448.
Varma, A., and Palsson, B. O. (1994). Stoichiometric flux balance models quantitatively predict growth and metabolic by-product secretion in wild-type Escherichia coli W3110. *Appl. Environ. Microbiol.* **60,** 3724–3731.
Veening, J. W., *et al.* (2008). Bistability, epigenetics, and bet-hedging in bacteria. *Annu. Rev. Microbiol.* **62,** 193–210.
Vemuri, G. N., *et al.* (2006). Overflow metabolism in Escherichia coli during steady-state growth: Transcriptional regulation and effect of the redox ratio. *Appl. Environ. Microbiol.* **72,** 3653–3661.
Visser, D., *et al.* (2004). Analysis of in vivo kinetics of glycolysis in aerobic Saccharomyces cerevisiae by application of glucose and ethanol pulses. *Biotechnol. Bioeng.* **88,** 157–167.
Wang, L., and Hatzimanikatis, V. (2006). Metabolic engineering under uncertainty. I: Framework development. *Metab. Eng.* **8,** 133–141.

CHAPTER TWENTY-FIVE

Supply–Demand Analysis: A Framework for Exploring the Regulatory Design of Metabolism

Jan-Hendrik S. Hofmeyr*,† *and* Johann M. Rohwer*

Contents

Abstract

The living cell can be thought of as a collection of linked chemical factories, a molecular economy in which the principles of supply and demand obtain. Supply–demand analysis is a framework for exploring and gaining an understanding of metabolic regulation, both theoretically and experimentally, where regulatory performance is measured in terms of flux control and homeostatic maintenance of metabolite concentrations. It is based on a metabolic control analysis of a supply–demand system in steady state in which the degree of flux and concentration control by the supply and demand blocks is related to their local properties, which are quantified as the elasticities of supply and demand. These elasticities can be visualized as the slopes of the log–log rate characteristics of supply and demand. Rate characteristics not only provide insight about

* Department of Biochemistry, University of Stellenbosch, Private Bag X1, Stellenbosch, South Africa
† Centre for Studies in Complexity, University of Stellenbosch, Private Bag X1, Stellenbosch, South Africa

Methods in Enzymology, Volume 500
ISSN 0076-6879, DOI: 10.1016/B978-0-12-385118-5.00025-6

system behavior around the steady state but can also be expanded to provide a view of the behavior of the system over a wide range of concentrations of the metabolic intermediate that links the supply and the demand. The theoretical and experimental results of supply–demand analysis paint a picture of the regulatory design of metabolic systems that differs radically from what can be called the classical view of metabolic regulation, which generally explains the role of regulatory mechanisms only in terms of the supply, completely ignoring the demand. Supply–demand analysis has recently been generalized into a computational tool that can be used to study the regulatory behavior of kinetic models of metabolic systems up to genome-scale.

1. Introduction

Supply–demand analysis (Hofmeyr, 2008; Hofmeyr and Cornish-Bowden, 2000) is a framework for exploring and gaining an understanding of metabolic regulation, both theoretically and experimentally, and is the culmination of a series of studies initiated in the early 1990s (Hofmeyr, 1995; Hofmeyr and Cornish-Bowden, 1991, 1996; Hofmeyr *et al.*, 1993). Rohwer and Hofmeyr (2008) recently generalized supply–demand analysis into a computational tool for analyzing regulation in kinetic models of cellular systems of any complexity.

The concept of metabolic regulation is tightly linked to metabolic function, such as, for example, the control of flux, the homeostatic maintenance of metabolite concentrations, the control of transition time between steady states, or the dynamic and structural stability of the steady state. Therefore, to say that a system is regulated is to mean that its intrinsic properties have been molded by evolution to fulfill specific functions (Hofmeyr, 1995; Hofmeyr and Cornish-Bowden, 1991). Because mass action is the intrinsic driving force for self-organization of reaction networks, we broadly define metabolic regulation as *the alteration of reaction properties to augment or counteract the mass-action trend in a network of reactions* (Hofmeyr, 1995; Reich and Sel'kov, 1981; Rohwer and Hofmeyr, 2010). In supply–demand analysis, we measure regulatory performance in terms of flux control and homeostatic maintenance of metabolite concentrations.

Whereas the classical view of metabolic regulation usually considers the biosynthetic pathway in isolation without considering what happens to the products—a situation that we have likened to an economic analysis of a factory that ignores the consumption of the products of the factory (Hofmeyr and Cornish-Bowden, 2000)—supply–demand analysis takes the demand into consideration. What results from this has led Oliver (2002) to remark that it "could mean that biologists in the twenty-first century need a rethink of their view of cellular economy that is every bit as radical as that initiated for political economy by John Stuart Mill and William Stanley Jevons in the nineteenth century."

One of the most powerful features of supply–demand analysis is a tool, the graph of combined rate characteristics, with which to visualize the regulatory design of a metabolic system. In the complex system of interacting processes that is the living cell, the properties of the agents that facilitate these processes, such as enzymes, transporters, and receptors, must be tuned to each other if the system is to behave harmoniously. To our knowledge, the graph of combined rate characteristics is the only way of visually combining all these properties in one picture.

2. The Functional Organization of Metabolism

The living cell can be thought of as a collection of linked chemical factories, a molecular economy in which the principles of supply and demand obtain. Biochemical textbooks and the ubiquitous wall charts depicting metabolic pathways tend to obscure this functional organization of the metabolism of all living organisms into blocks that produce and consume metabolic products (Fig. 25.1). These metabolic blocks communicate with each other through intermediates that couple the blocks by either the linear or the cyclic linkage types depicted in Fig. 25.2.

In the cyclic system, the sum of the concentrations of X and Y is constant; they form a moiety-conserving cycle (Hofmeyr *et al.*, 1986). X and Y can therefore vary only within this constraint, which, in turn, implies that the cycle can be reduced to a linear system with the concentration ratio y/x (or its inverse) as the coupling variable. If, for example, Y is ATP and X is ADP, ATP can be regarded as the phosphate-charged form of the ADP moiety, so that y/x is the ratio of charged over uncharged moiety. Hofmeyr (1997) has shown that even in more complex cycles, such as the ATP–ADP–AMP system in the presence of an active adenylate kinase, this principle still holds: here, the charged over uncharged ratio is ([ATP] + 0.5[ADP])/([AMP] + 0.5[ADP]).

3. Quantitative Analysis of Supply–Demand Systems

The question as to how the properties of the supply and demand blocks determine the behavior and control of the steady-state flux and concentration of the intermediate (p or y/x in Fig. 25.2) can be answered within the framework of metabolic control analysis (Heinrich and Rapoport, 1974; Kacser *et al.*, 1995). Here, we use combined log–log rate characteristics (Hofmeyr, 1995) as a powerful visual aid to explain the analysis.

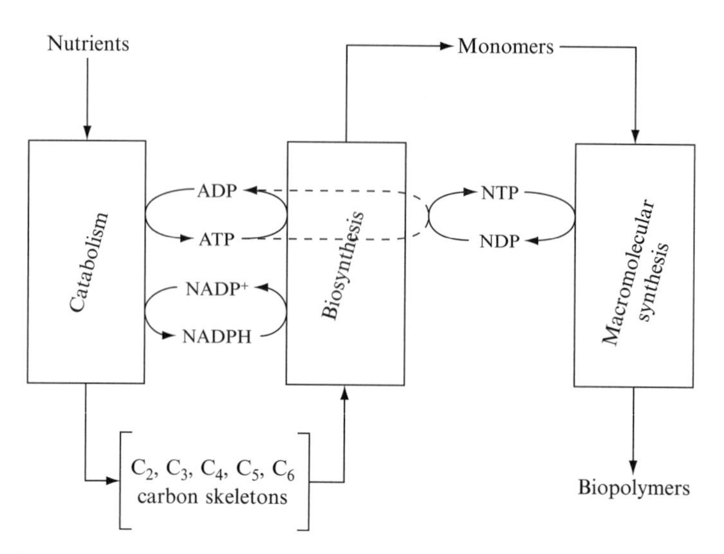

Figure 25.1 The functional organization of intermediary metabolism. The primary energy sources are degraded by catabolic pathways to form ATP, reducing equivalents (NADPH), and C_3—C_6 metabolic intermediates (e.g., sugar phosphates, activated CoA intermediates, PEP, pyruvate, oxaloacetate, and 2-oxoglutarate) that act as carbon skeletons for biosynthetic (anabolic) processes that produce monomers for the synthesis of biopolymers (proteins from amino acids, nucleic acids from nucleotides, lipids from fatty acids) and higher-order cellular structures; these processes also require an input of free energy (NTP, nucleotide triphosphates).

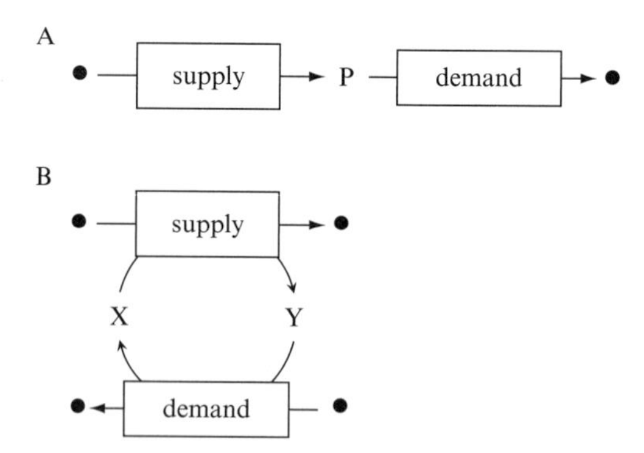

Figure 25.2 Two ways in which metabolic supply and demand reaction blocks are linked. (A) linear; (B) cyclic. In the linear system, the demand takes the linking metabolite P on to form a further product (here depicted by a solid circle). In the cyclic system, the demand reverses the action of the supply as far as the linking metabolite is concerned. For example, if the supply oxidizes X, the demand reduces it, or if the supply phosphorylates X to form Y, the demand dephosphorylates Y back to X.

Figure 25.3 shows how the steady state of a supply–demand system is formed and how the distribution of flux and concentration control depends on the properties of the supply and demand blocks. On the graph, the natural logarithms of the supply and demand rates are plotted as a function of the natural logarithm of the concentration variable that links them. If the supply and demand were catalyzed by single enzymes, these curves would represent, for example, the familiar Michaelis–Menten or Hill responses of a rate with respect to a product or a substrate. In general, however, the supply and demand are reaction blocks, so that the rate curves actually represent the variation in the local steady-state fluxes of the isolated supply and demand blocks as they respond to variation in the concentration of P. The use of logarithmic rather than linear scales has a number of advantages (Hofmeyr, 1995), the most important being that it allows direct comparison of the magnitude of steady-state responses to perturbations at different positions of the rate and concentration scale (equal distances on different parts of a logarithmic axis represent equal percentage change in the variable represented on that axis). Another advantage is that the curves retain their form when they vary with a parameter that is a multiplier of the rate equation, such as enzyme concentration.

The intersection of the supply and demand rate characteristic represents the steady state, which is characterized by a flux, J, and concentration of P, $\bar{p}$. From the graph, it should be clear that the response in the steady state to small perturbations in the activities of supply or demand depends completely on the *elasticity coefficients*, that is, the slopes of the tangents to the double logarithmic rate characteristics at the steady-state point.

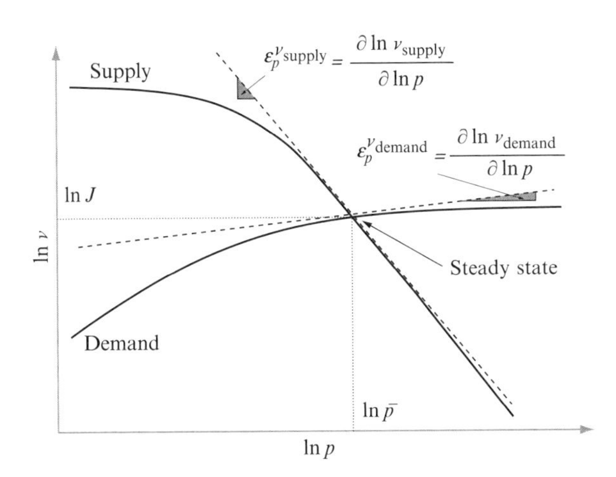

Figure 25.3 The rate characteristics of a supply–demand system plotted in double logarithmic space, showing the steady state where the two rate characteristics intersect; J is the steady-state flux, and $\bar{p}$ is the steady-state concentration of P. The slopes of the tangents to the rate characteristics at the steady-state point are the so-called elasticities of the supply and demand rates with respect to P (Hofmeyr and Cornish-Bowden, 2000). Reproduced with permission from Hofmeyr and Cornish-Bowden (2000).

The control of the supply and demand blocks over the flux and concentration of P is quantified in terms of the response of these steady-state variables to small perturbations, $d \ln v_{\text{supply}}$ or $d \ln v_{\text{demand}}$ in the activities of the supply or demand blocks. The degrees to which supply and demand control J and $\bar{p}$ are given by the *flux–control coefficients*:

$$C^J_{\text{supply}} = \frac{d \ln J}{d \ln v_{\text{supply}}}; \quad C^J_{\text{demand}} = \frac{d \ln J}{d \ln v_{\text{demand}}} \tag{25.1}$$

and the *concentration-control coefficients*

$$C^p_{\text{supply}} = \frac{d \ln p}{d \ln v_{\text{supply}}}; \quad C^p_{\text{demand}} = \frac{-d \ln p}{d \ln v_{\text{demand}}} \tag{25.2}$$

Hofmeyr and Cornish-Bowden (2000) provide a graphical thought experiment that illustrates these definitions.

Flux-control and concentration-control coefficients obey the following summation relationships:

$$C^J_{\text{supply}} + C^J_{\text{demand}} = 1 \tag{25.3}$$

$$C^p_{\text{supply}} + C^p_{\text{demand}} = 0 \tag{25.4}$$

These are specific cases of the so-called *summation theorems* of control analysis (Kacser *et al.*, 1995).

Further, using the definitions of the elasticities of supply and demand given in Fig. 25.3, the *connectivity theorems* (Kacser *et al.*, 1995) can also be derived:

$$C^J_{\text{supply}}\varepsilon^{v_{\text{supply}}}_p + C^J_{\text{demand}}\varepsilon^{v_{\text{demand}}}_p = 0 \tag{25.5}$$

$$C^p_{\text{supply}}\varepsilon^{v_{\text{supply}}}_p + C^p_{\text{demand}}\varepsilon^{v_{\text{demand}}}_p = -1 \tag{25.6}$$

The summation and connectivity theorems provide enough information to express the control coefficients in terms of elasticities of supply and demand (Hofmeyr and Cornish-Bowden, 1991). The flux-control coefficients are

$$C^J_{\text{supply}} = \frac{\varepsilon^{v_{\text{demand}}}_p}{\varepsilon^{v_{\text{demand}}}_p - \varepsilon^{v_{\text{supply}}}_p} \tag{25.7}$$

and

$$C^J_{\text{demand}} = \frac{-\varepsilon_p^{v_{\text{supply}}}}{\varepsilon_p^{v_{\text{demand}}} - \varepsilon_p^{v_{\text{supply}}}} \tag{25.8}$$

and the concentration-control coefficients:

$$C^p_{\text{supply}} = -C^p_{\text{demand}} = \frac{1}{\varepsilon_p^{v_{\text{demand}}} - \varepsilon_p^{v_{\text{supply}}}} \tag{25.9}$$

Note that $\varepsilon_p^{v_{\text{supply}}}$ is typically a negative quantity, that is, product inhibits supply through enzyme inhibition and thermodynamic back-pressure. The ratio of elasticities determines the distribution of flux-control between supply and demand (if $|\varepsilon_p^{v_{\text{supply}}}/\varepsilon_p^{v_{\text{demand}}}| > 1$, the demand has more control over the flux than the supply; if $|\varepsilon_{\bar{p}}^{v_{\text{supply}}}/\varepsilon_{\bar{p}}^{v_{\text{demand}}}| < 1$, the demand has less control over the flux than the supply). With regard to $\bar{p}$, it is not the distribution of $\bar{p}$-control that is of interest (C^p_{supply} always being equal to $-C^p_{\text{demand}}$ no matter what the values of the elasticities), but what determines the magnitude of the variation in $\bar{p}$ (and, therefore, its homeostatic maintenance): the larger $\varepsilon_p^{v_{\text{demand}}} - \varepsilon_p^{v_{\text{supply}}}$, the smaller the absolute values of both C^p_{supply} and C^p_{demand}, and the better the homeostatic regulation of $\bar{p}$. This algebraic analysis is clearly illustrated by the different configuration of rate characteristics around the steady state shown in Fig. 25.4.

Figure 25.4A shows a situation where the absolute values of the elasticities of supply and demand (the slopes of the lines) are equal, so that the functions of flux and concentration control are equally distributed: the same percentage change in the activity of either supply or demand causes the same change in the flux ($C^J_{\text{supply}} = C^J_{\text{demand}} = 0.5$). The magnitude of the variation in $\bar{p}$ is determined to the same degree by the supply and demand elasticities. We call this a *functionally undifferentiated system*. Similarly, Fig. 25.4B shows a functionally undifferentiated system in which the elasticities of supply and demand are much smaller (shallower slopes). This clearly illustrates the principle that the sum of supply and demand elasticities determines the degree of concentration control of $\bar{p}$: the larger the sum, the smaller the values of the control coefficients.

In Fig. 25.4C and D, the elasticity of demand is decreased considerably (the demand becomes more saturated with P). Comparing equal percentage changes in demand (in C) and supply (in D), it is clear that most of the flux control has been transferred to the demand. However, because $\varepsilon_p^{v_{\text{demand}}} \ll |\varepsilon_p^{v_{\text{supply}}}|$, the magnitude of the variation in $\bar{p}$ is now largely determined by the supply elasticity. This phenomenon is even more pronounced in Fig. 25.4E and F where the supply slope is very steep. Therefore, the steeper the slope of the supply rate characteristic, the narrower the band of variation in $\bar{p}$ and, therefore, the better the homeostatic maintenance of $\bar{p}$. The opposite would

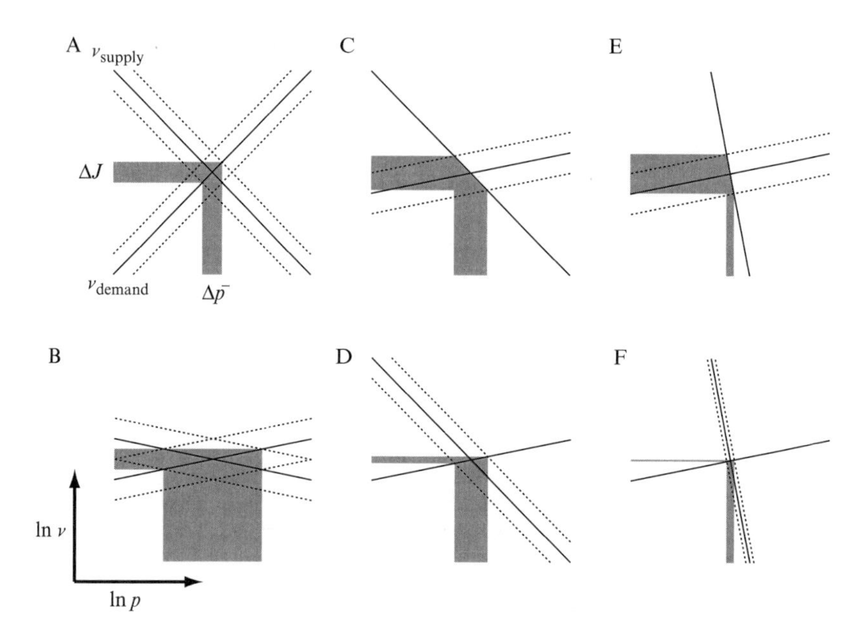

Figure 25.4 The effect on the steady state of varying the activity of the demand (C, E) or supply (D, F) or both (A, B). The slope of each line is an elasticity of either supply or demand at the steady state. The dotted lines show a set percentage increase or decrease in activity. The shaded regions show the magnitude of the response in the steady-state flux (horizontal) and concentration of P (vertical). See text for explanation.

obtain if the supply elasticity were small, whereas the demand elasticity were large: flux control would shift to the supply, while the elasticity of demand would determine the magnitude of variation in $\bar{p}$.

Supply–demand analysis therefore shows that the functions of flux and concentration control are mutually exclusive in the sense that if one block controls the flux, it loses any influence over the magnitude of variation in the concentration of the linking intermediate $\bar{p}$: this becomes the sole function of the other block. This finding has profound consequences for any view of metabolic regulation.

The above analysis has only considered the response of the steady state to small variations in the activity of supply or demand without considering either the form of the full rate characteristics or the position of the steady state in relation to equilibrium. Rate characteristics can also be used to obtain an overall view of the limits within which the system can fulfill its functions.

A supply pathway must be able to meet increasing demand for its product at least up to some limit and to cope with low demand in such a way that its product and intermediate metabolite concentrations do not tend toward their equilibrium concentrations (most biosynthetic pathways have huge equilibrium constants so that near-equilibrium conditions would cause fatally high accumulation of supply pathway intermediates and product (Atkinson, 1977). Textbook wisdom has it that allosteric feedback inhibition of supply by its product is responsible for satisfying demand, while it has little to say about low demand. Supply–demand analysis teaches us otherwise.

Figure 25.5 shows a set of supply–demand rate characteristics that span the full range of p to its equilibrium value (assuming that the substrate for the supply pathway is buffered and therefore constant). For the supply to be able to meet a specific range of variation in demand activity, it cannot have any flux control in that range. Focusing for the moment on the supply curve, it is clear that only in the shaded band between steady states 2 and 3 will the supply be able to meet the variation in demand while keeping $\bar{p}$ reasonably constant. When demand becomes higher than 2, it loses control over the flux (steady state 1) with a concomitant sharp decrease in $\bar{p}$. An increase in the maximal activity of the supply (the plateau at 1) would extend the range in which the supply can meet the demand. However, it is also clear that the presence of allosteric feedback inhibition is not a prerequisite for flux control by demand: in the shaded band on the right, demand also controls the flux in the absence of allosteric feedback (the dashed supply characteristic), and the supply is equally effective in keeping $\bar{p}$ homeostatic. The dramatic difference between the two situations is in the concentration where P is homeostatically maintained: without feedback inhibition, it can only be near equilibrium (with all the accompanying disadvantages),

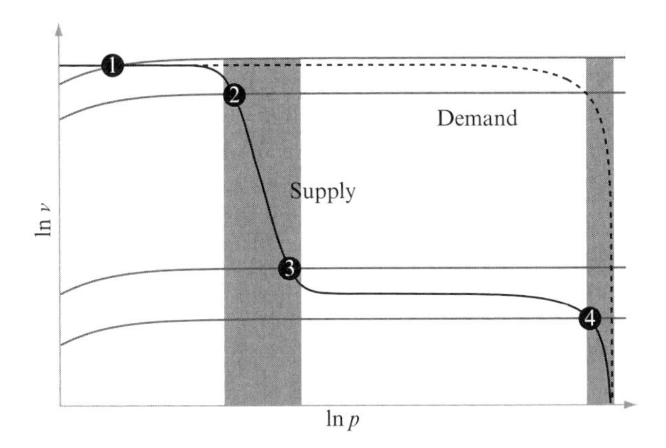

Figure 25.5 The steady-state behavior of a supply–demand system with (solid) and without (dashed) inhibition of supply by its product P. The gray lines represent different demand activities. The four marked steady states are discussed in the text. The rate characteristics were generated with PySCeS (Olivier *et al.*, 2005) for the supply–demand system described in Hofmeyr and Cornish-Bowden (1991) using the reversible Hill (Hofmeyr and Cornish-Bowden, 1997) and reversible Michaelis–Menten rate equations with realistic parameter values. Reproduced with permission from Hofmeyr and Cornish-Bowden (2000).

whereas with feedback inhibition, it can be maintained orders of magnitude away from equilibrium (at a concentration around the $p_{0.5}$ of the allosteric enzyme). Clearly, therefore, when demand controls flux, the functional role of feedback inhibition is homeostatic maintenance of $\bar{p}$ at a concentration far from equilibrium.

In general, each elasticity coefficient is the sum of a mass-action term that depends only on Γ/K_{eq} and a binding term that is determined by the binding properties of the enzyme. The mass-action term in the supply elasticity approaches 0 in conditions far from equilibrium and $-\infty$ near equilibrium, where it completely swamps the binding term, which typically varies between 0 and the Hill coefficient (Hofmeyr, 1995; Rohwer and Hofmeyr, 2010). Kinetic effects such as allosteric feedback inhibition can therefore only play a regulatory role far from equilibrium where the mass-action term is negligible. This is also shown by the solid curve in Fig. 25.5: there is a lower limit (around 3) to the range in which $\bar{p}$ can be kinetically regulated; below this limit, $\bar{p}$ jumps to the region where the mass-action term dominates the supply elasticity.

Hofmeyr (2008) considered a more complex version of the simple supply–demand system in Fig. 25.5, to which was added a catabolic demand for P and the induction of the synthesis of the first supply enzyme at low concentrations of P, with P acting as a corepressor. This allowed, on the one hand, for an increase in supply capacity at low P, thereby extending the range in which the supply could match variation in demand activity. However, the inclusion of a catabolic demand provided an overflow valve into which P could be channeled once the demand fell too low, thus preventing P to jump to near-equilibrium concentrations. However, for the system to function in this way, the parameters of all the various regulatory mechanisms had been carefully matched. First, the strength of binding of P to the repressor protein had to be matched to the strength of allosteric binding of P to the first supply enzyme in order to ensure a smooth transition in the supply rate characteristic from the regime of allosteric regulation to that of enzyme expression. Second, the binding properties of the catabolic demand had to be chosen so that it did not kick in at too low or too high a concentration of P; positive cooperative binding of P to the catabolic demand also increased the regulatory effectivity. This analysis showed in a particularly clear way the power of combined rate characteristics to visualize and understand the regulation design of metabolism in the way that, to our knowledge, no other method provides.

4. Generalized Supply–Demand Analysis

Ordinary supply–demand analysis of a kinetic model can be easily achieved *in silico* by making the intermediate around which the rate characteristic is to be constructed a fixed (clamped) species of the model, thus turning it into a model parameter. This parameter is then varied over a wide range through a parameter scan. An implicit assumption of this approach is that the system has been or can readily be partitioned into supply and demand; however, when faced with the complexity of cellular pathways or of large models of such pathways, the choice of intermediate around which to perform the supply–demand analysis is often far from obvious. This has hampered the application of supply–demand analysis to large kinetic models of cellular pathways.

To address this shortcoming, we have generalized supply–demand analysis in such a way that it can easily be performed on a kinetic models of any cellular system, large or small, without requiring prior knowledge of its regulatory structure (Rohwer and Hofmeyr, 2008). Generalized supply–demand analysis works in the following way: *each* of the variable intermediates is *clamped in turn* and thus made into a parameter of the system. Its concentration is then varied above and below the reference steady-state value in the original system through a parameter scan, and the fluxes through the supply and demand reactions that are directly connected to the intermediate are plotted on a log–log rate characteristic. Every flux that directly produces the intermediate is a separate supply flux, and likewise, each flux that directly consumes it is a separate demand flux. There will thus be as many rate characteristics as there are reactions that produce or consume the intermediate. It should be emphasized that this procedure is valid for arbitrary models and does not presuppose a subdivision of the system into supply and demand blocks.

Generalized supply–demand analysis yields as many combined rate characteristic graphs as there are variable species in the system. As will be shown below, the following important features about the regulation of the system can be identified from the shapes of the curves and associated elasticities and response coefficients:

1. Potential sites of regulation;
2. Regulatory metabolites;
3. The quantitative relative contribution of different routes of interaction from an intermediate to a supply or demand block;
4. Sites of functional differentiation where one of the supply or demand blocks predominantly controls the flux, and the other determines the degree of homeostatic buffering of the intermediate.

We exemplify generalized supply–demand analysis with a model of a linear five-enzyme pathway containing a feedback loop. The simulations use two variants of a kinetic model of the linear pathway in Fig. 25.6; detailed model descriptions and computational methods are provided in Rohwer and Hofmeyr (2008).

Model I This is the base-line, undifferentiated version in which all five enzymes have identical kinetic parameters and are modeled with reversible

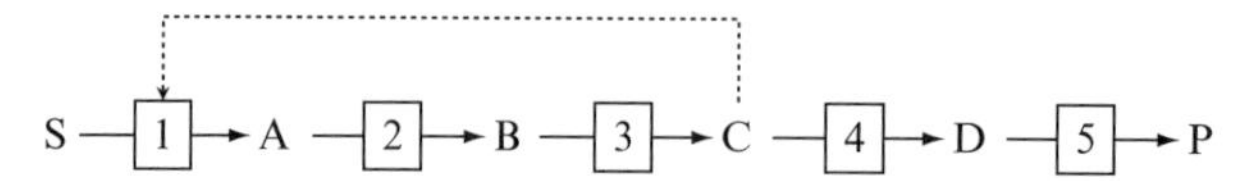

Figure 25.6 A five-enzyme linear pathway converting substrate S to product P. In one of the models, the first enzyme is allosterically inhibited by intermediate C (see main text). Reproduced with permission from Rohwer and Hofmeyr (2008).

Michaelis–Menten kinetics (with the exception of enzyme 5, which is modeled with irreversible Michaelis–Menten kinetics). There is no allosteric feedback from C to enzyme 1.

Model II In this model, enzyme 1 is inhibited allosterically by C and is modeled with reversible Hill kinetics (Hofmeyr and Cornish-Bowden, 1997). The limiting rates of enzymes 2 and 3 have been increased so that they are close to equilibrium. Enzymes 4 and 5 together have almost complete control over the flux through the pathway.

As explained above, a generalized supply–demand analysis is performed by clamping each variable species of the model, in turn, and varying its concentration to generate the supply and demand rate characteristics. This yields graphs such as in Fig. 25.7, which shows the generalized supply–demand analysis around metabolite B in model I. To facilitate the interpretation of such graphs, this specific case is discussed in detail. The intersection of the log–log rate characteristics of supply and demand marks the steady-state point. The supply rate characteristic is drawn in light gray and the demand rate characteristic in medium gray. The slopes of the tangents to the rate characteristics (indicated by dashed lines on the graph) equal the flux–response coefficients (see e.g., Kacser *et al.*, 1995) of supply and demand toward B (in Fig. 25.7, J_{12} signifies the flux through the supply block and J_{345} that through the demand block). These response coefficients quantify how sensitively the supply and demand fluxes respond toward changes in b and are equivalent to "block-elasticities" (Fell and Sauro, 1985) or coresponse coefficients (Hofmeyr and Cornish-Bowden, 1996; Hofmeyr *et al.*, 1993) in the complete system where B is not clamped.

Supply–demand analysis as described in Section 3 assumes that the only communication between supply and demand is via the linking intermediate. In this situation, the supply–demand block control coefficients of the complete pathway can be directly calculated from the supply and demand block elasticities (see Eqs. (25.7)–(25.9)), and the distribution of flux control is determined by the ratio of the block elasticities, while the magnitude of concentration control is determined by the difference $\varepsilon_b^{v_{345}} - \varepsilon_b^{v_{12}}$. Figure 25.7 also shows graphically the elasticities of the enzymes that produce or consume B. In this case, B is a product of v_2 and a substrate for v_3, so Fig. 25.7 shows $\varepsilon_b^{v_2}$ (thick light gray line) and $\varepsilon_b^{v_3}$ (thick medium gray line).

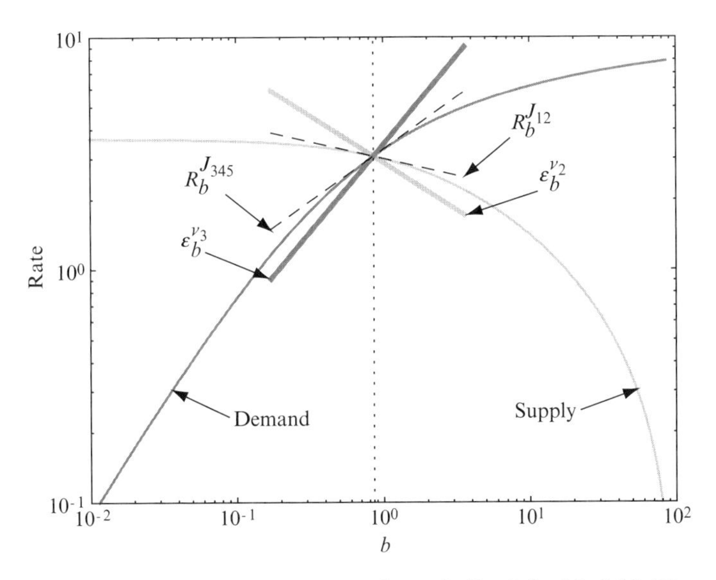

Figure 25.7 Supply–demand analysis around metabolite B for Model I. The concentration of B was clamped and varied to generate the supply and demand rate characteristics, as described in the text. The steady-state concentration is indicated by a vertical dotted line. The rate characteristics, response coefficients (tangents to the rate characteristics at the steady-state point), and elasticities of the supply and demand enzymes directly connected to B are labeled on the graph. Reproduced with permission from Rohwer and Hofmeyr (2008).

The crux of generalized supply–demand analysis now lies in the comparison of the values of the response coefficients with the elasticities of the enzymes that are directly connected to the clamped metabolite. In Fig. 25.7, these values differ, that is, $R_b^{J_{12}} \neq \varepsilon_b^{v_2}$ and $R_b^{J_{345}} \neq \varepsilon_b^{v_3}$. In other cases, they will be seen to agree. However, before comparing them in detail, first we have to present the generalized supply–demand analysis of all metabolites for both models.

The graphs in Fig. 25.8 present the results of the generalized supply–demand analysis on models I and II. To avoid clutter, the graphs are not annotated but they follow the same convention as Fig. 25.7. The only additional piece of information required is that of an allosteric modifier elasticity ($\varepsilon_c^{v_1}$ in Fig. 25.8B with c clamped, as only model II has the feedback loop). This is drawn in a dark gray line to set it apart from the supply and demand elasticities.

For larger models where a particular metabolite may be connected to a number of different supply and demand fluxes and the overall graph is too cluttered, the individual rate characteristics for each flux and their associated elasticity and response coefficients can be drawn on separate graphs. This choice depends on the particular model analyzed, but ultimately, it is only a

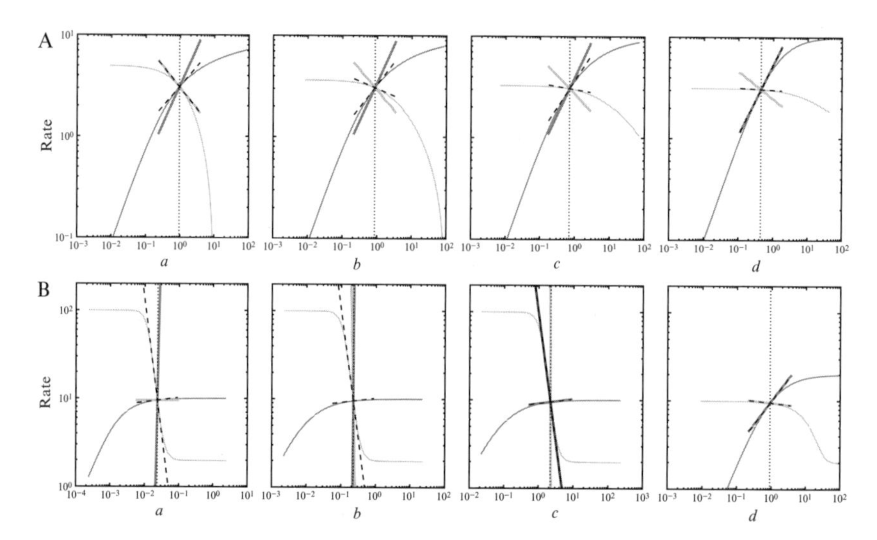

Figure 25.8 Generalized supply–demand analysis of the system depicted in Fig. 25.6. The concentrations *a*–*d* were clamped, in turn, and varied to generate the supply and demand rate characteristics, as described in the text. The supply rate characteristic is drawn in light gray, that for the demand in medium gray. The steady-state concentration of the clamped metabolite is indicated by a vertical dotted line. The response coefficients of the supply and demand blocks are indicated by black dashed lines. The elasticities of the supply and demand enzymes for the clamped intermediate they are directly connected to are indicated by thick lines of the same color as the rate characteristic. Model variants: (A) model I, (B) model II (see main text). In (B), the allosteric elasticity $\varepsilon_c^{v_1}$ is indicated by a dark gray thick line. Adapted from Rohwer and Hofmeyr (2008) with permission.

question of visualizing the data and in no way changes the principle of the analysis.

The graphs in Fig. 25.8 contain a wealth of information. As shown in Rohwer and Hofmeyr (2008), they can be interpreted on four levels, that is, differences in the rate characteristic shapes as one proceeds from one metabolite to the next in the pathway, comparison of elasticity and response slopes, identification of points of functional differentiation and homeostasis, and finally, refined analysis through partial response coefficients.

4.1. Differences in rate characteristic shapes

The first assessment criterion of generalized supply–demand analysis merely looks at the general shapes of the supply and demand rate characteristics and is not yet concerned with elasticities and response coefficients. In model I (Fig. 25.8A), all enzymes have identical kinetics and the overall shapes of the rate characteristics are similar for metabolites *A*–*D*. In model II (Fig. 25.8B), however, the pattern for D is different from those for A–C (which are still similar). This means that the kinetic properties of enzyme 4 are such that site of regulation has been introduced into the system. In this specific case, the reason is that enzyme 4 has been made insensitive to changes in the concentration of C ($\varepsilon_c^{v_4} \approx 0$). In general, such zero elasticities, whether toward substrate or product, induce a change in the rate characteristic shape because they shift the flux control to demand or supply, respectively. Overall, changes in the rate characteristic shapes thus pin-point potential sites of regulation.

4.2. Comparison of elasticities and response coefficients

Generalized supply–demand analysis can be extended to a second level by comparing the values of the elasticities and flux–response coefficients at the steady-state point for each metabolite. From the partitioned response property of control analysis (Kacser *et al.*, 1995),

$$R_p^J = \varepsilon_p^{v_i} \times C_{v_i}^J \tag{25.10}$$

it follows that $R_p^J = \varepsilon_p^{v_i}$ if $C_{v_i}^J = 1$. This means that the enzyme on which the intermediate acts directly must have full control over its own flux. Figure 25.8 shows that in general, response and elasticity coefficients differ. There are, however, a few notable exceptions. The first of these is the trivial case of the first and last metabolites in the chain (A and D): response coefficients and elasticities will generally agree because the supply block for A and demand block for D each consists only of a single enzyme. (The exception of $\varepsilon_a^{v_1} \neq R_a^{J_1}$ in Fig. 25.8B has to do with the feedback loop and is further discussed in Rohwer and Hofmeyr (2008).)

Aside from the trivial case, any agreement between elasticity and response coefficient points to a site of regulation. Eq. (25.10) shows that the response coefficient can equal the elasticity either if the control coefficient is one (as discussed above), or if the elasticity is zero (which effectively makes the value of the control coefficient irrelevant). The first case obtains, for example, in Fig. 25.8B with c clamped, where $\varepsilon_c^{v_1} = R_c^{J_{123}}$ (feedback loop with $C_{v_1}^{J_{123}} = 1$). Here, C can be classified as a "regulatory metabolite" with respect to its supply block because the flux response of this supply toward the clamped metabolite concentration is exactly the same as the activity response (i.e., elasticity) of the enzyme directly affected by the clamped metabolite. The flux-control coefficient of one causes the flux response to be transmitted fully through the block.

The second case (zero elasticity) obtains, for example, in Fig. 25.8B, where $\varepsilon_c^{v_4} = R_c^{J_{45}} \approx 0$. Such a zero elasticity confers flux control (in the complete system) to that particular block and results in functional differentiation of the system, which is further discussed below.

4.3. Functional differentiation and homeostasis

Functional differentiation has been discussed in Section 3 (see also Fig. 25.4) and is characterized by flux- and concentration control being functions of different blocks. Complete flux control by a supply or demand block (over the whole pathway) can easily be identified by a zero response coefficient (i.e., block elasticity) of that block toward the intermediate (e.g., $R_c^{J_{45}}$ in Fig. 25.8B). The response coefficient of the other block ($R_c^{J_{123}}$) will then determine the degree of homeostasis in the intermediate: the larger its numerical value, the better the homeostatic buffering. In model II (Fig. 25.8B), the properties of the feedback elasticity $\varepsilon_c^{v_1}$, which equals $R_c^{J_{123}}$ here, set the steady-state concentration of C and determine its degree of homeostatic buffering. In this sense, the steady-state concentration of C can be regarded as "regulated." Why C can be considered a "regulatory" metabolite when considering the supply block in isolation has been discussed above.

4.4. Multiple routes of interaction

When two or more direct routes of interaction exist from a clamped metabolite to a particular supply or demand block, generalized supply–demand analysis can be further refined by dissecting the response coefficient into partial response coefficients. An example is the generalized supply–demand analysis around metabolite C in Fig. 25.6, where C can affect both enzymes 1 and 3 directly (the former through allosteric inhibition, the latter through product inhibition). By calculating the total response coefficient as the sum of the two partial response coefficients referring to each of these routes of interaction, their individual contribution to the total response coefficient can be quantified and depicted graphically. This will not be further discussed here, and the reader is referred to Rohwer and Hofmeyr (2008).

4.5. Requirements for and limitations of the approach

Generalized supply–demand analysis, by virtue of being a computational method, requires a fully parameterized kinetic model. This is of course a limitation because such models do not exist for all pathways; however, their number and size are increasing steadily as any regular inspection of the JWS Online (Olivier and Snoep, 2004) or BioModels (le Novère *et al.*, 2006) databases will show. The application of the approach is thus limited by the size of available models, but this is not a limitation of the analysis *per se*. Indeed, efforts are under way toward building a genome-scale model of cellular metabolism (Smallbone *et al.*, 2010), and there is no reason in principle why generalized supply–demand analysis cannot be applied to such a model.

It could be argued naively that if a kinetic model of a pathway is available, then we know all the relevant regulatory mechanisms, so why is generalized supply–demand analysis necessary and what new insight can we gain from it? However, by way of example, the mere presence of a feedback loop does not mean that it is always active, and under some conditions, (such as very high demand in the pathway discussed by Hofmeyr and Cornish-Bowden, 2000) the regulation may actually follow a different route. Generalized supply–demand analysis thus identifies active routes of regulation under a *particular set of conditions*; these are dynamic properties of the model and depend on its particular state, they cannot be inferred from model structure alone. In addition, as the size of kinetic models increases (Smallbone *et al.*, 2010), their level of complexity approaches that of the systems they model. Tools are thus required for interrogating and analyzing such models, and generalized supply–demand analysis provides a first point of entry for such an analysis.

5. Experimental Applications of Supply–Demand Analysis

This section deals with methods for applying supply–demand analysis to experimental systems. A number of examples from the literature are discussed.

5.1. Double modulation

The classical method for performing a supply–demand analysis is based on the double-modulation method of metabolic control analysis, originally introduced by Kacser and Burns (1979). The requirements are that a system can be partitioned around an intermediary metabolite whose concentration can be readily determined, thus separating the pathway into a supply and a demand block. The pathway flux must also be measurable. Independent modulations are then made in the supply and demand blocks (through overexpression of an enzyme, adding an inhibitor, varying the initial pathway substrate concentration, or other means), and the resulting changes in pathway flux and intermediate concentration monitored. The variation of flux with intermediate concentration for the supply perturbation yields the demand rate characteristic, and vice versa.

The double-modulation method forms the basis for the "top-down" (Brown *et al.*, 1990) and "modular" (Schuster *et al.*, 1993) approaches to metabolic control analysis. Importantly, though, in these analyzes, the aim is to obtain elasticities that can be used for the calculation of control coefficients, and consequently only small perturbations around the reference steady state are required. By contrast, supply–demand analysis aims to paint a picture of the regulation of a system over a wide range (Fig. 25.5). Larger parameter variations are thus considered, but still using the general methodology of double modulation.

In our group, we have applied this method to perform a supply–demand analysis of fermentative anaerobic free-energy metabolism in *Saccharomyces cerevisiae* (Kroukamp *et al.*, 2002). The yeast was grown anaerobically in energy-limited low-glucose chemostat cultures; the demand for cellular free energy (ATP) was perturbed by adding an uncoupler (benzoic acid) at various concentrations, while the supply was perturbed by changing the dilution rate to give different residual glucose concentrations in the chemostat. The measured elasticities ($\varepsilon^{v_{\text{supply}}}_{\text{ATP/ADP}} = -0.84$, $\varepsilon^{v_{\text{demand}}}_{\text{ATP/ADP}} = 7.2$) were used to calculate the control coefficients ($C^{J}_{\text{supply}} = 0.9$, $C^{J}_{\text{demand}} = 0.1$), showing that the bulk of flux control resided in the supply under the conditions used in the investigation (Kroukamp *et al.*, 2002), and that the ATP/ADP concentration ratio was under strong homeostatic control. These results contrasted with those obtained by Schaaff *et al.* (1989), who overexpressed a series of glycolytic enzymes in yeast grown in batch culture, singly and in combination, and found that none of these manipulations affected the glycolytic flux. Possible explanations for the discrepancies between these two studies could be that the latter did not include overexpression of the glucose transporter, or that the different experimental conditions—glucose-limited chemostat versus glucose-excess batch culture—account for the observed differences. At low glucose concentration, the lower activity of the glucose transporter would increase its flux-control coefficient (Kroukamp *et al.*, 2002).

5.2. Selected examples of experimental supply–demand analysis

The first verification of the prediction by Hofmeyr and Cornish-Bowden (2000) that control of flux in many instances lies outside pathways, such as glycolysis, that form part of intermediary metabolism was provided by Koebmann *et al.* (2002) in their study of the control of glycolytic flux in *Escherichia coli* (see Oliver, 2002). They developed a molecular genetic tool that specifically induces ATP hydrolysis in living cells without interfering with other aspects of metabolism, with which they showed that the majority (>75%) of the control of growth rate resides in the demand for ATP, and not in the glycolytic supply. A later study by Causey *et al.* (2003) provided further support for these findings. The results from *E. coli* differ from those of yeast discussed in the previous section, but again it is difficult to compare the results because of differences in the experimental conditions, as the experiments by Koebmann *et al.* (2002) were also performed under glucose excess in batch culture.

Since these studies, a number of experimental investigations into the nature of metabolic regulation in a variety of organisms have explicitly used the supply–demand analytic approach. The study of the control of biosynthetic flux to glycogen in muscle by Schafer *et al.* (2004) is particularly interesting in that here flux control was shown to reside in the supply of glucose-6-phosphate (G6P), specifically in the GLUT4 glucose transporter (not, as previously thought (Roach, 2002), in the demand with glycogen synthase the purported rate-limiting step). They found that G6P homeostasis is achieved by insulin-dependent phosphorylation and allosteric activation of glycogen synthase sensitivity to the upstream G6P. This regulatory mechanism allows muscle cells to tolerate large flux increases across their transporters without significantly changing their own metabolite pools; it is another instance of the functional differentiation predicted by supply–demand analysis: when supply controls the flux, homeostatic maintenance of metabolite concentrations becomes the function of the demand. Supply–demand analysis was therefore instrumental in solving the apparent paradox of glycogen synthase being sensitive to allosteric and covalent regulation despite having a minimal role in flux control.

Mendoza-Cózatl and Moreno-Sánchez (2006) explicitly used supply–demand rate characteristics in their study of the control of glutathione (GSH) and phytochelatin synthetic flux under cadmium stress. Their kinetic model was based on data obtained from GHS synthesis in tobacco cell suspension cultures during illumination. They found that when the GSH supply system was studied in isolation, γ-glutamylcysteine synthetase, which is feedback-inhibited by GSH, is rate limiting. However, in the full supply–demand system that took into account the GSH-consuming reactions, low demand activity exerted most of the flux control, while at high demand, the supply and demand blocks shared the control of flux.

Another instance of flux-control by demand and homeostatic maintenance of concentration by supply was found by Jørgensen *et al.* (2004), who studied the regulatory role of CTP synthase in the biosynthesis of CTP and dCTP in *Lactococcus lactis*. They found that CTP synthase, a supply enzyme, has no control on the growth rate but a strong control on the CTP and dCTP concentrations, as predicted by supply–demand analysis.

These experimental studies are excellent examples of the explanatory power of supply–demand analysis. Space precludes the discussion of other studies, but the reader is encouraged to follow-up Aledo *et al.* (2008), Boada *et al.* (2004), de Atauri *et al.* (2005), McCormick *et al.* (2009), Meléndez-Hevia and Paz-Lugo (2008), Nazaret and Mazat (2008), Santos *et al.* (2006), and Yuan *et al.* (2009). Supply–demand analysis also holds important implications for biotechnology, especially with regard to designing strategies for manipulating the metabolic behavior of organisms (Cornish-Bowden *et al.*, 1995).

ACKNOWLEDGMENT

The authors acknowledge financial support from the South African National Research Foundation (NRF). Any opinion, findings, and conclusions or recommendations expressed in this material are those of the authors, and therefore the NRF does not accept any liability in regard thereto.

REFERENCES

Aledo, J. C., Jiménez-Rivérez, S., Cuesta-Munoz, A., and Romero, J. M. (2008). The role of metabolic memory in the ATP paradox and energy homeostasis. *FEBS J.* **275,** 5332–5342.

Atkinson, D. E. (1977). Cellular Energy Metabolism and Its Regulation. Academic Press, New York.

Boada, J., Cuesta, E., Perales, J. C., Roig, T., and Bermudez, J. (2004). Glutathione content and adaptation to endogenously induced energy depletion in Mv1Lu cells. *Free Radic. Biol. Med.* **36,** 1555–1565.

Brown, G. C., Hafner, R. P., and Brand, M. D. (1990). A 'top-down' approach to the determination of control coefficients in metabolic control theory. *Eur. J. Biochem.* **188,** 321–325.

Causey, T. B., Zhou, S., Shanmugam, K. T., and Ingram, L. O. (2003). Engineering the metabolism of *Escherichia coli* W3110 for the conversion of sugar to redox-neutral and oxidized products: Homoacetate production. *Proc. Natl. Acad. Sci. USA* **100,** 825–832.

Cornish-Bowden, A., Hofmeyr, J.-H. S., and Cárdenas, M. L. (1995). Strategies for manipulating metabolic fluxes in biotechnology. *Bioorg. Chem.* **23,** 439–449.

de Atauri, P., Orrell, D., Ramsey, S., and Bolouri, H. (2005). Is the regulation of galactose 1-phosphate tuned against gene expression noise? *Biochem. J.* **387,** 77–84.

Fell, D. A., and Sauro, H. M. (1985). Metabolic control and its analysis. Additional relationships between elasticities and control coefficients. *Eur. J. Biochem.* **148,** 555–561.

Heinrich, R., and Rapoport, T. A. (1974). A linear steady-state treatment of enzymatic chains: General properties, control and effector strength. *Eur. J. Biochem.* **42,** 89–95.

Hofmeyr, J.-H. S. (1995). Metabolic regulation: A control analytic perspective. *J. Bioenerg. Biomembr.* **27,** 479–490.

Hofmeyr, J.-H. S. (1997). Anaerobic energy metabolism in yeast as a supply-demand system. *In* "New Beer in an Old Bottle: Eduard Buchner and the Growth of Biochemical Knowledge," (A. Cornish-Bowden, ed.), pp. 225–242. Universitat de València, València, Col.lecció Oberta.

Hofmeyr, J.-H. S. (2008). The harmony of the cell: The regulatory design of cellular processes. *Essays Biochem.* **45,** 57–66.

Hofmeyr, J.-H. S., and Cornish-Bowden, A. (1991). Quantitative assessment of regulation in metabolic systems. *Eur. J. Biochem.* **200,** 223–236.

Hofmeyr, J.-H. S., and Cornish-Bowden, A. (1996). Co-response analysis: A new experimental strategy for metabolic control analysis. *J. Theor. Biol.* **182,** 371–380.

Hofmeyr, J.-H. S., and Cornish-Bowden, A. (1997). The reversible Hill equation: How to incorporate cooperative enzymes into metabolic models. *Comput. Appl. Biosci.* **13,** 377–385.

Hofmeyr, J.-H. S., and Cornish-Bowden, A. (2000). Regulating the cellular economy of supply and demand. *FEBS Lett.* **476,** 47–51.

Hofmeyr, J.-H. S., Kacser, H., and van der Merwe, K. J. (1986). Metabolic control analysis of moiety-conserved cycles. *Eur. J. Biochem.* **155,** 631–641.

Hofmeyr, J.-H. S., Cornish-Bowden, A., and Rohwer, J. M. (1993). Taking enzyme kinetics out of control; putting control into regulation. *Eur. J. Biochem.* **212,** 833–837.

Jørgensen, C. M., Hammer, K., Jensen, P. R., and Martinussen, J. (2004). Expression of the pyrG gene determines the pool sizes of CTP and dCTP in *Lactococcus lactis*. *Eur. J. Biochem.* **271,** 2438–2445.

Kacser, H., and Burns, J. A. (1979). Molecular democracy: Who shares the controls? *Biochem. Soc. Trans.* **7,** 1149–1160.

Kacser, H., Burns, J. A., and Fell, D. A. (1995). The control of flux: 21 years on. *Biochem. Soc. Trans.* **23,** 341–366.

Koebmann, B. J., Westerhoff, H. V., Snoep, J. L., Solem, C., Pedersen, M. B., Nilsson, D., Michelsen, O., and Jensen, P. R. (2002). The extent to which ATP demand controls the glycolytic flux depends strongly on the organism and conditions for growth. *Mol. Biol. Rep.* **29,** 41–45.

Kroukamp, O., Rohwer, J. M., Hofmeyr, J.-H. S., and Snoep, J. L. (2002). Experimental supply-demand analysis of anaerobic yeast energy metabolism. *Mol. Biol. Rep.* **29,** 203–209.

le Novère, N., Bornstein, B., Broicher, A., Courtot, M., Donizelli, M., Dharuri, H., Li, L., Sauro, H., Schilstra, M., Shapiro, B., Snoep, J. L., and Hucka, M. (2006). BioModels Database: A free, centralized database of curated, published, quantitative kinetic models of biochemical and cellular systems. *Nucleic Acids Res.* **34,** D689–D691.

McCormick, A. J., Watt, D. A., and Cramer, M. D. (2009). Supply and demand: Sink regulation of sugar accumulation in sugarcane. *J. Exp. Bot.* **60,** 357–364.

Meléndez-Hevia, E., and Paz-Lugo, P. D. (2008). Branch-point stoichiometry can generate weak links in metabolism: The case of glycine biosynthesis. *J. Biosci.* **33,** 771–780.

Mendoza-Cózatl, D. G., and Moreno-Sánchez, R. (2006). Control of glutathione and phytochelatin synthesis under cadmium stress. Pathway modeling for plants. *J. Theor. Biol.* **238,** 919–936.

Nazaret, C., and Mazat, J.-P. (2008). An old paper revisited: "A mathematical model of carbohydrate energy metabolism. interaction between glycolysis, the Krebs cycle and the H-transporting shuttles at varying ATPases load" by V. V. Dynnik, R. Heinrich and E.E. Sel'kov. *J. Theor. Biol.* **252,** 520–529.

Oliver, S. (2002). Metabolism: Demand management in cells. *Nature* **418,** 33–34.

Olivier, B. G., and Snoep, J. L. (2004). Web-based kinetic modelling using JWS Online. *Bioinformatics* **20,** 2143–2144.

Olivier, B. G., Rohwer, J. M., and Hofmeyr, J.-H. S. (2005). Modelling cellular systems with PySCeS. *Bioinformatics* **21,** 560–561.

Reich, J. G., and Sel'kov, E. E. (1981). Energy metabolism of the cell: A theoretical treatise. Vol. 29 of Koebmann *et al.* (2002).

Roach, P. J. (2002). Glycogen and its metabolism. *Curr. Mol. Med.* **2,** 101–120.

Rohwer, J. M., and Hofmeyr, J.-H. S. (2008). Identifying and characterising regulatory metabolites with generalised supply-demand analysis. *J. Theor. Biol.* **252,** 546–554.

Rohwer, J. M., and Hofmeyr, J.-H. S. (2010). Kinetic and thermodynamic aspects of enzyme control and regulation. *J. Phys. Chem.B* **114,** 16280–16289.

Santos, V., Galdeano, C., Gomez, E., Alcon, A., and Garcia-Ochoa, F. (2006). Oxygen uptake rate measurements both by the dynamic method and during the process growth of *Rhodococcus erythropolis* IGTS8: Modelling and difference in results. *Biochem. Eng. J.* **32,** 198–204.

Schaaff, I., Heinisch, J., and Zimmermann, F. K. (1989). Overproduction of glycolytic enzymes in yeast. *Yeast* **5,** 285–290.

Schafer, J. R. A., Fell, D. A., Rothman, D., and Shulman, R. G. (2004). Protein phosphorylation can regulate metabolite concentrations rather than control flux: The example of glycogen synthase. *Proc. Natl. Acad. Sci. USA* **101,** 1485–1490.

Schuster, S., Kahn, D., and Westerhoff, H. V. (1993). Modular analysis of the control of complex metabolic pathways. *Biophys. Chem.* **48,** 1–17.
Smallbone, K., Simeonidis, E., Swainston, N., and Mendes, P. (2010). Towards a genome-scale kinetic model of cellular metabolism. *BMC Syst. Biol.* **4,** 6.
Yuan, J., Doucette, C. D., Fowler, W. U., Feng, X.-J., Piazza, M., Rabitz, H. A., Wingreen, N. S., and Rabinowitz, J. D. (2009). Metabolomics-driven quantitative analysis of ammonia assimilation in *E. coli*. *Mol. Syst. Biol.* **5,** 302.

CHAPTER TWENTY-SIX

Modular Kinetic Analysis

Klaas Krab

Contents

Abstract

Modularization is an important strategy to tackle the study of complex biological systems. Modular kinetic analysis (MKA) is a quantitative method to extract kinetic information from such a modularized system that can be used to determine the control and regulatory structure of the system, and to pinpoint and quantify the interaction of effectors with the system. The principles of the method are described, and the relation with metabolic control analysis is discussed. Examples of application of MKA are given.

1. Introduction

To deal with complex biological systems, it is often necessary to find a middle way between studying the system as a whole, or taking into account all individual processes occurring in the system. One such

Department of Molecular Cell Physiology, IMC, Faculty of Earth and Life Sciences, Vrije Universiteit, De Boelelaan 1085, Amsterdam, The Netherlands

Methods in Enzymology, Volume 500
ISSN 0076-6879, DOI: 10.1016/B978-0-12-385118-5.00026-8

middle way is to divide the system conceptually in a few modules, each comprising a subset of the individual processes in the system (Brand, 1996; Kahn and Westerhoff, 1991; Schuster *et al.*, 1993). These modules interact with a number of well-characterized intermediates (e.g., see Fig. 26.6). Modular kinetic analysis (MKA) then is the determination of steady-state relationships between these connecting intermediates and the fluxes through the modules. To obtain these "module kinetics," module fluxes and intermediate concentrations are measured while the modules are individually perturbed (by inhibition or activation of processes making up these modules, or by introducing extra feeds into the connecting intermediates).

1.1. Nomenclature and link with metabolic control analysis

MKA is intimately linked to metabolic control analysis (MCA). This can be seen from the fact that when the method was first developed by Brand and coworkers, it was actually called "top-down control analysis" (Brown *et al.*, 1990; Hafner *et al.*, 1990) or "elasticity analysis" (Brand, 1998; Harper and Brand, 1995), the latter term after the elasticity coefficients from MCA. These coefficients quantify the local change in the rate of a reaction step when the concentration of a metabolite, acting as reactant, product, or modifier for that reaction, is uniquely perturbed (see Box 26.1). These coefficients can be obtained from the slopes of the module kinetic curves at their intersection at steady state (Brand *et al.*, 1988). In fact, as the name implies, top-down MCA actually is carried out not with individual enzyme-catalyzed processes, but with modules that can contain a number of coupled processes. A key feature of this type of MCA is the calculation of control of system variables from the elasticity coefficients (and fluxes) obtained from kinetic curves as in MKA. Another MCA method which is intimately linked with MKA is coresponse analysis (Cornish-Bowden and Hofmeyr, 1994). In this approach, elasticity coefficients are determined from coresponse coefficients (defined as the relative change in an intermediate, divided by the relative change in a flux$_i$ when the system is perturbed elsewhere, see Box 26.2 for an example). The link is here that the slopes of the MKA curves in fact are coresponse coefficients, with the only distinctive feature that modules are considered, and not necessarily individual reactions. Closely related is the double modulation method (Kacser and Burns, 1979), as elaborated by Giersch (1994).

The preferred name for the method is Modular kinetic analysis (Amo and Brand, 2007; Ciapaite *et al.*, 2005, 2009; Kikusato *et al.*, 2010; Krab *et al.*, 2000), because this name implies that the method is not limited to calculation of the MCA-derived values in the steady state.

BOX 26.1 MCA Definitions Relevant for MKA

The structure of the mathematical definitions of the coefficients is cause in the denominator and effect in the numerator. The exception is the coresponse coefficient, where both denominator and numerator refer to effects, and the cause is specified separately (as an index).

Flux control coefficient:

$$C_i^J = \frac{\lambda_i}{J}\frac{\partial J}{\partial \lambda_i} = \frac{\partial \ln J}{\partial \ln \lambda_i}$$

global; refers to difference between steady states. J is a flux, λ_i a parameter to which the rate of the process carried out by module i is proportional (for a module consisting of a single enzyme-catalyzed process, this could be an enzyme concentration, or a $V_{\max}$).

Concentration control coefficient:

$$C_i^x = \frac{\lambda_i}{X}\frac{\partial X}{\partial \lambda_i} = \frac{\partial \ln X}{\partial \ln \lambda_i}$$

global; refers to difference between steady states. X is the concentration of a linking intermediate, λ_i as above.

Response coefficient:

$$R_Y^Z = \frac{Y}{Z}\frac{\partial Z}{\partial Y} = \frac{\partial \ln Z}{\partial \ln Y}$$

global; Y is an external parameter (such as an effector concentration), Z is a system variable (such as a flux or an intermediate concentration).

Elasticity coefficient:

$$\varepsilon_X^i = \frac{X}{v_i}\frac{\partial v_i}{\partial X} = \frac{\partial \ln v_i}{\partial \ln X}$$

(applied to intermediates that link modules) local; refers to an infinitesimal change within a steady state. v_i is the rate of a module-catalyzed process, X as above.

Elasticity coefficient:

$$\varepsilon_Y^i = \frac{Y}{v_i}\frac{\partial v_i}{\partial Y} = \frac{\partial \ln v_i}{\partial \ln Y}$$

(applied to external parameters) local; refers to an infinitesimal change within a steady state. Y and v_i as above.

Coresponse coefficients:

$${}^{P}O_J^x = \left(\frac{J}{X}\frac{\partial X}{\partial J}\right)_{\text{change in } P} = \left(\frac{\partial \ln X}{\partial \ln J_i}\right)_{\text{change in } P}$$

(Continued)

BOX 26.1 (Continued)

(here between an intermediate concentration and a flux) global; P is a parameter affecting the system. If P affects module i, the coresponse coefficient is seen to be a ratio between two control coefficients:

$$ {}^{i}O_{J}^{x} = \frac{C_{i}^{x}}{C_{i}^{J}} $$

Partial flux control coefficient (actually, an internal partitioned response coefficient; Ainscow and Brand, 1999b):

$$ {}^{X}R_{i}^{J_k} = C_{i}^{X}\varepsilon_{X}^{k} \quad \text{with} \quad C_{i}^{J_k} = \delta_{ik} + \sum_{\text{all intermediates } X} {}^{X}R_{i}^{J_k} $$

these quantify the effect of modulation of module i on flux J_k as transmitted by intermediate X. δ_{ik} has the value 0, except when $k = i$, when it has the value 1 (the special case where the effect of modulation on the flux through the module itself is considered; Wanders *et al.*, 1984).

Regulatory strength (Kahn and Westerhoff, 1993), also known as internal partial or partitioned response coefficient (Kesseler and Brand, 1994b; Kacser and Burns, 1973):

$$ {}^{i}R_{X}^{Z} = C_{i}^{Z}\varepsilon_{X}^{i} \quad \text{with} \quad R_{X}^{Z} = \sum_{\text{all modules } i} {}^{i}R_{X}^{Z} $$

these quantify the effect of modulation of intermediate X on system variable Z as transmitted by module i.

External partial or partitioned response coefficient (Kacser and Burns, 1973; Kholodenko, 1988):

$$ {}^{i}R_{Y}^{Z} = \varepsilon_{Y}^{i}C_{i}^{Z} \quad \text{with} \quad R_{Y}^{Z} = \sum_{\text{all modules } i} {}^{i}R_{Y}^{Z} $$

these quantify the effect of modulation of external parameter Y on system variable Z as transmitted by module i. Note that for this relation it is not required to use the relative change in Y (one can assume that $\partial Y/Y = 1$ in the definitions of ε_{Y}^{i} and R_{Y}^{Z}), so in principle changes of Y from a value of 0 can be considered, see the example in Section 2.5.

A remark on the distinction between the use of MCA coefficients before and after modularization: in MKA modules are seen as units, so, for example, the definition of flux control coefficients refer to a situation where all fluxes in the module are modified to the same extent, and elasticity coefficients in MKA are "block elasticities," again referring to v of the module as a whole and not to v of the particular step in the module sensitive to the compound involved.

BOX 26.2 Coresponse Analysis

Generation of modular kinetic data as in Fig. 26.3 actually amounts to determination of coresponses between fluxes and intermediates. In the steady state, these are quantified by the coresponse coefficients (${}^{k}O_{J_i}^{X}$): the relative change in intermediate X, divided by the relative change in flux J_i through a module i when the system is perturbed elsewhere (at module $k \neq i$). In the scheme of Fig. 26.4, modulation of step 1 changes the flux through steps 2 and 3. These changes are mediated by changed levels of X and Y (to step 2) or Y alone (to step 3). Similarly, modulation of step 2 or step 3 is transmitted to the other steps via X and Y. This can be quantified with the help of control coefficients and elasticity coefficients according to the following relations:

Modulations resulting in changed flux through step 1:

$$C_2^{J_1} = C_2^{X} \varepsilon_X^1 \quad \text{or} \quad 1 = {}^{2}O_J^{X} \varepsilon_X^1$$

(when step 2 is modified)

$$C_3^{J_1} = C_3^{X} \varepsilon_X^1 \quad \text{or} \quad 1 = {}^{3}O_J^{X} \varepsilon_X^1$$

(when step 3 is modified)

Modulations resulting in changed flux through step 2:

$$C_1^{J_2} = C_1^{X} \varepsilon_X^2 + C_1^{Y} \varepsilon_Y^2 \quad \text{or} \quad 1 = {}^{1}O_J^{X} \varepsilon_X^2 + {}^{1}O_J^{Y} \varepsilon_Y^2$$

(when step 1 is modified)

$$C_3^{J_2} = C_3^{X} \varepsilon_X^2 + C_3^{Y} \varepsilon_Y^2 \quad \text{or} \quad 1 = {}^{3}O_J^{X} \varepsilon_X^2 + {}^{3}O_J^{Y} \varepsilon_Y^2$$

(when step 3 is modified)

Modulations resulting in changed flux through step 3:

$$C_1^{J_3} = C_1^{Y} \varepsilon_Y^3 \quad \text{or} \quad 1 = {}^{1}O_J^{Y} \varepsilon_Y^3$$

(when step 1 is modified)

$$C_2^{J_3} = C_2^{Y} \varepsilon_Y^3 \quad \text{or} \quad 1 = {}^{2}O_J^{Y} \varepsilon_Y^3$$

(when step 2 is modified)

Note that $J_1 = J_2 = J_3 = J$. The coresponses are the inverted slopes of the curves in Fig. 26.3 at their intersections at steady state.

This way sets of linear equations are generated that can be used to calculate the elasticity coefficients, and from these, the control and regulation structure of the system.

2. Description of the Method

2.1. Basic approach

The method starts from some knowledge of the layout of the system. In fact, this knowledge to a large extent determines the choice of modules, in addition to practical considerations such as whether fluxes and intermediates can actually be measured. Also important is the possibility to perturb modules individually. Note that this division of the system is determined by practical considerations and is different from the search of functional modules based on analysis of genome-scale networks (Bundy *et al.*, 2007; Guimera and Nunes Amaral, 2005; Pereira-Leal *et al.*, 2004).

After the choice of the modules and connecting intermediates, the system is brought into a steady state. Then, for example, in a series of titrations with effectors of individual modules, the steady state is shifted, and changes in fluxes and connecting intermediates are recorded. The response of the fluxes through those modules that are not titrated is analyzed as a function of intermediate concentration ("module kinetics").

A useful application of this method is to study in what way an effector with unknown site or sites of interaction with the system changes these module kinetics. Both these procedures (determination of module kinetics and of the site or sites of action of a modifier) will be illustrated in Section 2.2 for the simplest modular system consisting of two modules connected by a single intermediate.

2.2. The single-intermediate input–output system

The easiest way to visualize how module kinetics are obtained is with the help of the system shown in Fig. 26.1A. This system consists of two modules, linked by a single explicit intermediate X. One module (the input) produces X, the other module (the output) consumes X. All factors that determine the flux through input or output, except X, are assumed to be constant. Figure 26.1B sketches the way in which the input and output flux depend on the concentration of X; the intersection of these "module kinetics" determines the steady state values of flux and [X].

To determine the shapes of the kinetic curves, one interferes with one module, while the other module reacts to the changing steady-state value of [X]. This procedure is illustrated in Fig. 26.1C and D: changing the output generates a series of steady states that together determine the kinetic curve for the input (Fig. 26.1C) and *vice versa* (Fig. 26.1D; e.g., Brand *et al.*, 1988). Figure 26.1E shows a possible outcome of an experiment in which an unknown modulator of the system is present at fixed concentration: from

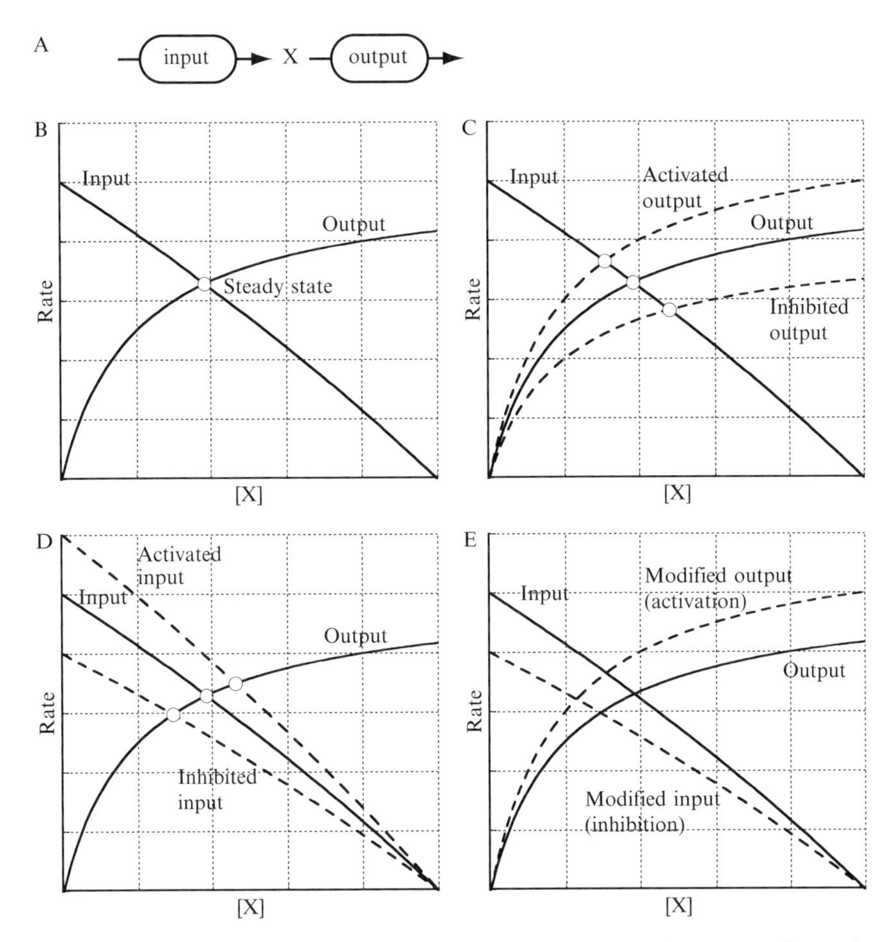

Figure 26.1 (A) A system consisting of two modules (input and output) and a single connecting intermediate. (B) Module kinetics of input and output, and position of the steady state. (C) Determination of input kinetics by variation of the output. (D) Determination of output kinetics by variation of the input. (E) Example of a possible effect of an external modifier on the kinetic curves.

the changes in the input- and output-curves in Fig. 26.1E, it can be concluded that the unknown modulator inhibits the input and activates the output.

2.3. Multi-intermediates, multimodule systems

In many cases, a biological system can be represented by two modules connected by a single intermediate (Rohwer and Hofmeyr, 2008). However, generally the questions we would like to ask about biological systems involve more than two modules, connected by a number of intermediates.

There are several approaches that can be followed to decompose the network into modules to analyze their kinetics.

a. The system can be divided in a linear sequence of modules and intermediates (Fig. 26.2A). In this case, separate analysis can be carried out for each intermediate. The system is divided into an input and output module for the intermediate of choice (see Fig. 26.3). This is a useful approach to identify the site of action of effectors: for example, in the system of Fig. 26.2A a specific inhibitor of step 2 would inhibit the output in MKA around X, and the input in MKA around Y (e.g., Krab *et al.*, 2000).
b. Multiple inputs and outputs for a single connecting intermediate (Fig. 26.2B). Here it is essential that the modules only interact via X. In this case, module kinetics can be determined individually by the strategy illustrated in Fig. 26.2B for a system consisting of a single input and two outputs. First, module kinetics are determined for the input (module 1) and the total output (module 2, incorporating both steps 2 and 3); then one of the outputs is disabled by inhibition or removal of a necessary substrate other than X, and the changed output module kinetics are determined in the usual manner (by gradual modification of the input). This way, the kinetics of module 2* (step 3) is

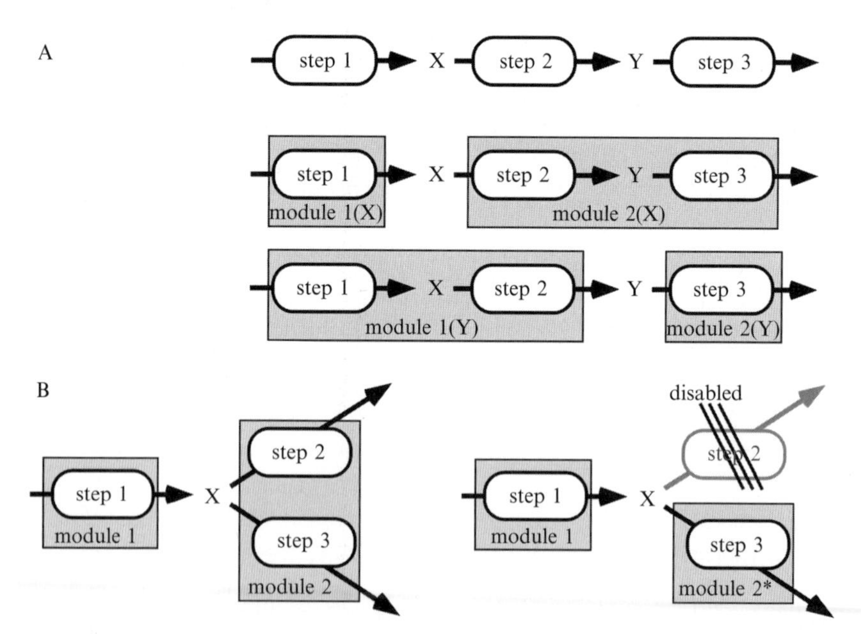

Figure 26.2 (A) Possible modularizations of a linear three-step system with two intermediates. (B) A simple branched three-step system with one intermediate; two experimental conditions in one of which one of the module is disabled.

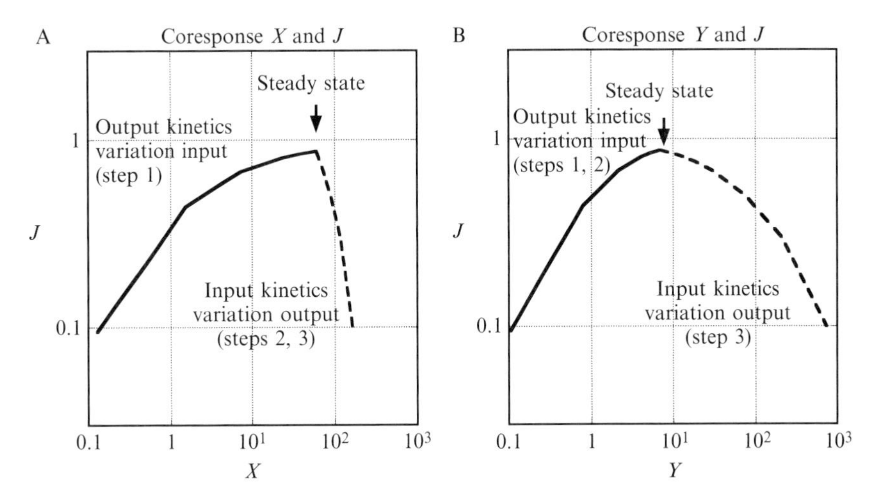

Figure 26.3 Simulation of module kinetics in the system of Fig. 26.2A. (A) Modularization around X, (B) Modularization around Y.

obtained: the kinetics of step 2 is found by subtracting the curve for step 3 from the curve for module 2 (examples of this procedure can be found in Hafner *et al.*, 1990; Van den Bergen *et al.*, 1994).

c. The system contains modules that interact via more than one intermediate. If the aim of the MKA is to determine the distribution of metabolic control at steady state, it is not always necessary to measure all intermediates. Measurement of an intermediate can be circumvented by including it in a module, provided that there is an independent means to determine control distribution within this module (an example of this specific methodology can be found in Ainscow and Brand, 1995).

d. Complex modular systems. The focus in more complex systems generally is on control analysis by MKA, although inspection of module kinetics still is useful in identifying sites of interaction with modifiers. Interaction with modifiers can still be quantified in terms of partitioned response coefficients (see Box 26.1; Ainscow and Brand, 1999c). General versions of methodology involve mathematical symbolisms which tend to be impenetrable to the impatient practical biochemist. For practical purposes, coresponse analysis (Cornish-Bowden and Hofmeyr, 1994) probably is the least daunting way in which elasticities, and from these, control distributions can be obtained.

Modularization is relatively unproblematic when the modules considered are monofunctional (Rohwer *et al.*, 1996). A module is monofunctional when three criteria are satisfied: (1) intermediates inside the module should not affect processes outside of the module; (2) intermediates inside the module should not be linked to intermediates outside the module by moiety conservation (as this would break the first rule via the back door); (3) fluxes crossing the border of the module should have only a single degree of freedom: if the value of one flux is known, all the others can be calculated.

When there is significant multifunctionality in the modules under consideration, sometimes it is advisable to use extra branches to linking intermediates as perturbations of the steady state rather than inhibition or activation of module components (to avoid simultaneous perturbations of several modules, cf. Ainscow and Brand, 1999a).

In MKA, coresponses between a flux and an intermediate are measured (e.g., in Fig. 26.3 the responses of J and X or Y to perturbations of the three modules). In the simple case of an input–output system, the two coresponse coefficients are identical to the elasticity coefficients of input and output with respect to the linking intermediate. In more complex systems, the coresponse coefficients between fluxes and intermediates take part in linear relationships with the elasticity coefficients as coefficients. See Box 26.1 for the definitions of the MCA terms used, and Box 26.2 for an example of calculation of elasticity coefficients from coresponses in a linear three-module system (Fig. 26.4).

Important for practical MKA is the selection of effectors (inhibitors and activators) that modulate the activity of modules, or extra branches that can independently change intermediate concentrations. This set of effectors and extra branches can be thought of as a toolkit for the determination of module kinetics and the control properties of the system. A very thorough motivation of the selection of such a toolkit can be found in Ainscow and Brand (1999a). It is here that prior knowledge about the system is essential. However, overlooking, for example, sites of interaction between system and toolkit modifiers is quickly revealed as inconsistent results, provided that sufficient coresponses are measured. An example of this can be found in Ainscow and Brand (1999a), where one of the modifiers (uncoupler) gave deviating results attributed to interaction with other modules than proton leak. This type of finding will lead to extrapolation from current understanding of system structure to improved understanding.

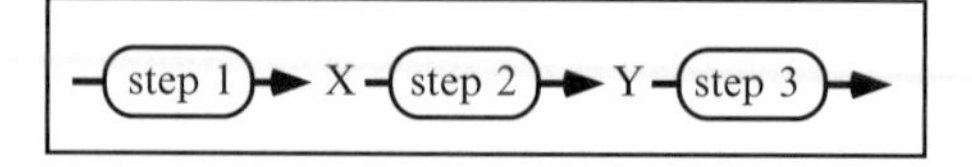

Figure 26.4 Three-step linear mechanism without feed-forward or feed-back. Each step is treated as a module.

2.4. Metabolic control analysis

MKA yields elasticity coefficients and fluxes at steady state; these can be used to calculate control distributions using the theorems of MCA. Once the control coefficients are known, the door is opened for detailed analysis of internal regulation (e.g., see Kesseler and Brand, 1994b; Ainscow and Brand, 1999b). Partial flux control coefficients and regulatory strengths (see Box 26.1) can be calculated to analyze regulation within the modular system in detail.

A note of caution here is that the control coefficients are those of the modules; so the view of control and regulatory structures is exactly as detailed as the modularization chosen.

2.5. Response analysis

A particularly useful application of MKA is the analysis of the responses of the system to external perturbations. In addition to the pattern of interaction between the effector and the system as revealed by the changes in module kinetics, the availability of the control structure of the system allows detailed analysis of partial responses to the effector.

The total response of a system variable (Z) to a change in an external parameter (Y, the effector) can be dissected as a sum of contributions by different modules in the system. These contributions are given by the partial external response coefficients (Box 26.1). Some care should be taken that the system variable Z is an exclusive function of the concentrations of the explicit intermediates. Variables that depend (additionally) on internal fluxes or intermediates within modules do not allow equating the total response to the effector as a sum of partial response coefficients.

The procedure is illustrated in Fig. 26.5. Once the module kinetics in the presence and absence of modifier Y have been determined, the finite changes indicated by c − a and d − a in the figure can be used to calculate approximate values of the elasticity coefficients $\varepsilon_Y^{\text{input}}$ and $\varepsilon_Y^{\text{output}}$. Multiplying these with control coefficients obtained from the situation in the absence of Y, external partial or partitioned response coefficients can be calculated. For the example of Fig. 26.5, the values for the response of the flux J are given in the legend, together with the total response of J toward Y. In this example, the calculated sum of the partial response coefficients (0.043) exceeds the actual response (0.033). The error in this analysis is very dependent on the size of the modifications (in Fig. 26.5, the output is activated by 10%, the input is inhibited by 5%). Unfortunately, experimental considerations generally prevent changes that are sufficiently small to give satisfactory agreement between the total response (between steady states) and the summed partial response coefficients. However, the relative contributions of the partial responses are still informative.

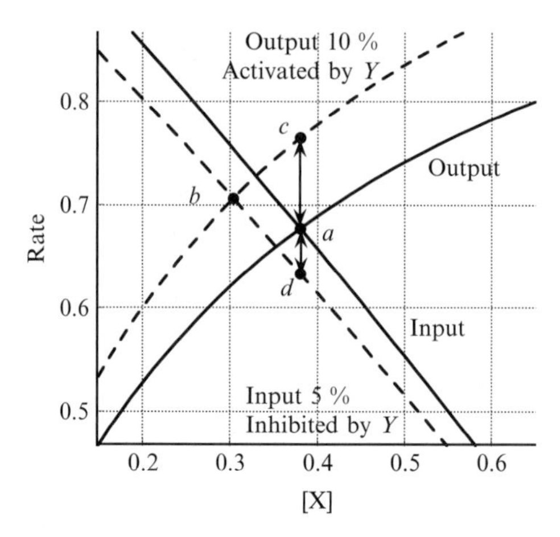

Figure 26.5 Effect of an external modifier Y which inhibits the input by 5% and activates the output by 10% in the system of Fig. 26.1A. The black dots indicate the steady states in the absence (a) and presence (b) of Y, and the hypothetical points on output (c) and input (d) kinetic curves that would occur if Y changes, but X would remain constant. For this example, the following MCA values have been calculated (steady state in the absence of Y): $\varepsilon_X^{\text{input}} = -0.556$, $\varepsilon_X^{\text{output}} = 0.342$, $C_{\text{input}}^J = 0.381$, and $C_{\text{output}}^J = 0.619$. $\varepsilon_Y^{\text{input}} = -0.05$ and $\varepsilon_Y^{\text{output}} = 0.1$ (assuming $\partial Y/Y = 1$, see Box 26.1), so that ${}^{\text{input}}R_Y^J = -0.019$, ${}^{\text{output}}R_Y^J = 0.062$. From the difference between the two steady states, $R_Y^J = 0.033$.

An experimental example of response analysis can be found in Kesseler and Brand (1994c).

3. Applications

3.1. Mitchondrial respiration and oxidative phosphorylation

Examples of MKA with modules connected by a single intermediate are abundant in the study of mitochondrial respiration and oxidative phosphorylation. Brand and coworkers analyzed in a number of publications a system consisting of mitochondrial membrane potential ($\Delta\psi$) as intermediate between a respiratory module (generating $\Delta\psi$), a phosphorylating module (dissipating $\Delta\psi$ in the synthesis and export of ATP) and a proton leak module (dissipating $\Delta\psi$). These studies resulted in determination of control distributions between these modules on $\Delta\psi$ and the different fluxes in this system, and included elucidation of the effect of factors such as Cd (Kesseler and Brand, 1994a,b,c), thyroid hormone (Lombardi *et al.*, 1998), genetic background (Amo and Brand, 2007), etc.

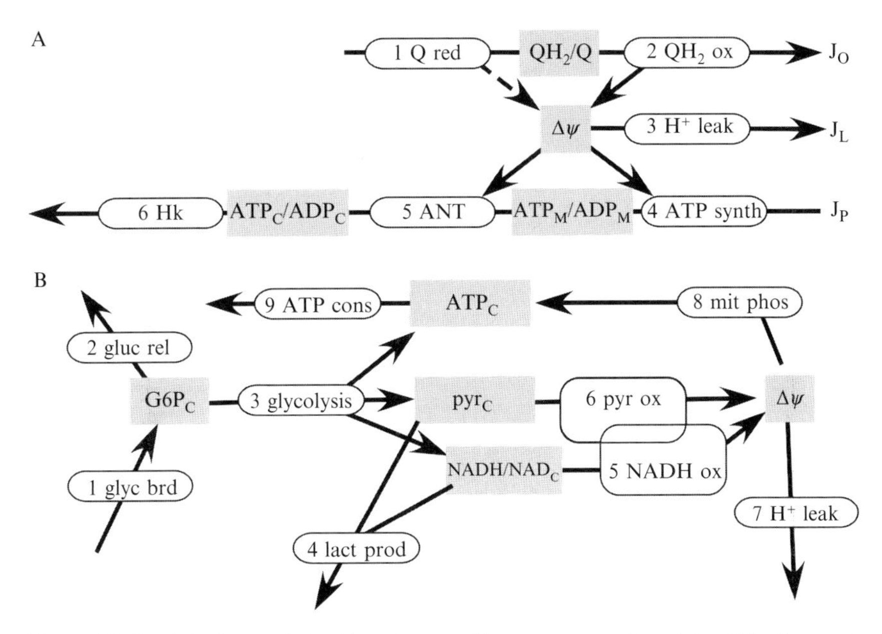

Figure 26.6 (A) Six module, four intermediate system analyzed by Ciapaite *et al.* (2006). Modules are Q red, coenzyme Q reducer; QH_2 ox, coenzyme Q oxidizer; H^+ leak, proton leak; ATP synth, phosphorylation; ANT, adenine nucleotide translocator; Hk, hexokinase (ATP consumer). J_O, J_L, J_P refer to independent fluxes. (B) Nine module, five intermediate system analyzed by Ainscow and Brand (1999a). Modules are glyc brd, glycogen breakdown; gluc rel, glucose release; glycolysis; lact prod, lactate production; NADH ox, NADH oxidation; pyr ox, pyruvate oxidation (overlapping with the NADH ox module); H^+ leak, proton leak; mit phos, phosphorylation; ATP cons, ATP consumption.

It is not always necessary to do a complete MKA analysis; in many studies the group of Brand has focused on the kinetics of the proton leak module. This has led to important insights about the importance of the proton leak process in scavenging of mitochondrially produced reactive oxygen species, thermogenesis, and the role of UCP proteins in this process (see Brand, 2005 for an overview).

In the same $\Delta\psi$-linked system other groups studied the effects of organic pollutants (Mildaziene *et al.*, 2002), toxic cations (Ciapaite *et al.*, 2009), or heat stress (Kikusato *et al.*, 2010) on electron transfer, ATP synthesis, and proton leak.

A structurally similar system was studied in plant mitochondria, where coenzyme Q redox state linked a number of Q-reducing and QH_2-oxidizing modules (Affourtit *et al.*, 2001; Van den Bergen *et al.*, 1994). Also here, it is possible to focus on the kinetics of a single module, as, for example, in a study of the kinetics of the alternative respiration module in *Arum maculatum* during flower development (Leach *et al.*, 1996).

A more extensive study of oxidative phosphorylation in mitochondria in relation to diabetes and obesity was carried out by Ciapaite *et al.* (2006). These authors considered the system of six modules and four intermediates (Q redox state, $\Delta\psi$ and ATP levels in the mitochondrial matrix, and the extramitochondrial space) illustrated in Fig. 26.6A. Their main goal was to identify the effect of long-chain fatty acyl-CoA esters in this system, but also the control structure of the system was obtained.

3.2. Other systems

A further extension of the study of oxidative phosphorylation in mitochondria is the study of ATP economics in hepatic cells by Ainscow and Brand (1999a,b). The authors considered a system of five intermediates and nine (partially overlapping) modules illustrated in Fig. 26.6B. The result of this study was a complete description of the control and regulation of the chosen modular system. Kroukamp *et al.* (2002) applied MKA around the intermediate ATP/ADP ratio in their study of anaerobic yeast energy metabolism.

ACKNOWLEDGMENTS

The author thanks Dr. F. J. Bruggeman for sharing his insights in the ramifications of MCA.

REFERENCES

Affourtit, C., Krab, K., Leach, G. R., Whitehouse, D. G., and Moore, A. L. (2001). New insights into the regulation of plant succinate dehydrogenase. On the role of the protonmotive force. *J. Biol. Chem.* **276,** 32567–32574.

Ainscow, E. K., and Brand, M. D. (1995). Top-down control analysis of systems with more than one common intermediate. *Eur. J. Biochem.* **231,** 579–586.

Ainscow, E. K., and Brand, M. D. (1999a). Top-down control analysis of ATP turnover, glycolysis and oxidative phosphorylation in rat hepatocytes. *Eur. J. Biochem.* **263,** 671–685.

Ainscow, E. K., and Brand, M. D. (1999b). Internal regulation of ATP turnover, glycolysis and oxidative phosphorylation in rat hepatocytes. *Eur. J. Biochem.* **266,** 737–749.

Ainscow, E. K., and Brand, M. D. (1999c). Quantifying elasticity analysis: How external effectors cause changes to metabolic systems. *Biosystems* **49,** 151–159.

Amo, T., and Brand, M. D. (2007). Were inefficient mitochondrial haplogroups selected during migrations of modern humans? A test using modular kinetic analysis of coupling in mitochondria from cybrid cell lines. *Biochem. J.* **404,** 345–351.

Brand, M. D. (1996). Top down metabolic control analysis. *J. Theor. Biol.* **182,** 351–360.

Brand, M. D. (1998). Top-down elasticity analysis and its application to energy metabolism in isolated mitochondria and intact cells. *Mol. Cell. Biochem.* **184,** 13–20.

Brand, M. D. (2005). The efficiency and plasticity of mitochondrial energy transduction. *Biochem. Soc. Trans.* **33,** 897–904.

Brand, M. D., Hafner, R. P., and Brown, G. C. (1988). Control of respiration in non-phosphorylating mitochondria is shared between the proton leak and the respiratory chain. *Biochem. J.* **255,** 535–539.

Brown, G. C., Hafner, R. P., and Brand, M. D. (1990). A 'top-down' approach to the determination of control coefficients in metabolic control theory. *Eur. J. Biochem.* **188,** 321–325.

Bundy, J. G., Papp, B., Harmston, R., Browne, R. A., Clayson, E. M., Burton, N., Reece, R. J., Oliver, S. G., and Brindle, K. M. (2007). Evaluation of predicted network modules in yeast metabolism using NMR-based metabolite profiling. *Genome Res.* **17,** 510–519.

Ciapaite, J., Van Eikenhorst, G., Bakker, S. J., Diamant, M., Heine, R. J., Wagner, M. J., Westerhoff, H. V., and Krab, K. (2005). Modular kinetic analysis of the adenine nucleotide translocator-mediated effects of palmitoyl-CoA on the oxidative phosphorylation in isolated rat liver mitochondria. *Diabetes* **54,** 944–951.

Ciapaite, J., Bakker, S. J., Diamant, M., van Eikenhorst, G., Heine, R. J., Westerhoff, H. V., and Krab, K. (2006). Metabolic control of mitochondrial properties by adenine nucleotide translocator determines palmitoyl-CoA effects. Implications for a mechanism linking obesity and type 2 diabetes. *FEBS J.* **273,** 5288–5302.

Ciapaite, J., Nauciene, Z., Baniene, R., Wagner, M. J., Krab, K., and Mildaziene, V. (2009). Modular kinetic analysis reveals differences in Cd^{2+} and Cu^{2+} ion-induced impairment of oxidative phosphorylation in liver. *FEBS J.* **276,** 3656–3668.

Cornish-Bowden, A., and Hofmeyr, J. H. (1994). Determination of control coefficients in intact metabolic systems. *Biochem. J.* **298,** 367–375.

Giersch, C. (1994). Determining elasticities from multiple measurements of steady-state flux rates and metabolite concentrations: Theory. *J. Theor. Biol.* **169,** 89–99.

Guimera, G., and Nunes Amaral, L. A. (2005). Functional cartography of complex metabolic networks. *Nature* **433,** 895–900.

Hafner, R. P., Brown, G. C., and Brand, M. D. (1990). Analysis of the control of respiration rate, phosphorylation rate, proton leak rate and protonmotive force in isolated mitochondria using the 'top-down' approach of metabolic control theory. *Eur. J. Biochem.* **188,** 313–319.

Harper, M. E., and Brand, M. D. (1995). Use of top-down elasticity analysis to identify sites of thyroid hormone-induced thermogenesis. *Proc. Soc. Exp. Biol. Med.* **208,** 228–237.

Kacser, H., and Burns, J. A. (1973). The control of flux. *Symp. Soc. Exp. Biol.* **27,** 65–104(see also: Biochem. Soc. Trans. 23, 341–366. (1995)).

Kacser, H., and Burns, J. A. (1979). Molecular democracy: Who shares the controls? *Biochem. Soc. Trans.* **7,** 1149–1160.

Kahn, D., and Westerhoff, H. V. (1991). Control theory of regulatory cascades. *J. Theor. Biol.* **153,** 255–285.

Kahn, D., and Westerhoff, H. V. (1993). The regulatory strength: How to be precise about regulation and homeostasis. *Acta Biotheor.* **41,** 85–96.

Kesseler, A., and Brand, M. D. (1994a). Localisation of the sites of action of cadmium on oxidative phosphorylation in potato tuber mitochondria using top-down elasticity analysis. *Eur. J. Biochem.* **225,** 897–906.

Kesseler, A., and Brand, M. D. (1994b). Effects of cadmium on the control and internal regulation of oxidative phosphorylation in potato tuber mitochondria. *Eur. J. Biochem.* **225,** 907–922.

Kesseler, A., and Brand, M. D. (1994c). Quantitative determination of the regulation of oxidative phosphorylation by cadmium in potato tuber mitochondria. *Eur. J. Biochem.* **225,** 923–935.

Kholodenko, B. N. (1988). How do external parameters control fluxes and concentrations of metabolites? An additional relationship in the theory of metabolic control. *FEBS Lett.* **232,** 383–386.

Kikusato, M., Ramsey, J. J., Amo, T., and Toyomizu, M. (2010). Application of modular kinetic analysis to mitochondrial oxidative phosphorylation in skeletal muscle of birds exposed to acute heat stress. *FEBS Lett.* **584,** 3143–3148.

Krab, K., Wagner, M. J., Wagner, A. M., and Møller, I. M. (2000). Identification of the site where the electron transfer chain of plant mitochondria is stimulated by electrostatic charge screening. *Eur. J. Biochem.* **267,** 869–876.

Kroukamp, O., Rohwer, J. M., Hofmeyr, J. H., and Snoep, J. L. (2002). Experimental supply-demand analysis of anaerobic yeast energy metabolism. *Mol. Biol. Rep.* **29,** 203–209.

Leach, G. R., Krab, K., Whitehouse, D. G., and Moore, A. L. (1996). Kinetic analysis of the mitochondrial quinol-oxidizing enzymes during development of thermogenesis in *Arum maculatum* L. *Biochem. J.* **317,** 313–319.

Lombardi, A., Lanni, A., Moreno, M., Brand, M. D., and Goglia, F. (1998). Effect of 3,5-di-iodo-L-thyronine on the mitochondrial energy-transduction apparatus. *Biochem. J.* **330,** 521–526.

Mildaziene, V., Nauciene, Z., Baniene, R., and Grigiene, J. (2002). Multiple effects of 2,2′,5,5′-tetrachlorobiphenyl on oxidative phosphorylation in rat liver mitochondria. *Toxicol. Sci.* **65,** 220–227.

Pereira-Leal, J. B., Enright, A. J., and Ouzounis, C. A. (2004). Detection of functional modules from protein interaction networks. *Proteins: Structure, function, and bioinformatics* **54,** 49–57.

Rohwer, J. M., and Hofmeyr, J. H. (2008). Identifying and characterising regulatory metabolites with generalised supply-demand analysis. *J. Theor. Biol.* **252,** 546–554.

Rohwer, J. M., Schuster, S., and Westerhoff, H. V. (1996). How to recognize monofunctional units in a metabolic system. *J. Theor. Biol.* **179,** 213–228.

Schuster, S., Kahn, D., and Westerhoff, H. V. (1993). Modular analysis of the control of complex metabolic pathways. *Biophys. Chem.* **48,** 1–17.

Van den Bergen, C. W., Wagner, A. M., Krab, K., and Moore, A. L. (1994). The relationship between electron flux and the redox poise of the quinone pool in plant mitochondria. Interplay between quinol-oxidizing and quinone-reducing pathways. *Eur. J. Biochem.* **226,** 1071–1078.

Wanders, R. J. A., Groen, A. K., Van Roermund, C. W. T., and Tager, J. M. (1984). Factors determining the relative contribution of the adenine-nucleotide translocator and the ADP-regenerating system to the control of oxidative phosphorylation in isolated rat-liver mitochondria. *Eur. J. Biochem.* **142,** 417–424.

CHAPTER TWENTY-SEVEN

Quantitative Analysis of Flux Regulation Through Hierarchical Regulation Analysis

Karen van Eunen,[*,†] Sergio Rossell,[‡] Jildau Bouwman,[§] Hans V. Westerhoff,[†,¶] *and* Barbara M. Bakker[*,†]

Contents

Abstract

Regulation analysis is a methodology that quantifies to what extent a change in the flux through a metabolic pathway is regulated by either gene expression or metabolism. Two extensions to regulation analysis were developed over the past years: (i) the regulation of V_{max} can be dissected into the various levels of the gene-expression cascade, such as transcription, translation, protein

* Department of Pediatrics, Center for Liver, Digestive and Metabolic Diseases, University Medical Center Groningen, University of Groningen, Groningen, The Netherlands
† Department of Molecular Cell Physiology, Faculty of Earth and Life Sciences, VU University, Amsterdam, The Netherlands
‡ Center for Molecular and Biomolecular Informatics, Nijmegen Center for Molecular Life Sciences, Radboud University Nijmegen Medical Center, Nijmegen, The Netherlands
§ Biosciences, TNO Quality of Life Zeist, The Netherlands
¶ Manchester Centre for Integrative Systems Biology, Manchester Interdisciplinary BioCentre, The University of Manchester, Manchester, United Kingdom

Methods in Enzymology, Volume 500
ISSN 0076-6879, DOI: 10.1016/B978-0-12-385118-5.00027-X

degradation, etc. and (ii) a time-dependent version allows following flux regulation when cells adapt to changes in their environment. The methodology of the original form of regulation analysis as well as of the two extensions will be described in detail. In addition, we will show what is needed to apply regulation analysis in practice.

Studies in which the different versions of regulation analysis were applied revealed that flux regulation was distributed over various processes and depended on time, enzyme, and condition of interest. In the case of the regulation of glycolysis in baker's yeast, it appeared, however, that cells that remain under respirofermentative conditions during a physiological challenge tend to invoke more gene-expression regulation, while a shift between respirofermentative and respiratory conditions invokes an important contribution of metabolic regulation. The complexity of the regulation observed in these studies raises the question what is the advantage of this highly distributed and condition-dependent flux regulation.

1. Introduction

In biology, the term regulation is used in different meanings. In the field of systems biology, regulation indicates which mechanisms the cell uses to accomplish changes in, or robustness of cellular functions (Bruggeman *et al.*, 2007). Regulation has various aspects: (i) the capacity of a living organism to respond to changes in its environment, (ii) the internal communication between different parts of cells or organisms, and (iii) the maintenance of homeostasis upon external perturbations (Kahn and Westerhoff, 1993).

The flux through a pathway or an enzyme can be changed by many different regulatory mechanisms. These mechanisms can be dissected crudely into three groups: that is, signal transduction, gene-expression, and metabolic regulation. As each of these can affect both others, the relation between them can be depicted as a triangle (Fig. 27.1). For instance, signal transduction can have either a direct effect on metabolism or an indirect effect via gene expression. As all these mechanisms work

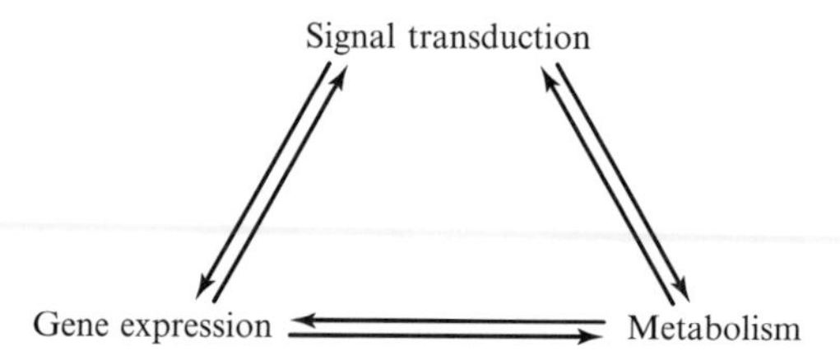

Figure 27.1 The triangle of regulation.

simultaneously and in interaction with each other, a quantitative approach is required to assess which are the important regulatory routes during specific adaptations of an organism. The goal of this chapter is to review recent advances in regulation analysis, which allows to dissect the relative importance of metabolic and gene-expression regulation and of the processes therein.

Paragraphs 1.1 and 1.2 describe the processes involved in gene-expression and metabolic regulation. In Section 2, we explain how these two levels of regulation can be quantitatively dissected via the hierarchical regulation analysis. We pay specific attention to the most recent extensions of regulation analysis, i.e. time-dependent regulation analysis and the dissection of regulation *within* the gene-expression cascade. Section 3 discusses experimental aspects of the method. Finally, in Section 4, we will give an overview of published results and discuss which regulatory patterns emerge from the experimental data.

1.1. Gene-expression regulation

The research fields of metabolism and gene expression have long been quite isolated from each other. However, a change in flux through a metabolic pathway is often regulated by the interplay between the two. Transcriptomics, proteomics, and metabolomics have brought the concerted regulation of gene expression and metabolism to light. To gain understanding of the mechanisms and strategies underlying this complex regulation, new theoretical tools are required. regulation analysis was developed to meet this need.

Expression of a gene can be regulated at various levels, including transcription, mRNA processing, transport and stability, translation, posttranslational modification, and protein stability. Nevertheless, the term "gene expression" is often used as a synonym of "transcription". This reflects that in the early years, research of gene expression focused almost exclusively on transcription (Jacob and Monod, 1961). In the past decades, however, posttranscriptional regulation mechanisms have been unraveled in ever increasing complexity. The observed poor correlations between mRNA and protein levels (Greenbaum *et al.*, 2003; Griffin *et al.*, 2002) suggest that translation and/or protein degradation play a quantitatively important role in the regulation of gene expression. Consequently, quantitative studies of gene expression should involve not only the regulation of transcription but also the posttranscriptional regulation events. When we talk about gene expression, we imply all the processes involved in the gene-expression cascade which lead to a pool of functional protein. Let us review these briefly.

The transcription process itself may be regulated on the one hand via the structure of the chromatin template and on the other hand through factors that affect RNA polymerase and its associated proteins (Myers and

Kornberg, 2000). The chromatin structure can be in a compact state in which the genes are silent (nonpermissive chromatin) and in a more relaxed state in which gene promoters are accessible to the transcription machinery (permissive chromatin). A number of molecular modifications of DNA, such as DNA methylation and histone modifications, are associated with the permissive or nonpermissive state of chromatin (Jaenisch and Bird, 2003; Zaina *et al.*, 2010). Transcription is further regulated by the binding of sequence-specific repressor and activator proteins (transcription factors) to DNA elements and/or to the transcription machinery at promoters.

The quantitative importance of posttranscriptional mechanisms in the regulation of protein concentrations and activities is becoming increasingly clear (Clark *et al.*, 2009; Daran-Lapujade *et al.*, 2004; Day and Tuite, 1998; Haanstra *et al.*, 2008; Kolkman *et al.*, 2006). In many eukaryotes, the first events after transcription are mRNA splicing and the transport of mRNA molecules from the nucleus to the cytosol, both of which can be regulated (Licatalosi and Darnell, 2010).

The second level of posttranscriptional regulation is the turnover of the mRNA molecules. This process is extensively regulated during development or in response to environmental changes (Newbury, 2006). This results in a wide variety of decay rates between different mRNAs, as well as in specific regulation of the decay rates of individual mRNA species as a function of the state of the cell (Jacobson and Peltz, 1996).

Translation regulation is the third level of posttranscriptional regulation. It involves mechanisms in which the regulators interact directly with the ribosome or with associated initiation factors, thus influencing the recognition of the translation-initiation region by the ribosome complex (Lindahl and Hinnebusch, 1992). In addition, regulatory factors may bind to untranslated regions of the mRNA and thereby change the translation rate. Such regulatory factors are often proteins, but also *trans*-acting RNAs have been described (Wilkie *et al.*, 2003). The regulation of mRNA decay and translation are connected by proteins involved in both processes simultaneously, notably ribosomal proteins. In general, mRNA molecules bound to the translation machinery are more stable. When translation is repressed and mRNA molecules enter into a ribosome-free state, they become prone to degradation due to enhanced decapping (Franks and Lykke-Andersen, 2008; Newbury, 2006). Nevertheless, mRNA decapping can also occur when the ribosomes are still attached to the mRNA molecule (Hu *et al.*, 2009).

The turnover of proteins is the fourth level of posttranscriptional regulation. Proteins can be degraded via two main mechanisms: (i) via selective degradation by the ubiquitin–proteasome system (Hilt, 2004) and (ii) via autophagy, characterized by nonselective bulk degradation (Klionsky, 2007; Mizushima and Klionsky, 2007; Nakatogawa *et al.*, 2009). The ubiquitin–proteasome system is required for rapid and selective degradation of proteins when fast adaptation to a changing environment is needed. Autophagy, however, is involved in bulk degradation of cytosolic proteins and even of entire organelles (Kraft *et al.*, 2008; Tolkovsky, 2009).

The final level of posttranscriptional regulation is the posttranslational modification of proteins. Posttranslational modification is the chemical modification of a protein after its translation. This can be done by (i) addition of functional groups to the protein (e.g., phosphorylation, adenylation, or glycosylation), (ii) changing the chemical nature of an amino acid (e.g., the conversion of asparagine into aspartic acid by deamidation), (iii) covalent linkage of proteins to other proteins or peptides (e.g., ubiquitination), or (iv) changing the structure of the protein by, for instance, proteolytic cleavage (Larsen *et al.*, 2006; Meri and Baumann, 2001).

Besides the processes discussed, cellular localization and complex formation of proteins and/or mRNAs also play an important role in regulation of protein function.

1.2. Metabolic regulation

At the metabolic level, fluxes can be regulated either by changes in the concentrations of enzyme substrate(s), product(s), and/or effectors or by changes in the affinities of the enzymes to these molecules. Enzyme effectors could be pH, other small ions, or metabolites that act as specific, allosteric regulators. The affinity of an enzyme toward a metabolite can be changed by the presence of competing metabolites, by binding of an allosteric regulator, by the modification of the enzyme, or by changes in the expression of the different isoenzymes. In the later case, metabolic and gene-expression regulation overlap. Sauro (1989) described a methodology that dissects the relative importance of the various components of metabolic regulation. He defined regulation as the response of the system to changes in the environment at a constant concentration of the enzyme. As not all enzymes interact directly with the environment, the metabolic regulation of the enzymes will depend mainly on the changes in the concentrations of their substrates, products, and/or effectors. For instance, for an enzyme i, which is inhibited by its product P according to a Michaelis–Menten mechanism, the rate (v) follows the equation:

$$v_i = e_i k_{\mathrm{cat},i} \frac{\frac{S}{K_S}}{1 + \frac{S}{K_S} + \frac{P}{K_P}}, \qquad (27.1)$$

in which e equals the enzyme concentration, k_{cat} is the catalytic rate constant, S is the substrate concentration, P is the product concentration, K_S is the affinity constant of the enzyme for its substrate, and K_P is the affinity constant of the enzyme for its product.

Since Sauro describes the regulation of the flux by changes in the concentrations of the substrate S and the product P, the total derivative of the enzyme rate is given by:

$$\mathrm{d}\log v_i = \varepsilon_S^{v_i}\mathrm{d}\log S + \varepsilon_P^{v_i}\mathrm{d}\log P \tag{27.2}$$

The elasticity coefficients (ε) reflect the fractional change in enzyme rates associated with the fractional change in concentrations of the substrate, the product, or certain effectors, at constant concentrations of all other effectors (Burns *et al.*, 1985). For instance, the elasticity of the rate v_i toward its substrate S is expressed by:

$$\varepsilon_S^{v_i} = \frac{\partial \log v_i}{\partial \log S} \tag{27.3}$$

If we divide both sides of Eq. (27.2) by d log v_i, we obtain the following summation law:

$$1 = \varepsilon_S^{v_i}\frac{\mathrm{d}\log S}{\mathrm{d}\log v_i} + \varepsilon_P^{v_i}\frac{\mathrm{d}\log P}{\mathrm{d}\log v_i} \tag{27.4}$$

The two terms dissect the metabolic regulation into a part that depends on the changes in S and a part that depends on the changes in P. The two ratios of differentials are not equal to the inverse of the elasticity coefficients but correspond to the coresponse coefficients for metabolic control analysis as defined by Hofmeyr and Cornish-Bowden (1996). In principle, this methodology can be extended to include multiple substrates, products, and effectors (Sauro, 1989). As it involves partial derivatives, however, it requires either infinitely small changes of metabolite concentrations or titration of the perturbing agent. This makes the method experimentally challenging and is probably the reason why—to our knowledge—it has hardly been applied to experimental data.

2. Theory of Regulation Analysis

The quantitative method of Sauro discussed above dissects the various components of the metabolic regulation of the flux. To include the regulation by gene expression, a complementary methodology was developed by Ter Kuile and Westerhoff (2001). This methodology, called hierarchical regulation analysis, dissects gene expression from metabolic regulation. Since it can handle large changes, it is relatively easily applied to experimental data. The adjective "hierarchical" was added to "regulation analysis" (but often dropped later, for brevity) to indicate that now different levels of the cellular hierarchy were addressed.

Regulation analysis allows unraveling how changes in the rate of an enzymatic reaction are brought about by the interplay of gene expression and metabolism (Rossell *et al.*, 2005; ter Kuile and Westerhoff, 2001). It dissects quantitatively to what extent the flux through the enzyme is regulated by changes in the interaction of the enzyme with its substrate(s) (S), product(s) (P), and/or effectors(s) (X), the so-called metabolic regulation, or by changes in gene expression, called hierarchical regulation (Fig. 27.2). The rate (v) through an enzyme i depends linearly on a function f that depends on enzyme concentration (e), and on a function g that depends on metabolic effectors (S, P, X, K).

$$v_i = f(e_i)g(S, P, X, K)_i \tag{27.5}$$

In most enzyme kinetics, the function $f(e_i)$ equals $k_{\text{cat},i}e_i$ (cf. Eq. (27.1)), which, in turn, equals $V_{\max,i}$. Changes in the function g are caused by changes in the concentrations of substrates (S), products (P), and effectors (X), or by changes in affinities ($1/K$) of enzyme i toward its substrates, products, and effectors. The function g can have many shapes, and the choice in Eq. (27.1) is just an example. If we project Eq. (27.5) into

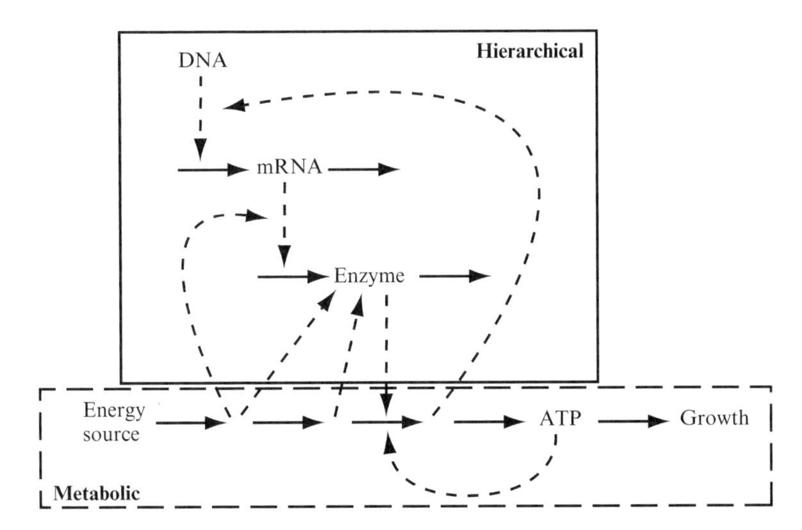

Figure 27.2 The distinction of hierarchical and metabolic regulation in regulation analysis. Production and consumption reactions are depicted by solid arrows, and regulatory events are depicted by dashed arrows. The hierarchical regulation refers to the regulation of the flux by changes in the gene-expression cascade, and the metabolic regulation refers to the regulation of the flux by changes in the metabolite concentration.

logarithmic space, the rate equation is dissected into a term that only depends on the enzyme concentration and a term that only depends on the concentrations of the metabolites and effectors.

$$\log v_i = \log f(e_i) + \log g(S, P, X, K)_i \tag{27.6}$$

This is equivalent to:

$$\log v_i = \log V_{\max,i} + \log g(S, P, X, K)_i \tag{27.7}$$

Considering the regulation that is responsible for the difference between two steady states (indicated by Δ), and dividing both sides of the equation by the logarithm of the relative change in the steady-state flux v_i through the enzyme, one obtains the following summation law:

$$1 = \frac{\Delta \log V_{\max,i}}{\Delta \log v_i} + \frac{\Delta \log g(S, P, X, K)_i}{\Delta \log v_i} = \rho_{h,i} + \rho_{m,i} \tag{27.8}$$

The hierarchical regulation coefficient ρ_h quantifies the relative contribution of changes in enzyme capacity ($V_{\max}$) to the regulation of the flux through the enzyme of interest and is associated with changes in the gene-expression cascade. The relative contribution of changes in the interaction of the enzyme with the rest of metabolism is reflected in the metabolic regulation coefficient ρ_m. Together, the two regulation coefficients should describe regulation completely, i.e. add up to 1.

2.1. Time-dependent regulation analysis

The original form of regulation analysis has been developed to compare two steady states. In order to obtain more insight into the adaptation strategies of organisms, it would be more informative to follow the patterns of regulation during the transition from one steady state to another. It is generally believed that metabolic regulation plays an important role at short-time scales, while regulation by gene expression kicks in later. This has, however, not been rigorously tested. Bruggeman *et al.* (2006) therefore derived regulation analysis as a function of time, which allows studying time-dependent regulation quantitatively. Two different methods of time-dependent regulation analysis were introduced. One is an integrative method over time and the second deals with instantaneous analysis of regulation over time (Bruggeman *et al.*, 2006).

For the integrative form of time-dependent regulation analysis, the regulation coefficients have been derived in a similar way as for the original form of regulation analysis. This form of regulation analysis integrates all the changes between t_0 and t according to the following equations:

$$1 = \frac{\log V_{\max,i}(t) - \log V_{\max,i}(t_0)}{\log v_i(t) - \log v_i(t_0)} + \frac{\log g(S,P,X,K)_i(t) - \log g(S,P,X,K)_i(t_0)}{\log v_i(t) - \log v_i(t_0)} = \rho_{\mathrm{h},i}(t) + \rho_{\mathrm{m},i}(t) \tag{27.9}$$

In this approach, the system is perturbed at time point t_0. When the new steady state is reached, the hierarchical and metabolic regulation coefficients have become equal to the values calculated with Eq. (27.8).

The instantaneous version of hierarchical regulation decomposes the time derivative of a rate v at each time point into a metabolic and a hierarchical contribution. Following the same strategy as above then leads to:

$$1 = \frac{\mathrm{d}\log V_{\max,\mathrm{i}}/\mathrm{d}t}{\mathrm{d}\log v_i/\mathrm{d}t}(t) + \frac{\mathrm{d}\log g(S,P,X,K)_i/\mathrm{d}t}{\mathrm{d}\log v_i/\mathrm{d}t}(t) = \pi_{\mathrm{h},i}(t) + \pi_{\mathrm{m},i}(t) \tag{27.10}$$

Dissecting $\pi_{\mathrm{m},i}(t)$ further into contributions by each individual metabolite leads to the alternative representation in Bruggeman *et al.* (2006). The difference between the two forms of time-dependent regulation analysis is that the integrative form determines the accumulated contributions of gene expression and metabolism between time point t_0 and t, while the instantaneous form expresses to which extent each mechanism is still active at time point t. In practice, this means that the time-dependent version requires accurate analysis of sufficient time points around t to determine the derivatives in Eq. (27.10). In contrast, in the integrative form, a single time point t can be compared to t_0. So far, there are only a limited number of experimental studies in which time-dependent regulation analysis was applied (van den Brink *et al.*, 2008; van Eunen *et al.*, 2009, 2010a), and in these cases, the integrative form was used.

2.2. Dissection of the regulation of V_{max}

Another extension made it possible to study in detail how the various processes involved in the gene-expression cascade regulate the flux. This was done by Daran-Lapujade *et al.* (2007) and Haanstra *et al.* (2008), who extended regulation analysis to dissect the hierarchical regulation into contributions by transcription, mRNA processing (splicing and degradation), translation, protein degradation, and posttranslational modification.

Here, we will show an example when dissecting the regulation of $V_{\max}$ into transcription, translation, protein degradation, and posttranslational modification. The V_{max} of an enzyme depends on the enzyme concentration e_i and its catalytic rate constant $k_{cat,i}$:

$$V_{max} = e_i k_{cat,i} \tag{27.11}$$

At steady state, the translation rate (v_{trans}) of an enzyme should equal its degradation rate (v_{deg}).

$$\frac{de_i}{dt} = v_{trans,i} - v_{deg,i} = 0 \tag{27.12}$$

The translation rate of a specific enzyme depends on the concentration of the ribosomes and of the mRNA coding for that specific enzyme. It is approximated by

$$v_{trans,i} = k_{trans,i}\text{ribosome}\cdot\text{mRNA}_i \tag{27.13}$$

in which k_{trans} is a first-order kinetic constant of the rate of translation. The degradation rate of the enzyme was taken proportional to the concentration of the enzyme and a phenomenological kinetic constant k_{deg}.

$$v_{deg,i} = k_{deg,i}e_i \tag{27.14}$$

Combining the Eqs. (27.11)–(27.14) results in:

$$V_{max} = \frac{k_{cat,i}k_{trans,i}\text{ribosome}\cdot\text{mRNA}_i}{k_{deg,i}} \tag{27.15}$$

If we project Eq. (27.15) into logarithmic space, consider the regulation that is responsible for the difference between two steady states (indicated by Δ) and divide both sides of the equation by the logarithm of V_{max}, the following summation law is obtained:

$$1 = \underbrace{\frac{\Delta\log k_{cat,i}}{\Delta\log V_{max}}}_{\rho_{PT,i}} + \underbrace{\frac{\Delta\log \text{mRNA}_i}{\Delta\log V_{max}}}_{\rho_{mRNA,i}} + \underbrace{\left(\frac{\Delta\log k_{trans,i}}{\Delta\log V_{max}} + \frac{\Delta\log \text{ribosome}}{\Delta\log V_{max}}\right)}_{\rho_{trans,i}} - \underbrace{\frac{\Delta\log k_{deg,i}}{\Delta\log V_{max}}}_{\rho_{deg,i}} \tag{27.16}$$

in which the regulation of V_{max} by posttranslational modification is reflected by ρ_{PT}, the regulation of V_{max} by the transcript concentration is reflected by ρ_{mRNA}, the regulation of V_{max} by the translation activity is reflected by

ρ_{trans}, and the ρ_{deg} quantifies the regulation by protein degradation. The sum of these four regulation coefficients add up to 1. With this extension, the hierarchical regulation of the flux through each enzyme can thus be dissected into the various levels of the gene-expression cascade. Depending on what is measurable, this methodology can be varied upon. Daran-Lapujade *et al.* (2007) included the specific growth rate of the cells, which leads to dilution of proteins. Haanstra *et al.* (2008) focused on the metabolism of mRNA in African trypanosomes and included mRNA precursor degradation, *trans*-splicing and degradation of mature mRNAs.

2.3. Possible outcomes of regulation analysis

For the original steady-state form of regulation analysis, Rossell *et al.* (2005) have classified the variety of numerical outcomes into five distinct categories. In the time-dependent regulation analysis, shifts from one category to another can be found. Below, the different categories of regulation will be discussed.

Purely hierarchical regulation In this category, the change in the enzyme rate is completely caused by the change in V_{max}, resulting in a hierarchical regulation coefficient ρ_h of 1 and a metabolic regulation coefficient ρ_m of 0. In this case, the metabolic function $g(S, P, X, K)$ does not change, implying that there is no metabolic regulation. This may be due to constant concentrations of all the metabolites that are involved. In general, however, the underlying molecular events can be more complex. In principle, changes in different K_m's and/or metabolite concentrations can occur simultaneously, but such that there is no net change of the metabolic function g. In conclusion, in this category, the metabolic changes do not make a net contribution to a change in the enzyme rate.

Purely metabolic regulation In this case, the change in enzyme rate results from a change in the metabolic function only. For example, the flux may be decreased by a decreased substrate concentration, while the V_{max} remains unchanged. Consequently, the ρ_m equals 1 and the ρ_h is 0. In general, not only the substrate but also the product or other metabolic effectors may be responsible for the observed metabolic regulation. Also shifts in isoenzyme expression may underlie the metabolic regulation if the isoenzymes have different affinity constants.

Cooperative regulation When changes in the V_{max} and the metabolic function affect the enzyme rate in the same direction, a ρ_h between 0 and 1 is obtained. From the summation law (Eq. (27.8)), it then follows that the ρ_m

is also in between 0 and 1. In this case, the increase of the rate is larger than the increase of the V_{max}; the remaining increment may be due to an increased substrate concentration, or in general to changes in any metabolic effectors.

Antagonistic regulation directed by V_{max} In this category, the changes in V_{max} and the metabolic function have antagonistic effects on the enzyme rate. Specifically, the increase in V_{max} is larger than the increase in rate because a decrease in substrate concentration (or an increase in product concentration) is counteracting the change in V_{max}. The net result is that the enzyme rate changes in the same direction as its V_{max}, but not quite as much as the V_{max}. This results in a ρ_h higher than 1 and consequently a ρ_m lower than 0. In this category, the hierarchical regulation is dominant over the metabolic regulation.

Antagonistic regulation directed by metabolism In this case, the V_{max} changes in the opposite direction as the enzyme rate, resulting in a ρ_h below 0 and a ρ_m higher than 1. As in the previous category, the changes in V_{max} and metabolism have antagonistic effects on the flux; however, now metabolism is dominant. Here, the change in the metabolic function g is larger than the overall change in rate because the change in V_{max} is counteracting the effect of the change in substrate concentration.

3. Experimental Tools for Regulation Analysis

Experimentally, the hierarchical regulation coefficient ρ_h is the one that is more readily determined, as it requires measurements of the V_{max} of the enzyme and the flux through it, under two conditions, according to:

$$\rho_h = \frac{\Delta \log V_{max,i}}{\Delta \log v_i} \tag{27.17}$$

This is often possible, although sometimes challenging. First, the V_{max} should be measured under physiologically relevant conditions. Second, when isoenzymes with different substrate affinities are active at the same time, independent analysis of each isoenzyme may be required. However, Rossell *et al.* (2005) demonstrated that this complication can often be overcome by a precise interpretation of the coefficients: if the overall V_{max} is measured, the metabolic regulation coefficient includes changes in K_m. Third, when the metabolic network is complex without a single major flux routing, it may be necessary to carry out more advanced metabolic flux analysis to resolve the intracellular fluxes (Christensen and Nielsen, 2000; Stephanopoulos, 1999).

3.1. Metabolic flux analysis

In order to be able to calculate the hierarchical regulation coefficient, determination of the V_{max} of the enzyme and flux through it are necessary. For complex metabolic pathways with side branches, the determination of the flux through the individual enzymes is not straightforward. If fluxes cannot be measured directly, they must be inferred from measured quantities through computer model-based interpretation. This is called metabolic flux analysis (Dauner, 2010; Sauer, 2006). We will discuss two methods of metabolic flux analysis, which are used in the field of systems biology.

The first method (stoichiometric flux analysis) is based on measurements of the rates of consumption of nutrients and the production of end products and on the known stoichiometry of the metabolic network. The method avoids difficulties of kinetic modeling (Llaneras and Pico, 2008; Stephanopoulos, 1999). It assumes that there is a metabolic quasi steady state for all internal metabolites. This assumption is, at the same time, one of the disadvantages of the method because it precludes the dynamic analysis of intracellular fluxes. The second method, based on isotope labeling, does allow dynamic analysis of the intracellular fluxes, including the flux through metabolic cycles (Wiechert, 2001). A growing cell culture is fed with isotope-labeled substrates until the isotope label is distributed throughout the network (Sauer, 2004; Wiechert, 2001). The particular distribution of the fluxes in the cell determines the specific labeling patterns in the metabolic intermediates. These specific labeling patterns are measured by NMR or mass spectrometry techniques. The flux data can be obtained from the labeling patterns by at least two different approaches. The first approach integrates simultaneously the ^{13}C data, extracellular fluxes, and biosynthetic requirements in computer models. The flux distribution is then obtained by fitting the intracellular fluxes to the experimental data by minimizing the differences between the observed and simulated isotopologue spectra (van Winden *et al.*, 2005; Wiechert, 2001). The second method is the flux ratio method, in which the relative distribution of converging pathways to the formation of a specific metabolite is quantified from a particular combination of NMR or mass pattern (Sauer, 2006). With the fitting method, absolute fluxes are obtained, whereas the flux ratio method results in relative values. However, the flux ratio approach is extended to the estimation of absolute fluxes by anchoring it to one flux, which is measured absolutely. In rapidly growing organisms, this may be the flux of biomass production. The relative fluxes then become that flux multiplied by the fractional composition of the biomass in terms of building blocks such as amino acids and nucleotides. The latter method is applicable at higher throughput (Fischer and Sauer, 2005; Fischer *et al.*, 2004).

3.2. Measuring V_{max}

The V_{max} is measured as the maximum rate of an enzyme in cell-free extracts at saturating substrate concentrations. There are a number of suitable methods available, which can be distinguished into "continuous" and "discontinuous" assays. In continuous assays, the disappearance or appearance of a given substrate or product, respectively, is continuously monitored. This is most often measured by spectroscopic techniques, such as ultraviolet-visible absorption and fluorescence emission (Harris and Keshwani, 2009). The frequently used NAD(P)H-linked assays belong to the group of "continuous" assays. These assays make use of the absorption and/or fluorescent properties of the NAD(P)H molecule. If the NAD(P)H is a product or substrate of the enzyme, the V_{max} can be directly measured by monitoring the (dis)appearance of NAD(P)H. However, when this is not the case, either one or more additional enzymes are included in the reaction mixture, in order to couple it to a reaction that does have NAD(P)H as substrate or product (Eisenthal and Danson, 2002).

Discontinuous enzyme assays are used when a given substrate–product pair exhibits either similar or no spectroscopic properties. In these assays, the reaction is stopped or "quenched" at different time points, in order to be able to measure the change in concentration of either the substrate or the product for each individual time point. An example of discontinuous assays is the radiometric assays by using substrates which contain a radioactive isotopic atom. These assays are usually extremely sensitive, which makes it possible to measure very low levels of enzyme activity (Eisenthal and Danson, 2002). A similar approach is to monitor the incorporation or release of stable isotopes when the substrate is converted into its product by using mass spectrometry (Honda *et al.*, 2007; Kranendijk *et al.*, 2009). Besides the use of isotopes, high-performance liquid chromatography could also be used to determine the concentrations of both substrates and products (Welling *et al.*, 1994). The V_{max} could then be calculated based on the consumption and/or production rates of the substrates and/or products of the enzyme.

4. Strategies of Flux Regulation

In theory, we may distinguish various flux regulation strategies: (i) modification of a single rate-limiting enzyme (single-site modulation), (ii) a simultaneous and proportional modulation of all the enzymes in the pathway (multisite modulation; Fell and Thomas, 1995), (iii) exclusively metabolic regulation, and (iv) different enzymes playing different regulatory roles, often even depending on the particular conditions and challenges. If we translate these strategies in terms of regulation analysis, single-site modulation means that one enzyme is regulated purely at the gene-expression level ($\rho_h = 1$), while all the others are regulated by changes in the metabolite concentrations ($\rho_h = 0$). Even though single rate-limiting enzymes may exist for some pathways (Bakker *et al.*, 1999; Diderich *et al.*, 1999; Reijenga *et al.*, 2001; Shulman *et al.*, 1995; Westerhoff and Arents, 1984), it is more common that the control of the flux is distributed over several enzymes (reviewed in Fell, 1992) and therefore, single-site modulation may be rather exceptional and certainly not the norm. Indeed, efforts to correlate a change in glycolytic flux to a change in the activity of any single enzyme have failed (Daran-Lapujade *et al.*, 2007; Even *et al.*, 2003; Haanstra *et al.*, 2008; Postmus *et al.*, 2008; Rossell *et al.*, 2006, 2008; ter Kuile and Westerhoff, 2001; van den Brink *et al.*, 2008; van Eunen *et al.*, 2009, 2010a). The opposite theory of multisite modulation proposes metabolite homeostasis, excluding the possibility of metabolic regulation ($\rho_h = 1$ for all enzymes). Examples of multisite modulation have been suggested, such as lipogenesis in mice, the urea cycle in rats, and photosynthesis in green plants (Fell, 1992). However, in none of these studies, the hierarchical regulation coefficients have been measured quantitatively, such that their equality has not been proven. Further, in regulation analysis of yeast glycolysis where this quantification was undertaken, this has never been found (see below). The third possible strategy, exclusive metabolic regulation, which corresponds to $\rho_h = 0$ for all enzymes in the pathway, has not yet been observed experimentally.

4.1. How are changes in fluxes regulated in practice?

So far hierarchical regulation analysis has been applied to quantify the regulation fluxes through glycolysis either in *Saccharomyces cerevisiae*, *Lactococcus lactis* or in *Trypanosoma brucei* (Daran-Lapujade *et al.*, 2007; Even *et al.*, 2003; Haanstra *et al.*, 2008; Postmus *et al.*, 2008; Rossell *et al.*, 2006, 2008; ter Kuile and Westerhoff, 2001; van den Brink *et al.*, 2008; van Eunen *et al.*, 2009, 2010a). The observed regulation of the flux through the glycolytic and fermentative pathways was extremely diverse. Different perturbations to the cells led to different patterns of distributed regulation (Daran-Lapujade *et al.*, 2007; Even *et al.*, 2003; Haanstra *et al.*, 2008; Postmus *et al.*, 2008; Rossell *et al.*, 2006, 2008; ter Kuile and Westerhoff, 2001; van den Brink *et al.*, 2008; van Eunen *et al.*, 2009, 2010a). Rossell *et al.* (2008, 2006) concluded that the flux through a pathway is regulated in a subtle way, that enzymes may play different regulatory roles and that the regulation of fluxes does not need to be governed by single drives or constraints.

In this paragraph, we will make an overview of the regulation data presented in literature so far. Since most studies were done in yeast, we will focus on which regulation strategy yeast glycolysis follows when adapting to environmental changes and if similar strategies are followed upon the different perturbations (Daran-Lapujade *et al.*, 2007; Postmus *et al.*, 2008; Rossell *et al.*, 2006, 2008; van den Brink *et al.*, 2008; van Eunen *et al.*, 2009, 2010a). Further, we will study if the growth conditions play a role in the regulation of the glycolytic flux.

The recent development of time-dependent regulation analysis made it possible to follow flux regulation in time when cells adapt to an environmental change. It also divided the regulation studies in two groups: on the one hand, those comparing two steady-state conditions and on the other hand, the time-dependent studies. The results obtained with the time-dependent form of regulation analysis can reveal shifts from one regulation category to another. For instance, it has been shown that during prolonged nitrogen starvation, the regulation of some enzymes shifted from hierarchical to metabolic, whereas the regulation of other enzymes shifted the other way around (van Eunen *et al.*, 2010a). This makes it difficult to compare regulation data obtained by the original steady-state regulation analysis to those from the time-dependent regulation analysis. For the time-dependent datasets, we have therefore used the last time point measured, which is probably close to the new steady state.

Figure 27.3 gives a complete overview of all regulation data that are currently available for yeast glycolysis. Overall, metabolic regulation dominated slightly over hierarchical regulation with 60% versus 40% (see Tables 27.1 and 27.2). The major exceptions were two nitrogen starvation experiments starting from respirofermentative growth (Table 27.2: N-starved versus nonstarved cells from either batch cultures or from aerobic glucose-limited chemostat cultures at a dilution rate of 0.35 h^{-1}) both showing dominant hierarchical regulation. Mean values over the glycolytic enzymes for the hierarchical regulation coefficients (ρ_h) were 0.9 and 1.6, respectively. Most of the enzymes showed a different regulation depending on the conditions that were studied. In the summary of Table 27.1, this becomes apparent from the large standard deviations which were taken over the various studies. Nevertheless, it is clear that GAPDH is mainly regulated by metabolism (90%) with a relatively small standard deviation (30%). Enolase showed cooperative regulation by both metabolism and gene expression under all conditions. Overall, this resulted in a distribution of 40% hierarchical versus 60% metabolic regulation with a small standard deviation (20%) (see Table 27.1).

Until now, it is impossible to draw general conclusions about the impact of the growth conditions prior to a perturbation, as only two studies were time resolved. Nevertheless, conclusions can be drawn about the influence of respiratory versus respirofermentative growth on the distribution of regulation. To this end, we have divided the experiments in two groups. The first group is composed of comparisons between conditions in which the cells exhibit (respiro-)fermentative metabolism. To this, first group belong the nitrogen starvation studies with cells from either a batch culture (Rossell *et al.*, 2006) or an aerobic glucose-limited chemostat culture at a dilution rate of 0.35 h^{-1} (van Eunen *et al.*, 2009), the shift from aerobic glucose-limited chemostat conditions at a growth rate of 0.35 h^{-1} to aerobic glucose-excess conditions (van Eunen *et al.*, 2009) and the comparison of the anaerobic glucose-limited chemostat culture in the presence and absence of benzoic acid (Daran-Lapujade *et al.*, 2007). The second group

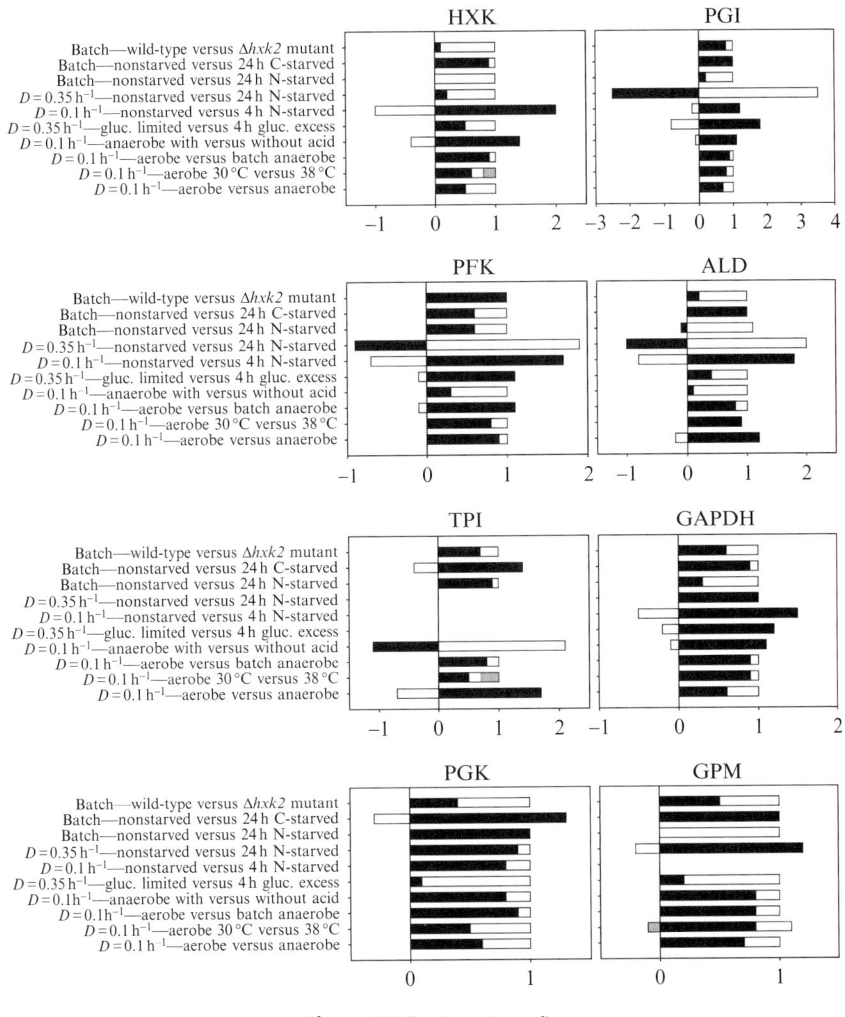

Figure 27.3 (continued)

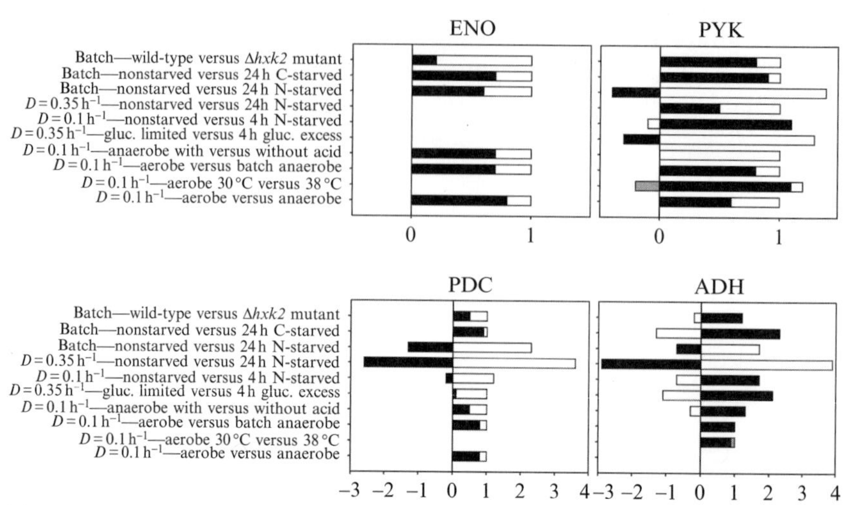

Figure 27.3 Overview of regulation analysis data. Black bars: metabolic regulation, ρ_m; white bars: hierarchical regulation, ρ_h; gray bars: regulation by temperature, ρ_T. Different studies presented are batch—wild-type versus Δ*hxk2* mutant (Rossell *et al.*, 2008), batch—nonstarved versus 24 h C-starved (Rossell *et al.*, 2006), batch—nonstarved versus 24 h N-starved (Rossell *et al.*, 2006), $D = 0.35\ h^{-1}$—nonstarved versus 24 h N-starved (van Eunen *et al.*, 2009), $D = 0.1\ h^{-1}$—nonstarved versus 4 h N-starved (van Eunen *et al.*, 2010a), $D = 0.35\ h^{-1}$—glc. limited versus 4 h glc. excess (van Eunen *et al.*, 2009), $D = 0.1\ h^{-1}$—anaerobe with versus without acid (Daran-Lapujade *et al.*, 2007), $D = 0.1\ h^{-1}$—aerobe versus batch anaerobe (van den Brink *et al.*, 2008), $D = 0.1\ h^{-1}$—aerobe 30 °C versus 38 °C (Postmus *et al.*, 2008), $D = 0.1\ h^{-1}$—aerobe versus anaerobe (Daran-Lapujade *et al.*, 2007).

contains studies that compared a condition in which the cells exhibit respiratory metabolism to a condition in which the cells exhibit (respiro-) fermentative metabolism. Studies belonging to the second group are the shift from an aerobic glucose-limited chemostat condition at a D of 0.1 h^{-1} to an anaerobic glucose-excess batch condition (van den Brink *et al.*, 2008), an aerobic glucose-limited chemostat culture grown at a growth rate of 0.1 h^{-1} at 30 °C compared to one at 38 °C (Postmus *et al.*, 2008), the glucose-limited chemostat cultures grown at a growth rate of 0.1 h^{-1} under aerobic versus anaerobic conditions (Daran-Lapujade *et al.*, 2007), the nitrogen-starved cells from an aerobic glucose-limited chemostat culture at a growth rate of 0.1 h^{-1} (van Eunen *et al.*, 2010a), the carbon-starved cells from a batch culture (Rossell *et al.*, 2006), and the comparison between the wild type and the Δ*hxk2* strain grown under batch conditions (Rossell *et al.*, 2008). In the studies from group 2 (respiratory vs. (respiro-)fermentative metabolism), the observed changes of flux were regulated for 90% by

Table 27.1 The average regulation coefficients per enzyme

Enzyme	ρ_h	Standard deviation	ρ_m
HXT	0.0	0.9	1.0
HXK	0.3	0.6	0.7
PGI	0.4	1.2	0.6
PFK	0.3	0.7	0.7
ALD	0.5	0.8	0.5
TPI	0.3	0.9	0.7
GAPDH	0.1	0.3	0.9
PGK	0.3	0.3	0.7
GPM	0.3	0.4	0.7
ENO	0.4	0.2	0.6
PYK	0.5	0.5	0.5
PDC	1.1	1.2	−0.1
ADH	0.2	1.6	0.8
Overall	*0.4*	*0.2*	*0.6*

The averages of all studies (Daran-Lapujade *et al.*, 2007; Postmus *et al.*, 2008; Rossell *et al.*, 2006, 2008; van den Brink *et al.*, 2008; van Eunen *et al.*, 2009, 2010a) were taken for each enzyme separately.

Table 27.2 The average regulation coefficients per physiological perturbation

Study	ρ_h	Standard deviation	ρ_m
Batch—wild-type versus Δ*hxk2* mutant	0.4	0.3	0.6
Batch—nonstarved versus 24 h C-starved	0.0	0.4	1.0
Batch—nonstarved versus 24 h N-starved	0.9	0.6	0.1
$D = 0.35\ h^{-1}$—nonstarved versus 24 h N-starved	1.6	1.6	−0.6
$D = 0.1\ h^{-1}$—nonstarved versus 4 h N-starved	−0.4	0.7	1.4
$D = 0.35\ h^{-1}$—glucose limited versus 4 h glucose excess	0.3	0.8	0.7
$D = 0.1\ h^{-1}$—anaerobe with versus without acid	0.4	0.7	0.6
$D = 0.1\ h^{-1}$—aerobe versus batch anaerobe	0.1	0.1	0.9
$D = 0.1\ h^{-1}$—aerobe 30 °C versus 38 °C	0.2	0.2	0.8
$D = 0.1\ h^{-1}$—aerobe versus anaerobe	0.2	0.3	0.8
Overall	*0.4*	*0.6*	*0.6*

The averages over all glycolytic enzymes were taken for each perturbation separately.

metabolism with a standard deviation of only 20% (Table 27.3). However, in group 1, in which always two conditions of (respiro-)fermentative metabolism were compared, regulation by gene expression was more important (80%) (Table 27.3).

Table 27.3 Average regulation coefficients after dividing the studies into two groups, that is, those comparing two conditions in which the cells exhibit (respiro-) fermentative metabolism (group 1) and those comparing a condition in which the cells exhibit respiratory metabolism to a condition in which the cells exhibit (respiro-) fermentative metabolism (group 2)

	Group 1			Group 2		
Enzyme	ρ_h	Standard deviation	ρ_m	ρ_h	Standard deviation	ρ_m
HXT	1.2	–[a]	−0.2	−0.3	0.8	1.3
HXK	0.5	0.6	0.5	0.1	0.6	0.9
PGI	0.9	1.9	0.2	0.1	0.2	0.9
PFK	0.7	0.9	0.3	0.0	0.4	1.0
ALD	1.2	0.6	−0.2	0.0	0.5	1.0
TPI	1.1	1.4	−0.1	−0.1	0.5	1.1
GAPDH	0.1	0.4	0.9	0.1	0.3	0.9
PGK	0.3	0.4	0.7	0.3	0.3	0.8
GPM	0.5	0.6	0.6	0.3	0.2	0.7
ENO	0.4	0.1	0.7	0.4	0.3	0.6
PYK	1.1	0.4	−0.1	0.1	0.2	0.9
PDC	1.8	1.4	−0.8	0.4	0.5	0.6
ADH	1.1	2.2	0.0	−0.4	0.6	1.4
Overall	*0.8*	*0.5*	*0.2*	*0.1*	0.2	*0.9*

The averages of all studies within the particular group were taken for each enzyme separately.
[a] Within this group the V_{max} of the hexose transporters was only measured in the study: batch—nonstarved versus 24 h N-starved.

Since in all individual studies a different type of regulation is found, none of the regulation strategies described at the beginning of this paragraph can be identified as the *generic* strategy of yeast glycolysis. However, on average group 1 ((respiro-)fermentative metabolism) approximates multisite modulation most closely. Yet, the large standard deviation (50%) in this group indicates that not all enzymes join in multisite modulation. Thereby, the proposed advantage of homogeneous multisite regulation that the metabolite concentrations can remain unaffected is lost. Indeed, metabolite concentrations were changed in the nitrogen starvation experiments carried out on the cells from chemostat cultures (van Eunen *et al.*, 2010a). In addition, also in this group, GAPDH is regulated metabolically. The third strategy of pure metabolic regulation is most closely approximated by group 2 (respiratory vs. (respiro-)fermentative metabolism). This dominant metabolic regulation suggests that respiratory cells, which typically have a low glycolytic flux, have a substantial overcapacity of enzyme activities. This allows an upregulation of glycolytic flux with minor induction of glycolytic gene expression.

5. Concluding Remarks

Regulation analysis has drastically changed our view on the dynamic regulation of metabolic fluxes. Regulation turned out to be distributed over various processes and depended on time, enzyme, and condition of interest. The observed complexity was beyond our expectations.

Until now, the focus of regulation analysis has been on the question *how* changes in specific fluxes were regulated? Technical advances to measure the processing and turnover of mRNAs and proteins in high throughput will allow to dissect the regulation within the gene-expression cascade in more detail and beyond specific metabolic pathways. Improvements in the analysis of enzyme activities *in vivo* or under physiological conditions (e.g., van Eunen *et al.*, 2010b) should pave the way for quantification of the role of individual metabolites in the overall regulation. It is not unlikely that a further dissection of regulation will reveal an every increasing complexity. The true challenge is therefore to discover meaningful underlying patterns. The question then becomes *why* fluxes (or at least the glycolytic flux) are regulated in such a complex manner. The *why* question assumes a strategy or a purpose behind this regulation, or in less teleological terms an evolutionary advantage in the context of limited resources. Since proposed strategies that make intuitively sense, like multisite regulation of single rate-limiting enzymes can explain only few experimental cases, it is likely that other evolutionary purposes are served than hitherto thought. For instance, an economic use of cellular space and biosynthetic capacity (Molenaar *et al.*, 2009) and the dynamics of cellular responses (Assmus *et al.*, 2006) should be taken into account in future studies.

In conclusion, the challenge for systems biology will not only be to understand the mechanisms of biological regulation but even more to understand how certain mechanisms could have evolved.

REFERENCES

Assmus, H. E., Herwig, R., Cho, K. H., and Wolkenhauer, O. (2006). Dynamics of biological systems: Role of systems biology in medical research. *Expert Rev. Mol. Diagn.* **6,** 891–902.

Bakker, B. M., Michels, P. A., Opperdoes, F. R., and Westerhoff, H. V. (1999). What controls glycolysis in bloodstream form Trypanosoma brucei? *J. Biol. Chem.* **274,** 14551–14559.

Bruggeman, F. J., de Haan, J., Hardin, H., Bouwman, J., Rossell, S., van Eunen, K., Bakker, B. M., and Westerhoff, H. V. (2006). Time-dependent hierarchical regulation analysis: Deciphering cellular adaptation. *Syst. Biol. (Stevenage)* **153,** 318–322.

Bruggeman, F. J., Rossell, S., Van Eunen, K., Bouwman, J., Westerhoff, H. V., and Bakker, B. (2007). Systems biology and the reconstruction of the cell: From molecular components to integral function. *Subcell. Biochem.* **43,** 239–262.

Burns, J. A., Cornishbowden, A., Groen, A. K., Heinrich, R., Kacser, H., Porteous, J. W., Rapoport, S. M., Rapoport, T. A., Stucki, J. W., Tager, J. M., Wanders, R. J. A., and Westerhoff, H. V. (1985). Control analysis of metabolic systems. *Trends Biochem. Sci.* **10,** 16.

Christensen, B., and Nielsen, J. (2000). Metabolic network analysis. A powerful tool in metabolic engineering. *Adv. Biochem. Eng. Biotechnol.* **66,** 209–231.

Clark, A., Dean, J., Tudor, C., and Saklatvala, J. (2009). Post-transcriptional gene regulation by MAP kinases via AU-rich elements. *Front. Biosci.* **14,** 847–871.

Daran-Lapujade, P., Jansen, M. L., Daran, J. M., van Gulik, W., de Winde, J. H., and Pronk, J. T. (2004). Role of transcriptional regulation in controlling fluxes in central carbon metabolism of Saccharomyces cerevisiae. A chemostat culture study. *J. Biol. Chem.* **279,** 9125–9138.

Daran-Lapujade, P., Rossell, S., van Gulik, W. M., Luttik, M. A., de Groot, M. J., Slijper, M., Heck, A. J., Daran, J. M., de Winde, J. H., Westerhoff, H. V., Pronk, J. T., and Bakker, B. M. (2007). The fluxes through glycolytic enzymes in Saccharomyces cerevisiae are predominantly regulated at posttranscriptional levels. *Proc. Natl. Acad. Sci. USA* **104,** 15753–15758.

Dauner, M. (2010). From fluxes and isotope labeling patterns towards in silico cells. *Curr. Opin. Biotechnol.* **21,** 55–62.

Day, D. A., and Tuite, M. F. (1998). Post-transcriptional gene regulatory mechanisms in eukaryotes: An overview. *J. Endocrinol.* **157,** 361–371.

Diderich, J. A., Teusink, B., Valkier, J., Anjos, J., Spencer-Martins, I., van Dam, K., and Walsh, M. C. (1999). Strategies to determine the extent of control exerted by glucose transport on glycolytic flux in the yeast Saccharomyces bayanus. *Microbiology* **145**(Pt. 12), 3447–3454.

Eisenthal, R., and Danson, M. (2002). Enzyme Assays: A Practical Approach. Oxford University Press, Inc., New York.

Even, S., Lindley, N. D., and Cocaign-Bousquet, M. (2003). Transcriptional, translational and metabolic regulation of glycolysis in Lactococcus lactis subsp. cremoris MG 1363 grown in continuous acidic cultures. *Microbiology* **149,** 1935–1944.

Fell, D. A. (1992). Metabolic control analysis: A survey of its theoretical and experimental development. *Biochem. J.* **286**(Pt. 2), 313–330.

Fell, D. A., and Thomas, S. (1995). Physiological control of metabolic flux: The requirement for multisite modulation. *Biochem. J.* **311**(Pt. 1), 35–39.

Fischer, E., and Sauer, U. (2005). Large-scale in vivo flux analysis shows rigidity and suboptimal performance of *Bacillus subtilis* metabolism. *Nat. Genet.* **37,** 636–640.

Fischer, E., Zamboni, N., and Sauer, U. (2004). High-throughput metabolic flux analysis based on gas chromatography-mass spectrometry derived 13C constraints. *Anal. Biochem.* **325,** 308–316.

Franks, T. M., and Lykke-Andersen, J. (2008). The control of mRNA decapping and P-body formation. *Mol. Cell* **32,** 605–615.

Greenbaum, D., Colangelo, C., Williams, K., and Gerstein, M. (2003). Comparing protein abundance and mRNA expression levels on a genomic scale. *Genome Biol.* **4,** 117.

Griffin, T. J., Gygi, S. P., Ideker, T., Rist, B., Eng, J., Hood, L., and Aebersold, R. (2002). Complementary profiling of gene expression at the transcriptome and proteome levels in Saccharomyces cerevisiae. *Mol. Cell. Proteomics* **1,** 323–333.

Haanstra, J. R., Stewart, M., Luu, V. D., van Tuijl, A., Westerhoff, H. V., Clayton, C., and Bakker, B. M. (2008). Control and regulation of gene expression: Quantitative analysis of the expression of phosphoglycerate kinase in bloodstream form Trypanosoma brucei. *J. Biol. Chem.* **283,** 2495–2507.

Harris, T. K., and Keshwani, M. M. (2009). Measurement of enzyme activity. *Methods Enzymol.* **463,** 57–71.

Hilt, W. (2004). Targets of programmed destruction: A primer to regulatory proteolysis in yeast. *Cell. Mol. Life Sci.* **61,** 1615–1632.

Hofmeyr, J. H., and Cornish-Bowden, A. (1996). Co-response analysis: A new experimental strategy for metabolic control analysis. *J. Theor. Biol.* **182,** 371–380.

Honda, A., Mizokami, Y., Matsuzaki, Y., Ikegami, T., Doy, M., and Miyazaki, H. (2007). Highly sensitive assay of HMG-CoA reductase activity by LC-ESI-MS/MS. *J. Lipid Res.* **48,** 1212–1220.

Hu, W., Sweet, T. J., Chamnongpol, S., Baker, K. E., and Coller, J. (2009). Co-translational mRNA decay in *Saccharomyces cerevisiae*. *Nature* **461,** 225–229.

Jacob, F., and Monod, J. (1961). Genetic regulatory mechanisms in the synthesis of proteins. *J. Mol. Biol.* **3,** 318–356.

Jacobson, A., and Peltz, S. W. (1996). Interrelationships of the pathways of mRNA decay and translation in eukaryotic cells. *Annu. Rev. Biochem.* **65,** 693–739.

Jaenisch, R., and Bird, A. (2003). Epigenetic regulation of gene expression: How the genome integrates intrinsic and environmental signals. *Nat. Genet.* **33**(Suppl.), 245–254.

Kahn, D., and Westerhoff, H. V. (1993). The regulatory strength: How to be precise about regulation and homeostasis. *Acta Biotheor.* **41,** 85–96.

Klionsky, D. J. (2007). Autophagy: From phenomenology to molecular understanding in less than a decade. *Nat. Rev. Mol. Cell Biol.* **8,** 931–937.

Kolkman, A., Daran-Lapujade, P., Fullaondo, A., Olsthoorn, M. M., Pronk, J. T., Slijper, M., and Heck, A. J. (2006). Proteome analysis of yeast response to various nutrient limitations. *Mol. Syst. Biol.* **2,** 0026.

Kraft, C., Deplazes, A., Sohrmann, M., and Peter, M. (2008). Mature ribosomes are selectively degraded upon starvation by an autophagy pathway requiring the Ubp3p/Bre5p ubiquitin protease. *Nat. Cell Biol.* **10,** 602–610.

Kranendijk, M., Salomons, G. S., Gibson, K. M., Aktuglu-Zeybek, C., Bekri, S., Christensen, E., Clarke, J., Hahn, A., Korman, S. H., Mejaski-Bosnjak, V., Superti-Furga, A., Vianey-Saban, C., *et al.* (2009). Development and implementation of a novel assay for L-2-hydroxyglutarate dehydrogenase (L-2-HGDH) in cell lysates: L-2-HGDH deficiency in 15 patients with L-2-hydroxyglutaric aciduria. *J. Inherit. Metab. Dis.* **32,** 713–719.

Larsen, M. R., Trelle, M. B., Thingholm, T. E., and Jensen, O. N. (2006). Analysis of posttranslational modifications of proteins by tandem mass spectrometry. *Biotechniques* **40,** 790–798.

Licatalosi, D. D., and Darnell, R. B. (2010). RNA processing and its regulation: Global insights into biological networks. *Nat. Rev. Genet.* **11,** 75–87.

Lindahl, L., and Hinnebusch, A. (1992). Diversity of mechanisms in the regulation of translation in prokaryotes and lower eukaryotes. *Curr. Opin. Genet. Dev.* **2,** 720–726.

Llaneras, F., and Pico, J. (2008). Stoichiometric modelling of cell metabolism. *J. Biosci. Bioeng.* **105,** 1–11.

Meri, S., and Baumann, M. (2001). Proteomics: Posttranslational modifications, immune responses and current analytical tools. *Biomol. Eng.* **18,** 213–220.

Mizushima, N., and Klionsky, D. J. (2007). Protein turnover via autophagy: Implications for metabolism. *Annu. Rev. Nutr.* **27,** 19–40.

Molenaar, D., van Berlo, R., de Ridder, D., and Teusink, B. (2009). Shifts in growth strategies reflect tradeoffs in cellular economics. *Mol. Syst. Biol.* **5,** 323.

Myers, L. C., and Kornberg, R. D. (2000). Mediator of transcriptional regulation. *Annu. Rev. Biochem.* **69,** 729–749.

Nakatogawa, H., Suzuki, K., Kamada, Y., and Ohsumi, Y. (2009). Dynamics and diversity in autophagy mechanisms: Lessons from yeast. *Nat. Rev. Mol. Cell Biol.* **10,** 458–467.

Newbury, S. F. (2006). Control of mRNA stability in eukaryotes. *Biochem. Soc. Trans.* **34,** 30–34.

Postmus, J., Canelas, A. B., Bouwman, J., Bakker, B. M., van Gulik, W., de Mattos, M. J., Brul, S., and Smits, G. J. (2008). Quantitative analysis of the high temperature-induced glycolytic flux increase in *Saccharomyces cerevisiae* reveals dominant metabolic regulation. *J. Biol. Chem.* **283,** 23524–23532.

Reijenga, K. A., Snoep, J. L., Diderich, J. A., van Verseveld, H. W., Westerhoff, H. V., and Teusink, B. (2001). Control of glycolytic dynamics by hexose transport in *Saccharomyces cerevisiae. Biophys. J.* **80,** 626–634.

Rossell, S., van der Weijden, C. C., Kruckeberg, A. L., Bakker, B. M., and Westerhoff, H. V. (2005). Hierarchical and metabolic regulation of glucose influx in starved *Saccharomyces cerevisiae. FEMS Yeast Res.* **5,** 611–619.

Rossell, S., van der Weijden, C. C., Lindenbergh, A., van Tuijl, A., Francke, C., Bakker, B. M., and Westerhoff, H. V. (2006). Unraveling the complexity of flux regulation: A new method demonstrated for nutrient starvation in *Saccharomyces cerevisiae. Proc. Natl. Acad. Sci. USA* **103,** 2166–2171.

Rossell, S., Lindenbergh, A., van der Weijden, C. C., Kruckeberg, A. L., van Eunen, K., Westerhoff, H. V., and Bakker, B. M. (2008). Mixed and diverse metabolic and gene-expression regulation of the glycolytic and fermentative pathways in response to a HXK2 deletion in *Saccharomyces cerevisiae. FEMS Yeast Res.* **8,** 155–164.

Sauer, U. (2004). High-throughput phenomics: Experimental methods for mapping fluxomes. *Curr. Opin. Biotechnol.* **15,** 58–63.

Sauer, U. (2006). Metabolic networks in motion: 13C-based flux analysis. *Mol. Syst. Biol.* **2,** 62.

Sauro, H. M. (1989). Regulatory responses and control analysis: Assessment of the relative importance of internal effectors. *In* "Control of Metabolic Processes," (A. Cornish-Bowden and M. L. Cardenas, eds.), Plenum Press, New York.

Shulman, R. G., Bloch, G., and Rothman, D. L. (1995). In vivo regulation of muscle glycogen synthase and the control of glycogen synthesis. *Proc. Natl. Acad. Sci. USA* **92,** 8535–8542.

Stephanopoulos, G. (1999). Metabolic fluxes and metabolic engineering. *Metab. Eng.* **1,** 1–11.

ter Kuile, B. H., and Westerhoff, H. V. (2001). Transcriptome meets metabolome: Hierarchical and metabolic regulation of the glycolytic pathway. *FEBS Lett.* **500,** 169–171.

Tolkovsky, A. M. (2009). Mitophagy. *Biochim. Biophys. Acta* **1793,** 1508–1515.

van den Brink, J., Canelas, A. B., van Gulik, W. M., Pronk, J. T., Heijnen, J. J., de Winde, J. H., and Daran-Lapujade, P. (2008). Dynamics of glycolytic regulation during adaptation of *Saccharomyces cerevisiae* to fermentative metabolism. *Appl. Environ. Microbiol.* **74,** 5710–5723.

van Eunen, K., Bouwman, J., Lindenbergh, A., Westerhoff, H. V., and Bakker, B. M. (2009). Time-dependent regulation analysis dissects shifts between metabolic and gene-expression regulation during nitrogen starvation in baker's yeast. *FEBS J.* **276,** 5521–5536.

van Eunen, K., Dool, P., Canelas, A. B., Kiewiet, J., Bouwman, J., van Gulik, W. M., Westerhoff, H. V., and Bakker, B. M. (2010a). Time-dependent regulation of yeast glycolysis upon nitrogen starvation depends on cell history. *IET Syst. Biol.* **4,** 157–168.

van Eunen, K., Bouwman, J., Daran-Lapujade, P., Postmus, J., Canelas, A. B., Mensonides, F. I., Orij, R., Tuzun, I., van den Brink, J., Smits, G. J., van Gulik, W. M., Brul, S., *et al.* (2010b). Measuring enzyme activities under standardized in vivo-like conditions for systems biology. *FEBS J.* **277,** 749–760.

van Winden, W. A., van Dam, J. C., Ras, C., Kleijn, R. J., Vinke, J. L., van Gulik, W. M., and Heijnen, J. J. (2005). Metabolic-flux analysis of Saccharomyces cerevisiae CEN. PK113-7D based on mass isotopomer measurements of (13)C-labeled primary metabolites. *FEMS Yeast Res.* **5,** 559–568.

Welling, G. W., Scheffer, A. J., and Welling-Wester, S. (1994). Determination of enzyme activity by high-performance liquid chromatography. *J. Chromatogr. B Biomed. Appl.* **659,** 209–225.

Westerhoff, H. V., and Arents, J. C. (1984). Two (completely) rate-limiting steps in one metabolic pathway? The resolution of a paradox using bacteriorhodopsin liposomes and the control theory. *Biosci. Rep.* **4,** 23–31.

Wiechert, W. (2001). 13C metabolic flux analysis. *Metab. Eng.* **3,** 195–206.

Wilkie, G. S., Dickson, K. S., and Gray, N. K. (2003). Regulation of mRNA translation by 5′- and 3′-UTR-binding factors. *Trends Biochem. Sci.* **28,** 182–188.

Zaina, S., Perez-Luque, E. L., and Lund, G. (2010). Genetics talks to epigenetics? The interplay between sequence variants and chromatin structure. *Curr. Genomics* **11,** 359–367.

CHAPTER TWENTY-EIGHT

Origins of Stochastic Intracellular Processes and Consequences for Cell-to-Cell Variability and Cellular Survival Strategies

A. Schwabe,[*,†] M. Dobrzyński,[†] K. Rybakova,[†,‡] P. Verschure,[†,§] *and* F. J. Bruggeman[*,†,‡,§]

Contents

[*] Life Sciences, Centre for Mathematics and Computer Science (CWI), Amsterdam, The Netherlands
[†] Netherlands Institute for Systems Biology (NISB), Amsterdam, The Netherlands
[‡] Molecular Cell Physiology, VU University, Amsterdam, The Netherlands
[§] Swammerdam Institute for Life Sciences, University of Amsterdam, Amsterdam, The Netherlands

Methods in Enzymology, Volume 500
ISSN 0076-6879, DOI: 10.1016/B978-0-12-385118-5.00028-1

Abstract

Quantitative analyses of the dynamics of single cells have become a powerful approach in current cell biology. They give us an unprecedented opportunity to study dynamics of molecular networks at a high level of accuracy in living single cells. Genetically identical cells, growing in the same environment and sharing the same growth history, can differ remarkably in their molecular makeup and physiological behaviors. The origins of this cell-to-cell variability have in many cases been traced to the inevitable stochasticity of molecular reactions. Those mechanisms can cause isogenic cells to have qualitatively different life histories. Many studies indicate that molecular noise can be exploited by cell populations to enhance survival prospects in uncertain environments. On the other hand, cells have evolved noise-suppression mechanisms to cope with the inevitable noise in their functioning so as to reduce the hazardous effects of noise. In this chapter, we discuss key experiments, theoretical results, and physiological consequences of molecular stochasticity to introduce this exciting field to a broader community of (systems) biologists.

1. Cell-to-Cell Heterogeneity and Measurement Techniques

Traditionally, cell biology studies cells in populations using harsh methods for cell lysis. Proteins were often purified to study their properties. Even though molecular stochasticity and cell-to-cell variability were anticipated already decades ago (Berg, 1978; Novick and Weiner, 1957; Spudich and Koshland, 1976), only after the development of fluorescent proteins, single-molecule measurements and new microscopy techniques have quantitative studies of (single) molecular processes at a single cell level become reality. What these studies show is that grinding cells and studying average cell compositions is a limited approach and only scratches the surface of the interesting biology that can be found at the level of single cells.

Early studies of single *Escherichia coli* cells used fluorescent proteins (GFP, YFP, etc.) to quantify the distribution of whole cell fluorescence across a cell population (Elowitz *et al.*, 2002; Ozbudak *et al.*, 2002). Single cell dynamics were measured early on using synthetic and natural gene networks in *E. coli* (Becskei and Serrano, 2000; Elowitz and Leibler, 2000; Gardner *et al.*, 2000). Those studies indicated that isogenic cells can differ markedly in their protein levels and that cellular dynamics often show fluctuations and cell-to-cell variability. At those times, an arbitrary fluorescence unit was used in the quantification of cellular variability. Even though this unit likely was linearly correlated with molecule abundance, it did not allow for straightforward counting of single molecules. Those experiments followed later, first for mRNA and then for protein. More recent experiments on signaling dynamics of single eukaryotic cells (Cohen-Saidon *et al.*, 2009; Nelson *et al.*, 2004; Shankaran *et al.*, 2009) have shown that decision making at the level of a single cell is highly heterogeneous. Goentoro *et al.* suggested that cells use fold-change detection to correct for fluctuations in baseline signaling activity. In those experiments, signaling molecules are tagged to fluorescent proteins to allow for real-time imaging of signaling events.

Single-molecule counting experiments of mRNA were first carried out using a bacteriophage-derived protein, MS2 (Chubb *et al.*, 2006; Golding *et al.*, 2005). This protein binds particular RNA sequences and, when fused to a fluorescent protein, allows for single-molecule counting of mRNA equipped with an array of 50–100 MS2-binding sequences in its untranslated region (Bertrand *et al.*, 1998). In this manner, mRNA was quantified in *E. coli* and mammalian cells. These studies indicated that transcription can occur in bursts; genes toggle between transcriptionally active and inactive states. Usage of such MS2-based methods allows for time-resolved monitoring of transcription at the single cell level. The disadvantage of this method lies in its requirement for genetic engineering of target genes. A more straightforward approach is the usage of RNA fluorescent in situ hybridization (Raj *et al.*, 2006; Youk *et al.*, 2010; Zenklusen *et al.*, 2008). This technique does not allow for time-resolved studies of transcription activity; it only gives snapshots of single cell transcript levels and simultaneous analysis of many cells in a population gives insight into population transcription-level variability. An alternative approach is single-cell qPCR, which is limited to mRNA with levels higher than tens per cell (Bengtsson *et al.*, 2005).

Single-molecule monitoring of fluorescent-protein tagged proteins is at the moment limited to slowly diffusing proteins (Yu *et al.*, 2006). This method already pushes current microscopy methods to its limits as the number of photons emitted per protein cellular location is so low that currently only slowly diffusing (membrane or DNA-bound) proteins can be visualized and counted. The development of single-molecule counting methods is an active research field and will certainly lead to improved methods in the next couple of years. For more in depth discussions on the current experimental capabilities, we refer the reader to two excellent reviews (Raj and van Oudenaarden, 2009; Xie *et al.*, 2008).

In the next sections, we will discuss the results of those experimental studies in the light of stochastic models of mRNA and protein turnover.

2. Theoretical Insights and Experimental Evidence

2.1. Fluctuations in molecule numbers are inevitable consequences of the nature of molecular reactions

The mRNA level at steady state is such that its averaged synthesis and degradation rates balance. This mRNA level would remain truly fixed over time only when individual synthesis and degradation events occur in synchrony. This is, however, not the case; transcription depends on the stochastic diffusive encounters of (often low-copy number) transcription factors, RNA polymerase, and cofactors at regulatory sites and promotors on the DNA. Degradation of mRNA, regardless of whether it is spontaneous or catalyzed, is also a stochastic process. The average mRNA lifetime is given by the inverse of its degradation constant but individual mRNA lifetimes are, in the simplest case, exponentially distributed. We will come back to those calculations later on. In other words, there is no reason to believe that synthesis and degradation occur in synchrony so the null-hypothesis from this argumentation is that mRNA should fluctuate at steady state around its average level. And indeed it does.

- But how large are those fluctuations? In other words, how large are the deviations in mRNA molecule number from its average?
- What sets the timescale for those fluctuations, that is, can we speak of average mRNA levels during a time interval as long as the generation time?
- Can we expect concentrations of mRNAs, proteins, and metabolites to fluctuate to the same extent?
- How do network structures such as cascades and regulatory circuitry influence fluctuations? Can mRNA fluctuations induce protein fluctuations?
- How does cell division distort molecule levels and how quickly do the correlations between the number of molecules in mother and daughter cells decay?

Those questions will be addressed in the sections by discussing stochastic model predictions and experimental results.

2.2. Noise in mRNA numbers at steady state

For some models, the distributions of mRNA across a cell population can be obtained in closed form, that is, the stochastic model can be analytically solved in terms of transcription and degradation kinetics. Such descriptions are particularly insightful as they allow us to investigate how the dispersion of mRNA distributions across a population of cells depends on the transcription and degradation kinetics. One such model is the simplest imaginable production–degradation scheme:

$$\emptyset \xrightarrow{k_{\mathrm{m}}^{+}} \mathrm{mRNA} \xrightarrow{k_{\mathrm{m}}^{-} n_{\mathrm{m}}} \emptyset \tag{28.1}$$

Here mRNA, measured in molecules per cell, is synthesized at a rate of k_{m}^{+} molecules per minute and degraded at a rate of $k_{\mathrm{m}}^{-} n_{\mathrm{m}}$ molecules per minute, where n_{m} denotes mRNA copy number (Fig. 28.1).

The macroscopic description of this system assumes that the amount of mRNA changes in a continuous manner. Therefore, the discrete nature of molecules is discarded, which is valid if the concentrations are *high*. The temporal evolution of the continuous variable quantifying the abundance of mRNA is given by the following differential equation:

$$\frac{\mathrm{d}\langle n_{\mathrm{m}}(t)\rangle}{\mathrm{d}t} = k_{\mathrm{m}}^{+} - k_{\mathrm{m}}^{-}\langle n_{\mathrm{m}}(t)\rangle \tag{28.2}$$

where $\langle n_{\mathrm{m}}(t)\rangle$ denotes the average number of mRNA in the system at a particular moment in time t, and the average is taken over an entire cell population.[1]

At steady state, the amount of mRNA synthesized at any given time interval is balanced by the degradation process. Hence, the familiar macroscopic result for the mean number of mRNA at steady state

$$\langle n_{\mathrm{m}}\rangle_{\mathrm{ss}} = \frac{k_{\mathrm{m}}^{+}}{k_{\mathrm{m}}^{-}} \tag{28.3}$$

The mean lifetime of a single mRNA molecule is given by $\tau_{\mathrm{m}} = 1/k_{\mathrm{m}}^{-}$.

In order to account for fluctuations inherent in synthesis and degradation we turn to a mesoscopic description in terms of the chemical master equation (CME) (Mcquarrie, 1967; Van Kampen, 2003). The CME deals with the dynamics of chemical transitions between separate states corresponding to discrete numbers of molecules of chemical species (Fig. 28.2). Hence, instead of the evolution of a continuous concentration as in Eq. (28.2), the quantity evolved in the CME is the probability distribution $p(n_{\mathrm{m}}, t)$, which in the case of a single chemical species, mRNA in our example, describes the probability of finding the system in a state with n_{m} mRNA molecules at time t. Gains and losses due to chemical

[1] More precisely, brackets denote the ensemble average. In our case, the ensemble includes an infinite number of mental copies of the system, each corresponding to a different number of mRNA molecules that the system might contain.

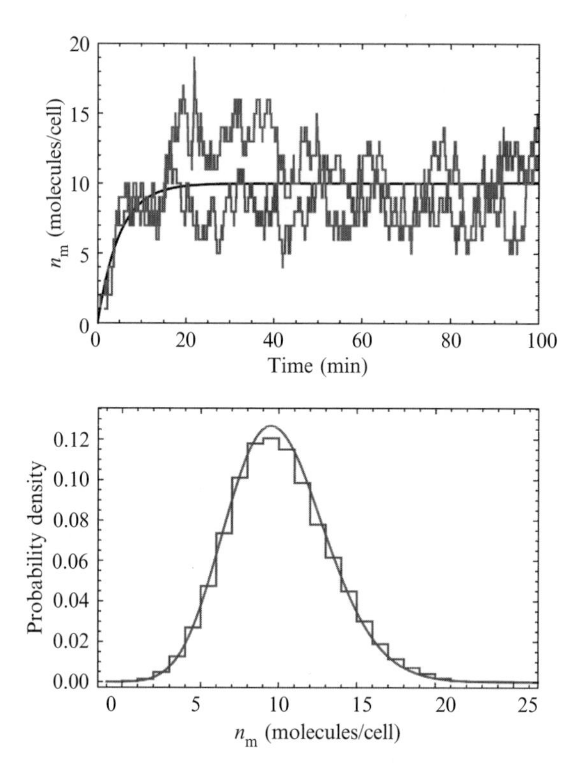

Figure 28.1 Simulation results for the transcription model shown in Eq. (28.1). The synthesis rate constant equals 2 mRNAs/(min · cell) and the degradation constant is 0.2 min^{-1}, this gives a steady state of 10 mRNAs/cell. The upper figure gives the result of the macroscopic dynamics described by Eq. (28.2) starting from 0 mRNAs/cell to the steady state (black) and two mesoscopic simulations that mimic single-cell behavior (blue, purple). Those simulations are sample trajectories that obey the master equation (Eq. (28.4)) obtained using the Gillespie algorithm (the direct method) (Gillespie, 2007). The lower figure shows the steady-state probability density for mRNA copy numbers; the blue line is determined from simulations and the red line from Eq. (28.7). (For interpretation of the references to color in this figure legend, the reader is referred to the Web version of this chapter.)

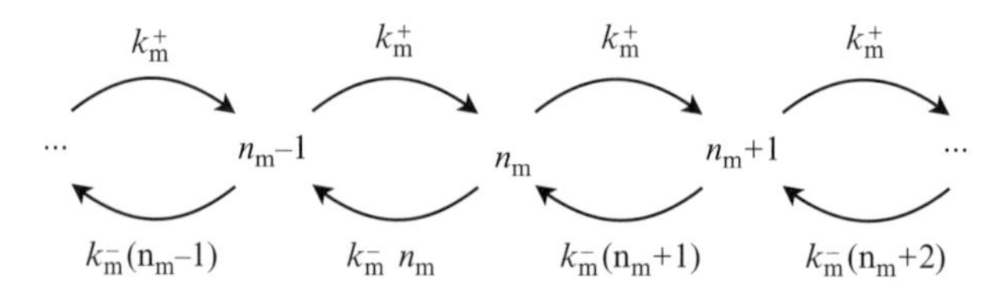

Figure 28.2 Discrete states in the production–degradation model. The illustration of chemical transitions between discrete states corresponding to the abundance of mRNA molecules, n_m, in the reaction scheme specified by Eq. (28.1). The chemical master equation (Eq. (28.4)) describes temporal changes of the probability for a single state n_m.

reactions on the right hand side of Eq. (28.2) are now written in the following form:[2]

$$\frac{\mathrm{d}p(n_\mathrm{m},t)}{\mathrm{d}t} = k_\mathrm{m}^+ p(n_\mathrm{m}-1,t) + k_\mathrm{m}^-(n_\mathrm{m}+1)p(n_\mathrm{m}+1,t) - \left(k_\mathrm{m}^+ p(n_\mathrm{m},t) + k_\mathrm{m}^- n_\mathrm{m} p(n_\mathrm{m},t)\right) \tag{28.4}$$

The first line of the right hand side includes "gains," that is, all chemical reactions that result in state n_m. These are the synthesis of a *single* mRNA molecule in state $n_\mathrm{m}-1$ and the degradation of a *single* mRNA molecule in state $n_\mathrm{m}+1$. Similarly, the second line describes, respectively, the synthesis and the degradation event that lead to abandonment of the state n_m—the "losses." (For $n_\mathrm{m}=0$ the $p(n_\mathrm{m}-1)$ becomes zero; the number of molecules cannot be negative.) We have in principle an infinite number of those equations, one for every discrete state n_m, which poses a significant challenge for obtaining closed analytical solutions for more complex reaction schemes.

At steady state, $\mathrm{d}p(n_\mathrm{m},t)/\mathrm{d}t=0$, by following all states starting from $n_\mathrm{m}=0$, one obtains a recurrence relation (the dependence on t is now omitted):

$$\frac{p(n_\mathrm{m})}{p(n_\mathrm{m}-1)} = \frac{k_\mathrm{m}^+}{k_\mathrm{m}^- n_\mathrm{m}} \tag{28.5}$$

All states are now related to $p(0)$ as

$$p(n_\mathrm{m}) = \langle n_\mathrm{m}\rangle^{n_\mathrm{m}} \frac{1}{n_\mathrm{m}!} p(0) \tag{28.6}$$

Since contributions from all probabilities $p(n_\mathrm{m})$ have to sum up to 1, that is, $\sum_{n_\mathrm{m}=0}^{\infty} p(n_\mathrm{m}) = 1$, $p(0)$ can be solved for and yields, $p(0) = \mathrm{e}^{-\langle n_\mathrm{m}\rangle}$. Putting this all together, we obtain the Poisson distribution for the steady-state distribution of the number of mRNA molecules:

$$p(n_\mathrm{m}) = \frac{\langle n_\mathrm{m}\rangle^{n_\mathrm{m}}}{n_\mathrm{m}!} \mathrm{e}^{-\langle n_\mathrm{m}\rangle} \tag{28.7}$$

[2] From this equation, we can recover the macroscopic solution by realizing that

$$\frac{\mathrm{d}\langle n_\mathrm{m}\rangle(t)}{\mathrm{d}t} = \sum_{n_\mathrm{m}=0}^{\infty} n_\mathrm{m} \frac{\mathrm{d}p(n_\mathrm{m},t)}{\mathrm{d}t}$$

This last result is not generally true; it is only valid for cases where the reaction rates depend linearly on molecular copy numbers, otherwise higher moments appear in the macroscopic description—which are typically negligible for large molecule copy numbers (Van Kampen, 2003).

We can obtain the first moment (the mean) and the second moment of mRNA level from $\langle n_\mathrm{m} \rangle = \sum_{n_\mathrm{m}}^{\infty} n_\mathrm{m} p(n_\mathrm{m})$ and $\langle n_\mathrm{m}^2 \rangle = \sum_{n_\mathrm{m}}^{\infty} n_\mathrm{m}^2 p(n_\mathrm{m})$. The variance is then given by $\langle \delta_{n_\mathrm{m}}^2 \rangle = \langle n_\mathrm{m}^2 \rangle - \langle n_\mathrm{m} \rangle^2$. The standard deviation is defined as the square root of the variance. For the Poisson distribution, the variance equals the mean. We now define mRNA copy number noise as the dimensionless ratio of the variance in mRNA copy number divided by the squared mean:

$$\frac{\langle \delta n_\mathrm{m}^2 \rangle}{\langle n_\mathrm{m} \rangle^2} = \frac{1}{\langle n_\mathrm{m} \rangle} \tag{28.8}$$

The ratio of the variance over the mean, the Fano factor, equals 1 in this case indicating that we are dealing with a Poisson distribution

$$\frac{\langle \delta n_\mathrm{m}^2 \rangle}{\langle n_\mathrm{m} \rangle} = 1 \tag{28.9}$$

Even though the theory in this section gives somewhat more insight into the sort of distributions we can expect, the model is too simple to capture experimental data. In the next section, we will use a stochastic model for transcription that has proven very useful for understanding experimental studies. One of the possible outcomes of this model is the Poisson distribution for mRNA as explained in this section.

We can draw a number of conclusions from this section:

- Fluctuations around an average occur even in the simplest reaction system that can attain a steady state.
- Noise in the copy number of a molecule reduces as the average copy number increases, Eq. (28.8). This is an important rule of thumb that is often used to argue that noise in high protein levels, such as metabolic enzymes, typically will not occur. Below we will show that this is not generally the case as noise in low-copy number mRNAs or due to slowly switching genes between ON and OFF states can propagate to protein levels.
- A property of the Poisson process is that it has a Fano factor of 1. Below we will discuss cases where the Fano factor is larger, which is more in agreement with experiments.

2.3. A switching-gene model that captures many experimental findings

The model described in the previous section considered a gene with a constant transcription activity. Such constitutively active genes are an exception; typically prokaryotic transcription activity depends on a number

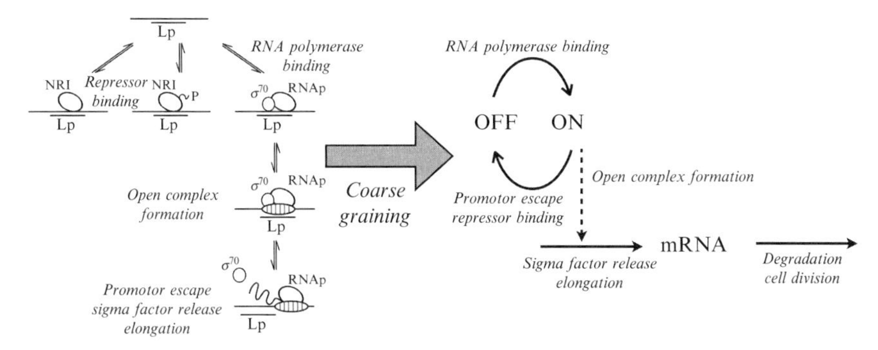

Figure 28.3 Core model for regulated transcription and mRNA degradation. The scheme on the right has been a frequently used model in the experimental analysis of stochastic single-cell transcription. Most of the findings of those studies can be explained in terms of this model. It is a simplification of the biochemical processes associated with gene regulation by an activating or inhibiting transcription factor. The scheme on the left describes the regulation of a promotor, L_p, involved in the regulation of ammonium uptake enzymes in *E. coli* by an inhibiting transcription factor, *NRI* (and its phosphorylated form *NRIP*). In prokaryotes, transcription is carried out by a RNA polymerase assisted by a sigma factor; σ^{70} in this case. It is questionable to what extent this model can be applied to eukaryotic genes as their mode of regulation is so much more complicated than prokaryotic gene regulation.

of transcription factors, RNA polymerase, and a sigma factor, whereas eukaryotic transcription depends on dozens of proteins with activities that depend on the gene's chromatin state dynamics. A coarse-grained model that describes these processes is shown in Fig. 28.3. It contains the following reaction events:[3]

$$\begin{aligned} &\mathrm{ON} \xrightarrow{k_{\mathrm{off}}\, n_{\mathrm{on}}} \mathrm{OFF} \\ &\mathrm{OFF} \xrightarrow{k_{\mathrm{on}}(1-n_{\mathrm{on}})} \mathrm{ON} \\ &\xrightarrow{k_{\mathrm{m}}^{+}\, n_{\mathrm{on}}} \mathrm{mRNA} \xrightarrow{k_{\mathrm{m}}^{-}\, n_{\mathrm{m}}} \end{aligned} \tag{28.11}$$

In Fig. 28.4, some of the dynamics of the model outlined in Eq. (28.11) is illustrated. This model can operate in three different noise regimes. The upper figures correspond to the case where the gene switches very rapidly relative to the dynamics of mRNA (characteristic timescale for this is $1/k_{\mathrm{m}}^{-}$). This means that mRNA cannot track individual switching events and the

[3] The analytical distribution of mRNA is known for this model (Shahrezaei and Swain, 2008)

$$p(n_{\mathrm{m}}) = \frac{\left(\frac{k_{\mathrm{m}}^{+}}{k_{\mathrm{m}}^{-}}\right)^{n_{\mathrm{m}}} e^{-\left(k_{\mathrm{m}}^{+}/k_{\mathrm{m}}^{-}\right)}}{n_{\mathrm{m}}!} \frac{\Gamma\left(\frac{k_{\mathrm{on}}}{k_{\mathrm{m}}^{-}} + n_{\mathrm{m}}\right)\Gamma\left(\frac{k_{\mathrm{on}}}{k_{\mathrm{m}}^{-}} + \frac{k_{\mathrm{off}}}{k_{\mathrm{m}}^{-}}\right)}{\Gamma\left(\frac{k_{\mathrm{on}}}{k_{\mathrm{m}}^{-}} + \frac{k_{\mathrm{off}}}{k_{\mathrm{m}}^{-}} + n_{\mathrm{m}}\right)\Gamma\left(\frac{k_{\mathrm{on}}}{k_{\mathrm{m}}^{-}}\right)} 1F1\left(\frac{k_{\mathrm{on}}}{k_{\mathrm{m}}^{-}}, \frac{k_{\mathrm{on}}}{k_{\mathrm{m}}^{-}} + \frac{k_{\mathrm{off}}}{k_{\mathrm{m}}^{-}} + n_{\mathrm{m}}, \frac{k_{\mathrm{m}}^{+}}{k_{\mathrm{m}}^{-}}\right) \tag{28.10}$$

Here $\Gamma(x)$ denotes the so-called gamma function defined as $\Gamma(x) = \int_0^\infty t^{x-1} e^{-t} dt$ and 1F1 is the Kummer confluent hypergeometric function. This equation was used in Fig. 28.4.

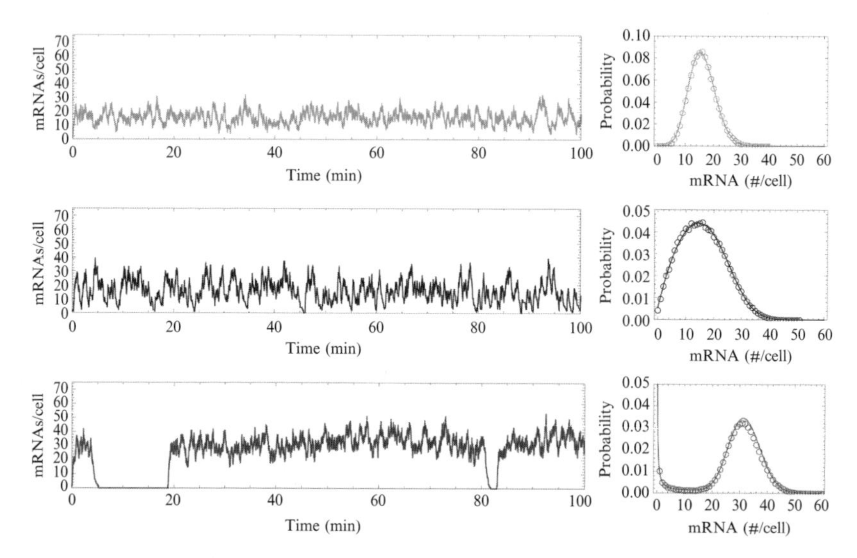

Figure 28.4 Different noise regimes for the switching-gene model. The network shown in Fig. 28.3 can generate operate at three different noise regimes (see main text). In the plots on the right, the circular data points are simulations results and the line corresponds to the analytical distribution, Eq. (28.10).

noise in mRNA is then approximately given by $1/\langle n_{\text{m}}\rangle$; the system operates in its Poisson regime. If the switch operates at the timescale of mRNA dynamics (middle figures), then mRNA levels respond to switches in the gene between ON and OFF such that during the ON and OFF phases significant synthesis and degradation events can occur leading to a broader mRNA distribution. The bottom figures show the situation where the gene switches very slowly; mRNA dynamics is in this case faster than characteristic timescales in the switch (i.e., $1/k_{\text{on}}$ and $1/k_{\text{off}}$). In this regime, mRNA can attain a steady state during the ON phase and all mRNA is degraded during the OFF phase. The probability density plots (on the right) indicate that noise transfer from the switch to mRNA (middle figure) leads to a higher dispersion in the mRNA distribution than when the system operates in its Poisson regime. In other words, in the blue simulations, the Fano factor is larger than 1, which indicates transcription bursts as we shall see in the next section. The bottom probability density plot shows a bimodal distribution, indicating that many cells either have no (or little) mRNA or have high mRNA levels. Bimodal distributions have been found experimentally as well (Blake *et al.*, 2003; Yu *et al.*, 2006). The origin of bimodal distributions here lies in timescale separation between the switching gene and mRNA turnover. This stochastic behavior cannot be captured by the macroscopic description of the dynamics. An alternative mechanism for bimodal distributions of cell states is bistable dynamics (we will discuss this shortly below) (Becskei *et al.*, 2001; Maamar *et al.*, 2007; Ozbudak *et al.*, 2004). This is a form of dynamics that can be recovered from the macroscopic descriptions. Dwell times of cells in each of the two stable steady states are again random variables and depend on the rate of fluctuation occurrence and the probability of having a fluctuation of a high enough size.

For this model (Eq. (28.11)), the noise in mRNA becomes (using, for instance, linear noise approximation; Elf and Ehrenberg, 2003; Paulsson, 2004; Bruggeman *et al.*, 2009)

$$\frac{\langle \delta^2 n_{\mathrm{m}} \rangle}{\langle n_{\mathrm{m}} \rangle^2} = \frac{1}{\langle n_{\mathrm{m}} \rangle} + \frac{\tau_{\mathrm{o}}}{\tau_{\mathrm{o}} + \tau_{\mathrm{m}}} \frac{\langle \delta^2 n_{\mathrm{on}} \rangle}{\langle n_{\mathrm{on}} \rangle^2} \tag{28.12}$$

Here, $\tau_{\mathrm{o}} = 1/(k_{\mathrm{off}} + k_{\mathrm{on}})$ and $\tau_{\mathrm{m}} = 1/k_{\mathrm{m}}^{-}$. This equation shows that if the turnover of mRNA is slow, relative to timescales in the switch, that is, $\tau_{\mathrm{m}} \gg \tau_{\mathrm{o}}$, the noise becomes approximately equal to the Poisson limit, $1/\langle n_{\mathrm{m}} \rangle$. In this limit, mRNA cannot track gene switching events. When mRNA is synthesized and degraded fast enough, the noise can be appreciably larger than $1/\langle n_{\mathrm{m}} \rangle$. In this case, mRNAs with high mean levels can become noisy, due to the term $\frac{\tau_o}{\tau_{\mathrm{o}}+\tau_{\mathrm{m}}} \frac{\langle \delta^2 n_{\mathrm{on}} \rangle}{\langle n_{\mathrm{on}} \rangle^2}$, which is maximally equal to $\frac{\langle \delta^2 n_{\mathrm{on}} \rangle}{\langle n_{\mathrm{on}} \rangle^2}$; the noise in the ON state, which we refer to as the noise in the switch. It can be shown that $\frac{\langle \delta^2 n_{\mathrm{on}} \rangle}{\langle n_{\mathrm{on}} \rangle^2} = \frac{k_{\mathrm{off}}}{k_{\mathrm{on}}}$. The gene operates in a bursting regime if on average more than 1 mRNA is made per ON state, that is, the mean mRNA burst size, $\langle b_{\mathrm{m}} \rangle = k_{\mathrm{m}}^{+}/k_{\mathrm{off}}$, is larger than 1.[4] Whether transcription bursts are significant depends on the duration of the OFF state (Dobrzynski and Bruggeman, 2009).

Using Eq. (28.12), we can express the Fano factor for mRNA in terms of the mean mRNA burst size (with $\tau_{\mathrm{on}} = 1/k_{\mathrm{off}}$ and $\tau_{\mathrm{off}} = 1/k_{\mathrm{on}}$)

$$\frac{\langle \delta^2 n_{\mathrm{m}} \rangle}{\langle n_{\mathrm{m}} \rangle} = 1 + \frac{\tau_{\mathrm{m}}}{\tau_{\mathrm{m}} + \tau_{\mathrm{o}}} \frac{\tau_{\mathrm{off}}}{\tau_{\mathrm{off}} + \tau_{\mathrm{on}}} \frac{\tau_{\mathrm{o}}}{\tau_{\mathrm{on}}} \langle b_{\mathrm{m}} \rangle \tag{28.13}$$

An important conclusion can be drawn from this equation. For illustrative purposes, we shall assume that the lifetime τ_{m} of mRNA is fixed, and so are the parameters of the switch, that is, τ_{on} and τ_{off}. Then the mean amount of mRNA, $\langle n_{\mathrm{m}} \rangle$, scales linearly with the mean burst size, $\langle b_{\mathrm{m}} \rangle$, just as the Fano factor. Therefore, despite the large mean of the mRNA abundance significant deviations from Poisson regime can occur (Fano factor > 2).

[4] The mean transcription rate is equal to $\langle b_{\mathrm{m}} \rangle / (\tau_{\mathrm{cycle}})$; the burst size divided by the gene cycle time $(1/k_{\mathrm{on}} + 1/k_{\mathrm{off}})$.

We quantify them with following conditions: (i) mRNAs live longer than the timescale of the switch ($\tau_m > \tau_o$), (ii) the gene spends most of the time in the OFF phase ($\tau_{off} > \tau_{on}$ and, therefore, $\tau_o > \tau_{on}$), and (iii) the burst size is larger than 1. Those timescale conditions correspond to leaky gene expression. Golding *et al.* (2005) measured the ON and OFF duration of their version of the *lac* operon and found 6 and 37 min, respectively.

Leaky gene expression conditions have indeed been found to give rise to transcription bursts in prokaryotes (Cai *et al.*, 2006; Golding *et al.*, 2005; Yu *et al.*, 2006). In those cases, the burst size of mRNA (Golding *et al.*, 2005) and protein (Cai *et al.*, 2006; Yu *et al.*, 2006) was found to be 4 and 5, respectively. Recently, a near genome-wide study of mRNA noise in *E. coli* indicated that the mean Fano factor for mRNA is 1.6; Golding *et al.* (2005) found a Fano factor of 4. Across all mRNAs, they found a different scaling for mRNA than for proteins, mRNA noise scaled inversely with the mean mRNA level. Burst sizes in eukaryotes are believed to be much higher (Raj and van Oudenaarden, 2008).

Models (Bruggeman *et al.*, 2009; Ozbudak *et al.*, 2002; Paulsson, 2004) predict that the Fano factor for a protein translated from mRNA that is constantly transcribed (highly active gene conditions) is given by[5]

$$\frac{\langle \delta^2 n_p \rangle}{\langle n_p \rangle} = 1 + \frac{\tau_m}{\tau_m + \tau_p} \frac{\langle n_p \rangle}{\langle n_m \rangle} = 1 + \frac{\langle b_p \rangle}{1 + \left(k_p^-/k_m^-\right)} \approx 1 + \langle b_p \rangle \qquad (28.15)$$

Here $\langle n_m \rangle = k_m^+/k_m^-$ (mean mRNA copy number), $\langle n_p \rangle = k_p^+ \langle n_m \rangle / k_p^-$ (mean protein copy number), $\tau_m = 1/k_m^-$ (mRNA lifetime), $\langle b_p \rangle = k_p^+/k_m^-$ (protein burst size), and $\tau_p = 1/k_p^-$ (protein lifetime). The simplification assumes that mRNA has a higher degradation rate than protein, $k_m^- \gg k_p^-$. For large mean protein copy numbers, the noise in protein becomes $\frac{\langle \delta^2 n_p \rangle}{\langle n_p \rangle^2} \approx \frac{\langle b_p \rangle}{\langle p \rangle}$.

Taniguchi *et al.* (2010) found that proteins with low levels per cells (<10) displayed noise in the Poisson limit ($1/\langle n_p \rangle$). Proteins with higher concentration displayed noise that was independent of the mean level and about two to three times higher than the intrinsic noise level, which was

[5] The distribution of protein molecules, if they are produced by translation bursts (at some mRNA copy number, n_m) and degraded by a first-order process, can be shown to follow a Gamma distribution (Friedman et al., 2006; Paulsson and Ehrenberg, 2000):

$$p(n_p | n_m) = \frac{\langle b_p \rangle^{n_p}}{\left(1 + \langle b_p \rangle\right)^{k_p^+/k_p^- n_m + n_p}} \frac{\Gamma\left(k_p^+/k_p^- n_m + n_p\right)}{\Gamma\left(k_p^+/k_p^- n_m\right) n_p!} \qquad (28.14)$$

with mean $\langle n_p \rangle = n_m k_p^+/k_p^- \langle b_p \rangle$ and the variance as $\langle n_p \rangle (1 + \langle b_p \rangle)$.

$\sim$0.1. They also confirmed that the mean levels of essential proteins are higher than the mean level determined for all proteins.

One of the earlier studies on protein noise tested a prediction of Eq. (28.15); that is, the Fano factor depends linearly on translation rate (or efficiency), k_{p}^{+}, and is independent of transcription rate (or efficiency), k_{m}^{+} in *E. coli* (Ozbudak *et al.*, 2002). Blake *et al.* (2003) did find a dependence of noise on transcription efficiency in *Saccharomyces cerevisiae* and argued that the more complex design of transcription regulation in yeast can account for these differences with *E. coli*. In a follow-up study, Blake *et al.* (2006) continued their analysis of transcription-mediated noise in yeast and found that transcription bursts can readily occur in eukaryotic transcription control mechanisms and lead to a noise increase. They then showed that the survival prospects of yeast cells when exposed to high antibiotic levels increases when the cell population is more variable and reduces with cell-to-cell variability at low levels of antibiotics. The level of transcription noise and response time were experimentally tuned by manipulating the DNA sequence of the TATA box; the consensus TATA box promotes transcription bursts. These experiments show how the quality of the TATA box contributes to survival prospects and indeed this correlates with the high occurrence of TATA boxes in stress-related genes.

Transcription bursts have been experimentally observed in a number of studies: in prokaryotes (Cai *et al.*, 2006; Golding *et al.*, 2005; Ozbudak *et al.*, 2002; Taniguchi *et al.*, 2010; Yu *et al.*, 2006) and eukaryotes (Chubb *et al.*, 2006; Youk *et al.*, 2010; Zenklusen *et al.*, 2008). Bursts do not have a constant size for every ON phase; the burst sizes also follow a distribution. This originates from the fact that the lifetime of the ON phase and the transcription initiation time are not fixed. They are each random variables; in model Eq. (28.11), they are both drawn from exponential distributions with means $1/k_{\mathrm{off}}$ and $1/k_{\mathrm{m}}^{+}$ (below more about such distributions). For the model shown in Eq. (28.11), the burst size distribution follows a geometric distribution, which models the distributions of trials, b_{m}, before success given the probability for success (i.e., the occurrence of a switch to the OFF state, with probability p_{on}):

$$p(b_{\mathrm{m}}) = (1 - p_{\mathrm{on}})^{b_{\mathrm{m}}} p_{\mathrm{on}} \tag{28.16}$$

with b_{m} as the mRNA burst size and p_{on} defined as $\frac{k_{\mathrm{off}}}{k_{\mathrm{off}}+k_{\mathrm{m}}^{+}}$. This discrete probability mass function has as mean $\langle b_{\mathrm{m}} \rangle = \frac{k_{\mathrm{m}}^{+}}{k_{\mathrm{off}}} = \frac{\tau_{\mathrm{off}}}{\tau_{\mathrm{ini}}}$ and noise $\frac{\langle \delta^2 b_{\mathrm{m}} \rangle}{\langle b_{\mathrm{m}} \rangle^2} = 1 + \frac{1}{\langle b_{\mathrm{m}} \rangle}$. Those distributions have been experimentally found by Yu *et al.* (2006) and Golding *et al.* (2005) in *E. coli*. These studies indicate that the noise in burst size typically is very large and that burst sizes in prokaryotes are small (see also Taniguchi *et al.*, 2010).

One other important source of noise in molecule number could be unequal partitioning of mRNAs and proteins (Berg, 1978; Kaufmann *et al.*, 2007; Sigal *et al.*, 2006), most noticeably those molecules occurring at low molecule numbers, such as transcription factors. At cell division, molecule numbers are distributed according to a binomial distribution with probabilities given by the fractional volumes of the mother and the daughter cell. The noise in a binomial distribution quickly drops with the number of molecules being partitioned; therefore, this effect is only minor for molecules with large copy numbers (more than 50 molecules/cell). A more important effect of cell division is that mother and daughter cells can remain correlated for some time in their expression states if molecule levels have lifetimes longer than the generation time. This effect has been found experimentally (Berg, 1978; Kaufmann *et al.*, 2007).

From this section, we can conclude:

- Transcription often occurs in bursts and typically a Fano factor larger than 1 indicates bursts
- Switching genes can function in different noise regimes: (1) in an intrinsic noise regime where the gene switches fast and individual switches cannot be tracked by transcription (a gene in an active state regulated by a highly abundant transcription factor); the noise in mRNA is Poissonian, (2) in a noise propagation regime where gene switches occur on a comparable timescale as mRNA turnover, gene switches do lead to changes in mRNA concentrations and the Fano factor for mRNA is larger than 1, and (3) very slow switch regimes where mRNA dynamics can attain a high-level mRNA steady state during the ON period and a null state during the OFF state. This model gives rise to a bimodal distribution of mRNA.
- The Fano factor of proteins equals $1 + \langle b_p \rangle$ if the mRNA lives short and translation bursts occur.
- In a near genome-wide study of *E. coli*, mRNA noise scales inversely with mean mRNA levels. Proteins with concentrations smaller than 10 per cell have noise equals $1/\langle n_p \rangle$ and proteins with higher concentrations have concentration-independent noise between 0.1 and 0.3.
- In yeast, protein noise scales inversely proportional to protein level (for 43 proteins).
- The burst size as predicted by the model of Eq. (28.11) is geometrically distributed and this has also been found experimentally for β-galactosidase levels and *lac* operon expression under repressed conditions (Golding *et al.*, 2005; Yu *et al.*, 2006).

2.4. Eukaryotic translation bursts and eukaryotic protein noise

The previous section already indicated that the noise in the protein level (in the simplest models) is the sum of an intrinsic noise term ($1/\langle n_p \rangle$) and an extrinsic noise term through which noise in other molecule levels propagate to protein. Depending on the magnitudes of timescale separation between protein turnover, fluctuations in the regulator, and noise in the regulator, the extrinsic noise terms can either be significant or not and cause the Fano factor to increase above 1. A model that can guide us in understanding experimental data on noise in eukaryotic and prokaryotic protein levels is one that considers a switching promotor activity and synthesis and degradation of protein, as shown in Fig. 28.5. For this cascade, the noise in protein can be shown to be sum of three contributions: intrinsic noise in protein, the intrinsic noise in mRNA ($1/\langle n_m \rangle$) propagating to protein, and the noise in gene ON state ($\langle \delta^2 n_{on} \rangle / \langle n_{on} \rangle^2 = k_{off}/k_{on}$) propagating via mRNA to protein (Bruggeman *et al.*, 2009). The magnitude of the latter two contributions depend on whether mRNA can track gene toggling and whether protein can track mRNA·fluctuations

$$\frac{\langle \delta^2 n_p \rangle}{\langle n_p \rangle^2} = \frac{1}{\langle n_p \rangle} + \frac{\tau_m}{\tau_m + \tau_p} \frac{1}{\langle n_m \rangle} + \frac{\tau_o \left(\tau_o \tau_p + \tau_m \left(\tau_o + \tau_p \right) \right)}{\left(\tau_o + \tau_m \right) \left(\tau_m + \tau_p \right) \left(\tau_o + \tau_p \right)} \frac{\langle \delta^2 n_{on} \rangle}{\langle n_{on} \rangle^2} \tag{28.17}$$

Typically, mRNA lives shorter than protein, that is, $\tau_m \ll \tau_p$ and then this equation can be simplified (by taking its limit to $\tau_m/\tau_p \rightarrow 0$) to give the

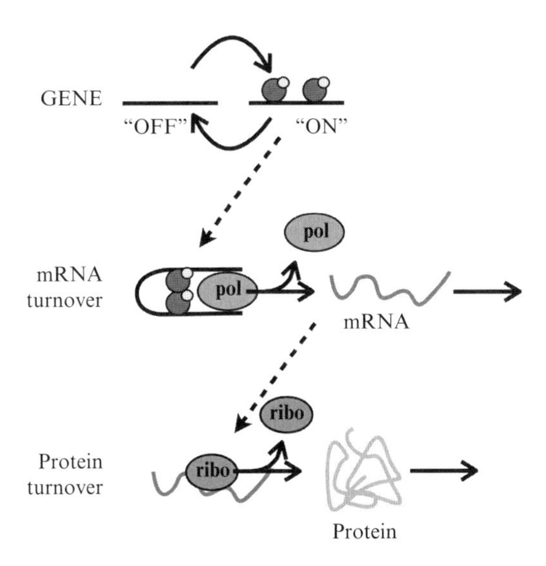

Figure 28.5 Gene activity toggling and a transcription–translation cascade. In this figure, a gene switching between an active and inactive state (induced by transcription factors or chromatin remodeling) is shown that in its active state induces transcription. The transcript is read in translation and both the mRNA and protein are degraded either by dilution due to cell growth or due to catalyzed degradation.

following Fano factor for protein (compare to Eq. (28.13) for the Fano factor in mRNA):

$$\frac{\langle \delta^2 n_{\mathrm{p}} \rangle}{\langle n_{\mathrm{p}} \rangle} = 1 + \frac{\tau_{\mathrm{p}}}{\tau_{\mathrm{o}} + \tau_{\mathrm{p}}} \frac{\tau_{\mathrm{off}}}{\tau_{\mathrm{on}} + \tau_{\mathrm{off}}} \frac{\tau_{\mathrm{o}}}{\tau_{\mathrm{on}}} \langle b_{\mathrm{m}} \rangle \langle b_{\mathrm{p}} \rangle \qquad (28.18)$$

Equation (28.18) illustrates that the Fano factor is larger than 1 in the case of (i) combined transcription and translation bursts, (ii) protein lives longer than the timescale of the gene switch, and (iii) the ON shorter than the OFF duration (and, therefore, $\tau_{\mathrm{on}} < \tau_{\mathrm{o}}$). So for a given average protein level, a Fano factor larger than one occurs for leaky gene expression as well as for highly active genes with pulsed expression events.

How close is Eq. (28.18) to actual measurements?[6] Division of Eq. (28.18) by the mean protein copy number gives a protein noise expression $\langle \delta^2 n_{\mathrm{p}} \rangle / \langle n_{\mathrm{p}} \rangle^2 = C/\langle n_{\mathrm{p}} \rangle$. In Bar-Even *et al.* (2006), this linear relationship between noise and the inverse mean protein level was observed and they estimated the constant C to be about 1200. This study considered 43 proteins in *S. cerevisiae*. Since, also high variability in signaling protein levels have been found in eukaryotic signaling (Cohen *et al.*, 2009; Cohen-Saidon *et al.*, 2009; Sigal *et al.*, 2006) and that proteins tend to live longer than mRNA (Cohen *et al.*, 2009). Equation (28.18) should be a promising relation. Chubb *et al.* (2006) and Raj *et al.* (2006) have found evidence for short and pulsed transcription activity in eukaryotes. Whether the constant C would also be large for prokaryotic systems remains to be seen. It is probably only significant in the case of leaky gene expression and then little mRNA is made. Golding *et al.* (2005) estimated an mRNA burst size of about 4 for an *E. coli* strain with reduced repressor levels under gene repression conditions (on average in the burst size 0.6 in *E. coli*; Taniguchi *et al.*, 2010). The differences between eukaryotic and prokaryotic noise are probably mostly due to the timescale of the fluctuations in gene activity, which is for prokaryotes predominantly deriving from a competition between repressors and RNA polymerases for binding at the promotor (these processes occur on seconds and minutes time-scales), whereas for eukaryotes switches between gene activity states are likely much slower (tens of minutes to hours) due to slow changes in chromatin states that modulate the affinity of chromatin for transcription regulators.

From this section we conclude that

- The Fano factor and noise of proteins can be large even when mRNA is fast when transcription and translation occur in bursts and ON durations

[6] There is one important limitation to this equation: it is not applicable in the bimodal regime of the model as it strictly models the dispersion of protein copy number distributions when they obey a Gaussian distribution (linear noise approximation conditions, Elf, Van Kampen). Bimodality is absent when $\tau_{\mathrm{o}}/(\tau_{\mathrm{o}} + \tau_{\mathrm{p}}) \approx < 0.5$.

are as long as or shorter than OFF durations. This has been found experimentally in yeast.
- Equation (28.18) applies to infrequent transcription bursts and those have indeed been found experimentally.
- Slow gene switching in eukaryotes due to for instance chromatin changes between repressive and permissive gene states would induce bimodal protein distributions depending on the lifetime of protein and gene switch rate.

2.5. Noise propagation in molecular networks

In the previous section, we found that the noise in mRNA (Eq. (28.11)) and protein (Eq. (28.15)) is given by the kinetics of their synthesis and degradation events besides noise cascading from molecular regulators of the synthesis event, that is, the switching gene and the template mRNA. These examples indicate noise can propagate through networks and that molecules with high mean numbers can still be noisy if their synthesis and degradation are regulated by "noisy" molecules.

Elowitz *et al.* (2002) have elegantly shown this phenomenon experimentally. They equipped *E. coli* with two gene constructs, each encoding a fluorescent protein under the control of the same promotor. Those copies are regulated by the same regulatory molecules. They reasoned that the noise in two fluorescent molecules should have two contributions: one deriving from fluctuations in their common regulators (extrinsic or global noise), which is equal for both genes, and another that derives from the individual synthesis and degradation events of the fluorescent proteins. To disentangle those contributions they plotted the fluorescence values of both reporters in a single cell as function of each other. If the found dependency is a straight line, the extrinsic fluctuations dominate, and otherwise, when a cloud of points is observed, the intrinsic noise reduces the correlations between fluorescence levels in a single cell. No matter how precisely the two gene copies are controlled, the latter process will always limit precision of gene regulation. They found that under full activation conditions (for strong promotors) their constructs were least noisy; the noise increased by about a factor of 5 at conditions of transcription repression. As a function of fluorescence level they found that the intrinsic noise reduces with higher mean fluorescence levels. Total noise was often due more to extrinsic than to intrinsic noise. The authors suggested that low transcription factor and/or mRNA levels would explain the relatively high extrinsic noise. Extrinsic and intrinsic noise have different origins: extrinsic fluctuations occur at timescale close to the generation time (about 40 min), whereas intrinsic fluctuations occur much faster (characteristic timescale 9 min; Rosenfeld *et al.*, 2005).

Pedraza and van Oudenaarden (2005) investigated noise propagation in a synthetic gene cascade composed out of three interlinked gene systems. The first gene was constitutively active and encoded the *lac* repressor, which under the influence of IPTG could variably induce the second operon, composed out of the *tet* repressor and a fluorescent protein (CFP). The third operon was downregulated by *TetR* and also encoded a fluorescent protein (YFP). Global fluctuations in expression level were monitored by a fourth gene encoding RFP (not linked to the cascade). Input–output characterization of the cascade, fluorescence as function of IPTG levels, showed that the cascade was indeed tunable. The dependency of CFP and YFP self-correlation (SD/mean; square root of noise) and cross-correlations as function of IPTG were determined for all the data (global fluctuations were IPTG independent). Then they tried to model this data using a stochastic model that was analytically trackable and gave expression for noise (and correlations) as shown in Eq. (28.17). Noise downstream of the cascade is then typically a function of intrinsic noise at this level of the cascade, time-averaging of input fluctuations from upstream, the sensitivity of this cascade level to upstream changes in molecules numbers (the logarithmic gain or local response coefficient) and global noise. In this manner, the noise in the output of the cascade, the YFP fluorescence, could be described as the sum intrinsic, transmitted, and global noise. Depending on the IPTG level either global or transmitted noise was highest. Intrinsic noise was negligible at low levels of IPTG (when YFP is high) and about one-third of the total noise at high levels. Transmitted noise was shown to be very dependent on the gain of YFP to CFP (maximally 3; corresponding to an input–output relationship modeled with a Hill equation with exponent 3). This study confirmed and quantified the terms in noise propagation equations such as Eq. (28.17). Becskei and Serrano (2000) showed that negative feedback could reduce cell-to-cell variability in gene expression using synthetic gene networks. Yu *et al.* (2008) showed a similar result for a naturally occurring system. Negative feedback can also lead to the enhancement of noise depending on how the timescale separation plays out in a gene cascade (Bruggeman *et al.*, 2009).

The main concepts of this section can be summarized as follows:

- Noise in molecule numbers can propagate from one level to the next.
- Steady-state noise in an mRNA has an intrinsic and extrinsic component; the former depends on individual mRNA synthesis and degradation events occurring out of sync whereas the latter originates from fluctuating levels of global transcription regulators, such as RNA polymerase and transcription factors.
- The extent of noise propagation depends on the sensitivity of molecular reactions to the fluctuating regulator, the timescale separation between the target and regulatory fluctuations and the noise in the regulator (Fig. 28.6).
- Experiments can be satisfactorily explained in terms of approximate stochastic theory.
- Noise transmission and total noise depend on gene expression activity.

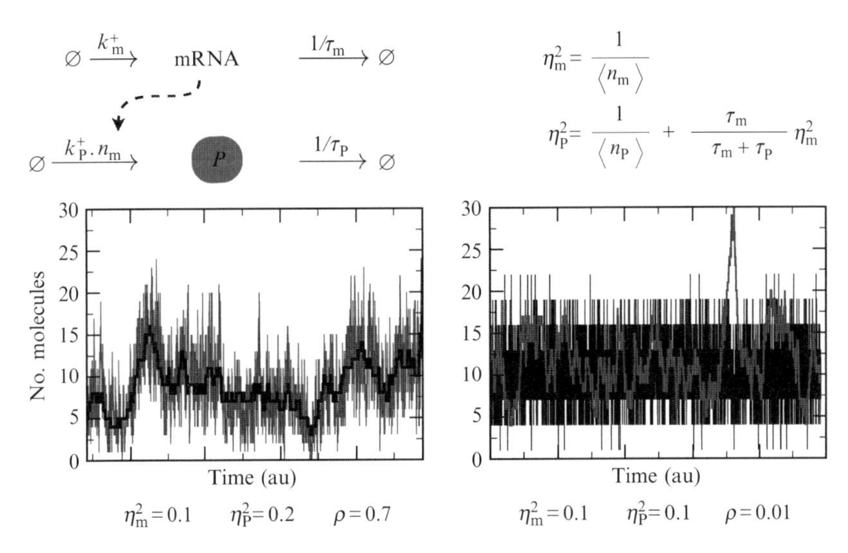

Figure 28.6 Timescale separation affects noise propagation in a simple gene expression model. The synthesis of protein P (red line) takes place only in the presence of mRNA (black); hence, the production rate equal to $k_P^+ n_m \tau_m = 1/k_m^-$ and $\tau_P = 1/k_P^-$ denotes the lifetime of mRNA and protein, respectively. Noise, variance divided by the mean squared, is denoted by η. (Left plot) The lifetime of the protein is short compared to the lifetime of mRNA; hence, the noise in mRNA propagates downstream of the network. Correlation ρ between mRNA and P concentration is high. (Right plot) Long-lived proteins "average out" fluctuations in mRNA; correlation is low. (For interpretation of the references to color in this figure legend, the reader is referred to the Web version of this chapter.)

3. Beneficial and Detrimental Effects of Molecular Noise

3.1. Changing and uncertain environments: Stochastic phenotype switching by microorganisms

As discussed in the previous sections, heterogeneity is an inevitable characteristic of cell populations. Here we discuss some physiological consequences of this. In the simplest case, expression levels of proteins fluctuate around the mean level giving rise to a unimodal distribution and theoretical considerations suggest that such gradual fluctuations can be advantageous to the growth of a population if there are large (temporal) fluctuations in the optimum expression level of the protein and strong selective pressures (Tănase-Nicola and ten Wolde, 2008; Zhang and Hill, 2005). Benefits arising from this type of heterogeneity have been demonstrated experimentally with yeast mutant strains displaying increased heterogeneity in stress response proteins which under extreme stress conditions have higher survival rates than the wild type (Bishop *et al.*, 2007; Smith *et al.*, 2007). However, environmental fluctuations in the optimum expression level need to be extremely high in this scenario. A different picture emerges when considering switch-like systems leading to multimodal protein distributions. In this case, different subpopulations are characterized by different expression levels of a set of proteins and by different growth rates which are environment-dependent. Examples for such switch systems include the competence pathway in *Bacillus subtilis* (Maamar *et al.*, 2007; Suel *et al.*, 2007), the mating pheromone pathway in yeast (Colman-Lerner *et al.*, 2005), viral latency (Weinberger *et al.*, 2005), bacterial persistence (Balaban *et al.*, 2004; Kussell and Leibler, 2005), and sporulation in *B. subtilis* (Veening *et al.*, 2008). Switching between different states can occur in either a purely stochastic manner or the switching rates can be dependent on the current environment or history of recently encountered environmental conditions. Theoretical considerations suggest that either strategy can be advantageous depending on the strength of selection pressure, duration of environmental states, etc. Purely stochastic switching can be the optimal strategy if some environments are lethal to a certain phenotype (e.g., growing vs. persister cells) or if the environment changes on timescales shorter than the response time of adaptive systems or if environmental changes are very rare so that the cost of maintaining a sensory system does not pay off (Kussell and Leibler, 2005; Leibler and Kussell, 2010; Thattai and van Oudenaarden, 2004; Wolf *et al.*, 2005). Experimentally the effects of switching on population fitness were tested with a synthetic genetic switch in *S. cerevisiae* with experimentally tunable switching rates. The two subpopulations resulting from the bistable switch were designed to have growth advantages under different conditions (Acar *et al.*, 2008). These experiments showed that population growth rate is optimized when the rates of transition between the two stable states are tuned to the rates of environmental change. That bet-hedging is indeed an evolvable strategy was demonstrated with a laboratory evolution experiment (Beaumont *et al.*, 2009). Bacterial populations were exposed to varying environments and several rounds of selections. From these experiments a strain that displayed phenotypic switching between different colony morphologies could be selected.

3.2. Bistable switches in cellular decision making

Often the cause for the presence of two or more distinct classes of phenotypes in a population of isogenetic cells is an underlying bi- or multistable gene regulatory network or signaling network. Two different network motifs have been described that can generate bistability: positive feedback loops (Thomas *et al.*, 1976) and multistep covalent modification (Markevich *et al.*, 2004; Ortega *et al.*, 2006; Saez-Rodriguez *et al.*, 2008). Networks with direct feedback (positive autoregulation), double negative feedback (a toggle switch), or more complex network structures with net positive feedback can all lead to multiple stable states for certain parameter ranges (Angeli *et al.*, 2004). Although the terms bistability of the network and bimodality of protein levels are often used almost synonymously, bistability is not a strict prerequisite for bimodality. Systems that have unstable states which can persist for prolonged times before the single stable state is reached can lead to bimodality. Further, stochasticity of reactions can also lead to bimodal distributions although a deterministic analysis of the reaction network predicts only one stable state (Fig. 28.7) (Artyomov *et al.*, 2007; To and Maheshri, 2010).

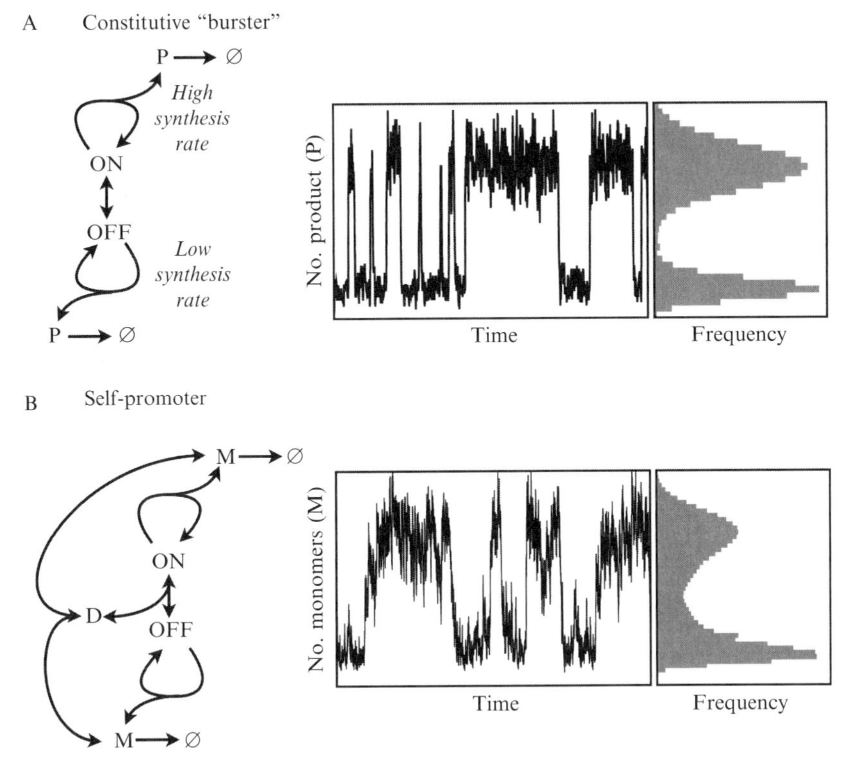

Figure 28.7 Two sources of bimodality in protein distribution. Bimodal steady-state protein distribution may arise purely due to stochastic effects in gene expression (A) or due to bistability in a nonlinear dynamical system (B). (A) A spontaneous stochastic switch, a "burster," with constitutive production of *P*. If ON and OFF states are long enough to establish their respective quasi steady states, the system will "flip" between two distinct states. (B) A classical mechanism leading to bimodality—a positive feedback. Monomer *M* promotes its own synthesis by forming a dimer *D*, which further induces the active ON state. Dimerization introduces nonlinearity which may lead to bistability if none of the processes acts too strongly (Isaacs *et al.*, 2003).

Switches between the (stable) states can arise from fluctuations in network components that cross the threshold of the positive feedback. In this way variations around an average level (a unimodal distribution of concentrations) can be amplified and give rise to a bimodal distribution of protein levels that can lead to qualitatively different phenotypes. The role of noise in switching between states has been analyzed for the competence pathway in *B. subtilis* (Maamar *et al.*, 2007; Suel *et al.*, 2007). In populations entering stationary phase, a certain fraction of cells express the key regulator of competence ComK at high levels and acquire competence for DNA uptake while the majority of cells remain in the low ComK expression state. The core regulatory motif is a positive feedback loop of ComK cooperatively binding to its own promoter and activating expression. ComK levels are low during exponential growth and transiently increase when cells reach stationary phase. To test whether fluctuations around this increased average level could explain why only a fraction of cells enter the competent state, Maamar *et al.* and Suel *et al.* engineered strains that have the same average ComK concentration as the wild type but with an increased or decreased magnitude of fluctuations around the average. Noise in basal ComK levels was reduced by changing transcription and translation efficiency (Maamar *et al.*, 2007) and by using a septation mutant (Suel *et al.*, 2007). In both cases, the reduction of noise in the basal ComK levels led to a decrease in the fraction of cells that switched to the competent state, consistent with the hypothesis that fluctuation induced crossing of the threshold for positive feedback causes switching to the high expression state.

3.3. Eukaryotic signaling and cell-to-cell variability

While noise and heterogeneity can be advantageous in changing and unpredictable environments, noise in signaling poses a limit to the transfer of information, of how much a cell or an organism can learn about its environment and consequently how accurately it can adapt. One fundamental limit to information transfer derives from diffusional noise in the input of signaling pathways (Berg and Purcell, 1977; Bialek and Setayeshgar, 2005): if the input signal is the concentration of some molecule (a ligand or a transcription factor, etc.), the number of complexes formed (receptor-ligand-complex, promoter-TF-complex) inevitably fluctuates over time and the minimum fractional accuracy with which a cell can determine this number is inversely proportional to the square root of the number of receptors and the time over which complex formation is monitored. Therefore, an increase in the accuracy of sensing at this level can only be achieved at the expense of either increased receptor concentrations or with longer averaging times resulting in a slower response to changes in the signal. Such fluctuations in receptor occupancy on the cell surface have been measured experimentally for the EGF receptor (Sako *et al.*, 2000) as well as a chemotaxis receptor in dictyostelium (Ueda *et al.*, 2001) with total internal reflection fluorescence microscopy. Besides fluctuations in the input of signaling pathways, also the concentrations of its components, i.e. the proteins involved in signal transmission, can be heterogenous in a cell population (Cohen-Saidon *et al.*, 2009; Colman-Lerner *et al.*, 2005) and fluctuate over time within a single cell (Sigal *et al.*, 2006). It then depends on the architecture of the signaling network how much such component level fluctuations affect the signaling output. A study of drug-induced apoptosis indicated that high levels of heterogeneity in the time between drug addition and apoptosis could be explained by a model taking into account the heterogeneity in the levels of five regulatory proteins (Spencer *et al.*, 2009). Other signaling pathways have been shown to be remarkably robust to fluctuations in protein levels. One mechanism by which this property can be achieved is fold-change-detection which has been found experimentally for the EGF/ERK2 pathway (Cohen-Saidon *et al.*, 2009) as well as the Wnt/β-catenin pathway (Goentoro and Kirschner, 2009). Since the output of such signaling pathways consists of the relative difference of the concentration of a protein before and after the signal rather than its absolute numbers, such mechanisms can buffer fluctuations in protein levels and allow precise signaling in the presence of noise. Interestingly, theoretical considerations suggest that fluctuations in protein levels could in principle also enhance signal transduction, a phenomenon called stochastic resonance (Mcdonnell *et al.*, 2009). It is defined as an increase in a metric of quality of signal transmission due to an increase in random noise. Prerequisites for this phenomenon are that the system operates at a nonoptimal parameter regime (nonoptimal with regard to signal transmission) and that it contains nonlinearities. Since any network within a cell is subject to various constraints (on component numbers, achievable timescales, etc.) and since nonlinear regulation is commonly found in signaling pathways, it is well possible that these two conditions are true for signaling network. However, the relevance of this phenomenon for biochemical signaling networks needs to be explored experimentally.

3.4. Noisy decision making in eukaryotic development

The development of multicellular organisms requires the generation of multitudes of specialized cell types in a defined spatiotemporal order. Nevertheless, a number of cell fate decisions in eukaryotes seem to exploit noise in the levels of transcription factors. Mouse embryonic stem cells display a bimodal distribution of the transcription factor Nanog, with the cells in the low Nanog state being predisposed for differentiation. Cells can switch between the high and low expression states so that at any time a certain fraction of cells is poised for differentiation while without external cues all cells remain pluripotent. The bimodal distribution of Nanog is consistent with a core regulatory circuit between Nanog and Oct4 involving both positive and negative feedback (Kalmar *et al.*, 2009). A similar observation, that fluctuations in the level of a transcription factor can make subpopulations more prone for further differentiation without yet committing them to the differentiated cell fate, has been made in hematopoietic progenitor cells for the factor Sca1 (Chang *et al.*, 2008). The distribution of Sca1 and the dynamics of its fluctuations could best be described by a model for the progenitor cell population that includes multiple discrete states differing in their transcriptomes and their predisposition for differentiation, and stochastic transitions between these states. It has been hypothesized that the noise driven generation of subpopulations which are predisposed for differentiation and subsequent fixation into differentiated cells is a very general mechanism for eukaryotic cell fate decisions that is repeated several times in the course of development (Arias and Hayward, 2006; Kilfoil *et al.*, 2009). The fixation of predisposed cells to certain cell fates can either be guided by signals from other cells or be purely stochastic generating a certain fraction of differentiated cells. The latter mechanism has been proposed for the early development of mice embryo, where stochastic decisions on a cellular level lead to differentiation into three different cell types in a mosaic pattern, followed by cell sorting/migration to establish defined spatial pattern (Dietrich and Hiiragi, 2007).

4. Conclusion

Inside cells, it is a noisy business: diffusion-driven reactive collisions of molecules, probabilistic spontaneous decay of molecules, binomial partitioning of molecules between daughter cells at variable cell division times, and short (and imprecise) transcription bursts are key determinants of cell-to-cell variability in process rates, response times, and molecule levels. At a higher level, this molecular noise gives rise to variability in physiological properties.

On the one hand, noise is a nuisance for cells; sophisticated regulatory mechanisms are inevitably distorted by molecular noise. Hereby, information transfer is limited by noise preventing accurate cellular decision making. On the other hand, noise offers an opportunity for cells in uncertain and dynamic environments; it allows for randomization of states to enhance their survival chances (bet-hedging). Studies of how cells cope and exploit noise in terms of molecular mechanisms and how associated trade-offs and physicochemical constraints delimit cell behavior are therefore highly relevant.

Most studies so far dealt with quantification of noise—early studies—or with the evaluation of how cells exploit or are constrained by noise—more recent studies—at the level of small molecular networks. How molecular noise propagates through larger segments of a cell's molecular networks is unexplored: is there noise buffering or amplification? Another unexplored aspect is how microbial cells deal with noise when they adapt to changes in conditions. In those situations, noise reduces performance as a cell needs to distinguish changes induced by the environment from those induced by noise. How can microbes cope with this? Perhaps they tune their own noise to different frequencies to enhance identification of environmental changes?

For eukaryotes, many questions are still open as well. The noise in proteins turns out to be likely much higher than in prokaryotes, presumably due to slow gene switching induced by chromatin dynamics. It has been shown that this leads to large variability in signaling protein levels causing difference in single-cell dynamics and drug sensitivity. Again, how can cells have reliable signal transfer when their protein levels are unreliable? Fold-change detection has been suggested to be a solution to this problem and it has shown to be operational in a number of different systems in mammalian cells and *E. coli* chemotaxis. It is such universal mechanisms that systems biology is searching for.

In cell biology, we are slowly gaining more insight into the functioning and constraints of small molecular networks through more advanced experimentation and mathematical modeling. The capabilities have advanced to a stage that we can start thinking about studying larger molecular networks. Then we will see how noisy cells really are and whether the noise at the level of single genes and small networks propagates and pushes cells to functional regimes where it needs sophisticated control circuitry to sustain itself and adapt.

ACKNOWLEDGMENTS

We thank L. Anink, J. Blom, T. Hellings, M. Mandjes, D. Piebes, and R. van Driel for discussions. A. S. and F. J. B. thank the Netherlands Institute for Systems Biology (www.sysbio.nl), the Centre for Mathematics and Computer Science, and the Dutch Organization for Scientific Research (NWO; Grant number 837.000.001) for funding. K. R. and F. J. B. thank the Marie Curie Network *NucSys* for funding.

REFERENCES

Acar, M., Mettetal, J. T., and Van Oudenaarden, A. (2008). Stochastic switching as a survival strategy in fluctuating environments. *Nat. Genet.* **40**(4), 471–475.

Angeli, D., Ferrell, J. E., and Sontag, E. D. (2004). Detection of multistability, bifurcations, and hysteresis in a large class of biological positive-feedback systems. *Proc. Natl. Acad. Sci. USA* **101,** 1822–1827.

Arias, A. M., and Hayward, P. (2006). Filtering transcriptional noise during development: Concepts and mechanisms. *Nat. Rev. Genet.* **7,** 34–44.

Artyomov, M. N., Das, J., Kardar, M., and Chakraborty, A. K. (2007). Purely stochastic binary decisions in cell signaling models without underlying deterministic bistabilities. *Proc. Natl. Acad. Sci. USA* **104,** 18958–18963.

Balaban, N. Q., Merrin, J., Chait, R., Kowalik, L., and Leibler, S. (2004). Bacterial persistence as a phenotypic switch. *Science* **305,** 1622–1625.

Bar-Even, A., Paulsson, J., Maheshri, N., Carmi, M., O'Shea, E., Pilpel, Y., and Barkai, N. (2006). Noise in protein expression scales with natural protein abundance. *Nat. Genet.* **38** (6), 636–643.

Beaumont, H. J., Gallie, J., Kost, C., Ferguson, G. C., and Rainey, P. B. (2009). Experimental evolution of bet hedging. *Nature* **462,** 90–93.

Becskei, A., and Serrano, L. (2000). Engineering stability in gene networks by autoregulation. *Nature* **405**(6786), 590–593.

Becskei, A., Séraphin, B., and Serrano, L. (2001). Positive feedback in eukaryotic gene networks: Cell differentiation by graded to binary response conversion. *EMBO J.* **20**(10), 2528–2535.

Bengtsson, M., Ståhlberg, A., Rorsman, P., and Kubista, M. (2005). Gene expression profiling in single cells from the pancreatic islets of Langerhans reveals lognormal distribution of mRNA levels. *Genome Res.* **15**(10), 1388–1392.

Berg, O. G. (1978). A model for the statistical fluctuations of protein numbers in a microbial population. *J. Theor. Biol.* **71**(4), 587–603.

Berg, H. C., and Purcell, E. M. (1977). Physics of chemoreception. *Biophys. J.* **20,** 193–219.

Bertrand, E., Chartrand, P., Schaefer, M., Shenoy, S. M., Singer, R. H., and Long, R. M. (1998). Localization of ash1 mRNA particles in living yeast. *Mol. Cell* **2**(4), 437–445.

Bialek, W., and Setayeshgar, S. (2005). Physical limits to biochemical signaling. *Proc. Natl. Acad. Sci. USA* **102,** 10040–10045.

Bishop, A. L., Rab, F. A., Sumner, E. R., and Avery, S. V. (2007). Phenotypic heterogeneity can enhance rare-cell survival in 'stress-sensitive' yeast populations. *Mol. Microbiol.* **63,** 507–520.

Blake, W. J., Kaern, M., Cantor, C. R., and Collins, J. J. (2003). Noise in eukaryotic gene expression. *Nature* **422**(6932), 633–637.

Blake, W. J., Balázsi, G., Kohanski, M. A., Isaacs, F. J., Murphy, K. F., Kuang, Y., Cantor, C. R., Walt, D. R., and Collins, J. J. (2006). Phenotypic consequences of promoter-mediated transcriptional noise. *Mol. Cell* **24**(6), 853–865.

Bruggeman, F. J., Blüthgen, N., and Westerhoff, H. V. (2009). Noise management by molecular networks. *PLoS Comput. Biol.* **5**(9).

Cai, L., Friedman, N., and Xie, X. S. (2006). Stochastic protein expression in individual cells at the single molecule level. *Nature* **440**(7082), 358–362.

Chang, H. H., Hemberg, M., Barahona, M., Ingber, D. E., and Huang, S. (2008). Transcriptome-wide noise controls lineage choice in mammalian progenitor cells. *Nature* **453,** 544–547.

Chubb, J. R., Trcek, T., Shenoy, S. M., and Singer, R. H. (2006). Transcriptional pulsing of a developmental gene. *Curr. Biol.* **16**(10), 1018–1025.

Cohen, A. A., Kalisky, T., Mayo, A., Geva-Zatorsky, N., Danon, T., Issaeva, I., Kopito, R. B., Perzov, N., Milo, R., Sigal, A., and Alon, U. (2009). Protein dynamics in individual human cells: Experiment and theory. *PLoS One* **4**(4).

Cohen-Saidon, C., Cohen, A. A., Sigal, A., Liron, Y., and Alon, U. (2009). Dynamics and variability of ERK2 response to EGF in individual living cells. *Mol. Cell* **36**(5), 885–893.

Colman-Lerner, A., Gordon, A., Serra, E., Chin, T., Resnekov, O., Endy, D., Pesce, C. G., and Brent, R. (2005). Regulated cell-to-cell variation in a cell-fate decision system. *Nature* **437**(7059), 699–706.

Dietrich, J.-E., and Hiiragi, T. (2007). Stochastic patterning in the mouse pre-implantation embryo. *Development* **134**(23), 4219–4231.

Dobrzynski, M., and Bruggeman, F. J. (2009). Elongation dynamics shape bursty transcription and translation. *Proc. Natl. Acad. Sci. USA* **106**(8), 2583–2588.
Elf, J., and Ehrenberg, M. (2003). Fast evaluation of fluctuations in biochemical networks with the linear noise approximation. *Genome Res.* **13**(11), 2475–2484.
Elowitz, M. B., and Leibler, S. (2000). A synthetic oscillatory network of transcriptional regulators. *Nature* **403**(6767), 335–338.
Elowitz, M. B., Levine, A. J., Siggia, E. D., and Swain, P. S. (2002). Stochastic gene expression in a single cell. *Science* **297**(5584), 1183–1186.
Friedman, N., Cai, L., and Xie, X. S. (2006). Linking stochastic dynamics to population distribution: An analytical framework of gene expression. *Phys. Rev. Lett.* **97**(16), 168302–168306.
Gardner, T. S., Cantor, C. R., and Collins, J. J. (2000). Construction of a genetic toggle switch in Escherichia coli. *Nature* **403**(6767), 339–342.
Gillespie, D. T. (2007). Stochastic simulation of chemical kinetics. *Annu. Rev. Phys. Chem.* **58,** 35–55.
Goentoro, L., and Kirschner, M. W. (2009). Evidence that fold-change, and not absolute level, of beta-catenin dictates wnt signaling. *Mol. Cell* **36**(5), 872–884.
Golding, I., Paulsson, J., Zawilski, S. M., and Cox, E. C. (2005). Real-time kinetics of gene activity in individual bacteria. *Cell* **123**(6), 1025–1036.
Isaacs, F. J., Hasty, J., Cantor, C. R., and Collins, J. J. (2003). Prediction and measurement of an autoregulatory genetic module. *Proc. Natl. Acad. Sci. USA* **100**(13), 7714–7719.
Kalmar, T., Lim, C., Hayward, P., Munoz-Descalzo, S., Nichols, J., Garcia-Ojalvo, J., and Martinez Arias, A. (2009). Regulated fluctuations in Nanog expression mediate cell fate decisions in embryonic stem cells. *PLoS Biol.* **7,** e1000149.
Kaufmann, B. B., Yang, Q., Mettetal, J. T., and van Oudenaarden, A. (2007). Heritable stochastic switching revealed by single-cell genealogy. *PLoS Biol.* **5**(9), e239.
Kilfoil, M. L., Lasko, P., and Abouheif, E. (2009). Stochastic variation: From single cells to superorganisms. *HFSP J.* **3,** 379–385.
Kussell, E., and Leibler, S. (2005). Phenotypic diversity, population growth, and information in fluctuating environments. *Science* **309,** 2075–2078.
Leibler, S., and Kussell, E. (2010). Individual histories and selection in heterogeneous populations. *Proc. Natl. Acad. Sci. USA* **107**(29), 13183–13188.
Maamar, H., Raj, A., and Dubnau, D. (2007). Noise in gene expression determines cell fate in Bacillus subtilis. *Science* **317**(5837), 526–529.
Markevich, N. I., Hoek, J. B., and Kholodenko, B. N. (2004). Signaling switches and bistability arising from multisite phosphorylation in protein kinase cascades. *J. Cell Biol.* **164,** 353–359.
Mcdonnell, M. D., Abbott, D., and Friston, K. J. (2009). What is stochastic resonance? Definitions, misconceptions, debates, and its relevance to biology. *PLoS Comput. Biol.* **5** (5), e1000348.
Mcquarrie, D. A. (1967). Stochastic approach to chemical kinetics. *J. Appl. Prob.* **4,** 413–478.
Nelson, D. E., Ihekwaba, A. E., Elliott, M., Johnson, J. R., Gibney, C. A., Foreman, B. E., Nelson, G., See, V., Horton, C. A., Spiller, D. G., Edwards, S. W., McDowell, H. P., *et al.* (2004). Oscillations in NF-kappaB signaling control the dynamics of gene expression. *Science* **306**(5696), 704–708.
Novick, A., and Weiner, M. (1957). Enzyme induction as an all-or-none phenomenon. *Proc. Natl. Acad. Sci. USA* **43**(7), 553–566.
Ortega, F., Garces, J. L., Mas, F., Kholodenko, B. N., and Cascante, M. (2006). Bistability from double phosphorylation in signal transduction. Kinetic and structural requirements. *FEBS J.* **273,** 3915–3926.
Ozbudak, E. M., Thattai, M., Kurtser, I., Grossman, A. D., and van Oudenaarden, A. (2002). Regulation of noise in the expression of a single gene. *Nat. Genet.* **31**(1), 69–73.

Ozbudak, E. M., Thattai, M., Lim, H. N., Shraiman, B. I., and Van Oudenaarden, A. (2004). Multistability in the lactose utilization network of Escherichia coli. *Nature* **427** (6976), 737–740.

Paulsson, J. (2004). Summing up the noise in gene networks. *Nature* **427**(6973), 415–418.

Paulsson, J., and Ehrenberg, M. (2000). Random signal fluctuations can reduce random fluctuations in regulated components of chemical regulatory networks. *Phys. Rev. Lett.* **84**(23), 5447–5450.

Pedraza, J. M., and van Oudenaarden, A. (2005). Noise propagation in gene networks. *Science* **307**(5717), 1965–1969.

Raj, A., and van Oudenaarden, A. (2008). Nature, nurture, or chance: Stochastic gene expression and its consequences. *Cell* **135**(2), 216–226.

Raj, A., and van Oudenaarden, A. (2009). Single-molecule approaches to stochastic gene expression. *Annu. Rev. Biophys.* **38,** 255–270.

Raj, A., Peskin, C. S., Tranchina, D., Vargas, D. Y., and Tyagi, S. (2006). Stochastic mRNA synthesis in mammalian cells. *PLoS Biol.* **4**(10), e309.

Rosenfeld, N., Joung, J. W., Alon, U., Swain, P. S., and Elowitz, M. B. (2005). Gene regulation at the single-cell level. *Science* **307**(5717), 1962–1965.

Saez-Rodriguez, J., Hammerle-Fickinger, A., Dalal, O., Klamt, S., Gilles, E. D., and Conradi, C. (2008). Multistability of signal transduction motifs. *IET Syst. Biol.* **2,** 80–93.

Sako, Y., Minoghchi, S., and Yanagida, T. (2000). Single-molecule imaging of EGFR signalling on the surface of living cells. *Nat. Cell Biol.* **2,** 168–172.

Shahrezaei, V., and Swain, P. S. (2008). Analytical distributions for stochastic gene expression. *Proc. Natl. Acad. Sci. USA* **105**(45), 17256–17261.

Shankaran, H., Ippolito, D. L., Chrisler, W. B., Resat, H., Bollinger, N., Opresko, L. K., and Wiley, H. S. (2009). Rapid and sustained nuclear-cytoplasmic ERK oscillations induced by epidermal growth factor. *Mol. Syst. Biol.* **5,** 332.

Sigal, A., Milo, R., Cohen, A., Geva-Zatorsky, N., Klein, Y., Liron, Y., Rosenfeld, N., Danon, T., Perzov, N., and Alon, U. (2006). Variability and memory of protein levels in human cells. *Nature* **444**(7119), 643–646.

Smith, M. C., Sumner, E. R., and Avery, S. V. (2007). Glutathione and Gts1p drive beneficial variability in the cadmium resistances of individual yeast cells. *Mol. Microbiol.* **66,** 699–712.

Spencer, S. L., Gaudet, S., Albeck, J. G., Burke, J. M., and Sorger, P. K. (2009). Non-genetic origins of cell-to-cell variability in TRAIL-induced apoptosis. *Nature* **459,** 428–432.

Spudich, J. L., and Koshland, D. E. (1976). Non-genetic individuality: Chance in the single cell. *Nature* **262**(5568), 467–471.

Suel, G. M., Kulkarni, R. P., Dworkin, J., Garcia-Ojalvo, J., and Elowitz, M. B. (2007). Tunability and noise dependence in differentiation dynamics. *Science* **315,** 1716–1719.

Tănase-Nicola, S., and ten Wolde, P. R. (2008). Regulatory control and the costs and benefits of biochemical noise. *PLoS Comput. Biol.* **4**(8), e1000125.

Taniguchi, Y., Choi, P. J., Li, G. W., Chen, H., Babu, M., Hearn, J., Emili, A., and Xie, X. S. (2010). Quantifying e. coli proteome and transcriptome with single-molecule sensitivity in single cells. *Science* **329**(5991), 533–538.

Thattai, M., and van Oudenaarden, A. (2004). Stochastic gene expression in fluctuating environments. *Genetics* **167,** 523–530.

Thomas, R., Gathoye, A. M., and Lambert, L. (1976). A complex control circuit. Regulation of immunity in temperate bacteriophages. *Eur. J. Biochem.* **71,** 211–227.

To, T. L., and Maheshri, N. (2010). Noise can induce bimodality in positive transcriptional feedback loops without bistability. *Science* **327,** 1142–1145.

Ueda, M., Sako, Y., Tanaka, T., Devreotes, P., and Yanagida, T. (2001). Single-molecule analysis of chemotactic signaling in Dictyostelium cells. *Science* **294,** 864–867.

Van Kampen, N. G. (2003). Stochastic processes in physics and chemistry. North-Holland.

Veening, J. W., Stewart, E. J., Berngruber, T. W., Taddei, F., Kuipers, O. P., and Hamoen, L. W. (2008). Bet-hedging and epigenetic inheritance in bacterial cell development. *Proc. Natl. Acad. Sci. USA* **105,** 4393–4398.

Weinberger, L. S., Burnett, J. C., Toettcher, J. E., Arkin, A. P., and Schaffer, D. V. (2005). Stochastic gene expression in a lentiviral positive-feedback loop: Hiv-1 tat fluctuations drive phenotypic diversity. *Cell* **122**(2), 169–182.

Wolf, D. M., Vazirani, V. V., and Arkin, A. P. (2005). Diversity in times of adversity: Probabilistic strategies in microbial survival games. *J. Theor. Biol.* **234,** 227–253.

Xie, X. S., Choi, P. J., Li, G. W., Lee, N. K., and Lia, G. (2008). Single-molecule approach to molecular biology in living bacterial cells. *Annu. Rev. Biophys.* **37,** 417–444.

Youk, H., Raj, A., and van Oudenaarden, A. (2010). Imaging single mRNA molecules in yeast. *Methods Enzymol.* **470,** 429–446.

Yu, J., Xiao, J., Ren, X., Lao, K., and Xie, X. S. (2006). Probing gene expression in live cells, one protein molecule at a time. *Science* **311**(5767), 1600–1603.

Yu, R. C., Pesce, C. G., Colman-Lerner, A., Lok, L., Pincus, D., Serra, E., Holl, M., Benjamin, K., Gordon, A., and Brent, R. (2008). Negative feedback that improves information transmission in yeast signalling. *Nature* **456**(7223), 755–761.

Zenklusen, D., Larson, D. R., and Singer, R. H. (2008). Single-RNA counting reveals alternative modes of gene expression in yeast. *Nat. Struct. Mol. Biol.* **15**(12), 1263–1271.

Zhang, X. S., and Hill, W. G. (2005). Evolution of the environmental component of the phenotypic variance: Stabilizing selection in changing environments and the cost of homogeneity. *Evolution* **59,** 1237–1244.

SECTION NINE

MANAGING SYSTEMS BIOLOGY DATA, MODELS AND RESEARCH

CHAPTER TWENTY-NINE

The SEEK: A Platform for Sharing Data and Models in Systems Biology

Katy Wolstencroft,* Stuart Owen,* Franco du Preez,[†] Olga Krebs,[‡] Wolfgang Mueller,[‡] Carole Goble,* *and* Jacky L. Snoep[†,§,||]

Contents

* School of Computer Science, University of Manchester, Manchester, United Kingdom
[†] Manchester Interdisciplinary Biocentre, University of Manchester, Manchester, United Kingdom
[‡] Heidelberg Institute for Theoretical Studies (Hits), gGmbH, Schloss-Wolfsbrunnenweg 35, Heidelberg, Germany
[§] Department of Biochemistry, Stellenbosch University, Private Bag X1, Matieland, South Africa
[||] Molecular Cell Physiology, VU Amsterdam, The Netherlands

Methods in Enzymology, Volume 500
ISSN 0076-6879, DOI: 10.1016/B978-0-12-385118-5.00029-3

Abstract

Systems biology research is typically performed by multidisciplinary groups of scientists, often in large consortia and in distributed locations. The data generated in these projects tend to be heterogeneous and often involves high-throughput "omics" analyses. Models are developed iteratively from data generated in the projects and from the literature. Consequently, there is a growing requirement for exchanging experimental data, mathematical models, and scientific protocols between consortium members and a necessity to record and share the outcomes of experiments and the links between data and models. The overall output of a research consortium is also a valuable commodity in its own right. The research and associated data and models should eventually be available to the whole community for reuse and future analysis.

The SEEK is an open-source, Web-based platform designed for the management and exchange of systems biology data and models. The SEEK was originally developed for the SysMO (systems biology of microorganisms) consortia, but the principles and objectives are applicable to any systems biology project. The SEEK provides an index of consortium resources and acts as gateway to other tools and services commonly used in the community. For example, the model simulation tool, JWS Online, has been integrated into the SEEK, and a plug-in to PubMed allows publications to be linked to supporting data and author profiles in the SEEK.

The SEEK is a pragmatic solution to data management which encourages, but does not force, researchers to share and disseminate their data to community standard formats. It provides tools to assist with management and annotation as well as incentives and added value for following these recommendations. Data exchange and reuse rely on sufficient annotation, consistent metadata descriptions, and the use of standard exchange formats for models, data, and the experiments they are derived from.

In this chapter, we present the SEEK platform, its functionalities, and the methods employed for lowering the barriers to adoption of standard formats. As the production of biological data continues to grow, in systems biology and in the life sciences in general, the need to record, manage, and exploit this wealth of information in the future is increasing. We promote the SEEK as a data and model management tool that can be adapted to the specific needs of a particular systems biology project.

1. Introduction

The number of systems biology research projects has grown rapidly over the past decade. Some of these projects are very large, for instance, SysMO (http://www.sysmo.net), a European project studying the systems biology of microorganisms, consists of over 320 scientists working in more than 120 research groups, organized into 13 distributed projects across Europe. Typically, such systems biology projects contain a heterogeneous group of scientists with a variety of life science, informatics, and computational modeling backgrounds. In addition to heterogeneity in research background of the scientists, there can also be a great diversity between research projects, with large differences in data types, experimental procedures, and models.

The multidisciplinary nature of systems biology projects necessitates a good exchange of data and models, such that an effective iterative cycle between experiment and model can take place. To make such an exchange possible, it is necessary that the data and models are described according to community accepted standards, and that sufficient annotation and metadata is available. In this regard, data sharing in systems biology faces the same issues as any data sharing in science. Reuse and future interpretation relies on common naming schemes and reporting standards and understanding the data in the context of the experiment(s) that created it. Conforming to these common standards, however, can be time-consuming and complicated, so the challenge for data management systems is to achieve this with minimal disruption to the daily activities of scientists by providing tooling, expertise, and best practice guidelines.

Classic data management systems have focused on prescriptive database and warehouse solutions for storing data. Such solutions are not always useful for the researchers however, as it would take a long time before the databases are developed and available. By that time, researchers may have large collections of unstructured legacy data. These solutions also require researchers to understand and adhere to rigid data structures and upload data in unfamiliar environments. For example, large-scale scientific data-sharing projects, such as the BIRN (http://www.birncommunity.org/), caBIG (https://**cabig**.nci.nih.gov/), or GridPP (http://www.gridpp.ac.uk/), insist that each participant agrees to specific formats and model specifications and adapts to a common infrastructure. If data management resources have been budgeted for, the conversion of data to the prescribed standards is possible and such approaches can be successful, but in the general case, resources are limited and such solutions are too heavy weight for many consortia. In addition, each individual must understand the standards and the data model in the new system in order to participate and must conform fully to this model. The resulting data is uniform and of a high standard, but the time required for submission of data may result in low user participation with only small amounts of data being deposited.

An important aspect of data management is therefore a cost–benefit analysis. Here costs would not only be the development and maintenance of the infrastructure (software development and hardware) but would also include effort of researchers in the projects to make the data and models available. The benefit would be the availability and reusability of data and the availability of tools to work with the data. A good balance between costs and benefit must be found, and is not necessarily static. The greater the standardization, the more reusable and comparable the data becomes, but there is a limit to the time and effort that can be expected from individual researchers without added benefits and incentives for their own work.

A more difficult aspect of data management is the reluctance of researchers to make their data available, especially before publication. Clearly, if data is only submitted to central repositories after publication, the members of the consortia do not have full benefit from the available resources produced throughout the projects, which can hinder collaborations. Therefore, it is essential that control over sharing individual data items and models remains with the researchers and encourages incrementally sharing with colleagues and the wider community. In contrast, funding bodies are now making much clearer demands on researchers to share their results more quickly and many publicly funded initiatives must adhere to new data-sharing policies. In SysMO, researchers are expected to pool their research capacities and know-how, and strongly promote the sharing of data, methods, models, and results within the consortium and with the systems biology community.

To meet these data management challenges, technical as well as social, the SysMO-DB project has designed, developed, and deployed a Web-based infrastructure (the SEEK) and a methodology to overcome these barriers and enable sharing and exchange in systems biology. Although developed for the SysMO consortium, the SEEK platform addresses general issues in systems biology data sharing and is applicable and adaptable to other consortia. It is available as open-source software and is designed for easy installation (http://www.sysmo-db.org/). The SEEK platform is consequently spreading. The Virtual Liver (http://www.sysmo-db.org/), EraSysBio+ (http://www.erasysbio.net), and UniCellSys (http://www.unicellsys.eu/) consortia are all examples of large systems biology networks that have adopted the SEEK.

In this chapter, we describe the SEEK and illustrate its functionality with examples from the SysMO consortium. We start with an overview of the SEEK platform and an outline of its design principles. Next, we discuss data management issues in more depth and show how the SEEK and associated tools assist scientists with the above-mentioned problems. We finish the chapter with a more general discussion about the state of data sharing in the life sciences, and how suitable incentives can be found to encourage individuals and institutions to become more open.

2. The SEEK Platform

The SEEK is the name given to the whole SysMO-DB data-sharing platform. Its development follows a rapid and incremental cycle with new functionality becoming available with each release (approximately every 2 months). As a result, the first version of the SEEK was deployed to the SysMO consortium within a year.

2.1. The PALs

A rapid and user-oriented development of the SEEK is ensured by frequent interactions in site visits and workshops with a focus group of users, the SysMO-PALs. PALs (Project Area Liaisons) are representatives from each SysMO project who are a mixture of experimentalists, modelers, and informaticians. PALs are Ph.D. students and postdocs, but not group leaders or project managers. This was a conscious decision to connect with people who could meet regularly and would be responsible for daily data generation and curation. The SysMO-PALs are an extension of the SysMO-DB design team. New developments in SEEK are trialed with the PALs before release to the rest of the community and the PALs describe new requirements and request possible new features. For example, they recommended we provide a directory of SysMO consortium members, they highlighted the importance of standard operating procedures (SOPs) and protocols in understanding experimental context, and they raised sensitive issues surrounding data security and access control. PALs also have extra responsibility in managing project membership and metadata. Consequently, the PALs gather intelligence from their projects and act as a dissemination mechanism for new developments with SysMO-DB.

2.2. Access to the SEEK

The SEEK Web interface is the main user access point to the system and provides a gateway to all other SysMO-DB resources. The SEEK comprises of the SysMO Yellow Pages, an Assets Catalog, a model simulation environment, and links to external resources. Figure 29.1 shows a screenshot of the SEEK, showing a summary of a collection of experiments and their associations. In addition to the Web interface, the SEEK also has a REST interface, allowing programmatic access to resources and allowing future federations of SEEK instances from different communities. The following sections describe the different elements of the SEEK and how they interact.

2.3. The Yellow Pages

The Yellow Pages list the members of the consortium, their projects, institutes, and expertise. This information helps foster links between SysMO projects and individual scientists, allowing people with the correct skill sets to be identified for collaborations. The Yellow Pages also links data and models and other SysMO assets to the scientists that produced them. Each asset is owned and controlled by the person registering it in SEEK. Individuals can decide whether to share immediately with the whole consortium, with their own project, or to restrict access to a few close collaborators.

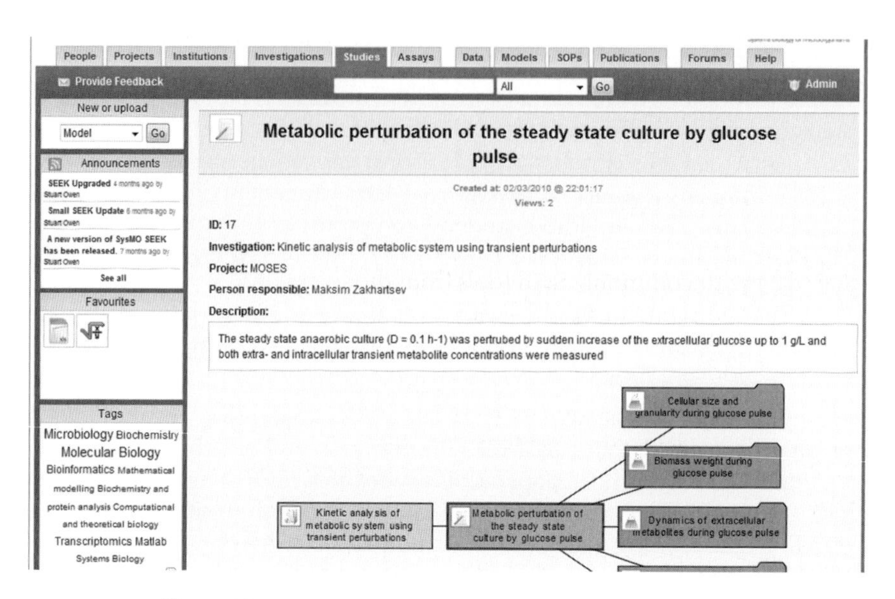

Figure 29.1 A screenshot of the SysMO–SEEK interface.

2.4. The SEEK assets catalog

The Assets catalog is a registry of who owns what resources and where they can be found. SysMO assets include data, models, protocols, SOPs, workflows, and publications. These assets may be held centrally in the SEEK, or they may be held in local project repositories. If they are held elsewhere, the SEEK indexes descriptions of the assets and can extract them from these external sources on demand, but it does not store them. Each asset (local or remote) is managed by the individual scientist who created and uploaded it. If assets are based on previously registered assets, an attribution system allows this to be recorded, which ensures scientists retain the credit for all their work. SysMO assets are registered with persistent URIs to allow stable referencing from publications.

Assets are associated with one another using the ISA hierarchy (Investigation, Study, Assay; Sansone *et al.*, 2008). ISA provides a framework for pooling data files and models in their experimental context. For example, data files can be associated with the SOPs used to create them; models can be associated with files containing construction data and validation data. Individual experiments (assays) and any associated assets can be grouped into larger studies and investigations, where the results of a combination of assays are required for biological interpretation.

ISA-TAB is being developed as a community standard and is a general tabular format for describing data from different types of omics experiments. By following such a community initiative, we enable future exchange of data with other public resources. In SEEK, we have extended the ISA omics concept to encompass mathematical models, to allow all SEEK assets to be described in the same ISA format. Figure 29.2 shows an ISA description of a set of SysMO experiments.

2.5. Access to external resources

The SysMO–SEEK is a gateway to other resources. SysMO users can analyze their data with commonly used tools from the community (e.g., JWS Online (http://jjj.mib.ac.uk), a model simulation environment; Olivier and Snoep, 2004), or they can explore descriptions of asset in a community context (e.g., using the BioPortal ontology repository; Noy *et al.*, 2009). By providing these services, we encourage uptake in the consortium and transform SysMO–SEEK from a static repository to an active, dynamic resource. Links to publications prepare the way for dissemination to the wider community. Direct links between the publications and associated data and models will allow the SEEK to become a supplementary material store for published work.

In the near future, the ability to run analyses through the SEEK will be implemented, driven by Taverna Workflows (Hull *et al.*, 2006; http://www.taverna.org.uk). A collection of workflows will assist in the meta-analysis of data registered in the SEEK.

Figure 29.3 shows the architecture of the SEEK, demonstrating how the different components interact. The SEEK adopts a modular approach, so access to other external resources can be added incrementally. The central

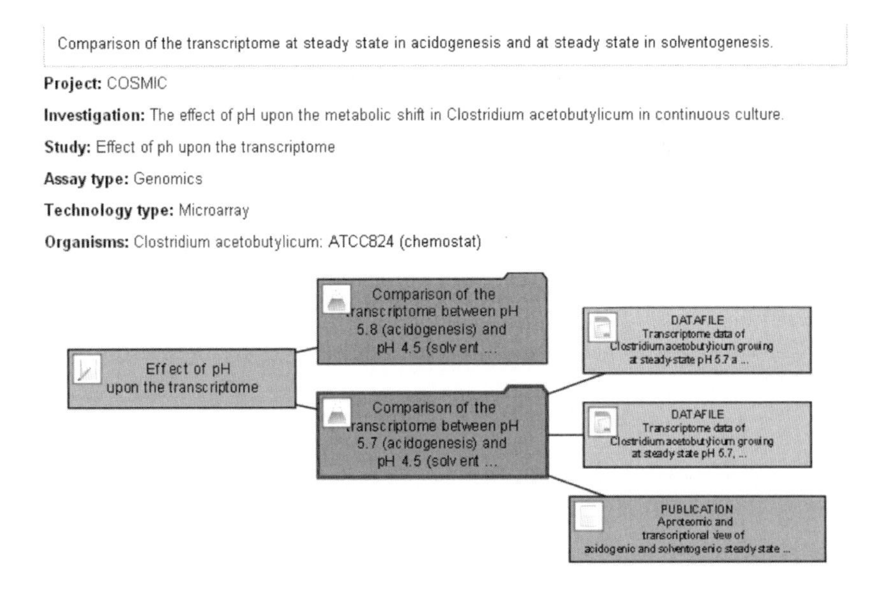

Figure 29.2 A screenshot of the ISA structure in SEEK and the interconnection of data and other assets in context of the experiments that created them.

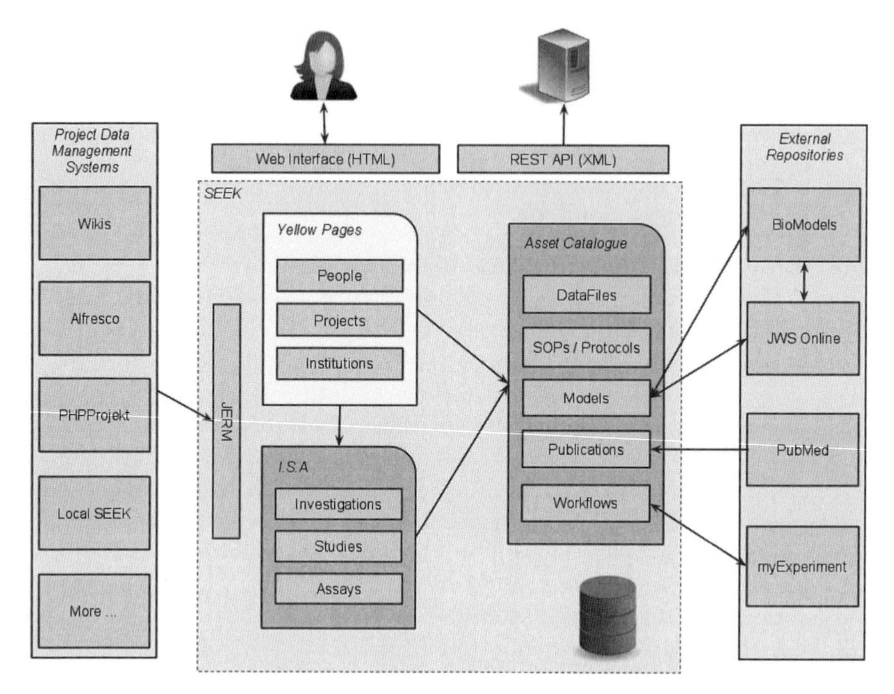

Figure 29.3 The architecture of the SEEK.

piece of the architecture is the JERM (Just Enough Results Model), which allows the interpretation of relationships between assets and the understanding of their contents. The JERM is fully described in Section 4.

The link to the JWS Online simulation environment is also a crucial part of the SEEK architecture. It provides a suite of tools for model management, annotation, and simulation. JWS Online can be accessed via SEEK, via a Web browser or via Web services. The interface gives access to the model parameters and initial conditions and allows the user to select between a number of functionalities such as time integration, steady state analysis, parameter scanning, and metabolic control analysis. In addition, an SBGN (Le Novère *et al.*, 2009) reaction schema is linked to the model interface via which the user can display the rate equations, view the annotation, and access external links to the models.

3. The Challenges of Data Management

The architecture of the SEEK platform allows for a flexible approach to uploading and linking assets. Such a record of data and models from a large research initiative is important in its own right, but the real challenge lies in being able to interpret the contents of the assets, which is necessary for comparison with other data sets and for further analysis.

In this section, we discuss the current issues with identifying and interpreting biological data and describe some community initiatives that are attempting to resolve and standardize data descriptions.

3.1. Biological object identity

Combining different types of biological data hinges on understanding exactly which biological objects interact and also on being able to identify the same biological objects in different data sets. Public data repositories often contain overlapping sets of information with the same biological objects having different names and identifiers. For example, Table 29.1 shows the different names for fructose-bisphosphate aldolase A and different identifiers for this protein. This protein intersects several central metabolic pathways, including glycolysis, the pentose phosphate pathway, and the fructose and mannose metabolism pathway. Consequently, it features in many protein and pathway resources. It is therefore essential that we are able to map synonyms back to official names of genes and proteins to enable the integration of information and data relating to this biological object from multiple sources.

The need for consistency in naming biological objects and the use of unique identifiers is well recognized (Howe *et al.*, 2008), but many scientists

Table 29.1 The names and synonyms of a gene and its product in different life science databases

	Preferred name	Synonyms	ID
Gene name	ALDOA	ALDA GSD12 MGC10942 MGC17716 GC17767	NCBI-GI: 4557305 NCBI-GeneID: 226 HGNC: 414 HPRD: 00070 Ensembl: ENSG00000149925 KEGG: hsa226
Protein name	Fructose-bisphosphate aldolase A	Lung cancer antigen NY-LU-1 Muscle-type aldolase	Uniprot: P04075 PIR: S14084
Enzyme classification	Aldolase A, fructose-bisphosphate		EC:4.1.2.13

still use colloquial names to refer to genes, proteins, and metabolites, for example, in daily practice. For their own use, this may not be problem, but for publishing results, or for querying across multiple data files, it can be impossible to determine if the same biological object is being referred to. It is also vital that biological objects are annotated with the most accurate description possible. For example, in the case of metabolites such as glucose, it may be necessary to define which isomer is being referred to. For example, if you know you are only measuring the concentration of D-glucose, you should annotate your data with the identifier for the D-glucose isomer. If you do not know if it is D-glucose, or a mixture of D- and L-glucose, you should annotate your data with the identifier for glucose to avoid adding inaccurate annotation.

Public databases and commonly used resources provide a collection of "official" names for biological entities. For example, UniProt IDs (The UniProt Consortium, 2010) can be used to refer to specific proteins, or ChEBI (Chemical Entities of Biological Interest) IDs (Degtyarenko *et al.*, 2008) can be used to refer to chemical entities. In SysMO-DB, we recommend the consistent use of these common vocabularies for all data values, and we have made such a list available for the most common data types in the SysMO projects. This allows identical biological objects to be identified and more easily compared across data sets.

3.2. Data in context

Problems with keeping track of data extend beyond the use of biological identifiers. The flexibility and adaptability of biological systems to varying external conditions make the experimental context in which data are obtained of high importance. Data must be recorded with enough description to enable others to understand how it was generated and for what purpose. Descriptions about the data are often referred to as metadata (data about the data).

If, for example, the experimental protocol for preparing a particular biological sample is unknown, or the methods used for collecting particular measurements are not recorded, it is not possible to compare those results with others. If a model contains parameters that cannot be traced back to any source or to the conditions under which the parameters were determined, it is not possible to validate it, as model behavior might be very dependent on these conditions.

Metadata is an important aspect of data management and data sharing. Annotating experimental results with a consistent set of information allows for easier discovery of relevant data as well as enabling others to potentially reuse it.

Metadata ranges from simple descriptions (e.g., who performed the experiment and when), to more detailed descriptions of growth conditions of biological species, sampling and preparation of samples, and description of the experimental conditions. All of this information is typically available if the work is published, but it is not computationally accessible. Also, if data featured in published articles is submitted to databases or supplementary data stores, it should ideally contain enough metadata for interpretation without a reliance on the publication.

For many types of biological data, there are already community agreed standards for metadata reporting. These standards are often termed *minimum information models*. These models aim to describe the least required for others to interpret and reuse data in the future. The MIBBI portal (Minimum Information about Biological and Biomedical Investigations; http://mibbi.org; Taylor *et al.*, 2008) is a collection of all current minimum information models in the life sciences. To date, there are over 30 in use in the community.

MIBBI models are a pragmatic solution to metadata collection. They recognize data annotation is a time-consuming and under-valued activity. By defining a minimum set of required metadata, they encourage more cooperation and buy-in from the community. In SysMO-DB, we also adopt the minimum information model idea with our JERM (described fully in Section 4).

As well as the MIBBI models, some communities also specify that metadata must be recorded using common vocabularies and ontologies. This makes querying the data computationally more straightforward, but many laboratory scientists have no experience or expertise in using such resources. Vocabularies are numerous and can be complex and difficult to navigate. In SysMO-DB, we try and combat these problems by providing extra tooling to help understand which minimum information models, vocabularies, and standards should be used in which circumstances (Table 29.2).

Despite the provision of standards and vocabularies, data annotation is still time-consuming. It can add a huge overhead to the workload of the scientist, so extra incentives to stimulate an adequate annotation of data are being used. In fields such as transcriptomics, it is often a prerequisite to submit data to public repositories before publications are accepted. In other fields, journals are specifying requirements for supplementary data submissions.

In both of these cases, data is shared at the point of publication but not before. Scientists are reluctant to release their data until they are able to use it in a publication. Any data management system designed for scientists must respect this publication lifecycle and allow scientists to remain in control of data release and dissemination. In large consortia like SysMO, however, sharing within the consortium before publication is desirable. Making access control simple whilst ensuring consortium members are working under a common data-sharing policy can encourage sharing. Our aim is to provide good software tools to assist with annotation and to enhance the analysis capabilities (such as model simulations, integrative workflows, visualization

Table 29.2 A selection of minimum information models and their accompanying biological ontologies

Data	MIBBI model	Ontologies	Standards body
Microarray	MIAME: Minimum Information about a Microarray Experiment (Brazma *et al.*, 2001)	MGED (Whetzel *et al.*, 2006)	Functional Genomics Data Society
Proteomics	MIAPE: Minimum Information about a Proteomics Experiment (Taylor *et al.*, 2007)	PSI-MI, PSI-MS, PSI-MOD	Proteomics Standards Initiative
Interaction experiments	MIMIX: Minimum Information about a Molecular Interaction Experiment (Orchard *et al.*, 2007)	PSI-MI Protein–Protein Interaction	Proteomics Standards Initiative
Systems biology models	MIRIAM: Minimal Information Required In the Annotation of biochemical Models (Le Novère *et al.*, 2005)	SBO: Systems Biology Ontology	BioModels.net
Systems biology model simulation	MIASE: Minimum Information About a Simulation Experiment	KISAO: Kinetic Simulation Algorithm Ontology	BioModels.net

of results, versioning, and recording of the scientific process), to stimulate rather than force scientists to annotate and upload their data.

As more data is produced in the public domain, the importance of making it available and reusable increases. There must, however, be clear incentives for scientists to describe and annotate their data sufficiently to make sharing possible and good practice. The software developed by SysMO-DB provides some of these incentives.

4. The JERM Infrastructure

The JERM is the central organizational framework for the SEEK. It allows the exchange, interpretation, and comparison between different types of data and results files. The JERM describes the minimum information required to identify and interpret assets. For example, for experimental data, the JERM describes what type of experiment was performed, who performed it, what was measured, and what the values in the data sets mean. It also allows for linkage between data, SOPs, and models and therefore helps retain the context of the original experiment.

The JERM addresses the questions:

- What is the minimum information required to find the data?
- What is the minimum amount of information required to interpret the data?

The JERM follows the same principles as MIBBI. It is a minimum metadata specification to reduce the overheads of the scientists describing their data. Each asset has a title, a SEEK ID, and an upload date. It is also associated with its creator(s) and a project. Other common elements help place the asset in context; for example, each asset should be associated with an assay and an assay type. If it is a data asset, it should be associated with SOPs, environmental conditions, and factors studied. However, different types of data will require different JERMs at a more detailed level. The minimum information required to describe a microarray experiment, for example, is not the same as the minimum information required to describe a proteomics experiment using NMR. To make the data reusable for other scientist/studies, it must be clear how the samples were prepared and what samples were used in the experiments.

In the SEEK, we promote the use of JERM compliance by providing JERM templates. The majority of SysMO scientists use Excel as a primary data management tool, so we provide JERM templates as spreadsheets to further encourage compliance and scientists upload their data in this format. It is also possible to acquire the same templates in alternative formats (e.g., XML schemas) for scientists using relational databases for storing local data.

4.1. JERM harvesters and extractors

The SysMO SEEK is an assets catalog. It is a registry of assets stored in distributed project resources as well as assets stored centrally. In order to make use of all assets, wherever they are stored, the SEEK uses harvesters for gathering data and extractors for interpreting their contents.

The retrieval and extraction of assets from the SysMO–SEEK is therefore a two-stage process. Assets are registered in SEEK and searched over using their metadata. They are not retrieved from distributed project resources until required. If they are JERM compliant, further metadata can be indexed from within the asset using the SysMO extractors.

When assets match search results, they are retrieved on demand, provided the user has permission to view and download them. The SEEK harvesters control this process. They connect to a variety of local project resources, including wikis, content management systems, and relational databases. Harvesters can also be instructed to trawl over distributed resources at regular intervals in order to identify new SysMO assets automatically.

JERM compliance is optional. If data is uploaded in a JERM-compliant format, querying over that data is easier, and more tools are available for using that data in analyses. Data can be uploaded in a noncompliant format, but there will be no attempt to parse or understand the contents. Adding incentives for data discovery and reuse encourages JERM compliance. This makes exchange and the eventual dissemination and export to other resources much more straight forward. It also helps with the registration of SysMO assets when they are stored in distributed project resources. JERM harvesters and extractors can be used to connect to these distributed assets on demand (Fig. 29.4).

4.2. The SEEK and data management

Data in systems biology projects range from those produced by traditional molecular biology experiments through to the outputs from the latest techniques in omics high-throughput experiments. Typically, in these projects, transcriptomics, proteomics, and metabolomics analyses are conducted on the same samples, often alongside enzymatic activity analyses, protein–protein interaction studies, and network analyses. Consequently, data can be large, complex, and in a variety of formats. The SEEK must cater for all these types of data, allowing storage (in the SEEK or at remote sites), and searches and comparisons between data sets. Consequently, JERM-compliant templates for different types of experimental data are being produced in collaboration with SysMO researchers. These templates are potentially useful to other communities, so the collection will be made available as a SEEK resource.

4.3. The SEEK and model management

Model management in SEEK includes storage, annotation, and simulation. Mathematical models play an important role in systems biology projects. They are crucial for understanding the behavior of systems components, the description of experimental data, and the analysis and understanding of the systems under study.

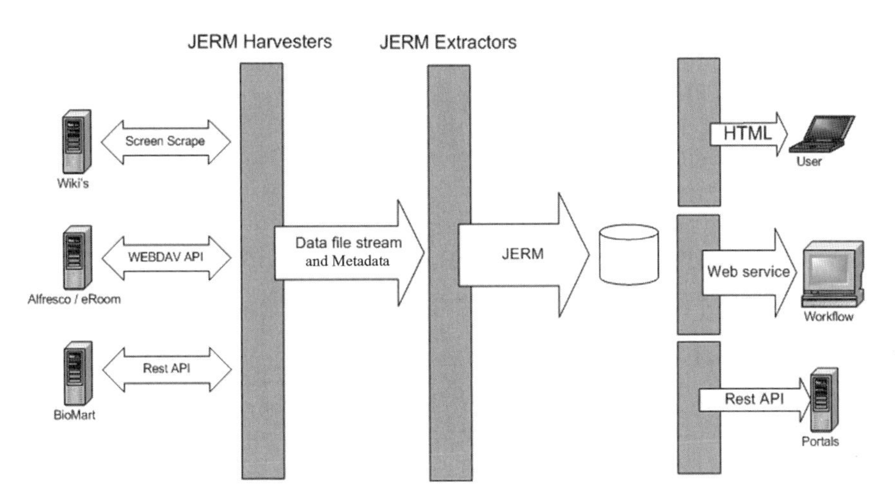

Figure 29.4 JERM Harvesters and Extractors are used in combination to retrieve SysMO assets held in distributed locations. The data is normally returned to the SEEK interface via the HTML pages, but can also be returned via REST services to enable computational access to the data for analyses.

Model management can be divided into a number of different tasks; model construction, simulation, validation, annotation, storage, and dissemination. Although model construction and validation would largely fall under the responsibility of the respective scientists, SysMO-DB provides tools to facilitate these steps, which would usually involve links with experimental data and models. In the case of a mathematical model that is constructed using a *bottom-up* approach, such tools should enable visualization of data sets used for the parameterization of the individual rate equations together with its goodness of fit. For model validation, a different data set, for instance, a time trace for model variables obtained on the complete system, could be used and then a visualization of the complete model together with the validation set should be possible. This example represents an idealized situation. There are not many models available in the scientific literature that show all data sets used for model construction and model validation. However, the SEEK provides the possibility to present large data sets along with the models and therefore promote these good modeling practices. Importantly, such practices make the complete model building process transparent and reproducible. They would remove any uncertainties on how model parameters were derived.

In the SEEK, SBML (Hucka *et al.*, 2003) is the recommended model format. Scientists are free to upload models in other formats, but the extra tools and functionality provided by JWS Online require SBML for use.

4.4. The SEEK and process management

Process management in the SEEK encompasses SOPs and protocols from laboratory investigations as well as data analysis protocols and model building methods. Conceptually, there should be no difference between these different types of protocol. Each SOP describes the necessary conditions to understand and interpret the resulting data. This can then be reused by other members of the consortium to perform the same experiments. In large, diverse consortia, where scientists are studying different organisms and different biological systems, the greatest added value can come from sharing methods and techniques rather than directly comparing data.

The multidisciplinary nature of systems biology projects means that cutting-edge technologies are often adapted and employed. Some require the development of new protocols. Sharing such protocols allows fast emergence of best practice and knowledge transfer between consortium partners. Unlike data, scientists are often willing to share protocols before their results are published, specifically if those protocols are obtained or modified from the literature.

As with data and models, standards for SOPs are recommended but SysMO-DB does not enforce them. Consortium members are free to upload or register SOPs in any format, but the *Nature Protocols* format is recommended (http://www.nature.com/nprot).

4.4.1. SOPs and protocols

The distinction between SOPs and protocols is important in distributed projects. A SOP is a protocol that has been agreed upon by a whole project or consortia. SOPs are essential for any part of the work that depends on standardizing practices across the board. For example, when preparing cultures and samples that will be used in all subsequent experiments, it is important that they are prepared in exactly the same way to allow effective data comparisons. In SysMO, each project has a set of SOPs for the growth of cultures, which have typically been optimized over several iterations. For other experimental work, some protocols are used by the whole consortia, and some are only used by individuals.

4.4.2. Protocols for informatics experiments

In the SEEK, we make no distinction between laboratory experiments and informatics experiments. The bioinformatics analysis of data is simply considered to be another kind of experiment. Therefore, data and results should be recorded along with the SOPs and protocols used to produce it. In certain cases, however, the bioinformatics protocol may actually be executable. If the analysis was performed using a scientific workflow (e.g., in Taverna) (Hull *et al.*, 2006), the workflow itself contains the protocol for the experiment and can therefore be shared and run again with the same data for verification or with new data to perform related analyses. In the next phase of development, common data analysis tasks will be made available as Taverna workflows through the SEEK interface.

4.4.3. Protocols for models

It is not yet common practice to write SOPs and protocols for modeling work, but capturing the process and the context of assumptions in the model is important, so we encourage the recording and storage of SOPs for modeling in SEEK. An important aspect of this work is identifying data that has been used for model construction and data that has been used for model validation.

4.5. Publications

The primary method for sharing scientific research remains the scientific publication. Publications can be registered in SEEK via a PubMed ID or a DOI. SEEK automatically matches author names to SEEK profiles and registers the publication abstract for searching. Any other asset can also be linked to a publication, which means that supplementary material for the paper can be associated directly from the SEEK.

5. The SEEK Functionalities: Annotating and Linking Assets

Annotation of assets, be it data or models is time-consuming and difficult. Scientists tend to start with annotation as and when they must do so for publication. For effective collaboration across distributed researchers, however, this practice has to be encouraged earlier.

For data annotation, the JERM templates provide a mechanism to help with this process. By using the JERM templates or schemas provided, SysMO scientists can produce JERM-compliant data. However, the templates only address the structure of the data; we must also consider the content. For mathematical models, a MIRIAM annotation standard has been published (Le Novère *et al.*, 2005), and we have implemented a tool in SEEK, OneStop to annotate models according to this standard, and in the same time adhere to SBML (Hucka *et al.*, 2003) and SBGN (Le Novère *et al.*, 2009) model and network description standards as well.

In this section, we introduce these tools, show how they are used in the SEEK, and illustrate the strength of annotation in linking assets.

5.1. Data annotation and RightField

Typically, MIBBI standards dictate that particular values in a minimum information model should be annotated with terms from a particular domain-specific ontology. For example, when referring to the name of a chemical entity, that entity should be identified by its ChEBI entry (Degtyarenko *et al.*, 2008), or when annotating SBML models, annotation terms should be taken from the SBO (Systems Biology Ontology; http://www.ebi.ac.uk/sbo/). This is effectively another layer of annotation that is expected from SysMO scientists, but many are not familiar with the ontologies, or the advantages of uniform annotation for search and retrieval. Therefore, the approach we have adopted in SysMO-DB is to provide tools to make this process more accessible and straightforward.

RightField (Wolstencroft *et al.*, 2011) is an open-source application that provides a mechanism for embedding ontology annotation support in Excel spreadsheets. Individual cells, columns, or rows in spreadsheets can be restricted to particular ranges of allowed classes or instances from chosen ontologies. Bioinformaticians, with experience in ontologies and data annotation, can prepare RightField-enabled spreadsheets with embedded ontology term selection support for distribution across the consortium.

RightField supports the loading of ontologies (in OWL, OBO, or RDF format; http://www.w3.org/standards/) directly from the BioPortal ontology repository, or from a local machine. When a spreadsheet has been marked up with terms from selected ontologies, they are embedded into the Excel file. Once marked-up and saved, the RightField-enabled spreadsheet contains embedded worksheets with information concerning the origins and versions of ontologies, used in the annotation. This encapsulation stage is crucial. With everything embedded in the spreadsheet, scientists do not require any new applications to use it and they can complete annotation offline should they wish. This also makes the spreadsheets readily exchangeable and enables a series of experiments to be annotated with the same versions of the same marked-up ontologies, even if the live ontologies change during this time. Ontology versions can be updated if the spreadsheet is opened again in RightField, but it is not automatic.

The RightField-enabled spreadsheet presents selected ontology terms to the users as a simple drop-down list, enabling scientists to consistently annotate their data without the need to understand the numerous metadata standards and ontologies available to them. The result is semantic annotation by stealth. RightField facilitates an annotation process that is quicker, less error-prone, and more efficient. By combining JERM templates and embedded ontology terms with RightField, we provide an infrastructure that promotes and encourages compliance and standardization. The result is a collection of data files with consistent annotation that is consequently easier to search and compare. Examples of these can be downloaded from the RightField Web site (http://www.rightfield.org.uk).

Figure 29.5 shows RightField being used to mark up a transcriptomics data template with terms from the MGED ontology.

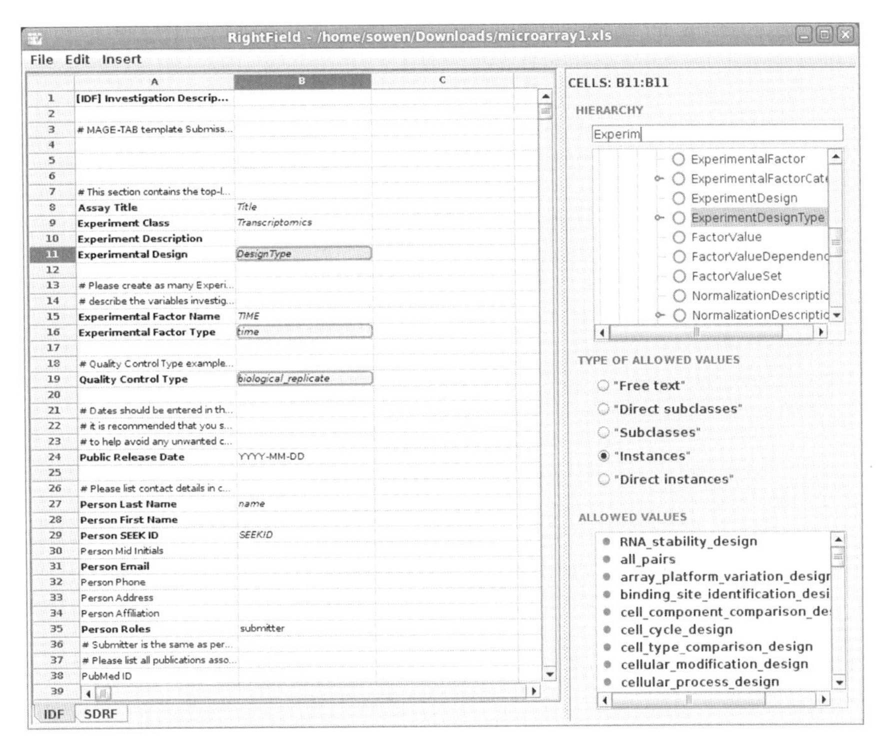

Figure 29.5 The RightField application showing the embedding of ontology terms into a spreadsheet template from SysMO.

5.2. Tools for model annotation

The recommended standard for exchanging systems biology models is SBML, but SBML alone is not sufficient for a comprehensive understanding of the model. In systems biology, models are often used to simulate a specific system and contain variables and parameters which represent physical biological entities. Annotating the model with unique identifiers for molecular species (e.g., ChEBI), reaction steps (e.g., KEGG), and enzyme species (e.g., EC numbers), for example, allows model simulation results to be analyzed in the context of experimental data and enables others to interpret the model more effectively. MIRIAM annotation and MIASE (Waltemath *et al.*, 2011) can be used for this contextual understanding and description of simulation experiments, respectively.

For small models, a standardized model description format might not appear to be that important. For example, the formulation of ordinary differential equations (ODEs) with parameter values and initial conditions seems simple enough. However, when screening the scientific literature, it quickly becomes evident that few models are described in sufficient detail that they can be reconstructed and simulations be repeated. This might reflect the aims of the scientists to illustrate a principle more than to build a realistic model, but still it is disconcerting that most of these models can never be used again. The JWS Online database was created to address this issue. It is both a curated models repository and a simulation environment. The OneStop tool is an extension to the JWS Online environment to allow MIRIAM annotation and the construction and editing of models.

OneStop provides an interface where users can define their model in a number of text fields in a Web browser. Subsequently, the model can be simulated using the JWS interface. Models can be defined from scratch, but the user can also upload SBML files or any of the models from the JWS Online or Biomodels databases (Le Novère et al., 2006). Models can be saved in SBML format and an automated SBGN schema generator and a tool for MIRIAM annotation are available. The annotation tool makes use of Web services from semanticSBML (Krause et al., 2010).

Examples of text fields for model description are shown in Fig. 29.6 and the model annotation field is shown in Fig. 29.7. These tools are part of OneStop and integrated into the SEEK environment.

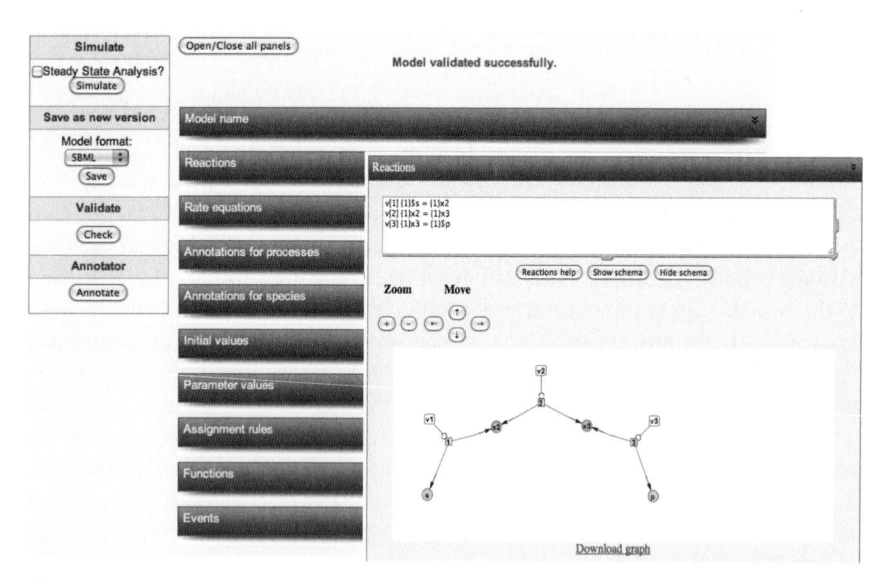

Figure 29.6 The OneStop model constructor. Via a number of text files a user can define a mathematical model, which can subsequently be simulated via the JWS interface. Models can be defined from scratch or uploaded from the JWS Online database or Biomodels database in SBML format. Good error catching and graphical displays of reaction networks (in SBGN format) and rate equations enhance the functionality of the model constructor.

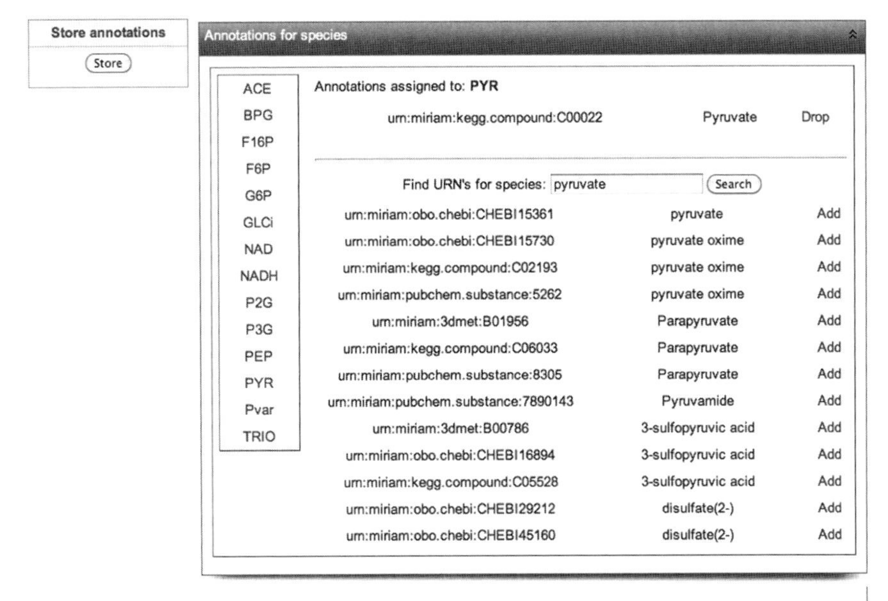

Figure 29.7 The OneStop model annotator. Using the semanticSBML (Krause *et al.*, 2010) Web services, unique identifiers can be given to model variables and model reactions. OneStop makes it possible to annotate the model according to the MIRIAM specification (Le Novère *et al.*, 2005).

5.3. Linking data and models

Linking data and models relies crucially on the annotation of both the data sets and the model components. There is currently no community standard or fixed structure for expressing the connections between them. In SysMO-DB, we have been working on a number of scenarios for how data and models could be linked. At a basic level, SEEK users can specify that a particular data set was used either in the construction or validation of a model. If data is the result of a model simulation run, we can also draw this distinction. However, much greater detail is needed for comprehensive integration.

For a number of metabolic models, we have illustrated how data can be linked to the individual processes. For instance, for *bottom-up* models, a user could have an experimental data set for each of the model processes and on the basis of the data, the user would formulate a mathematical equation.

In JWS Online, it is possible to work with an isolated rate equation. Users can plot the rate equation with the parameter values used in the model (and he/she can change these parameter values). In addition, experimental data sets can be uploaded (for instance, as excel files), and plotted together with the rate equation used in the model. In the SEEK, we are developing mechanisms for easily importing/exporting data for plotting against models (e.g., see Fig. 29.8). Typically, such data sets would be used for model construction. For model validation, one could use data sets obtained with the complete organism/pathway, and such data sets would be linked to complete models.

6. Incentives for Sharing Data

The SEEK is a sharing initiative driven by funding councils in Europe, as a platform to assist the SysMO consortia members but also to ensure that the ever-increasing amounts of scientific data generated by public funding are made available to the community for further analysis and reuse. The SEEK provides a repository for all data and models from one funding

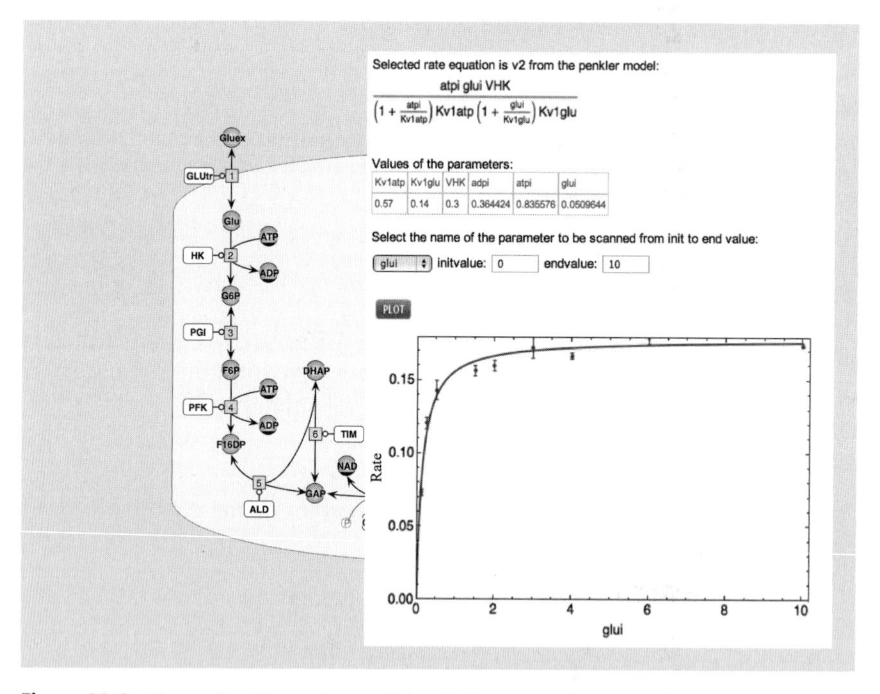

Figure 29.8 Example of experimental data linking to individual rate equation. Saturation of the hexokinase reaction with internal glucose is shown as an example for the linking of individual rate equations with experimental data. When the user clicks the reaction process (v2) in the SBGN schema, the rate equation of the reaction is loaded from the model package. The user can select to plot the rate equation as a function of its parameters (here glui for internal glucose was selected). If data for the specific model and rate equation are available in the JWS database, these are selected and plotted together with the rate equation.

initiative, creating a central focus for scientists involved in the initiative as well as a record of the research developed from it. It allows researchers to search and compare results or experimental techniques and include data from earlier work in wider studies. These outcomes are of benefit to the wider systems biology community, but there must be incentives for the SysMO scientists to spend time on data curation, annotation, and sharing. The SEEK encourages participation by providing such incentives, which are the provision of a safe haven for data and other assets, a set of tools for further analysis of these assets and for easy implementation of data and model standards in annotation, the opportunity for individuals to receive credit and attribution for their data contributions, and the ability to easily export assets to other public repositories. The following section describes these advantages for SysMO consortium members to adopt the SEEK.

6.1. Secure and continuous storage

Consortium members are obliged to make SysMO assets available to the community for 10 years as a condition of funding. If scientists opt to retain assets locally, the responsibility of ensuring they are available for others remains with them. During the day-to-day running of a project, this is often the case. However, when projects finish, the individuals responsible for local upkeep and maintenance may move to new institutions.

If all SysMO assets are not uploaded to SEEK at the end of a project, the responsibility to make the data available long-term also remains with the scientists. SysMO-DB provides an archiving service to allow SysMO projects to publish all assets centrally at the end of their funding period, providing a guaranteed safe haven for assets for an initial period of 10 years. This releases scientists from the overheads associated with maintaining individual resources and enables the whole consortium and others to benefit from the pooling of SysMO assets.

6.2. Credit and attribution

Biological data can take months to collect and longer to analyze and publish. Traditionally, this data has only been used as evidence in the resulting publication, but data reuse is becoming more common as a result of large-scale analyses and the emergence of public repositories.

If data is adequately annotated and documented, it is potentially useful for future experiments and some data sets can even become widespread "reference" sets that are reused in multiple investigations.

In the life sciences, scientists are credited for their publications, but not traditionally for the actual data. Obtaining data of a good quality that can be used in multiple analyses is an advantage to the whole community. Therefore, the concept of data citation is becoming more popular

(Editorial, 2009) and mechanisms to enable this are now being proposed (http://www.datacite.org).

Ensuring SysMO scientists gain maximum credit for their work is an important incentive for registering and sharing. SysMO assets are associated with the profiles of their creators and registered with a unique and persistent URL to allow direct links to be made from publications and other online sources.

Attribution is an equally important issue. Experiments are often based on other experiments. SOPs are modified to improve efficiency, for example, or raw data is normalized or analyzed. In these cases, the same scientist may not be responsible for the original and subsequent work, so it must be made clear which parts belong to which individuals. In the SEEK, credit and attribution are clearly visible. It is possible to credit other people for any work being shared and any asset can be attributed to any other, to signify that it was based on earlier work.

6.3. Export to public repositories

The SEEK is a unified interface to the outcomes of SysMO, but many journals require data to be submitted to public repositories before papers can be published. This is particularly true with omics data. For example, microarray data must be submitted to ArrayExpress (Parkinson *et al.*, 2005) or GEO (Barrett *et al.*, 2009) before any paper is published relating to it. Such public repositories require data in particular formats to comply with community metadata standards. In SEEK, we plan to offer conversion services to allow one-click export, either by making use of tools from the ISA Infrastructure (Rocca-Serra *et al.*, 2010; in this case, the ISA Converter; http://isatab.sourceforge.net/converter.html) or by directly mapping from the SysMO-JERM models. The advantage for SysMO users is that data annotation and formatting only needs to be done once, at the initial registration with SEEK. In addition, we provide tools that makes adhering to such standards easier, for instance, for mathematical models, we have the OneStop tool for generating MIRIAM annotated SBML models, together with networks schema drawn according to SBGN standards.

7. The SEEK: Experiences

Since the initial release of the SEEK in SysMO, we have seen a gradual rise in uptake and use. There are already over 1700 assets registered in the SEEK and over 200 active users. As expected, we see a spectrum of compliance levels with registered assets. Some are registered with sparse metadata and remain unchanged, whilst others are richly described, or have incremental metadata additions to conform to the JERM.

We have, however, observed a much lower uptake of recommended formats and standards than we expected. For example, for models, SysMO-DB recommends SBML (which is also the community standard), but many in the consortium do not use it for the following reasons:

- It is not seen as fit for purpose.
- It is still under constant development, and therefore is viewed as too unstable.
- A lack of specific tooling support means it is difficult to import and export from applications already in use.

These issues can, to some extent, be surmounted by simple interventions once they have been identified. For example, more tooling can be provided to help with format exchange from common applications (as in OneStop, the JWS Online Model Constructor), and the consortium can officially propose new directions to the standards developers to address shortcomings in the specification. In the meantime, the "Just Enough" principles of SysMO-DB ensure that consortium members are already free to share in any format until these new developments can be implemented.

The "Just Enough" design in SysMO-DB is the most fundamental part of the System. It is essential to provide a flexible model which users are free to interact with at different levels of compliance and detail, and at different times.

The flexibility of using a minimum model like the JERM encourages uptake and encourages social connections between consortium members. For assets that are poorly annotated, discussing their meaning and contents with the creator is the most efficient route to a contextual understanding. Therefore, data sharing and integration can be achieved through automated methods with the JERM extractors and harvesters, or through dialog between consortium members with the SysMO–SEEK Yellow Pages. The social connections also tie individuals' reputations to their assets. This encourages the addition of more metadata to prevent misinterpretation.

The next steps for SysMO-DB involve a greater focus on data analysis. Data exchange and sharing is becoming more popular, but the primary concern has been to encourage this behavior and ensuring assets are recorded and archived. Now we have a growing collection of data and models, we need to provide more sophisticated ways of exploring and comparing them.

The overall SysMO-DB design methodology has been successful because we have focused on the specific concerns of the user community and built a solution that fits in with existing practices. Everything is designed in consultation with the SysMO-PALs focus group, so they can help us identify bottlenecks and essential new features. Within the consortia, the PALs have formed their own network of young systems biology researchers with experience in data management and close collaborations between modelers and experimentalists.

The emergence of large-scale scientific consortia, in systems biology and other areas, coupled with the rapid development of more high-throughput experimental techniques, is driving changes in the way we record, reuse, and reward data. Data management is consequently becoming more complex both locally and at a community level. To properly pool research outcomes and promote reuse, it must be easier for scientists to manage and publish data. This means providing tools for data storage and for data standardization and analysis. The SysMO-DB project offers a suite of tools for a pragmatic data management solution to allow sharing in a large consortium and beyond with minimum impact on the daily work of researchers.

ACKNOWLEDGMENTS

Funding: This work was funded by the BBSRC and the BMBF. SysMO-DB: Supporting Data Access and Integration (BBG0102181).

REFERENCES

Barrett, T., Troup, D. B., Wilhite, S. E., Ledoux, P., Rudnev, D., Evangelista, C., Kim, I. F., Soboleva, A., Tomashevsky, M., Marshall, K. A., Phillippy, K. H., Sherman, P. M., *et al.* (2009). NCBI GEO: Archive for high-throughput functional genomic data. *Nucleic Acids Res.* **37**(Database issue), D885–D890.

Brazma, A., Hingamp, P., Quackenbush, J., Sherlock, G., Spellman, P., Stoeckert, C., Aach, J., Ansorge, W., Ball, C. A., Causton, H. C., Gaasterland, T., Glenisson, P., *et al.* (2001). Minimum information about a microarray experiment (MIAME)-toward standards for microarray data. *Nat. Genet.* **29**(4), 365–371.

Degtyarenko, K., de Matos, P., Ennis, M., Hastings, J., Zbinden, M., McNaught, A., Alcántara, R., Darsow, M., Guedj, M., and Ashburner, M. (2008). ChEBI: A database and ontology for chemical entities of biological interest. *Nucleic Acids Res.* **36**(Database issue), D344–D350.

Editorial (2009). Data producers deserve citation credit. *Nat. Genet.* **41**(10), 1045.

Howe, D., Costanzo, M., Fey, P., Gojobori, T., Hannick, L., Hide, W., Hill, D. P., Kania, R., Schaeffer, M., St Pierre, S., Twigger, S., White, O., *et al.* (2008). Big data: The future of biocuration. *Nature* **455**(7209), 47–50.

Hucka, M., Finney, A., Sauro, H. M., Bolouri, H., Doyle, J. C., Kitano, H., Arkin, A. P., Bornstein, B. J., Bray, D., Cornish-Bowden, A., Cuellar, A. A., Dronov, S., *et al.* (2003). SBML Forum. THE systems biology markup language (SBML): A medium for representation and exchange of biochemical network models. *Bioinformatics* **19**(4), 524–531.

Hull, D., Wolstencroft, K., Stevens, R., Goble, C., Pocock, M. R., Li, P., and Oinn, T. (2006). Taverna: A tool for building and running workflows of services. *Nucleic Acids Res.* **34**(Web server issue), W729–W732.

Krause, F., Uhlendorf, J., Lubitz, T., Schulz, M., Klipp, E., and Liebermeister, W. (2010). Annotation and merging of SBML models with semanticSBML. *Bioinformatics* **26,** 421–422.

Le Novère, N., Le Finney, A., Hucka, M., Bhalla, U., Campagne, F., Collado-Vides, J., Crampin, E., Halstead, M., Klipp, E., Mendes, P., Nielsen, P., Sauro, H., *et al.* (2005).

Minimal information requested in the annotation of biochemical models (MIRIAM). *Nat. Biotechnol.* **23,** 1509–1515.

Le Novère, N., Bornstein, B., Broicher, A., Courtot, M., Donizelli, M., Dharuri, H., Li, L., Sauro, H., Schilstra, M., Shapiro, B., Snoep, J. L., and Hucka, M. (2006). BioModels database: A free, centralized database of curated, published, quantitative kinetic models of biochemical and cellular systems. *Nucleic Acids Res.* **34,** D689–D691.

Le Novère, N., Hucka, M., Mi, H., Moodie, S., Schreiber, F., Sorokin, A., Demir, E., Wegner, K., Aladjem, M., Wimalaratne, S. M., Bergman, F. T., Gauges, R., *et al.* (2009). The systems biology graphical notation. *Nat. Biotechnol.* **27,** 735–741.

Noy, N. F., Shah, N. H., Whetzel, P. L., Dai, B., Dorf, M., Griffith, N., Jonquet, C., Rubin, D. L., Storey, M. A., Chute, C. G., and Musen, M. A. (2009). BioPortal: Ontologies and integrated data resources at the click of a mouse. *Nucleic Acids Res.* **37** (Web server issue), W170–W173.

Olivier, B. G., and Snoep, J. L. (2004). Web-based kinetic modelling using JWS Online. *Bioinformatics* **20,** 2143–2144.

Orchard, S., Salwinski, L., Kerrien, S., Montecchi-Palazzi, L., Oesterheld, M., Stümpflen, V., Ceol, A., Chatr-Aryamontri, A., Armstrong, J., Woollard, P., Salama, J. J., Moore, S., *et al.* (2007). The minimum information required for reporting a molecular interaction experiment (MIMIx). *Nat. Biotechnol.* **25,** 894–898.

Parkinson, H., Sarkans, U., Shojatalab, M., Abeygunawardena, N., Contrino, S., Coulson, R., Farne, A., Lara, G. G., Holloway, E., Kapushesky, M., Lilja, P., Mukherjee, G., *et al.* (2005). ArrayExpress—A public repository for microarray gene expression data at the EBI. *Nucleic Acids Res.* **33**(Database issue), D553–D555.

Rocca-Serra, P., Brandizi, M., Maguire, E., Sklyar, N., Taylor, C., Begley, K., Field, D., Harris, S., Hide, W., Hofmann, O., Neumann, S., Sterk, P., *et al.* (2010). ISA software suite: Supporting standards-compliant experimental annotation and enabling curation at the community level. *Bioinformatics* **26**(18), 2354–2356.

Sansone, S. A., Rocca-Serra, P., Brandizi, M., Brazma, A., Field, D., Fostel, J., Garrow, A. G., Gilbert, J., Goodsaid, F., Hardy, N., Jones, P., Lister, A., *et al.* (2008). The first RSBI (ISA-TAB) workshop: "Can a simple format work for complex studies?" *Omics* **12**(2), 143–149.

Taylor, C. F., Paton, N. W., Lilley, K. S., Binz, P. A., Julian, R. K., Jones, A. R., Zhu, W., Apweiler, R., Aebersold, R., Deutsch, E. W., Dunn, M. J., Heck, A. J., *et al.* (2007). The minimum information about a proteomics experiment (MIAPE). *Nat. Biotechnol.* **25,** 887–893.

Taylor, C. F., Field, D., Sansone, S. A., Aerts, J., Apweiler, R., Ashburner, M., Ball, C. A., Binz, P. A., Bogue, M., Booth, T., Brazma, A., Brinkman, R. R., *et al.* (2008). Promoting coherent minimum reporting guidelines for biological and biomedical investigations: The MIBBI project. *Nat. Biotechnol.* **26**(8), 889–896.

The UniProt Consortium (2010). The Universal Protein Resource (UniProt) in 2010. *Nucleic Acids Res.* **38**(Database issue), D142–D148.

Waltemath, D., Adams, R., Beard, D. A., Bergmann, F. T., Bhalla, U. S., *et al.* (2011). Minimum information about a simulation experiment (MIASE). *PLoS Comput. Biol.* **7**(4).

Whetzel, P. L., Parkinson, H., Causton, H. C., Fan, L., Fostel, J., Fragoso, G., Game, L., Heiskanen, M., Morrison, N., Rocca-Serra, P., Sansone, S. A., Taylor, C., *et al.* (2006). The MGED ontology: A resource for semantics-based description of microarray experiments. *Bioinformatics* **22**(7), 866–873.

Wolstencroft, K., Owen, S., Horridge, M., Krebs, O., Mueller, W., du Snoep, J. L., Preez, F., and Goble, C. (2011). RightField: Embedding ontology annotation in spreadsheets. *Bioinformatics* 10.1093/bioinformatics/btr312.

CHAPTER THIRTY

Crossing the Boundaries: Delivering Trans-disciplinary Science in a Disciplinary World

Elizabeth A. Elliot* *and* Neil W. Hayes†

Contents

Abstract

Major research initiatives are increasingly drawing on multiple disparate disciplines and systems biology is a key exemplar. Trans-disciplinary research occurs where individual disciplinary traditions combine to create new shared knowledge that cannot be said to fit within the domain of any single discipline. Generation of new understanding of biological systems at the cell, organ, or organism level clearly meets these criteria, and we therefore consider systems biology research a truly trans-disciplinary undertaking. Aside from the technological challenges of combining research outcomes of the contributing disciplines, directing and managing the overall research program also presents a significant challenge. In this chapter, we discuss the challenges of and enablers

* Centre for Systems Biology at Edinburgh, The University of Edinburgh, Edinburgh, United Kingdom
† Manchester Centre for Integrative Systems Biology, Manchester Interdisciplinary Biocentre, The University of Manchester, Manchester, United Kingdom

Methods in Enzymology, Volume 500
ISSN 0076-6879, DOI: 10.1016/B978-0-12-385118-5.00030-X

to working across the broad range of disciplines that contribute to systems biology research; we discuss potential management models that may be adopted and the features, benefits, and drawbacks of each, introducing examples of management models adopted at two UK Systems Biology Centres.

1. Introduction

Systems biology is a classic example of a field of study that crosses traditional boundaries. Relevant scientific disciplines include informatics, engineering, physics, mathematics, and biology; however, systems biology also looks to the social sciences and humanities where issues such as organization, ethics, philosophy, and law are encountered. Trans-disciplinary research may be described as the coordinated interaction and integration across multiple disciplines resulting in the restructuring of disciplinary knowledge and the creation of new shared knowledge (Jakobsen *et al.*, 2004), whereas interdisciplinary research involves coordinated interaction across multiple disciplines which at best may result in potential integration of knowledge centered around the issue under study. The ability of systems biology to create new understanding and consequently new shared knowledge at the level of the cell, tissue, or organism drawing inputs from diverse disciplines spanning engineering and the physical sciences and the humanities clearly meets the criteria for classification as trans-disciplinary research. This broad remit is where many of the challenges in managing and organizing such research arise; the different disciplines each bring their own perspective on what constitutes knowledge and different methodological traditions. As Charles Handy noted, "God, it has been said, did not see fit to divide up the world to accord with the faculties of universities" (Handy, 1999).

Systems biology focuses on the systematic study of complex interactions in biological systems, representing a move away from reductionism and toward integration. Seeking to integrate the physical with the life sciences is a fundamental aspect of all systems biology research, whether a top-down, middle-out, or bottom-up approach is adopted. It can be applied to developing understanding on various scales from the subcellular, for example, studying metabolic or regulatory pathways, to the whole organism with concepts such as the silicon human (Westerhoff *et al.*, 2011) which envisages a hierarchy of interacting models capable of describing function at the cell, tissue, organ, and organism levels. Whilst management of the former example could be considered relatively straightforward, the latter is a very different matter and in between the two a whole spectrum of possibilities exists. Systems biology has therefore the potential to deliver an enormous increase in understanding and lead to major progress in medicine, agriculture, and industry and as such the economic impact is expected to be substantial.

Relevant to the profile of this emergent field, significant public sector investment has also been made both internationally and nationally. The United States, Japan, and the countries of the European Union have all implemented mechanisms to fund systems biology research. The establishment of centers in systems biology, designed to integrate multidisciplinary teams with sufficient stability and critical mass, has become a major theme. Here in the United Kingdom, since 2005 a joint funding initiative between the BBSRC and EPSRC has invested in excess of £47M in the creation of six dedicated Centres for Integrative Systems Biology (CISBs). One of the unique aspects of this BBSRC/EPSRC funding initiative was a dual investment in both the science and the management of the science.

Whilst the array of disciplines that contribute to systems biology have the potential to achieve a deeper and more comprehensive understanding of the subject, they also bring greater potential for conflict and misunderstanding between the participants. This complexity has driven development and adoption of new and unfamiliar organization structures and management approaches more commonly associated with interdisciplinary and transdisciplinary projects than traditional scientific research. This chapter examines the challenges of managing systems biology within an institutional context and, with reference to theoretical management strategies, the best practice approaches to management taken by the Manchester Centre for Integrative Systems Biology (MCISB; http://www.mcisb.org/) and the Centre for Systems Biology at Edinburgh (CSBE; www.csbe.ed.ac.uk). The authors consider that no organizational structure will suit all and no single management strategy is the key to success, but rather that an awareness of a structure's intrinsic strengths and weaknesses and the willingness to share best practice will support the most effective delivery of any organizations' broader goals.

2. Theoretical Management Strategies

Organizational structure defines the culture, relationships, and ultimately performance of all workplaces. A succinct description could be the units and workflows within an organization that makes the organization capable of achieving its higher-level business goals. Within a higher education institution, this type of structure encompasses all units; examples could range from Estates & Buildings departments to world-leading research groups. Likewise example, workflows encompass all activity streams from teaching to research, to knowledge exchange. The organization structure acts to align these diverse groups and activities to ensure delivery of broader strategic aims. Whilst a range of organizational structures exist, their presence *per se* is conserved across all employment sectors.

2.1. Functional organizations

Earlier in the twentieth century, functional organization structures prevailed across all employment sectors and today they still predominate in the public sector. They represent a traditional, vertical hierarchy structure familiar to many working in the higher education sector. Functional organization can bring a number of business advantages. Individual specialist departments (functions) can deliver individual contributions of excellence and operational efficiency through clear line management structures. The grouping of specialists within functional groups provides clear career development paths and a professional base for expertise and continuing professional development (CPD). Equally, the same structure can bring a number of disadvantages. Working successfully across several functions demands a level of coordination. Individual functions can become overspecialized leading to extended and inefficient communication lines between functions. Inappropriate weighting may be attached to functional priorities at the expense of broader organizational goals. Finally, whilst staff within a function benefit from a clear professional base, those staff operating across functions may lack the authority to lead work for which they are responsible and can be hindered by the absence of a single base for their professional development (Fig. 30.1).

2.2. Project-based organizations

At the other end of spectrum from the more traditional organizational structures lie the project-based organizational structures. In this scenario, organization activities are divided into individual projects. One could imagine a project to build a road, publish a book, or supervise a Ph.D. Functional departments do not exist and instead teams of staff with diverse expertise collaborate to deliver a collective goal. This type of structure can also be variously described as "divisional" or "product" but all require self-contained teams with diverse skills represented. A project structure brings with it total resource control for project teams, distinct project identities, and clear targets and responsibilities. Within a team, information can be easily shared and communication is simplified. Project structures also open up possibilities for portfolio management assessing whole organization activities considering resourcing, global strategy alignment, realizing the benefits, risks, and interdependencies. And yet just as with functional organization structures, project-based organizations can encounter disadvantages. Project teams can develop overly strong identities with a tendency to collect staff and resources and increasing isolation from broader organizational goals. Individual projects cannot offer continuity of employment; for example, when the book is published, or the road is built, staff need to be reallocated to new projects and such temporarily unassigned staff can lose direction and motivation. Project-based organizational structures are common in the aerospace, construction, and engineering industries, and it

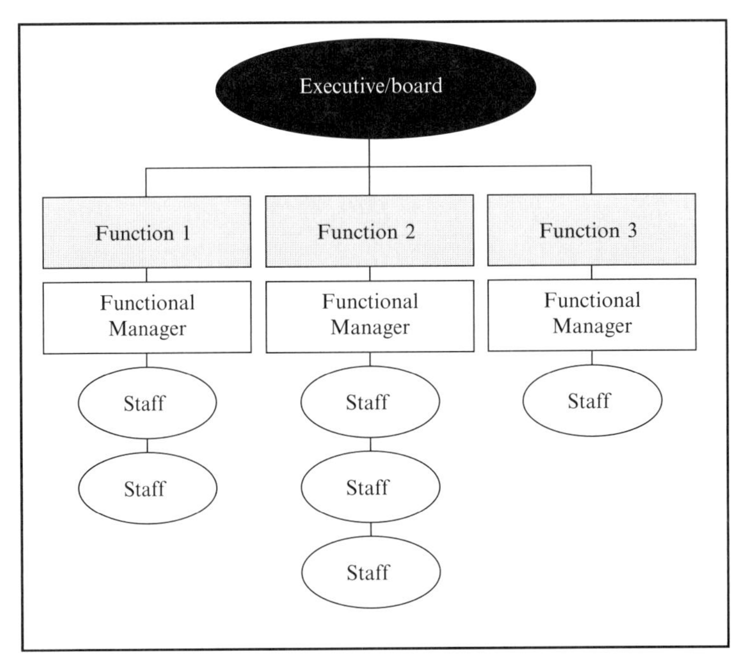

Figure 30.1 Functional organization structure. This vertical structure is represented with staff (white ellipse) assigned to function groups (shaded rectangle) with clear line management and reporting through functional managers (white rectangle) ultimately to the executive/board (black ellipse).

could be argued that in such structures a project team has simply replaced a functional department (Fig. 30.2).

2.3. Matrix organization structure

A matrix structure combines elements of both the functional and the project-based organizational structures described above. Within the organization, functional departments are retained and in parallel project teams can also be established with both reporting to the organization executive. The matrix organization originated from a need to integrate customer requirements with organizational priorities for economies of scale and development of specialist skills (Handy, 1999). A matrix structure allows the co-coordinated and tailored deployment of functional department expertise across all of the organizations projects. Individual projects can draw on the resources of the entire parent organization. The arrangement facilitates the consistent application of corporate strategy, policy, and procedures. The comparable weighting given to both functional and project teams provides useful bases for professional staff working "within discipline"

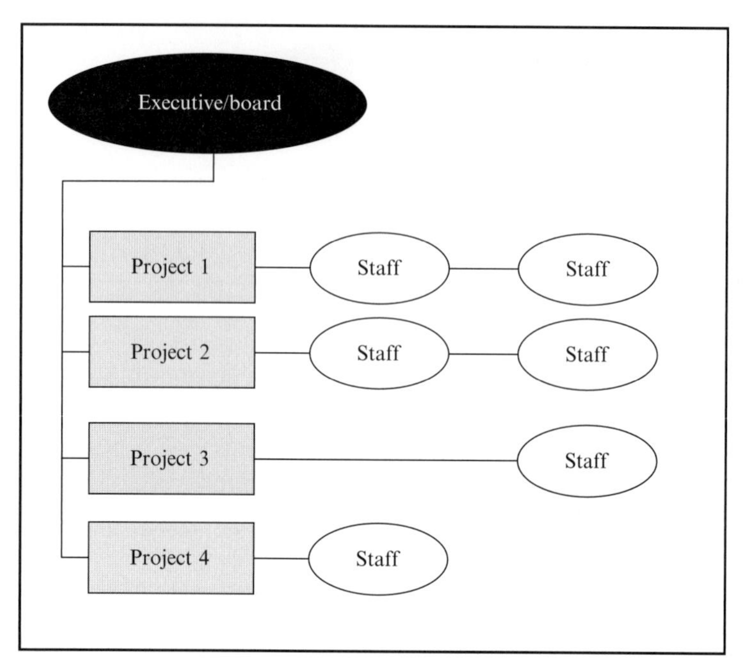

Figure 30.2 Projectized organization structure. This structure is represented with staff (white ellipse) assigned to project groups (shaded rectangle) with clear line management and reporting through project managers ultimately to the executive/board (black ellipse).

and those with an interdisciplinary remit. However, matrix structures are not without challenges. Line management can be blurred between functional and project groupings risking separation of authority and responsibility. Individual staff may experience divided loyalties, to the project versus the function and the necessity to share resources carries a risk of resource conflict. Communications are necessarily more demanding in a matrix structure and yet poor communication contributes to poor coordination and inefficient problem solving (Fig. 30.3).

3. Consideration of Real-World Examples

3.1. The challenge of knowledge creation and management in systems biology

The main challenges associated with managing trans-disciplinary research, of which systems biology is a classic example, have been previously identified as issues relating to knowledge exchange and the formation of shared goals (Hollaender *et al.*, 2008). Knowledge production and communication

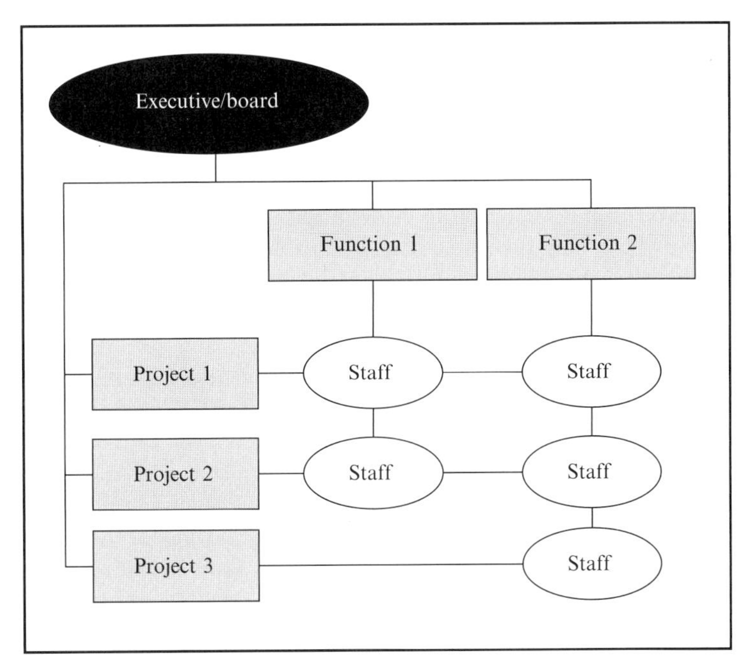

Figure 30.3 Matrix organization structure. This structure is represented with staff (white ellipse) assigned to function and project groups (shaded rectangle). Management and reporting lines flow both vertically and horizontally and senior management staff within both project and function teams report ultimately to the executive/board (black ellipse).

between disciplines has also been identified as crucial to the management of systems biology (Allarakhia and Wensley, 2007). Contributing factors influencing trans-disciplinary research including geographical location, cross-disciplinary literacy, organizational structure, different methodological traditions, diverse range of stakeholders, and the different views regarding what constitutes knowledge (Hollaender *et al.*, 2008; Jakobsen *et al.*, 2004; Monteiro and Keating, 2009) have been reported. Whilst the formation of interdisciplinary research centers has gone some way to addressing the issues associated with geographical location and organizational structure by providing a framework for cross-disciplinary interaction and more matrix style project management, there are still some significant challenges for systems biology to address. It has been suggested that for effective knowledge exchange and mutual learning between disciplines the tendency of scientists to organize a project to avoid conflicting overlap needs to be resisted (Hollaender *et al.*, 2008) and that through controlled conflict new joint knowledge is produced and consensus achieved. This is particularly important for systems biology where no single discipline operating in isolation can define a solution at the systems level. One of the major potential frustrations facing systems biology researchers is this interdependence on others and a perception of a lack of control over the knowledge creation process. This is particularly evident in large-scale research undertakings where the different disciplines interact in series and the research outcomes are only realized in the final publication of a validated model. Those involved in early stage data gathering may become disillusioned by the delay between completion of their input and recognition of their contribution in the form of the final published manuscript or frustrated because discoveries made later in the knowledge creation process require their work to be revisited often after a significant amount of time has passed since the original undertaking. Similarly, those involved in the later stages may be frustrated by a perceived lack of progress and a concern that the assumptions upon which they base their work may be undermined by new discoveries in the data gathering phase. The management of this process needs to be carefully construed to ensure maximum benefits are realized without becoming destructive to the functioning of the team. How different disciplines evaluate what constitutes valid results and how best to present data (Monteiro and Keating, 2009) also forms part of the challenge in terms of finding an acceptable common scientific language by which team members can communicate effectively at the system level is another important factor in knowledge management. By the very nature of systems biology research, the high-level project goals are broadly defined compared to those associated with disciplinary research topics. When these are translated into more specific goals and objectives to be worked on by individual team members, it can result in a high degree of uncertainty and risk as a development or breakthrough in another area may have significant implications. There is a danger this may lead to disillusionment if the individual is not connected to the project through shared goals and a robust system of change management is not implemented. An important factor in the management of systems biology is the recognition of different stakeholder goals regarding the development and dissemination of new knowledge. Academics seek publication in peer-reviewed journals and presentation at scientific conferences, funding bodies want evidence of beneficial societal impact, and industry seeks intellectual property rights and commercial exploitation. Amalgamating these differing requirements into a common set of goals is important to the overall success of the project; it has been previously reported (Monteiro and Keating, 2009) that participation in medical projects can serve as an incentive for scientists to collaborate where the goals are contextualized in terms of saving lives or promoting health.

3.2. Application to CISBs—Mixed organizational structures

Organizational structures can be considered in terms of the parent organization in its entirety. An underlying logic relating to the choice of appropriate organizational structure is based on the concept of "fit"

(Roberts, 2004). Experience teaches that specific organizational structures align with contextual factors to deliver good performance, others do not. Increasing project scale and complexity favor a move away from functional organization and toward a projectized or matrix structure.

However, within large institutions, it is also possible to employ a range of different organizational structures in a composite model. An institution could consider a broadly functional parent structure but this is entirely consistent with individual high-profile projects selected as needing increased autonomy. Projects that are determined as business critical by the institution may necessitate at least a balanced matrix structure. The option to deploy tailored management structures offers institutions a number of advantages. Critical areas of work can be delivered in the "best fit" organizational context and the journey along the organizational continuum can be piloted at lower risk than complete organizational culture change. Organizational structure, either at the parent institution level or at the individual project level, should not be considered as static and appropriateness can be reassessed as part of management good practice.

Where individual institutions are considering a composite model, supporting some activities as project initiatives overlying existing functional domains, due thought should be given to the interface. Although it is possible to use this composite structure with minimal adjustments to overall hierarchy, the result can be an artificial condition where power still resides in the functions and leadership within project teams has minimal authority. Delivering against a trans-disciplinary agenda, for example, leading the CISB to an international research profile in this context, demands skilled situational leadership.

3.3. MCISB

The management approach adopted by MCISB is designed primarily to address the key challenges of managing trans-disciplinary research projects namely; knowledge production, the formation of shared goals and communication between disciplines (Allarakhia and Wensley, 2007; Hollaender *et al.*, 2008). But also addresses other factors influencing trans-disciplinary research (Hollaender *et al.*, 2008; Jakobsen *et al.*, 2004; Monteiro and Keating, 2009) which include

- Geographical location
- Cross-disciplinary literacy
- Organizational structure
- Different methodological traditions
- Diverse range of stakeholders
- Different views regarding what constitutes knowledge
- Broadly defined goals compared to mono- or interdisciplinary research

The MCISB is located in an interdisciplinary research center, The Manchester Interdisciplinary Biocentre (MIB), which promotes several key benefits; colocation of researchers from the various disciplines in a shared office space encourages communication on a social level and facilitates effective implementation of a matrix management model. This structure offers the researchers both the benefit of being able to participate in cutting-edge trans-disciplinary research whilst remaining within their functional disciplines for line management and career development. In order for this model to work effectively, it is important that clear shared goals are established not only amongst the researchers but also amongst their line managers. Conflicts between functional and Centre objectives are avoided through engagement of the functional line managers as key members of the Centre management board with a vested interest in the success of the initiative. The overall Centre mission is governed by the management board which is advised by an international scientific advisory board and an industrial partnership committee to ensure external stakeholder interests are represented. The management board determines and reviews project goals and sets the strategic plan for the Centre. The overall activity is coordinated by a project manager who manages execution of the Centre strategy and communication between parties.

One of the early successes of MCISB was the development of a workflow for bottom-up systems biology, which mapped interdependencies between the various disciplinary contributions onto a cycle of continuous improvement through a process of hypothesis, experimental design, biochemical assay, modeling, and validation. This workflow has proven a valuable asset in improving integration between different disciplines and forging the development of shared goals. A recent strategy adopted to further facilitate development of shared goals has been the concept of Centre papers, these are targeted publications which draw on inputs from the entire team and reflect overall progress made on the Centre's core research program. They also provide a useful vehicle for knowledge production and the promotion of cross-disciplinary literacy as the focus is on the creation of new knowledge and understanding at the systems level. As a necessary part of the knowledge creation process, there needs to be a degree of controlled conflict between the different disciplinary traditions in order for new shared knowledge to be generated; the Centre papers provide a safe environment for this conflict to occur through debate focused on the problem (communicating a systems level understanding of the issue under study) rather than on the validity of methodologies used by the different disciplinary traditions. If understanding at the systems level can be achieved, this can provide the lingua franca by which the different disciplines communicate. The MCISB also utilizes discussion meetings where each discipline takes turn to lead the discussion and present a particularly significant result which is discussed in the context of the overall program as another means of generating controlled conflict and stimulating debate. The combination of these two approaches helps the team develop a deeper appreciation of their individual contribution to the shared goals and a better understanding of the role of other disciplines in the knowledge creation process.

Communication plays a key role in the management of systems biology both within the Centre and through its interaction with external stakeholders. The MCISB utilizes various communication tools to engage and manage its stakeholder relationships. Some of the internal procedures such as the use of discussion meetings and Centre papers have already been described; however, other tools include the use of wikis, systems biology themed seminars, Web pages, social networking sites, school visits and internships, formal project meetings and reports. Both the wiki and the Web site provide valuable tools for sharing data with internal and external stakeholders, serving as repositories for project data, historical archives, bulletin boards for news and forthcoming events, links to other sites of interest to the systems biology community and providing access to the systems biology toolsets developed by MCISB. The management reporting structure operates at both operational and strategic levels; operational issues such as progress of the core research program and routine financial and administration matters are dealt with through monthly program reviews, whereas strategic issues such as long-term scientific direction, future funding, staffing levels, and facilities requirements are addressed at monthly management board meetings. Program reviews engage researchers employed within the Centre, the Centre's Director, Deputy Director, and Project Manager, providing the opportunity to reflect on progress in the core research activities and take appropriate management action where required. These meetings also represent the means by which change and risk management are implemented, ensuring that all changes to the scientific program and the reasons for them are clearly communicated to the members of the team on whom they will have the greatest impact (Fig. 30.4).

3.4. Centre for Systems Biology at Edinburgh

A system can be considered as a set of interacting components that form the integrated whole. Using a broad definition like this immediately illustrates the applicability across a range of scales and range of subject matter. One logical extension is the consideration of organizations as complex social systems and the application of systems management theory.

The management and organizational structure of CSBE is entirely consistent with a basic model of an open system; inputs are received from the environment, converted to outputs and outputs are released into the environment. This includes the inputs of staff, existing data, financial resources, and experimental equipment; inputs are then converted into new data, new experimental procedures, and new research collaborations.

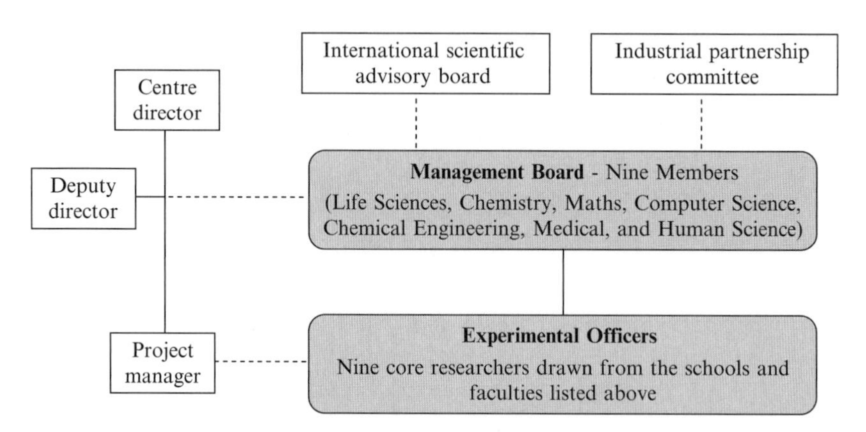

Figure 30.4 MCISB management structure.

Outputs are the products of such a system, and these include, amongst other things, new knowledge, intellectual property, and peer-reviewed publications. Within this open system are feedback interactions and component subsystems; examples could include research projects, staff training activities, business development, finance, and human resources. Some subsystems are internal to the Centre, others demand integration with external functions and the CSBE Centre Management function is integral to this interface. Practical examples include the smooth operation of diverse financial management and contractual and intellectual property systems across and within all participating institutions. A parallel analysis is to consider the Centre Management function as the interface between the more projectized CSBE organizational structure and the functional parent institution organizational structure.

CSBE Centre Management also takes the lead on delivering internal management and communication processes. We operate a structure of tiered management meetings, recognizing the importance of appropriate communication without an overburdened meeting schedule. Overall scientific governance and direction is provided by the CSBE Management Board, and a parallel Senior Systems Biology Group oversees integration of the Centre with the broader national and international research communities. Annual meetings of the International Scientific Advisory Board provide vital external context and insight, as well as leveraging strategically valuable cross-sector relationships (Fig. 30.5).

Research within CSBE is organized by project team with each project, from RNA metabolism in yeast to Process Algebra representing a subsystem. Projects illustrated in Fig. 30.5 fall into two main categories, "Theory & Tools" and "Biological Systems." CSBE also has two overarching platform research projects, the Kinetic Parameter Facility and the Systems Biology Software Infrastructure. CSBE was designed to include a project

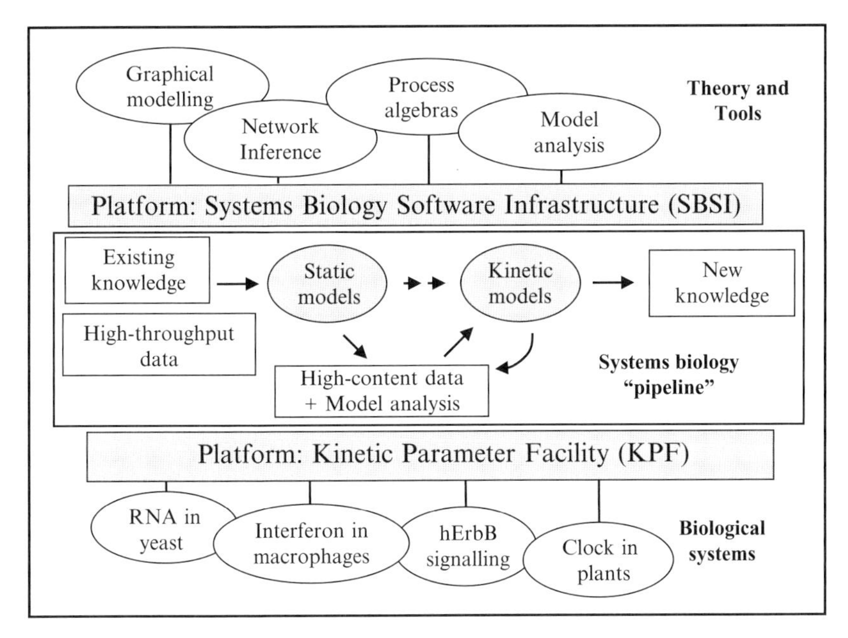

Figure 30.5 CSBE research project organization. The core systems biology "pipeline" is represented at the center of this diagram, illustrating the iterative cycle of new data generation, existing knowledge integration to static and kinetic system models, and the ultimate generation of new knowledge. The platform technology SBSI sits across computational research within the Centre and individual research projects are indicated in upper ellipses. The Kinetic Parameter Facility sits across all biological research within the Centre and individual research projects are indicated in lower ellipses.

portfolio with each bringing a target system of different scale, complexity, and preferred modeling platform but conserved across all projects is the central activity of dynamic systems modeling, the provision of input data, an iterative cycle of model development and analysis, and the output of new knowledge.

In all cases, research line management is delivered within the subsystem through the project team. All research projects also benefit from planned additional investment in project management time. Interestingly, project management staff are line managed ultimately through the Centre Director illustrating the unique overview aspects of these roles. The projects' staff body represent previously disparate disciplines and all absolutely require the establishment and maintenance of effective relationships with colleagues working in neighboring subsystems. Project management staff members are responsible for integrating the activities of adjacent subsystems toward the collective goal or "pipeline," and it is increasingly recognized that boundary management is of vital importance to the effectiveness of those in systems biology supervisory roles.

A range of boundary management techniques were deployed within the CSBE open system. For example, the use of wikis for informal communication providing a single and secure base for research collaboration data, a portal for routine administration and shared access to status reports and financial information. At the research level, the use of theme-specific laboratory meetings brought broader communities of researchers together than standard single discipline laboratory meetings. Effort was also directed to lowering perceived barriers to participation, with a regular seminar series including a mix of invited external speakers and regular representation of internal work-in-progress, to informal learning lunches that provided a more relaxed learning environment.

Infrastructure can also impact significantly on the successful delivery of trans-disciplinary research and CSBE is housed in the new CH Waddington building. Joining CSBE in this purpose-built facility are several major research leaders and their groups brought to Scotland by the Scottish Universities Life Sciences Alliance (SULSA). The CH Waddington building brings together a diverse array of researchers from mathematicians to biologists, physicists, and software developers and, with its cutting-edge laboratories, integrated offices, and social spaces provides a real hub for trans-disciplinary science in Scotland. Importantly, the provision of pilot project funding was then able to pump prime these interactions and foster a complementary new generation of systems biology research projects.

4. Conclusions

Systems biology shares a set of management challenges with other trans-disciplinary research initiatives. The principal challenges relate to the generation and exchange of knowledge and the formation of shared goals between researchers of different if not opposed (e.g., mathematics and molecular biology) disciplinary traditions. In this chapter, we have described the different approaches taken by the CSBE and MCISB to developing management structures to address these challenges. Whilst the two Centres have arrived at very different models for the implementation of best practice, CSBE has adopted an open system approach, whereas MCISB has utilized a matrix management model, there are a number of common elements at the core of each approach. Both are housed within buildings designed to promote interdisciplinary research allowing researchers of different disciplines to share office space and interact on both a social and a professional level. Both utilize theme-specific scientific meetings to engage a wider range of disciplines to help generate new knowledge at the systems level through increased interaction and providing a format for controlled conflict. Owing to the diverse range of stakeholders associated with systems biology initiatives, both Centres have employed a wide range of communication tools and boundary management techniques to build and maintain stakeholder relationships. The importance of the formation of shared goals is also recognized in both approaches; CSBE achieves this by utilizing boundary management techniques for the integration of adjacent subsystems, whereas MCISB achieves this through the focus on generation of new knowledge at the systems level and utilization of the workflow as a framework on which such knowledge is developed. Finally both approaches rely on strong project management to function effectively; due to the complexity of the work being undertaken, the broadly defined goals, the different disciplines and methodological traditions involved, and the diverse range of stakeholders that need to be engaged, an effective means of change and risk management is required to avoid individuals becoming disillusioned and maintain motivation when the inevitable changes occur.

No single organizational structure or management strategy can be applied that will provide the optimal solution in all circumstances; however, it is important to be aware of the influence such a choice may have. The approach selected will need to consider a number of factors such as the maturity and stability of the research group, the nature and complexity of the question being investigated, and political and cultural factors dominant within the organization. Regardless of the organizational structure or management strategy adopted, central to achieving a successful outcome is implementation of a mechanism for effective cross-disciplinary communication and the establishment of shared goals. The experiences of the MCISB and CSBE highlight the importance of a professional approach to the management of systems biology research and the value of and case for continued investment by funding bodies in support of management of complex trans-disciplinary research.

ACKNOWLEDGMENTS

The authors would like to acknowledge support from BBSRC and EPSRC: The Centre for Systems Biology at Edinburgh is a Centre for Integrative Systems Biology (CISB) funded by BBSRC and EPSRC, reference BB/D019621/1. The Manchester Centre for Integrative Systems Biology (MCISB) is a Centre for Integrative Systems Biology (CISB) funded by the BBSRC and EPSRC, reference BB/C008219/1.

REFERENCES

Allarakhia, M., and Wensley, A. (2007). Systems biology: A disruptive biopharmaceutical research paradigm. *Technol. Forecast. Soc. Change* **74,** 1643–1660.

Handy, C. (1999). Understanding Organisations. 4th edn. Penguin Group, London, UK.

Hollaender, K., *et al.* (2008). Management. *In* "Handbook of Transdisciplinary Research" (G. Hirsch Hadorn, *et al.*, eds.). Springer, Dordecht, Netherlands (chapter 25).

Jakobsen, C. H., *et al.* (2004). Barriers and facilitators to integration among scientists in transdisciplinary landscape analyses: A cross-country comparison. *Forest Policy Econ.* **6,** 15–31.

Monteiro, M., and Keating, E. (2009). Managing misunderstandings: The role of language in interdisciplinary scientific collaboration. *Sci. Commun.* **31,** 6–28.

Roberts, J. (2004). The Modern Firm: Organisational Design for Performance and Growth. Oxford University Press, UK.

Westerhoff, H. V., Verma, M., Bruggeman, F. J., Kolodokin, A., Swat, M., Hayes, N., Nardelli, M., Bakker, B. M., and Snoep, J. L. (2011). Chapter 19 from silicon cell to silicon human. *In* "BetaSys" (B. Booß-Bavnbek, *et al.*, eds.). Springer, New York.

Author Index

B

C

E

H

L

M

S

T

U

V

W

Subject Index

T

U

Y

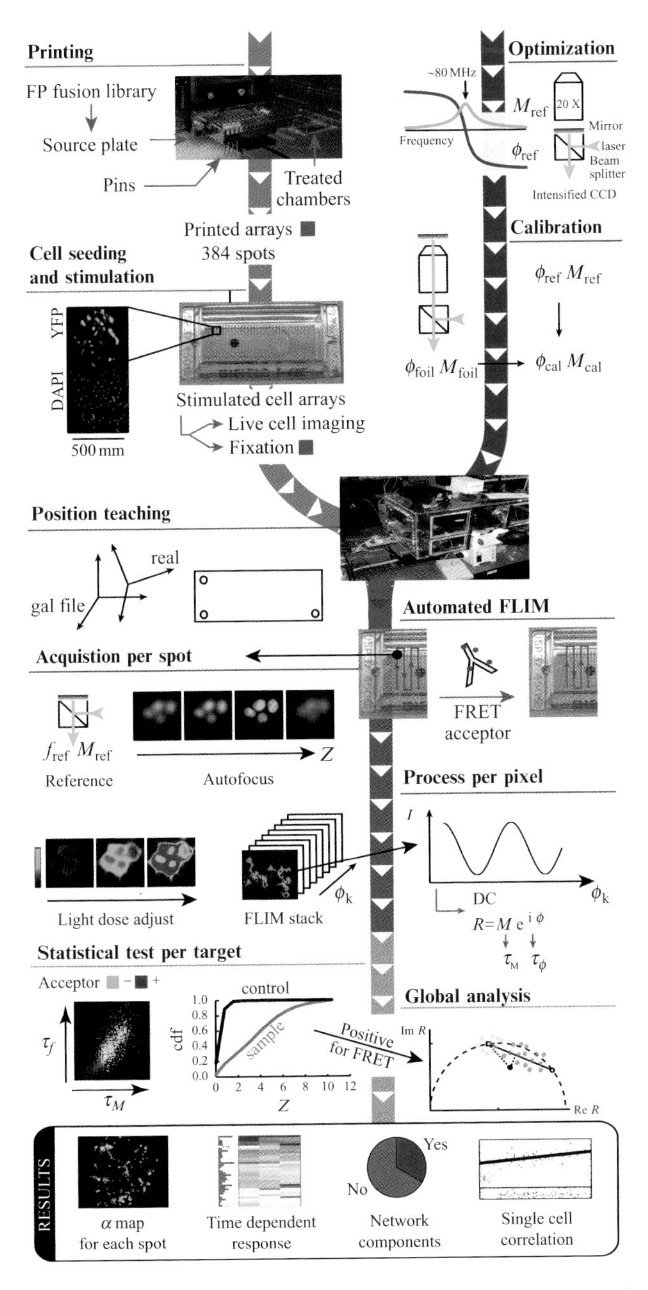

Hernán E. Grecco *et al.*, Figure 3.1 CA-FLIM visual workflow. The blue lane indicates the sample preparation steps to obtain a cell array. Green squares indicate pause points as indicated in the procedure section. The red lane indicates the microscopy preparation workflow. Both lanes converge in the automated FLIM violet lane, where after teaching the positions the spots are sequentially scanned. In each spot, optimization preparation steps can be performed before the FLIM stack is acquired. The yellow lane indicates the postprocessing of the data, in which targets showing FRET are selected using an statistical test and the fraction of posttranslationally modified protein per pixel is obtained. The resulting dataset can be used for different analysis.

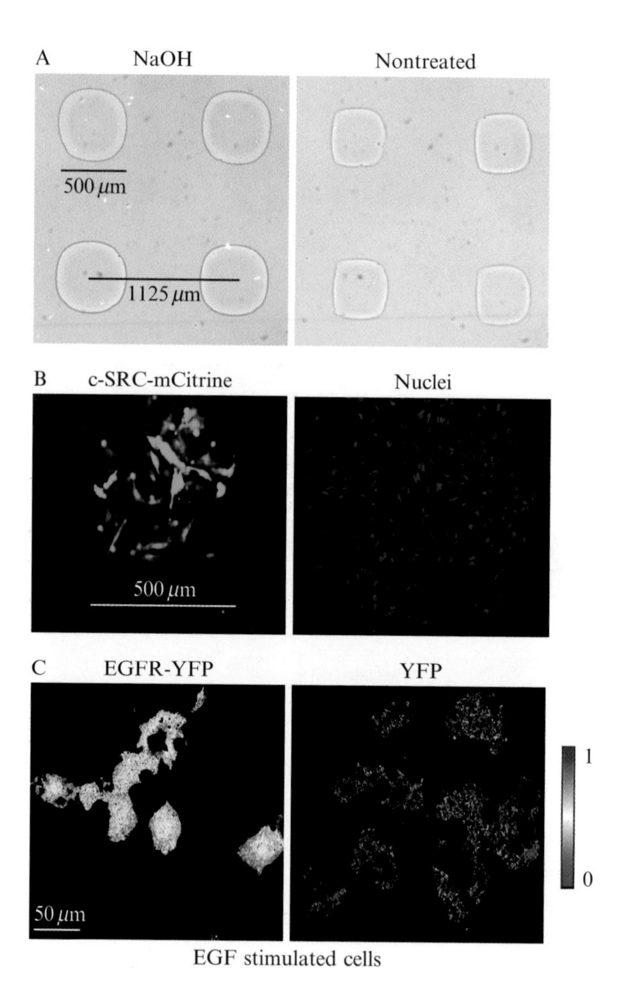

Hernán E. Grecco *et al.*, Figure 3.2 Quality control. (A) Spot quality. Comparison of spot shape and size between NaOH treated and nontreated LabTek chambers. (B) Transfection efficiency. Representative example of the transfection efficiency. Total number of cells is counted by staining nuclei (left panel). (C) Spots containing plasmids encoding for EGFR-YFP and YFP alone are used as a positive and negative controls respectively in cell arrays stimulated with EGF.

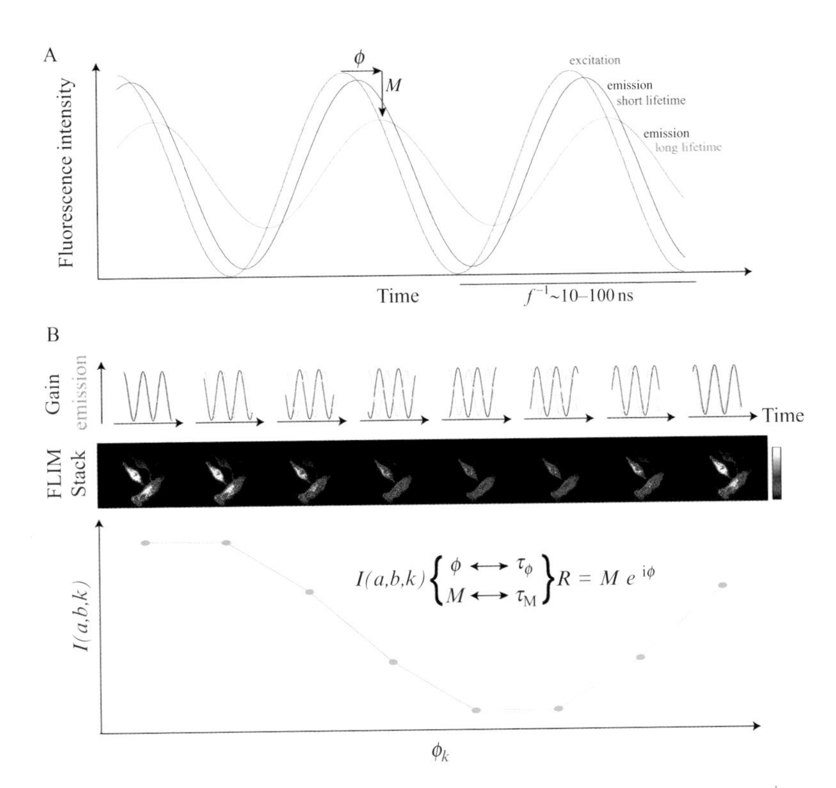

Hernán E. Grecco *et al.*, Figure 3.3 FD-FLIM. (A) Sinusoidally modulated excitation and emission of two fluorophores with two different lifetimes (short and long). The phase shift, demodulation, and the period of the wave are indicated. (B) Acquisition of the FLIM stack by homodyne detection. Upper panels show the modulated emission along with the different phases of the modulated gain of the intensifier. Middle panel shows the different phases of the acquired FLIM stack. Plot in the lower panel shows the change in $I(i, j, k)$ in the different phases of the FLIM stack.

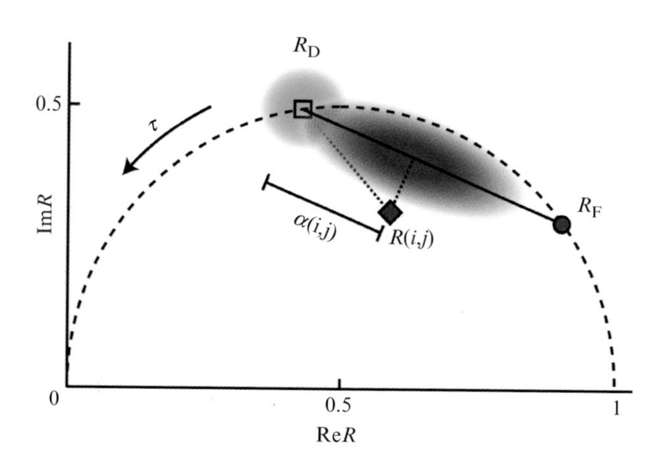

Hernán E. Grecco *et al.*, Figure 3.4 Global analysis in the complex plane. Complex plane schematic representation of phase shift and modulation data before (green cloud) and after (red cloud) the addition of the FRET acceptor. The dashed semicircle represents those points characterized by a single fluorescence lifetime (τ). R_D and R_F are the values for the donor alone and in complex with the acceptor. A mixture will be a linear combination of those. The fraction of donor in complex with the acceptor (α) is the length of the projection of $R - R_D$ onto the vector $R_F - R_D$.